T0319179

MOLECULAR ASPECTS OF ALCOHOL AND NUTRITION

MOLECULAR ASPECTS OF ALCOHOL AND NUTRITION

A VOLUME IN THE MOLECULAR NUTRITION SERIES

Edited by

VINOOD B. PATEL, PHD

Department of Biomedical Sciences, Faculty of Science and Technology, University of Westminster, London, UK

AMSTERDAM • BOSTON • HEIDELBERG • LONDON
NEW YORK • OXFORD • PARIS • SAN DIEGO
SAN FRANCISCO • SINGAPORE • SYDNEY • TOKYO
Academic Press is an imprint of Elsevier

Academic Press is an imprint of Elsevier
125, London Wall, EC2Y 5AS, UK
525 B Street, Suite 1800, San Diego, CA 92101-4495, USA
225 Wyman Street, Waltham, MA 02451, USA
The Boulevard, Langford Lane, Kidlington, Oxford OX5 1GB, UK

British Library Cataloguing-in-Publication Data
A catalogue record for this book is available from the British Library

Library of Congress Cataloging-in-Publication Data
A catalog record for this book is available from the Library of Congress

ISBN: 978-0-12-800773-0

For information on all Academic Press publications
visit our website at http://store.elsevier.com/

Publisher: Mica Haley
Senior Acquisitions Editor: Tari K. Broderick
Editorial Project Manager: Jeff Rossetti
Production Project Manager: Chris Wortley
Designer: Victoria Pearson

Typeset by Thomson Digital

Printed and bound in the United States of America

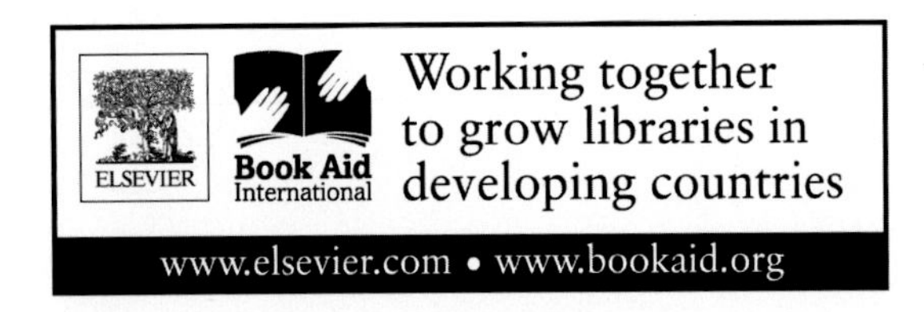

Table of Contents

List of Contributors ix
Series Preface xiii
Preface xv

I
GENERAL AND INTRODUCTORY ASPECTS

1. Nutrients and Liver Metabolism
DANIEL GYAMFI, KWABENA OWUSU DANQUAH

Introduction 3
Hepatic Carbohydrate Metabolism 3
Hepatic Fat Metabolism 7
Hepatic Amino Acid Metabolism 11
Conclusions 13
References 14

2. Alcohol Metabolism: General Aspects
REEM GHAZALI, VINOOD B. PATEL

Introduction 17
Alcohol Metabolism 17
The Epigenetic Changes of Alcohol 18
Conclusions 20
References 20

II
MOLECULAR BIOLOGY OF THE CELL

3. Alcohol and Aldehyde Dehydrogenases: Molecular Aspects
KWABENA OWUSU DANQUAH, DANIEL GYAMFI

Overview of Alcohol and Aldehyde Dehydrogenases 25
Human ALDH Superfamily 25
Modifications of ALDH and Biomedical Implications 30
Genomics of the ALDH Superfamily 31
ALDH Polymorphism in Alcohol-Related Diseases 39
Conclusions 40
References 41

4. Alcohol Intake and Apoptosis: A Review and Examination of Molecular Mechanisms in the Central Nervous System
MARIA D. CAMARGO, CHERRY IGNACIO, PATRICK BURKE, FRANK A. MIDDLETON

Introduction 45
Molecular Profiling of the Effects of Ethanol Consumption in Adolescent Rats 53
Conclusions 58
References 59

5. Pathogenic Mechanisms in Alcoholic Liver Disease (ALD): Emerging Role of Osteopontin
JASON D. COOMBES, WING-KIN SYN

Introduction 63
General Mechanisms of ALD 63
ALD Fibrosis – Pathogenic Mechanisms 64
Role of OPN in ALD 66
Conclusions 68
References 68

6. The Role of CD36 in the Pathogenesis of Alcohol-Related Disease
CALEB T. EPPS, ROBIN D. CLUGSTON, AMIT SAHA, WILLIAM S. BLANER, LI-SHIN HUANG

Introduction 71
The Structure-Function Relationship of CD36 71
CD36 and FA Uptake 73
CD36 and Ca^{2+} Signaling 76
CD36 as a Pattern Recognition Receptor 77
CD36 in the Brain 78
Conclusions 79
References 81

7. Thiamine Deficiency and Alcoholism Psychopathology
ANN M. MANZARDO

Introduction 85
Biochemistry of Thiamine Pyrophosphate 85
Secondary Effects of Thiamine Deficiency 87
Thiamine Deficiency, Mood and Behavior 87
Thiamine Analogs 88
Benfotiamine Trial 89
Conclusions 91
References 92

8. Vitamin B Regulation of Alcoholic Liver Disease
CHARLES H. HALSTED, VALENTINA MEDICI

Effects of Chronic Alcoholism on the Availability of Folate, Vitamin B12, and Vitamin B6 95
Interactions of Selected B Vitamins in Hepatic Methionine Metabolism with Implications for Alcoholic Liver Injury 98

Effects of Chronic Alcohol Exposure on Methionine Metabolic Pathways 99
Epidemiology and Pathogenesis of Alcoholic Liver Disease 99
Regulatory Effects of Altered Reactions in Methionine Metabolism and its Metabolites on Mechanisms of Alcoholic Liver Disease 100
The Ethanol-Fed Micropig Model for the Interaction of Ethanol with Methionine Metabolism, and Induction of Alcoholic Liver Disease 101
Effects of SAM in Clinical Trials of Treatment of ALD 101
Relationships of Altered Methionine Metabolism to Epigenetic Regulation of Gene Expression in Alcoholic Liver Injury 102
Conclusion 103
References 103

9. Interactions Vitamin D – Bone Changes in Alcoholics

EMILIO GONZÁLEZ-REIMERS, FRANCISCO SANTOLARIA-FERNÁNDEZ, GERALDINE QUINTERO-PLATT, ANTONIO MARTÍNEZ-RIERA

Introduction 107
Vitamin D Metabolism 107
Vitamin D Among Alcoholics 108
Vitamin D Deficiency and Bone Metabolism 109
Effects of Therapy 114
Conclusion 114
References 115

10. Antioxidant Treatment of Alcoholism

CAMILA S. SILVA, GUILHERME V. PORTARI, HELIO VANNUCCHI

Importance of the Antioxidant Systems to the Organism 119
Antioxidant Status in Alcoholism 120
Antioxidant Treatment of Alcoholism 121
Conclusions 125
References 128

11. Selenium Dietary Supplementation and Oxidative Balance in Alcoholism

OLIMPIA CARRERAS, MARÍA LUISA OJEDA, FÁTIMA NOGALES

Introduction 133
Alcohol and Malnutrition: Selenium Supplementation 134
Alcohol and Selenium Tissue Distribution: Selenium Supplementation 135
Alcohol and Oxidative Balance: Selenium Supplementation 137
Conclusion 139
References 139

12. Role of Zinc in Alcoholic Liver Disease

WEI ZHONG, QIAN SUN, ZHANXIANG ZHOU

Introduction 143
Zinc Metabolism and Function 143
Occurrence of Zinc Deficiency in ALD 144
Mechanisms of Zinc Dyshomeostasis in ALD 145
Beneficial Effects of Zinc on ALD 147
Molecular Mechanisms of Zinc Protection Against ALD 149
Zinc Supplementation in Human ALD 151
Conclusions 152
References 152

13. Interactions Between Alcohol and Folate

BOGDAN CYLWIK, LECH CHROSTEK, EWA GRUSZEWSKA

Introduction 157
Conclusions 166
References 167

14. Effects of Acetaldehyde on Intestinal Barrier Function

ELHASEEN E. ELAMIN, AD A. MASCLEE, DAISY M. JONKERS

Introduction 171
The Intestinal Epithelial Barrier and Apical Junctional Complex 171
Generation and Metabolism of Acetaldehyde in the Gastrointestinal Tract 172
Role of Intestinal Microbiota in Production and Metabolism of Acetaldehyde 175
Effects of Acetaldehyde on Intestinal Barrier Epithelial Integrity 175
Mechanisms of Acetaldehyde-Induced Intestinal Barrier Dysfunction 177
Modulation of Acetaldehyde-Induced Intestinal Epithelial Barrier Function 178
Conclusions 180
References 181

15. Cholesterol Regulation by Leptin in Alcoholic Liver Disease

BALASUBRAMANIYAN VAIRAPPAN

Introduction 187
Alcohol Metabolism 187
Effects of Alcohol on Lipid Metabolism 187
Triglycerides 188
Fatty Acids 189
Hepatic Lipid Metabolism is Controlled by Several Master Transcription Factors 189
Cholesterol 191
Obesity, Leptin, and ALD 192
Conclusions 196
References 197

16. The Corticotropin Releasing Factor System and Alcohol Consumption

ANDREY E. RYABININ, WILLIAM J. GIARDINO

Introduction: Alcohol and the Brain 201
CRF, Urocortins, and the HPA-Axis 201
Function of CRF System Components Revealed by Genetic Knockouts 202
Effects of EtOH on the HPA-Axis 203
Effects of EtOH on the Extrahypothalamic CRF System 203
Effects of CRF Receptors on EtOH Drinking 204

Ucn1 and EtOH Drinking 205
Interactions of CRF System, Stress, EtOH Drinking and Dependence 206
Conclusions 206
References 207

17. Metabolic Profiling Approaches for Biomarkers of Ethanol Intake
H.G. GIKA, I.D. WILSON

Introduction 213
Metabolic Profiling 213
Analytical Technologies for Global Metabolic Profiling 213
Ethanol Exposure Studies on Animal Models 215
Ethanol Exposure Studies in Man 219
Conclusions 220
References 221

III
GENETIC MACHINERY AND ITS FUNCTION

18. Gene Expression in Alcoholism: An Overview
REEM GHAZALI, VINOOD B. PATEL

Introduction 225
Brain Serotonin and Alcoholism 225
NMDA Receptor and Alcoholism 226
Fetal Alcohol Syndrome 226
Sweet Preference and Alcohol Dependence 226
Stem Cell Therapy and Alcoholism 227
Genetic Risks of Alcoholism 227
MicroRNA and Alcoholism 227
Conclusions 228
References 228

19. *Cytochrome P4502E1* Gene Polymorphisms and the Risks of Ethanol-Induced Health Problems in Alcoholics
TAO ZENG, KE-QIN XIE

The Polymorphisms of *CYP2E1* Gene and the Nomenclatures of *CYP2E1* Alleles 231
The Polymorphisms of *CYP2E1* Gene and Individual Susceptibility to Alcohol Dependence (AD) 233
CYP2E1 Gene Polymorphisms and Susceptibility to Alcoholic Liver Disease (ALD) 235
CYP2E1 Gene Polymorphism and Risks of Alcoholic Pancreatitis (AP) 239
Conclusions 242
References 243

20. Genes Associated with Alcohol Withdrawal
KESHENG WANG, LIANG WANG

Introduction 247
Candidate Gene Studies 248
Genome-Wide Association Study 253
Limitations 254
Conclusions 254
References 255

21. Alcohol and Epigenetic Modulations
CLAUDIO D'ADDARIO, MAURO MACCARRONE

Introduction 261
Epigenetics 262
Alcohol Effects on Epigenetic Mechanisms 264
Alcohol Metabolism Influences Nutritional Status via Epigenetic Mechanisms 268
Future Directions 269
Conclusions 270
References 270

22. The miRNA and Extracellular Vesicles in Alcoholic Liver Disease
FATEMEH MOMEN-HERAVI, SHASHI BALA

Introduction 275
Extracellular Vesicle-Associated miRNAs as Potential Biomarkers for ALD 278
miRNA-Targeted Therapies for ALD 281
Conclusions 284
References 285

23. Alcohol Metabolism and Epigenetic Methylation and Acetylation
MARISOL RESENDIZ, SHERRY CHIAO-LING LO, JILL K. BADIN, YA-JEN CHIU, FENG C. ZHOU

Introduction 287
Alcohol Metabolism and Epigenetic Acetylation 288
Alcohol and Methyl Metabolism 289
Alcohol Affects the Epigenetics Converting Enzymes 295
Epigenetic Footprint of Alcohol Induced Dysfunction and Disease 296
Treatment and Biomarkers of Alcohol Diseases Through Epigenetic Pathway 299
Conclusions 299
References 301

24. Molecular Mechanisms of Alcohol-Associated Carcinogenesis
HELMUT K. SEITZ, SEBASTIAN MUELLER

Introduction 305
Effect of Ethanol on DNA 306
Effect of Ethanol on Epigenetics 308
Alcohol, Retinoids, and Cancer 309
The Role of Estrogens in Ethanol Mediated Breast Cancer 310
Summary and Conclusions 311
References 312

25. Molecular Link Between Alcohol and Breast Cancer: the Role of Salsolinol

MARIKO MURATA, KAORU MIDORIKAWA, SHOSUKE KAWANISHI

Introduction 315
Salsolinol Derived from Alcohol, as a Novel Causative Substance of Breast Cancer 315
DNA Damage Capability of Salsolinol 316
Proposed Mechanisms of Oxidative DNA Damage Induced by Salsolinol 318
Cell Proliferating Ability of Salsolinol 320
Conclusion 321
References 322

26. Ethanol Impairs Phospholipase D Signaling in Astrocytes

UTE BURKHARDT, JOCHEN KLEIN

Astrocytes 325
Phospholipase D 325
Transphosphatidylation 325
Isoforms of Phospholipase D 326
Regulation of Phospholipase D 326
Functions of Phospholipase D 328
Interruption of the PLD Signaling Pathway by Ethanol 329
Conclusions 332
Summary 333
References 334

27. Metabolic Changes in Alcohol Gonadotoxicity

GANNA M. SHAYAKHMETOVA, LARYSA B. BONDARENKO

Introduction 337
Metabolic Changes in Males' Alcohol Gonadotoxicity 339
Metabolic Changes in Females' Alcohol Gonadotoxicity 346
Conclusions 350
References 351

28. Molecular Effects of Alcohol on Iron Metabolism

KOSHA MEHTA, SEBASTIEN FARNAUD, VINOOD B. PATEL

Introduction 355
Overview of Iron Metabolism 355
Alcohol and Iron: Clinical Observations and Molecular Events 360
Exploring the Iron-Alcohol Link for Diagnosis and Therapeutics 362
Conclusions 365
References 365

Index 369

List of Contributors

Jill K. Badin, BS Department of Cellular & Integrative Physiology, Stark Neuroscience Research Institute, Indiana University-Purdue University at Indianapolis, Indianapolis, IN, USA

Shashi Bala, PhD Department of Medicine, University of Massachusetts Medical School, Worcester, MA, USA

William S. Blaner, PhD Department of Medicine, Division of Preventive Medicine & Nutrition, Columbia University, New York, NY, USA

Larysa B. Bondarenko, PhD, DSc General Toxicology Department, Institute of Pharmacology and Toxicology of National Academy of Medical Sciences of Ukraine, Kyiv, Ukraine

Patrick Burke, MS Department of Neuroscience & Physiology, SUNY Upstate Medical University, Syracuse, NY, USA

Ute Burkhardt, PhD Department of Pharmacology, Goethe University Frankfurt, Frankfurt, Hessen, Germany

Maria Camargo Moreno, MPH Department of Biochemistry & Molecular Biology, SUNY Upstate Medical University, Syracuse, NY, USA; Department of Neuroscience & Physiology, SUNY Upstate Medical University, Syracuse, NY, USA; Developmental Exposure Alcohol Research Center, Binghamton University, Binghamton, NY, USA

Olimpia Carreras, PhD Department of Physiology, Faculty of Pharmacy, Seville University, Seville, Spain

Ya-Jen Chiu, BS Department of Anatomy and Cell Biology, Indiana University-Purdue University at Indianapolis, Indianapolis, IN, USA; Department of Life Science, National Taiwan Normal University, Taipei, Taiwan

Lech Chrostek, PhD Department of Biochemical Diagnostics, Medical University of Bialystok, Bialystok, Poland

Robin D. Clugston, PhD Department of Medicine, Division of Preventive Medicine & Nutrition, Columbia University, New York, NY, USA

Jason D. Coombes, PhD Regeneration and Repair Group, The Institute of Hepatology, London, UK; Transplant Immunology and Mucosal Biology, Kings College London, UK

Bogdan Cylwik, MD Department of Paediatric Laboratory Diagnostics, Medical University of Bialystok, Bialystok, Poland

Claudio D'Addario, PhD Faculty of Bioscience and Technology for Food, Agriculture and Environment, University of Teramo, Teramo, Italy; Department of Clinical Neuroscience, Center for Molecular Medicine, Karolinska Institutet, Stockholm, Sweden

Kwabena Owusu Danquah, MSc, PhD Department of Medical Laboratory Technology, College of Health Sciences, Kwame Nkrumah University of Science & Technology, Kumasi, Ghana

Elhaseen E. Elamin, MD, MSc, PhD Division of Gastroenterology-Hepatology, School for Nutrition, Toxicology and Metabolism, Maastricht University Medical Center, Maastricht, the Netherlands

Caleb T. Epps, MS Department of Medicine, Division of Preventive Medicine & Nutrition, Columbia University, New York, NY, USA

Sebastien Farnaud, MSc, PhD Department of Life Sciences, University of Bedfordshire, Luton, Bedfordshire, UK

Reem Ghazali, MSc Department of Clinical Biochemistry, King Abdul Aziz University, Jeddah, Saudi Arabia; Department of Biomedical Sciences, Faculty of Science & Technology, University of Westminster, London, UK

William J. Giardino, PhD Department of Psychiatry & Behavioral Sciences, Stanford University, Stanford, CA, USA

Helen G. Gika, PhD Department of Chemical Engineering, Aristotle University of Thessaloniki, Thessaloniki, Greece

Emilio González-Reimers, PhD Servicio de Medicina Interna, Hospital Universitario de Canarias, Universidad de La Laguna, Tenerife, Canary Islands, Spain

Ewa Gruszewska Department of Biochemical Diagnostics, Medical University of Bialystok, Bialystok, Poland

Daniel Gyamfi, MSc, PhD Department of Medical Laboratory Technology, College of Health Sciences, Kwame Nkrumah University of Science & Technology, Kumasi, Ghana

Charles H. Halsted, MD University of California Davis, The Genome and Biomedical Sciences Facility, Davis, California, USA

Li-Shin Huang, PhD Department of Medicine, Columbia University, New York, NY, USA; Department of Medicine, Division of Preventive Medicine & Nutrition, Columbia University, New York, NY, USA

Cherry Ignacio, PhD Department of Biochemistry & Molecular Biology, SUNY Upstate Medical University, Syracuse, NY, USA; Department of Neuroscience & Physiology, SUNY Upstate Medical University, Syracuse, NY, USA; Developmental Exposure Alcohol Research Center, Binghamton University, Binghamton, NY, USA

Daisy M. Jonkers, PhD Division of Gastroenterology-Hepatology, School for Nutrition, Toxicology and Metabolism, Maastricht University Medical Center, Maastricht, the Netherlands

Shosuke Kawanishi, PhD Department of Environmental and Molecular Medicine, Mie University Graduate School of Medicine, Tsu, Japan; Laboratory of Public Health, Department of Health Sciences, Faculty of Pharmaceutical Sciences, Suzuka University of Medical Science, Suzuka, Japan

Jochen Klein, PhD Department of Pharmacology, Goethe University Frankfurt, Frankfurt, Hessen, Germany

Sherry Chiao-Ling Lo, PhD Department of Anatomy and Cell Biology, Indiana University-Purdue University at Indianapolis, Indianapolis, IN, USA; Indiana Alcohol Research Center, Indiana University-Purdue University at Indianapolis, Indianapolis, IN, USA

Mauro Maccarrone, MS, PhD School of Medicine and Center of Integrated Research, Campus Bio-Medico University of Rome, Rome, Italy; European Center for Brain Research (CERC)/Santa Lucia Foundation, Rome, Italy

Ann M. Manzardo, PhD Department of Psychiatry and Behavioral Sciences, University of Kansas Medical Center, Kansas City, Kansas, USA

Antonio Martínez-Riera, MD Servicio de Medicina Interna, Hospital Universitario de Canarias, Universidad de La Laguna, Tenerife, Canary Islands, Spain

Ad A. Masclee, MD, PhD Division of Gastroenterology-Hepatology, School for Nutrition, Toxicology and Metabolism, Maastricht University Medical Center, Maastricht, the Netherlands

Valentina Medici, MD Division of Gastroenterology and Hepatology, University of California Davis, Sacramento, California, USA

Kosha Mehta, MSc, PhD Department of Biomedical Sciences, Faculty of Science and Technology, University of Westminster, London, UK

Frank A. Middleton, PhD Department of Biochemistry & Molecular Biology, SUNY Upstate Medical University, Syracuse, NY, USA; Department of Neuroscience & Physiology, SUNY Upstate Medical University, Syracuse, NY, USA; Department of Psychiatry & Behavioral Sciences, SUNY Upstate Medical University, Syracuse, NY, USA; Developmental Exposure Alcohol Research Center, Binghamton University, Binghamton, NY, USA

Kaoru Midorikawa, PhD Department of Environmental and Molecular Medicine, Mie University Graduate School of Medicine, Tsu, Japan

Fatemeh Momen-Heravi, DDS, MPH Department of Medicine, University of Massachusetts Medical School, Worcester, MA, USA

Sebastian Mueller, MD, PhD Centre of Alcohol Research, University of Heidelberg, Heidelberg, Baden-Württemberg, Germany; Department of Medicine (Gastroenterology and Hepatology), Salem Medical Centre, Heidelberg, Baden - Württemberg, Germany

Mariko Murata, MD, PhD Department of Environmental and Molecular Medicine, Mie University Graduate School of Medicine, Tsu, Japan

Fátima Nogales, MD Department of Physiology, Faculty of Pharmacy, Seville University, Seville, Spain

María Luisa Ojeda, PhD Department of Physiology, Faculty of Pharmacy, Seville University, Seville, Spain

Kwabena Owusu Danquah, MSc, PhD Department of Medical Laboratory Technology, College of Health Sciences, Kwame Nkrumah University of Science & Technology, Kumasi, Ghana

Vinood B. Patel, PhD Department of Biomedical Sciences, Faculty of Science & Technology, University of Westminster, London, UK

Guilherme V. Portari, PhD Department of Nutrition, Federal University of Triângulo Mineiro, Uberaba, MG, Brazil

Geraldine Quintero-Platt, MD Servicio de Medicina Interna, Hospital Universitario de Canarias, Universidad de La Laguna, Tenerife, Canary Islands, Spain

Marisol Resendiz, BA, BS Department of Cellular & Integrative Physiology, Stark Neuroscience Research Institute, Indiana University-Purdue University at Indianapolis, Indianapolis, IN, USA

Andrey E. Ryabinin, PhD Department of Behavioral Neuroscience, Oregon Health and Science University, Portland, OR, USA

Amit Saha, BA Department of Medicine, Division of Preventive Medicine & Nutrition, Columbia University, New York, NY, USA

Francisco Santolaria-Fernández, MD Servicio de Medicina Interna, Hospital Universitario de Canarias, Universidad de La Laguna, Tenerife, Canary Islands, Spain

Helmut K. Seitz, MD, PhD Centre of Alcohol Research, University of Heidelberg, Heidelberg, Baden-Württemberg, Germany; Department of Medicine (Gastroenterology and Hepatology), Salem Medical Centre, Heidelberg, Baden - Württemberg, Germany

Ganna M. Shayakhmetova, PhD General Toxicology Department, Institute of Pharmacology and Toxicology of National Academy of Medical Sciences of Ukraine, Kyiv, Ukraine

Camila S. Silva, PhD Division of Biochemical Toxicology, National Center for Toxicological Research, US Food and Drug Administration, Jefferson, AR, USA

Qian Sun, PhD Center for Translational Biomedical Research, University of North Carolina at Greensboro, Kannapolis, NC, USA

Wing-Kin Syn, MBChB Regeneration and Repair Group, The Institute of Hepatology, London, UK; Liver Unit, Barts Health NHS Trust, London, UK; Department of Physiology, University of the Basque Country, Leioa, Spain; Department of Surgery, Loyola University, Chicago, IL, USA; Transplant Immunology and Mucosal Biology, Kings College London, UK

Balasubramaniyan Vairappan, PhD Department of Biochemistry, Jawaharlal Institute of Postgraduate Medical Education and Research (JIPMER), Dhanvantari Nagar, Pondicherry, India

Helio Vannucchi, MD, PhD Department of Internal Medicine, Division of Nutrition, University of São Paulo, Ribeirão Preto, Brazil

Kesheng Wang, PhD Department of Biostatistics and Epidemiology, College of Public Health, East Tennessee State University, Johnson City, USA

Liang Wang, MD, PhD, MPH Department of Biostatistics and Epidemiology, College of Public Health, East Tennessee State University, Johnson City, USA

Ian D. Wilson, DSc Department of Surgery & Cancer, Imperial College, London, UK

Ke-Qin Xie, PhD Institute of Toxicology, School of Public Health, Shandong University, Jinan City, China

Tao Zeng, PhD Institute of Toxicology, School of Public Health, Shandong University, Jinan City, China

Wei Zhong, DVM, PhD Center for Translational Biomedical Research, University of North Carolina at Greensboro, Kannapolis, NC, USA

Feng C. Zhou, PhD Department of Cellular & Integrative Physiology, Stark Neuroscience Research Institute, Indiana University-Purdue University at Indianapolis, Indianapolis, IN, USA; Department of Anatomy and Cell Biology, Indiana University-Purdue University at Indianapolis, Indianapolis, IN, USA; Indiana Alcohol Research Center, Indiana University-Purdue University at Indianapolis, Indianapolis, IN, USA; Indiana University School of Medicine, and Department of Psychology, Indiana University-Purdue University at Indianapolis, Indianapolis, IN, USA

Zhanxiang Zhou, PhD Center for Translational Biomedical Research, University of North Carolina at Greensboro, Kannapolis, NC, USA

Series Preface

In this series on Molecular Nutrition, the editors of each book aim to disseminate important material pertaining to molecular nutrition in its broadest sense. The coverage ranges from molecular aspects to whole organs, and the impact of nutrition or malnutrition on individuals and whole communities. It includes concepts, policy, preclinical studies, and clinical investigations relating to molecular nutrition. The subject areas include molecular mechanisms, polymorphisms, SNPs, genomic wide analysis, genotypes, gene expression, genetic modifications, and many other aspects. Information given in the Molecular Nutrition series relates to national, international, and global issues.

A major feature of the series that sets it apart from other texts is the initiative to bridge the transintellectual divide so that it is suitable for novices and experts alike. It embraces traditional and nontraditional formats of nutritional sciences in different ways. Each book in the series has both overviews and detailed and focused chapters.

Molecular Nutrition is designed for nutritionists, dieticians, educationalists, health experts, epidemiologists, and health-related professionals such as chemists. It is also suitable for students, graduates, postgraduates, researchers, lecturers, teachers, and professors. Contributors are national or international experts, many of whom are from world-renowned institutions or universities. It is intended to be an authoritative text covering nutrition at the molecular level.

V.R. Preedy
Series Editor

Preface

Molecular Nutrition is designed for nutritionists, dietitians, educationalists, health experts, and epidemiologists, as well as for health-related professionals such as chemists. It is also suitable for students, graduates, postgraduates, researchers, lecturers, teachers, and professors. Contributors are national or international experts, many of whom are from world-renowned institutions or universities. It is intended to be a *one-stop-shop* of everything to do with nutrition at the molecular level.

In this book, *Molecular Aspects of Alcohol and Nutrition*, we focus on alcohol. Readers must be mindful that, for many, alcohol is an important part of the diet. For example, in the USA, the NIAAA has reported that in the last year, 60% of women and 72% of men are drinkers. Approximately 20–35% of women and men drink daily. Thus, alcohol is an important nutrient in its own right and contributes significantly to overall dietary intake. In some individuals, however, ethanol may comprise about three quarters of total energy intake. Therefore, in this book, alcohol is considered a component in the diet rather than a mood-altering psychoactive agent.

Alcohol is an important field to study within *Molecular Nutrition* as both alcohol and its toxic metabolite acetaldehyde disrupts and impairs normal macro- and micronutrient regulation at the molecular level. For example, alcohol affects many gene pathways involved in fat and glucose metabolism such as fatty acid oxidation and synthesis, the SREBP pathway, the ketogenic pathway, pathways involved in the regulation of methionine and glutathione synthesis, Vitamin A regulation, and mitochondrial genes involved in energy production. Alcohol also disrupts iron homeostatic control from intestinal iron absorption to iron response genes in the liver. Alcohol also affects and interferes with cell signaling, cell death pathways, calcium homeostasis leading to osteoporosis, and oxygen balance.

This book is relevant to nutritionists and nutrition researchers as alcohol disrupts normal fat, carbohydrate, and protein metabolic processes occurring in the liver, as well as other parts of the body. An example of this disturbance is the disorder, alcoholic liver disease (ALD), in which fat deposition occurs. Alcohol is also a major contributing factor to diabetes and the metabolic syndrome. In the United States alone there are 18 million people with alcoholism and 26 million people who have diabetes; nondiabetic alcoholics are at risk of hypoglycemia. Alcohol also affects nutrient absorption in the gastrointestinal tract, which can lead to anemia and reduced amounts of fat-soluble vitamins. Secondary consequences of alcohol consumption are reduced levels of minerals such as magnesium and calcium and trace elements such as zinc.

Alcohol also affects many extrahepatic organs and tissues. For example, thiamine deficiency can cause the memory loss seen in Wernicke/Korsakoff syndrome; acute and chronic pancreatitis can lead to disruption in fat and glucose regulation; fetal alcohol syndrome itself is a serious consequence of maternal alcohol consumption, which is also a major contributing factor to the lack of nutrients consumed by the alcoholic mother. In terms of the treatment of alcoholics, an area of continued research important for nutritionists is studying the effect of parenteral nutrition, high-protein diets, and antioxidant therapy.

Molecular Aspects of Alcohol and Nutrition is divided into three sections. Section I covers nutrition and alcohol in terms of general aspects and contribution to the diet. In Section II, there are chapters on alcohol and aldehyde dehydrogenases, metabolic profiling, apoptosis, CRF1 receptor signaling, nutrient pathways, choline supplementation, and Vitamins B and D. Section III includes discussion of enzymes, gene expression, microRNAs, epigenetic modulations, DNA damage, repair proteins, DNA methylation, sweet preference genes, ALDH2, iron regulation, carcinogenesis, and DNA adducts.

Vinood B. Patel

SECTION I

GENERAL AND INTRODUCTORY ASPECTS

1 *Nutrients and Liver Metabolism* 3
2 *Alcohol Metabolism: General Aspects* 17

CHAPTER

1

Nutrients and Liver Metabolism

Daniel Gyamfi, MSc, PhD, Kwabena Owusu Danquah, MSc, PhD

Department of Medical Laboratory Technology, College of Health Sciences, Kwame Nkrumah University of Science & Technology, Kumasi, Ghana

INTRODUCTION

The liver, considered to be the largest and heaviest organ (1.0–1.5 kg) in the human body, consists of two lobes: a larger right lobe, and a smaller left lobe. It is located beneath the diaphragm, and in the upper right quadrant of the abdominal cavity. The role of the liver in nutrients' metabolism cannot be overlooked, since the liver is positioned strategically in the human body, such that almost all absorbed nutrients, after digestion in the gastrointestinal tract, pass through the liver via the hepatic portal vein, before they are distributed to other tissues; though some nutrients can also get to the liver via the systemic circulation from other tissues, such as skeletal muscles (lactate, amino acids, etc.) and adipose tissues (fatty acids).

The liver is involved in metabolism of the major nutrients: carbohydrate, lipid, and protein (Figure 1.1). It can take up glucose (after a meal), and oxidizes it (via glycolysis), to produce energy, and at high blood glucose concentration, the liver stores glucose as glycogen (glycogenesis), and converts glucose into fat (lipogenesis). When blood glucose concentration is low (during fasting), the liver releases glucose via glycogenolysis, and can synthesize glucose (gluconeogenesis) from noncarbohydrate precursors (lactate, pyruvate, glycerol, and certain amino acids). Formerly, regulation of these pathways particularly at the molecular level was mainly assigned to hormones, such as insulin and glucagon, but recent advances indicate that nutrients themselves play an important role in the transcriptional regulation of most of these pathways, independent of hormones. This chapter describes the major pathways involved in hepatic energy metabolism, and seeks to tackle some of the current advances made in biochemical and, especially, transcriptional regulations of these pathways by carbohydrate responsive element binding protein (ChREBP), peroxisomal proliferator-activated receptors (PPARs), and sterol-regulatory-element-binding protein 1c (SREBP-1c).

HEPATIC CARBOHYDRATE METABOLISM

Carbohydrate is one of the major sources of energy for the human body. Glucose (a monosaccharide) is considered as the central molecule in carbohydrate metabolism, since the metabolism of other important monosaccharides (such as fructose, galactose, etc.), which are commonly found in diet, is linked virtually to glucose metabolic pathways. The liver plays an essential role in the carbohydrate metabolism, by monitoring and stabilizing blood glucose levels in two major conditions: absorptive (feeding) state when the liver stores glucose as glycogen (via glycogenesis); and postabsorptive (fasting) state when the liver releases glucose (via glycogenolysis), or synthesizes glucose from noncarbohydrate sources (via gluconeogenesis).

Glycogenesis and Glycogenolysis

Absorptive State

Intestinal glucose that is absorbed after the digestion of a carbohydrate-containing meal is carried by hepatic portal vein to the liver. Within the liver, glucose crosses the membrane of hepatocytes by way of facilitated diffusion, using principally GLUT 2 glucose transporters, that are membrane bound transporters with very high K_m, ranging from 15 mmol/L to 20 mmol/L, and their expression, and activities are independent of insulin signaling. As a result, the rate at which glucose is taken up by the hepatocytes is proportional to the concentration of glucose in the blood.

Once taken up by the hepatocytes, the enzyme glucokinase (GK) (the rate-limiting enzyme for hepatic glucose utilization) phosphorylates glucose to

Molecular Aspects of Alcohol and Nutrition. http://dx.doi.org/10.1016/B978-0-12-800773-0.00001-X

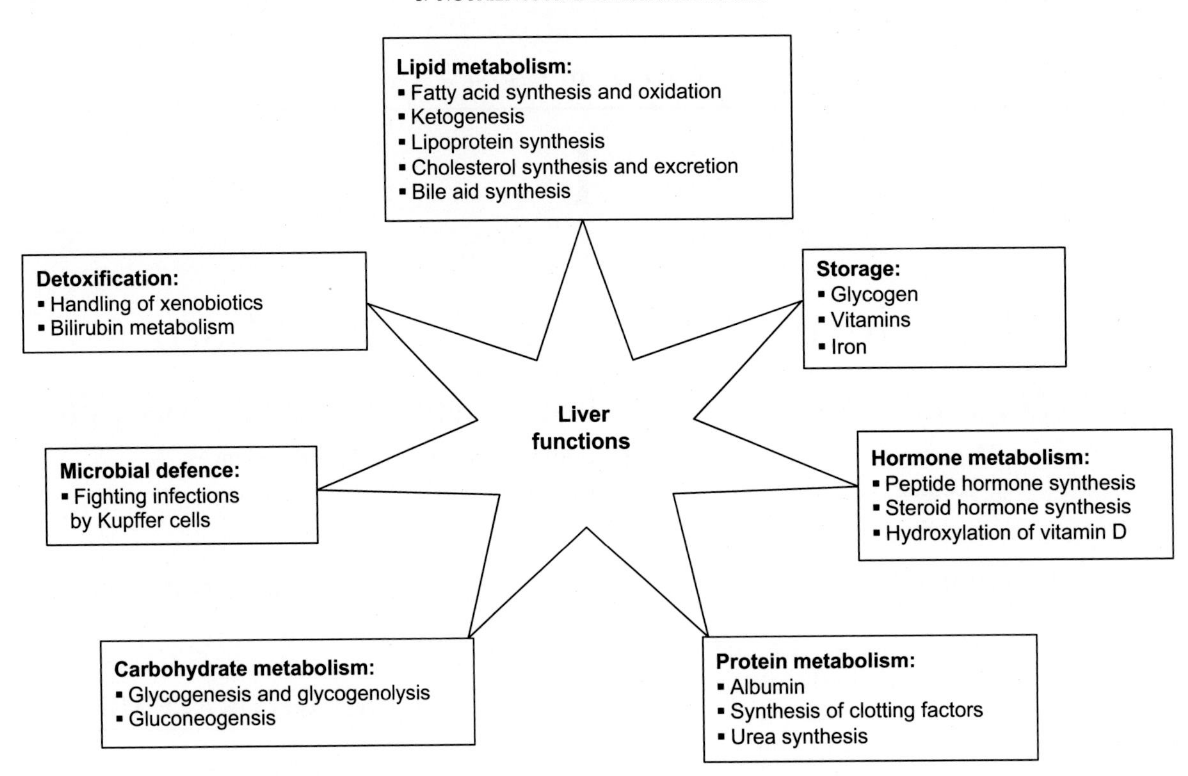

FIGURE 1.1 **The main functions of the liver in categories.**

glucose-6-phosphate. Glucokinase (hexokinase IV) belongs to the hexokinase family with a high K_m for glucose (10 mmol/L), and under physiological conditions, it is not affected by its product, glucose-6-phosphate, which allows for postprandial glycogen storage within the hepatocytes (Figure 1.2). Glucokinase (inactive in the fasting state, since it is bound to glucokinase regulatory protein (GKR) within the nucleus), is active in postprandial state, since abundance of glucose, and insulin-action synergistically cause its rapid dissociation from GKR with subsequent translocation into the cytoplasm.[1,2] All these characteristic features exhibited by the liver help in the rapid uptake, and phosphorylation of glucose during high blood glucose concentration.

Depending on the systemic metabolic state, the glucose-6-phosphate synthesized within the hepatocytes may either be further processed in glycolysis, or be utilized for glycogen synthesis (glycogenesis). Liver glycogen formation is catalyzed by glycogen synthase (GS) after conversion of glucose-6-phosphate to uridine diphospate-glucose (UDP-glucose) via glucose-1-phosphate. GS is activated allosterically by glucose-6-phosphate, but inactive when phosphorylated. Protein kinases such as, AMP-activated protein kinase (AMPK), protein kinase A (PKA), and glycogen synthase kinase 3 (GSK3) phosphorylate GS. Insulin suppresses PKA leading to dephosphorylation of GS, and activation of glycogen synthesis. The creation of branches in glycogen molecule, to increase nonreducing ends for chain extension or degradation, is catalyzed by branching enzyme. However, recent research indicates that glycogen formation after meals is not only from the pathway that directly synthesizes glycogen from glucose, but also from an indirect pathway, where glycogen can be synthesized using three-carbon gluconeogenic substrates such as glycerol, lactate, etc.[3,4]

On the other hand, through glycolysis (a 10-step process that occurs in the cytoplasm with or without the presence of oxygen), glucose-6-phosphate synthesized is converted to pyruvate with a net gain of two ATP and two NADH molecules per glucose molecule. Pyruvate is further either converted to lactate (in the absence of oxygen) or decarboxylated to acetyl-CoA that can either be processed in the tricarboxylic acid (TCA) cycle (to produce energy, reducing equivalents (NADH and $FADH_2$) and carbon dioxide) in the mitochondria (in the presence of oxygen), or be utilized for *de novo* lipogenesis (DNL) in the cytosol. The reducing equivalents generated in the TCA cycle (NADH and $FADH_2$), and glycolysis (NADH) enter the mitochondrial electron transport chain to synthesize ATP via oxidative phosphorylation. Glycolysis is regulated by GK, phosphofructokinase 1 (inhibited by its product fructose-1,6-bisphosphate), AMP, and pyruvate kinase (that catalyzes the final reaction, phosphoenolpyruvate to pyruvate). Insulin, epinephrine, and glucagon also regulate pyruvate kinase via the phosphoinositide 3-kinase (PI3K) pathway. Another alternative pathway

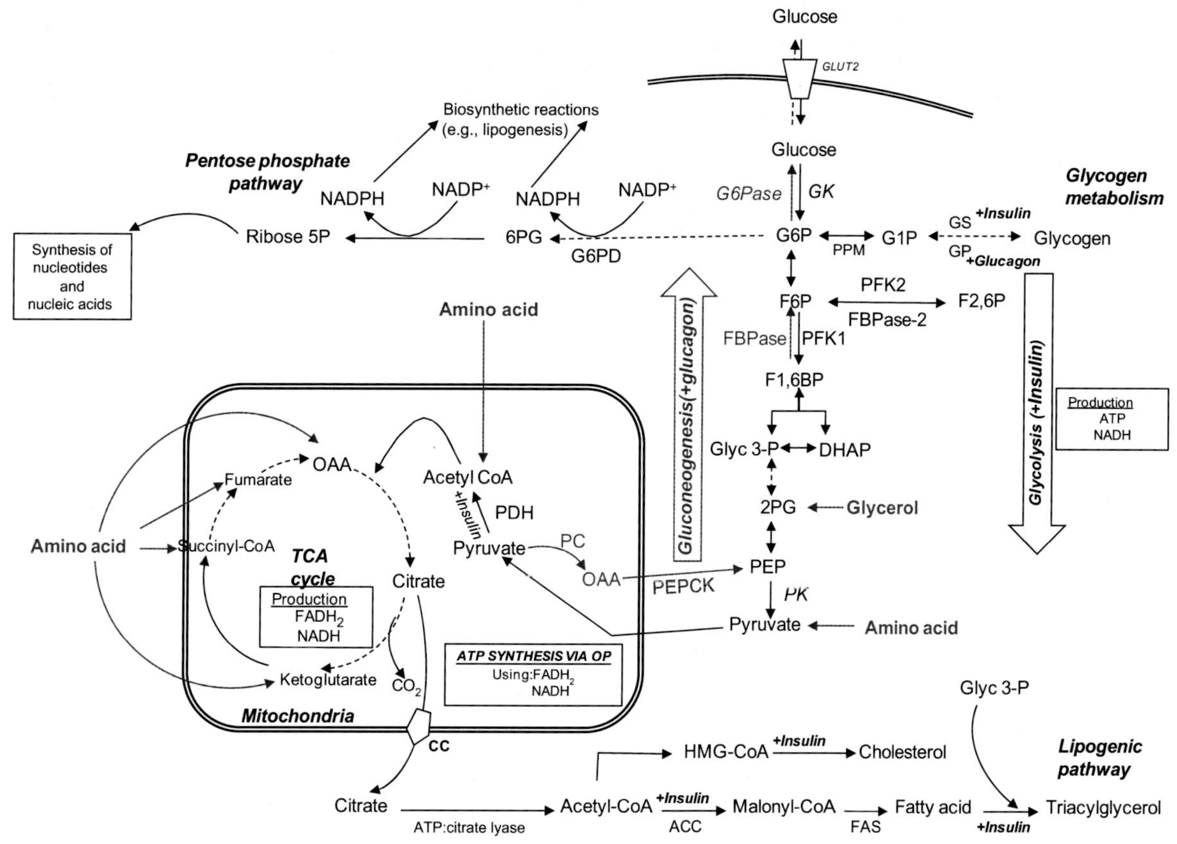

FIGURE 1.2 Overview of hepatic carbohydrate metabolism. After a meal, glucose is taken up by hepatocytes via glucose transporter 2 (GLUT2), and oxidizes via glycolysis in the cytosol to form pyruvate that produces more ATP through tricarboxylic acid (TCA) cycle, and mitochondrial oxidative phosphorylation pathway. The glucose 6-phosphate (G6P) formed in glycolysis can also be utilized to synthesize glycogen by glycogen synthase (GS), or can enter pentose phosphate pathway. In the excess intake of dietary carbohydrate, glucose is converted to fatty acids (triacylglycerol) through lipogenesis where pyruvate is converted to acetyl-CoA, and then citrate that exits the mitochondria via citrate carrier (CC) into the cytosol. During low cellular energy condition (e.g., fasting), the stored liver glycogen is broken down by the enzyme glycogen phosphorylase (GP) to G6P that can join the glycolytic and pentose phosphate pathways, or be converted by the enzyme glucose 6-phosphatase (G6Pase) to glucose. When glycogen becomes depleted (e.g., in starvation), hepatocyte synthesize glucose via gluconeogenesis using pyruvate, glycerol, certain amino acids and the intermediates of TCA cycle. These processes are regulated by hormones, such as, insulin and glucagon. Dashed arrows indicate multiple enzymatic steps. The blue portions represent gluconeogenesis. ACC, acetyl-CoA carboxylase; FAS, fatty acid synthase; G1P, glucose 1-phosphate; PPM, phosphoglucomutase; GK, glucokinase; F6P, fructose 6-phosphate; F1,6P, fructose 1,6-bisphosphate; FBPase, fructose 1,6-bisphosphatase; Glyc 3-P, glyceraldehydes 3-phosphate; G6PD, glucose 6-phosphate dehydrogenase; 6PG, 6-phosphogluconate; Ribose 5P, ribose 5-phosphate; OAA, oxaloacetate; PK, pyruvate kinase; PEP, phosphoenolpyruvate; PC, pyruvate carboxylase; PEPCK, phosphoenolpyruvate carboxykinase; PDH, pyruvate dehydrogenase; 2PG, 2-phosphoglycerate; DHAP, dihydroxyacetone phosphate; PFK1, phosphofructokinase 1; PFK2, phosphofructokinase 2; F2,6P, fructose 2,6-bisphosphate; FBPase2, fructose 2,6-bisphosphatase.

for degradation of glucose-6-phosphate in hepatocytes is the pentose phosphate pathway that provides the cell with NADPH, an important antioxidant and cosubstrate for DNL, and cholesterol synthesis.[5]

Postabsorptive State

When the concentration of glucose in the blood falls (for instance, between meals, or during starvation/fasting), the glycogen stored in the liver is broken down to release glucose (glycogenolysis) to restore the blood glucose concentration. The breakdown of glycogen is catalyzed by glycogen phosphorylase (GP) that cleaves glucose via phosphorolysis from the nonreducing ends of glycogen polymer to produce glucose-1-phosphate, which is converted to glucose-6-phosphate by phosphoglucomutase. Glycogen phosphorylase action stops at four glucose monomers to a branched point, and debranching enzyme cleaves the last four glucose monomers. The glucose-6-phosphate synthesized is transported finally into the endoplasmic reticulum (ER), and dephosphorylated by glucose-6-phosphatase, mainly synthesized in the liver, to produce glucose (Figure 1.2) that is released into the circulation.

The regulation of glycogenolysis is the reverse of glycogenesis, thus glycogen phosphorylase is activated, while glycogen synthase is inhibited.[4] Glycogen phosphorylase is activated allosterically by AMP, and is active when phosphorylated. Activated PKA causes phosphorylation of GP. Hormones such as glucagon and catecholamines including epinephrine and norepinephrine (under stressful condition), phosphorylates glycogen phosphorylase via PKA activation, which is inhibited by insulin, to cause glycogenolysis.[4]

Gluconeogenesis

During prolonged fasting or starvation, or restriction dietary carbohydrate, liver glycogen store is depleted, and gluconeogenesis, the synthesis of glucose from noncarbohydrate sources, occurs. The liver is the major organ involved in this process. The main gluconeogenic substrates are pyruvate, lactate, amino acids (e.g., alanine, aspartate), intermediate metabolites of the TCA cycle and glycerol.[3] These substrates are either generated in the liver, or transported from extrahepatic tissues to the liver through the systemic circulation. Gluconeogenesis is not the direct reverse of glycolysis since significant differences exist at irreversible steps in glycolysis where regulation occurs.

Gluconeogenesis is regulated by both gluconeogenic substrates availability, and hormones (especially glucagon) that affect the expression/activation of enzymes involved in the pathway.[6] The rate of gluconeogenesis is increased when gluconeogenic substrates are elevated. For example, the liver gluconeogenic activity is increased during physical exercise, since there is rise in blood lactate (a major gluconeogenic substrate) concentration that is converted subsequently into glucose by the hepatocytes and then released into circulation to support muscular activities. Gluconeogenesis is stimulated by glucagon whereas insulin prevents gluconeogenesis. Glucagon inactivates pyruvate kinase via cyclic AMP mediation thereby channeling phosphoenolpyruvate into gluconeogenic pathway. Also, glucagon reduces the concentration of fructose 2,6-bisphosphate (signal molecule which levels increase postabsorptively). Decreased fructose 2,6-bisphosphate concentration allosterically inhibits phosphofructokinase, and activates fructose 1,6-bisphosphatase favoring increased gluconeogenesis.[7] Thus glucagon predominance leads to the generation of glucose-6-phosphate in the liver, which is subsequently released as glucose into the bloodstream. Thus, glucagon dominance resembles glycogenolysis. It means, therefore, that within a short period, the two metabolic processes occur almost at the same time. A good example is that during an overnight fast, gluconeogenesis and glycogenolysis contribute, roughly, an equal proportion to the overall glucose production.[8] However, it is believed that, in prolonged fasting/starvation, gluconeogenesis is the predominant source of glucose production.[4,8]

Regulation of Blood Glucose Levels

Human plasma glucose concentration is maintained stable in a narrow range between minimum values of around 3 mmol/L after prolonged fasting or vigorous muscular exercise, and maximum values of around 9 mmol/L acquired after feeding.[9] This is achieved by carefully regulated homeostatic mechanisms within the body, since serious metabolic adverse effects result if these processes are not controlled tightly. For example, hyperglycemia causes glucose induced oxidative damage, and also nonenzymatic glycosylation of proteins leading to protein dysfunction.[3,9] On the other hand, hypoglycemia results in insufficient supply of glucose to obligatory glycolytic tissues such as neuronal cells, erythrocytes, and fibroblasts.[9] The liver plays a central role in buffering the plasma glucose concentration with the help of hormones, and glucose nutrient itself. Insulin, the only known blood glucose lowering hormone, is secreted by the pancreas (β-cells of the islets of Langerhans). Hyperglycemia stimulates insulin secretion as well as amino acids, free fatty acids,[8] and ketone bodies. Insulin acts by decreasing blood glucose levels through increasing of glucose entry into insulin-sensitive tissues in addition to enhancing glucose metabolism (glycolysis and glycogenesis). The primary counter-regulatory hormone to hypoglycemia, glucagon (secreted by the α-cells of the islets of Langerhans of pancreas) promotes the release of glucose into circulation by stimulating glycogenolysis and gluconeogenesis (Figure 1.2). Other counterregulatory hormones include thyroid hormones, adrenaline, growth hormones, and glucocorticoids.

Under longer-term regulation, insulin has been detected to have effects on glucose (and lipid) metabolism by modulating the expression of specific genes that are involved in the metabolic pathways of glucose.[4,10] Insulin-sensitive tissues (especially, the liver) contain transcription factor sterol regulatory element-binding protein 1c (SREBP-1c) that has been identified recently to mediate the insulin signaling. Insulin acts by stimulating the transcription, and the proteolytic maturation of SREBP-1c, which increases sequentially the expression of a family of genes involved in glucose utilization and also fatty acid synthesis as expanded on in Section "Regulation of Hepatic Fatty Acid Synthesis," and as a result, SREBP-1c is considered as a thrifty gene.[11] In addition to insulin acting through the transcription factor SREBP-1c, glucose has been identified also to cause increased utilization of glucose by stimulating the transcription factor carbohydrate responsive element-binding protein (ChREBP), which in turn activates the expression of genes of most the enzymes involved in glucose metabolism and lipogenesis (see the expansion in Section "Regulation of Hepatic Fatty Acid Synthesis").[4,10] For example, the activity of pyruvate kinase has been shown to be wholly under the regulation of glucose, and not under SREBP-1c regulation.[12]

HEPATIC FAT METABOLISM

Fats, consisting of fatty acids and glycerol, are the largest energy reserve in mammals.[13] The liver plays an essential role in lipid metabolism, since the liver can both synthesize, and oxidize fatty acids. The body fatty acid pool consists of typically dietary fatty acids, and to a lesser extent, fatty acids synthesized by the liver from nonfat sources like carbohydrate, in addition to plasma nonesterified fatty acids (NEFA) released from adipose tissues via lipolysis.[14,15] After a meal, to aid intestinal lipid absorption, lipid droplets are emulsified by bile acids, synthesized within the hepatocytes and secreted into the bile duct, making them accessible to lipase hydrolyzation. Thus, a healthy liver is crucial for intestinal lipid absorption. Over 95% of dietary fat (85–90% in infants) is absorbed by the small intestine after a meal, and either transported by the hepatic portal vein to the liver (i.e., fatty acids with less than 10 carbon units),[4] or processed by the enterocytes into nascent chylomicrons (the largest group of the circulating lipoproteins) that are released into the lymphatic systems, and transported, finally, via the blood to the liver, and adipose tissues.[16] Besides these chylomicrons, the liver packages endogenously synthesized triacylglycerol (TAG) as very low density lipoproteins (VLDLs) that are secreted into circulation, contributing to the plasma fatty acid pool. Recently, autophagy (a lysosomal pathway in which dispensable cellular constituents are recycled into important energy sources during fasting state) has been implicated to play a role in hepatic lipid homeostasis.[17] Thus, macroautophagy occurs during starvation, which leads to the fusion of lysosomes, and lipid droplets into autophagosomes, which are then degraded to release fatty acids.[5] However, fatty acids that enter or are synthesized within the liver can undergo either oxidation or can be stored as triacylglycerol (Figure 1.3).

Hepatic Fatty Acid Uptake and Oxidation

Fatty acids in circulation (that are mostly bound to albumin) are transported across the membrane of hepatocytes into the cytosol by passive diffusion, or by facilitated transporters such as fatty acid transport proteins (FATP) (FATP2 and FATP5 commonly expressed in hepatocytes), fatty acid translocase (CD36/FAT), and caveolin-1.[5,18] The breakdown of fatty acids by β-oxidation (oxidation of fatty acids on the β-carbon atom) in the human liver occurs in the mitochondria (responsible for the oxidation of short chain (<4 carbon), medium chain (4–12 carbon) and long chain (10–20 carbon) fatty acids), and other intracellular sites such as peroxisomes, involved in the oxidation of toxic, very long chain fatty acids (VLCFAs), and branched-chain fatty acids.[5] However, peroxisomal β-oxidation results in hydrogen peroxide production, and also low ATP synthesis since peroxisomes do not contain an electron transport chain.[15] It is now believed that the initial oxidation of VLCFAs by the peroxisomes shortens them prior to mitochondrial β-oxidation. The estimated contribution of peroxisomal β-oxidation toward total rate of hepatic fatty acid oxidation ranges from 5% to 30%. Aside β-oxidation, other types of fatty acid oxidation include α-oxidation and ω-oxidation carried out by cytochrome P450 4A in the endoplasmic reticulum (microsomes). During periods of increased influx of fatty acids to the liver, these extramitochondrial fatty acid oxidation systems especially the peroxisomes become of great importance. Thus, peroxisomes proliferation is stimulated by agents known as peroxisome proliferators, which include fatty acids and fatty-acid-derived molecules, leading to increased expression of genes involved in peroxisomal β-oxidation.[4]

Considering mitochondrial β-oxidation, the fatty acids, after entering the hepatocytes, are transported in the cytosol by specific fatty acid binding proteins (FABP). The fatty acids are then activated by esterification – forming thioester link with coenzyme-A (CoA) catalyzed by the enzyme acyl-CoA synthase (ACS) to produce acyl-CoA derivatives of the fatty acids. ACS enzyme is present in the outer mitochondrial membrane. The small and medium chain fatty acyl-CoA molecules (usually up to 10 carbon atoms) are able to cross the mitochondrial membrane into the mitochondrial matrix by diffusion. On the other hand, the longer fatty acyl-CoAs cannot cross the mitochondrial membrane by diffusion, but require to be linked to a polar molecule called carnitine catalyzed by the enzyme carnitine palmitoyl transferase 1 (CPT-1) (found in the outer face of the inner mitochondrial membrane) to form fatty acylcarnitine (i.e., the substitution of the CoA group with carnitine molecule) (Figure 1.3). The acylcarnitine is then transported across the inner mitochondrial membrane by the enzyme acylcarnitine translocase into the mitochondrial matrix where the acylcarnitine is reconverted to fatty acyl-CoA by the enzyme carnitine acyltransferase 2 that is found in the matrix side of the inner mitochondrial membrane. The intramitochondrial fatty acyl-CoA is oxidized through β-oxidation to produce acetyl-CoA, a two-carbon unit that is sequential removed from the end of the fatty acid chain. During β-oxidation process, electrons are transferred to cause the reduction of flavin-adenine dinucleotide (FAD) and oxidized nicotinamide adenine dinucleotide (NAD^+) to produce coenzymes $FADH_2$ and NADH, respectively, which in turn donate electrons to the electron transport chain to drive ATP synthesis via oxidative

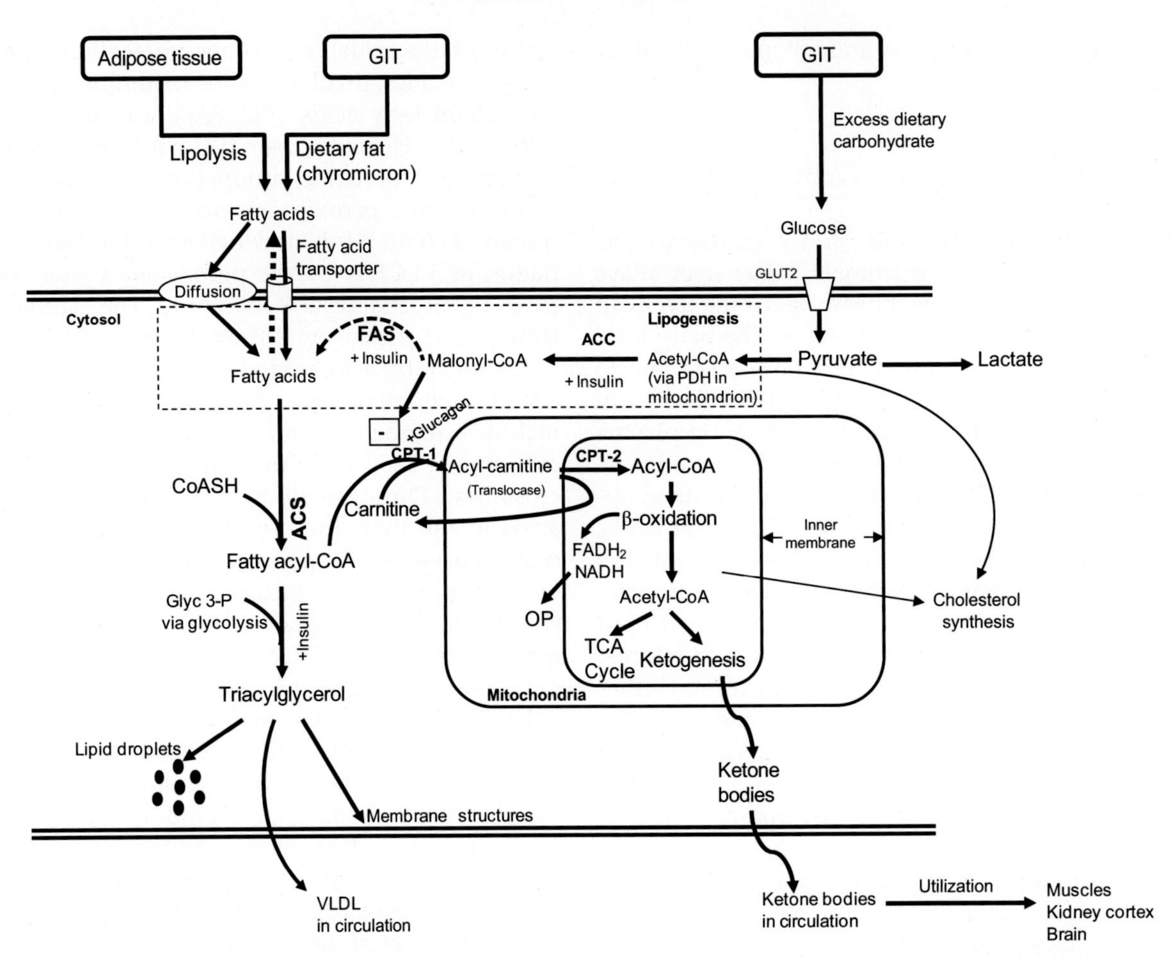

FIGURE 1.3 Overview of hepatic lipid metabolism. Fatty acids (from diet or lipolysis) enter the hepatocytes via diffusion, or fatty acid transporters. The fatty acids are then esterified with coenzyme-A (CoASH) catalyzed by the enzyme acyl-CoA synthase (ACS) to form fatty acyl-CoA. In order to be oxidized, the small- to medium-sized acyl-CoAs enter the mitochondria by simple diffusion, but longer ones are aided by polar molecule carnitine to form acyl-carnitine derivatives catalyzed by carnitine-palmitoyl transferase-1 (CPT-1) with a subsequent reconversion to acyl-CoA by the inner mitochondrial membrane enzyme carnitine-palmitoyl transferase-2 (CPT-2). The acyl-CoA derivatives then undergo β-oxidation leading to acetyl-CoA production. Depending upon the energy status of the cell, the acetyl-CoA can: enter the TCA cycle (for more ATP production); or be used for ketogenesis (e.g., during starvation) to produce ketone bodies, which are subsequently released into the circulation, and utilized by the skeletal muscle, heart, and brain. The oxidation of fatty acids generates NADH and $FADH_2$ that are used to synthesize ATP via the oxidative phosphorylation (OP) pathway. Excess intake of dietary carbohydrates leads to conversion of glucose into fatty acids via lipogenesis through acetyl-CoA with the help of the enzymes acetyl-CoA carboxylase (ACC), and fatty acid synthase (FS). Hormonal regulation occurs by insulin inhibiting fatty acid oxidation by activating ACC to increase the production of malonyl-CoA (to inhibit CPT-1) leading to increased fatty acid synthesis, whereas glucagon increases fatty acid oxidation through its action on CPT-1. Fatty acids in the hepatocytes can also be converted to triacylglycerol, and stored as lipid droplets, incorporated into membrane structures or secreted as into circulation as very low density lipoprotein (VLDL). ACS, acyl-CoA synthase; CoASH, coenzyme-A; Glyc 3-P, glyceraldehyde 3-phosphate.

phosphorylation. In fact, it is a rapid and effective way of energy production, especially during fasting/starvation, since, for instance, the oxidation of one molecule of palmitate produces up to 129 ATP equivalents.[5,7] For this reason, under extreme conditions, humans are able to fast and survive for 60–90 days without food, and it is even longer in obese persons (from 6 months to 1 year).[7,19]

The acetyl-CoA synthesized can be oxidized completely to carbon dioxide in the tricarboxylic acid cycle (TCA). However, in abundance of fatty acids, the acetyl CoA is converted via ketogenesis (Figure 1.3) to acetoacetate, and 3-hydroxybutyrate. These products 3-hydroxybutyrate, acetoacetate, and acetone (a decarboxylated form of acetoacetate) are known as ketone bodies. Though the liver synthesizes these ketone bodies, it cannot reutilize them as fuel. However, they are important energy source for certain extrahepatic tissues, such as skeletal muscle, cardiac muscle, or renal cortex, etc.[19] In addition, the brain can utilize 3-hydroxybutyrate as a major source of energy during starvation, and in diabetes mellitus.[19,20]

Regulation of Hepatic Fatty Acid Oxidation

The regulation of β-oxidation of fatty acids is achieved mainly by availability of fatty acids, and also by hormones. The nonesterified fatty acids in the blood are mostly from the breakdown of stored triacylglycerol in adipose tissues, controlled by hormone-sensitive triacylglycerol lipase. Fatty acids β-oxidation is regulated through the inhibition of the enzyme CPT-1 by malonyl-CoA, an intermediate of *de novo* lipogenesis. Thus, insulin secreted after a meal hinders fatty acids β-oxidation through activation of acetyl-CoA carboxylase, which in turn, increases the production of malonyl-CoA leading to inhibition of CPT-1. On the other hand, low energy situation (such as fasting) – less insulin secretion – also results in a decrease in malonyl-CoA production, and an increase in fatty acid oxidation. Thus, the inhibition of CPT-I by malonyl-CoA prevents simultaneous oxidation and synthesis of fatty acids within the hepatocytes, a potential futile cycle. The role of malonyl-CoA provides a good link between carbohydrate, and fat metabolism. Glucagon has been proposed to increase fatty acid oxidation besides increasing gluconeogenesis, and glycogenolysis during fasting.[21]

In terms of long-term modulation, nuclear receptors known as peroxisomal proliferator-activated receptors (PPARs) have been identified recently to play an important role. PPARs are activated by peroxisome proliferators, such as fatty acids and fatty-acid derivatives. Different types of these receptors exist: PPARα, PPARβ, and PPARγ (for a comprehensive review, see Refs [22,23]). The liver is among the tissues that highly express PPARα. PPARα controls the expression of genes involved in mitochondrial and extramitochondrial fatty acid β-oxidation in the liver. This implies that any abnormality in the expression of these genes can adversely affect hepatic fatty acid oxidation. Thus, increased plasma fatty acids causes the activation of PPARα leading to increased peroxisomal proliferation, and expression of genes encoding for enzymes involved in fatty acid oxidation.[23]

Hepatic Fatty Acid Synthesis

Depending on the metabolic state, an alternative fate of fatty acids taken up by the liver is their conversion via esterification to triacylglycerol (TAG) that is stored as cytosolic lipid droplets, or incorporated into membrane structures of the hepatocytes. The TAG can be utilized by the liver to synthesize very low–density lipoprotein (VLDL) that is secreted into the bloodstream.

However, fatty acids can also be produced in the cytosol of the hepatocytes from dietary carbohydrates, and amino acids, when consumed in excess, through acetyl-CoA in a process called *de novo* lipogenesis (Figure 1.3). In this process, acetyl-CoA provides the carbon atoms, while NADPH and ATP supply reducing equivalents and energy, respectively. Acetyl-CoA can be synthesized in the mitochondria from the oxidation of pyruvate (e.g., from glycolysis), the catabolism of carbon skeleton of certain amino acids (e.g., lysine, leucine, isoleucine), and the degradation of ketone bodies. However, the inner mitochondrial membrane is not permeable to acetyl-CoA, therefore, a bypass arrangement is made for the transfer of acetyl-CoA from the mitochondria to the cytosol. It condenses with oxaloacetate catalyzed by the enzyme citrate synthase to form citrate, which moves freely across the inner mitochondrial membrane via citrate carrier into the cytosol.[24] In the cytosol, the citrate is cleaved by ATP:citrate lyase to liberate acetyl-CoA and oxaloacetate. The cytosolic acetyl-CoA is carboxylated by the enzyme acetyl-CoA carboxylase to form malonyl CoA. Fatty acids are, then, synthesized from the intermediates of acetyl-CoA and malonyl CoA by successive addition of two-carbon units catalyzed by fatty acid synthase (FAS) complex, an enzyme complex with seven different functional activities in a single polypeptide chain.[7] The fatty acids synthesized may combine with glyceraldehydes 3-phosphate (an intermediate found in the glycolytic pathway) to form TAG and phospholipids (i.e., glycerolipids) (Figure 1.3).

Regulation of Hepatic Fatty Acid Synthesis

Acetyl-CoA carboxylase (ACC), the key enzyme implicated in the regulation of fatty acid synthesis, controls the committed step in fatty acid synthesis. ACC is inactive when phosphorylated by the enzyme AMP-activated protein kinase (inhibited by ATP and activated by AMP), and acetyl-CoA carboxylase is activated by dephosphorylation via the enzyme protein phosphatase 2A (PP2A). Under low cellular energy (high AMP–ATP ratio), the protein kinase is activated that then inactivates acetyl-CoA carboxylase, and fatty acids synthesis is, therefore, inhibited.

ACC is also regulated hormonally. Glucagon, epinephrine, and norepinephrine cause inactivation of the enzyme by cyclic AMP-dependent phosphorylation. However, insulin dephosphorylates, and activates the enzyme. Thus, after intake of a diet rich in carbohydrates, insulin secretion is increased, which in turn stimulates ACC leading to increased synthesis of malonyl CoA (an inhibitor of fatty acids oxidation) (Figure 1.2), which promotes subsequently fatty acid synthesis and storage. In energy-requiring conditions (such as, fasting), glucagon and epinephrine inhibit fatty acids synthesis by blocking the activity of protein phosphatase 2A leading to inactivation of ACC. This means that the activity of the lipogenic pathway is dependent strongly upon the nutritional conditions.

Under longer-term regulation, *de novo* lipogenesis is regulated transcriptionally by both glucose, and insulin signaling pathways. This happens in response to dietary carbohydrates and both glycolytic and lipogenic gene expressions are synergistically induced by both

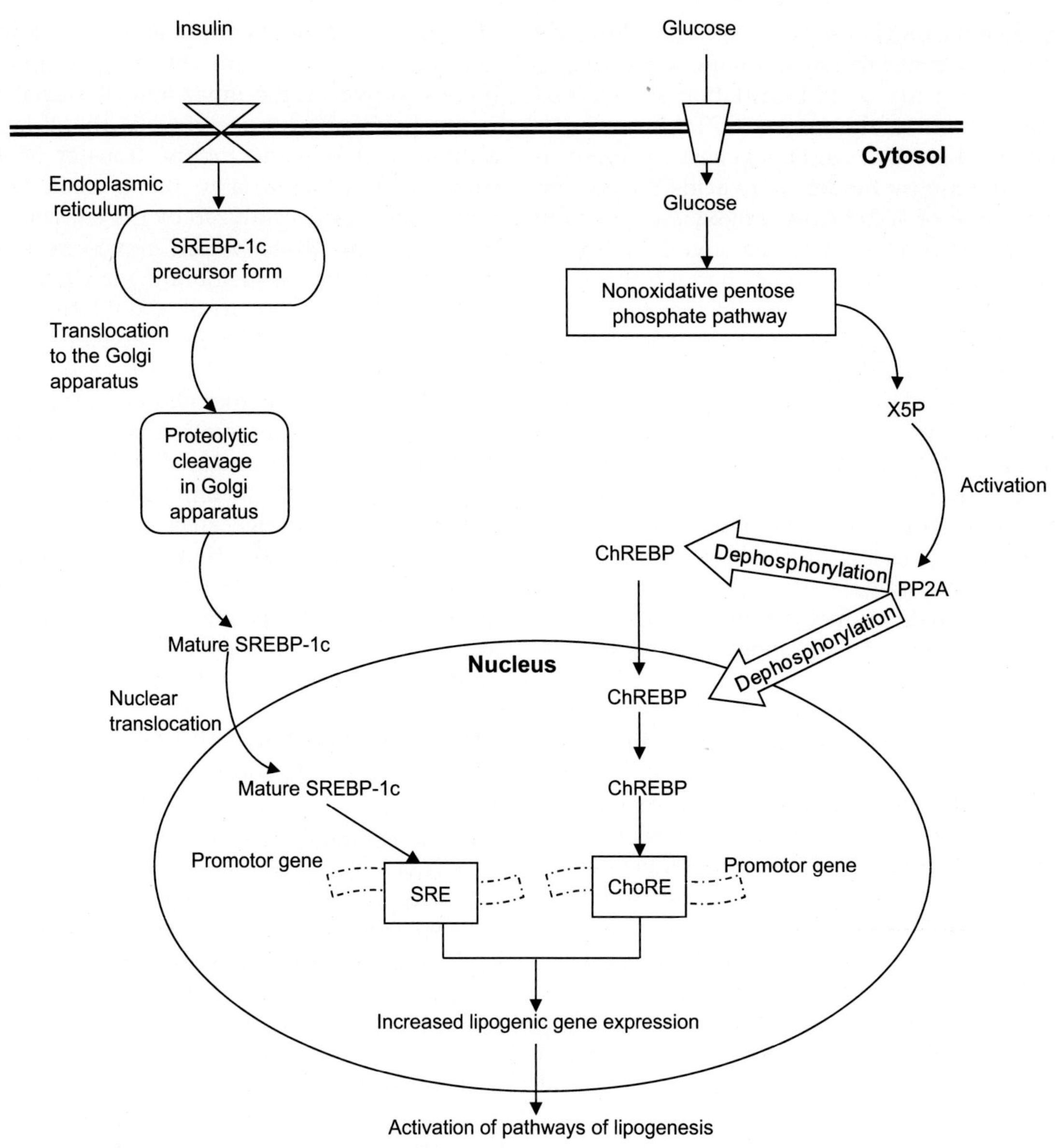

FIGURE 1.4 Activation of transcription factors ChREBP and SREBP-1c. During high concentration of glucose, glucose stimulates the enzyme protein phosphatase 2A (PP2A) via xylulose 5-phosphate (X5P), an intermediate of the nonoxidative pentose phosphate pathway, to activate ChREBP via double dephosphorylation (i.e., before and after entry into the nucleus). Activated ChREBP binds to the carbohydrate responsive element (ChoRE) binding site of the promoter gene of the target lipogenic genes causing transcriptional activation of these lipogenic genes, and ultimately lead to increased gene expression. On the other hand, insulin acts by activating the proteolytic maturation of SREBP-1c in both the endoplasmic reticulum, and Golgi apparatus to form mature SREBP-1c that translocates into the nucleus, and attaches to the sterol regulatory element (SRE) binding site of its promoter gene causing an increase in lipogenic gene expression. ChREBP, carbohydrate responsive element-binding protein; SREBP-1c, sterol regulatory element-binding protein-1c.

high concentrations of insulin and glucose, and not insulin alone as previously known[11] (Figure 1.4). Insulin activates SREBP-1c, an essential transcription factor involved in the regulation of several lipogenic genes.[25] SREBP-1c is synthesized as an inactive precursor bound to the membranes of the ER, and undergoes posttranslational modification (via double proteolytic cleavage to liberate the N-terminal domain in the Golgi apparatus) to produce a transcriptional active mature nuclear form.[14,18] Insulin, being one of the most potent activators of SREBP-1c, promotes the expression, and posttranslational maturation of SREBP-1c that then stimulates the expression of lipogenic genes, such as fatty acid synthase, stearoyl-CoA desaturase-1, and ACC.[14,25,26] However, it has been detected that the SREBP-1c activity alone cannot entirely explain the stimulation of lipogenic gene expression by dietary carbohydrate since SREBP-1c gene deletion in mice only resulted in a 50% decline in fatty acid synthesis.[14] The glucose-signaling transcription factor has been recently discovered as carbohydrate-responsive element-binding protein (ChREBP).[10,27] ChREBP is identified as a large protein (approximately 100 kDa) consisting of 864 amino acids with several domains, notably, a polyproline domain, a nuclear localization signal domain (near the N-terminus), a leucine-zipper-like domain, and a basic loop–helix–leucine-zipper domain.[4,11,28] ChREBP also has several potential phosphorylation sites for PKA and AMP-activated protein kinase (AMPK).[27] Thus, at low glucose concentrations or during starvation, ChREBP is phosphorylated on serine (ser)196, ser626 and threonine666 by PKA, on Ser568 by AMPK, and then, translocated from the nucleus into the cytosol.[29] However, during high glucose concentrations, glucose acts through xylulose 5-phosphate (X5P), an intermediate of the pentose phosphate pathway, which activates protein phosphatase 2A (PP2A). PP2A in turn activates ChREBP via first dephosphorylation, which allows its translocation into the nucleus followed by another dephosphorylation by the same enzyme PP2A. The activated ChREBP binds to the carbohydrate responsive element (ChoRE) present in the promoter regions of the target genes – glycolytic and lipogenic genes – to cause increased expression of these genes (e.g., liver pyruvate kinase, ACC, and FAS genes),[28,29] leading ultimately to increased lipogenesis (Figure 1.4). This means that during excess dietary carbohydrate (and amino acids) intake, the liver converts glucose to fat (i.e., when liver glycogen storage is full to capacity), to stabilize blood glucose concentration, and prevent against hyperglycemia.[4]

Dietary Regulation of Hepatic Lipid Metabolism

It is widely recognized that the composition of fatty acids in a diet is an important factor that can influence the hepatic lipid metabolism. Dietary fatty acids can control various metabolic pathways involved in lipid metabolism, essentially, through the regulation of gene transcription of specific enzymes.[24] With the importance attached to the key role played by the liver in lipid metabolism, the evidence indicates that dietary fats and their oxidized metabolites may not only influence the pathogenesis of liver diseases, but may also prevent and/or reverse disease manifestations.[24,30] The intake of diet containing short-chain and medium-chain fatty acids (SCFAs and MCFAs) results in an increased production of energy, which could be attributed to the direct movement of SCFAs and MCFAs from the intestine to the liver through the hepatic portal vein. SCFAs are quickly oxidized due to the easy transport of their fatty acyl-CoAs by diffusion into the mitochondria.[4] On the other hand, long-chain fatty acids (LCFAs), for example, polyunsaturated fatty acids (PUFAs), especially n-6 and n-3 series, are oxidized preferentially for energy than saturated fatty acids,[31] and are potent inhibitors of hepatic lipogenesis.[24] PUFAs have been found to decrease significantly the activity and the expression of the mitochondrial citrate carrier (transporter of citrate outside the mitochondria for cytosolic fatty acid biosynthesis), and cytosolic lipogenic enzymes.[24] However, the intake of a diet enriched in monounsaturated fatty acids (MUFA) or SFA did not cause any substantial effect on mitochondrial citrate carrier activity and expression, and as a result, did not affect *de novo* fatty acid synthesis.[24,32]

HEPATIC AMINO ACID METABOLISM

The liver is a key organ involved in amino acid metabolism. It is the first organ to receive intestinally absorbed dietary amino acids after digestion via hepatic portal vein. It is also the only organ with a complete urea-cycle synthesizing enzymes to eliminate the nitrogen from amino acids to form urea (a nontoxic compound, mainly excreted by the kidneys) during amino acid catabolism. It is capable of synthesizing certain amino acids and proteins – both proteins required within the liver itself, and proteins exported into circulation such as albumin, acute phase proteins, coagulation factors, etc. It utilizes amino acids to synthesize certain essential substances (e.g., glutathione, taurine). Finally, the liver is responsible for the production of glucose via gluconeogenesis from muscle-derived amino acids during starvation, and in certain conditions, such as, diabetes mellitus.

Hepatic Amino Acid Catabolism

In relation to energy metabolism, almost all amino acid catabolism takes place predominantly in the liver, with the exception of the branched chain amino acids (catabolized initially in the skeletal muscles), and amino acid catabolism, in general, contributes to about 10–15% of body daily energy requirements under normal physiological condition. About half of the liver energy requirement is met from hepatic amino acid catabolism.[6] However, amino acids are not stored, especially in mammals, simply for energy production. Thus, all proteins have their biological functions in the body, and the rate of protein turnover is maintained under normal conditions.[6] However, due to the ability of amino acids to be converted into glucose unlike fatty acids, they provide energy needs in drastic conditions, such as, starvation when the circulating blood glucose levels must be preserved in the absence of dietary supply of carbohydrate, and glycogen depletion, for the sustenance of obligatory glycolytic tissues, such as, the brain, erythrocytes, etc.

The intracellular concentration of amino acids is considerably higher than extracellular concentration. This means that active transporters are required to move amino acids, against the concentration gradient, into the

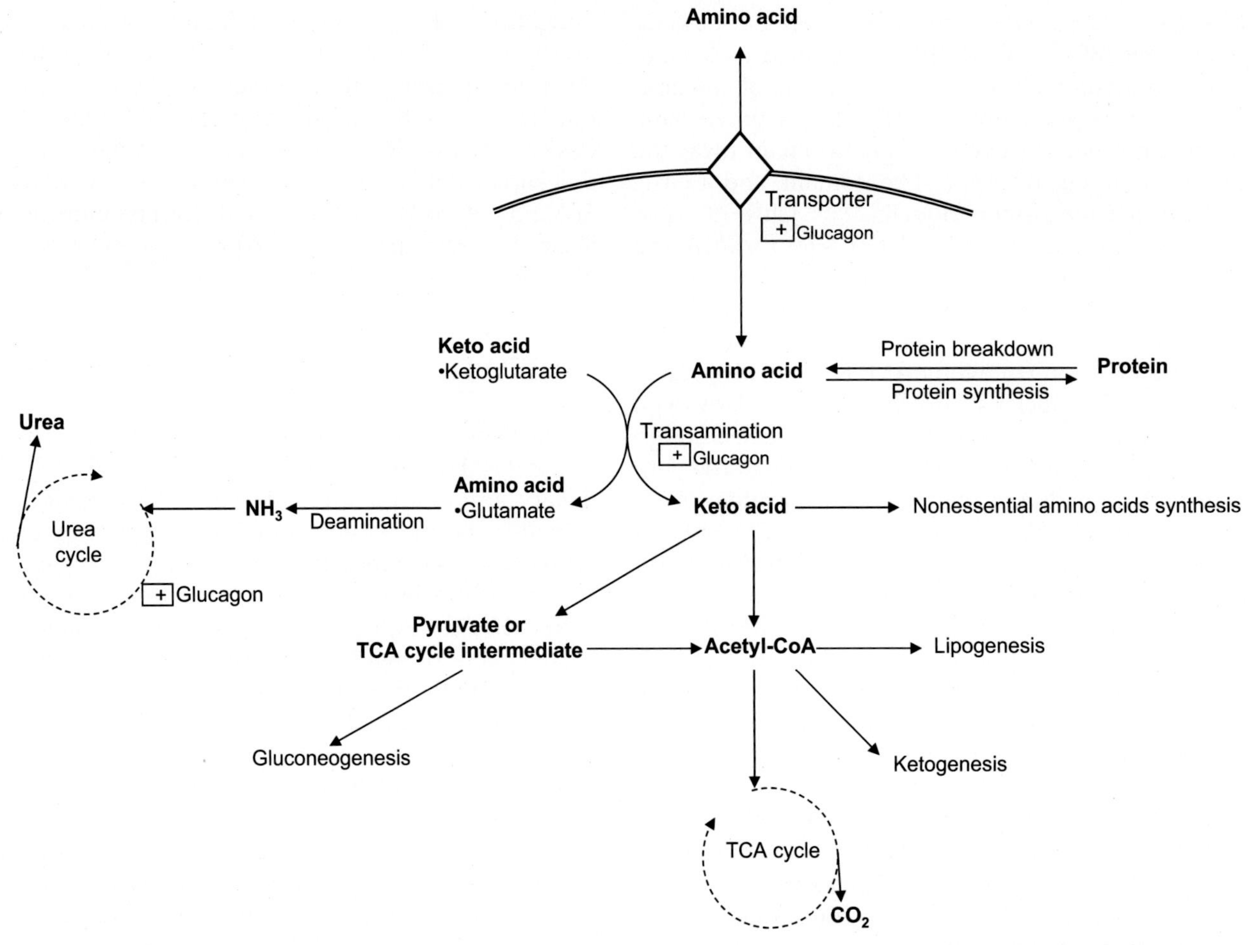

FIGURE 1.5 Overview of hepatic amino acid metabolism. Amino acids enter the hepatocytes via amino acid transporters. They can undergo protein synthesis or be catabolized via initial transamination reaction to form their respective keto acids. Depending upon the metabolic state of the body and the type of amino acid involved, these keto acids can be oxidized in the TCA cycle, or be utilized for gluconeogenesis (e.g., pyruvate during starvation), ketogenesis (e.g., lysine during starvation), lipogenesis (e.g., during excess dietary protein intake), or the synthesis of nonessential amino acids. Glutamate, acting as the central collection of amino group, undergoes oxidative deamination to remove the amino group to form ammonia (NH_2), which is converted to urea via the urea cycle.

hepatocytes. There are sodium-dependent and sodium-independent transporters (normally referred to as system; e.g., systems *L*, *A*, *ASC*, $N^{(m)}$, etc.) involved in amino acid transport. After a meal (containing proteins), almost all the absorbed dietary amino acids (with the exception of branched chain amino acids) in hepatic portal vein are taken up, and metabolized by the hepatocytes. Also, hepatic protein degradation is a source of amino acids within the hepatocytes.

An important general reaction involved in the amino acid catabolism in the hepatocytes is the transamination reaction, the transfer of the amino group from an amino acid to a keto acid with the amino acid, which lost the amino group forming a corresponding keto acid, and the keto acid that accepted the amino group forms a corresponding amino acid. The hepatocytes are relatively rich in the enzymes that catalyze transamination reactions known as aminotransferases, (e.g., alanine aminotransferase and aspartate aminotransferase). Depending upon the metabolic state, the keto acid formed can either enter an anabolic or catabolic pathway (Figure 1.5). For example, during starvation or after an overnight fast, alanine and glutamine are the predominant amino acids released from muscles, and other peripheral tissues into circulation. Of these two amino acids, alanine, and not glutamine (a good substrate for kidney and intestinal mucosal cells), is a good substrate for hepatic metabolism. Alanine is catabolized in the liver to form glucose via initial transamination reaction to produce the keto acid, pyruvate, followed by the conversion of pyruvate via gluconeogenesis to produce glucose. The glucose synthesized is released by the liver into circulation, which may be utilized by the skeletal muscles as fuel, through glycolysis producing pyruvate, which in turn may be transaminated to form alanine (i.e., glucose–alanine cycle). This cycle that occurs between the liver and the skeletal muscles forms a link between glucose and the amino acid metabolism. Other glucogenic amino acids, such as, aspartate (keto acid, oxaloacetate), glutamate (keto acid, α-ketoglutarate), etc., can all, in principle, be converted to glucose via gluconeogenesis, so that the body's protein reserves – especially, the bulk of skeletal muscle proteins – could help in sustaining glucose production for considerable time.

Regulation of Amino Acid Metabolism

In the short-term, liver amino acid catabolism is regulated primarily by substrate availability. During feeding state, the substrate supply is mainly from dietary amino acids, but in starvation, mainly from the breakdown of body proteins that is hormonally controlled.

In long-term, both substrate supply and hormonal control mechanisms are involved. The two main hormones that play an important role in the control of amino acid catabolism are glucagon, and cortisol. These hormones act by stimulating the main enzymes involved in amino acid catabolism and urea synthesis. Also, glucagon has been detected to stimulate amino acid transporters to enhance the uptake of amino acids, particularly, alanine. In the case of substrate supply, when the dietary protein intake is low, the expression of hepatic enzymes involved in amino acid catabolism is suppressed. However, in the presence of high dietary protein intake, there is increased expression of these enzymes. Also, substrate (amino acids) availability is proposed to increase the expression of amino acid transporters to enhance amino acids uptake. Thus, not much has been achieved in terms of molecular regulation of amino acids metabolism; an area that needs more investigation if glucose and fatty acids have been detected to regulate gene expression to affect hepatic energy metabolism.

CONCLUSIONS

The liver plays essential role in the nutrient metabolism, more particularly, in respect of energy metabolism. It stores glucose as glycogen (via glycogenesis), and converts glucose into fatty acids (via lipogenesis), under high blood glucose concentration (e.g., after meals) stimulated by insulin, and by both insulin (via SREBP-1c) and glucose (via ChREBP), respectively. During low blood glucose concentration (e.g., fasting/starvation), the liver releases glucose from glycogen (via glycogenolysis), and can synthesize glucose from noncarbohydrate sources such as lactate, amino acids, and glycerol via gluconeogenesis to restore blood glucose concentration under the influence of glucagon, and other stress hormones. Reliance of amino acids as energy source becomes prominent during starvation facilitated by glucagon. In the liver, fatty acids can either be oxidized to generate energy or stored as triacylglycerol, a major reserved energy source. Fatty acid oxidation is regulated by fatty acid availability through activation of PPARα that increases the expression of genes encoding for enzymes involved in fatty acid oxidation, possibly under the influence of glucagon, but inhibited by insulin.

Key Facts

Glucose

- Glucose is an organic compound made up of hydrogen, carbon and oxygen.
- It was first discovered by Andreas Marggraf, a German Scientist, in 1747.
- It is classified as a monosaccharide, and is also referred to as dextrose.
- Glucose is the main energy source for the human body.
- Excess glucose in blood is stored in the body, mainly in the liver, and in the muscles as glycogen.

Liver

- The liver is the second largest organ (to the skin) in the human body, and located under the diaphragm on the right side of the body.
- It is the only organ in the body that has the ability to regenerate after injury.
- The liver is divided into three zones, each with a specific function and susceptibility to particular illness. Blood with nutrients and oxygen enters the first zone, travels to the second zone, and third zone, before exiting the liver.
- It plays numerous metabolic roles including nutrients' metabolism, detoxification, bile synthesis, etc., in addition to the storage of nutrients, such as, iron, glycogen, copper, vitamin A, etc.
- It can be adversely affected by certain drugs (e.g., high dose of acetaminophen), excessive alcohol and fat intake, excess iron and copper deposition, viral infections, etc.

Summary Points

- The liver is involved centrally in the energy metabolism from the major food nutrients – carbohydrate, lipids, and proteins.
- The liver oxidizes glucose to produce energy (via glycolysis, TCA cycle, and oxidative phosphorylation), and stores glucose as glycogen (via glycogenesis) when blood glucose concentration is high (after carbohydrate-containing meals) under the influence of insulin. The liver releases glucose (via glycogenolysis) through glucagon to buffer the blood glucose concentration when low; this is a process inhibited by insulin.

- When liver glycogen is depleted (during starvation), the liver produces glucose from noncarbohydrate sources (such as, lactate and glycerol) via gluconeogenesis under the influence of glucagon and substrate availability.
- The liver can either oxidize fatty acids to produce energy via β-oxidation enhanced by fatty acid availability, and activated transcriptional factor PPARα but inhibited by insulin; or store fatty acids as triacylglycerol under the influence of insulin. It can also synthesize fatty acids from glucose via *de novo* lipogenesis activated by glucose via ChREBP, and insulin via SREBP-1c during excess carbohydrate.
- Amino acids, not utilized purposefully to form proteins for energy production since proteins have their biological functions, can be catabolized by the liver under drastic conditions, such as starvation (no dietary carbohydrate supply and glycogen depletion) to produce energy via gluconeogenesis with the help of glucagon.

Abbreviations

ACC	Acetyl-CoA carboxylase
ACS	Acyl-CoA synthase
AMP	Adenosine 5′-monophosphate
ATP	Adenosine 5′-triphosphate
ChoRE	Carbohydrate responsive element
ChREBP	Carbohydrate responsive element-binding protein
CPT-1	Carnitine-palmitoyl transferase-1
CPT-2	Carnitine-palmitoyl transferase-2
F6P	Fructose 6-phosphate
$FADH_2$	Flavin adenine dinucleotide (reduced form)
FAS	Fatty acid synthase
FFA	Free fatty acids
G6P	Glucose 6-phosphate
G6Pase	Glucose 6-phosphatase
G6PD	Glucose 6-phosphate dehydrogenase
GK	Glucokinase
GKR	Glucokinase regulatory protein
GLUT2	Glucose transporter 2
Glyc 3-P	Glyceraldehydes 3-phosphate
GP	Glycogen phosphorylase
GS	Glycogen synthase
GSK3	Glycogen synthase kinase 3
HMG-CoA	3-Hydroxy-3-methylglutaryl-CoA
K_m	Michaelis constant
LCFAs	Long-chain fatty acids
MCFAs	Medium-chain fatty acids
NADH	Nicotinamide adenine dinucleotide (reduced form)
$NADP^+$, NADPH	Nicotinamide adenine dinucleotide phosphate ($^+$, oxidized form; H, reduced form)
NEFAs	Nonesterified fatty acids
OAA	Oxaloacetate
PDH	Pyruvate dehydrogenase
PEP	Phosphoenol pyruvate
PK	Pyruvate kinase
PP2A	Protein phosphatase 2A
PPARs	Peroxisomal proliferator-activated receptors
PUFAs	Polyunsaturated fatty acids
Ser	Serine
SRE	Sterol regulatory element
SREBP-1c	Sterol regulatory element-binding protein-1c
TAG	Triacylglycerol
TCA cycle	Tricarboxylic acid cycle
VLDL	Very low–density lipoprotein
X5P	Xylulose 5-phosphate

References

1. Cullen KS, Al-oanzi ZH, Harte FPMO, Agius L, Arden C. Glucagon induces translocation of glucokinase from the cytoplasm to the nucleus of hepatocytes by transfer between 6-phosphofructo 2-kinase/fructose 2, 6-bisphosphatase-2 and the glucokinase regulatory protein. *Biochim Biophys Acta* 2014;**1843**:1123–34.
2. Agius L. Glucokinase and molecular aspects of liver glycogen metabolism. *Biochem J* 2008;**414**:1–18.
3. Nuttall FQ, Ngo A, Gannon MC. Regulation of hepatic glucose production and the role of gluconeogenesis in humans: is the rate of gluconeogenesis constant? *Diab Metab Res Rev* 2008;**24**:438–58.
4. Gyamfi D, Patel V. Liver metabolism: biochemical and molecular regulations. In: Preedy VR, Lakshman R, Srirajaskanthan R, Watson RR, editors. *Nutrition, diet therapy, and the liver*. USA: CRC Press; 2009. p. 3–15.
5. Bechmann LP, Hannivoort RA, Gerken G, Hotamisligil GS, Trauner M, Canbay A. The interaction of hepatic lipid and glucose metabolism in liver diseases. *J Hepatol* 2012;**56**:952–64.
6. Frayn KN. *Metabolic regulation: a human perspective*. 3rd ed. UK: Wiley-Blackwell; 2010.
7. Satyanarayana U, Chakrapani U. *Biochemistry*. 3rd ed. Kolkata: Books and Allied (P) Ltd; 2006.
8. Rui L. Energy metabolism in the liver. *Compr Physiol* 2014;**4**:177–97.
9. König M, Bulik S, Holzhütter H-G. Quantifying the contribution of the liver to glucose homeostasis: a detailed kinetic model of human hepatic glucose metabolism. *PLoS Comput Biol* 2012;**8**:e1002577.
10. Dentin R, Girard J, Postic C. Carbohydrate responsive element binding protein (ChREBP) and sterol regulatory element binding protein-1c (SREBP-1c): two key regulators of glucose metabolism and lipid synthesis in liver. *Biochimie* 2005;**87**:81–6.
11. Denechaud P-D, Dentin R, Girard J, Postic C. Role of ChREBP in hepatic steatosis and insulin resistance. *FEBS Lett* 2008;**582**:68–73.
12. Stoeckman AK, Towle HC. The role of SREBP-1c in nutritional regulation of lipogenic enzyme gene expression. *J Biol Chem* 2002;**277**:27029–35.
13. Frayn KN, Arner P, Yki-Järvinen H. Fatty acid metabolism in adipose tissue, muscle and liver in health and disease. *Essays Biochem* 2006;**42**:89–103.
14. Foufelle F, Ferr P. Hepatic steatosis: a role for *de novo* lipogenesis and the transcription factor SREBP-1c. *Diab Obes Metab* 2010;**12**:83–92.
15. Nguyen P, Leray V, Diez M, et al. Liver lipid metabolism. *J Anim Physiol Anim Nutr (Berl)* 2008;**92**:272–83.
16. Sampath H, Ntambi JM. Polyunsaturated fatty acid regulation of genes of lipid metabolism. *Annu Rev Nutr* 2005;**25**:317–40.
17. Czaja MJ. Autophagy in health and disease. 2. Regulation of lipid metabolism and storage by autophagy: pathophysiological implications. *Am J Physiol Cell Physiol* 2010;**298**:C973–8.
18. Canbay A, Bechmann L, Gerken G. Lipid metabolism in the liver. *Z Gastroenterol* 2007;**45**:35–41.
19. Cahill GF, Veech RL. Ketoacids? Good medicine? *Trans Am Clin Climatol Assoc* 2003;**114**:149–61 discussion 162–163.
20. Veech RL, Chance B, Kashiwaya Y, Lardy HA, Cahill GF. Ketone bodies, potential therapeutic uses. *IUBMB Life* 2001;**51**:241–7.

21. Longuet C, Sinclair EM, Maida A, et al. The glucagon receptor is required for the adaptive metabolic response to fasting. *Cell Metab* 2008;**8**:359–71.
22. Pawlak M, Lefebvre P, Staels B. Molecular mechanism of PPARα action and its impact on lipid metabolism, inflammation and fibrosis in non-alcoholic fatty liver disease. *J Hepatol* 2015;**62**:720–33.
23. Reddy JK, Hashimoto T. Peroxisomal beta-oxidation and peroxisome proliferator-activated receptor alpha: an adaptive metabolic system. *Annu Rev Nutr* 2001;**21**:193–230.
24. Ferramosca A, Zara V. Modulation of hepatic steatosis by dietary fatty acids. *World J Gastroenterol* 2014;**20**:1746–55.
25. Hegarty BD, Bobard A, Hainault I, Ferré P, Bossard P, Foufelle F. Distinct roles of insulin and liver X receptor in the induction and cleavage of sterol regulatory element-binding protein-1c. *Proc Natl Acad Sci USA* 2005;**102**:791–6.
26. Foufelle F, Ferré P. New perspectives in the regulation of hepatic glycolytic and lipogenic genes by insulin and glucose: a role for the transcription factor sterol regulatory element binding protein-1c. *Biochem J* 2002;**366**:377–91.
27. Dentin R, Denechaud P-D, Benhamed F, Girard J, Postic C. Hepatic gene regulation by glucose and polyunsaturated fatty acids: a role for ChREBP. *J Nutr* 2006;**136**:1145–9.
28. Postic C, Dentin R, Denechaud P-D, Girard J. ChREBP, a transcriptional regulator of glucose and lipid metabolism. *Annu Rev Nutr* 2007;**27**:179–92.
29. Xu X, So J-S, Park J-G, Lee A-H. Transcriptional control of hepatic lipid metabolism by SREBP and ChREBP. *Semin Liver Dis* 2013;**33**:301–11.
30. Masterton GS, Plevris JN, Hayes PC. Review article: omega-3 fatty acids – a promising novel therapy for non-alcoholic fatty liver disease. *Aliment Pharmacol Ther* 2010;**31**:679–92.
31. Jones P, Kubow S. Lipids, sterols, and their metabolites. In: Shils M, Shike M, Ross A, Caballero B, Cousins RJ, editors. *Modern nutrition in health and disease*. Philadelphia: Lippincott Williams and Wilkins; 2006.
32. Ferramosca A, Conte A, Burri L, et al. A krill oil supplemented diet suppresses hepatic steatosis in high-fat fed rats. *PLoS One* 2012;**7**:e38797.

CHAPTER

2

Alcohol Metabolism: General Aspects

Reem Ghazali, MSc,***, *Vinood B. Patel, PhD***

*Department of Clinical Biochemistry, King Abdul Aziz University, Jeddah, Saudi Arabia

**Department of Biomedical Sciences, Faculty of Science & Technology, University of Westminster, London, UK

INTRODUCTION

Heavy and prolonged alcohol consumption plays a major role in the prevalence of alcoholic liver disease (ALD) and many other diseases, like some malignant neoplasms and heart diseases in Western countries.[1] About 76 million, out of 2 billion people worldwide, who regularly consume alcohol with an average of 13 g/day are diagnosed with disorders in alcohol use. In England, a high average consumption of alcohol is associated highly with a range of 200 conditions, acute and chronic, such as cardiovascular disease and cancers, with an estimated cost of £ 2.7 billion of the direct annual health care expenditure by the National Health Services in the United Kingdom, which soared up to £3.5 between 2011 and 2012.[2–5]

Beer, wine, as well as spirits, are the most commonly consumed types of alcohol, in ascending order. Regardless of the type of alcohol drink, ALD is considered as the prime cause of chronic liver disease, and has a central causative role leading to increased mortality rates worldwide, due to its ability to develop into fibrosis and cirrhosis. In 2004, around 3.8% of all deaths were related to alcohol consumption. Yet, according to a report published by the National Institute of Alcohol Abuse and Alcoholism, liver cirrhosis was ranked as the twelfth major cause of mortality in the United States, with a total deaths number of 29,925 in 2007, of which 48% were alcohol related.[2,3]

It seems that alcohol intake elevates with the urbanization and industrialization of the country. As stated by the World Health Organization (WHO), the population of Eastern Europe are the highest alcohol consumers in the world, and the average of pure alcohol intake per capita is 12.2 L/year, whereas, the eastern region of the Mediterranean, and the Middle East showed the lowest percentage with an average of 0.7 L/year/capita.[6]

ALCOHOL METABOLISM

Alcohol is an organic compound that can be metabolized in the body by two different pathways: oxidative and nonoxidative. Oxidative alcohol metabolism occurs mainly in the liver, in the cytosol of the hepatocyte. In this pathway, alcohol is metabolized by an enzyme called alcohol dehydrogenase (ADH) into acetaldehyde, which is considered a highly toxic and reactive molecule, and is accompanied by nicotinamid adenine dinucliotide (NAD) reduction into NAD hydrogenase (NADH). Other enzymes are also involved in the oxidation of alcohol, such as the microsomal ethanol oxidizing system (MEOS), known as cytochrome P450 isoenzymes. Cytochrome P450 comprises a group of enzymes that includes CYP2E1, 3A4, and 1A2, and is predominantly found in the endoplasmic reticulum, and contributes to alcohol metabolism. Generally, with chronic alcohol consumption and/or when the concentration of alcohol is elevated, CYP2E1 is induced, resulting in oxidizing the excessive amount of alcohol to acetaldehyde, as well as generating reactive oxygen species (ROS) that contributes to endoplasmic reticulum (ER) stress.[7,8] Catalase, another enzyme located in the peroxisomes, has a minor role in metabolizing ethanol to acetaldehyde. Finally, acetaldehyde oxidation occurs in the mitochondria, by the enzyme aldehyde dehydrogenase (ALDH2) that has the highest affinity to acetaldehyde among the 19 ALDH mammalian genes, producing acetate and another NADH molecule that in turn can be oxidized by the mitochondrial electron transport chain (ETC).[3,9,10] On the other hand, acetate is released into the bloodstream, and can thus be metabolized further by peripheral tissues to form carbon dioxide, water or fatty acids.[11]

Nonoxidative ethanol metabolism is mediated by the esterification of ethanol and fatty acids, or fatty acyl-CoA into fatty acid ethyl esters (FAEE). Nonoxidative ethanol

Molecular Aspects of Alcohol and Nutrition. http://dx.doi.org/10.1016/B978-0-12-800773-0.00002-1

TABLE 2.1 Key Molecular Alterations Following Alcohol Metabolism

Metabolite or molecule	Mechanism	Effects
Acetaldehyde	Binding to membrane protein, DNA or lipid	Changing the fluidity of the membrane by changing the cellular homeostasis. Inducing ER stress causing steatohepatitis
FAEE	Causing oxidative phosphorylation uncoupling ATP synthesis reduction	Destabilization of the cellular membrane. Increasing the permeability of the membrane
NADH/NAD ratio	ETC disruption resulting in electron leakage and ATP synthesis reduction	Increasing the cellular membrane permeability
ROS	Production of reactive lipid molecules	Damage the membrane of the cell and changing the membrane stabilization

ER, endoplamic reticulum; NADH/NAD, nicotinamide adenine dinucleotide (reduced/oxidized form); FAEE, fatty acid ethyl ester; ROS, reactive oxygen species; ETC, electron transport chain; ATP, adinosin triphosphate.

metabolism only oxidizes about 1% of ethanol by FAEE synthase enzymes found in the cytosol and microsomes of the liver, pancreas, heart, and brain. These organs can be easily damaged by ethanol and its nonoxidative (FAEE) metabolite, as well as oxidative metabolites (acetaldehyde), and generated ROS. FAEE can destabilize the cellular and lysosomal membranes by increasing the fluidity of the membrane, consequently leading to mitochondrial oxidative phosphorylation uncoupling in the ETC, followed by a reduction in the adenosine triphosphate (ATP) produced (Table 2.1).[3]

Alcohol metabolism can exert a significant fluctuation and elevation in the ratio of NADH/NAD through the production of acetaldehyde and acetic acid, causing an extravagant reduction in the level of the NAD. Under normal conditions, the approximate ratios of NADH/NAD in the hepatic cytosol and mitochondria are 700:1 and 7–8:1, respectively. However, chronic alcohol consumption leads to an elevation of this ratio, and disturbs the metabolism of several pathways, such as carbohydrate, uric acid, lipids, and protein metabolism. Elevated NADH/NAD ratios also influence the modulation of gene expression, cell death, and mitochondrial permeability transition opening (MPT); the latter leads to mitochondrial swelling, loss of the mitochondrial membrane potential, and subsequently reduced ATP synthesis (Table 2.1). Acute or chronic alcohol consumption can increase the availability of NADH that drives the mitochondrial ETC, leading to increased electron leakage at complexes I and III, and subsequent elevation in ROS formation.[3,10,12]

ROS include molecules such as hydrogen peroxide, hydroxyl radicals, hypochlorite and superoxide radicals, and have the oxidative ability to damage DNA, as well as the membrane of the cell. The latter, induced through the production of the lipid diffusible molecules known as reactive aldehydes, such as malondialdehyde and 4-hydroxynonenal, in the peroxidation process of the lipid (Table 2.1).[13] The formation of ROS, mediated by the CYP2E1 isoenzyme, is also generated by various reactions, such as lipid metabolism and NADPH oxidase; however, the majority is via the mitochondrial ETC.[3,10,14] Epigenetic changes caused by oxidative alcohol metabolism can be induced by several mechanisms, such as high NADH/NAD ratio, and via the production of acetate, acetaldehyde, and ROS.[10] However, due to the electrophilic nature of acetaldehyde, it can covalently bind to molecules, such as protein, lipids, and DNA, changing the homeostasis of the cell, by altering protein structure and promoting DNA mutations, as well as inducing ER stress (Table 2.1).[8,15]

THE EPIGENETIC CHANGES OF ALCOHOL

Chronic alcohol consumption has a central causative role in altering epigenetic mechanisms. These alterations include the modification of histones, such as the acetylation or phosphorylation of histones and DNA methylation changes; that in turn contributes to diseases, such as alcoholic liver disease and cancer.[16,17] Ethanol affects negatively methionine metabolism, and reduces the availability of *S*-adenosylmethionine (SAMe) that is considered the major methyl donor in this pathway. SAMe is formed mainly in the liver by the methylation of homocysteine into methionine, and ultimately, into SAMe. The latter has a key role in inducing histone and DNA methylation; thus, a reduction in the level of hepatic SAMe results in lowering the methylation of DNA and histones, thereby causing epigenetic changes.[2] Ethanol also has a direct or indirect effect on the regulation or expression of many transcriptional factors. For instance, sterol regulatory element-binding protein 1c (SREBP-1c) is up-regulated by high alcohol consumption levels, resulting in increased synthesis of fat in the hepatocytes. Thus, through its metabolites (acetaldehyde), ethanol can directly increase the gene transcription of SREBP-1c, or can activate indirectly some factors, such as tumor

TABLE 2.2 Important Genes Involved in ALD Development

Gene	Definition	Function	Ethanol's effects
SREBP-1c	Sterol regulatory element-binding protein 1c	Stimulates fat synthesis	Upregulating the expression of SREBP-1c, and increasing the synthesis of lipids
PPAR-α	Peroxisome proliferator-activated receptor	Inducing fat oxidation	Downregulating PPAR-α expression, and inhibiting lipids oxidation
TGF-β	Transforming growth factor	Inducing fibrogenesis	Stimulates glycogen synthesis by inducing TGF-β, and small mother against decapentaplegic (SMAD) signaling, as well as activating TGF-β II receptor
PNPLA3	Patatin-like phospholipase domain-containing 3	Regulating lipid homeostasis and accelerating fibrogenesis	Upregulating *PNPLA3* in liver fibrosis
OPN	Osteopontin	Profibrogeneic protein. Activating the HSC and ECM component deposition	Activation of OPN. Induces plasminogen activator liver matrix remodeling
P90RSK	Member of the ribosomal S6 kinase family	Participation in HSC activation and the development of liver fibrosis	P90RSK activation causing EBP-β phosphorylation. Inducing liver fibrosis

HSC, hepatic stellate cells; ECM, extracellular components; EBP-β, transcriptional factor C.

necrosis factor alpha (TNF-α), adenosine, and endocannabinoid, all of which can increase the expression of the SREBP-1c gene, and induce fatty liver (Table 2.2).[2,15]

Other transcriptional genes that can be affected by the chronic consumption of alcohol are the genes encoding enzymes involved in the oxidation of fatty acids, such as peroxisome proliferator-activated receptor (PPAR)-α. This hormone receptor is a gene regulator that can be inactivated directly by acetaldehyde that inhibits the transcriptional activity and DNA-binding ability of PPAR-α, resulting in an accumulation of fatty acids in the hepatocyte (Table 2.2).[2]

There are various genes that are highly expressed in the liver of patients with excessive alcohol consumption, in which some genes are related to hepatitis and fibrosis. Transforming growth factor (TGF)-β is a secreted protein that is encoded by a high gene expression in many tissues. It is involved in alcoholic liver fibrosis, and can be activated by acetaldehyde production that can also induce the expression of TFG-β II receptor, and subsequently, increasing the synthesis of collagen, a major component of the extracellular matrix, as well as up-regulating the expression of collagen α1(I) mRNA (Table 2.2).[15]

Patatin-like phospholipase domain-containing 3 (*PNPLA3*) rs738409 is another gene that has a role in lipid homeostasis, and contributes in accelerating the rate of fibrosis in alcoholic liver disease. PNPLA3 is highly expressed in liver and adipose tissues, and a single polymorphism in the nucleotide of *PNPLA3* gene, with an isoleucine-methionine substitution in the 148 M variant, is also highly correlated with the risk of alcoholic steatohepatitis, severity of cirrhosis, and hepatocellular carcinoma. However, this gene is only highly expressed in alcoholics and nonalcoholic cirrhotic livers, and not in cirrhotic hepatitis C, as it can be mainly stimulated by alcohol and liver accumulated fat (Table 2.2).[18–20]

Osteopontin (OPN) is a gene with various functions. It contributes to the pathogenesis of many conditions involving autoimmunity, inflammation, cancer in many tissues, and fibrosis. It has been found that the level of this protein in the liver is correlated highly with the incidence of fibrosis in ALD. OPN acts as a profibrogenic protein, and promotes the activation of hepatic stellate cells (HSC), and the deposition of ECM components. Its expression along with the level of serum OPN were also markedly elevated in alcoholic hepatitis that reflects its implication in the rise of alcoholic fibrosis and hepatitis (Table 2.2).[15,21] Plasminogen activation by HSC along with remodeling of the ECM have a key role in fibrinolysis. Thus, chronic alcohol consumption activates OPN that subsequently induces plasminogen activators, a process that in turn induces the remodeling of the liver matrix by converting plasminogen into plasmin.[21]

P90RSK is an enzymatic protein known as a member of the ribosomal S6 kinase family. It is involved in various cellular processes, such as cellular proliferation, synthesis of the cytockines and collagens, migration, apoptosis, as well as signaling pathways, and may mediate the repair of damaged tissue caused by chronic liver injury. This protein is a serine/threonine kinase that plays a vital role in the extracellular signal-regulated kinase (ERK), and performs as downstream mediator. Furthermore, p90RSK contributes in the activation of HSC and consequently in the development of liver fibrosis.[22,23] In their study, Morales-Ibanez et al., found that p90RSK is highly up-regulated in the advanced stage of hepatocellular fibrosis in patients with alcoholic hepatitis and other chronic liver injuries, due to its activation in the HSC, leading to the phosphorylation of the transcriptional factor C/EBP-β that has a key role in the synthesis of the collagen and, consequently, causing liver fibrinogensis.[23]

CONCLUSIONS

Alcohol metabolism is a central causative component of risk, associated with many diseases occurring through two different pathways. The alcohol metabolite, acetaldehyde, is a crucial toxic factor that has a major role, along with ethanol, in inducing epigenetic changes via various mechanisms, such as elevated NADH/NAD ratios and ROS levels. These changes can cause DNA and histone modifications by reducing their methylation. PPAR-α and SREBP-1c are transcriptional factors that are also affected by the epigenetic changes caused by alcohol. Acute and chronic alcohol consumption can cause severe tissue injury, and there are also many other genes that are highly upregulated and expressed in advanced stage of liver alcoholic fibrosis, cirrhosis and hepatitis, such as p90RSK, osteopontin, TGF-β, and *PNPLA3*. Understanding the nature of these genetic changes may help in the discovery of a reliable treatment.

Key Facts of Alcohol

- Pure alcohol, as a form, was first discovered by Mohammed Al-Razi in Persia.
- Alcohol, as a term, is commonly referred to ethanol that is considered as the predominant constituent of alcohol beverages.
- Alcohol is mainly metabolized in the liver.
- The liver can be damaged by heavy alcohol consumption.
- Alcoholic liver disease is one of the major causes of death worldwide.

Summary Points

- Ethanol is an organic compound that can be metabolized naturally in the body.
- Ethanol is mainly metabolized by oxidative alcohol metabolism into acetaldehyde and acetate.
- About 1% of ethanol is metabolized by nonoxidative alcohol metabolism, producing fatty acid ethyl esters (FAEE).
- Ethanol metabolites can affect the fluidity of the cellular membrane by increasing its permeability.
- Reactive oxygen species and high NADH/NAD ratio that are generated during alcohol metabolism, due to heavy alcohol consumption, can cause epigenetic changes.
- Histones acetylation or phosphorylation, as well as DNA methylation, are induced by *S*-adenosylmethionine (SAMe) that acts as a methyl donor for these two processes.
- *S*-adenosylmethionine (SAMe) reduction that is caused by excessive alcohol consumption reduces DNA and histones methylation, causing epigenetic changes.
- Ethanol has a negative effect on the regulation of gene expression.
- PPAR-α, SREBP-1c, and PNPLA3 are some genes affected by chronic alcohol consumption.

References

1. Popova S, Rehm J, Patra J, Zatonski W. Comparing alcohol consumption in central and eastern Europe to other European countries. *Alcohol Alcohol* 2007;**42**(5):465–73.
2. Gao B, Bataller R. Alcoholic liver disease: pathogenesis and new therapeutic targets. *Gastroenterol* 2011;**141**(5):1572–85.
3. Waszkiewicz N, Szajda S, Zalewska A, Szulc A, Kępka A, Minarowska A, Wojewódzka-Żelezniakowicz M, Konarzewska B, Chojnowska S, Ladny J, Zwierz K. Alcohol abuse and glycoconjugate metabolism. *Folia Histochem Cytobiol* 2012;**50**(1):1–11.
4. Knott C, Coombs N, Stamatakis E, Biddulph J. All cause mortality and the case for age specific alcohol consumption guidelines: pooled analyses of up to 10 population based cohorts. *BMJ* 2015;**350**:h384.
5. British Medical Association Board of Science. *Alcohol misuse: tackling the UK epidemic*. London: British Medical Association; 2008.
6. Roswall N, Weiderpass E. Alcohol as a risk factor for cancer: existing evidence in a global perspective. *J Prevent Med Public Health* 2015;**48**(1):1–9.
7. Leung T, Nieto N. CYP2E1 and oxidant stress in alcoholic and non-alcoholic fatty liver disease. *J Hepatol* 2013;**58**(2):395–8.
8. Ji C. Dissection of endoplasmic reticulum stress signalling in alcoholic and non-alcoholic liver injury. *J Gastroenterol Hepatol* 2008;**1**:S16–24.
9. Lemasters J, Holmuhamedov L, Czerny C, Zhong Z, Maldonado N. Regulation of mitochondrial function by voltage dependent anion channels in ethanol metabolism and the Warburg effect. *Biochim Biophys Acta* 2012;**1818**(6):1536–44.
10. Zakhari S. Alcohol metabolism and epigenetics changes. *Alcohol Res Current Rev* 2013;**35**(1):6–16.
11. Caballería J. Current concepts in alcohol metabolism. *Ann Hepatol* 2003;**2**(2):60–8.
12. Cunningham C, Bailey M. Ethanol consumption and liver mitochondria function. *Neurosignals* 2001;**10**:271–82.
13. Seitz H, Becker P. Alcohol metabolism and cancer risk. *Alcohol Res Health* 2007;**30**(1):38–47.
14. Leung T, Rajendran R, Singh S, Garva R, Krstic-Demonacos M, Demonacos C. Cytochrome P4502E1 (CYP2E1) regulates the response to oxidative stress and migration of breast cancer cells. *Breast Cancer Res* 2013;**15**(6):R107.
15. Ceni E, Mello T, Galli A. Pathogenesis of alcoholic liver disease: role of oxidative metabolism. *World J Gastroenterol* 2014;**20**(47):17756–72.
16. Shukla S, Velazquez J, French S, Lu S, Ticku M, Zakhari S. Emerging role of epigenetics in the actions of alcohol. *Alcohol Clin Exp Res* 2008;**32**(9):1525–34.
17. Mandrekar P. Epigenetic regulation in alcoholic liver disease. *World J Gastroenterol* 2011;**17**(20):2456–64.

18. Tian C, Stokowski R, Kershenobich D, Ballinger D, Hinds D. Variant in PNPLA3 is associated with alcoholic liver disease. *Nature Genet* 2010;**42**(1):21–3.
19. Nischalke H, Berger C, Luda C, Berg T, Müller T, Grünhage F, Lammert F, Coenen M, Krämer B, Körner C, Vidovic N, Oldenburg J, Nattermann J, Sauerbruch T, Spengler U. The PNPLA3 rs738409 148M/M genotype is a risk factor for liver cancer in alcoholic cirrhosis but shows no or weak association in hepatitis C cirrhosis. *PLOS One* 2011;**6**(11):e27087.
20. Burza M, Molinaro A, Attilia M, Rotondo C, Attilia F, Ceccanti M, Ferri F, Maldarelli F, Maffongelli A, De Santis A, Attili A, Romeo S, Ginanni Corradini S. PNPLA3 I148M (rs738409) genetic variant and age at onset of at-risk alcohol consumption are independent risk factors for alcoholic cirrhosis. *Liver Internat* 2015;**34**(4):514–20.
21. Seth D, Duly A, Kuo P, McCaughan G, Haber P. Osteopontin is an important mediator of alcoholic liver disease via hepatic stellate cell activation. *World J Gastroenterol* 2014;**20**(36):13088–104.
22. Armstrong S. Protein kinase activation and myocardial ischemia/reperfusion injury. *Cardiovasc Res* 2004;**61**(3):427–36.
23. Morales-Ibanez O, Affò S, Rodrigo-Torres D, Blaya D, Millán C, Coll M, Perea L, Odena G, Knorpp T, Templin M, Moreno M, Altamirano J, Miquel R, Arroyo V, Ginès P, Caballería J, Sancho-Bru P, Bataller R. Kinase analysis in alcoholic hepatitis identifies p90RSK as a potential mediator of liver fibrogenesis. *Gut* 2015; Available from: http://gut.bmj.com/content/early/2015/02/04/gutjnl-2014-307979.short..

SECTION II

MOLECULAR BIOLOGY OF THE CELL

3 *Alcohol and Aldehyde Dehydrogenases: Molecular Aspects* 25

4 *Alcohol Intake and Apoptosis: A Review and Examination of Molecular Mechanisms in the Central Nervous System* 45

5 *Pathogenic Mechanisms in Alcoholic Liver Disease (ALD): Emerging Role of Osteopontin* 63

6 *The Role of CD36 in the Pathogenesis of Alcohol-Related Disease* 71

7 *Thiamine Deficiency and Alcoholism Psychopathology* 85

8 *Vitamin B Regulation of Alcoholic Liver Disease* 95

9 *Interactions Vitamin D – Bone Changes in Alcoholics* 107

10 *Antioxidant Treatment and Alcoholism* 119

11 *Selenium Dietary Supplementation and Oxidative Balance in Alcoholism* 133

12 *Role of Zinc in Alcoholic Liver Disease* 143

13 *Interactions Between Alcohol and Folate* 157

14 *Effects of Acetaldehyde on Intestinal Barrier Function* 171

15 *Cholesterol Regulation by Leptin in Alcoholic Liver Disease* 187

16 *The Corticotropin Releasing Factor System and Alcohol Consumption* 201

17 *Metabolic Profiling Approaches for Biomarkers of Ethanol Intake* 213

CHAPTER

3

Alcohol and Aldehyde Dehydrogenases: Molecular Aspects

Kwabena Owusu Danquah, MSc, PhD, Daniel Gyamfi, MSc, PhD

Department of Medical Laboratory Technology, College of Health Sciences, Kwame Nkrumah University of Science & Technology, Kumasi, Ghana

OVERVIEW OF ALCOHOL AND ALDEHYDE DEHYDROGENASES

The superfamily of aldehyde dehydrogenase (ALDH) is a group of oxidizing enzymes that are well recognized for playing crucially important roles, not only in the detoxification of endogenous and exogenous aldehydes,[1] but also in formation of essential molecules, such as retinoic acid, betaine, and gamma-aminobutyric acid of cellular relevance.[2] Besides the enzymatic roles of ALDH, nonenzymatic functions of ALDH include binding to some hormones and other small molecules, and diminishing the hazardous effect of UV in the cornea of the eyes.[3] The far-ranging functions of ALDH indubitably suggest that any mutations in ALDH genes would have broader medical implications, especially in relation to alcohol consumption. This chapter seeks to review the molecular functions of ALDH members, modifications of ALDH, and mutational implications of ALDH in relation of alcohol.

HUMAN ALDH SUPERFAMILY

The ALDH superfamily gene is present in all the three taxonomic domains: the Archaea, Eubacteria, and Eukarya, strongly suggesting the essential roles ALDHs play in cellular processes in these organisms. To date, 19 human ALDH superfamily members have been identified, and the genes are located on different chromosomes.[4] The number of ALDH superfamily will increase as more human sequenced genomes are carried out and annotated. The 19-superfamily members of human ALDH can be broadly and loosely grouped into those that are involved in alcohol metabolism pathway (ALDH1A1, ALDH1B1, ALDH2, ALDH3B1, ALDH3B2, and ALDH3A2, (Table 3.1)) and those that are not. Table 3.1 summarizes the general features of all the 19 members of the ALDH superfamily, in terms of substrate specificity, natural variants, etc. Figure 3.1 shows the evolutionary pattern of the 19-superfamily members, indicating how one ALDH protein shares protein similarity with another.

Human Alcohol-Metabolizing ALDH

The human alcohol metabolizing ALDH are involved in alcohol metabolism pathway by converting aldehydes to acetate, and they include ALDH1A1, ALDH1B1, ALDH2, ALDH3B1, ALDH3B2, and ALDH3A2. Among the members that metabolize alcohol-derived acetaldehyde, ALDH2 and ALDH1B1 show a great protein similarity (Figure 3.1). Interestingly, these are the two main enzymes that are greatly involved in alcohol-derived acetaldehyde metabolism, though ALDH1A1 metabolizes acetaldehyde with poor performance. ALDH1A1 seems to be closer to the ancestral gene from which others evolved. Further, there are some conserved domains among these ALDH, as indicated in the open boxes of Figure 3.2. These domains are evolutionarily conserved because of unique and relevant roles they play in the respective ALDH proteins.

The ALDH1A1 protein is made up of 500 amino acid residues, of which 11 residues are cysteine. They are located in the liver, brain, lungs, eye, breast, and ovary. Different isoforms have been identified: the liver cytosolic isoform, and the cytosolic form of erythrocytes that is at a very low concentration.[5] Three natural variants exist so far in the forms of Asn121Ser, Gly125Arg, and Ile77Phe; they have normal catalytic activities.

Molecular Aspects of Alcohol and Nutrition. http://dx.doi.org/10.1016/B978-0-12-800773-0.00003-3

TABLE 3.1 Structural and Functional Features of Human ALDH Proteins

Protein name	AA length	Organ locations	Enzyme substrate	PTM	Natural variant	Alcohol pathway
Aldehyde dehydrogenase 1 family, member A1 (ALDH 1A1)	500	Liver, brain, lungs, eye, breast, ovary	Retinaldehyde	Acetylation	121-121(N-S), 125-125 (G-R), 177-177 (I-F)	No
Aldehyde dehydrogenase 1 family, member A2 (ALDH 1A2)	518	Testis, prostate, lungs, kidney	Octanal and decanal		50-50 (E-G), 110-110 (A-V), 348-348 (V-I), 436-436 (E-K)	No
Aldehyde dehydrogenase 1 family, member A3 (ALDH 1A3)	512	Salivary gland, stomach, kidney, eye	Retinaldehyde	Acetylation	89-89 (R-C), 145-145 (A-V), 369-369 (I-F), 493-493(A-P)	No
Aldehyde dehydrogenase 1 family, member B1 (ALDH 1B1)	517	Testis, liver, pancreas, lung, thyroid, kidney	Acetaldehyde	Acetylation	86-86 (A-V), 107-107 (L-R), 202-202 (T-I), 253-253 (V-M)	Yes
Aldehyde dehydrogenase 1 family, member L1 (ALDH 1L1)	902	Liver, brain, prostate, bladder, kidney	10-formyltetrahydrofolate	Phosphorylation	254-254 (L-P), 330-330 (V-F), 429-429 (E-A), 436-436 (A-T), 436-436 (A-V), 448-448 (S-N), 481-481 (S-G), 511-511 (A-V), 793-793 (D-G), 803-803 (E-K), 812-812 (I-V)	No
Aldehyde dehydrogenase 1 family, member L2 (ALDH 1L2)	923	Skin, heart,	10-formyltetrahydrofolate			No
Aldehyde dehydrogenase 2 family (ALDH2)	517	Liver	Aldehyde			Yes
Aldehyde dehydrogenase family 3 member B1 (ALDH3A1)		Stomach, esophagus, lung; liver and kidney	Benzaldehyde, saturated and unsaturated aldehydes			No
Aldehyde dehydrogenase 3 family, member A2 (ALDH3A2)	508	Liver, heart, esophagus, pancreas, trachea	Aliphatic aldehydes			Yes
Aldehyde dehydrogenase 3 family, member B1 (ALDH3B1)	468	Liver, salivary gland, cervix, kidney, lung	Saturated and unsaturated aldehydes			Yes
Aldehyde dehydrogenase 3 family, member B2 (ALDH3B2)	385	Salivary gland, parotid	Aldehyde			Yes
Aldehyde dehydrogenase 4 family, member A1 (ALDH4A1)	563	Liver, pancreas, lung, kidney	Δ-1-pyrroline-5-carboxylate	Acetylation	16-16 (P-L), 352-352 (S-L), 470-470 (V-I), 473-473 (T-A)	No
Aldehyde dehydrogenase 5 family, member A1 (ALDH5A1)	535	Brain, liver, heart, thymus	Succinate semialdehyde	Acetylation and disulfide bond	93-93 (C-F), 176-179 (G-R), 223-223 (C-Y), 255-255 (N-S), 268-268 (G-E), 335-335 (N-K), 382-382 (P-L), 382-382 (P-Q), 409-409 (G-D), 487-487 (V-E), 533-533 (G-R)	No

Aldehyde dehydrogenase 6 family, member A1 (ALDH6A1)	535	Kidney, prostate, liver, heart	Oxopropanoate	Acetylation	446-446 (G-R)	No
Aldehyde dehydrogenase 7 family, member A1 (ALDH7A1)	539	Kidney, cervix, heart, lungs	(S)-2-amino-6-oxohexanoate betainealdehyde	Acetylation	199-199 (A-V), 202-202 (G-V), 291-291 (G-E), 301-301 (N-I), 335-335 (R-Q), 395-395 (V-G), 412-412 (E-Q), 439-439 (K-Q), 458-458 (S-N)	No
Aldehyde dehydrogenase 8 family, member A1 (ALDH8A1)	487	Liver, kidney, spleen, intestine,	Retinaldehyde			No
Aldehyde dehydrogenase 9 family, member A1 (ALDH9A1)	494	Liver, kidney, muscle	γ-Trimethylaminobutyraldehyde	Acetylation	116-116 (C-S)	No
Aldehyde dehydrogenase 16 family, member A1 (ALDH16A1)	802	Pancreas, lymph node, placenta		Phosphorylation	110-110 (E-K), 227-227 (L-V)	??
Aldehyde dehydrogenase 18 family, member A1 (ALDH18A1)	795	Intestine, pancreas, thyroid, kidney	Glutamate	Phosphorylation	84-84 (R-Q), 299-299 (T-I), 372-372 (S-Y), 784-784 (H-Y)	No

AA, amino acid; PTM, posttranslational modification; A, alanine; R, arginine; N, asparagine; D, aspartic acid; C, cysteine; Q, glutamine; E, glutamic acid; G, glycine; H, histidine; I, isoleucine; L, leucine; K, lysine; M, methionine; F, phenylalanine; P, proline; S, serine; T, threonine; W, tryptophan; Y, tyrosine; V, valine.

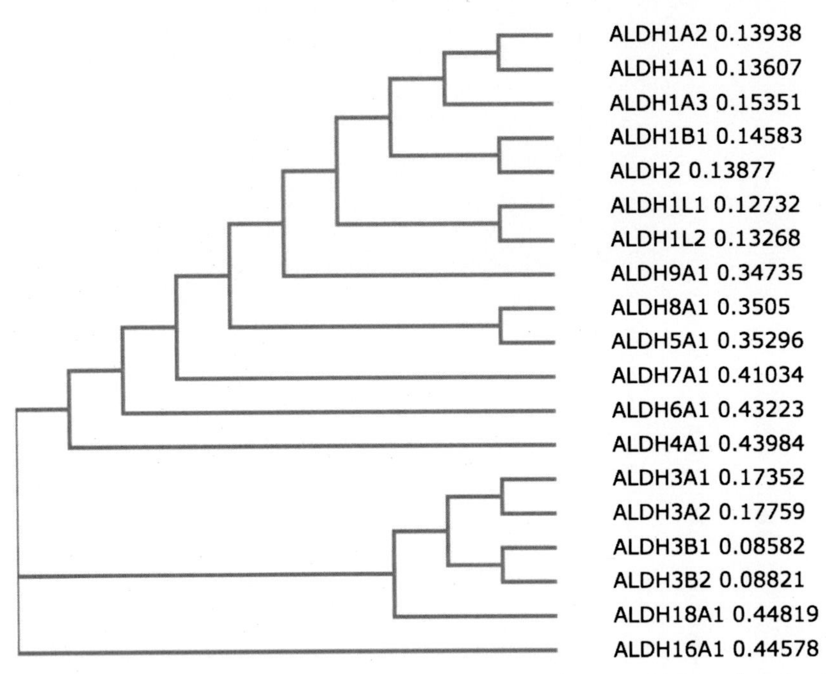

FIGURE 3.1 A neighbor-joining dendrogram of 19 human ALDH protein sequences (among which are those involved in alcohol metabolism: ALDH1A1, ALDH1B1, ALDH2, ALDH3). The branch length represents relative protein sequence similarity.

ALDH1B1 is mainly involved in the detoxification of alcohol-derived acetaldehyde, aside other roles in the metabolism of corticosteroids, biogenic amines, neurotransmitters, and lipid peroxidation. Bioinformatics analysis suggests that about 16 amino acid residues of this protein are extremely hyperacetylated.

ALDH2 is mainly mitochondrial acetaldehyde dehydrogenase and plays a crucial role in ethanol metabolism.[6] The protein has 17-amino acid signal peptide.[7]

ALDH3A1 is expressed and localized in the stomach, esophagus, lung, liver, and kidney. It plays a major role in detoxifying alcohol-derived acetaldehyde. They are also involved in the metabolism of corticosteroids, biogenic amines, neurotransmitters, and lipid peroxidation. ALDH3A1 preferentially oxidizes aromatic aldehyde substrates (Table 3.1). It has a high K_m for acetaldehyde.[8] Unlike other ALDH, no natural variants have been identified associated with ALDH3A1. However, a mutation of Cys244Ser has been found to ablate completely the catalytic activity of the protein.[9]

ALDH3A2 is a 508-amino acid protein expressed in the liver, heart, esophagus, pancreas, and trachea (Table 3.1). It catalyzes the oxidation of long-chain aliphatic aldehydes to fatty acids. Further, it also breaks down a variety of saturated and unsaturated aliphatic aldehydes between 6 and 24 carbons in length, as well as converting sphingosine 1-phosphate (SIP) degradation product hexadecenal into hexadecenoic acid.[10] There are about 24 existing natural ALDH3A2 variants, and almost all of the variants have severe loss of activity, and are associated with Sjoegren–Larsson syndrome (SLS).[11]

Nonalcohol-Metabolizing Human ALDH

The ALDH1A2 is located in the testis, prostate, lungs, and kidney; it metabolizes octanal and decanal, but does not metabolize citral, benzaldehyde, acetaldehyde, and propanal efficiently.

The 512-amino acid ALDH1A3 is expressed in the salivary gland, stomach, and kidney. Four natural variants have currently been identified (Table 3.1). At the 5′ prime upstream of the gene are TATA and CCAAT boxes, and Sp1 binding sites.[12] Mutations in this gene result in a condition called microphthalmia isolated 8 (MCOP8).[13]

The ALDH1L1 protein is a cytosolic protein consisting of 1-310 amino acid residue N-terminal domain that shares sequence homology and structural topology with other enzymes that utilize 10-formyltetrahydrofolate. The linker sequence consists of a 400–902 amino acid residues, and originates from an aldehyde dehydrogenase related gene; the linker sequence joins the N-terminal to the C-terminal domain. Bioinformatics analysis indicates potential sites for N-glycosylation, phosphorylation, and N-myristoylation. ALDH1L1, in the presence of $NADP^+$, oxidizes 10-formyltetrahydrofolate to tetrahydrofolate and CO_2. However, in the absence of $NADP^+$, 10-formyltetrahydrofolate is hydrolyzed into tetrahydrofolate and formic acid. ALDH1L1 induces phosphorylation of TP53 at Ser6, a required step in the activation of apoptosis,[14] induces phosphorylation of MAPK8 and MAPK9,[14] and folates enzyme with tumor suppressor-like properties that inhibits cell motility.[15]

In contrast, ALDH1L2 is a mitochondrial form of 10-formyltetrahydrofolate dehydrogenase, and it converts 10-formyltetrahydrofolate to tetrahydrofolate, in the presence of $NADP^+$. The ALDH1L2 protein shares 72% identity with human ALDH1L1, and shares three distinct domains. However, the 22-amino acid N-terminal mitochondrial-targeting signal of ALDH1L2 is not present in the ALDH1L1. The extra N-terminal sequence of 22 amino acids residues that is predicted to be a mitochondrial translocation signal lacks negatively charged amino acids but is enriched with positively charged residues, including five arginines and two lysines.[16]

ALDH6A1 is involved in valine and pyrimidine metabolism. It also binds fatty acyl-CoA. A mutated Gly446Arg natural variant is associated with methylmalonate semialdehyde dehydrogenase deficiency (MMSDHD), a metabolic disorder characterized by elevated beta-alanine, 3-hydropropionic acid, 3-amino and 3-hydroxyisobutyric acids in urine.[17]

ALDH7A1 performs several essential functions, mediating important protective effects. It achieves this via metabolism of betaine aldehyde to betaine, an important cellular osmolyte and methyl donor. It also protects cells from oxidative stress by metabolizing a number of lipid peroxidation-derived aldehydes.[18]

ALDH8A1 converts 9-*cis*-retinal to 9-*cis*-retinoic acid, but has a lower activity towards 13-*cis*-retinal. However, it has highest activity with benzaldehyde and decanal *in vitro*. Even though it has a preference for NAD, it shows considerable activity with NADP *in vitro*.[19]

```
ALDH2      MLRAAARFGPRLGRRLLSAAATQAVPAPNQQPEVFCNQIFINNEWHDAVSRKTFPTVNPS 60
ALDH1B1    MLRFLAPRLLSLQGRTARYSSAAALPSPILNPDIPYNQLFINNEWQDAVSKKTFPTVNPT 60
ALDH1A1    ----------------MSSSGTPDLPVLLTDLKIQYTKIFINNEWHDSVSGKKFPVFNPA 44
ALDH3A1    ------------------------------------------------------------
ALDH3A2    ------------------------------------------------------------
ALDH3B1    ------------------------------------------------------------
ALDH3B2    ------------------------------------------------------------

ALDH2      TGEVICQVAEGDKEDVDKAVKAARAAFQLGSPWRRMDASHRGRLLNRLADLIERDRTYLA 120
ALDH1B1    TGEVIGHVAEGDRADVDRAVKAAREAFRLGSPWRRMDASERGRLLNLLADLVERDRVYLA 120
ALDH1A1    TEEELCQVEEGDKEDVDKAVKAARQAFQIGSPWRTMDASERGRLLYKLADLIERDRLLLA 104
ALDH3A1    ------------MSKISEAVKRARAAFSSG---RTRPLQFRIQQLEALQRLIQEQEQELV 45
ALDH3A2    ---------------MELEVRRVRQAFLSG---RSRPLRFRLQQLEALRRMVQEREKDIL 42
ALDH3B1    ------------MDPLGDTLRRLREAFHAG---RTRPAEFRAAQLQGLGRFLQENKQLLH 45
ALDH3B2    ------------------------------------------------------------

ALDH2      ALETLDNGKPYVISYLVDLDMVLKCLRYYAG----WADKYHGKTIPIDGDFFSYTRHEPV 176
ALDH1B1    SLETLDNGKPFQESYALDLDEVIKVYRYFAG----WADKWHGKTIPMDGQHFCFTRHEPV 176
ALDH1A1    TMESMNGGKLYSNAYLNDLAGCIKTLRYCAG----WADKIQGRTIPIDGNFFTYTRHEPI 160
ALDH3A1    GALAADLHKNEWNAYYEEVVYVLEEIEYMIQKLPEWAADEPVEKTPQTQQDELYIHSEPL 105
ALDH3A2    TAIAADLCKSEFNVYSQEVITVLGEIDFMLENLPEWVTAKPVKKNVLTMLDEAYIQPQPL 102
ALDH3B1    DALAQDLHKSAFESEVSEVAISQGEVTLALRNLRAWMKDERVPKNLATQLDSAFIRKEPF 105
ALDH3B2    ------------------------------------MKDEPRSTNLFMKLDSVFIWKEPF 24
                                                      .         :   :*.

ALDH2      GVCGQIIPWNFPLLMQAWKLGPALATGNVVVMKVAEQTPLTALYVANLIKEAGFPPGVVN 236
ALDH1B1    GVCGQIIPWNFPLVMQGWKLAPALATGNTVVMKVAEQTPLSALYLASLIKEAGFPPGVVN 236
ALDH1A1    GVCGQIIPWNFPLVMLIWKIGPALSCGNTVVVKPAEQTPLTALHVASLIKEAGFPPGVVN 220
ALDH3A1    GVVLVIGTWNYPFNLTIQPMVGAIAAGNSVVLKPSELSENMASLLATIIPQY-LDKDLYP 164
ALDH3A2    GVVLIIGAWNYPFVLTIQPLIGAIAAGNAVIIKPSELSENTAKILAKLLPQY-LDQDLYI 161
ALDH3B1    GLVLIIAPWNYPLNLTLVPLVGALAAGNCVVLKPSEISKNVEKILAEVLPQY-VDQSCFA 164
ALDH3B2    GLVLIIAPWNYPLNLTLVLLVGALAAGSCVVLKPSEISQGTEKVLAEVLPQY-LDQSCFA 83
           *:   * .**:*: :    :  *:: *. *::* :* :      :* :: :  .  .

ALDH2      IVPGFGPTAGAAIASHEDVDKVAFTGSTEIGRVIQVAAGSSNLKRVTLELGGKSPNIIMS 296
ALDH1B1    IITGYGPTAGAAIAQHVDVDKVAFTGSTEVGHLIQKAAGDSNLKRVTLELGGKSPSIVLA 296
ALDH1A1    IVPGYGPTAGAAISSHMDIDKVAFTGSTEVGKLIKEAAGKSNLKRVTLELGGKSPCIVLA 280
ALDH3A1    VINGGVPETTELLKER--FDHILYTGSTGVGKIIMTAAAK-HLTPVTLELGGKSPCYVDK 221
ALDH3A2    VINGGVEETTELLKQR--FDHIFYTGNTAVGKIVMEAAAK-HLTPVTLELGGKSPCYIDK 218
ALDH3B1    VVLGGPQETGQLLEHR--FDYIFFTGSPRVGKIVMTAAAK-HLTPVTLELGGKNPCYVDD 221
ALDH3B2    VVLGGPQETGQLLEHK--LDYIFFTGSPRVGKIVMTAATK-HLTPVTLELGGKNPCYVDD 140
           :: *    :   :  :  .* : :**.. :*:::  ** . :*. ********.*  :
```

FIGURE 3.2 Alignment of seven human ALDH proteins (including alcohol-metabolizing ALDH) created by ClustalW. Dashes (–) represent sequence gaps, asterisks (*) represent identical amino acids (AAs), colons (:) represent very similar AAs, periods (.) represent less similar AAs, whereas spaces ( ) represent dissimilar AAs. Highly conserved domains are in open boxes.

```
ALDH2      DADMDWAVEQAHFALFFNQGQCCCAGSRTFVQEDIYDEFVERSVARAKSRVVGNPFDSKT 356
ALDH1B1    DADMEHAVEQCHEALFFNMGQCCCAGSRTFVEESIYNEFLERTVEKAKQRKVGNPFELDT 356
ALDH1A1    DADLDNAVEFAHHGVFYHQGQCCIAASRIFVEESIYDEFVRRSVERAKKYILGNPLTPGV 340
ALDH3A1    NCDLDVACRRIAWGKFMNSGQTCVAPDYILCDPSIQNQIVE-KLKKSLKEFYGEDAKKSR 280
ALDH3A2    DCDLDIVCRRITWGKYMNCGQTCIAPDYILCEASLQNQIVW-KIKETVKEFYGENIKESP 277
ALDH3B1    NCDPQTVANRVAWFRYFNAGQTCVAPDYVLCSPEMQERLLP-ALQSTITRFYGDDPQSSP 280
ALDH3B2    NCDPQTVANRVAWFCYFNAGQTCVAPDYVLCSPEMQERLLP-ALQSTITRFYGDDPQSSP 199
           :.* : . .      : : ** * * .  : . .: :.::   :  :     *:

ALDH2      EQGPQVDETQFKKILGYINTGKQEGAKLLCGGGIAADRGYFIQPTVFGDVQDGMTIAKEE 416
ALDH1B1    QQGPQVDKEQFERVLGYIQLGQKEGAKLLCGGERFGERGFFIKPTVFGGVQDDMRIAKEE 416
ALDH1A1    TQGPQIDKEQYDKILDLIESGKKEGAKLECGGGPWGNKGYFVQPTVFSNVTDEMRIAKEE 400
ALDH3A1    DYGRIISARHFQRVMGLIE-----GQKVAYGGT-GDAATRYIAPTILTDVDPQSPVMQEE 334
ALDH3A2    DYERIINLRHFKRILSLLE-----GQKIAFGGE-TDEATRYIAPTVLTDVDPKTKVMQEE 331
ALDH3B1    NLGRIINQKQFQRLRALLG-----CGRVAIGGQ-SDESDRYIAPTVLVDVQEMEPVMQEE 334
ALDH3B2    NLGHIINQKQFQRLRALLG-----CSRVAIGGQ-SNESDRYIAPTVLVDVQETEPVMQEE 253
              :. ::.::  :          ::  **        :: **:: .*      : :**

ALDH2      IFGPVMQILKFKTIEEVVGRANNSTYGLAAAVFTKDLDKANYLSQALQAGTVWVN--CYD 474
ALDH1B1    IFGPVQPLFKFKKIEEVVERANNTRYGLAAAVFTRDLDKAMYFTQALQAGTVWVN--TYN 474
ALDH1A1    IFGPVQQIMKFKSLDDVIKRANNTFYGLSAGVFTKDIDKAITISSALQAGTVWVN--CYG 458
ALDH3A1    IFGPVLPIVCVRSLEEAIQFINQREKPLALYMFSSNDKVIKKMIAETSSGGVAANDVIVH 394
ALDH3A2    IFGPILPIVPVKNVDEAINFINEREKPLALYVFSHNHKLIKRMIDETSSGGVTGNDVIMH 391
ALDH3B1    IFGPILPIVNVQSLDEAIEFINRREKPLALYAFSNSSQVVKRVLTQTSSGGFCGNDGFMH 394
ALDH3B2    IFGPILPIVNVQSVDEAIKFINRQEKPLALYAFSNSSQVVNQMLERTSSGSFGGNEGFTY 313
           ****:  :. .:.:::.:     *.     *:    *: . .     .     .:* .  *

ALDH2      VFGAQSPFGGYKMSGSGRELGEYGLQAYTEVKTVTVKVPQKNS----------------- 517
ALDH1B1    IVTCHTPFGGFKESGNGRELGEDGLKAYTEVKTVTIKVPQKNS----------------- 517
ALDH1A1    VVSAQCPFGGFKMSGNGRELGEYGFHEYTEVKTVTVKISQKNS----------------- 501
ALDH3A1    ITLHSLPFGGVGNSGMGSYHGKKSFETFSHRRSCLVRPLMNDEGLKVRYPPSPAKMTQH- 453
ALDH3A2    FTLNSFPFGGVGSSGMGAYHGKHSFDTFSHQRPCLLKSLKREGANKLRYPPNSQSKVDWG 451
ALDH3B1    MTLASLPFGGVGASGMGRYHGKFSFDTFSHHRACLLRSPGMEKLNALRYPPQSPRRLRM- 453
ALDH3B2    ISLLSVPFGGVGHSGMGRYHGKFTFDTFSHHRTCLLAPSGLEKLKEIHYPPYTDWNQQL- 372
           .       ****    ** *     *:    :. ::. :.   :       :

ALDH2      ----------------------------------
ALDH1B1    ----------------------------------
ALDH1A1    ----------------------------------
ALDH3A1    ----------------------------------
ALDH3A2    KFFLLKRFNKEKLGLLLLTFLGIVAAVLVKAEYY 485
ALDH3B1    ---LLVAMEAQGCSCTLL---------------- 468
ALDH3B2    ---LRWGMGSQ--SCTLL---------------- 385
```

FIGURE 3.2 *(cont.)*

ALDH16A1 exists in two isoforms; isoform 1 and isoform 2 consist of 802 and 751 amino acid residues, respectively.[20]

ALDH18A1 is also called delta-1-pyrroline-5-carboxylate synthetase (P5CS), and it is a bifunctional enzyme, converting L-glutamate to glutamic δ-semialdehyde, a metabolic precursor for proline biosynthesis.[21] ALDH18A1 has two isoforms: short and long isoforms that are produced through alternative splicing. These isoforms differ by a 2-amino acid insert at the N-terminal region of the gamma-glutamyl kinase active site. The long isoform is expressed in many tissues, and is required for proline synthesis. The short isoform, however, is highly expressed in the gut, where it is involved in arginine biosynthesis. The long isoform is insensitive to ornithine, but the short isoform is sensitive to orinithine due to lack of two amino acids.[22] Two variants (Arg84Glu and His784Tyr) of ALDH18A are associated with autosomal recessive cutis laxa type 3A (ARCL3A), a condition characterized by facial dysmorphism, with a properoid appearance, growth retardation, and intellectual difficulties.[21,23]

MODIFICATIONS OF ALDH AND BIOMEDICAL IMPLICATIONS

Intracellular modifications of proteins, including ALDH, during protein translation, or after translation as a response to stimuli, enrich these proteins for a plethora of functions that the nonmodified counterpart cannot do; or even if it can do them, it will be done inefficiently. Among the posttranslational modifications (PTM) are acetylation, phosphorylation, glycosylation, nitration, and S-nitrosylation. Some of the human ALDH superfamily members that are modified after translation are summarized in Table 3.1.

Acetylation

Cellular addition of acetyl groups to proteins has been reported to play significant roles, influencing the behavior of cells. Acetyl modification of proteins, such microtubules, enhances kinesin-1 binding, and subsequent cargo transport.[24] Acetylation of GD3 in glioma cells has also been shown to impact invasiveness of glioma cells,[25,26] whereas the control of the rate of cell growth and normal development is regulated by cyclical acetylation and deacetylation of p53.[27] Not only acetylation of proteins influences biological functions, but also hyperacetylation of lysine residues dramatically effect cellular behavior – for example, a long chain acyl-CoA dehydrogenase is inactivated via excessive acetylation of its lysine residues, affecting its role in mitochondrial fatty acid β-oxidation pathway, as demonstrated in Sirtuin 3 gene-knockout mice.[28]

A proteomic survey revealed that ALDH2, ALDH4A1, and ALDH6A1 were acetylated under physiological conditions,[29] and that several hepatic proteins are hyperacetylated in chronic alcohol consumption in rats,[29] suggesting a vital role of acetylated mitochondrial proteins in alcohol-related liver diseases.

Phosphorylation

Mitochondrial activity of ALDH2 is considerably inhibited in oxidative or nitrosative stress environment, induced by acute exposure of ethanol, acetaminophen, or CCl_4. This leads to mitochondrial dysfunction and severe liver injury in the pericentral regions, through apoptosis or necrosis.[30,31] Further, the oxidative stress environment induced by these compounds can simultaneously decrease the levels of antioxidants, as well as activate c-Jun N-terminal protein kinase (JNK)-mediated cell death signaling pathway,[32] and stimulate STAT-1 mediated proinflammatory signaling pathway.[33]

Activation of JNK by CCl_4 leads to irreversible phosphorylation of the serine residues of hepatic ALDH2, resulting in the inactivation of mitochondrial ALDH2.[34–36] However, cytosolic ALDH1A1 is not inhibited in the presence of CCl_4.[37] The inhibition of ALDH2 via JNK activation could result in decreased cellular defense strength, and increased lipid peroxidation – a likely facilitator in pericentral hepatic necrosis.[38] Phosphorylation of mitochondrial ALDH2 can, however, be induced in the presence of alcohol that stimulates protein kinase Cε (PKCε).[39] The stimulated PKCε is translocated to mitochondria in a time-dependent manner to bind to ALDH2 protein.[40] In experimental models, alcohol-induced phosphorylation, PKCε-medicated, of ALDH2 correlates strongly with decreased levels of 4-HNE, and a significant improvement of ischemia cardiac damage; it has been concluded, therefore, that the degree of the myocardial infarct size from experimental ischemia injury is inversely related to the activities of ALDH2.[41,42] Indeed, a recent proteomic analysis indicates that a cascade of molecular events, in both the mitochondria and the cytosol, defines the mechanism underpinning heart failure. Down-regulation of mitochondrial ALDH2 induces an increase in 4-hydroxynonenal (4-HNE), in both the mitochondria and the cytosol that results, in turn, in the inhibition of HSP70, phosphorylation of JNK, and later activation of p53, leading to cardiomyocyte apoptosis.[43]

S-Nitrosylation

In oxidative stress milieu, some proteins undergo modifications at the active sites – cysteine residues, including nitric oxide (NO), or peroxynitrite-dependent S-nitrosylation.[44,45] The active site cysteine is highly conserved in all ALDH members, and that cysteine 302 is the most residue that is modified via S-nitrosylation. In alcoholic persons, ALDH2 and ALDH1 are suppressed, and that oxidative stress microenvironment will lead to reversible modification of the cysteine residues in these proteins. It has been reported in chronic alcohol use or treatment that cytosolic ALDH1 is oxidatively nitrosylated in a reversible manner, leading to ALDH1 inactivation.[46] Similarly, mitochondrial ALDH2 that has a higher affinity for NO than cytosolic ALDH1 is also inactivated via S-nitrosylation pathway.[47]

Nitration

Addition of nitrogen to ALDH is another PTM that modifies the biological functions of ALDH isozymes. Unlike other PTMs of ALDH, such as acetylation, ALDH nitration is not alcohol induced. However, chronic usage of glyceryl nitrinitrate, GTN (an anti-ischemic drug for the treatment of angioplasty and cardiovascular diseases),[48,49] leads to the inhibition of ALDH2 via nitration signaling pathway.[50] The tyrosine residue of ALDH2 is nitrated in the presence of GTN.[45] Further, ALDH2 and ALDH7A1 have been confirmed, through proteomics and mass spectrometric analysis, to be nitrated in the kidneys of spontaneously hypertensive rats.[51]

GENOMICS OF THE ALDH SUPERFAMILY

Several variants have been identified in human ALDH. ALDH variants are a product of molecular mechanisms, such as gene duplication, alternative splicing, deletion of genes, etc. These genomic changes result in unique features of ALDH genes, leading to functional variations, compared to the parental or wild type gene.

Alternative Splicing of Human ALDHs

Sequencing of ALDH transcript has revealed that several ALDH genes encode multiple mRNA splice variants. Following splicing, ALDH exons are rearranged, such that some exons are fused in such a manner as to give different ALDH variants. With the exception of ALDH1B1, ALDH2, ALDH7A1, and ALDH9A1 that do not have alternative splice transcripts (perhaps not identified so far), the rest of the 19 ALDH members do (Table 3.2).

The human ALDH1 family has six member genes, including *ALDH1A1*, *ALDH1A2*, *ALDH1A3*, *ALDH1B1*, *ALDH1L1*, and *ALDH1L2*.

Of the two transcriptional variants of human ALDH1A1 gene (ALDH1A1_v1 and ALDH1A1_v2) that have been identified, ALDH1A1_v1 contains 13, and ALDH1A1_v2 has 11 exons. Compared to the native ALDH1A1_v1, the human ALDH1A1_v2 lacks the 3′ end of exon 7 and exon 9, a portion of exon 7, as well as missing exons 8, 10, 11, 12, and 13. Translation of ALDH1A1_v2 gives a truncated protein that misses 271 amino acid residues at the C-terminal region. Pfam analysis of ALDH1A1_v2 indicates the absence of Cys303 and Glu269

TABLE 3.2 Human ALDH Transcripts

Gene name (size (bp))	Transcript	Transcript size (bp)	Exons	Chromosomal location
ALDH1A1 (52.38)	ALDH1A1	2107	13	9q21.13
	ALDH1A1_v2	822	8	
ALDH1A2 (112.28)	ALDH1A2	3606	13	15q21.3
	ALDH1A2_v2	3492	12	
	ALDH1A2_v3	3210	11	
	ALDH1A2_v4	3234	12	
ALDH1A3 (36.77)	ALDH1A3	3622	13	15q26.3
	ALDH1A3_v2	3104	10	
ALDH1B1 (5.96)	ALDH 1B1	3088	2	9q11.1
ALDH1L1 (5.96)	ALDH 1L1	3306	23	3q21.3
	ALDH 1L1_v2	3179	23	
	ALDH 1L1_v3	2797	21	
	ALDH 1L1_v4	3399	7	
	ALDH 1L1_v5	1152	6	
ALDH1L2 (64.78)	ALDH 1L2	7555	23	12q23.3
	ALDH 1L2_v2	7860	22	
	ALDH 1L2_v3	2925	22	
ALDH2 (43.44)	ALDH2	2076	13	12q24.2
ALDH3A1 (10.31)	ALDH3A1	1794	11	17p11.2
	ALDH3A1_v2	2140	9	
	ALDH3A1_v3	1658	11	
	ALDH3A1_v4	1351	9	
	ALDH3A1_v5	2097	8	
	ALDH3A1_v6	1944	10	
	ALDH3A1_v7	1567	10	
ALDH3A2 (28.81)	ALDH3A2	3702	10	17p11.2
	ALDH3A2_v2	3823	11	
	ALDH3A2_v3	3627	11	

TABLE 3.2 Human ALDH Transcripts *(cont.)*

Gene name (size (bp))	Transcript	Transcript size (bp)	Exons	Chromosomal location
	ALDH3A2_v4		10	
	ALDH3A2_v5	2351	7	
	ALDH3A2_v6	1170	3	
ALDH3B1 (18.92)	ALDH3B1	2856	10	11q13
	ALDH3B1_v2	2745	9	
	ALDH3B1_v3	5879	9	
	ALDH3B1_v4	1408	7	
	ALDH3B1_v5	1686	9	
ALDH3B2 (12.43)	ALDH3B2	2660	10	11q13
	ALDH3B2_v2	2540	10	
	ALDH3B2_v3	1463	9	
ALDH4A1 (31.15)	ALDH4A1	3399	15	1p36
	ALDH4A1_v2	2386	16	
	ALDH4A1_v3	3074	14	
	ALDH4A1_v4	785	8	
	ALDH4A1_v5	839	9	
ALDH5A1 (42.24)	ALDH5A1	5131	10	6p22
	ALDH5A1_v2	5170	11	
	ALDH5A1_v3	1034	4	
ALDH6A1 (24.33)	ALDH6A1	4701	12	4q24.3
	ALDH6A1_v2	1627	7	
	ALDH6A1_v3	991	5	
	ALDH6A1_v4	724	4	
ALDH7A1 (50.39)	ALDH7A1	4953	18	5q31
ALDH8A1 (32.72)	ALDH8A1	2567	7	6q23.2
	ALDH8A1_v2	2405	6	
ALDH9A1 (36.56)	ALDH9A1	2500	11	1q23.1
ALDH16A1 (17.83)	ALDH16A1	3119	17	9q13.33
	ALDH16A1_2	2966	15	
ALDH18A1 (50.88)	ALDH18A1	3470	18	10q24.3
	ALDH18A1_v2	3464	18	

within the active sites, as it is in the native ALDH1A1_v1, leading to a possibility that ALDH1A1_v2 might not have enzymatic activity.[52]

There are four human ALDH1A2 transcriptional variants: ALDH1A2_v1, ALDH1A2_v2, ALDH1A2_v3, and ALDH1A2_v4. ALDH1A2_v1 is the consensus variant, and has the most prevalent transcript. Even though both ALDH1A2_v1 and ALDH1A2_v2 have 51.4 kb intron 1, ALDH1A2_v2 lacks exon 7. The exon 7 is within the coding region of the transcript, thus the ALDH1A2_v2 protein will be shorter than ALDH1A2_v1. Variant ALDH1A2_v3 that is derived from ALDH1A2_v2 lacks exons 1 and 2 of ALDH1A2_v2. ALDH1A2_v3 exon 1 has a distinct 5′-untranslated region (UTR), comprising of additional 15-bp upstream of exon 3. The translated ALDH1A2_v3 has a shorter N-terminal region, compared to ALDH1A2_v1. Finally, the ALDH1A2_v4 is derived from ALDH1A2_v2, but lacks 114-bp exon 7 of

ALDH1A2_v1, and uses an alternate exon 1, leading to a modified 5′ coding region.[52,53]

Three transcripts of ALDH1A3 have been identified in humans so far. The 3510-bp native variant (ALDH1A3_v1) has 13 exons.[54] ALDH1A3_v2 has 10 exons, due to the lack of exon 4 through 6 of ALDH1A3_v1. The encoded ALDH1A3_v2 lacks an internal segment within the ALDH peptide domain 5′ to the predicted cysteine and glutamate residues in the active site.[52] ALDH1A3_v3 has 11 exons, lacking exon 5, and part of exon 6 of ALDH1A3_v1.[54]

There are five transcriptional ALDH1L1 variants that have been identified. The native and major variant, ALDH1L1_v1, spans 3125 bp, and has 23 exons, just as ALDH1L1_v2.[55] Both ALDH1L1_v1 and ALDH1L1_v2 differ by an alternative exon 1. This results in varied translation initiation points on exons 2 in ALDH1L1_v1, and exon 1 in ALDH1L1_v1. The ALDH1L1_v3 contains 22 exons, but lacks 151-bp exon 13 that is present in ALDH1L1_v1 and ALDH1L1_v2, resulting in a truncation of a domain, due to a frame-shift that introduces an early termination signal. ALDH1L1_v4 and ALDH1L1_v5 are truncated transcripts that have no peptide domain in the translated proteins.[52]

Of the three transcriptional variants of ALDH1L2, ALDH1L2_v2 uses an alternate exon 1, a 5′ extended derivative of ALDH1L2_v1 exon 13. However, it lacks exons 1–12 of the ALDH1L2_v1. Translated ALDH1L2_v2 therefore retains the central portion of ALDH peptide domain, but the N-terminal and C-terminal formyl transferase peptide domains are missing. The third variant, ALDH1L2_v3, lacks a 70-bp exon 1 of ALDH1L2_v1.[52]

Two members, ALDH3A1 and ALDH3A2, make up the ALDH3 family. The ALDH3A family members have quite a number of variants relative to other ALDH families. Seven transcriptional variants of ALDH3A1 have been identified. The consensus variant, ALDH3A1_v1, is an 11-exon variant that encodes for 50.4 kDa. However, 9-exon ALDH3A1_v2 that has only 9 exons mainly encodes for a larger protein product, 61.6 kDa because its second exon is a fusion of exon 3, intron 3, and exon 4. The ALDH3A1_v3 also has 11 exons, but has a 5′ truncation of trinucleotide GAG of exon 7 in the coding region. ALDH3A1_v4 has 9 exons, but lacks exons 2 and 9 found in the consensus variant. An 8-exon ALDH3A1_v5 resembles ALDH3A1_v3 with regards to fusion of exons. Nonetheless, it does not have exon 1, and the fused exon has a 5′ truncation of the 88-bp exon 3 of ALDH3A1_v1. The 10-exon ALDH3A1_v6 lacks exon 7 of ALDH3A1_v1, and a truncation of 50 bp from the 5′ portion of exon 8. Like ALDH3A1_v6, the variant ALDH3A1_v7 is a 10-exon variant that lacks exon 2 of ALDH3A1_v1. The translated ALDH3A1_v7 is thus a functional peptide domain.[52]

Six transcriptional variants of human ALDH3A2 have been identified. The native variant, ALDH3A2_v1, has 10 exons, encoding for 54.9 kDa protein (expressed in microsomes). Relative to ALDH3A2_v1, the variant ALDH3A2_v2 has an additional 125-bp exon between exons 9 and 10; the translated protein has instead a larger molecular weight (57.5 kDa, expressed in peroxisomes).[56] The interesting features of ALDH3A2_v3 and ALDH3A2_v4 are that their coding regions are identical to those of ALDH3A2_v1 and ALDH3A2_v2, respectively. ALDH3A2_v5 uses an alternative exon 1 beginning upstream, and includes exon 4 of ALDH3A2_v1.[52]

This ALDH3B1 family member has five transcriptional variants. The primary variant ALDH3B1_v1 is a 10-exon transcript. Relative to the primary variant, ALDH3B1_v2 lacks exon 3. This variant encodes for a shorter protein with a complete ALDH peptide domain, even though exon 3 is within the coding region of the peptide; this is because its translation is not associated with the ALDH peptide domain. The exon 2 of ALDH3B1_v3 variant is a fusion of exon 2, intron 2, and exon 3 of the ALDH3B1_v1 transcript. The formation of exon 2 results in a 3′ shift in the transcript coding sequence, and later N-terminal truncation of the peptide. The 7-exon ALDH3B1_v4 transcript lacks exons 1 and 2, a 54-bp fragment from the 5′ end of exon 3, compared to ALDH3B1_v1. The ALDH3B1_v5 lacks exon 6. The ALDH3B1_v6 transcript is a 2516-bp fusion of intron 2 and exon 3 of ALDH3B1_v1.[52]

Of the three transcripts, ALDH3B2_v1 and ALDH3B2_v2 differ by an alternative exon 1, whereas ALDH3B2_v3 lacks exon 9 present in ALDH3B2_v1. The ALDH3B2_v3 transcript therefore has a shorter protein truncated at the C-terminus region of the peptide domain.[52]

The 3160-bp primary variant ALDH4A1 has 15 exons with identical coding regions to that of ALDH4A1_v2 (a 16-exon transcript). Thus, both transcripts encode the same protein. The ALDH4A1_v1 contains 1 kb insert at the 3′ UTR. The ALDH4A1_v3 variant lacks exon 4 of ALDH4A1_v1, resulting in a truncation of the protein at the 5′ of the peptide domain. The fourth and the fifth transcripts are shorter ones, thus appears a truncation at the C-terminus with partial domains, and no apparent active site residues. The ALDH4A1_v6 lacks exon 12, in comparison to ALDH4A1_v1, but is translated as a splice variant that is missing an internal 51-amino-acid segment.[52]

The 5131-bp ALDH5A1 primary variant is a 10-exon transcript, whereas the second transcript that spans 5170bp has 11 exons[57,58]: the extra 39-bp is transcribed from within intron 4, accounting for additional 13 amino acid residues. Both variants lack an in-frame coding region. A third variant, ALDH5A1_v3, lacks both 3′ and 5′ exon segments, resulting in a translated product that has truncated N- and C-termini, with a partial peptide domain, and no apparent active site residues.[52]

The primary transcript of ALDH6A1 is 12-exon, and the second transcript lacks exon 1 through 6. ALDH6A1_v2 begins 6-bp upstream from exon 7, in relation to ALDH6A1_v1. The 12th exon of ALDH6A1_v1 is transcribed as two separate exons; there is a 442-bp exon 6 and a 404-bp exon 7 in ALDH6A1_v2; both exons are separated by a 2237-bp intron. The coding sequence for this transcript ends within exon 6, at the same stop codon as the ALDH6A1_v1, thereby making exon 7 irrelevant to the protein's amino-acid sequence. The third transcript is a 5-exon, and the fourth transcript a 4-exon; they are both truncated at their 3′ ends. Their resultant translated proteins are truncated proteins at their COOH-termini. However, they retain a 5′ portion of the ALDH.[52]

A 2549-bp of ALDH8A1, as well as 2387-bps transcripts containing exons 7 and 6, respectively, has been identified.[19] The second variant lacks an in-frame segment within the coding region; that is, exon 6 of the primary transcript. When this is translated into a 433-amino acid splice variant, there are no apparent active site residues within the peptide domain.[52]

Two transcripts of ALDH16A1 have been identified so far, consisting of a 3119-bp that contains 17 exons, and a 2966-bp that has 15 exons.[20] The 6th exon of ALDH16A1_v2 is a fusion of exon 6, intron 6, and exon 7; its 15th exon is a fusion of exons 16 and 17, of ALDH16A1_v1. These fusions change the open reading frame, and introduce an early stop codon with a subsequent truncated peptide domain.[52]

This ALDH has 2 transcriptional variants, and both transcripts have 18 exons. ALDH18A1 and ALD18A1_v2 have 3470 and 3464 bps, respectively. The 6-bp difference is due to exon 6 that is 159 bp in the native transcript, and 153 bp in the second transcript.[52]

Copy-Number Variation, InDel, and Insertions

Copy-number variations (CNV) arise when there is a gain or loss of more than 1 kb DNA sequence; this can result in gene duplication, where a functional gene is produced, or a pseudogene is formed. InDels (insertion and deletion) variations are gain and loss of DNA sequence of 999–1000 bps.

Using the genomic variants in human genome (Build GRCh38: Dec. 2013, hg38), a query of database indicated the several CNVs identified in ALDH, as at January 2015. Hits from the query indicated the existence of about 157 ALDH variants identified from the database of genome variants (Table 3.3). Of these, CNV accounts

TABLE 3.3 Copy-Number Variation of Human ALDH Superfamily

ALDH gene	Accession no.	CNV	Deletion	Insertion	Duplication	MEI	Inversion
ALDH1A1	esv3370794			Gain			
	esv3423724			Gain			
	esv991190			Gain			
ALDH1A2	dgv12e194		Loss				
	esv1602894		Loss				
	esv2533857		Loss				
	esv2658819	Loss					
	esv2658819		Loss				
	esv2749738		Loss				
	esv2749739		Loss				
	esv3191417		Loss				
	esv4077		Loss				
	nsv569593	Loss					
	nsv569594	Loss					
	nsv569594	Loss					
	nsv569595	Loss					
	nsv569596	Loss					
	nsv833023	Gain					
	nsv976939				Gain		

(Continued)

TABLE 3.3 Copy-Number Variation of Human ALDH Superfamily *(cont.)*

ALDH gene	Accession no.	CNV	Deletion	Insertion	Duplication	MEI	Inversion
ALDH1A3	dgv4781n54	Gain					
	dgv4782n54	Gain + loss					
	dgv4783n54	Gain + loss					
	dgv4784n54	Gain					
	dgv4785n54	Gain + loss					
	dgv4786n54	Gain					
	dgv4787n54	Loss					
	esv1234590		Loss				
	esv1234590		Loss				
	esv1276733		Loss				
	esv1484966		Loss				
	esv2046145		Loss				
	esv2547886		Loss				
	esv26584	Loss					
	esv2750165		Loss				
	esv2750166		Loss				
	esv2750167		Loss				
	esv2750168		Loss				
	esv2750169		Loss				
	esv4123		Loss				
	esv4921	Loss					
	esv9086	Loss					
	esv993186		Loss				
	nsv457288	Loss					
	nsv469824	Loss					
	nsv469824	Loss					
	nsv476137			Novel insert			
	nsv570802	Gain					
	nsv570804	Gain + loss					
	nsv570816	Gain					
	nsv570825	Loss					
	nsv570826	Loss					
	nsv570827	Loss					
	nsv827467	Gain					
	nsv957971		Loss				
ALDH1B1	esv34172	Loss					
	nsv527063	Loss					
	nsv614271	Loss					
	nsv8445	Gain + loss					

TABLE 3.3 Copy-Number Variation of Human ALDH Superfamily *(cont.)*

ALDH gene	Accession no.	CNV	Deletion	Insertion	Duplication	MEI	Inversion
ALDH1L1	dgv8545n54	Loss					
	dgv8546n54	Loss					
	esv23630	Loss					
	esv23630	Loss					
	esv23981	Loss					
	esv2678139		Loss				Yes
	esv2725864		Loss				
	esv2725865		Loss				
	esv3305167					Gain	
	esv3308825					Gain	
	esv3405792			Gain			
	esv3442040			Gain			
	esv7912						
	nsv3985		Loss				
	nsv470871	Loss					
	nsv520037	Gain + loss					
	nsv524971	Loss					
	nsv591465	Gain					
	nsv591496	Loss					
	nsv829716	Gain + loss					
	nsv955334		Loss		Gain		
	nsv955335						
ALDH1L2	esv1002429			Loss			
	esv1010041		Loss				
	esv1010041		Loss				
	esv3306561					Gain	
	esv3309843					Gain	
	esv3442903			Gain			
	nsv832505	Gain					
	nsv832506	Gain					Yes
	nsv832506	Gain					
ALDH2	dgv2874n54	Gain					
	esv2760710	Gain					
	nsv455715	Gain					
	nsv470317	Gain					
	nsv515716	Gain					
ALDH3A1	nsv457704	Loss					
	nsv470582	Loss					

(Continued)

TABLE 3.3 Copy-Number Variation of Human ALDH Superfamily *(cont.)*

ALDH gene	Accession no.	CNV	Deletion	Insertion	Duplication	MEI	Inversion
	nsv526290	Loss					
	nsv526290	Loss					
	nsv574580	Loss					
ALDH3A2	esv2659432		Loss				
	esv2758679	Loss					
	esv4947	Loss					
	nsv428337	Loss					
	nsv953846		Loss				
ALDH3B1	dgv156n27	Loss					
	dgv1981n54	Gain					
	dgv1983n54	Loss					
	esv2759835	Gain + loss					
	nsv428260	Gain					
	nsv468606	Gain					
	nsv468615	Loss					
	nsv469856	Gain					
	nsv469964	Loss					
	nsv475863		Novel insert (gain)				
	nsv555277	Loss					
	nsv555278	Loss					
	nsv825963	Gain					
	nsv8836	Gain					
	nsv983037						
ALDH4A1	nsv479029						
	nsv545719	Loss					
ALDH5A1	esv2648573		Loss		Gain		
	esv2763542	Gain					
	esv34020	Loss					
	nsv519552	Gain					
	nsv965688						
ALDH6A1	esv22046	Loss					
	esv2614022		Loss				
	esv2660368	Loss					
	esv2759997	Gain					
ALDH7A1	esv2677758		Loss				
	esv3400328			Gain			
	nsv462442	Gain					
	nsv464063	Gain					
	nsv5489		Loss				

TABLE 3.3 Copy-Number Variation of Human ALDH Superfamily *(cont.)*

ALDH gene	Accession no.	CNV	Deletion	Insertion	Duplication	MEI	Inversion
	nsv599662	Gain					
	nsv604722	Gain					
	nsv604723	Gain					
ALDH9A1	esv1624307			Gain			
	esv2555363			Gain			
	esv33411	Loss					
	esv33567	Loss					
	nsv946481				Gain		
ALDH16A1	dgv52n68	Loss					
	esv1208352			Gain			
	esv1213251		Loss				
	esv1432834			Gain			
	esv1526018			Gain			
	esv2674843		Loss				
	esv2718721		Loss				
	esv3384892				Gain		
	nsv2518			Gain			
	nsv499659	Gain					
	nsv833862	Loss					
	nsv953599		Loss				
	nsv9739	Gain + loss					
ALDH18A1	nsv526180	Gain					

CNV, copy-number variations; MEI, mobile element insertion.

for 91, consisting of variants with loss, gain, and gain-and-loss of gene functions. Only nine variants of the 91 CNV have both gain-and-loss of gene function. There are 37 deletion variants, and all the deletion variants have loss of gene function. There are 17 insertion variants, out of which three have novel insertion: ALDH4A1, ALDH3B1, and ALDH1A3. There are six variants due to gene duplication (all the six variants have gain of function). Gene duplication variants are identified in ALDH1A2, ALDH1L1, ALDH3B1, ALDH5A1, ALDH9A1, and ALDH16A1. Mobile element insertions are found in four variants, mainly in ALDH1B2 (2) and ALDH1L2 (2). Among the 157 variants, there are two inversions found only in ALDH1B2 and ALDH1L2.

ALDH POLYMORPHISM IN ALCOHOL-RELATED DISEASES

The crucial roles of human ALDH in the metabolism of alcohol can never be underestimated. Alcohol consumption can be either detrimental or beneficial, depending on several factors, such as the nature of ALDH enzymes that metabolize acetaldehyde to acetate. Hence, any genetic variations that affect the catalytic functions and/or other protein domains of ALDH may also have other consequences for the drinkers.

The human ALDH1B1 has four variants (Arg86Val, Leu107Lys, Thr202Ile, and Val253Met). The Arg86Val, Leu107Lys, Thr202Ile, and Val253Met correspond to ALDH1B1*2, ALDH1B1*3, ALDH1B1*4, and ALDH1B1*5, respectively. A computational modeling suggests that ALDH1B1*2 is catalytically inactive, probably due to poor NAD^+ binding. Both ALDH1B1*3 and ALDH1B1*5 are catalytically active, and ALDH1B1*3 may be less able to metabolize all-trans retinaldehyde whereas ALDH1B1*5 may bind NAD^+ poorly.[59]

Human ALDH1B1 single nucleotide polymorphism (SNP) (C/T; rs2228093) results in the change of alanine to valine, at position 143. This polymorphism, found in Scandinavian individuals, has been strongly associated with alcohol hypersensitivity in these individuals.[60] Another SNP of the ALDH1B1 (G/T; rs2073478), resulting in leucine instead of arginine, at position 170, is reported to play a significant role in the development of coronary artery disease, a complex disease that has risk factors such as high alcohol consumption.[61]

There are currently no evidences suggesting that ALDH variants belonging to human ALDH3 family members are associated with any alcohol related disorders. It is reasonable to suggest that, unlike ALDH2 – a major player in alcohol metabolism, the rest play minimal roles in alcohol metabolism.

Mitochondrial ALDH2 is the most important protein in acetaldehyde removal in the liver, after alcohol dehydrogenase conversion of alcohol to acetaldehyde. The major polymorphism of ALDH2 is the ALDH2*2 (rs671) that has glutamate replaced with lysine, at amino acid position 504. ALDH2*2 protein product is catalytically inactive, with greatly reduced capability of metabolizing acetaldehyde.[62] The ALDH2*2 allele distribution among race is varying; it is most prevalent in East Asians, but very rare among Africans or Caucasians.[63] It is estimated that approximately 40% of Japanese have inactive forms of ALDH2, due to ALDH2*2 polymorphism.[64] Heterozygous individuals, with ALDH2*1/2*2 genotype, have only 6.25% of the normal ALDH2*1 protein, whereas the ALDH2*2 subunits are considered inactive. This explains the dominant effect of ALDH2*2.[65] Individuals with different ALDH2*2 genotypes will respond differently to alcohol challenges. For example, ALDH2*2/2*2 homozygous individuals who were given 0.1 g ethanol per kilogram body weight had 18 times blood acetaldehyde concentration, compared to heterozygous ALDH2*1/2*2 individuals on a similar treatment that had five times blood acetaldehyde concentration.[66] Accumulation of acetaldehyde in individuals with ALDH2*2 mutants leads to a "flushing response" that includes facial flushing, palpitations, drowsiness, and some more severe symptoms.[67] These responses that are due to the inactive form of ALDH2 deter many Japanese from drinking heavily and developing alcoholism.[68] ALDH2*2 is associated with alcoholic liver disease, cirrhosis, or pancreatitis in the alcoholic patients.[69]

The associated complications of ALDH2*2 are due to the accumulation of acetaldehyde, a carcinogen. Tumors of the respiratory tract, particularly adenocarcinomas and squamous cell carcinomas of the nasal mucosa in rats,[70] are produced after inhalation of acetaldehyde, and laryngeal carcinomas appear in hamsters after similar treatments.[71] Another report indicates that acetaldehyde induces chromosomal aberrations, micronuclei and sister chromatid exchanges in cultured mammalian cells.[72] These results and others[73] strongly lay the elucidative basis for the cancer-causing nature of acetaldehyde. A meta-analysis indicates a strong correlation with an increased risk of gastric cancer and ALDH2*2 (rs671).[74] Yokoyama and Omori have provided an elegant comprehensive report on the association of esophageal and head and neck cancer with ALDH2* allele.[75]

CONCLUSIONS

ALDH, together with alcohol dehydrogenase, plays a critical role in alcohol metabolism. There are, so far, 19 members of the ALDH superfamily, consisting of alcohol metabolizing and nonalcohol metabolizing ALDHs. The latter are involved in several essential cellular processes, including the metabolism of valine and pyrimidine by ALDH6A1, and protection of cells from oxidative stress by ALDH7A1.

Interestingly, the roles played by ALDHs are invariably influenced by modifications, such as an addition of phosphate, acetyl, nitric oxide, and nitrogen groups to the protein, during either physiological or stress conditions. Hyperacetylation of hepatic proteins in chronic alcohol consumption in rats is a typical example of the role of acetylated proteins in alcohol liver diseases. Further, alternative splicing, mutations, and deletions can hamper or enhance the ALDH functions. Several ALDH variants have been identified among a majority of ALDH members in humans. These variants are products of alternative splicing, a mechanism that results in the rearrangement of exons. Some ALDH variants, for example, one belonging to ALDH16A1, have exon and intron fused, leading to a truncated protein. Deletion and/or insertion of ALDH also result in a copy-number variation (CNV) of ALDH. Several ALDH CNVs have been identified, with ALDH functions enhanced and/or lost.

The ALDH2 is the major player in the metabolism of alcohol-derived acetaldehyde, a carcinogenic whose accumulation is detrimental to cellular processes. Individuals with mutant ALDH2 (ALDH2*2) have been identified to manifest a range of diseases and cancers, including liver and head and neck cancers.

Key Facts on Aldehyde Dehydrogenase, Member 2 (ALDH2)

- ALDH2 is the most researched aldehyde dehydrogenase.
- ALDH2 is mainly expressed in the mitochondrial of the liver and is the major player in alcohol metabolism.
- A major single base mutation of ALDH2 (ALDH2*2) results in a replacement of glutamate with lysine at amino acid position 504.
- ALDH2*2 mutation varies among populations, such that it is very common in East Asians, but very rare among Africans or Caucasians.

- Individuals with different forms of ALDH2*2 mutations respond quite differently following alcohol consumption, and have alcohol intolerance leading to facial flushing, palpitations, drowsiness.
- ALDH2*2 is associated with alcoholic liver disease, cirrhosis, or pancreatitis in alcoholic patients.

Summary Points

- Aldehyde dehydrogenases (ALDHs) are oxidizing enzymes that are involved in detoxification of both exogenous and endogenous aldehydes.
- Nineteen members of the human aldehyde dehydrogenase superfamily have been identified so far, and only a few are involved in the metabolism of alcohol-derived aldehydes, for example, ALDH2.
- ALDHs are modified by addition of acetyl and phosphate groups under certain conditions: for example, phosphorylation of ALDH is induced in the presence of alcohol, and results in apoptosis.
- A rich variety of ALDHs are produced through alternative splicing and mutations (deletion and gene duplication), leading to myriad functions.
- Single nucleotide polymorphism of ALDHs could result in alcohol dependence or cancer of the head and neck, etc.

Abbreviations

4-HNE	4-Hydroxynonenal
ALDH	Aldehyde dehydrogenase
ARCL3A	Autosomal recessive cutis laxa, type 3A
CCl_4	Carbon tetrachloride
CO_2	Carbon dioxide
CNV	Copy-number variations
DNA	Deoxyribonucleic acid
GTN	Glyceryl nitrinitrate
HSP70	Heat shock protein 70
JNK	c-Jun N-terminal protein kinase
MAPK8/9	Mitogen-activated protein kinase 8/9
MCOP8	Microphthalmia isolated 8
MMSDHD	Methylmalonate semialdehyde dehydrogenase deficiency
$NADP^+$	Nicotinamide adenine dinucleotide phosphate
PKCε	Protein kinase Cε
PTM	Posttranslational modification
SIP	Sphingosine 1-phosphate
SLS	Sjoegren–Larsson syndrome
SNP	Single nucleotide polymorphism
TP53	Tumor protein 53

References

1. Vasiliou V, Pappa A, Petersen DR. Role of aldehyde dehydrogenases in endogenous and xenobiotic metabolism. *Chem Biol Interact* 2000;**129**(1–2):1–19.
2. Dockham PA, Lee MO, Sladek NE. Identification of human liver aldehyde dehydrogenases that catalyze the oxidation of aldophosphamide and retinaldehyde. *Biochem Pharmacol* 1992;**43**(11):2453–69.
3. Chen Y, Thompson DC, Koppaka V, Jester JV, Vasiliou V. Ocular aldehyde dehydrogenases: protection against ultraviolet damage and maintenance of transparency for vision. *Prog Retin Eye Res* 2013;**33**:28–39.
4. Jackson B, Brocker C, Thompson DC, et al. Update on the aldehyde dehydrogenase gene (ALDH) superfamily. *Hum Genom* 2011;**5**(4):283–303.
5. Inoue K, Nishimukai H, Yamasawa K. Purification and partial characterization of aldehyde dehydrogenase from human erythrocytes. *Biochim Biophys Acta* 1979;**569**(2):117–23.
6. Perez-Miller S, Younus H, Vanam R, Chen CH, Mochly-Rosen D, Hurley TD. Alda-1 is an agonist and chemical chaperone for the common human aldehyde dehydrogenase 2 variant. *Nat Struct Mol Biol* 2010;**17**(2):159–64.
7. Braun T, Bober E, Singh S, Agarwal DP, Goedde HW. Evidence for a signal peptide at the amino-terminal end of human mitochondrial aldehyde dehydrogenase. *FEBS Lett* 1987;**215**(2):233–6.
8. Eckey R, Timmann R, Hempel J, Agarwal DP, Goedde HW. Biochemical, immunological, and molecular characterization of a "high K_m" aldehyde dehydrogenase. *Adv Experim Med Biol* 1991;**284**:43–52.
9. Khanna M, Chen CH, Kimble-Hill A, et al. Discovery of a novel class of covalent inhibitor for aldehyde dehydrogenases. *J Biol Chem* 2011;**286**(50):43486–94.
10. Nakahara K, Ohkuni A, Kitamura T, et al. The Sjogren–Larsson syndrome gene encodes a hexadecenal dehydrogenase of the sphingosine 1-phosphate degradation pathway. *Mol Cell* 2012;**46**(4):461–71.
11. Tsukamoto N, Chang C, Yoshida A. Mutations associated with Sjogren–Larsson syndrome. *Ann Hum Gen* 1997;**61**(Pt. 3):235–42.
12. Hsu LC, Chang WC, Hiraoka L, Hsieh CL. Molecular cloning, genomic organization, and chromosomal localization of an additional human aldehyde dehydrogenase gene, ALDH6. *Genomics* 1994;**24**(2):333–41.
13. Aldahmesh MA, Khan AO, Hijazi H, Alkuraya FS. Mutations in ALDH1A3 cause microphthalmia. *Clin Genet* 2013;**84**(2):128–31.
14. Oleinik NV, Krupenko NI, Krupenko SA. Cooperation between JNK1 and JNK2 in activation of p53 apoptotic pathway. *Oncogene* 2007;**26**(51):7222–30.
15. Oleinik NV, Krupenko NI, Krupenko SA. ALDH1L1 inhibits cell motility via dephosphorylation of cofilin by PP1 and PP2A. *Oncogene* 2010;**29**(47):6233–44.
16. Krupenko NI, Dubard ME, Strickland KC, Moxley KM, Oleinik NV, Krupenko SA. ALDH1L2 is the mitochondrial homolog of 10-formyltetrahydrofolate dehydrogenase. *J Biol Chem* 2010;**285**(30):23056–63.
17. Chambliss KL, Gray RG, Rylance G, Pollitt RJ, Gibson KM. Molecular characterization of methylmalonate semialdehyde dehydrogenase deficiency. *J Inherit Metab Dis* 2000;**23**(5):497–504.
18. Brocker C, Lassen N, Estey T, et al. Aldehyde dehydrogenase 7A1 (ALDH7A1) is a novel enzyme involved in cellular defense against hyperosmotic stress. *J Biol Chem* 2010;**285**(24):18452–63.
19. Lin M, Napoli JL. cDNA cloning and expression of a human aldehyde dehydrogenase (ALDH) active with 9-*cis*-retinal and identification of a rat ortholog, ALDH12. *J Biol Chem* 2000;**275**(51):40106–12.
20. Hanna MC, Blackstone C. Interaction of the SPG21 protein ACP33/maspardin with the aldehyde dehydrogenase ALDH16A1. *Neurogenetics* 2009;**10**(3):217–28.
21. Baumgartner MR, Hu CA, Almashanu S, et al. Hyperammonemia with reduced ornithine, citrulline, arginine and proline: a new inborn error caused by a mutation in the gene encoding delta(1)-pyrroline-5-carboxylate synthase. *Hum Mol Genet* 2000;**9**(19):2853–8.
22. Hu CA, Lin WW, Obie C, Valle D. Molecular enzymology of mammalian Delta1-pyrroline-5-carboxylate synthase. Alternative

splice donor utilization generates isoforms with different sensitivity to ornithine inhibition. *J Biol Chem* 1999;**274**(10):6754–62.
23. Bicknell LS, Pitt J, Aftimos S, Ramadas R, Maw MA, Robertson SP. A missense mutation in ALDH18A1, encoding Delta1-pyrroline-5-carboxylate synthase (P5CS), causes an autosomal recessive neurocutaneous syndrome. *Eur J Hum Gen* 2008;**16**(10):1176–86.
24. Reed NA, Cai D, Blasius TL, et al. Microtubule acetylation promotes kinesin-1 binding and transport. *Curr Biol* 2006;**16**(21):2166–72.
25. Kniep B, Kniep E, Ozkucur N, et al. 9-O-acetyl GD3 protects tumor cells from apoptosis. *Int J Cancer* 2006;**119**(1):67–73.
26. Birks SM, Danquah JO, King L, Vlasak R, Gorecki DC, Pilkington GJ. Targeting the GD3 acetylation pathway selectively induces apoptosis in glioblastoma. *Neuro Oncol* 2011;**13**(9):950–60.
27. Fu M, Wang C, Zhang X, Pestell RG. Acetylation of nuclear receptors in cellular growth and apoptosis. *Biochem Pharmacol* 2004;**68**(6):1199–208.
28. Hirschey MD, Shimazu T, Goetzman E, et al. SIRT3 regulates mitochondrial fatty-acid oxidation by reversible enzyme deacetylation. *Nature* 2010;**464**(7285):121–5.
29. Shepard BD, Tuma DJ, Tuma PL. Chronic ethanol consumption induces global hepatic protein hyperacetylation. *Alcohol Clin Exp Res* 2010;**34**(2):280–91.
30. Recknagel RO, Glende Jr EA, Dolak JA, Waller RL. Mechanisms of carbon tetrachloride toxicity. *Pharmacol Therap* 1989;**43**(1):139–54.
31. Song BJ, Suh SK, Moon KH. A simple method to systematically study oxidatively modified proteins in biological samples and its applications. *Methods Enzymol* 2010;**473**:251–64.
32. Lee YJ, Aroor AR, Shukla SD. Temporal activation of p42/44 mitogen-activated protein kinase and c-Jun N-terminal kinase by acetaldehyde in rat hepatocytes and its loss after chronic ethanol exposure. *J Pharmacol Exp Therap* 2002;**301**(3):908–14.
33. Yoo SH, Park O, Henderson LE, Abdelmegeed MA, Moon KH, Song BJ. Lack of PPARalpha exacerbates lipopolysaccharide-induced liver toxicity through STAT1 inflammatory signaling and increased oxidative/nitrosative stress. *Toxicol Lett* 2011;**202**(1):23–9.
34. Hjelle JJ, Grubbs JH, Beer DG, Petersen DR. Inhibition of rat liver aldehyde dehydrogenase by carbon tetrachloride. *J Pharmacol Exp Therap* 1981;**219**(3):821–6.
35. Hjelle JJ, Grubbs JH, Beer DG, Petersen DR. Time course of the carbon tetrachloride-induced decrease in mitochondrial aldehyde dehydrogenase activity. *Toxicol Appl Pharmacol* 1983;**67**(2):159–65.
36. Song BJ, Moon KH, Upreti VV, Eddington ND, Lee IJ. Mechanisms of MDMA (ecstasy)-induced oxidative stress, mitochondrial dysfunction, and organ damage. *Curr Pharmaceut Biotechnol* 2010;**11**(5):434–43.
37. Song BJ, Abdelmegeed MA, Yoo SH, et al. Post-translational modifications of mitochondrial aldehyde dehydrogenase and biomedical implications. *J Proteom* 2011;**74**(12):2691–702.
38. Mendelson KG, Contois LR, Tevosian SG, Davis RJ, Paulson KE. Independent regulation of JNK/p38 mitogen-activated protein kinases by metabolic oxidative stress in the liver. *Proc Natl Acad Sci USA* 1996;**93**(23):12908–13.
39. Chen CH, Gray MO, Mochly-Rosen D. Cardioprotection from ischemia by a brief exposure to physiological levels of ethanol: role of epsilon protein kinase C. *Proc Natl Acad Sci USA* 1999;**96**(22):12784–9.
40. Churchill EN, Disatnik MH, Mochly-Rosen D. Time-dependent and ethanol-induced cardiac protection from ischemia mediated by mitochondrial translocation of varepsilonPKC and activation of aldehyde dehydrogenase 2. *J Mol Cell Cardiol* 2009;**46**(2):278–84.
41. Budas GR, Disatnik MH, Chen CH, Mochly-Rosen D. Activation of aldehyde dehydrogenase 2 (ALDH2) confers cardioprotection in protein kinase C epsilon (PKCvarepsilon) knockout mice. *J Mol Cell Cardiol* 2010;**48**(4):757–64.
42. Chen CH, Budas GR, Churchill EN, Disatnik MH, Hurley TD, Mochly-Rosen D. Activation of aldehyde dehydrogenase-2 reduces ischemic damage to the heart. *Science* 2008;**321**(5895):1493–5.
43. Sun A, Zou Y, Wang P, et al. Mitochondrial aldehyde dehydrogenase 2 plays protective roles in heart failure after myocardial infarction via suppression of the cytosolic JNK/p53 pathway in mice. *J Am Heart Assoc* 2014;**3**(5):e000779.
44. Hess DT, Matsumoto A, Kim SO, Marshall HE, Stamler JS. Protein S-nitrosylation: purview and parameters. *Nat Rev Mol Cell Biol* 2005;**6**(2):150–66.
45. Radi R, Beckman JS, Bush KM, Freeman BA. Peroxynitrite oxidation of sulfhydryls. The cytotoxic potential of superoxide and nitric oxide. *J Biol Chem* 1991;**266**(7):4244–50.
46. Moon KH, Abdelmegeed MA, Song BJ. Inactivation of cytosolic aldehyde dehydrogenase via S-nitrosylation in ethanol-exposed rat liver. *FEBS Lett* 2007;**581**(21):3967–72.
47. Moon KH, Kim BJ, Song BJ. Inhibition of mitochondrial aldehyde dehydrogenase by nitric oxide-mediated S-nitrosylation. *FEBS Lett* 2005;**579**(27):6115–20.
48. Chen Z, Foster MW, Zhang J, et al. An essential role for mitochondrial aldehyde dehydrogenase in nitroglycerin bioactivation. *Proc Natl Acad Sci USA* 2005;**102**(34):12159–64.
49. Sydow K, Daiber A, Oelze M, et al. Central role of mitochondrial aldehyde dehydrogenase and reactive oxygen species in nitroglycerin tolerance and cross-tolerance. *J Clin Invest* 2004;**113**(3):482–9.
50. Oelze M, Knorr M, Schell R, et al. Regulation of human mitochondrial aldehyde dehydrogenase (ALDH-2) activity by electrophiles *in vitro*. *J Biol Chem* 2011;**286**(11):8893–900.
51. Tyther R, Ahmeda A, Johns E, Sheehan D. Proteomic identification of tyrosine nitration targets in kidney of spontaneously hypertensive rats. *Proteomics* 2007;**7**(24):4555–64.
52. Black WJ, Stagos D, Marchitti SA, et al. Human aldehyde dehydrogenase genes: alternatively spliced transcriptional variants and their suggested nomenclature. *Pharmacogenet Genom* 2009;**19**(11):893–902.
53. Ono Y, Fukuhara N, Yoshie O. TAL1 and LIM-only proteins synergistically induce retinaldehyde dehydrogenase 2 expression in T-cell acute lymphoblastic leukemia by acting as cofactors for GATA3. *Mol Cell Biol* 1998;**18**(12):6939–50.
54. Koenig U, Amatschek S, Mildner M, Eckhart L, Tschachler E. Aldehyde dehydrogenase 1A3 is transcriptionally activated by all-trans-retinoic acid in human epidermal keratinocytes. *Biochem Biophys Res Commun* 2010;**400**(2):207–11.
55. Krupenko SA, Oleinik NV. 10-formyltetrahydrofolate dehydrogenase, one of the major folate enzymes, is down-regulated in tumor tissues and possesses suppressor effects on cancer cells. *Cell Growth Different* 2002;**13**(5):227–36.
56. Rogers GR, Markova NG, De Laurenzi V, Rizzo WB, Compton JG. Genomic organization and expression of the human fatty aldehyde dehydrogenase gene (FALDH). *Genomics* 1997;**39**(2):127–35.
57. Chambliss KL, Hinson DD, Trettel F, et al. Two exon-skipping mutations as the molecular basis of succinic semialdehyde dehydrogenase deficiency (4-hydroxybutyric aciduria). *Am J Hum Genet* 1998;**63**(2):399–408.
58. Blasi P, Palmerio F, Caldarola S, et al. Succinic semialdehyde dehydrogenase deficiency: clinical, biochemical and molecular characterization of a new patient with severe phenotype and a novel mutation. *Clin Genet* 2006;**69**(3):294–6.
59. Jackson BC, Reigan P, Miller B, Thompson DC, Vasiliou V. Human ALDH1B1 Polymorphisms may affect the metabolism of acetaldehyde and all-trans retinaldehyde-*in vitro* studies and computational modeling. *Pharmaceut Res* 2015;**32**(5):1648–62.
60. Linneberg A, Gonzalez-Quintela A, Vidal C, et al. Genetic determinants of both ethanol and acetaldehyde metabolism influence alcohol hypersensitivity and drinking behaviour among Scandinavians. *Clin Experim Allerg* 2010;**40**(1):123–30.

61. Wang Y, Du F, Zhao H, et al. Synergistic association between two alcohol metabolism relevant genes and coronary artery disease among Chinese hypertensive patients. *PloS One* 2014;**9**(7):e103161.
62. Yoshida A, Hsu LC, Yasunami M. Genetics of human alcohol-metabolizing enzymes. *Prog Nucleic Acid Res Mol Biol* 1991;**40**:255–87.
63. Goedde HW, Agarwal DP, Fritze G, et al. Distribution of ADH2 and ALDH2 genotypes in different populations. *Hum Genet* 1992;**88**(3):344–6.
64. Higuchi S, Matsushita S, Murayama M, Takagi S, Hayashida M. Alcohol and aldehyde dehydrogenase polymorphisms and the risk for alcoholism. *Am J Psych* 1995;**152**(8):1219–21.
65. Crabb DW, Edenberg HJ, Bosron WF, Li TK. Genotypes for aldehyde dehydrogenase deficiency and alcohol sensitivity. The inactive ALDH2(2) allele is dominant. *J Clin Invest* 1989;**83**(1):314–6.
66. Enomoto N, Takase S, Yasuhara M, Takada A. Acetaldehyde metabolism in different aldehyde dehydrogenase-2 genotypes. *Alcohol Clin Experim Res* 1991;**15**(1):141–4.
67. Muramatsu T, Higuchi S, Shigemori K, et al. Ethanol patch test – a simple and sensitive method for identifying ALDH phenotype. *Alcohol Clin Experim Res* 1989;**13**(2):229–31.
68. Harada S, Agarwal DP, Goedde HW, Tagaki S, Ishikawa B. Possible protective role against alcoholism for aldehyde dehydrogenase isozyme deficiency in Japan. *Lancet* 1982;**2**(8302):827.
69. Li D, Zhao H, Gelernter J. Strong protective effect of the aldehyde dehydrogenase gene (ALDH2) 504lys (*2) allele against alcoholism and alcohol-induced medical diseases in Asians. *Hum Genet* 2012;**131**(5):725–37.
70. Woutersen RA, Appelman LM, Van Garderen-Hoetmer A, Feron VJ. Inhalation toxicity of acetaldehyde in rats. III. Carcinogenicity study. *Toxicology* 1986;**41**(2):213–31.
71. Feron VJ, Kruysse A, Woutersen RA. Respiratory tract tumours in hamsters exposed to acetaldehyde vapour alone or simultaneously to benzo(a)pyrene or diethylnitrosamine. *Eur J Can Clin Oncol* 1982;**18**(1):13–31.
72. Dellarco VL. A mutagenicity assessment of acetaldehyde. *Mut Res* 1988;**195**(1):1–20.
73. Vaca CE, Fang JL, Schweda EK. Studies of the reaction of acetaldehyde with deoxynucleosides. *Chem Biol Interact* 1995;**98**(1):51–67.
74. Wang HL, Zhou PY, Liu P, Zhang Y. ALDH2 and ADH1 genetic polymorphisms may contribute to the risk of gastric cancer: a meta-analysis. *PloS One* 2014;**9**(3):e88779.
75. Yokoyama A, Omori T. Genetic polymorphisms of alcohol and aldehyde dehydrogenases and risk for esophageal and head and neck cancers. *Alcohol* 2005;**35**(3):175–85.

CHAPTER

4

Alcohol Intake and Apoptosis: A Review and Examination of Molecular Mechanisms in the Central Nervous System

Maria Camargo Moreno, MPH[*,**,‡], *Cherry Ignacio, PhD*[*,**,‡], *Patrick Burke, MS*[**], *Frank A. Middleton, PhD*[*,**,†,‡]

*Department of Biochemistry & Molecular Biology, SUNY Upstate Medical University, Syracuse, NY, USA
**Department of Neuroscience & Physiology, SUNY Upstate Medical University, Syracuse, NY, USA
†Department of Psychiatry & Behavioral Sciences, SUNY Upstate Medical University, Syracuse, NY, USA
‡Developmental Exposure Alcohol Research Center, Binghamton University, Binghamton, NY, USA

INTRODUCTION

The focus of this review is on the effect of alcohol consumption on cell death in the brain and central nervous system (CNS). Although some level of cell death is considered normal and, indeed, necessary for proper organ and tissue development, widespread increases in cell death, following chronic ethanol exposure, can be harmful or lethal to the organism. This is particularly true for cell death in the CNS. Unlike many of the other cells and tissues affected by ethanol, cells in the CNS are unable to regenerate or be replaced if they are lost. Thus, regulating cell death within the CNS is critically important to cell and tissue viability.

Cell Death Through Necrosis and Apoptosis

One way that cells regulate cell death is through the activation or inhibition of two specific cell-signaling pathways, termed necrosis and apoptosis. Before discussing the evidence that either of these processes occurs as a consequence of ethanol exposure, it is first helpful to briefly distinguish them from each other.

Necrosis

The process of necrosis is considered a somewhat disorganized cellular or tissue response to a direct injury or major insult, such as physical trauma, infection, or anoxia. The steps involved in necrosis include release of lysosomal enzymes, fragmentation of the cell membrane, release of cytosolic contents to the extracellular space, and initiation of a potent inflammatory response to the local area.

Apoptosis

In contrast to necrosis, when cells experience more subtle challenges, or simply as a consequence of normal changes in tissue development, they often initiate the process of programmed cell death, or apoptosis.[1,2] Apoptosis is distinguished from necrosis in that cells undergoing apoptosis rarely cause an inflammatory response in the surrounding cells and tissue. However, as necessary as apoptosis is to multicellular organisms (and to some single cell organisms like yeast), it can also lead to detrimental effects, if it occurs in excess. In the case of ethanol exposure to humans, the well-known increase in apoptosis is highly detrimental to the organism, regardless of the organ in which the apoptosis has been measured.[3–9]

In this review, we consider much of the evidence for apoptosis in the CNS as a possible consequence of exposure to nutritionally relevant (i.e., physiological) concentrations of ethanol. Indeed, there is ample evidence that apoptosis can and does occur in this context. The specifics of how ethanol damages cells have been extensively

Molecular Aspects of Alcohol and Nutrition. http://dx.doi.org/10.1016/B978-0-12-800773-0.00004-5

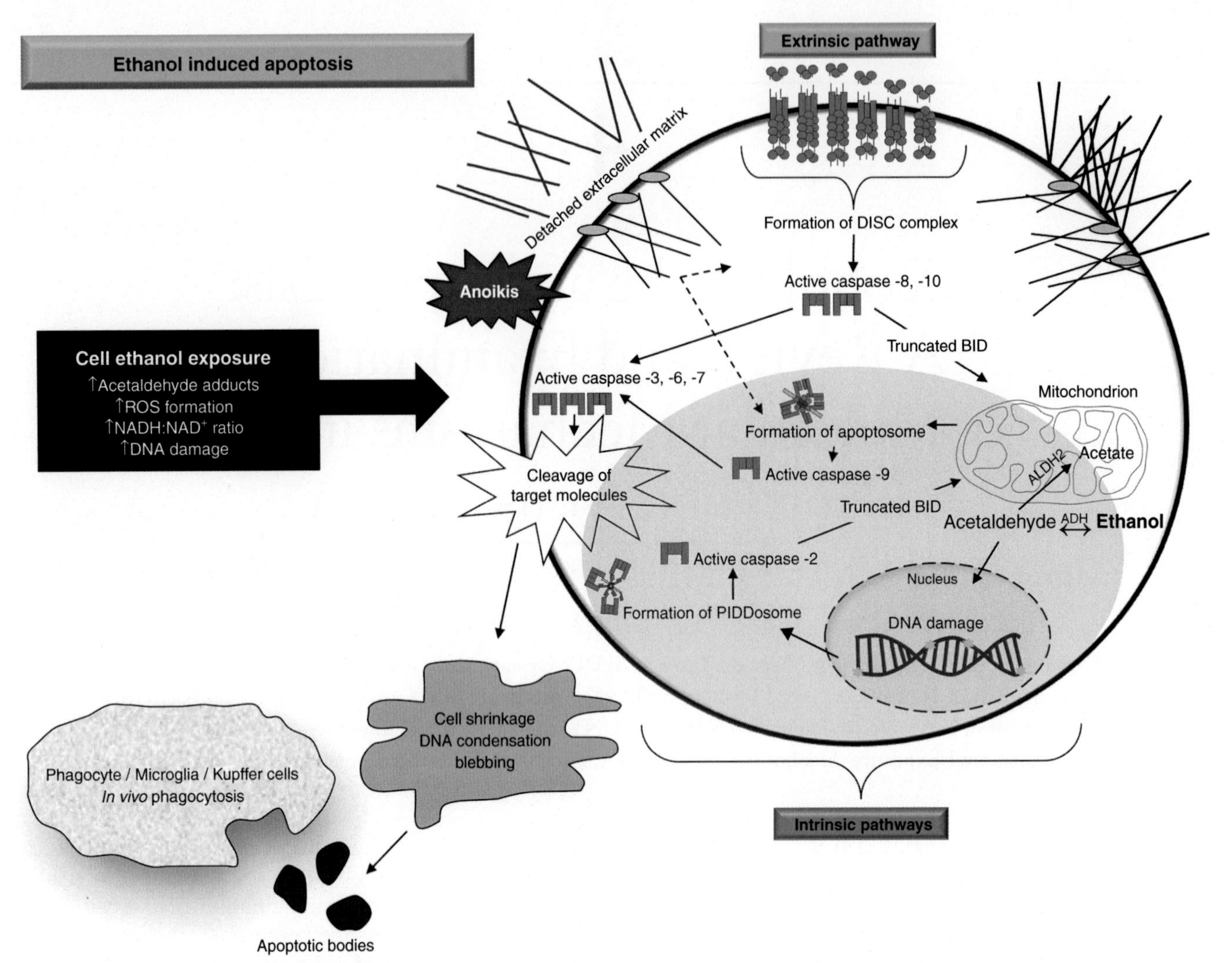

FIGURE 4.1 **Cellular apoptosis induced by exposure to ethanol.** Intrinsic and extrinsic pathways may be activated, as well as anoikis. Although different events can initiate each, there are points of convergence at the level of effectors caspases-3, -6, and -7, and both culminate with formation of apoptotic bodies and eventual phagocytosis.

investigated and are now recognized as including single and double strand DNA breaks, and DNA adduct formation, acetaldehyde formation, reactive oxygen species (ROS) formation, and increased NADH:NAD$^+$ ratio formation.[10–12] Despite general agreement about the consistency of these particular events, the cellular pathways activated in cell death are still being determined. Thus, as will become evident, the rate at which apoptosis occurs, and whether there is also any accompanying necrosis, depends on many variables and experimental conditions.

There are three main pathways that can be activated for a cell to reach an apoptotic state: intrinsic, extrinsic, and granzyme/perforin. There is also a fourth noncaspase pathway, mediated by the release of the mitochondrial apoptosis inducing factor, but this has been much less studied in regards to ethanol exposure.[13–15] Importantly, the three main pathways each have their own subpathways, and specific positive and negative regulators, but share the final steps in common: activation of proteases (specifically the effector caspases-3, -6, and -7), activation of endonucleases, chromatin condensation and marginalization, cell shrinkage, plasma blebbing, DNA fragmentation, flipping of phosphatidylserine to expose the negatively charged hydrophilic head to the extracellular space, and formation of apoptotic bodies that are engulfed by neighboring cells and by specialized macrophages, such as Kupffer cells in the liver, or microglia in the CNS (Figure 4.1).[2]

Although they share many later events in common, the onset of apoptosis is actually distinct in the three main pathways. The extrinsic pathway is distinguished by activation of initiator caspases-8 and -10. These caspases are themselves activated by the formation of the death-inducing signaling complex (DISC). DISC is turned on by the binding of extracellular cytokines FAS L, TNF-α,

TWEAK, TRAIL, and APP to their respective death receptors: FAS, TNF-R1, TRAMP, TRAIL-R1, TRAIL-R2, and DR6.[4] Binding of the ligand to one of these death receptors produces a conformational change in the cytosolic part of the complex that promotes procaspase-8 and -10 binding, along with the adaptor proteins FADD or TRADD.[16] This binding is what leads to the formation of DISC and ensuing activation of caspase-8 and -10 (Figure 4.1).[2]

In contrast to the extrinsic pathway, one of the two intrinsic pathways is distinguished by the activation of caspase-9, when it forms part of the apoptosome. The two other proteins that comprise the apoptosome are the mitochondria-released proteins APAF1 and cytochrome c.[2,17] The intrinsic pathway is notable because of the mitochondrial outer membrane permeabilization (that allows for cytochrome c and APAF1 release, along with calcium). Nonetheless, this feature alone is not unique to the intrinsic pathway, since the other two main apoptosis pathways can also lead to mitochondrial outer membrane permeabilization, via the truncation of the BID protein (Figure 4.1).[18]

The second intrinsic pathway is triggered by the activation of caspase-2, and several subpathways have been identified that can promote this activation, including the controversial formation of a PIDDosome by PIDD and RAIDD proteins.[19,20] Finally, the last main apoptosis pathway, termed the granzyme/perforin pathway, is activated by cytotoxic T lymphocytes, a type of white blood cell. This pathway has been extensively characterized in peripheral tissues. However, this immune-cell triggered apoptosis is not seen as the starting point of ethanol induced apoptosis.

In addition to the three main pathways, of particular interest is the less characterized process of anoikis. It is defined as apoptosis induced by cell signals that occur as a result of detachment from the extracellular matrix. Integrins, transmembrane proteins, and cytoskeletal rearrangements all transduce information from the extracellular matrix to both the intrinsic and extracellular apoptotic pathways, in order to induce anoikis.[21] It is important to note that the type of anchorage dependence to the extracellular matrix varies considerably in different cells and tissues. Thus, anoikis uses different signal transduction pathways, depending on the local environment.[22]

Ethanol Activated Apoptosis in the CNS

Downstream of the initiation events, there is more consensus as to the apoptotic pathways activated by ethanol. In the CNS, it has traditionally been thought that the intrinsic apoptotic pathway is primarily activated by ethanol exposure. In contrast, in the gastrointestinal tract and liver, studies largely support activation of the extrinsic pathway after ethanol exposure.[4,8] Notably, the evidence for the intrinsic pathway is mainly based on increases in active caspase-3, and the absence of increases in active caspase-8, following exposures.[23] However, because caspase-3 is an effector caspase that is activated by the extrinsic, intrinsic, or granzyme/perforin pathways, its presence alone cannot be viewed as solely supporting the involvement of the intrinsic pathway. Along these lines, it is important to point out that there are *in vitro* studies that have reported greater extrinsic pathway activation, based on Fas and FasL expression,[24–27] as well as at least one study reporting increased caspase-2, -3, -8, and -9 activity following ethanol exposure.[28] Thus, at least some data support activation of both intrinsic and extrinsic pathways, and the differences in outcomes may depend on the tissue and cell types, as well as timing, dose, and routes of exposure.

The upstream signaling mechanisms that lead to increased apoptosis following ethanol exposure are still being debated. Some of the candidate signaling pathways that have been closely examined in this regard include neurotrophins, PI3K, p53, PTEN, GABA, and glutamate receptors and recycling machinery, among others.[1,29] While it is not possible to review all of this evidence here, we will highlight the major findings relevant to ethanol exposures covering one or more developmental periods: gestation, adolescence, and adulthood (Table 4.1).

Fetal Alcohol Exposure

The etiology of fetal alcohol spectrum disorder (FASD), though variable, is well known. However, our understanding of the molecular mechanisms through which alcohol exposure produces the FASD phenotype is incomplete, as is the explanation for why maternal alcohol consumption exerts its most detrimental effects on brain development, and subsequent measures of brain function, despite the fact that the entire embryo is exposed.[67] Moreover, within the CNS, although there are indications that the teratogenic effects of ethanol are widespread and affect "almost the entire brain" in both humans and animal models,[68,69] there appears to be an enhanced vulnerability to cell death and apoptosis in certain brain areas. These more sensitive areas include the anterior vermis, hippocampus, corpus callosum, caudate, and regions of the frontal, parietal, and temporal lobes.[68,69] Notably, however, the timing and duration of the exposure clearly influences these findings. Consequently, identifying possible explanations for the regional and temporal gradients in vulnerability to cell death, following ethanol exposure, has been a subject of considerable interest and study. Several mechanisms, already mentioned, have been implicated in such findings, including differences in cell–cell interactions, gene expression and epigenetic responses, relative differences in oxidative stress, DNA damage and cell cycle regulation,

TABLE 4.1 Apoptotic and Cell Stress Markers Studied in Alcohol-Exposed Neural Tissue From Different Aged Models

Marker	Fetal	Adolescent	Adult (includes postmortem)
Pro-caspase-3		Decrease,[30] increase[31]	Increase[32]
Active caspase-3	Increase[33–36]	Increase,[30,37] not detected[31]	Increase,[14,38,39] decrease[15]
Caspase-3 activity	Increase[13,28]	Increase[30,40]	Decrease[41]
Caspase-3 mRNA			Increase[38]
Bax	Apoptosis dependent on Bax presence[36]		Decrease in mitochondria,[30] increase[32]
Active caspase-8	No change[36]		
p53	Apoptosis independent of p53 presence[34]		No change[15]
PUMA	Apoptosis dependent on PUMA presence[34]		
NOXA	Apoptosis independent of NOXA[34]		
Bid	Apoptosis independent of Bid[34]		Decrease[15]
Bad	Apoptosis independent of Bad[34]		Decrease in mitochondria[14]
Bim	Apoptosis independent of Bim[34]		
Bcl2			Decrease in mitochondria,[14] decrease[32,42]
Hrk/Dp5	Apoptosis independent of Hrk/Dp5[34]		
TUNEL	Increase[43]	Small increase,[44] increase[45]	Increase,[46,47] not detected[48]
Pyknotic cells	Increase[43,49]	Increase[40,44]	Increase[14,47,48]
Lipid peroxidation	Increase[50]	Increase[51]	
Superoxide anion formation	Increase[50]		
Reactive oxygen species (ROS) generation	Increase[52,53]		
Dibutyryl cAMP	Decrease[54]		
Blockade of NMDA	Increase[55–57]		Decrease[58]
Activation of GABA receptors	Increase[55]		
Activation of PI 3-kinase	Decrease[59]		
Tyrosine autophosphorylation of IGF-I receptor	Decrease[60]		
Apoptotic ultrastructure	Increase[36]		Increase[47]
Intracellular Ca^{2+}	Decrease[56,57]		
Cytochrome c release	Increase[13]	Increase[30]	Increase[14,32]
AIF release	Increase[13]		Increase,[14] decrease[15]
DNA fragmentation	Increase[13,61]	Increase[30,40]	
4-Hydroxynonenal	Increase[13]		Increase[62]
F-Actin cortical ring structure	Increase[63,64]		
Cell adhesion molecules	Change in expression,[65] increase in expression[66]		
Cleavage of poly(ADP-ribose) polymerase-1		Increase[30]	Increase[14,32]

TABLE 4.1 Apoptotic and Cell Stress Markers Studied in Alcohol-Exposed Neural Tissue From Different Aged Models *(cont.)*

Marker	Fetal	Adolescent	Adult (includes postmortem)
PKC-δ		Decrease[30]	
Apaf-1		No change[30]	
Procaspase-9		Decrease[30]	Increase[32]
Active caspase-9		Increase[30]	Increase[14]
Active caspase-7			Increase[14]
IAP2			Decrease[15]
Caspase-2 activity	Increase[28]		
Caspase-8 activity	Increase[28]		
Caspase-9 activity	Increase[28]		
Fas FasL mRNA	Increase[24–26]		

and differences in the availability for trophic support from growth factors.[52]

Third Trimester Models

As noted previously, two of the brain regions most affected in human FASD are the cerebellum, particularly the anterior vermis, and the hippocampus. For this reason, most of the early research work using FASD models focused on these structures. That research also importantly identified the third trimester as one of heightened vulnerability to ethanol.[70] In rodent models, early studies by Bauer-Moffat and Altman[71] showed that there was enhanced reduction in cerebellar weight, compared to whole brain weight, reduced thickness of the cerebellar cortex, and decreases in both Purkinje cell number and granule cell number, due to alcohol exposure. Subsequent research by West and coworkers confirmed that high blood alcohol levels (BALs), especially from small concentrated doses (i.e., binge drinking), lead to decreased numbers of granule cells and Purkinje cells,[72] as well as decreased neurons in two major sources of input to the cerebellum – the inferior olive and trigeminal nucleus – but no decreases in hippocampal cell number, except in the CA1 region, at the highest BAL.[73,74]

A pathway for the observed decreases in cell number and volume measures is apoptosis, as detected by the identification of active caspase-3 immunohistostaining. Following up on these seminal studies, single binge third trimester alcohol exposure models in rodents have been developed that use two 2.5 g/kg subcutaneous injections given 2 h apart on the day postnatal day 7 (P7). This corresponds to the third trimester in humans, based on the equivalent patterns of brain and physiological development. In these single day exposures, widespread increases in silver staining of neurons are reported throughout the brain, indicating compromised cell membranes and neuronal degeneration.[33,55] Increased caspase-3 was also observed in the rostral forebrain and the cerebral cortex in the parietal lobe immediately superior to the hippocampus (i.e., where the scientists focused their analysis).[33,34]

Some of the more recent work on apoptosis due to fetal alcohol exposure has been done using third trimester binge drinking models in nonhuman primates. Burke et al.[75] found that the offspring of Vervet monkeys that binge-drank four times per week during the third trimester had decreased numbers of neurons in the frontal cortex, at 2 years of age. The authors interpreted their data as supporting the evidence for impairments in frontal lobe functions, such as executive function, in children with FASD. Building on these findings, Farber et al.[35] looked at whether single 8-h exposures to ethanol, at doses producing BALs of 300–400 mg/dL in pregnant Cynomologous monkeys (equivalent to binge drinking levels seen in humans), was sufficient to cause apoptosis at various time-points throughout the third trimester (gestational days 105–155 in macaque). In this case, apoptosis was assessed immediately after the end of the 8-h alcohol exposure. The authors found that the alcohol-exposed fetal brains had more than 60-fold higher active caspase-3 staining than the saline-exposed controls, and this was observed in various regions throughout the brain.[35] Moreover, despite the generally widespread nature of the increased apoptosis, it did appear that different brain regions were affected during the earlier part of the third trimester (G105–G135), compared to the later part of the third trimester (G140–G155). The areas with earlier susceptibility included the cerebellum, subiculum and entorhinal cortex, as well as the striatum, thalamus, and inferior colliculus. In contrast, superficial cortical layers (that are the last to develop) appeared to be more susceptible to later exposures.[35]

As already mentioned, numerous molecules and processes have been found to contribute to apoptosis in the CNS, following ethanol exposure, including BAX, GABA agonists, glutamate antagonists, insulin like growth factor 1, superoxide anion generation, lipid peroxidation, acetaldehyde, cyclic AMP, and 4-hydroxynonenal.[13,36,50,52–56,59,60,76,77] One way to identify whether the involvement of the molecules is essential to ethanol-induced apoptosis is to knock-down or knock-out that gene in mice, and observe if apoptosis levels decrease. This is what Young et al.[36] did using Bax-null (homozygous Bax −/−) and Bax-deficient (Bax −/+) mice in a third trimester exposure paradigm. Using active caspase-3 staining, they reported a complete lack of apoptosis in the null Bax −/− mice, but no change in apoptosis in deficient (Bax+/−) mice, compared to wild-type mice, following ethanol exposure. This is striking evidence for the critical involvement of Bax and the mitochondria in ethanol-induced apoptosis, since Bax translocates from the cytosol to the outer membrane of the mitochondria, and increases mitochondrial permeability, as part of cellular apoptosis.

Additional molecules specific to neurons have also been identified that play a role in neuronal ethanol induced apoptosis. Ethanol is a well-characterized GABA agonist, and glutamate antagonist. Since excessive amounts of GABA agonists and glutamate antagonists have been found to cause widespread apoptosis in the brain,[55,76] these types of compounds can be compared with ethanol to test the critical involvement of their signaling pathways in ethanol-induced apoptosis. It was already known that the addition of *N*-methyl-D-aspartate (NMDA), glycine, or glutamate (at nonexcitotoxic levels) had trophic effects on cultures of postnatal day 6–8 (P6–P8) cerebellar granule neurons. However, Wegelius and Korpi,[57] and Bhave et al.[56] both correlated ethanol induced apoptosis to a decrease in the effectiveness of NMDA to serve as a trophic factor. These authors proposed that the effect of ethanol on apoptosis was, therefore, "mediated, at least in part, by inhibition of NMDA receptor function."[56]

In similar studies, ethanol exposure was also determined to contribute to apoptosis through inhibition of another molecule with trophic effects: Insulin-like growth factor 1 (IGF-1).[78] Addition of 100 mM ethanol for 24-h exposure to primary cultured P7–P8 cerebellar granule neurons inhibited the neurotrophic action of IGF-1, with cell viability decreasing from 90% to 35%, as measured by trypan blue staining, and confirmation of apoptotic morphology with Hoechst staining. Notably, however, the tropic effects of IGF-1 were subsequently proposed by Zhang et al.[59] to be dependent on PI3-kinase, since inhibition of PI3-kinase using LY294002 also lead to increases in apoptosis, regardless of the presence of IGF-1.

Additional *in vitro* studies of third trimester models of ethanol exposure have also coupled increased apoptosis following ethanol or ethanol metabolite exposure with DNA damage. Holownia et al.[77] used primary cultures of rat astrocytes from newborn rat pups to show that there was more DNA fragmentation after exposure to ethanol (4.1%), but even greater increases in DNA fragmentation after exposure to the ethanol metabolite acetaldehyde (17.9%). Notably, in this study, the authors quantified DNA fragmentation using the random oligonucleotide primed synthesis (ROPS) assay, and compared 20 mM ethanol exposure with 0.448 mM acetaldehyde concentration, as measured by HPLC. They also showed that maintaining levels of acetaldehyde (by coculturing with ADH transfected CHO cells) caused increased DNA fragmentation greater than a 4-day exposure to ethanol alone, and thus could represent the putative cause of apoptosis in ethanol exposures.

First Trimester Models

In addition to third trimester models of FASD, there have also been several studies of alcohol-induced apoptosis during the first trimester and second trimester, including work in nonrodent species.[43,49,79] For example, Cartwright et al.[43] used a chick embryo model of FASD to determine ethanol increased apoptosis in neural crest cells, when ethanol was applied at the gastrulation stage (HH4), but not when applied after neural crest migration commenced (HH12). Such findings were thought to provide a possible mechanistic explanation for early ethanol exposures to give rise to the facial phenotype of FASD (flat groove between nose and upper lip, thin upper lip, and small eye width) through effects on facial progenitor cells. Other studies using first and second trimester ethanol exposure models in mice also provided clear evidence of increased apoptosis, but demonstrated that the sites most affected were distinct to the timing of ethanol injections.[49,79] For example, loss of premigratory cranial neural crest cell populations, affecting the hindbrain and trigeminal, sensory, and motor nuclei, was detected when mice were exposed to ethanol at GD8.[49]

In addition to measuring total levels of apoptosis, some first trimester fetal alcohol exposure studies have also sought to identify the specific molecules involved in increasing apoptosis. Kotch et al. reported increased superoxide formation coinciding with increased apoptosis in first trimester mouse embryo explants exposed to ethanol versus controls or versus embryos exposed to ethanol and the antioxidant superoxide dismutase.[50] De et al.[54] noted that ethanol is known to reduce endogenous levels of dibutyryl cyclic AMP (Bt_2cAMP), and tested whether adding it to cell cultures could be neuroprotective. They reported that addition of Bt_2cAMP reduced levels of ethanol-induced cell death (from ~46% to ~36%) in primary fetal (G19–G21) hypothalamic cell cultures, although it was still greater than that of controls (~18%).[54] While their data supported a connection between ethanol-induced neuronal apoptosis and cAMP, those authors also pointed out that increased cAMP levels have been shown to promote apoptotic cell death in other cell types, including thymocytes and lymphoid cell lines.

Other notable molecular events implicated in early gestational FASD models include direct DNA damage, and formation of 4-hydroxynonenal (HNE), a byproduct of ethanol-enhanced lipid oxidation. For example, using centrifugation and fluorometric HPLC techniques, Holownia et al.[77] and Ramachandran et al.[13] reported 25% increased DNA fragmentation, and 23% increased HNE formation, after in utero exposure of G17 and G18 embryos to ethanol versus an isocaloric dextrose solution. Ramachandran et al.[80] subsequently demonstrated that *in vitro* addition of HNE increased cytochrome c and AIF release from fetal mitochondria. Thus, both DNA damage and HNE formation are implicated in ethanol-induced neuronal apoptosis, and could be triggers for other molecular events.

Stem Cell and Slice Culture Models

There has been considerable interest in evaluating the possible effects of early gestational exposures of ethanol on cell proliferation, and cell migration, in the developing cerebral cortex. Indeed, the precise coordination of these two processes is critical to normal brain development, and directly affects cell fate. During corticogenesis, cells that will become projection neurons have to exit the cell cycle at a precise time, and attach to a radial glial fiber that they use to translocate to the outermost lamina. Once the neurons arrive, they then have to detach from the radial glia, as they begin to fully differentiate. Ethanol exposures during these events appear to disrupt cell cycle timing, altering both migration, and final cell fate determination.[81,82] In an attempt to model the events that take place in the developing brain as a result of ethanol exposure, studies have been done using organotypic slice cultures exposed to ethanol, or the antiproliferative (and anti-inflammatory) growth factor TGF β1 that stimulates precursor cell differentiation. Siegenthaler[66] showed that addition of 400 mg/dL ethanol or TGF β1 increased the expression of the cell adhesion molecule nCAM, and the integrins α3, αv, and β1. Interestingly, however, the combined application of TGF β1 and ethanol led to decreased expression of the same molecules. The potential ability of both ethanol and TGF β1 to profoundly influence cell adhesion was further supported by some of our own previous work, examining the effects of these compounds on rat B104 neuroblastoma cells, where significant changes in more than 20 cell adhesion genes and proteins were observed.[65]

It is quite possible that the aberrant cell migration reported after ethanol exposures, as well as the alteration in cell proliferation and increased apoptosis, are all related to the process of anoikis that promotes apoptosis, as a result of detachment from the extracellular matrix. Previous research makes such a connection possible. Using human pluripotent cells, Wang et al. reported increased caspase-dependent anoikis when the proliferative growth factor FGF was removed from culture, as measured by the presence of stained F-Actin structures that resembled cortical rings indicative of anoikis.[63] Miñambres et al.[64] subsequently showed that the same peripheral membrane associated actin rings formed in cultured astrocytes exposed to ethanol. Notably, in studies that will be presented later in this chapter, we have observed additional evidence for the potential involvement of anoikis in the brain's response to ethanol intake.

Adolescent Ethanol Models

Although there is more attention given to fetal exposure than to adolescent exposure, ethanol exposure during either time can have adverse effects on the developing brain, and predispose newborns and adolescents to an alcohol use disorder (AUD) in adulthood.[83–85] Moreover, chronic postnatal exposure to ethanol also causes significant neuronal death in the brain. Studies of adolescent alcohol exposures have been performed in rodent, primates, and humans. The adolescence studies typically model the human age range from 10 years to 20 years of age. Notably, this is a critical time period in which there is pruning of 40–50% of synapses, extensive myelination, alterations in neurotransmitter receptor levels and sensitivity, and in general, significant developmental changes in the brain.[37]

Perhaps because it is in such a dynamic state, one of the major findings regarding the effects of ethanol exposure during adolescence is that it predisposes to increased alcohol consumption in adulthood.[86] Moreover, adolescent ethanol exposure in humans is also associated with decreased brain volume in young adults and adolescents, as measured by MRI.[87] An obvious mechanism for reduced brain volume is the loss of cells. In a rat model of chronic alcohol consumption that began in adolescence (P35) and lasted for 8 weeks, reductions in cortical neurons, astrocytes, and microglia were all reported.[51] Unlike fetal alcohol exposure studies, however, direct apoptosis measures have not commonly been used to determine if such cell losses are due to apoptosis or other processes, although some studies have examined activated (cleaved) caspase-3 immunohistochemistry, with variable results.

Some studies examining caspase-3 immunohistochemistry have reported increased staining after ethanol exposure. In a 5-week ethanol vapor exposure starting at weaning (P23) in rats, one group reported significantly increased caspase-3 staining at 2 weeks, and 8 weeks postwithdrawal, compared to controls.[37] Another group showed that rats given 10% (v/v) ethanol as their only fluid source for 8 weeks, beginning in adolescence, showed increased caspase-3 activity in both the cerebellum and neocortex. Likewise, Pascual et al.[40] also found an increase in caspase-3 activity and DNA fragmentation in the cerebellum, neocortex, and hippocampus of adolescent rats that had been administered 3 g/kg ethanol by i.p. injection for 2 consecutive days, at 48-h intervals, over 2 weeks. Similarly, Jang et al.[45] reported increased caspase-3 and TUNEL staining in the dentate gyrus, after administering P30 adolescent rats 2 g/kg ethanol by i.p. injection, once a day, for 3 days. A more invasive, but perhaps more biologically appropriate, model of ethanol consumption used intragastric gavage to administer 25% (w/v) ethanol every 8 h, for 4 days (called the Majchrowicz method),[88] and also reported significantly increased TUNEL staining.[44] Thus, several studies have documented increased caspase-3 activity and TUNEL staining following adolescent ethanol exposure. However, results from other studies argue against increased apoptosis following ethanol exposure, and instead propose that only an inflammatory response is taking place,[40,44] For example, one study reported increased inactive caspase-3, but not active caspase-3, along with changes in phospholipase A and Parp-1, specifically in the hippocampus in a rat model that used combined 1-week exposures during adolescence (starting at P37), and again in adulthood (starting at P68).[31] Nonetheless, whether it occurs by apoptosis or necrosis, there is a general consensus that neural degeneration is increased due to adolescent ethanol exposure, mainly identified by cupric silver staining.[89]

Adult Models

As in prenatal and adolescent studies, adult alcohol exposure studies for the most part identify neurodegeneration in the brain, as well as volume decreases in the brain, after various types of ethanol exposure.[42,58,62,90–93] The frontal cortex, hippocampus, cerebellum, and white matter have all been highlighted as areas with possibly greater ethanol vulnerability, although some papers have reported no neurodegeneration in these areas, as well.[94–96] The positive findings have led to the development of a frontal lobe theory of ethanol susceptibility that coincides with frontal lobe-dependent neuropsychological deficits in adults with AUD.[92]

The mechanisms that produce neurodegeneration in the adult brain, after ethanol exposure, are an area of active investigation. When specifically looking for apoptotic markers, several studies have identified an increase in apoptosis in the adult model of ethanol exposure. For example, in a 12-week study of adult male Wistar rats, consumption of 10% (v/v) ethanol in water was found to increase caspase-3 and TUNEL staining.[14] Those investigators further proposed that hippocampal apoptosis was dependent on increases in caspase-9, but frontal cortex apoptosis was caspase-independent, and instead mediated by AIF. In contrast to these assertions, results from a 5-month study in C57BL/6 female mice, consuming 10% (v/v) ethanol in water, showed increased active caspase-3 staining, and caspase-3 mRNA in the medial frontal cortex.[38] In addition, an 11-day study in male Sprague–Dawley rats administered 2 g/kg ethanol by i.p. injection also reported increased caspase-3 and caspase-9 in the neocortex.[32] Besides an increase in caspase-3, and -9, some studies in adult humans and rodent models have also found increased TUNEL staining. In post mortem human brains of adults with a positive history of AUD (defined by intake of >80 g of alcohol per day), increased TUNEL positive cells were reported, although the authors also noted that the cells did not possess "typical morphological features of apoptosis."[46] In contrast, in a 6-week study of male Sprague–Dawley rats, involving consumption of 6.4% (v/v) ethanol liquid diet, researchers reported an increase in TUNEL staining, with a concomitant increase in pyknotic staining in the dentate gyrus of the hippocampus, particularly the proliferative subgranular zone.[47]

As in the studies of adolescent ethanol exposures, the adult apoptosis question becomes highly complicated to address because there are several studies that have reported no increases or actual decreases in apoptosis, depending on the paradigm and region of study. For example, one study that used a 3-week ethanol self-administration paradigm, followed by 6 weeks of ethanol vapor exposure in adult male Wistar rats, reported increased hippocampal apoptosis, but decreased mPFC apoptosis, as measured by active caspase-3 immunohistochemistry. Another study that assessed total brain homogenates of adult male Sprague–Dawley rats, after a single dose of 2.5 g/kg ethanol, showed no difference in caspase-3 activity, compared to controls, and reported that ethanol exposure after brain trauma actually decreased relative caspase-3 activity.[41] Likewise, a post mortem human study of male AUD subjects also reported decreases in activated caspase-3 in the prefrontal cortex.[97] And finally, using the Majchrowicz method of ethanol exposure,[88] intragastric infusion of 5 g/kg 25% (w/v) ethanol to adult male Sprague–Dawley did not increase TUNEL labeling, and demonstrated no morphological changes in cells indicative of apoptosis, using transmission electron microscope, although increases in pyknotic cells were reported.[48]

MOLECULAR PROFILING OF THE EFFECTS OF ETHANOL CONSUMPTION IN ADOLESCENT RATS

To evaluate the evidence for ethanol-induced apoptotic mechanisms in one particularly relevant developmental age (adolescence), we have recently been performing comprehensive genome-wide transcriptional profiling of the brain and peripheral blood leukocytes, using a 3-week daily drinking paradigm. Our first studies utilized a total of 24 (12 male and 12 female) Long–Evans adolescent rats (Harlan Labs). At the time of weaning, these animals were all housed in individual cages, fed *ad lib* chow, and exposed to a 12-h reverse light/dark cycle. At postnatal day 29 (the beginning of adolescence), the animals were split into two groups, with one group fed exclusively an *ad lib* liquid diet (from OpenSource Research Diets™) that we modified to contain 6.7% ethanol (accounting for 35% of the total caloric content). Another group of rats (termed "pair-fed," or PF) was weight-matched with the ET-fed rats, and fed aliquots of a control liquid diet (with maltose replacing ethanol), defined by the volume of liquid ethanol diet the matched ET-fed rats consumed. Throughout the course of the study, body weight and alcohol consumption of the animals was measured. The ethanol diet consumption data were normalized to body weight and reported in milliliter per kilogram (kg). At the study-midpoint, approximately 500 μL of venous tail blood was obtained within 4 h of the peak feeding time, and blood ethanol concentrations (BECs) were determined using an analox analyzer. Then, after 3 weeks, all the rats were euthanized, and we collected blood samples for gene expression profiling, as well as a liver function test (LFT) enzyme panel. The LFT indicators included alanine aminotransferase (ALT), aspartate aminotransferase (AST), alkaline phosphatase (ALP), total and direct bilirubin, and albumin. In addition to the blood sampling, we also dissected the hippocampus and cerebellar vermis for molecular profiling. RNA samples from the blood, hippocampus, and vermis were analyzed for ethanol-induced changes in expression, in using standard protocols on the rat gene 1.0 ST array (Affymetrx). This analysis was performed using a 3-way ANOVA to probe for effects of the diet (ethanol vs. control), or tissue (blood, hippocampus, vermis), while using the pair factor to mitigate the influence of gender and caloric consumption.

Using this treatment paradigm, we first observed that all the rats consumed relatively stable amounts of the liquid diet, and gained weight during the course of the study, with males and females consuming the same volume of ethanol, adjusted per body weight (Figure 4.2). Furthermore, when we measured the BEC levels at the midpoint of the study, we found that ethanol-consuming rats exceeded 150 mg/dL, with female rats showing a nonsignificant trend for higher levels, than their male adolescent counterparts (220.3 mg/dL vs. 151.3 mg/dL), despite consumption of the same quantity of ethanol per body weight. The results from the liver function tests also confirmed that the ethanol consuming rats were experiencing signs of liver damage, with consistent and significant elevations in AST, ALP, and the AST/ALT ratio in ethanol versus PF rats (Figure 4.2).

We next examined the gene expression data to reveal the most robustly affected genes across tissue types (Table 4.2), after correcting for multiple testing. Notably, most of these genes showed robust changes in more than one tissue type, with two genes (Fos and Plekho1) showing significant changes in all three tissues. Validation of selected changes in expression was then performed using real-time quantitative RT-PCR, for six genes of interest that were significantly changed in the array data – Atf5, Plcl2, Fos, Npas4, Dusp1, Nr4a2, and FosB. Each of these genes was independently confirmed as changed by these assays, in at least one tissue type, with evidence of a high correlation (R = 0.91) in the changes seen by PCR and array data for these genes across the tissues.

Finally, in order to determine the possible relevance of some of our most changed genes to cellular apoptosis and necrosis pathways, we performed bioinformatic analysis of the most robustly changed genes, using the Gene Ontology tool in Partek Genomics Suite software, as well as the Core Analysis and Molecular Activity Prediction tools in Ingenuity® IPA software (Qiagen, Valencia, CA). This analysis revealed that, among the top 40 enriched gene networks identified as enriched in our significant genes, there were five related to positive regulation of apoptosis and anoikis, two related to oxidative stress, 11 related to general transcriptional regulation, and five related to cellular differentiation and synaptic plasticity (Table 4.3). Because we were specifically interested in apoptosis-related networks, we further examined the genes related to apoptosis and anoikis, and found several that belonged to multiple gene networks (Table 4.4). Notably, three of the genes that we identified (Sik1, Aes, and Ptrh2) are specifically involved in the positive regulation of anoikis, with two (Sik1 and Aes) showing robustly increased expression, and the other (Ptrh2) showing robustly decreased expression, following ethanol consumption. The complementary bioinformatic analysis we completed used Ingenuity IPA software to map the functions of a more complete set of transcripts (n = 118) that showed FDR-corrected p-values less than 0.05 for the main effect of ethanol consumption, across the three tissues. This revealed robust mapping of genes to both pro- and antiapoptosis related networks. Moreover, the Molecular Activity Predictor tool also showed bidirectional changes in several specific subpathways, with a net effect of inhibiting apoptosis related genes, and activating prosurvival genes. When visualized across the

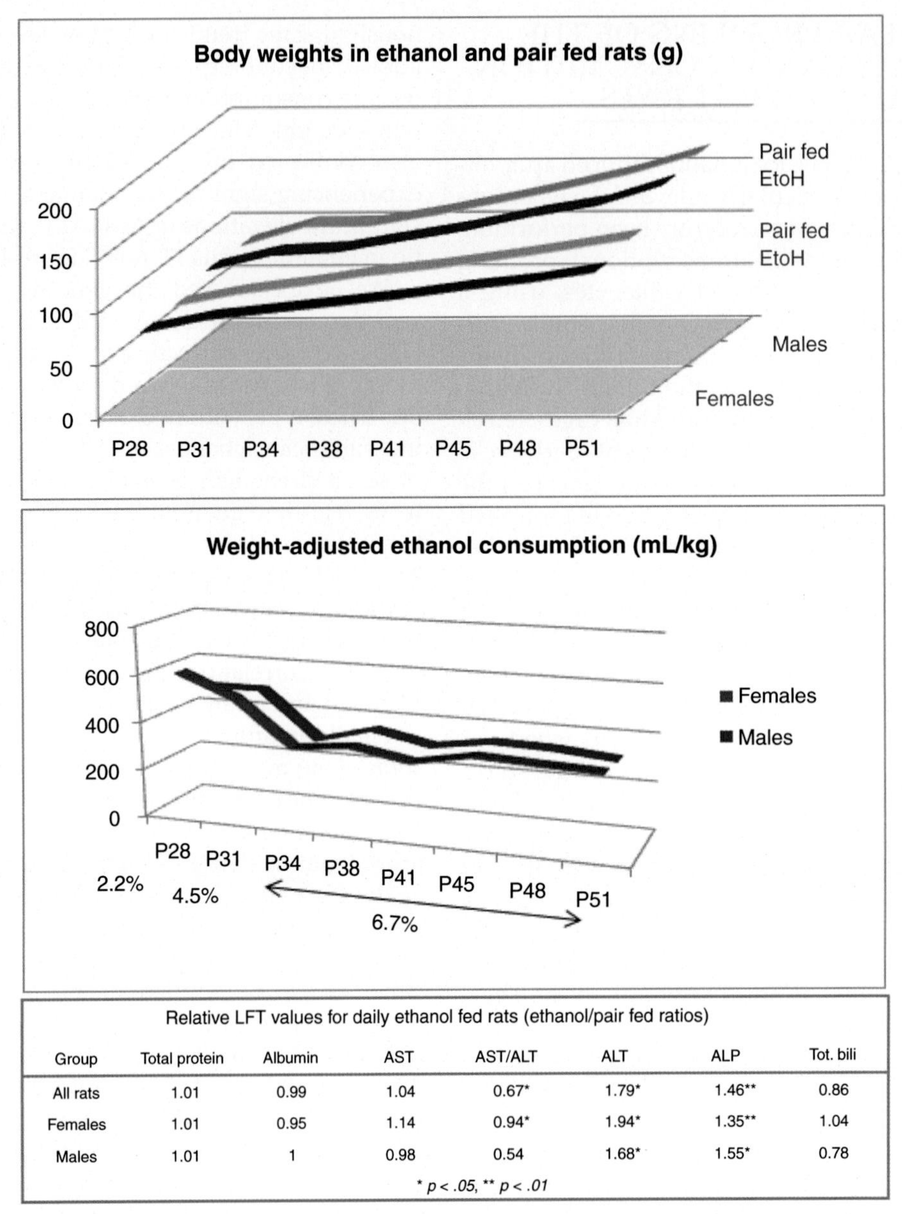

Relative LFT values for daily ethanol fed rats (ethanol/pair fed ratios)

Group	Total protein	Albumin	AST	AST/ALT	ALT	ALP	Tot. bili
All rats	1.01	0.99	1.04	0.67*	1.79*	1.46**	0.86
Females	1.01	0.95	1.14	0.94*	1.94*	1.35**	1.04
Males	1.01	1	0.98	0.54	1.68*	1.55*	0.78

* $p < .05$, ** $p < .01$

FIGURE 4.2 **Body weight, ethanol consumption, and liver function tests (LFT) in ethanol-consuming rats.** Note that both males and females consuming ethanol, and their pair-fed controls that received isocaloric nonalcoholic liquid diet, all gained weight during the 3-week daily consumption period, and that the ethanol-consuming rats showed signs of significant liver damage, demonstrating a toxic level of ethanol exposure was produced. Abbreviations: ALP, alkaline phosphatase; ALT, alanine aminotransferase; AST, aspartate aminotransferase; Tot. bili., total bilirubin.

three tissues, these predicted responses were most evident in the two CNS regions versus blood (Figure 4.3). It is also worthwhile to point out that, although the net direction of change tended to be unidirectional, there were several genes in these apoptosis-related networks, whose direction of change was not consistent with the predicted overall change.

Thus, from our own studies, we have obtained clear molecular evidence that, overall, there is evidence of both proapoptosis and antiapoptosis processes going on, following ethanol consumption during adolescence. Moreover, although we focused this review on studies that considered specific apoptotic measures in their outcomes, our data demonstrated that other molecules that are not necessarily widely studied in apoptosis research may in fact be more important to consider than some of the traditional apoptotic markers. So, the disparities in the literature might be best explained by careful consideration of our results. There are clearly robust changes caused by ethanol, but depending on the timing of when you look, the paradigm, and the tissue you might see changes in different directions that could involve different sets of genes and proteins than you originally envisioned (Figure 4.3).

TABLE 4.2 Top 25 Significantly Changed Genes Following 3-Week Daily Ethanol Consumption

		3-way ANOVA	Blood (ethanol vs pair fed)		Hippocampus		Vermis	
Gene	**Symbol**	**Corrected *p*-value**	**Fold change**	***p*-Value**	**Fold change**	***p*-Value**	**Fold change**	***p*-Value**
FBJ osteosarcoma oncogene	Fos	3.4E-09	1.5	0.00290	1.9	0.00001	3.3	2.4E-12
Activating transcription factor 5	Atf5	2.8E-07	2.0	1.4E-14	1.1	0.13163	1.2	0.02262
Neuronal PAS domain protein 4	Npas4	2.8E-07	1.0	0.94849	3.0	4.1E-08	4.0	1.2E-10
Nuclear receptor subfamily 4, group A, member 1	Nr4a1	4.2E-06	−1.1	0.47043	2.1	1.3E-06	3.0	4.5E-11
Pleckstrin homology domain containing, family O member 1	Plekho1	2.4E-05	1.2	0.00115	1.1	0.02221	1.4	8.2E-08
Nuclear receptor subfamily 4, group A, member 2	Nr4a2	0.00015	1.1	0.57388	1.4	0.00350	2.1	7.5E-10
Early growth response 1	Egr1	0.00016	1.1	0.34583	2.0	6.3E-07	1.7	0.00006
Nuclear receptor subfamily 4, group A, member 3	Nr4a3	0.00025	1.0	0.87099	1.3	0.01918	2.3	7.5E-11
Fos-like antigen 2	Fosl2	0.00025	1.1	0.15229	1.1	0.12513	1.5	3.5E-10
Metallothionein 1a	Mt1a	0.00036	2.6	2.4E-11	−1.1	0.51231	1.4	0.00880
Regulatory factor X, 5 (influences HLA class II expression)	Rfx5	0.00047	−1.3	0.00008	−1.1	0.23119	−1.3	0.00001
Metallothionein 1a	Mt1a	0.00082	2.3	1.2E-10	−1.1	0.54899	1.3	0.01159
Lymphatic vessel endothelial hyaluronan receptor 1	Lyve1	0.00099	2.6	3.3E-10	1.1	0.69291	1.2	0.08978
FBJ osteosarcoma oncogene B	Fosb	0.00099	1.0	0.47546	1.1	0.05818	1.5	2.4E-09
Salt-inducible kinase 1	Sik1	0.00200	1.0	0.63604	1.4	0.00032	1.6	4.5E-06
Dual specificity phosphatase 1	Dusp1	0.00267	1.1	0.72991	1.9	0.00008	2.0	0.00002
RNA binding motif, single stranded interacting protein 2	Rbms2	0.00293	1.7	1.8E-09	1.1	0.07964	1.0	0.85257
Similar to RIKEN cDNA 6430548M08	RGD1304884	0.00449	1.5	1.3E-07	1.1	0.31642	1.1	0.06697
VGF nerve growth factor inducible	Vgf	0.00596	1.1	0.56178	1.4	0.00613	1.8	0.00000
Cytohesin 1	Cyth1	0.00697	−1.4	4.7E-06	−1.1	0.24574	−1.2	0.01881
PC4 and SFRS1 interacting protein 1	Psip1	0.00740	−1.4	3.8E-07	−1.1	0.10079	−1.1	0.24210
Similar to RIKEN cDNA 1110007C09	RGD1306058	0.00790	1.5	4.1E-09	1.0	0.78355	1.1	0.07815
1026-1094, tRNA-Val	tRNA-Val	0.00790	1.9	7.5E-09	1.1	0.26975	1.1	0.58552
Cd63 molecule	Cd63	0.00790	1.7	2.1E-08	1.0	0.86206	1.2	0.03663
jun B proto-oncogene	Junb	0.00790	1.0	0.89278	1.6	0.00091	1.9	0.00001

TABLE 4.3 Top 40 Significantly Enriched Gene Ontologies Following 3-Weeks Daily Ethanol Consumption

Gene ontology	GO ID	Enrich score	Fold change	*p*-Value	# Genes
Positive regulation of anoikis	2000210	14.2	6.1	7.0E-07	3
Response to organophosphorus	46683	13.2	16.1	1.9E-06	8
Response to purine-containing compound	14074	12.2	16.0	5.2E-06	8
Response to cAMP	51591	12.0	14.1	5.9E-06	7
Core promoter sequence-specific DNA binding	1046	12.0	10.1	6.0E-06	5
Positive regulation of apoptotic process	43065	10.8	19.8	2.0E-05	10
Positive regulation of programmed cell death	43068	10.7	19.8	2.3E-05	10
RNA polymerase II core promoter sequence-specific DNA binding	979	10.1	8.1	3.9E-05	4
Positive regulation of cell death	10942	9.9	19.8	4.9E-05	10
Core promoter binding	1047	9.8	10.1	5.8E-05	5
Protein heterodimerization activity	46982	9.0	19.7	0.00012	10
Regulation of anoikis	2000209	8.7	6.1	0.00016	3
RNA polymerase II core promoter proximal region sequence-specific DNA binding transcription factor activity	982	8.2	10.1	0.00027	5
Response to oxidative stress	6979	8.0	15.9	0.00034	8
Transcription factor complex	5667	7.6	14.0	0.00051	7
Transcription regulatory region sequence-specific DNA binding	976	7.4	12.0	0.00064	6
RNA polymerase II regulatory region sequence-specific DNA binding	977	7.3	10.0	0.00065	5
RNA polymerase II core promoter proximal region sequence-specific DNA binding transcription factor activity involved in positive regulation of transcription	1077	7.3	8.1	0.00067	4
Sequence-specific DNA binding RNA polymerase II transcription factor activity	981	7.1	13.9	0.00082	7
RNA polymerase II regulatory region DNA binding	1012	7.1	10.0	0.00086	5
Intermediate filament cytoskeleton	45111	6.7	6.1	0.00119	3
Ligand-activated sequence-specific DNA binding RNA polymerase II transcription factor activity	4879	6.6	6.1	0.00139	3
Direct ligand regulated sequence-specific DNA binding transcription factor activity	98531	6.5	6.1	0.00150	3
RNA polymerase II transcription regulatory region sequence-specific DNA binding transcription factor activity involved in positive regulation of transcription	1228	6.5	8.1	0.00157	4
Transcription from RNA polymerase II promoter	6366	6.4	13.9	0.00164	7
Positive regulation of proteasomal ubiquitin-dependent protein catabolic process	32436	6.3	6.1	0.00185	3
Negative regulation of kinase activity	33673	6.2	10.0	0.00199	5
Positive regulation of proteasomal protein catabolic process	1901800	6.0	6.1	0.00239	3
Response to reactive oxygen species	302	6.0	10.0	0.00254	5
Negative regulation of transferase activity	51348	6.0	10.0	0.00254	5
Skeletal muscle cell differentiation	35914	6.0	6.1	0.00254	3
Response to peptide hormone	43434	5.9	13.9	0.00265	7
Steroid hormone receptor activity	3707	5.9	6.1	0.00270	3

TABLE 4.3 Top 40 Significantly Enriched Gene Ontologies Following 3-Weeks Daily Ethanol Consumption *(cont.)*

Gene ontology	GO ID	Enrich score	Fold change	*p*-Value	# Genes
Formation of primary germ layer	1704	5.9	6.1	0.00286	3
Response to peptide	1901652	5.6	13.9	0.00379	7
Regulation of neuronal synaptic plasticity	48168	5.5	6.1	0.00414	3
Monooxygenase activity	4497	5.5	8.0	0.00415	4
Response to insulin	32868	5.4	10.0	0.00462	5
Positive regulation of cAMP metabolic process	30816	5.3	6.1	0.00501	3

TABLE 4.4 Common Apoptosis-Related Genes Altered by Ethanol Consumption

Gene	Symbol	FDR	Fold change (blood)	Fold change (hippocampus)	Fold change (vermis)	Positive regulation of apoptotic process	Positive regulation of programmed cell death	Positive regulation of cell death	Regulation of anoikis	Positive regulation of anoikis
Nuclear receptor subfamily 4, group A, member 1	Nr4a1	4.2E-06	0.91	2.07	3.00	✓	✓	✓		
Early growth response 1	Egr1	0.00016	1.13	2.01	1.72	✓	✓	✓		
Nuclear receptor subfamily 4, group A, member 3	Nr4a3	0.00025	1.02	1.29	2.31	✓	✓	✓		
Salt-inducible kinase 1	Sik1	0.00200	1.05	1.45	1.63	✓	✓	✓	✓	✓
Dual specificity phosphatase 1	Dusp1	0.00267	1.05	1.92	2.04	✓	✓	✓		
TCF3 (E2A) fusion partner	Tfpt	0.01061	1.60	1.01	1.10	✓	✓	✓		
Amino-terminal enhancer of split	Aes	0.01410	1.20	1.10	1.07	✓	✓	✓	✓	✓
Peptidyl-tRNA hydrolase 2	Ptrh2	0.03653	0.88	0.83	0.88	✓	✓	✓	✓	✓
Guanine nucleotide binding protein (G protein), beta polypeptide 2 like 1	Gnb2l1	0.04221	0.77	0.97	0.98	✓	✓	✓		
RNA binding motif protein 5	Rbm5	0.04283	0.81	0.93	0.78	✓	✓	✓		

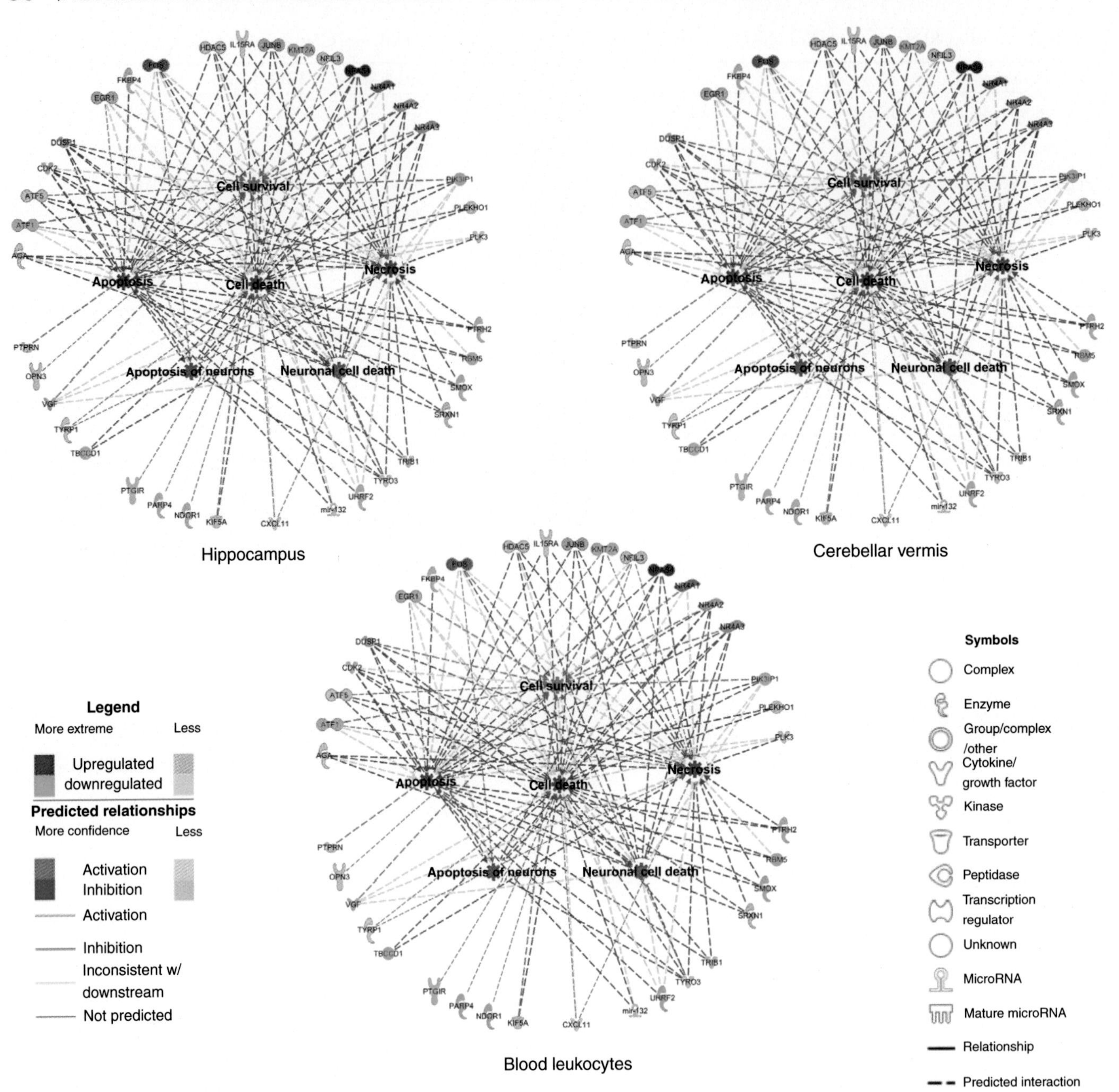

FIGURE 4.3 Molecular activity prediction of the transcriptional effects of ethanol consumption. The Ingenuity IPA Molecular Activity Predictor tool was used to analyze all of the genes identified as significantly changed, following 3 weeks ethanol consumption across three tissue types (hippocampus, cerebellar vermis, and blood leukocytes). This analysis revealed that, overall, there is a net attempt to inhibit proapoptotic networks (blue), and activate prosurvival networks (orange). These likely represent conserved compensatory responses to ethanol intake.

CONCLUSIONS

Ethanol is a potent central nervous system (CNS) depressant, with a wide range of side effects on the body. For example, cell membranes are highly permeable to alcohol. Therefore, once alcohol enters the bloodstream, it rapidly diffuses into nearly every tissue and organ. However, food stores, hydration, gender, body mass, and many other variables, all affect the absorbency and metabolism of alcohol.[98] The short-term and long-term effects of moderate alcohol consumption can include changes in metabolism, dehydration, blurred vision, ataxia, cognitive impairment, a reduced ability to evaluate the consequences of behavior, and even anterograde amnesia.[97,99] Chronic effects of excessive consumption can include damage to virtually every organ in the body, through apoptosis and/or necrosis.

Apoptosis is a necessary biological process. However, an unregulated increase in apoptosis can be detrimental to the body or, in the case of ethanol-associated cancers, a complete lack of apoptosis can be lethal. There are three main apoptotic pathways, one extrinsic and two intrinsic, as well as a granzyme/perforin pathway, and an additional noncaspase dependent pathway. The literature regarding the CNS indicates that it is particularly vulnerable to the damaging effects of ethanol, with general agreement that prenatal ethanol exposures increase neuronal apoptosis, but disagreement regarding the importance of apoptosis following adolescent and adult exposures. Here, we have reviewed studies that considered specific apoptotic measures in their outcomes. This evidence tells us apoptosis cannot be discounted as being highly activated in ethanol exposures, at any of the three developmental life stages (fetal, adolescent, adult). Examination of our own molecular data on the effects of daily ethanol consumption in adolescent drinking rats indicates that proapoptotic responses may be the most obvious or easiest to detect, but that a more inclusive analysis reveals that cells in the CNS and periphery also may attempt to inhibit apoptosis, and promote survival. Such responses may be the most critical in an organ or tissue such as the CNS, where the capacity for cellular renewal and replacement is limited. However, our data also showed the same trend in the highly renewable leukocyte population in the peripheral blood. Thus, we cannot avoid the conclusion that the studies supporting increases in apoptosis-related responses, and the studies failing to demonstrate these increases, may both have validity. Explanations for any apparent disparities may be found after careful consideration of the specific experimental paradigms and ages of the animals, tissues and cell types that are studied, and the molecular markers examined.

In summary, FASD and AUD remain as preventable disorders that have lifelong health consequences for individuals. Future research should consider more comprehensive approaches regarding the molecular determinants of ethanol-induced brain and tissue damage, if we are to make progress to redress the toll alcohol abuse and misuse take on our society. What can be concluded at this point is that ethanol neurotoxicity involves the apoptotic process, but there is a need to further elucidate the precise mechanisms involved.

Key Facts

- Chronic or high levels of ethanol exposure lead to apoptosis in widespread tissues, but the brain in particular.
- Apoptosis in the brain is most consistently detected following fetal exposures, but also seen to varying degrees following adolescent and adult exposures.
- Ethanol damages cells through acetaldehyde formation, increased reactive oxygen species formation, DNA adduct formation, and single and double strand break formation in DNA.
- Along with protein and DNA markers of apoptosis, molecular profiling reveals a diverse set of molecular adaptations that are both pro- and antiapoptotic in nature, indicating that many cells may attempt to actively inhibit this process. following ethanol exposure.

References

1. Kuan C-Y, Roth KA, Flavell RA, Rakic P. Mechanisms of programmed cell death in the developing brain. *Trends Neurosci* 2000;**23**(7):291–7.
2. Danial NN, Korsmeyer SJ. Cell death: critical control points. *Cell* 2004;**116**(2):205–19.
3. Dunty WC, Chen S-y, Zucker RM, Dehart DB, Sulik KK. Selective vulnerability of embryonic cell populations to ethanol-induced apoptosis: implications for alcohol-related birth defects and neurodevelopmental disorder. *Alcohol Clin Exp Res* 2001;**25**(10):1523–35.
4. Guicciardi ME, Gores GJ. Apoptosis: a mechanism of acute and chronic liver injury. *Gut* 2005;**54**(7):1024–33.
5. Nowoslawski L, Klocke BJ, Roth KA. Molecular regulation of acute ethanol-induced neuron apoptosis. *J Neuropathol Exp Neurol* 2005;**64**(6):490–7.
6. Seo JB, Gowda GAN, Koh D-S. Apoptotic damage of pancreatic ductal epithelia by alcohol and its rescue by an antioxidant. *PLoS ONE* 2013;**8**(11):e81893.
7. Lui S, Jones RL, Robinson NJ, Greenwood SL, Aplin JD, Tower CL. Detrimental effects of ethanol and its metabolite acetaldehyde, on first trimester human placental cell turnover and function. *PLoS ONE* 2014;**9**(2):e87328.
8. Wang K. Molecular mechanisms of hepatic apoptosis. *Cell Death Dis* 2014;**5**:e996.
9. Singhal PC, Reddy K, Ding G, et al. Ethanol-induced macrophage apoptosis: the role of tgf-β. *J Immunol* 1999;**162**(5):3031–6.
10. Cederbaum AI, Lu Y, Wu D. Role of oxidative stress in alcohol-induced liver injury. *Arch Toxicol* 2009;**83**(6):519–48.
11. Higuchi H, Kurose I, Kato S, Miura S, Ishii H. Ethanol-induced apoptosis and oxidative stress in hepatocytes. *Alcohol Clin Exp Res* 1996;**20**(9 Suppl.):340A–6A.
12. Maneesh M, Jayalekshmi H, Dutta S, Chakrabarti A, Vasudevan DM. Role of oxidative stress in ethanol induced germ cell apoptosis – an experimental study in rats. *Indian J Clin Biochem* 2005;**20**(2):62–7.
13. Ramachandran V, Perez A, Chen J, Senthil D, Schenker S, Henderson GI. In utero ethanol exposure causes mitochondrial dysfunction, which can result in apoptotic cell death in fetal brain: a potential role for 4-hydroxynonenal. *Alcohol Clin Exp Res* 2001;**25**(6):862–71.
14. Sunkesula SRB, Swain U, Babu PP. Cell death is associated with reduced base excision repair during chronic alcohol administration in adult rat brain. *Neurochem Res* 2008;**33**(6):1117–28.
15. Johansson S, Ekstrom TJ, Marinova Z, et al. Dysregulation of cell death machinery in the prefrontal cortex of human alcoholics. *Int J Neuropsychopharmacol* 2009;**12**(1):109–15.
16. Wehrli P, Viard I, Bullani R, Tschopp J, French LE. Death receptors in cutaneous biology and disease. *J Invest Dermatol* 2000;**115**(2): 141–8.
17. Green DR. Apoptotic pathways: ten minutes to dead. *Cell* 2005; **121**(5):671–4.

18. Benn SC, Woolf CJ. Adult neuron survival strategies-slamming on the brakes. *Nat Rev Neurosci* 2004;**5**(9):686–700.
19. Janssens S, Tinel A. The piddosome, DNA-damage-induced apoptosis and beyond. *Cell Death Differ* 2012;**19**(1):13–20.
20. Tinel A, Tschopp J. The piddosome, a protein complex implicated in activation of caspase-2 in response to genotoxic stress. *Science* 2004;**304**(5672):843–6.
21. Guadamillas MC, Cerezo A, Del Pozo MA. Overcoming anoikis – pathways to anchorage-independent growth in cancer. *J Cell Sci* 2011;**124**(Pt. 19):3189–97.
22. Taddei ML, Giannoni E, Fiaschi T, Chiarugi P. Anoikis: an emerging hallmark in health and diseases. *J Pathol* 2012;**226**(2):380–93.
23. Olney JW, Ishimaru MJ, Bittigau P, Ikonomidou C. Ethanol-induced apoptotic neurodegeneration in the developing brain. *Apoptosis* 2000;**5**(6):515–21.
24. De la Monte S, Derdak Z, Wands JR. Alcohol, insulin resistance and the liver-brain axis. *J Gastroenterol Hepatol* 2012;**27**:33.
25. Hicks SD, Miller MW. Effects of ethanol on transforming growth factor β1-dependent and -independent mechanisms of neural stem cell apoptosis. *Exp Neurol* 2011;**229**(2):372–80.
26. Cheema ZF, West JR, Miranda RC. Ethanol induces fas/apo [apoptosis]-1 mrna and cell suicide in the developing cerebral cortex. *Alcohol Clin Exp Res* 2000;**24**(4):535–43.
27. Mooney SM, Lein PJ, Miller MW. Fetal alcohol spectrum disorder: targeted effects of ethanol on cell proliferation and survival. In: Rakic P, Rubenstein JLR, editors. *Neural circuit development and function in the brain*. Oxford: Academic Press; 2013. p. 521–37.
28. Vaudry D, Rousselle C, Basille M, et al. Pituitary adenylate cyclase-activating polypeptide protects rat cerebellar granule neurons against ethanol-induced apoptotic cell death. *Proc Natl Acad Sci USA* 2002;**99**(9):6398–403.
29. Nikolić M, Gardner HAR, Tucker KL. Postnatal neuronal apoptosis in the cerebral cortex: physiological and pathophysiological mechanisms. *Neuroscience* 2013;**254**:369–78.
30. Rajgopal Y, Chetty CS, Vemuri MC. Differential modulation of apoptosis-associated proteins by ethanol in rat cerebral cortex and cerebellum. *Eur J Pharmacol* 2003;**470**(3):117–24.
31. Tajuddin NF, Przybycien-Szymanska MM, Pak TR, Neafsey EJ, Collins MA. Effect of repetitive daily ethanol intoxication on adult rat brain: significant changes in phospholipase a2 enzyme levels in association with increased parp-1 indicate neuroinflammatory pathway activation. *Alcohol* 2013;**47**(1):39–45.
32. Badshah H, Kim TH, Kim MJ, et al. Apomorphine attenuates ethanol-induced neurodegeneration in the adult rat cortex. *Neurochem Int* 2014;**74**:8–15.
33. Olney JW. Fetal alcohol syndrome at the cellular level. *Addict Biol* 2004;**9**(2):137–49 discussion 51.
34. Ghosh AP, Walls KC, Klocke BJ, Toms R, Strasser A, Roth KA. The proapoptotic bh3-only, bcl-2 family member, puma is critical for acute ethanol-induced neuronal apoptosis. *J Neuropathol Exp Neurol* 2009;**68**(7):747–56.
35. Farber NB, Creeley CE, Olney JW. Alcohol-induced neuroapoptosis in the fetal macaque brain. *Neurobiol Dis* 2010;**40**(1):200–6.
36. Young C, Klocke BJ, Tenkova T, et al. Ethanol-induced neuronal apoptosis *in vivo* requires bax in the developing mouse brain. *Cell Death Differ* 2003;**10**(10):1148–55.
37. Ehlers CL, Liu W, Wills DN, Crews FT. Periadolescent ethanol vapor exposure persistently reduces measures of hippocampal neurogenesis that are associated with behavioral outcomes in adulthood. *Neuroscience* 2013;**244**:1–15.
38. Alfonso-Loeches S, Pascual-Lucas M, Blanco AM, Sanchez-Vera I, Guerri C. Pivotal role of tlr4 receptors in alcohol-induced neuroinflammation and brain damage. *J Neurosci* 2010;**30**(24):8285–95.
39. Richardson HN, Chan SH, Crawford EF, et al. Permanent impairment of birth and survival of cortical and hippocampal proliferating cells following excessive drinking during alcohol dependence. *Neurobiol Dis* 2009;**36**(1):1–10.
40. Pascual M, Blanco AM, Cauli O, Miñarro J, Guerri C. Intermittent ethanol exposure induces inflammatory brain damage and causes long-term behavioural alterations in adolescent rats. *Eur J Neurosci* 2007;**25**(2):541–50.
41. Kanbak G, Kartkaya K, Ozcelik E, et al. The neuroprotective effect of acute moderate alcohol consumption on caspase-3 mediated neuroapoptosis in traumatic brain injury: the role of lysosomal cathepsin l and nitric oxide. *Gene* 2013;**512**(2):492–5.
42. Moselhy HF, Georgiou G, Kahn A. Frontal lobe changes in alcoholism: a review of the literature. *Alcohol Alcohol* 2001;**36**(5):357–68.
43. Cartwright MM, Tessmer LL, Smith SM. Ethanol-induced neural crest apoptosis is coincident with their endogenous death, but is mechanistically distinct. *Alcohol Clin Exp Res* 1998;**22**(1):142–9.
44. Morris SA, Eaves DW, Smith AR, Nixon K. Alcohol inhibition of neurogenesis: a mechanism of hippocampal neurodegeneration in an adolescent alcohol abuse model. *Hippocampus* 2010;**20**(5): 596–607.
45. Jang MH, Shin MC, Jung SB, et al. Alcohol and nicotine reduce cell proliferation and enhance apoptosis in dentate gyrus. *Neuroreport* 2002;**13**(12):1509–13.
46. Ikegami Y, Goodenough S, Inoue Y, Dodd PR, Wilce PA, Matsumoto I. Increased tunel positive cells in human alcoholic brains. *Neurosci Lett* 2003;**349**(3):201–5.
47. Herrera DG, Yagüe AG, Johnsen-Soriano S, et al. Selective impairment of hippocampal neurogenesis by chronic alcoholism: protective effects of an antioxidant. *Proc Natl Acad Sci USA* 2003;**100**(13):7919–24.
48. Obernier JA, Bouldin TW, Crews FT. Binge ethanol exposure in adult rats causes necrotic cell death. *Alcohol Clin Exp Res* 2002;**26**(4):547–57.
49. Kotch LE, Sulik KK. Patterns of ethanol-induced cell death in the developing nervous system of mice; neural fold states through the time of anterior neural tube closure. *Int J Dev Neurosci* 1992;**10**(4):273–9.
50. Kotch LE, Chen SY, Sulik KK. Ethanol-induced teratogenesis: free radical damage as a possible mechanism. *Teratology* 1995;**52**(3): 128–36.
51. Teixeira FB, Santana LNdS, Bezerra FR, et al. Chronic ethanol exposure during adolescence in rats induces motor impairments and cerebral cortex damage associated with oxidative stress. *PLoS ONE* 2014;**9**(6):e101074.
52. Goodlett CR, Horn KH, Zhou FC. Alcohol teratogenesis: mechanisms of damage and strategies for intervention. *Exp Biol Med* 2005;**230**(6):394–406.
53. Brocardo PS, Gil-Mohapel J, Christie BR. The role of oxidative stress in fetal alcohol spectrum disorders. *Brain Res Rev* 2011;**67** (1–2):209–25.
54. De A, Boyadjieva NI, Pastorcic M, Reddy BV, Sarkar DK. Cyclic amp and ethanol interact to control apoptosis and differentiation in hypothalamic beta-endorphin neurons. *J Biol Chem* 1994;**269**(43):26697–705.
55. Ikonomidou C, Bittigau P, Ishimaru MJ, et al. Ethanol-induced apoptotic neurodegeneration and fetal alcohol syndrome. *Science* 2000;**287**(5455):1056–60.
56. Bhave SV, Hoffman PL. Ethanol promotes apoptosis in cerebellar granule cells by inhibiting the trophic effect of nmda. *J Neurochem* 1997;**68**(2):578–86.
57. Wegelius K, Korpi ER. Ethanol inhibits nmda-induced toxicity and trophism in cultured cerebellar granule cells. *Acta Physiol Scand* 1995;**154**(1):25–34.
58. Farber NB, Heinkel C, Dribben WH, Nemmers B, Jiang X. In the adult cns, ethanol prevents rather than produces nmda antagonist-induced neurotoxicity. *Brain Res* 2004;**1028**(1):66–74.

59. Zhang FX, Rubin R, Rooney TA. *N*-methyl-D-aspartate inhibits apoptosis through activation of phosphatidylinositol 3-kinase in cerebellar granule neurons: a role for insulin receptor substrate-1 in the neurotrophic action of *n*-methyl-D-aspartate and its inhibition by ethanol. *J Biol Chem* 1998;**273**(41):26596–602.
60. Cui S-J, Tewari M, Schneider T, Rubin R. Ethanol promotes cell death by inhibition of the insulin-like growth factor i receptor. *Alcohol Clin Exp Res* 1997;**21**(6):1121–7.
61. Holownia A, Ledig M, Braszko JJ, Ménez JF. Acetaldehyde cytotoxicity in cultured rat astrocytes. *Brain Res* 1999;**833**(2):202–8.
62. Tajuddin N, Moon K-H, Marshall SA, et al. Neuroinflammation and neurodegeneration in adult rat brain from binge ethanol exposure: abrogation by docosahexaenoic acid. *PLoS ONE* 2014;**9**(7):e101223.
63. Wang X, Lin G, Martins-Taylor K, Zeng H, Xu R-H. Inhibition of caspase-mediated anoikis is critical for basic fibroblast growth factor-sustained culture of human pluripotent stem cells. *J Biol Chem* 2009;**284**(49):34054–64.
64. Miñambres R, Guasch RM, Perez-Aragó A, Guerri C. The rhoa/rock-i/mlc pathway is involved in the ethanol-induced apoptosis by anoikis in astrocytes. *J Cell Sci* 2006;**119**(2):271–82.
65. Miller MW, Mooney SM, Middleton FA. Transforming growth factor β1 and ethanol affect transcription and translation of genes and proteins for cell adhesion molecules in b104 neuroblastoma cells. *J Neurochem* 2006;**97**(4):1182–90.
66. Siegenthaler JA, Miller MW. Transforming growth factor β1 modulates cell migration in rat cortex: effects of ethanol. *Cerebral Cortex* 2004;**14**(7):791–802.
67. Riley EP, Infante MA, Warren KR. Fetal alcohol spectrum disorders: an overview. *Neuropsychol Rev* 2011;**21**(2):73–80.
68. Roebuck TM, Mattson SN, Riley EP. A review of the neuroanatomical findings in children with fetal alcohol syndrome or prenatal exposure to alcohol. *Alcohol Clin Exp Res* 1998;**22**(2):339–44.
69. Lebel C, Roussotte F, Sowell ER. Imaging the impact of prenatal alcohol exposure on the structure of the developing human brain. *Neuropsychol Rev* 2011;**21**(2):102–18.
70. Marcussen BL, Goodlett CR, Mahoney JC, West JR. Developing rat purkinje cells are more vulnerable to alcohol-induced depletion during differentiation than during neurogenesis. *Alcohol* 1994;**11**(2):147–56.
71. Bauer-Moffett C, Altman J. Ethanol-induced reductions in cerebellar growth of infant rats. *Exp Neurol* 1975;**48**(2):378–82.
72. Bonthius DJ, West JR. Alcohol-induced neuronal loss in developing rats: increased brain damage with binge exposure. *Alcohol Clin Exp Res* 1990;**14**(1):107–18.
73. Napper RM, West JR. Permanent neuronal cell loss in the inferior olive of adult rats exposed to alcohol during the brain growth spurt: a stereological investigation. *Alcohol Clin Exp Res* 1995;**19**(5):1321–6.
74. Miller MW. Effect of pre- or postnatal exposure to ethanol on the total number of neurons in the principal sensory nucleus of the trigeminal nerve: cell proliferation and neuronal death. *Alcohol Clin Exp Res* 1995;**19**(5):1359–63.
75. Burke MW, Palmour RM, Ervin FR, Ptito M. Neuronal reduction in frontal cortex of primates after prenatal alcohol exposure. *Neuroreport* 2009;**20**(1):13–7.
76. Ikonomidou C, Bosch F, Miksa M, et al. Blockade of nmda receptors and apoptotic neurodegeneration in the developing brain. *Science* 1999;**283**(5398):70–4.
77. Holownia A, Ledig M, Braszko JJ, Ménez J-F. Acetaldehyde cytotoxicity in cultured rat astrocytes. *Brain Res* 1999;**833**(2):202–8.
78. Zhang FX, Rubin R, Rooney TA. Ethanol induces apoptosis in cerebellar granule neurons by inhibiting insulin-like growth factor 1 signaling. *J Neurochem* 1998;**71**(1):196–204.
79. Lipinski RJ, Hammond P, O'Leary-Moore SK, et al. Ethanol-induced face-brain dysmorphology patterns are correlative and exposure-stage dependent. *PLoS ONE* 2012;**7**(8):e43067.
80. Ramachandran V, Watts LT, Maffi SK, Chen J, Schenker S, Henderson G. Ethanol-induced oxidative stress precedes mitochondrially mediated apoptotic death of cultured fetal cortical neurons. *J Neurosci Res* 2003;**74**(4):577–88.
81. Komatsu S, Sakata-Haga H, Sawada K, Hisano S, Fukui Y. Prenatal exposure to ethanol induces leptomeningeal heterotopia in the cerebral cortex of the rat fetus. *Acta Neuropathol* 2001;**101**(1):22–6.
82. Mooney SM, Siegenthaler JA, Miller MW. Ethanol induces heterotopias in organotypic cultures of rat cerebral cortex. *Cerebral Cortex* 2004;**14**(10):1071–80.
83. Clapper RL, Buka SL, Goldfield EC, Lipsitt LP, Tsuang MT. Adolescent problem behaviors as predictors of adult alcohol diagnoses. *Int J Addict* 1995;**30**(5):507–23.
84. Miller MW, Spear LP. The alcoholism generator. *Alcohol Clin Exp Res* 2006;**30**(9):1466–9.
85. Grant BF. The impact of a family history of alcoholism on the relationship between age at onset of alcohol use and dsm-iv alcohol dependence: results from the national longitudinal alcohol epidemiologic survey. *Alcohol Health Res World* 1998;**22**(2):144–7.
86. Barr CS, Schwandt ML, Newman TK, Higley JD. The use of adolescent nonhuman primates to model human alcohol intake: neurobiological, genetic, and psychological variables. *Ann NY Acad Sci* 2004;**1021**(1):221–33.
87. De Bellis MD, Clark DB, Beers SR, et al. Hippocampal volume in adolescent-onset alcohol use disorders. *Am J Psych* 2000;**157**(5):737–44.
88. Majchrowicz E. Induction of physical dependence upon ethanol and the associated behavioral changes in rats. *Psychopharmacologia* 1975;**43**(3):245–54.
89. Crews FT, Braun CJ, Hoplight B, Switzer III RC, Knapp DJ. Binge ethanol consumption causes differential brain damage in young adolescent rats compared with adult rats. *Alcohol Clin Exp Res* 2000;**24**(11):1712–23.
90. Collins MA, Moon KH, Tajuddin N, Neafsey EJ, Kim HY. Docosahexaenoic acid (dha) prevents binge ethanol-dependent aquaporin-4 elevations while inhibiting neurodegeneration: experiments in rat adult-age entorhino-hippocampal slice cultures. *Neurotox Res* 2013;**23**(1):105–10.
91. Pfefferbaum A, Lim KO, Zipursky RB, et al. Brain gray and white matter volume loss accelerates with aging in chronic alcoholics: a quantitative mri study. *Alcohol Clin Exp Res* 1992;**16**(6):1078–89.
92. Kril JJ, Halliday GM, Svoboda MD, Cartwright H. The cerebral cortex is damaged in chronic alcoholics. *Neuroscience* 1997;**79**(4): 983–98.
93. Cippitelli A, Damadzic R, Frankola K, et al. Alcohol-induced neurodegeneration, suppression of transforming growth factor-β, and cognitive impairment in rats: prevention by group ii metabotropic glutamate receptor activation. *Biol Psychiatry* 2010;**67**(9):823–30.
94. Harding AJ, Wong A, Svoboda M, Kril JJ, Halliday GM. Chronic alcohol consumption does not cause hippocampal neuron loss in humans. *Hippocampus* 1997;**7**(1):78–87.
95. Jensen GB, Pakkenberg B. Do alcoholics drink their neurons away? *Lancet* 1993;**342**(8881):1201–4.
96. Korbo L. Glial cell loss in the hippocampus of alcoholics. *Alcohol Clin Exp Res* 1999;**23**(1):164–8.
97. Grattan KE, Vogel-Sprott M. Maintaining intentional control of behavior under alcohol. *Alcohol Clin Exp Res* 2001;**25**(2):192–7.
98. Ramchandani VA, Kwo PY, Li T-K. Effect of food and food composition on alcohol elimination rates in healthy men and women. *J Clin Pharmacol* 2001;**41**(12):1345–50.
99. Goodwin DW, Crane JB, Guze SB. Alcoholic "blackouts": a review and clinical study of 100 alcoholics. *Am J Psych* 1969;**126**(2):191–8.

CHAPTER

5

Pathogenic Mechanisms in Alcoholic Liver Disease (ALD): Emerging Role of Osteopontin

Jason D. Coombes, PhD[*,§], *Wing-Kin Syn, MBChB*[*,**,†,‡,§]

*Regeneration and Repair Group, The Institute of Hepatology, London, UK
**Liver Unit, Barts Health NHS Trust, London, UK
†Department of Physiology, University of the Basque Country, Leioa, Spain
‡Department of Surgery, Loyola University, Chicago, IL, USA
§Transplant Immunology and Mucosal Biology, Kings College London, UK

INTRODUCTION

Alcoholic liver disease (ALD) is a leading cause of morbidity and mortality worldwide, and a major indication for liver transplantation. ALD encompasses a spectrum of clinicopathological states, ranging from steatosis, steatohepatitis, and fibrosis-cirrhosis (liver scar accumulation, with or without the presence of architecture distortion and organ dysfunction), and these features often coexist in the ALD liver. Significantly, individuals who develop fibrosis-cirrhosis are at an increased risk of developing liver complications, such as portal hypertension, liver cancer, and liver failure.[1]

While the majority of heavy drinkers develop liver steatosis, only a minority develop the more advanced stages of ALD (i.e., steatohepatitis and fibrosis-cirrhosis stage).[2] These suggest that additional host or environmental factors are important modulators of disease outcomes.[3,4] Our ability to identify individuals at risk of progressive liver disease, however, is limited, in part because of the incomplete understanding of pathogenic mechanisms.[5] Future studies are necessary to unravel the molecular and cellular signals that underpin this complex disease, and to identify targets for antifibrotic therapy.

In this chapter, we will first provide a brief overview of known pathogenic pathways. We will then focus on the emerging role of osteopontin (OPN) in liver disease, and ALD fibrosis.

GENERAL MECHANISMS OF ALD

We recognize that excessive alcohol consumption is a leading cause of chronic liver disease, but only about a third of heavy drinkers actually develop clinically advanced ALD such as fibrosis-cirrhosis, steatohepatitis, and liver failure.[6] The vast majority will develop liver steatosis (fatty liver) that is generally considered "benign," and potentially reversible on cessation of alcohol intake. Understanding how and why only some individuals develop advanced ALD is clinically important, as it would enable clinicians to identify and focus on those at the highest risk of disease progression. Despite extensive research, our understanding of pathogenic mechanisms remains incomplete.

Current dogma suggests that oxidative stress (from alcohol metabolism and reactive oxygen species formation), immune dysregulation, and endotoxins from the gut, in concert, promote the development of steatohepatitis, a more advanced stage of ALD.[7] Fibrosis and cirrhosis ensues when liver repair mechanisms become overwhelmed, and scar tissue accumulates.[8]

Alcoholic Liver Steatosis

Hepatic steatosis develops in nearly 90% of heavy drinkers, and results from an imbalance between lipid synthesis and breakdown. Alcohol disrupts the NADH: NAD ratio that impairs fatty acid oxidation such as,

Molecular Aspects of Alcohol and Nutrition. http://dx.doi.org/10.1016/B978-0-12-800773-0.00005-7

repression of peroxisome proliferator-activated- (PPAR) and adenosine monophosphate-activated protein kinase (AMPK), but enhances the influx of free fatty acids from adipose stores to the liver, and upregulates lipogenic enzymes (i.e., sterol regulatory element binding protein 1c, SREBP-1c).[9–11] Mitochondria damage during lipid peroxidation also leads to the loss of ApoB100, resulting in the accumulation of very low-density lipoprotein (VLDL), while glycosylation of lipoproteins in the Golgi apparatus leads to macrovesicular fat accumulation.[12,13]

Alcoholic Steatohepatitis and Fibrosis

Alcohol Metabolism and Oxidative Stress

The liver is the major site for alcohol metabolism (~90%). Oxidation by alcohol dehydrogenase, a cytosolic enzyme in hepatocytes, leads to the formation of acetaldehyde. In turn, acetaldehyde is converted to acetate by acetaldehyde dehydrogenase. Acetaldehyde mediates hepatocyte injury and inflammation, and covalently binds to DNA and protein to form adducts; in turn, these protein-adducts amplify injury and stimulate fibrogenesis.[14,15] The production of acetate also enhances liver inflammation and injury through the generation of acetyl-coenzyme A (acetyl-CoA) that modulates the production of the proinflammatory cytokines, TNF-α and IL-8.[16] When in excess, alcohol can also be oxidized by the microsomal ethanol oxidizing system (MEOS), also known as cytochrome P450 2E1 (CYP2E1).[17] This pathway leads to additional production of reactive oxygen species (ROS) that further accelerates ALD progression through the activation of nuclear factor kappa B (NFκB) (a key transcription factor), and increased expression of proinflammatory cytokines and chemokines.[18] Chemokines recruit neutrophils and effector lymphocytes that perpetuate liver injury and fibrogenesis. When alcohol is consumed in excess, levels of (protective) antioxidants (such as glutathione) are also reduced, thereby enhancing liver injury.

Gut Permeability and Lipopolysaccharides (LPS)

In addition to alcohol metabolism and ROS formation, endotoxins (or lipopolysaccharide, LPS) from the gut are a key driver of ALD progression.[19] Alcohol increases gut permeability, and results in the translocation of LPS bacterial products from the gut into the portal circulation.[20] The portal circulation then delivers LPS to the liver, where it binds to toll-like receptor 4 (TLR4) expressed on liver macrophages, activates downstream NFκB signaling (via MyD88-independent pathways), and upregulates transcription of proinflammatory cytokines, such as IL-1 and TNF-α.[21] This autoregulatory feedback loop leads to an enhanced inflammatory response, and promotes liver injury. Mice genetically deficient in TLR4 are protected from alcohol-induced liver injury,[21,22] and mice treated with oral antibiotics also exhibit reduced level of injury.[23] In humans, elevated levels of LPS are detectable in serum of individuals who drink excessively, and among those with ALD.[24,25]

ALD FIBROSIS – PATHOGENIC MECHANISMS (FIGURE 5.1)

Individuals who develop steatohepatitis are at risk of liver fibrosis and cirrhosis.

Overview of Liver Repair

The liver normally regenerates itself after an acute injury (i.e., injurious stimulus is removed, hepatocytes replicate, and liver progenitor cells (or stem cell) differentiate into new cells), and fibrosis does not occur.[8,26]

The ability of a normal hepatocyte to replicate is lost during chronic injury, such as alcoholic liver disease, nonalcoholic fatty liver disease, chronic viral hepatitis (B and C), and genetic metabolic diseases. Cumulative evidence show that, under such conditions, resurrection of morphogenic signals (in addition to other signaling pathways), and an exaggerated liver progenitor cell response occur. *Liver fibrosis* develops when elevated levels of profibrogenic cytokines (such as transforming growth factor- and osteopontin) promote a dysregulated progenitor cell response (i.e., fibrogenic liver repair), and activate liver pericytes (also known as the hepatic stellate cell, HSC) into collagen-producing myofibroblasts.[7]

The mechanisms that regulate liver repair remain unclear, and studies have only begun to unravel this. Clinical and experimental data suggest that hepatocyte injury (or death), immune and cytokine imbalance, and extrahepatic signals, are important modulators of disease outcomes.

Role of Hepatocyte Injury-Death (Damage Associated Molecular Patterns, DAMPs/Other Signals)

Individuals who develop progressive chronic liver disease invariably exhibit ongoing liver injury or death, as reflected by abnormal liver biochemistry tests (i.e., ALT or AST levels), elevated serum levels of cleaved keratin 18 (K18) (a marker of caspase activation), and/or necrotic-apoptosis on liver histology.[27] The association between epithelial cell death and fibrosis came to fore in studies where mice with increased hepatocyte apoptosis developed more fibrosis,[28] while administration of caspase inhibitors (that block apoptosis) abrogated fibrotic outcomes in mice fed the methionine-choline deficient diet,[29] or subjected to bile-duct ligation.[30] In a more recent study, the loss of RIP3 protected mice from necroptosis

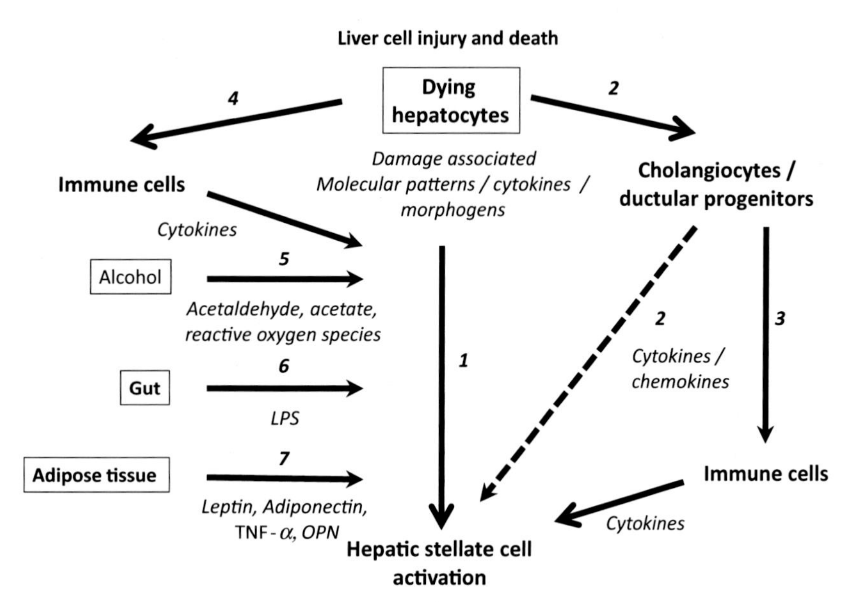

FIGURE 5.1 **Summary of pathogenic mechanisms.** Hepatocyte injury and death (apoptosis, necrosis, necroapoptosis) is characteristic of chronic liver disease, including alcoholic liver disease (ALD). Released damage associated molecular patterns (DAMP) (including ATP, HMGB1), cytokines (such as IL-33), and morphogens (such as hedgehog ligands) activate hepatic stellate cells into collagen-secreting myofibroblasts.[1] These mediators also stimulate proliferation of liver progenitors (liver stem cells), and promote their differentiation (i.e., epithelial-mesenchymal transition) into fibroblastic cells.[2] Liver progenitor cells are highly reactive, and secrete large amounts of cytokines and chemokines (including CXCL16, CCL20) that recruit inflammatory cells.[3] Immune cells may also be activated directly by DAMPs and cytokines/morphogens.[4] In concert, activated immune cell subsets accumulate, and secrete factors that amplify the injury and fibrogenic responses.

In ALD, alcohol metabolites (acetaldehyde and acetate) can directly mediate hepatocyte injury, inflammation, and stimulate hepatic stellate cell activation.[5] Alcohol increases gut permeability, and promotes the translocation of bacterial products (i.e., endotoxin/lipopolysaccharide, LPS) from the gut, into the portal circulation. LPS engages receptors expressed on liver macrophages and hepatic stellate cells, and triggers the secretion of proinflammatory cytokines such as TNF-α; in turn, TNF-α enhances liver injury.[6] The adipose tissue secretes high levels of cytokines (osteopontin, TNF-α) and adipokines (leptin, adiponectin), and is a key modulator of liver disease outcome.[7] Leptin directly activates hepatic stellate cells, while adipose-derived TNF-α levels correlate with severity of alcoholic hepatitis.

and attenuated NASH-fibrosis, additionally highlighted the link between death and fibrosis.[31,32]

Nevertheless, the mechanistic link between injury-death and fibrosis remains poorly understood. Previously, others proposed that dying hepatocytes were engulfed by HSC and macrophages that consequently became activated and enhanced fibrosis.[33] We and others reported that stressed and dying hepatocytes release factors such as IL-33 and hedgehog (Hh) ligands (such as sonic Hh) that activate neighboring hepatic stellate cells, immune cells, and ductular/liver progenitor cells.[34–36] Hh ligands, for example, stimulate proliferation of liver progenitor cells, and their reprogramming to a fibroblastic phenotype (i.e., epithelial-mesenchymal transition),[37] and activate HSC to collagen-producing myofibroblasts.[38] Activation of the Hh pathway also stimulates the secretion of chemokines that recruit immune cell subsets into the liver; in turn, immune cells respond to Hh ligands and secrete even more inflammatory and fibrogenic cytokines that amplify the fibrotic outcome.[39] Interestingly, inhibiting hepatocyte apoptosis with a pan-caspase inhibitor reduced expression of Hh ligands, and attenuated Hh pathway activity.[29] These observations in cell culture and animal models have been recapitulated in human liver disease: activation of the Hh pathway occurs in ALD, and correlates with disease severity and outcome.[40]

Recent studies show that dying cells can release danger signals (i.e., damage-associated molecular patterns), such as the high mobility groupbox 1 (HMGB1) and ATP that trigger immune cell recruitment, and exacerbate acute liver injury (sterile inflammation).[41,42] HMGB1 binds to TLR4 receptors on liver macrophages and HSC, and binds to RAGE on liver progenitors, providing credence that HMGB1 may contribute to ALD progression. Indeed, hepatocyte necrosis and apoptosis are characteristic features of progressive ALD (i.e., alcoholic steatohepatitis).[27,43]

Immune Cells and Chemokines/Cytokines

Liver disease progression is associated intimately with inflammatory cell infiltration.[44]

The liver sinusoidal endothelium is the site of immune cell trafficking into the liver, and this recruitment process is regulated tightly by the coordinated expression of adhesion molecules, integrins, and chemokines (also known as chemotactic cytokines).[45] During chronic liver injury, induction of ICAM1, VCAM1, VAP1 (adhesion molecules), CXCL16, and CXCR3 chemokines (CXCL9-11) promote the recruitment of immune subsets (such as effector T cells, T regulatory cells, natural killer T cells (NKT), B cells, monocytes, and dendritic cells). In turn, recruited immune cells secrete pro- and anti-inflammatory mediators (including cytokines) that modulate disease outcome.[7] Compared with ALD-steatosis that is not associated with any significant increase in adhesion molecule expression, ALD-steatohepatitis is characterized by an overexpression of VCAM1, ICAM1, and VAP1.[46,47] Consistently, ALD livers are enriched with large numbers of CD8 and NKT cells that perpetuate and amplify liver injury and fibrogenesis.

CCL20 (also known as macrophage inflammatory protein, MIP-3) is one of the most upregulated chemokine in patients with ALD-steatohepatitis, and correlates with ALD fibrosis stage, degree of portal hypertension, clinical severity and mortality.[48] Interestingly, CCL20 can be induced by LPS (high levels of LPS in human ALD), and was found to exert a direct, profibrogenic effect on HSC, while loss of CCL20 abrogated LPS-induced hepatic injury.

Increased levels of proinflammatory and profibrogenic cytokines are detected in patients with progressive ALD. TNF-α, IL-22, Osteopontin (OPN), and CCL20 are examples of key mediators that are dysregulated in human and rodent models of ALD. TNF-α is the prototypical proinflammatory cytokine that is upregulated early in disease, and secreted by macrophages following LPS activation.[49] Binding of TNF-α to its receptors leads to the recruitment of downstream adaptor proteins and associated signaling that target the mitochondria, and mitochondria-derived ROS are critical to TNF-α induced hepatocyte injury. These observations in mice led to subsequent clinical trials that evaluated the role of etanercept in patients with severe ALD-steatohepatitis. Unfortunately, no clinical benefit was observed with etanercept,[50] and further studies on the utility and safety of cell death inhibitors are needed.

In contrast to TNF-α IL-22 (a member of the IL-10 family) exhibits antioxidant, antisteatotic, and anti-apoptotic effects, and plays an important role in tissue repair.[51] It is secreted by T cells, NK cells, and γδT cells, and binds to IL-22R1 receptors expressed on hepatocytes, leading to activation of STAT3.[52] Studies show that IL-22 inhibits hepatocyte death, and is protective in models of ischemic-reperfusion injury, and ALD. Forced overexpression of IL-22 reduced hepatic steatosis, and attenuated injury in mice, in part through decreased TNF-α expression, ROS, and lipid peroxidation. Importantly, IL-22 repressed fibrogenesis by promoting HSC senescence via a STAT3-SOCS3 pathway.[53] Additional studies will be necessary to evaluate the potential therapeutic benefit of exogenous IL-22 in a more physiological model of ALD.

The role of OPN in ALD will be discussed in the following sections.

Contributions From Adipose Tissues and the Gut

Adipose-derived cytokines (TNF-α, IL-6, IL-8, IL-10, transforming growth factor-) and adipokines (leptin, adiponectin, resistin) are important modulators of liver injury and fibrosis.[54] For example, leptin is a profibrogenic adipokine that directly activates HSC, and promotes progression of nonalcoholic steatohepatitis (NASH), while adiponectin exerts a protective, anti-inflammatory, and antifibrotic effect. Accumulating data suggest that the adipose tissue may also be involved in the pathogenesis of ALD. Among those with ALD, excess weight for over 10 years is an independent risk factor for steatohepatitis and cirrhosis,[55,56] and body mass index (BMI) is an independent risk factor for fibrosis.[57] In a study of individuals with ALD, adipose tissue inflammation correlates with ALD severity: adipose IL10 positively correlated with fibrosis stage, while adipose TNF-α correlated with the Maddrey's prognosis score.[55] Interestingly, the withdrawal of alcohol for just one week reduces macrophage infiltration in subcutaneous adipose tissue, and switches macrophage toward a M2-anti-inflammatory phenotype,[58] reattesting the intricate link between adipose tissue and ALD liver.

The gut plays a key role in shaping homeostatic responses, and modulating disease outcomes. Commensal bacteria maintain integrity of the gut mucosa, and changes to the gut microbiota are associated with the metabolic syndrome and NASH.[59] In addition to changes in gut permeability and LPS (discussed earlier), alcohol can directly alter the composition of the gut microbiome.[60] In mice, the administration of alcohol increases the populations of Actinobacteria and Firmicutes (compared with pair-fed controls),[61] and leads to a fall in gut branched chain amino acids.[62]. Similar changes to the gut bacterial populations are detected among patients with advanced ALD, and selective gut decontamination with nonabsorbable antibiotics reduces ALD complications.[63] Further studies are needed to understand if changes to the gut microbiota directly modulate the natural history of ALD.

ROLE OF OPN IN ALD

Introduction to OPN and its Role in Liver Disease

Osteopontin (OPN) (also known as secreted phosphoprotein-1, Spp1) is a proinflammatory cytokine and matrix protein (matricellular protein) that is upregulated in human chronic liver disease (viral hepatitis B and C, NASH, PBC),[64] and in various murine models of liver injury (bile-duct ligation (biliary fibrosis), methionine-choline deficient diet (NASH-fibrosis), and carbon tetrachloride injection (advanced fibrosis)).[65] It is expressed normally by epithelial cells (particularly cholangiocytes and liver progenitors), immune cells (macrophages, dendritic cells, B and T cell subsets), HSC, and endothelial cells, and its expression is regulated by cytokines (TNF-α, IL-6), growth factors (platelet-derived growth factor, PDGF), morphogens (hedgehog and Wnt signals), adipokines (leptin), and sympathetic hormones (epinephrine).[66]

When secreted, OPN engages with integrins (including α_v, α4, α9) and the CD44 receptor, leading to the activation of diverse signaling pathways (PI3K, MAPK, ERK, Src). In addition, OPN exists as multiple isoforms (intracellular and extracellular), and undergoes post-translational modifications (i.e., phosphorylation, glycosylation, and proteolytic cleavage), adding to the complexity of OPN regulation and function.[67,68]

Studies in cell cultures show that OPN:

1. Upregulates α-smooth muscle actin and collagen 1α1 in HSC (markers of HSC activation)[69]
2. Stimulates the proliferation and wound healing responses of liver progenitors[65]
3. Activates immune cells and modulates immune cell functions (e.g., switching dendritic cell phenotype)[70]

Role of OPN in Chronic Liver Diseases

Human

Liver and blood levels of OPN are elevated significantly in patients with NASH, chronic biliary diseases (PBC and PSC), and chronic viral hepatitis (B and C). Interestingly, expression of OPN correlates with the degree of Hh pathway activation, and correlates with liver fibrosis stage.[69] A recent report further demonstrated that OPN directly promotes HCV viral replication *in vitro*, and was associated with lower rates of sustained virological response, following treatment with pegylated interferon and ribavirin.[64]

Models

Levels of OPN in blood and liver are also elevated after bile-duct ligation, the methionine-choline deficient diet (and the Western diet), and chronic carbon tetrachloride administration.[65] The highest expressions of OPN are seen in liver progenitors and cholangiocytes, but are also expressed by HSC, endothelial cells, and immune cells. OPN is an important viability factor, inhibits cell death, and enhances liver fibrogenesis. Treating liver progenitors with recombinant OPN promotes their reprogramming into fibroblast-like cells, via modulation of the TGF-β signaling pathway, while OPN blockade inhibits this process. Importantly, OPN neutralization with OPN-specific antibodies or aptamers, during liver injury, significantly abrogates the liver progenitor cell response and attenuates fibrogenesis.

Accumulating data suggest that OPN isoforms exhibit differential cellular effects.[67,68] Intracellular OPN is important for cell viability and is intricately linked to cytoskeletal proteins (such as ezrin, radixin, moesin),[71] while extracellular OPN (the isoform targeted by OPN antibodies or aptamers) modulates downstream transduction signals (integrin-PI3K/AKT, MAPK; Ski/SnoN-Smad).[65] OPN knockout mice are deficient in both intra- and extracellular OPN, and, as such, do not necessarily recapitulate neutralization studies *in vitro* or *in vivo*. Future studies will be needed to elucidate the roles of other OPN isoforms in liver injury and fibrosis.

Outside of the liver, studies show that OPN is a key regulator of cell metabolism.[72] OPN is overexpressed in inflamed adipose tissues, in inflamed vascular endothelium, and is upregulated in patients with the metabolic syndrome.[73] Individuals with the metabolic syndrome are at increased risk of NASH and NASH-fibrosis,[74] and adipose tissue inflammation correlates with ALD severity,[55] implicating a potential role for OPN in adipose tissue-liver tissue (fibrosis) crosstalk. This is supported by observations that OPN neutralization improves peripheral insulin sensitivity and metabolic profiles, and reduces NASH severity.[75]

Role of OPN in ALD

Human

As anticipated, OPN is one of the most highly upregulated genes, in patients with ALD. In the first study (from France), OPN was found to be elevated in serum and livers of patients with ALD, and serum OPN levels correlated with fibrosis stage.[76] Liver OPN correlated with serum OPN levels, and also reflected the degree of liver inflammation and fibrosis. Interestingly, adipose tissue OPN was also upregulated in these individuals, and correlated with liver fibrosis stage. In a recent study (from Spain), investigators confirmed that liver and serum OPN were significantly increased in patients with alcoholic hepatitis, and correlated with clinical severity (positive) and short-term outcomes (negative). Individuals with alcoholic hepatitis also tendered to have a lower frequency of the CC genotype of the +1239C SNP of the OPN gene.[77]

Models

Comparable with other models of liver injury, OPN expression is significantly upregulated, following chronic alcohol binge. Unlike wild-type (control) mice, OPN knockout mice were protected against ALD, and expressed repressed levels of inflammatory cytokines.[77] OPN is a well-recognized chemoattractant for neutrophils, and rats fed the Lieber–Decarli ethanol diet with LPS (another model of ALD) also upregulated OPN, and accumulated large numbers of neutrophils. Conversely, treatment with an OPN neutralizing antibody prevented the accumulation of neutrophils, and attenuated liver injury.[78]

Surprisingly, however, there have been several recent reports that OPN is protective in models of ALD. In the first, Nieto and coworkers used an OPN-(hepatocyte) transgenic mouse, and demonstrated that overexpression of OPN in hepatocytes protected mice from inflammation, namely macrophage numbers and TNF-α (+) cells; from injury (ALT levels); from steatosis; and reduced LPS levels.[79] They proposed that OPN binds to LPS, and prevents LPS-induced macrophage activation, thereby reducing ROS/TNF-α secretion. They further reported that supplementation with milk OPN attenuates liver injury, by limiting the translocation of gut-LPS into the portal circulation, through an increase in intestinal gland height, mucin content, and crypt cell proliferation. In another study, Tsukamoto and coworkers observed that OPN knockout mice developed alcoholic hepatitis, without the need for an additional weekly binge (acute on chronic alcoholic hepatitis model).[80] These findings appear contradictory to the reported outcomes of other liver injury models, and may be explained by differences in injury models (i.e., acute versus chronic, type of injury), and roles of specific OPN isoforms (i.e., use of OPN knockout mice versus administration of OPN neutralization antibodies or aptamers that target extracellular OPN).

CONCLUSIONS

ALD is a leading cause of chronic liver disease, but treatment options remain limited to steroids, for those with severe alcoholic hepatitis. Recent studies implicate OPN as a key player in ALD pathogenesis and outcome. Future studies will be needed to elucidate the roles of OPN isoforms across ALD-states (i.e., steatosis, alcoholic steatohepatitis, and fibrosis-cirrhosis), with models that recapitulate human disease in a better manner.

Key Facts in ALD and Osteopontin

- Heavy drinking is associated with steatosis.
- Oxidative stress, immune disruption and toxins arising from alcohol processing in the liver lead to inflammation and cell death.
- Individuals with steatohepatitis are at risk of developing liver fibrosis and cirrhosis.
- Advanced progression of ALD requires activation of signaling pathways leading to inflammation and fibrosis cellular responses.
- Osteopontin is a proinflammatory cytokine and matricellular protein.
- Osteopontin is highly elevated in ALD patient serum and liver tissue. Osteopontin levels correlate with fibrosis stage.
- Osteopontin binds with integrins and CD44 receptor, leading to activation of diverse signaling pathways such as PI3K, MAPK, and SMAD.
- OPN promotes immune, inflammatory, and myofibroblast activity.

Summary Points

- To halt the progression of ALD, new therapies are required.
- Novel therapeutic strategies could target molecular and cellular mechanisms of fibrosis.
- Neutralization of osteopontin is a potential antifibrotic strategy in ALD.

Abbreviations

ALD	Alcoholic liver disease
OPN	Osteopontin
Steatosis	Imbalance between lipid synthesis and breakdown within the liver, resulting in intracellular accumulation of triglyceride in hepatocytes
Steatohepatitis	Steatosis accompanied by liver injury and inflammation
Fibrosis	Excessive deposition of connective tissue, evident by histopathology. Represents scarring formed in uncontrolled response to chronic injury
Cirrhosis	Advanced injury to the liver whereby the normal architecture is disrupted and replaced by excessive accumulation of scarring. Causes potentially irreversible loss of liver function

References

1. Gao B, Bataller R. Alcoholic liver disease: pathogenesis and new therapeutic targets. *Gastroenterol* 2011;**141**(5):1572–85.
2. Teli MR, Day CP, Burt AD, et al. Determinants of progression to cirrhosis or fibrosis in pure alcoholic fatty liver. *Lancet* 1995;**346**: 987–90.
3. Tsukamoto H, Machida K, Dynnyk A, et al. "Second hit" models of alcoholic liver disease. *Semin Liver Dis* 2009;**29**:178–87.
4. Wilfred de Alwis NM, Day CP. Genetics of alcoholic liver disease and nonalcoholic fatty liver disease. *Semin Liver Dis* 2007;**27**:44–54.
5. Altamirano J, Bataller R. Alcoholic liver disease: pathogenesis and new targets for therapy. *Nat Rev Gastroenterol Hepatol* 2011;**8**: 491–501.
6. Chedid A, Mendenhall CL, Gartside P, et al. Prognostic factors in alcoholic liver disease. *Am J Gastroenterol* 1991;**86**:210–6.
7. Seth D, Haber PS, Syn WK, et al. Pathogenesis of alcohol-induced liver disease: classical concepts and recent advances. *J Gastroenterol Hepatol* 2011;**26**(7):1089–105.
8. Diehl AM, Chute J. Underlying potential: cellular and molecular determinants of adult liver repair. *J Clin Invest* 2013;**123**(5):1858–60.

9. Baraona E, Lieber CS. Effects of ethanol on lipid metabolism. *J Lipid Res* 1979;**20**:289–315.
10. You M, Fischer M, Deeg MA, et al. Ethanol induces fatty acid synthesis pathways by activation of sterol regulatory element-binding protein (SREBP). *J Biol Chem* 2002;**277**:29342–7.
11. You M, Matsumoto M, Pacold CM, et al. The role of AMP-activated protein kinase in the action of ethanol in the liver. *Gastroenterol* 2004;**127**:1798–808.
12. Hoek JB, Cahill A, Pastorino JG. Alcohol and mitochondria: a dysfunctional relationship. *Gastroenterol* 2002;**122**:2049–63.
13. Sozio MS, Liangpunsakul S, Crabb D. The role of lipid metabolism in the pathogenesis of alcoholic and nonalcoholic hepatic steatosis. *Semin Liver Dis* 2010;**30**:378–90.
14. Setshedi M, Wands JR, Monte SM. Acetaldehyde adducts in alcoholic liver disease. *Oxid Med Cell Longev* 2010;**3**:178–85.
15. Mottaran E, Stewart SF, Rolla R, et al. Lipid peroxidation contributes to immune reactions associated with alcoholic liver disease. *Free Radic Biol Med* 2002;**32**:38–45.
16. Orman ES, Odena G, Bataller R. Alcoholic liver disease: pathogenesis, management, and novel targets for therapy. *J Gastroenterol Hepatol* 2013;**28**(Suppl. 1):77–84.
17. Lieber CS. Alcoholic fatty liver: its pathogenesis and mechanism of progression to inflammation and fibrosis. *Alcohol* 2004;**34**(1):9–19.
18. Albano E, Vidali M. Immune mechanisms in alcoholic liver disease. *Genes Nutr* 2010;**5**(2):141–7.
19. Purohit V, Brenner DA. Mechanisms of alcohol-induced hepatic fibrosis: a summary of the Ron Thurman Symposium. *Hepatol* 2006;**43**:872–8.
20. Rao R. Endotoxemia and gut barrier dysfunction in alcoholic liver disease. *Hepatol* 2009;**50**:638–44.
21. Hritz I, Mandrekar P, Velayudham A, et al. The critical role of toll-like receptor (TLR) 4 in alcoholic liver disease is independent of the common TLR adapter MyD88. *Hepatol* 2008;**48**:1224–31.
22. Uesugi T, Froh M, Arteel GE, et al. Toll-like receptor 4 is involved in the mechanism of early alcohol-induced liver injury in mice. *Hepatol* 2001;**34**(1):101–8.
23. Adachi Y, Moore LE, Bradford BU, et al. Antibiotics prevent liver injury in rats following long-term exposure to ethanol. *Gastroenterol* 1995;**108**(1):218–24.
24. Parlesak A, Schafer C, Schutz T, et al. Increased intestinal permeability to macromolecules and endotoxemia in patients with chronic alcohol abuse in different stages of alcohol-induced liver disease. *J Hepatol* 2000;**32**(5):742–7.
25. Lin RS, Lee FY, Lee SD, et al. Endotoxemia in patients with chronic liver diseases: relationship to severity of liver diseases, presence of esophageal varices, and hyperdynamic circulation. *J Hepatol* 1995;**22**(2):165–72.
26. Best J, Dollé L, Manka P, et al. Role of liver progenitors in acute liver injury. *Front Physiol* 2013;**4**:258.
27. Nanji AA, Hiller-Sturmhofel S. Apoptosis and necrosis: two types of cell death in alcoholic liver disease. *Alcohol Health Res World* 1997;**21**:325–30.
28. Canbay A, Higuchi H, Bronk SF, et al. Fas enhances fibrogenesis in the bile duct ligated mouse: a link between apoptosis and fibrosis. *Gastroenterol* 2002;**123**(4):1323–30.
29. Witek RP, Stone WC, Karaca FG, et al. Pan-caspase inhibitor VX-166 reduces fibrosis in an animal model of nonalcoholic steatohepatitis. *Hepatol* 2009;**50**(5):1421–30.
30. Canbay A, Feldstein A, Baskin-Bey E, et al. The caspase inhibitor IDN-6556 attenuates hepatic injury and fibrosis in the bile duct ligated mouse. *J Pharmacol Exp Ther* 2004;**308**(3):1191–6.
31. Gautheron J, Vucur M, Reisinger F, et al. A positive feedback loop between RIP3 and JNK controls nonalcoholic steatohepatitis. *EMBO Mol Med* 2014;**6**:1062–74.
32. Petrasek J, Iracheta-Vellve A, Csak T, et al. STINGIRF3 pathway links endoplasmic reticulum stress with hepatocyte apoptosis in early alcoholic liver disease. *Proc Natl Acad Sci USA* 2013;**110**:16544–9.
33. Canbay A, Feldstein AE, Higuchi H, et al. Kupffer cell engulfment of apoptotic bodies stimulates death ligand and cytokine expression. *Hepatol* 2003;**38**(5):1188–98.
34. Li J, Razumilava N, Gores GJ, et al. Biliary repair and carcinogenesis are mediated by IL-33-dependent cholangiocyte proliferation. *J Clin Invest* 2014;**124**:3241–51.
35. Luedde T, Kaplowitz N, Schwabe RF. Cell death and cell death responses in liver disease: mechanisms and clinical relevance. *Gastroenterol* 2014;**147**(4):765–83.
36. Jung Y, Witek RP, Syn WK, et al. Signals from dying hepatocytes trigger growth of liver progenitors. *Gut* 2010;**59**(5):655–65.
37. Syn WK, Jung Y, Omenetti A. Hedgehog-mediated epithelial-to-mesenchymal transition and fibrogenic repair in nonalcoholic fatty liver disease. *Gastroenterol* 2009;**137**(4):1478–88.
38. Choi SS, Omenetti A, Witek RP, et al. Hedgehog pathway activation and epithelial-to-mesenchymal transitions during myofibroblastic transformation of rat hepatic cells in culture and cirrhosis. *Am J Physiol Gastrointest Liver Physiol* 2009;**297**(6):G1093–106.
39. Omenetti A, Syn WK, Jung Y, et al. Repair-related activation of hedgehog signaling promotes cholangiocyte chemokine production. *Hepatol* 2009;**50**(2):518–27.
40. Jung Y, Brown KD, Witek RP, et al. Accumulation of hedgehog-responsive progenitors parallels alcoholic liver disease severity in mice and humans. *Gastroenterol* 2008;**134**(5):1532–43.
41. Chen GY, Nunez G. Sterile inflammation: sensing and reacting to damage. *Nat Rev Immunol* 2010;**10**:826–37.
42. Lotze MT, Zeh HJ, Rubartelli A, et al. The grateful dead: damage-associated molecular pattern molecules and reduction/oxidation regulate immunity. *Immunol Rev* 2007;**220**:60–81.
43. Ziol M, Tepper M, Lohez M, et al. Clinical and biological relevance of hepatocyte apoptosis in alcoholic hepatitis. *J Hepatol* 2001;**34**:254–60.
44. Shetty S, Lalor PF, Adams DH. Lymphocyte recruitment to the liver: molecular insights into the pathogenesis of liver injury and hepatitis. *Toxicol* 2008;**254**(3):136–46.
45. Lalor PF, Faint J, Aarbodem Y, et al. The role of cytokines and chemokines in the development of steatohepatitis. *Semin Liver Dis* 2007;**27**(2):173–93.
46. Haydon G, Lalor PF, Hubscher SG, et al. Lymphocyte recruitment to the liver in alcoholic liver disease. *Alcohol* 2002;**27**(1):29–36.
47. Afford SC, Fisher NC, Neil DA, et al. Distinct patterns of chemokine expression are associated with leukocyte recruitment in alcoholic hepatitis and alcoholic cirrhosis. *J Pathol* 1998;**186**(1):82–9.
48. Affò S, Morales-Ibanez O, Rodrigo-Torres D, et al. CCL20 mediates lipopolysaccharide induced liver injury and is a potential driver of inflammation and fibrosis in alcoholic hepatitis. *Gut* 2014;**63**(11):1782–92.
49. Kitazawa T, Nakatani Y, Fujimoto M, et al. The production of tumor necrosis factor-alpha by macrophages in rats with acute alcohol loading. *Alcohol Clin Exp Res* 2003;**27**(8 Suppl.):72S–5S.
50. Boetticher NC, Peine CJ, Kwo P, et al. A randomized, double-blinded, placebo-controlled multicenter trial of etanercept in the treatment of alcoholic hepatitis. *Gastroenterol* 2008;**135**:1953–60.
51. Ki SH, Park O, Zheng M, et al. Interleukin-22 treatment ameliorates alcoholic liver injury in a murine model of chronic-binge ethanol feeding: role of signal transducer and activator of transcription 3. *Hepatol* 2010;**52**:1291–300.
52. Wolk K, Witte E, Witte K, et al. Biology of interleukin-22. *Semin Immunopathol* 2010;**32**:17–31.
53. Kong X, Feng D, Wang H, et al. Interleukin-22 induces hepatic stellate cell senescence and restricts liver fibrosis in mice. *Hepatol* 2012;**56**:1150–9.
54. Gerner RR, Wieser V, Moschen AR, et al. Metabolic inflammation: role of cytokines in the crosstalk between adipose tissue and liver. *Can J Physiol Pharmacol* 2013;**91**(11):867–72.

55. Naveau S, Cassard-Doulcier AM, Njiké-Nakseu M, et al. Harmful effect of adipose tissue on liver lesions in patients with alcoholic liver disease. *J Hepatol* 2010;**52**(6):895–902.
56. Naveau S, Giraud V, Borotto E, et al. Excess weight risk factor for alcoholic liver disease. *Hepatol* 1997;**25**(1):108–11.
57. Raynard B, Balian A, Fallik D, et al. Risk factors of fibrosis in alcohol-induced liver disease. *Hepatol* 2002;**35**(3):635–8.
58. Voican CS, Njiké-Nakseu M, Boujedidi H, et al. Alcohol withdrawal alleviates adipose tissue inflammation in patients with alcoholic liver disease. *Liver Int* 2015;**35**(3):967–78.
59. Vajro P, Paolella G, Fasano A. Microbiota and gut-liver axis: their influences on obesity and obesity-related liver disease. *J Pediatr Gastroenterol Nutr* 2013;**56**(5):461–8.
60. Szabo G. Gut-liver axis in alcoholic liver disease. *Gastroenterol* 2015;**148**(1):30–6.
61. Bull-Otterson L, Feng W, Kirpich I, et al. Metagenomic analyses of alcohol induced pathogenic alterations in the intestinal microbiome and the effect of Lactobacillus rhamnosus GG treatment. *PLoS One* 2013;**8**:e53028.
62. Xie G, Zhong W, Zheng X, et al. Chronic ethanol consumption alters mammalian gastrointestinal content metabolites. *J Proteome Res* 2013;**12**:3297–306.
63. Bass NM, Mullen KD, Sanyal A, et al. Rifaximin treatment in hepatic encephalopathy. *N Engl J Med* 2010;**362**:1071–81.
64. Choi SS, Claridge LC, Jhaveri R, et al. Osteopontin is up-regulated in chronic hepatitis C and is associated with cellular permissiveness for hepatitis C virus replication. *Clin Sci (Lond)* 2014;**126**(12): 845–55.
65. Coombes JD, Swiderska-Syn M, Dollé L, et al. Osteopontin neutralisation abrogates the liver progenitor cell response and fibrogenesis in mice. *Gut* 2015;**64**(7):1120–31.
66. Uede T. Osteopontin, intrinsic tissue regulator of intractable inflammatory diseases. *Pathol Int* 2011;**61**(5):265–80.
67. Inoue M, Shinohara ML. Intracellular osteopontin (iOPN) and immunity. *Immunol Res* 2011;**49**(1–3):160–72.
68. Gimba ER, Tilli TM. Human osteopontin splicing isoforms: known roles, potential clinical applications and activated signaling pathways. *Cancer Lett* 2013;**331**(1):11–7.
69. Syn WK, Choi SS, Liaskou E, et al. Osteopontin is induced by hedgehog pathway activation and promotes fibrosis progression in nonalcoholic steatohepatitis. *Hepatol* 2011;**53**(1):106–15.
70. Shinohara ML, Kim HJ, Kim JH, et al. Alternative translation of osteopontin generates intracellular and secreted isoforms that mediate distinct biological activities in dendritic cells. *Proc Natl Acad Sci USA* 2008;**105**(20):7235–9.
71. Zohar R, Suzuki N, Suzuki K, et al. Intracellular osteopontin is an integral component of the CD44-ERM complex involved in cell migration. *J Cell Physiol* 2000;**184**(1):118–30.
72. Shi Z, Mirza M, Wang B, et al. Osteopontin-a alters glucose homeostasis in anchorage-independent breast cancer cells. *Cancer Lett* 2014;**344**(1):47–53.
73. Musso G, Paschetta E, Gambino R, et al. Interactions among bone, liver, and adipose tissue predisposing to diabesity and fatty liver. *Trends Mol Med* 2013;**19**(9):522–35.
74. Armstrong MJ, Adams LA, Canbay A, et al. Extrahepatic complications of nonalcoholic fatty liver disease. *Hepatol* 2014;**59**(3): 1174–97.
75. Kiefer FW, Zeyda M, Gollinger K, et al. Neutralization of osteopontin inhibits obesity-induced inflammation and insulin resistance. *Diabetes* 2010;**59**(4):935–46.
76. Patouraux S, Bonnafous S, Voican CS, et al. The osteopontin level in liver, adipose tissue and serum is correlated with fibrosis in patients with alcoholic liver disease. *PLoS One* 2012;**7**(4):e35612.
77. Morales-Ibanez O, Domínguez M, Ki SH, et al. Human and experimental evidence supporting a role for osteopontin in alcoholic hepatitis. *Hepatol* 2013;**58**(5):1742–56.
78. Banerjee A, Apte UM, Smith R, et al. Higher neutrophil infiltration mediated by osteopontin is a likely contributing factor to the increased susceptibility of females to alcoholic liver disease. *J Pathol* 2006;**208**:473–85.
79. Ge X, Leung TM, Arriazu E, et al. Osteopontin binding to lipopolysaccharide lowers tumor necrosis factor-α and prevents early alcohol-induced liver injury in mice. *Hepatol* 2014;**59**(4):1600–16.
80. Lazaro R, Wu R, Lee S, et al. Osteopontin deficiency does not prevent but promotes alcoholic neutrophilic hepatitis in mice. *Hepatology* 2015;61(1):129–40.

CHAPTER

6

The Role of CD36 in the Pathogenesis of Alcohol-Related Disease

Caleb T. Epps, MS, Robin D. Clugston, PhD, Amit Saha, BA, William S. Blaner, PhD, Li-Shin Huang, PhD

Department of Medicine, Division of Preventive Medicine & Nutrition, Columbia University, New York, NY, USA

INTRODUCTION

Cluster of differentiation 36 (CD36) is a class B scavenger receptor expressed in a variety of cell types that binds to multiple ligands, including but not limited to: long-chain fatty acids (FAs), native and oxidized low density lipoproteins (ox-LDL), phospholipids, thrombospondins (TSPs), the malaria parasite, bacteria, and β-amyloid.[1–3] CD36 has been shown to play different roles in numerous physiological processes, as well as in disease susceptibility.[1–7] These include facilitating FA transport in adipose tissue, and heart and skeletal muscle;[4] lipid sensing in taste buds;[5] dietary fat absorption;[5] angiogenesis;[6] susceptibility to atherosclerosis;[1,2] severity of malaria infection;[7] and susceptibility to Alzheimer's disease.[3] Elevated expression of CD36 has been observed in the livers of rodents fed alcohol,[8,9] and a role for CD36 in the development of alcohol-related disease has been demonstrated by Clugston et al.,[10] who showed that whole-body CD36-deficient mice are resistant to alcoholic steatosis, and have a reduced mortality rate related to alcohol consumption. Although little else is currently known regarding how CD36 functions in the body upon chronic alcohol consumption, much has been described about the receptor's actions on various physiological and metabolic processes.[1,2,4–6] This chapter reviews the known functions of CD36 that may play a role in the development of alcohol-related disease, particularly in alcoholic liver disease. After a brief review of the structure–function relationship of CD36, this chapter will discuss CD36's role as a receptor for FA transport, a signaling molecule for cellular Ca^{2+} fluxes, and a pattern recognition receptor for pathogen-associated molecular patterns (PAMPs). In this chapter, we will also hypothesize how these known functions of CD36 could be linked to alcohol-related disease, particularly in the liver and the brain. The chapter concludes by posing key questions that, if answered, would further clarify the interaction between CD36 and alcohol.

THE STRUCTURE-FUNCTION RELATIONSHIP OF CD36

CD36 was discovered in the 1970s and was described as "glycoprotein IV" on platelet membranes.[11] Since then, much has been learned about the structure and function of CD36.[2,5] Human CD36 protein is a glycoprotein with 472 amino acids, and a molecular mass of approximately 88 kDa. It belongs to the class B scavenger receptor family that also includes scavenger receptor-B1, and the lysosomal protein LIMP2.[3] All members of this protein family share structural similarities, and have two transmembrane domains, as shown schematically for human CD36 in Figure 6.1. CD36 also has a large extracellular domain that includes a hydrophobic region creating a loop toward the plasma membrane, and two smaller cytoplasmic terminals. The carboxyl half of the extracellular domain also contains three disulfide bridges.[5]

Functional Binding Domains of CD36

CD36 has numerous functional binding domains, particularly in the amino-terminal half of the extracellular domain that has binding motifs for multiple ligands. For example, the FA binding domain of CD36 has been

Molecular Aspects of Alcohol and Nutrition. http://dx.doi.org/10.1016/B978-0-12-800773-0.00006-9

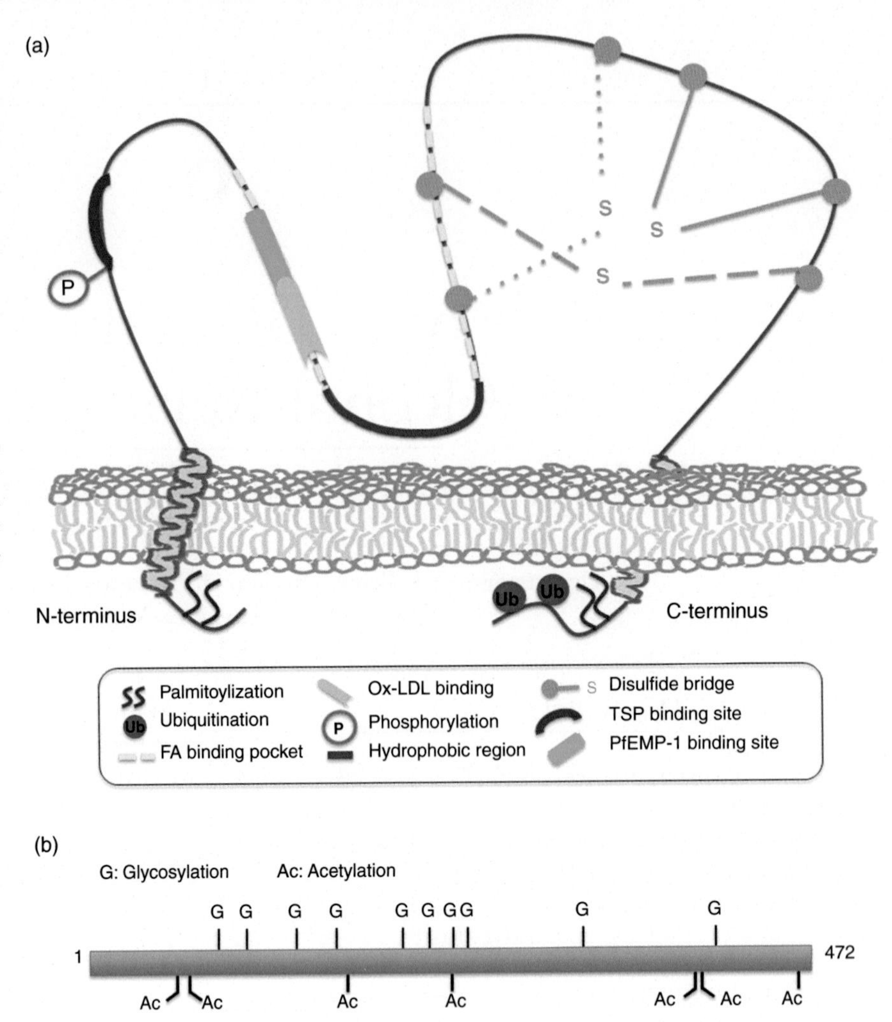

FIGURE 6.1 Schematic presentation of human CD36 protein. (a) The proposed topology shows that CD36 has two transmembrane regions, a large extracellular domain with three disulfide bridges, and two small cytoplasmic tails, with one on the amino terminus and one on the carboxyl-terminus of the protein. Functional binding domains include binding sites for thrombospondin (TSP, amino acids 93–120), PfEMP-1 (a protein derived from malaria parasite *P. falciparum*, amino acids 145–171), oxidized LDL (ox-LDL, amino acids 153–183), and the proposed fatty acid (FA) binding pocket (amino acids 127–279). Posttranslational modifications such as palmitoylation, phosphorylation, and ubiquitination are also shown. (b) Glycosylation (G) is found on asparagine at residues 79, 102, 134, 163, 205, 235, 247, 321, and 417. Acetylation (Ac) is found on lysine at residues 52, 56, 166, 231, 394, 398, and 463.

proposed to comprise amino acid residues 127–279, with a hydrophobic region at residues 186 to 204 forming a loop in the extracellular domain (Figure 6.1) to serve as part of a binding pocket for long-chain FAs, and other lipid ligands.[5,12] Binding of sulfo-*N*-succinimidyl oleate to lysine 164 irreversibly inhibits CD36-mediated FA uptake, and signaling in a number of cell types.[12,13]

TSPs bind to amino acids 93–120 of CD36 expressed on endothelial cells to inhibit angiogenesis, thereby playing a role in processes that require neovascularization, such as tumor growth and wound healing.[14] Amino acids 145–171 contain binding sites for proteins in the *Plasmodium falciparum* erythrocyte membrane binding protein 1 (PfEMP1) family, found in *P. falciparum*, a protozoan parasite that causes severe malaria.[15] PfEMP1 is present on the cell surface of infected host red blood cells, and its binding to CD36 on the surface of endothelial cells causes sequestration of the infected host red blood cells to various tissues, such as the heart and brain, resulting in severe tissue damage.[16] The binding domain for ox-LDL is at amino acid residues 153 through 183.[17] CD36-mediated internalization of ox-LDL in macrophages promotes the formation of foam cells and atherosclerotic plaques,[18] stimulates the release of cytokines,[19] and initiates the formation of reactive oxygen species (ROS).[20,21] Other ligands, such as advanced glycated products and hexarelin, a member of the growth-hormone-releasing peptide family, also bind to this domain.[2,22] Hexarelin and its derivative EP 80317, which has no growth hormone-releasing properties, have been found to compete with ox-LDLs for CD36 binding, leading to reduced atherosclerotic lesions in apoE-deficient mice.[23]

Functional domains are also found in the carboxyl-terminal cytoplasmic tail that is required for CD36-mediated phagocytosis of *Staphylococus aureus*, a Gram-positive bacterium.[24] Mutations at tyrosine 463 or cysteine 464 abolish CD36-mediated internalization of *S. aureus* and its cell wall component lipoteichoic acid in CD36-expressing cell lines and macrophages.[24] Also, internalization of ox-LDL is found to require the last six amino acids of this C-terminal tail.[25]

Posttranslational Modification of CD36

Posttranslational modifications can regulate the function of CD36. These include glycosylation, phosphorylation, ubiquitination, palmitoylation, and acetylation (Figure 6.1). For example, glycosylation of the extracellular domain is required for the trafficking of CD36 from the cytosol to the plasma membrane.[26] Phosphorylation at threonine 92 blocks TSP-1 binding to CD36, and its inhibitory effects on angiogenesis.[27] Dephosphorylation of CD36 by intestinal alkaline phosphatase has been shown to reduce intestinal FA uptake, and fat absorption.[28] Ubiquitination is found on two lysine residues (469 and 472) at the carboxyl cytoplasmic tail, and plays a role in intracellular CD36 trafficking.[5] Both cytoplasmic tails are palmitoylated at cysteines 3, 7 (amino terminus), 464, and 466 (carboxyl terminus).[29] Inhibition of palmitoylation has been shown to reduce ox-LDL uptake.[30] Finally, acetylation is found at several lysine residues, including lysine 166, the acetylation of which is thought to play a role in regulating FA uptake, and subsequent signaling.[5,31]

CD36 AND FA UPTAKE

Long-chain FAs are a critical source of energy and signaling molecules, and are required for membrane synthesis.[4] Excess FA uptake, however, can cause the accumulation of triglycerides (TGs), and toxic lipid intermediates that are seen frequently in diseased hearts or fatty livers.[32,33] Long-chain FAs can pass through phospholipid bilayers, via diffusion or facilitated transport by integral or membrane-associated proteins, such as CD36.[4] CD36's function as an FA transporter in adipocytes was first observed in 1993.[34] With the creation of animal models of CD36 deletion and overexpression, CD36 is well studied for its role in facilitating FA uptake in adipose tissue, heart, and skeletal muscle.[4]

CD36 Trafficking is Required for CD36-Facilitated FA Uptake

CD36 resides on the cell surface, but also localizes on intracellular vesicles (e.g., endosomes, the endoplasmic reticulum (ER), and lysosomes), as well as on mitochondria.[4,35] CD36-mediated FA uptake requires the translocation of CD36 from intracellular organelles to the plasma membrane, a process tightly regulated by nutritional demands and hormonal states.[4,36] For example, muscular contraction and insulin acutely promote the translocation of CD36 from intracellular organelles to the plasma membrane, and enhance FA uptake.[4] Muscular contraction also increases CD36 on the outer membrane of mitochondria. This has been associated with increased FA oxidation.[35] Furthermore, FoxO1, a transcription factor activated by fasting, is found to stimulate CD36 translocation to the plasma membrane, to enhance FA uptake and oxidation in order to meet the energy demands of the fasting state.[4,37]

On the cell surface, CD36 is found in lipid rafts; these are detergent-resistant microdomains, rich in cholesterol and sphingolipids.[38] Caveolae are a type of lipid raft coated by proteins, such as caveolin-1, linked functionally to CD36.[39] Embryonic fibroblasts from caveolin-1-deficient mice do not have CD36 on the cell surface, and show reduced FA uptake. Adenoviral expression of caveolin-1 restores the translocation of CD36 to plasma membrane, and normalizes FA uptake in caveolin-1-deficient embryonic fibroblasts.[39] Caveolin-1 expression, therefore, is required for membrane expression of CD36.

Intracellular CD36 trafficking is regulated by posttranslational modifications. It has been proposed that ubiquitinated CD36 in early endosomes are targeted to lysosomes for degradation, instead of recycling back to the plasma membrane.[4] Importantly, ubiquitination of CD36 is stimulated by the presence of long-chain FAs, but is inhibited by insulin,[40] indicating physiological mechanisms for regulating membrane recruitment of CD36. It is also known that inhibition of palmitoylation delays the processing of CD36 at the ER and trafficking through the secretory pathway, hampering the incorporation efficiency of CD36 into lipid rafts on the plasma membrane, and reducing the half-life of the CD36 protein.[30]

Metabolic Consequences of Impaired CD36-Mediated FA Uptake

CD36 deficiency in mice reduces FA uptake in various tissues, while increasing plasma FA and TG levels.[41] Reduced FA uptake in the heart and skeletal muscle results in a compensatory increase in glucose uptake, leading to an increase in insulin sensitivity, in both tissues.[41] Conversely, CD36 overexpression in the muscle increases FA oxidation while simultaneously decreasing plasma lipids.[42] Reduced FA uptake in adipose tissue protects CD36-deficient mice from diet-induced weight gain.[43] However, leptin secretion from CD36-deficient adipocytes is doubled, suggesting CD36-mediated FA uptake plays a role in regulating leptin production.[43] The reverse also holds true, with CD36 overexpression leading to reduced levels of plasma TG and FA, accompanied by hyperglycemia and hyperinsulinemia.[42]

CD36 deficiency also impairs FA uptake in the proximal end of the small intestine, where CD36 is abundantly expressed.[44] This defect, however, is compensated for by increased FA uptake in the distal intestine, via either passive diffusion, or an increase in other FA transporters present in the small intestine.[41,44] Although intestinal lipid absorption is normal in CD36-deficient mice, these animals have significant reductions in chylomicron secretion, likely due to impaired CD36-mediated signaling.[5,45] In addition, intestinally derived lipoprotein particles from these animals are smaller than the chylomicron particles from WT mice, and have a prolonged life in the circulation, resulting in postprandial hypertriglyceridemia.[46]

In humans, mutations have been found on the CD36 gene, located on chromosome 7q. CD36 deficiency occurs in 3–10% of African and Asian populations but is less common in European populations.[7,47] CD36-deficient subjects have been reported to exhibit a defect in myocardial FA uptake, delayed clearance of plasma FA after an oral meal, and abnormal formation of chylomicrons.[48–51] Studies also showed that increased plasma FA and TG levels are associated with the risk of Type II diabetes in CD36-deficient subjects.[48] Although CD36-deficiency and overexpression of CD36 are both considered metabolically detrimental, the literature from human and animal studies suggests a threshold or a range for CD36 expression to be "metabolically protective."[52] The readers are referred to published reviews, for details in the genetics of human CD36.[52,53]

Cardiac CD36 Expression and Alcoholic Cardiomyopathy

CD36-mediated FA uptake is a major pathway in the heart for acquiring circulating free FAs, for energy utilization.[32] CD36 deficiency ameliorates cardiac lipotoxicity, and restores cardiac function in transgenic mice overexpressing peroxisome proliferator-activated receptor (PPAR) α in the heart.[54] Conversely, elevated expression of cardiac CD36 mRNA and protein is associated with increased cardiac TG content, and reduced left ventricular function in diet-induced or genetically obese mice.[55] In addition, FA uptake rates in cardiomyocytes isolated from the genetically obese mice are positively correlated with cardiac TG content and CD36 expression levels, suggesting a role for excess CD36-mediated FA uptake in cardiac TG accumulation and dysfunction, associated with cardiomyopathy.[55]

Alcoholic cardiomyopathy is characterized by left ventricular dilation, and is associated with chronic and heavy alcohol consumption.[56] One potential cause is the dysregulation of fatty acid metabolism, and subsequent TG accumulation in the heart. Hu et al.[57] explored this possibility by chronically exposing C57BL/6J mice to alcohol in their drinking water, with different concentrations of ethanol (10, 14, and 18% v/v) for various durations (12, 8, and 4 weeks, respectively). The 14% alcohol-water was given after 4 weeks of 10% alcohol-water, and the 18% alcohol-water was given following 4 weeks of 10% and then 4 weeks of 14% alcohol-water. Echocardiography showed that all alcohol groups had significant decreases in fractional shortening and ejection fraction.[57] Isolated cardiomyocytes from the alcohol-fed mice showed a dose-dependent increase in FA uptake, compared to the controls. In addition, CD36 expression in cardiomyocytes was upregulated among the alcohol-fed mice, and correlated with the V_{max} for facilitated FA uptake. Furthermore, stearoyl-CoA desaturase 1 expression was upregulated, while genes involved in mitochondrial biogenesis and metabolism were downregulated. Overall, the studies by Hu et al.[57] suggest that CD36-mediated FA uptake, increased FA synthesis, and reduced FA oxidation may all contribute to TG accumulation in the heart, and the consequent cardiac dysfunction associated with chronic alcohol consumption.

Although these studies demonstrate a correlation between CD36 expression and CD36-mediated FA uptake in isolated cardiomyocytes from alcohol-fed mice, it was not known whether this could be demonstrated *in vivo* in alcohol-fed, CD36-deficient mice. In this regard, Clugston et al.[10] showed an expected decrease in cardiac FA uptake in CD36-deficient mice; however, they did not observe an alcohol effect on cardiac FA uptake in wild-type (WT) mice. The lack of an alcohol effect on cardiac FA uptake *in vivo* may be explained by differences in the experimental protocols of the two studies;[10,57] in the latter,[10] mice were fed a liquid Lieber–DeCarli diet, with a lower concentration of alcohol (5.1%, v/v) for a shorter duration (6 weeks). Thus, the effect of CD36 deficiency on the development of alcoholic cardiomyopathy warrants further investigation.

Hepatic CD36 Expression is Associated with Nonalcoholic Fatty Liver Disease (NAFLD)

The significance of CD36 in hepatic FA uptake has been controversial, due to its relatively low expression level in the liver. In CD36-deficient mice, hepatic FA uptake is unchanged in the fed state, and increased in the fasting state.[58,59] The latter is thought to be due to a compensatory increase in FA uptake, via other hepatic FA transporters, in response to increased circulating plasma FA levels in these animals. Similarly, hepatic FA uptake was not altered in human patients with CD36 deficiency, compared to normal control subjects.[60] However, CD36 is induced in the liver under nonphysiological conditions, and is associated with the development of hepatic steatosis, in both animal and human studies. For example, hepatic CD36 mRNA and proteins are increased in mice with high-fat diet-induced fatty livers, suggesting CD36's role in hepatic FA uptake in NAFLD.[61] Furthermore, adenoviral overexpression of CD36 in mice increases hepatic FA uptake and TG accumulation in the livers of WT and CD36-deficient mice, demonstrating CD36's ability to mediate FA uptake in the liver.[61] Interestingly, in human patients with NAFLD, Greco et al.[62] observed a positive correlation between liver fat content and hepatic CD36 expression. Furthermore, in addition to increased gene expression, immunostained liver biopsies show that CD36 proteins are located predominantly on the plasma membranes of hepatocytes in patients with NAFLD, whereas they are found more in the cytoplasm of nondiseased liver sections.[63]

Another line of evidence that suggests a role of CD36 in hepatic TG accumulation is that CD36 is a target gene of several transcription factors that regulate lipogenesis. For example, activation of CD36 by PPARγ in the liver has been reported, along with increased hepatic expression of PPARγ in diet-induced hepatic steatosis.[64] In this regard, the murine CD36 gene is known to be a direct target of PPARγ, as well as liver X receptor (LXR), and pregnane X receptor, all of which are nuclear receptors that promote hepatic steatosis.[64] Furthermore, Zhou et al.[64] showed that LXR ligand-induced hepatic steatosis is largely abolished in CD36-deficient mice, suggesting that CD36 plays a role in LXR-mediated steatosis in the liver.

In addition to being a target of nuclear receptors, both human and mouse CD36 genes are direct targets of the aryl hydrocarbon receptor (AhR), a ligand-activated transcription factor that belongs to the basic helix-loop-helix/period-AhR nuclear translocator-single-minded family of proteins.[65] Mice constitutively expressing an AhR-transgene developed hepatic steatosis that was associated with increased hepatic CD36 gene expression, and FA uptake in primary hepatocytes isolated from these transgenic mice was increased, compared to WT mice.[65] In addition, AhR-induced hepatic steatosis is largely abrogated in CD36-deficient mice, indicating a critical role of CD36 in AhR-mediated steatosis in the liver.[65]

Taken together, CD36 upregulation in the liver, and the consequent increase in hepatic FA uptake is a plausible mechanism for the development of NAFLD.

CD36 Deficiency Protects Against Alcoholic Steatosis in Mice

NAFLD and alcoholic fatty liver disease (AFLD), though different in etiology, share many common pathogenic characteristics: both diseases start as simple steatosis, characterized by TG accumulation and followed by inflammation of the liver, known as steatohepatitis.[66] As observed in NAFLD, increased hepatic CD36 mRNA and protein expression levels are associated with hepatic steatosis in mice upon chronic alcohol consumption.[8,9] Clugston et al.,[10] therefore, tested the hypothesis that CD36 deficiency protects against alcoholic steatosis, by reducing hepatic FA uptake in alcohol-fed mice. In this study, CD36-deficient mice and WT mice were fed either alcohol-containing (5.1%, v/v) or isocaloric control Lieber–DeCarli liquid diets, for a 6-week period. Interestingly, CD36-deficient mice had a lower basal TG

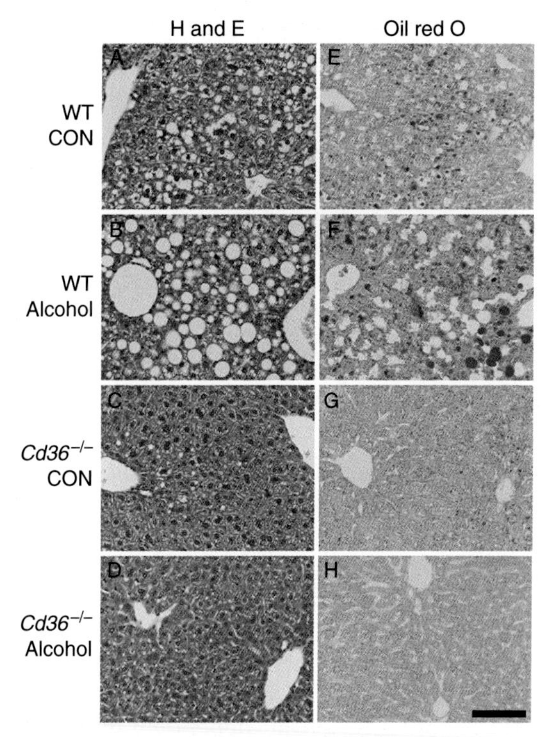

FIGURE 6.2 **Alcohol-fed WT mice develop pronounced steatosis, but CD36-deficient mice do not.** Age-matched male WT and CD36-deficient (*Cd36*−/−) mice were fed a Lieber–DeCarli liquid diet for 6 weeks, either a 5.1% (v/v) alcohol-containing (Alcohol) or an isocaloric control (CON) diet. Liver sections were stained with Hemotoxylin and Eosin (H and E, left column) for morphology, or Oil red O (right column) for neutral lipids. Scale bar = 100 μm. This research was originally published in *The Journal of Lipid Research. Adapted from Ref. [10].*

content, compared to WT mice. However, alcohol did not exacerbate the accumulation of TG, as observed in the WT mice.[10] Representative liver sections stained with hemotoxylin and eosin, and with Oil red O (for neutral lipids), are shown in Figure 6.2. This finding was striking, as CD36-deficient mice were reported to have increased hepatic TG, when fed solid diets such as rodent chow, high-fat, or high-fructose diets.[67] To determine if this protective effect was due to impaired FA uptake, each mouse was intravenously administered radiolabeled FA for 5 min, and the radioactivity present in the liver was then measured. Using this approach, hepatic FA uptake was not altered by CD36 deficiency or alcohol consumption.[10] However, using a ^3H_2O labeling technique, the authors did find that the lower TG in CD36-deficient mice was due to a significant decrease in *de novo* lipogenesis, that corresponded with lower expression levels of lipogenic genes, including PPARγ and its downstream target genes, such as fatty acid synthase, acetyl-CoA carboxylase 1, and stearoyl-CoA desaturase 1.[10] One possible explanation, as proposed by the authors, is that CD36-deficient mice have reduced glucose substrate for *de novo* lipogenesis because of an increase in hepatic glucose output, driven by the energy demand in tissues such as the heart, and skeletal muscles, where FA supplies were insufficient due to impaired CD36-mediated FA uptake.[10] An alternative explanation is that CD36 exerts other actions unrelated to FA uptake in alcohol-exposed livers. Accordingly, the following sections will review CD36's role as a signaling molecule for cellular Ca^{2+} homeostasis, and subsequent prostaglandin production, and as a pattern recognition receptor. These functions may alter lipid and glucose metabolism, and cause inflammation, all of which have been observed in alcohol-exposed livers.

CD36 AND Ca^{2+} SIGNALING

CD36 can function as a signaling molecule, and mediates various cellular and metabolic pathways. For example, CD36-dependent Ca^{2+} signaling is involved in the uptake of ox-LDL particles by macrophages, and this could have implications for understanding the pathogenesis of atherosclerosis.[68] CD36-deficient mice exhibit contraction anomalies in the myocardium that are associated with disturbed Ca^{2+} homeostasis, suggesting a critical role for CD36 in Ca^{2+}-dependent muscle contraction.[69] In taste bud cells, binding of long-chain FAs to CD36 alters cellular Ca^{2+} homeostasis, with an eventual release of the neurotransmitter serotonin, leading to a preference of "fatty" food via the gustatory pathway.[5,70,71] Unlike WT mice, CD36-deficient mice lose their preference for linoleic acid in a two-bottle preference test, indicating a role for CD36 in lipid sensing.[70] The following section will highlight the mechanism underlying CD36-mediated Ca^{2+} influx, via store-operated calcium channels (SOCs) upon ER Ca^{2+} depletion, a process also observed in CD36-mediated lipid sensing.[5]

CD36-Mediated Eicosanoid Production and Inflammation

Under physiological conditions, sarco/endoplasmic reticulum Ca^{2+}-ATPase (SERCA), functioning as a Ca^{2+} pump, transports cytosolic Ca^{2+} to the sarcoplasmic reticulum or ER, for utilization and storage.[72] In response to stimuli, such as muscle contraction, Ca^{2+} is released from the ER that in turn, stimulates mitochondrial ATP production, leading to increased SERCA activity.[72] Simultaneously, depletion of ER Ca^{2+} stores activates ER stromal interaction molecules, resulting in the influx of extracellular Ca^{2+} via SOCs on the plasma membrane, to maintain ER Ca^{2+} homeostasis.[73] Using thapsigargin, a SERCA inhibitor, Kuda et al.[74] showed that ER Ca^{2+} depletion, and the consequent increase of cytosolic Ca^{2+} concentration, promote the translocation of cytosolic CD36 to the plasma membrane in macrophages, and CD36-expressing Chinese hamster ovary cells (Figure 6.3). The persistent presence of CD36 in the plasma membrane

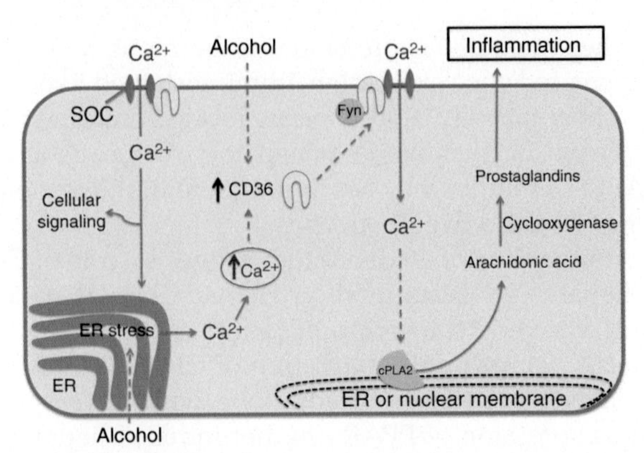

FIGURE 6.3 **A proposed model of CD36 and Ca^{+2}-signaling interaction in alcohol-exposed cells.** Under ER stress (e.g., induced by alcohol or other stimuli described in the text), Ca^{2+} from ER stores is released into the cytoplasm that, in turn, stimulates the translocation of CD36 from the cytosol to the plasma membrane. Plasma membrane CD36 then mediates the influx of extracellular Ca^{2+} via store-operated calcium channels (SOC). Increased cytoplasmic Ca^{2+} activates cytosolic Ca^{2+}-dependent phospholipase A2 (cPLA2) that mobilizes polyunsaturated fatty acids from membrane phospholipids to generate arachidonic acid and subsequent production of inflammatory prostaglandins, resulting in inflammation. This figure also shows that alcohol increases CD36 gene expression and ER stress. A link between CD36-mediated Ca^{2+} signaling and inflammation is proposed as a possible underlying mechanism for alcoholic liver disease.

of these cells causes Ca^{2+} influx via SOCs, activating Ca^{2+}-dependent cytosolic phospholipase A2 (cPLA2). Activated cPLA2 mobilizes polyunsaturated FAs from membrane phospholipids to generate arachidonic acids, and subsequent production of proinflammatory prostaglandins (PGs) via cyclooxygenases, resulting in inflammation.[74] Plasma membrane CD36 can also interact with Fyn, a Src-family tyrosine kinase, and activate extracellular-signal-regulated kinase (ERK) 1 and ERK2, members of the mitogen-activated protein kinase family that, in turn, phosphorylate cPLA2, and lead to further activation of cPLA2.[74] Unlike WT macrophages, CD36-deficient macrophages fail to mediate extracellular Ca^{2+} influx, and release less arachidonic acid into the medium, after thapsigargin treatment.[74] CD36-mediated Ca^{2+} influx in these cells is also diminished after UTP treatment[74] that causes ER Ca^{2+} depletion, via a different mechanism involving membrane purinergic receptors, and the inositol triphosphate pathway. It has also been implicated in the activation of cPLA2 and PGE2 in other cell types.[75,76] Thus, Kuda et al.[74] have identified a mechanism by which CD36 modulates cellular Ca^{2+} homeostasis in the event of ER Ca^{2+} depletion, potentially explaining the pleiotropic proinflammatory effects of CD36.

A Link between CD36-Mediated Ca^{2+} Signaling and Alcoholic Liver Disease?

Chronic alcohol consumption elevates expression of CD36 mRNA and protein.[8–10] CD36's action on cellular

Ca^{2+} homeostasis correlates well with many of alcohol's known effects on the liver (Figure 6.3). For example, alcohol induces ER stress that has been associated with impaired Ca^{2+} homeostasis.[77] It is known that both chronic and binge alcohol consumption activate cPLA2-mediated neuroinflammatory pathways, releasing arachidonic acid, and elevating oxidative stress adducts that contribute to brain damage.[78] Chronic alcohol consumption is also known to activate ERK1/2 signaling pathways.[79] Increased cellular Ca^{2+} can also increase Ca^{2+} levels in mitochondria, and lead to the production of ROS, and mitochondrial dysfunction.[80] Thus, it is tempting to speculate that elevated expression of hepatic CD36 contributes to impaired Ca^{2+} homeostasis in alcohol-exposed tissues.

ER stress and impaired Ca^{2+} homeostasis are associated with steatosis in NAFLD[81] and AFLD,[82] and many mechanistic pathways have been proposed. One possible mechanism for CD36-mediated hepatic steatosis in alcohol-exposed livers is the generation of proinflammatory prostaglandins, known to suppress VLDL secretion.[83,84] Surprisingly, Nassir et al.[85] showed that CD36 deficiency reduces VLDL secretion, and exacerbates hepatic steatosis in leptin-deficient mice. It was found that CD36 deficiency upregulates cyclooxygenase-1 expression in Kupffer cells (resident liver macrophages), resulting in increases in PG levels, and reduced VLDL secretion in these mice.[85] Despite this paradoxical result, it remains to be determined whether impaired VLDL secretion is a plausible mechanism for CD36's action in alcohol-exposed liver tissue.

CD36 AS A PATTERN RECOGNITION RECEPTOR

The innate immune system is the body's primary defense against pathogens, or injured host cells and tissues. CD36 and toll-like receptors (TLRs) are capable of recognizing PAMPs and damage-associated molecular patterns (DAMPs) found on these agents, and have been implicated in the innate immune response.[3,86] Upon ligand binding, CD36 and TLRs initiate a series of inflammatory responses, leading to resolution of infection, or wound healing. Sustained or chronic inflammation could lead to pathological conditions and diseases, including steatosis and steatohepatitis.[87]

TLR4 Plays a Key Role in Alcoholic Liver Disease

TLR4, one of the 13 TLRs identified to date, is a cell surface pattern recognition receptor found on a variety of cell types, including macrophages. It is well characterized for its ability to recognize lipopolysaccharide (LPS), an important membrane component of Gram-negative bacteria, and plays a key role in the development of both NAFLD and AFLD.[86,87] Alternatively, TLR2 recognizes membrane components of Gram-positive bacteria, such as lipoteichoic acid, and peptidoglycan.[86] Detailed reviews on TLRs and liver diseases have been previously published.[86–89] In general, activated TLRs, either in homodimeric or heterodimeric form, mediate signals through myeloid differentiation factor 88 (MyD88)-dependent and/or MyD88-independent pathways.[86] CD14 and myeloid differentiation protein 2 are both known to function as coreceptors that facilitate LPS-induced TLR4 signaling.[90,91] The activated TLR4 complex is subsequently bound to a MyD88 adaptor protein, and activates the transcription factor NF-κB that, in turn, activates the transcription and production of inflammatory cytokines, such as tumor necrosis factor (TNF)-α.[86] Alternatively, the TLR4 complex binds to Toll/IL-1 receptor domain-containing adaptor inducing interferon (IFN)-β (TRIF) instead of MyD88, ultimately leading to the activation of NF-κB, and consequent inflammation.[86] The TRIF pathway, however, is also able to activate the interferon regulator transcription factor 3 that leads to the production of IFN-β, an anti-inflammatory cytokine.[86]

LPS levels in the circulation are elevated by acute or chronic alcohol consumption in human and animal models.[92–94] This elevation is caused by the disruption of the intestinal epithelial barrier by toxic alcohol metabolites, leading to increased gut permeability, and thus allowing LPS and/or other microbial materials to enter the liver via the portal vein,[95–97] as shown schematically in Figure 6.4. Gut-derived LPS has been shown to be a crucial mediator of alcohol-induced liver injury, that is attenuated with the elimination of Gram-negative bacteria, in antibiotic-treated rats.[98]

Kupffer cells express TLR4, and are the first line of defense against gut-derived pathogens and PAMPs in the liver.[86,87] Depletion of Kupffer cells by gadolinium chloride markedly reduces alcohol-induced liver injury in rats, indicating a key role of these cells in the development of alcoholic liver disease.[99] Kupffer cells isolated from alcohol-fed rats also show more sensitivity to LPS-induced TNF-α production, than those from control-fed rats.[100] TLR4-deficient, but not TLR2- or MyD88-deficient, mice are resistant to alcoholic liver disease.[101] Thus, LPS-induced activation of Kupffer cells, via the Myd88-independent TLR4-mediated signaling pathway, plays a major role in the development of alcoholic liver disease.

A Role for CD36-Mediated Pattern Recognition and Signaling in Alcoholic Liver Disease?

CD36 functions as an essential coreceptor for the activation of the TLR2/TLR6 dimer in the phagocytosis

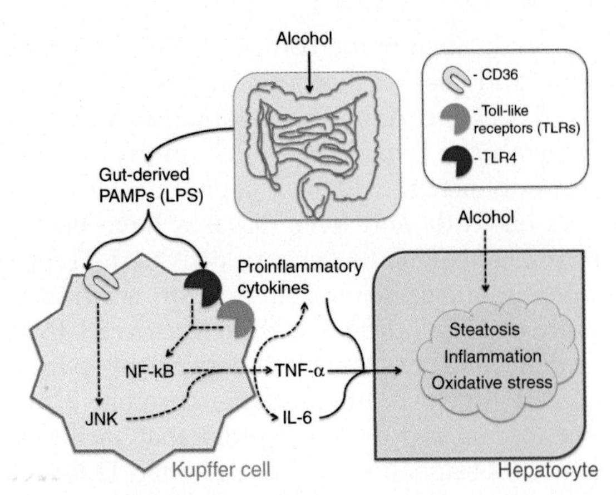

FIGURE 6.4 A proposed model showing alcohols effect on the liver via PAMP-induced inflammation. Alcohol increases gut permeability, allowing gut-derived pathogen-associated molecular patterns (PAMPs), such as lipopolysaccharide (LPS), to enter the circulation. Kupffer cells, the resident macrophage of the liver, recognizes PAMPs via TLRs, in particular TLR4. The activated TLR4 complex, as described in the text, upregulates transcription factor NF-κB that, in turn, upregulates the production of proinflammatory cytokines (e.g., TNF-α and IL-6), leading to inflammation. CD36 is also proposed to function as a pattern recognition receptor for gut-derived PAMPs that mediates LPS-induced inflammation, possibly via the JNK pathway, in alcohol-exposed livers.

of *S. aureus*.[24,102–104] CD36-deficient mice are unable to clear initial bacteremia efficiently, when infected intravenously with *S. aureus*, and have a high fatality rate, compared to WT mice, all of which survive.[24] Using human embryonic kidney (HEK) 293T cells expressing TLRs, and transfected with murine CD36, Stuart et al.[24] showed that CD36 also mediates uptake of *Escherichia coli*, a Gram-negative bacterium, but to a lesser extent compared to *S. aureus*. However, using HeLa cells transfected with human CD36, Baranova et al.[105] found that CD36 shows no preference in the uptake of various bacteria, including *S. aureus* and *E. coli*. In addition, LPS markedly induces the production of the chemokine interleukin 8 in CD36-expressing HEK293 cells that do not express TLR2 and TLR4, demonstrating that CD36 alone is able to elicit LPS-induced signaling, and the subsequent production of inflammatory cytokines and chemokines.[105] Inhibitor experiments in these cells further demonstrated that CD36 mediates LPS responsiveness via the JNK1/JNK2 pathway, and that this process is unaffected by NF-κB inhibitors.[105] Finally, CD36 is also implicated in adipose tissue inflammation, and diet-induced obesity, as demonstrated by Cai et al.[106] This study showed an attenuated response to LPS-induced cytokine production in primary adipocytes, and peritoneal macrophages, isolated from CD36-deficient mice fed a high-fat diet, compared to those from WT mice, on an identical diet.

CD36 can also recognize DAMPs as "altered self" components, and can trigger a sustained sterile inflammatory response.[3] Stewart et al.[107] showed that, upon binding to ox-LDL and β-amyloid, CD36 functions as a coreceptor for the formation of a TLR4/TLR6 heterodimer that triggers inflammatory responses, and ultimately results in symptoms found in atherosclerosis and Alzheimer's disease, respectively. Seimon et al.[108] identified a CD36-TLR2-dependent pathway for macrophage apoptosis, triggered by ER stress that can be induced by ox-LDL, oxidized phospholipids, and saturated FAs.

Taken together, the literature shows that CD36 can function alone as a pattern recognition receptor, or as a coreceptor for TLRs, for both PAMPs and DAMPs in various cell types. It is tempting to speculate, as shown in Figure 6.4, that elevated levels of CD36 in Kupffer cells from alcohol-exposed livers are capable of recognizing LPS, other PAMPs, or even DAMPs derived from damaged liver cells, and can trigger, possibly via the JNK pathway, the production of inflammatory cytokines and chemokines that lead to alcoholic liver injuries. Alternatively, CD36 could function as a coreceptor for the TLR4 homodimer complex in mediating alcohol-induced liver injuries. Overall, the literature supports a strong possibility for a CD36-mediated inflammatory response in alcohol-exposed livers. This possibility can be tested experimentally, by using mice lacking CD36 specifically in Kupffer cells, or in cells of myeloid origin.

CD36 IN THE BRAIN

Clugston et al.[10] reported that, during the course of their studies, CD36-deficient mice were completely protected from alcohol-induced mortality, whereas there was an alcohol-associated dose-dependent mortality rate in WT mice, as shown in Figure 6.5. It was unclear how CD36 deficiency exerted its protective effect from alcohol-induced death in these mice. Since CD36 is expressed in a wide variety of cell types, and exerts ligand-dependent pleiotropic effects, mouse models with cell-specific CD36 deficiency would be needed to address this question. This section will review the known actions of CD36 in the brain that may provide potential roles for CD36 in alcohol-exposed brains.

Chronic alcohol consumption can cause significant tissue damage through inflammation, oxidative stress, and hypoxia.[109–112] Several pathways for alcohol-induced neuroinflammation have been proposed, including disruption of the liver's detoxifying mechanisms, that would release LPS into the circulation, compromising brain immune privilege, and initiating inflammation;[113] upregulation of TLR4-mediated signaling;[110] upregulation of microglia (brain resident macrophages);[114] and oxidative stress in the vessels of the blood-brain barrier.[115,116]

CD36 is found on microglia and on microvascular endothelial cells in the brain.[117,118] CD36 on endothelial

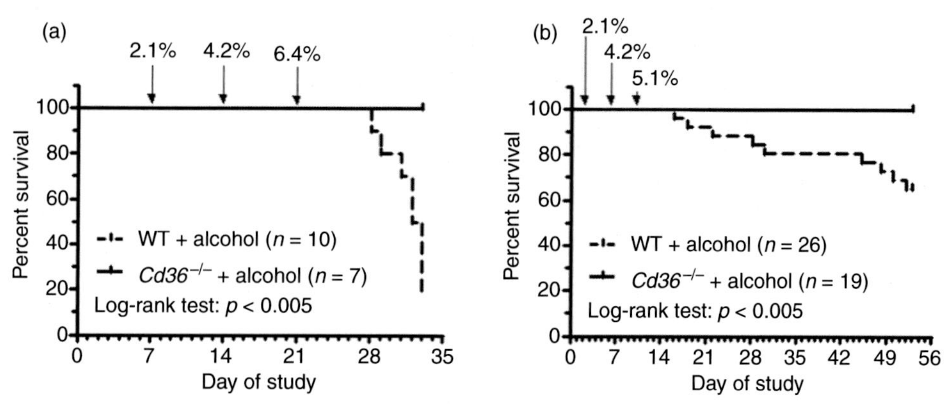

FIGURE 6.5 CD36-deficient mice are protected from mortality associated with feeding high levels of alcohol to WT mice. Kaplan–Meier survival curves are shown for two alcohol-feeding experiments. (a) A high mortality rate was observed in WT mice fed 6.4% alcohol, though no deaths were observed in the group of CD36-deficient ($Cd36^{-/-}$) mice. (b) Modification of the alcohol feeding protocol to contain less alcohol (5.1% alcohol) was associated with a lower mortality rate in the WT group fed alcohol. No deaths were observed in the group of CD36-deficient mice fed alcohol. No deaths were observed in the groups of WT and CD36-deficient mice fed the control liquid diet (data not shown). This research was originally published (as Supplemental Data) in *Journal of Lipid Research. Adapted from Ref. [10].*

cells may function as an FA transporter, and facilitate FA uptake into the brain.[117] On the other hand, the role of CD36 in microglial cells likely parallels those described for macrophages, and the proposed CD36-mediated LPS response in alcohol-exposed Kupffer cells, in the earlier section, may also apply to alcohol-exposed microglial cells, resulting in liver or brain damage, respectively.

CD36 is also known to mediate ROS production in response to β-amyloid binding, leading to the brain damage seen in Alzheimer's disease.[119] Furthermore, microglial CD36 protein expression is increased post-cerebral ischemia, associated with inflammation and subsequent tissue damage. ROS production, ischemic brain injury, and motor deficits are all reduced in CD36-deficient mice, indicating a role of CD36-mediated ROS production and tissue injury, in cerebral ischemia.[20] Finally, CD36 is induced by hypoxia inducible factor-1α and has been associated with the inflammatory process during hypoxia,[120] a condition associated with ischemia. Taken together, the literature supports a potential role of CD36-mediated inflammatory processes in alcohol-induced damage in the brain.

CONCLUSIONS

Recent studies[8–10,57] have implicated CD36 in the pathogenesis of alcohol-related diseases; however, its exact mechanism of action has not been elucidated. CD36 is a class B scavenger receptor with the capacity to bind diverse ligands, thereby implicating CD36 in a variety of physiological functions and processes. This chapter focused on the role of CD36 in FA uptake, Ca^{2+} signaling, and PAMP recognition, all of which have been proposed to contribute to the pathogenesis of alcohol-related diseases, particularly alcoholic liver disease. The potential role of CD36 in alcohol-induced neuroinflammation was also discussed. The questions below are posed to highlight the likely processes through which CD36 may contribute to the pathogenesis of alcohol-related disease, and to stimulate research in these areas.

First, what is the mechanism underlying the protective effect of CD36 deficiency against alcohol-induced steatosis? Clugston et al.[10] established that CD36-deficient mice are protected against alcohol-induced hepatic steatosis; however, it was also shown that this protective effect was not due to a decrease in hepatic FA uptake. To what, then, may this protective effect be attributed? Alternative possibilities discussed above include CD36's role in Ca^{2+} signaling, and recognition of gut-derived PAMPs. Given that CD36 is expressed in multiple cell types within the liver (including hepatocytes, Kupffer cell, stellate cells, and endothelial cells), alcohol feeding studies in cell type-specific CD36 knockouts would be advantageous for addressing this question. This approach would allow for the effects of CD36-deficiency in a single cell type to be elucidated, while eliminating the metabolic complications observed in whole-body CD36 knockouts. The availability of transgenic mice expressing Cre-recombinase driven by cell-specific promoters makes these experiments entirely feasible, and should be a priority for future studies.

Second, if CD36 deficiency protects against alcohol-induced hepatic steatosis, via its role in Ca^{2+} signaling, what is the mechanism by which this occurs? Because Ca^{2+} signaling has been implicated in a variety of processes, including obesity, inflammation, lipid perception, atherosclerosis development, and cardiovascular health, it would be necessary to tease out the specific relationship between CD36, alcohol, and Ca^{2+}. The use of inhibitors specific to potential factors underlying the CD36-dependent Ca^{2+} signaling pathway, cell-specific CD36 knockout mice, and alcohol-feeding protocols, would aid in answering this question.

Third, if CD36 deficiency protects against alcohol-induced hepatic steatosis, via its anti-inflammatory effects, what is the mechanism by which this occurs? Is it because CD36 plays a role in promoting inflammation by acting as a pattern recognition receptor to alcohol-induced, gut derived PAMPs and DAMPs? There are several steps between CD36 recognition of PAMPs and the production of proinflammatory molecules. Therefore, if the protective effect is due to this pathway, is it dependent on CD36 or another molecule in the pathway, such as TLR4? Primary Kupffer cells deficient of CD36 could be used to address some of these questions *in vitro*. For example, a reduction in alcohol-induced and LPS-mediated production of cytokines (e.g., TNF-α) in CD36-deficient Kupffer cells would confirm a role of CD36 in this process. Again, inhibitors and/or other biochemical approaches could be used to further identify the mechanisms underlying the proposed inflammatory processes mediated by CD36 in alcohol-exposed livers.

Fourth, the role CD36 plays in innate immunity is primarily associated with its presence on macrophages. For example, CD36 promotes inflammation in the liver, likely via its expression on the surface of Kupffer cells. Likewise, other tissues express CD36 in their resident macrophages; microglia in the brain express CD36 and, as discussed earlier in this chapter, may contribute to alcohol-induced brain damage and fatality. However, CD36 is also expressed in the microvascular epithelium of the brain. Therefore, what combination of alcohol and CD36 causes the alcohol-induced mortality, observed in WT mice but not in CD36-deficient mice? Is it due to CD36's role in microglial cells of the brain, or its role in the microvascular epithelium, or is death due to damage to another organ system altogether? Macrophage-specific knockouts of CD36 may aid in teasing out the role of macrophage CD36 in alcohol-related disease. However, additional tools will be required to distinguish the role of Kupffer CD36 versus microglia CD36, in the pathogenesis of alcohol-related disease.

In summary, there are several points of potential interaction between the effects of alcohol in multiple tissues and CD36. As pointed out by other reviewers, CD36 appears to be a Jack-of-all-trades[3] that wears many hats, some good and some bad.[1] Data from CD36-deficienct mice clearly show that these animals are protected from alcohol-induced hepatic steatosis and lethality, associated with consuming high concentrations of alcohol; however, the precise protective mechanism remains to be elucidated. Further research is required to define fully the significant role that CD36 plays in the pathogenesis of alcohol-related disease.

Key Facts of CD36

- CD36 is a class B scavenger receptor expressed in a variety of cell types that binds to multiple ligands.
- Ligands for CD36 include long-chain fatty acids, native and oxidized low-density lipoproteins, phospholipids, TSPs, the malaria parasite, bacteria, and β-amyloid.
- CD36 has pleiotropic functions, such as facilitating fatty acid uptake in adipose tissue, heart, and skeletal muscle; lipid sensing in taste buds; dietary fat absorption; and angiogenesis.
- CD36 can function as a signaling molecule, and mediates various cellular and metabolic pathways, including cellular calcium homeostasis and inflammation.
- CD36 mutation and altered CD36 expression have been implicated in susceptibility to atherosclerosis, severity of malaria infection, and susceptibility to Alzheimer's disease.
- Expression of CD36 is upregulated, in both alcoholic and nonalcoholic fatty livers.

Summary Points

- The chapter focuses on CD36, a class B scavenger receptor, and its potential role in alcohol-related disease.
- CD36 deficiency protects against alcoholic steatosis in mice consuming a low-fat/high carbohydrate Lieber–DeCarli diet, containing alcohol (5.1%, v/v). However, hepatic fatty acid uptake is not altered by CD36 deficiency, or chronic alcohol consumption.
- CD36-deficient mice have lower rates of *de novo* lipogenesis that corresponds with lower expression levels of genes involved in lipogenesis.
- Increased CD36 expression and fatty acid uptake in cardiomyocytes are positively associated with alcoholic cardiomyopathy in mice chronically consuming alcohol in their drinking water, suggesting a role of CD36-mediated fatty acid uptake in the development of alcoholic cardiomyopathy.
- In macrophages, CD36-mediated calcium signaling leads to an increase in the production of proinflammatory prostgalandins, and subsequent alterations in lipid metabolism, providing a possible role of CD36 in alcoholic steatosis.
- CD36 and TLRs are capable of recognizing pathogen-associated molecular patterns and damage-associated molecular patterns found on pathogens or injured host cells and tissues, and have been implicated in the innate immune response.
- Upon ligand binding, CD36 and TLRs initiate a series of inflammatory responses, leading to resolution of infection, or wound healing. Sustained or chronic inflammation could lead to pathological conditions

and diseases, including alcoholic steatosis and steatohepatitis, suggesting a possible role of CD36, as a pattern recognition receptor, in the development of alcoholic liver disease.

- CD36 is found on both microglia (brain macrophage) and microvascular endothelial cells in the brain. CD36 on endothelial cells may function as a fatty acid transporter, and facilitate fatty acid uptake into the brain. On the other hand, CD36 may function as a pattern recognition receptor in alcohol-exposed microglial cells. The literature supports a potential role of CD36 in alcohol-induced neuroinflammation, and oxidative stress.

Acknowledgment

This work was supported, in part, by the following NIH grants: K99AA022652 (RDC), R01DK068437 and R21AA021336 (WSB), and R21AA020561 (LSH).

Abbreviations

AFLD Alcoholic fatty liver disease
AhR Aryl hydrocarbon receptor
CD36 Cluster of differentiation 36
cPLA2 Cytosolic phospholipase A2
DAMPs Damage-associated molecular patterns
ER Endoplasmic reticulum
ERK Extracellular-signal-regulated kinase
FA Fatty acid
HEK Human embryonic kidney
IFN Interferon
LDL Low density lipoproteins
LPS Lipopolysaccharide
LXR Liver X receptor
MyD88 Myeloid differentiation factor 88
NAFLD Nonalcoholic fatty liver disease
Ox-LDL Oxidized-LDL
PAMPs Pathogen-associated molecular patterns
PfEMP1 *Plasmodium falciparum* erythrocyte membrane binding protein 1
PG Prostaglandin
PPAR Peroxisome proliferator-activated receptor
ROS Reactive oxygen species
SERCA Sarco/endoplasmic reticulum Ca^{2+}-ATPase
SOC Store-operated Ca^{2+} channel
TG Triglyceride
TLR Toll-like receptor
TNF-α Tumor necrosis factor α
TRIF Toll/IL-1 receptor domain-containing adaptor inducing interferon (IFN)-β
TSP Thrombospondin
WT Wild-type

References

1. Febbraio M, Silverstein RL. CD36: implications in cardiovascular disease. *Int J Biochem Cell Biol* 2007;**39**(11):2012–30.
2. Collot-Teixeira S, Martin J, McDermott-Roe C, Poston R, McGregor JL. CD36 and macrophages in atherosclerosis. *Cardiovascular Res* 2007;**75**(3):468–77.
3. Canton J, Neculai D, Grinstein S. Scavenger receptors in homeostasis and immunity. *Nat Rev Immunol* 2013;**13**(9):621–34.
4. Su X, Abumrad NA. Cellular fatty acid uptake: a pathway under construction. *Trends Endocrinol Metab* 2009;**20**(2):72–7.
5. Pepino MY, Kuda O, Samovski D, Abumrad NA. Structure-function of CD36 and importance of fatty acid signal transduction in fat metabolism. *Annu Rev Nutr* 2014;**34**:281–303.
6. Armstrong LC, Bornstein P. Thrombospondins 1 and 2 function as inhibitors of angiogenesis. *Matrix Biol* 2003;**22**(1):63–71.
7. Aitman TJ, Cooper LD, Norsworthy PJ, et al. Malaria susceptibility and CD36 mutation. *Nature* 2000;**405**(6790):1015–6.
8. Clugston RD, Jiang H, Lee MX, et al. Altered hepatic lipid metabolism in C57BL/6 mice fed alcohol: a targeted lipidomic and gene expression study. *J Lipid Res* 2011;**52**(11):2021–31.
9. Zhong W, Zhao Y, Tang Y, et al. Chronic alcohol exposure stimulates adipose tissue lipolysis in mice: role of reverse triglyceride transport in the pathogenesis of alcoholic steatosis. *Am J Pathol* 2012;**180**(3):998–1007.
10. Clugston RD, Yuen JJ, Hu Y, et al. CD36-deficient mice are resistant to alcohol- and high-carbohydrate-induced hepatic steatosis. *J Lipid Res* 2014;**55**(2):239–46.
11. Clemetson KJ, Pfueller SL, Luscher EF, Jenkins CS. Isolation of the membrane glycoproteins of human blood platelets by lectin affinity chromatography. *Biochim Biophys Acta* 1977;**464**(3):493–508.
12. Harmon CM, Abumrad NA. Binding of sulfosuccinimidyl fatty acids to adipocyte membrane proteins: isolation and amino-terminal sequence of an 88-kD protein implicated in transport of long-chain fatty acids. *J Membr Biol* 1993;**133**(1):43–9.
13. Coort SL, Willems J, Coumans WA, et al. Sulfo-N-succinimidyl esters of long chain fatty acids specifically inhibit fatty acid translocase (FAT/CD36)-mediated cellular fatty acid uptake. *Mol Cell Biochem* 2002;**239**(1–2):213–9.
14. Silverstein RL, Febbraio M. CD36, a scavenger receptor involved in immunity, metabolism, angiogenesis, and behavior. *Sci Signal* 2009;**2**(72):re3.
15. Baruch DI, Ma XC, Pasloske B, Howard RJ, Miller LH. CD36 peptides that block cytoadherence define the CD36 binding region for Plasmodium falciparum-infected erythrocytes. *Blood* 1999;**94**(6):2121–7.
16. Pasternak ND, Dzikowski R. PfEMP1: an antigen that plays a key role in the pathogenicity and immune evasion of the malaria parasite Plasmodium falciparum. *Int J Biochem Cell Biol* 2009;**41**(7):1463–6.
17. Puente Navazo MD, Daviet L, Ninio E, McGregor JL. Identification on human CD36 of a domain (155-183) implicated in binding oxidized low-density lipoproteins (Ox-LDL). *Arterioscler Thromb Vasc Biol* 1996;**16**(8):1033–9.
18. Podrez EA, Poliakov E, Shen Z, et al. A novel family of atherogenic oxidized phospholipids promotes macrophage foam cell formation via the scavenger receptor CD36 and is enriched in atherosclerotic lesions. *J Biol Chem* 2002;**277**(41):38517–23.
19. Martin-Fuentes P, Civeira F, Recalde D, et al. Individual variation of scavenger receptor expression in human macrophages with oxidized low-density lipoprotein is associated with a differential inflammatory response. *J Immunol* 2007;**179**(5):3242–8.
20. Cho S, Park EM, Febbraio M, et al. The class B scavenger receptor CD36 mediates free radical production and tissue injury in cerebral ischemia. *J Neurosci* 2005;**25**(10):2504–12.
21. Park YM, Febbraio M, Silverstein RL. CD36 modulates migration of mouse and human macrophages in response to oxidized LDL and may contribute to macrophage trapping in the arterial intima. *J Clin Invest* 2009;**119**(1):136–45.
22. Demers A, McNicoll N, Febbraio M, et al. Identification of the growth hormone-releasing peptide binding site in CD36: a photoaffinity cross-linking study. *Biochem J* 2004;**382**(Pt. 2):417–24.

23. Marleau S, Harb D, Bujold K, et al. EP 80317, a ligand of the CD36 scavenger receptor, protects apolipoprotein E-deficient mice from developing atherosclerotic lesions. *FASEB J* 2005;**19**(13):1869–71.
24. Stuart LM, Deng J, Silver JM, et al. Response to Staphylococcus aureus requires CD36-mediated phagocytosis triggered by the COOH-terminal cytoplasmic domain. *J Cell Biol* 2005;**170**(3):477–85.
25. Sun B, Boyanovsky BB, Connelly MA, Shridas P, van der Westhuyzen DR, Webb NR. Distinct mechanisms for OxLDL uptake and cellular trafficking by class B scavenger receptors CD36 and SR-BI. *J Lipid Res* 2007;**48**(12):2560–70.
26. Hoosdally SJ, Andress EJ, Wooding C, Martin CA, Linton KJ. The Human Scavenger Receptor CD36: glycosylation status and its role in trafficking and function. *J Biol Chem* 2009;**284**(24):16277–88.
27. Chu LY, Silverstein RL. CD36 ectodomain phosphorylation blocks thrombospondin-1 binding: structure-function relationships and regulation by protein kinase C. *Arterioscler Thromb Vasc Biol* 2012;**32**(3):760–7.
28. Lynes M, Narisawa S, Millan JL, Widmaier EP. Interactions between CD36 and global intestinal alkaline phosphatase in mouse small intestine and effects of high-fat diet. *Am J Physiol Regul Integr Comp Physiol* 2011;**301**(6):R1738–47.
29. Tao N, Wagner SJ, Lublin DM. CD36 is palmitoylated on both N- and C-terminal cytoplasmic tails. *J Biol Chem* 1996;**271**(37):22315–20.
30. Thorne RF, Ralston KJ, de Bock CE, et al. Palmitoylation of CD36/FAT regulates the rate of its post-transcriptional processing in the endoplasmic reticulum. *Biochim Biophys Acta* 2010;**1803**(11):1298–307.
31. Kuda O, Pietka TA, Demianova Z, et al. Sulfo-N-succinimidyl oleate (SSO) inhibits fatty acid uptake and signaling for intracellular calcium via binding CD36 lysine 164: SSO also inhibits oxidized low density lipoprotein uptake by macrophages. *J Biol Chem* 2013;**288**(22):15547–55.
32. Goldberg IJ, Trent CM, Schulze PC. Lipid metabolism and toxicity in the heart. *Cell Metab* 2012;**15**(6):805–12.
33. Fabbrini E, Sullivan S, Klein S. Obesity and nonalcoholic fatty liver disease: biochemical, metabolic, and clinical implications. *Hepatology* 2010;**51**(2):679–89.
34. Abumrad NA, el-Maghrabi MR, Amri EZ, Lopez E, Grimaldi PA. Cloning of a rat adipocyte membrane protein implicated in binding or transport of long-chain fatty acids that is induced during preadipocyte differentiation. Homology with human CD36. *J Biol Chem* 1993;**268**(24):17665–8.
35. Smith BK, Bonen A, Holloway GP. A dual mechanism of action for skeletal muscle FAT/CD36 during exercise. *Exerc Sport Sci Rev* 2012;**40**(4):211–7.
36. Schwenk RW, Luiken JJ, Bonen A, Glatz JF. Regulation of sarcolemmal glucose and fatty acid transporters in cardiac disease. *Cardiovasc Res* 2008;**79**(2):249–58.
37. Bastie CC, Hajri T, Drover VA, Grimaldi PA, Abumrad NA. CD36 in myocytes channels fatty acids to a lipase-accessible triglyceride pool that is related to cell lipid and insulin responsiveness. *Diabetes* 2004;**53**(9):2209–16.
38. Ehehalt R, Sparla R, Kulaksiz H, Herrmann T, Fullekrug J, Stremmel W. Uptake of long chain fatty acids is regulated by dynamic interaction of FAT/CD36 with cholesterol/sphingolipid enriched microdomains (lipid rafts). *BMC Cell Biol* 2008;**9**:45.
39. Ring A, Le Lay S, Pohl J, Verkade P, Stremmel W. Caveolin-1 is required for fatty acid translocase (FAT/CD36) localization and function at the plasma membrane of mouse embryonic fibroblasts. *Biochim Biophys Acta* 2006;**1761**(4):416–23.
40. Smith J, Su X, El-Maghrabi R, Stahl PD, Abumrad NA. Opposite regulation of CD36 ubiquitination by fatty acids and insulin: effects on fatty acid uptake. *J Biol Chem* 2008;**283**(20):13578–85.
41. Hajri T, Abumrad NA. Fatty acid transport across membranes: relevance to nutrition and metabolic pathology. *Annu Rev Nutr* 2002;**22**:383–415.
42. Ibrahimi A, Bonen A, Blinn WD, et al. Muscle-specific overexpression of FAT/CD36 enhances fatty acid oxidation by contracting muscle, reduces plasma triglycerides and fatty acids, and increases plasma glucose and insulin. *J Biol Chem* 1999;**274**(38):26761–6.
43. Hajri T, Hall AM, Jensen DR, et al. CD36-facilitated fatty acid uptake inhibits leptin production and signaling in adipose tissue. *Diabetes* 2007;**56**(7):1872–80.
44. Nassir F, Wilson B, Han X, Gross RW, Abumrad NA. CD36 is important for fatty acid and cholesterol uptake by the proximal but not distal intestine. *J Biol Chem* 2007;**282**(27):19493–501.
45. Nauli AM, Nassir F, Zheng S, et al. CD36 is important for chylomicron formation and secretion and may mediate cholesterol uptake in the proximal intestine. *Gastroenterol* 2006;**131**(4):1197–207.
46. Goudriaan JR, den Boer MA, Rensen PC, et al. CD36 deficiency in mice impairs lipoprotein lipase-mediated triglyceride clearance. *J Lipid Res* 2005;**46**(10):2175–81.
47. Yanai H, Chiba H, Fujiwara H, et al. Phenotype-genotype correlation in CD36 deficiency types I and II. *Thromb Haemost* 2000;**84**(3):436–41.
48. Furuhashi M, Ura N, Nakata T, Tanaka T, Shimamoto K. Genotype in human CD36 deficiency and diabetes mellitus. *Diab Med* 2004;**21**(8):952–3.
49. Kamiya M, Nakagomi A, Tokita Y, et al. Type I CD36 deficiency associated with metabolic syndrome and vasospastic angina: a case report. *J Cardiol* 2006;**48**(1):41–4.
50. Nagasaka H, Yorifuji T, Takatani T, et al. CD36 deficiency predisposing young children to fasting hypoglycemia. *Metabolism* 2011;**60**(6):881–7.
51. Yasunaga T, Koga S, Ikeda S, et al. Cluster differentiation-36 deficiency type 1 and acute coronary syndrome without major cardiovascular risk factors: case report. *Circ J* 2007;**71**(1):166–9.
52. Love-Gregory L, Abumrad NA. CD36 genetics and the metabolic complications of obesity. *Curr Opin Clin Nutr Metab Care* 2011;**14**(6):527–34.
53. Rac ME, Safranow K, Poncyljusz W. Molecular basis of human CD36 gene mutations. *Mol Med* 2007;**13**(5–6):288–96.
54. Yang J, Sambandam N, Han X, et al. CD36 deficiency rescues lipotoxic cardiomyopathy. *Circ Res* 2007;**100**(8):1208–17.
55. Ge F, Hu C, Hyodo E, et al. Cardiomyocyte triglyceride accumulation and reduced ventricular function in mice with obesity reflect increased long chain Fatty Acid uptake and de novo Fatty Acid synthesis. *J Obes* 2012;**2012**:205648.
56. Piano MR, Phillips SA. Alcoholic cardiomyopathy: pathophysiologic insights. *Cardiovasc Toxicol* 2014;**14**(2):291–308.
57. Hu C, Ge F, Hyodo E, et al. Chronic ethanol consumption increases cardiomyocyte fatty acid uptake and decreases ventricular contractile function in C57BL/6J mice. *J Mol Cell Cardiol* 2013;**59**:30–40.
58. Coburn CT, Knapp Jr FF, Febbraio M, Beets AL, Silverstein RL, Abumrad NA. Defective uptake and utilization of long chain fatty acids in muscle and adipose tissues of CD36 knockout mice. *J Biol Chem* 2000;**275**(42):32523–9.
59. Goudriaan JR, Dahlmans VE, Teusink B, et al. CD36 deficiency increases insulin sensitivity in muscle, but induces insulin resistance in the liver in mice. *J Lipid Res* 2003;**44**(12):2270–7.
60. Hames KC, Vella A, Kemp BJ, Jensen MD. Free fatty acid uptake in humans with CD36 deficiency. *Diabetes* 2014;**63**(11):3606–14.
61. Koonen DP, Jacobs RL, Febbraio M, et al. Increased hepatic CD36 expression contributes to dyslipidemia associated with diet-induced obesity. *Diabetes* 2007;**56**(12):2863–71.
62. Greco D, Kotronen A, Westerbacka J, et al. Gene expression in human NAFLD. *Am J Physiol Gastrointest Liver Physiol* 2008;**294**(5):G1281–7.

63. Miquilena-Colina ME, Lima-Cabello E, Sanchez-Campos S, et al. Hepatic fatty acid translocase CD36 upregulation is associated with insulin resistance, hyperinsulinaemia and increased steatosis in non-alcoholic steatohepatitis and chronic hepatitis C. *Gut* 2011;**60**(10):1394–402.
64. Zhou J, Febbraio M, Wada T, et al. Hepatic fatty acid transporter Cd36 is a common target of LXR, PXR, and PPARgamma in promoting steatosis. *Gastroenterol* 2008;**134**(2):556–67.
65. Lee JH, Wada T, Febbraio M, et al. A novel role for the dioxin receptor in fatty acid metabolism and hepatic steatosis. *Gastroenterol* 2010;**139**(2):653–63.
66. Sozio MS, Liangpunsakul S, Crabb D. The role of lipid metabolism in the pathogenesis of alcoholic and nonalcoholic hepatic steatosis. *Semin Liver Dis* 2010;**30**(4):378–90.
67. Hajri T, Han XX, Bonen A, Abumrad NA. Defective fatty acid uptake modulates insulin responsiveness and metabolic responses to diet in CD36-null mice. *J Clin Invest* 2002;**109**(10):1381–9.
68. Rahaman SO, Zhou G, Silverstein RL. Vav protein guanine nucleotide exchange factor regulates CD36 protein-mediated macrophage foam cell formation via calcium and dynamin-dependent processes. *J Biol Chem* 2011;**286**(41):36011–9.
69. Pietka TA, Sulkin MS, Kuda O, et al. CD36 protein influences myocardial Ca^{2+} homeostasis and phospholipid metabolism: conduction anomalies in CD36-deficient mice during fasting. *J Biol Chem* 2012;**287**(46):38901–12.
70. Laugerette F, Passilly-Degrace P, Patris B, et al. CD36 involvement in orosensory detection of dietary lipids, spontaneous fat preference, and digestive secretions. *J Clin Invest* 2005;**115**(11):3177–84.
71. Gaillard D, Laugerette F, Darcel N, et al. The gustatory pathway is involved in CD36-mediated orosensory perception of long-chain fatty acids in the mouse. *FASEB J* 2008;**22**(5):1458–68.
72. Mekahli D, Bultynck G, Parys JB, De Smedt H, Missiaen L. Endoplasmic-reticulum calcium depletion and disease. *Cold Spring Harb Perspect Biol* 2011;**3**(6).
73. Soboloff J, Rothberg BS, Madesh M, Gill DL. STIM proteins: dynamic calcium signal transducers. *Nat Rev Mol Cell Biol* 2012;**13**(9): 549–65.
74. Kuda O, Jenkins CM, Skinner JR, et al. CD36 protein is involved in store-operated calcium flux, phospholipase A2 activation, and production of prostaglandin E2. *J Biol Chem* 2011;**286**(20): 17785–95.
75. Lin WW, Chen BC. Pharmacological comparison of UTP- and thapsigargin-induced arachidonic acid release in mouse RAW 264.7 macrophages. *Br J Pharmacol* 1998;**123**(6):1173–81.
76. Degagne E, Grbic DM, Dupuis AA, et al. P2Y2 receptor transcription is increased by NF-kappa B and stimulates cyclooxygenase-2 expression and PGE2 released by intestinal epithelial cells. *J Immunol* 2009;**183**(7):4521–9.
77. Ji C. Mechanisms of alcohol-induced endoplasmic reticulum stress and organ injuries. *Biochem Res Int* 2012;**2012**:216450.
78. Collins MA, Tajuddin N, Moon KH, Kim HY, Nixon K, Neafsey EJ. Alcohol, phospholipase A-associated neuroinflammation, and omega3 docosahexaenoic acid protection. *Mol Neurobiol* 2014;**50**(1): 239–45.
79. Thakur V, Pritchard MT, McMullen MR, Wang Q, Nagy LE. Chronic ethanol feeding increases activation of NADPH oxidase by lipopolysaccharide in rat Kupffer cells: role of increased reactive oxygen in LPS-stimulated ERK1/2 activation and TNF-alpha production. *J Leukoc Biol* 2006;**79**(6):1348–56.
80. Gaspers LD, Memin E, Thomas AP. Calcium-dependent physiologic and pathologic stimulus-metabolic response coupling in hepatocytes. *Cell Calcium* 2012;**52**(1):93–102.
81. Fu S, Yang L, Li P, et al. Aberrant lipid metabolism disrupts calcium homeostasis causing liver endoplasmic reticulum stress in obesity. *Nature* 2011;**473**(7348):528–31.
82. Bartlett PJ, Gaspers LD, Pierobon N, Thomas AP. Calcium-dependent regulation of glucose homeostasis in the liver. *Cell Calcium* 2014;**55**(6):306–16.
83. Bjornsson OG, Sparks JD, Sparks CE, Gibbons GF. Prostaglandins suppress VLDL secretion in primary rat hepatocyte cultures: relationships to hepatic calcium metabolism. *J Lipid Res* 1992;**33**(7): 1017–27.
84. Perez S, Aspichueta P, Ochoa B, Chico Y. The 2-series prostaglandins suppress VLDL secretion in an inflammatory condition-dependent manner in primary rat hepatocytes. *Biochim Biophys Acta* 2006;**1761**(2):160–71.
85. Nassir F, Adewole OL, Brunt EM, Abumrad NA. CD36 deletion reduces VLDL secretion, modulates liver prostaglandins, and exacerbates hepatic steatosis in ob/ob mice. *J Lipid Res* 2013;**54**(11): 2988–97.
86. Mencin A, Kluwe J, Schwabe RF. Toll-like receptors as targets in chronic liver diseases. *Gut* 2009;**58**(5):704–20.
87. Szabo G, Petrasek J, Bala S. Innate immunity and alcoholic liver disease. *Dig Dis* 2012;**30**(Suppl. 1):55–60.
88. Petrasek J, Csak T, Szabo G. Toll-like receptors in liver disease. *Adv Clin Chem* 2013;**59**:155–201.
89. Kesar V, Odin JA. Toll-like receptors and liver disease. *Liver Int* 2014;**34**(2):184–96.
90. Zanoni I, Ostuni R, Marek LR, et al. CD14 controls the LPS-induced endocytosis of Toll-like receptor 4. *Cell* 2011;**147**(4): 868–80.
91. Schromm AB, Lien E, Henneke P, et al. Molecular genetic analysis of an endotoxin nonresponder mutant cell line: a point mutation in a conserved region of MD-2 abolishes endotoxin-induced signaling. *J Exp Med* 2001;**194**(1):79–88.
92. Fukui H, Brauner B, Bode JC, Bode C. Plasma endotoxin concentrations in patients with alcoholic and non-alcoholic liver disease: reevaluation with an improved chromogenic assay. *J Hepatol* 1991;**12**(2):162–9.
93. Mathurin P, Deng QG, Keshavarzian A, Choudhary S, Holmes EW, Tsukamoto H. Exacerbation of alcoholic liver injury by enteral endotoxin in rats. *Hepatol* 2000;**32**(5):1008–17.
94. Parlesak A, Schafer C, Schutz T, Bode JC, Bode C. Increased intestinal permeability to macromolecules and endotoxemia in patients with chronic alcohol abuse in different stages of alcohol-induced liver disease. *J Hepatol* 2000;**32**(5):742–7.
95. Bjarnason I, Peters TJ, Wise RJ. The leaky gut of alcoholism: possible route of entry for toxic compounds. *Lancet* 1984;**1**(8370): 179–82.
96. Draper LR, Gyure LA, Hall JG, Robertson D. Effect of alcohol on the integrity of the intestinal epithelium. *Gut* 1983;**24**(5):399–404.
97. Worthington BS, Meserole L, Syrotuck JA. Effect of daily ethanol ingestion on intestinal permeability to macromolecules. *Am J Dig Dis* 1978;**23**(1):23–32.
98. Adachi Y, Moore LE, Bradford BU, Gao W, Thurman RG. Antibiotics prevent liver injury in rats following long-term exposure to ethanol. *Gastroenterol* 1995;**108**(1):218–24.
99. Adachi Y, Bradford BU, Gao W, Bojes HK, Thurman RG. Inactivation of Kupffer cells prevents early alcohol-induced liver injury. *Hepatol* 1994;**20**(2):453–60.
100. Su GL, Klein RD, Aminlari A, et al. Kupffer cell activation by lipopolysaccharide in rats: role for lipopolysaccharide binding protein and toll-like receptor 4. *Hepatol* 2000;**31**(4):932–6.
101. Hritz I, Mandrekar P, Velayudham A, et al. The critical role of toll-like receptor (TLR) 4 in alcoholic liver disease is independent of the common TLR adapter MyD88. *Hepatol* 2008;**48**(4): 1224–31.
102. Philips JA, Rubin EJ, Perrimon N. Drosophila RNAi screen reveals CD36 family member required for mycobacterial infection. *Science* 2005;**309**(5738):1251–3.

103. Triantafilou M, Gamper FG, Haston RM, et al. Membrane sorting of toll-like receptor (TLR)-2/6 and TLR2/1 heterodimers at the cell surface determines heterotypic associations with CD36 and intracellular targeting. *J Biol Chem* 2006;**281**(41):31002–11.
104. Hoebe K, Georgel P, Rutschmann S, et al. CD36 is a sensor of diacylglycerides. *Nature* 2005;**433**(7025):523–7.
105. Baranova IN, Kurlander R, Bocharov AV, et al. Role of human CD36 in bacterial recognition, phagocytosis, and pathogen-induced JNK-mediated signaling. *J Immunol* 2008;**181**(10):7147–56.
106. Cai L, Wang Z, Ji A, Meyer JM, van der Westhuyzen DR. Scavenger receptor CD36 expression contributes to adipose tissue inflammation and cell death in diet-induced obesity. *PloS One* 2012;**7**(5): e36785.
107. Stewart CR, Stuart LM, Wilkinson K, et al. CD36 ligands promote sterile inflammation through assembly of a Toll-like receptor 4 and 6 heterodimer. *Nat Immunol* 2010;**11**(2):155–61.
108. Seimon TA, Nadolski MJ, Liao X, et al. Atherogenic lipids and lipoproteins trigger CD36-TLR2-dependent apoptosis in macrophages undergoing endoplasmic reticulum stress. *Cell Metabol* 2010;**12**(5):467–82.
109. Bailey SM. A review of the role of reactive oxygen and nitrogen species in alcohol-induced mitochondrial dysfunction. *Free Radic Res* 2003;**37**(6):585–96.
110. Pascual M, Balino P, Alfonso-Loeches S, Aragon CM, Guerri C. Impact of TLR4 on behavioral and cognitive dysfunctions associated with alcohol-induced neuroinflammatory damage. *Brain Behav Immun* 2011;**25**(Suppl. 1):S80–91.
111. Reddy VD, Padmavathi P, Kavitha G, Saradamma B, Varadacharyulu N. Alcohol-induced oxidative/nitrosative stress alters brain mitochondrial membrane properties. *Mol Cell Biochem* 2013;**375**(1–2):39–47.
112. Potula R, Haorah J, Knipe B, et al. Alcohol abuse enhances neuroinflammation and impairs immune responses in an animal model of human immunodeficiency virus-1 encephalitis. *Am J Pathol* 2006;**168**(4):1335–44.
113. Wang HJ, Zakhari S, Jung MK. Alcohol, inflammation, and gut-liver-brain interactions in tissue damage and disease development. *World J Gastroenterol* 2010;**16**(11):1304–13.
114. Riikonen J, Jaatinen P, Rintala J, Porsti I, Karjala K, Hervonen A. Intermittent ethanol exposure increases the number of cerebellar microglia. *Alcohol Alcohol* 2002;**37**(5):421–6.
115. Alikunju S, Abdul Muneer PM, Zhang Y, Szlachetka AM, Haorah J. The inflammatory footprints of alcohol-induced oxidative damage in neurovascular components. *Brain Behav Immun* 2011;**25**(Suppl. 1):S129–36.
116. Haorah J, Ramirez SH, Schall K, Smith D, Pandya R, Persidsky Y. Oxidative stress activates protein tyrosine kinase and matrix metalloproteinases leading to blood-brain barrier dysfunction. *J Neurochem* 2007;**101**(2):566–76.
117. Abumrad NA, Ajmal M, Pothakos K, Robinson JK. CD36 expression and brain function: does CD36 deficiency impact learning ability? *Prostaglandins Other Lipid Mediat* 2005;**77**(1–4): 77–83.
118. Husemann J, Loike JD, Anankov R, Febbraio M, Silverstein SC. Scavenger receptors in neurobiology and neuropathology: their role on microglia and other cells of the nervous system. *Glia* 2002;**40**(2):195–205.
119. Coraci IS, Husemann J, Berman JW, et al. CD36, a class B scavenger receptor, is expressed on microglia in Alzheimer's disease brains and can mediate production of reactive oxygen species in response to beta-amyloid fibrils. *Am J Pathol* 2002;**160**(1): 101–12.
120. Ortiz-Masia D, Diez I, Calatayud S, et al. Induction of CD36 and thrombospondin-1 in macrophages by hypoxia-inducible factor 1 and its relevance in the inflammatory process. *PloS One* 2012;**7**(10):e48535.

CHAPTER

7

Thiamine Deficiency and Alcoholism Psychopathology

Ann M. Manzardo, PhD

Department of Psychiatry and Behavioral Sciences, University of Kansas Medical Center, Kansas City, Kansas, USA

INTRODUCTION

Chronic severe alcoholism affects diet, as well as nutrient absorption, metabolism, and utilization, and increases the risk of severe nutritional deficiency, particularly for the B vitamins, due to physical properties that limit their bioavailability and cellular storage.[1–6] Vitamin B_1 (thiamine) is an essential nutrient commonly deficient in alcoholism, and a critical cofactor for a diverse array of biochemical reactions mediating essential biological functions in humans, including carbohydrate metabolism, cellular energy production, lipid, and amino acid metabolism that broadly impact the functioning of all cells.[7–10] Chronic thiamine deficiency impairs oxidative metabolism, increasing oxidative stress that may evoke an inflammatory neurodegenerative process modeling several common neurodegenerative diseases (e.g., Alzheimer's disease, Parkinson's disease), including alcoholism.[11–13] These neurodegenerative processes involve the induction of inflammatory mediators, also disturbed in alcoholism (e.g., MCP-1, TNF-α) and selectively undermine white matter integrity that is necessary for cortical communication among reward-related brain structures.[13–19] Erosion of these neural pathways through alcoholism-related thiamine deficiency may influence disease course, and response to alcoholism treatment.[14,17,18,20] Elevated energy demands associated with selected tissues (e.g., cardiac, nervous) increase their sensitivity to thiamine deficiency, leading to cardiomyopathy, and central and peripheral neuropathy that may progress to acute disease (e.g., Wernicke–Korsakoff syndrome) in alcoholism, and possibly death, if untreated.[21–24] The role of thiamine as a cofactor for key enzymes involved in oxidative metabolism (e.g., Kreb's cycle, Pentose-phosphate shunt) has been well-documented, and the neurological complications and comorbidities associated with alcoholism-related thiamine deficiency are described in depth in the medical literature.[7–9,21–24] Current alcoholism treatment protocols call for oral vitamin supplementation in outpatient settings to offset deficiency, and intravenous vitamins, including thiamine, for inpatients admitted for alcoholism treatment.[24,25] Nevertheless, evidence of chronic thiamine deficiency is still common in alcoholism, even in the absence of clinically recognizable symptoms of deficiency (e.g., disturbances in gate or vision).[23,24,26–29]

The broad effects of such chronic "subclinical" thiamine deficiency on protein synthesis, cell growth, signaling and function, and inflammatory response selectively impacts brain centers that regulate psychiatric traits, such as mood and impulsiveness that may directly or indirectly influence alcohol and other drug use and recovery.[30–39] Preclinical studies have shown that a deficiency of certain B vitamins, particularly thiamine deficiency, can modulate alcohol consumption in rat models of alcoholism,[40–47] and may represent a predisposing factor in the development of alcoholism.[7,48,49] However, there are a paucity of human studies considering the effects of thiamine deficiency and supplementation on alcoholism recovery, and other psychiatric outcomes.[48–50] The present chapter reviews the effects of thiamine deficiency on cellular and molecular pathways pertinent to addiction pathophysiology, and discusses the results of a randomized trial of thiamine supplementation in chronic severe alcoholism.

BIOCHEMISTRY OF THIAMINE PYROPHOSPHATE

Thiamine is a molecular catalyst for a large and diverse family of enzymes found in all life-forms that can be broadly classified into nine superfamilies, based upon sequence similarity and genetic characteristics of the active

Molecular Aspects of Alcohol and Nutrition. http://dx.doi.org/10.1016/B978-0-12-800773-0.00007-0

site that consists of two highly conserved catalytic domains (pyrimidine-binding, and pyrophosphate-binding domains).[10] The diversity and unique versatility of this enzyme family result from the stabilizing effects of the active diphosphate ion, thiamine pyrophosphate (TPP$^+$), on the reaction mechanics that enable the formation or cleavage of multiple types of covalent bonds involving elemental carbon (carbon—carbon, carbon—oxygen, carbon—nitrogen, and carbon—sulfur).[10] The thiazole ring structure and cationic form of TPP$^+$ may also stabilize molecular charge and membrane conductance, impacting current and voltage dependence of certain ion channels.[51] This includes ion channels implicated in alcoholism pathology (e.g., acetylcholine, GABA), encompassing a broad spectrum of biological processes and brain function, including emotion and memory formation.[52] The structural and catalytic architecture of thiamine-dependent enzymes in nature is characterized by Vogel and Pleiss.[10] Human thiamine-dependent enzymes fall into three of the nine superfamilies: decarboxylase (DC), transketolase (TK), and 2-ketoacid dehydrogenase (2K) superfamilies, and includes several key enzymes involved with carbohydrate, and lipid metabolism (see Table 7.1).[53]

The glycolytic metabolic pathway is linked to the tricarboxylic acid cycle (TCA) via a decarboxylation reaction that converts pyruvate to acetyl-CoA, using a multi-subunit enzyme complex that incorporates TPP$^+$ as a cofactor.[7–9,12] This initial and rate-limiting step of the TCA cycle establishes the metabolic efficiency of oxidative metabolism, during the production of adenosine triphosphate (ATP), the primary energy source for all cells, and underlies the basis for the global reliance upon thiamine for all complex organisms. A second, reversible decarboxylation reaction within the TCA cycle catalyzes the conversion of α-ketoglutarate to succinyl-CoA, through a similar multisubunit enzyme complex (alpha-ketoglutarate dehydrogenase complex), also reliant upon TPP$^+$ as a cofactor.[7–9,12] Disruption of this core metabolic function is an important component of the physiological effects observed in thiamine deficiency that arises from the reduction or loss of oxidative capacity, and increased reliance upon anaerobic metabolic pathways (i.e., lactic acid cycle) for energy production (ATP).[7–9,12]

The hexose monophosphate shunt (HMS) is an important secondary pathway for carbohydrate metabolism, with an oxidative and nonoxidative arm that incorporates another thiamine dependent enzyme, transketolase.[7,54] HMS enzymes generate several intermediate metabolic compounds with important biological functions, including nicotine adenine dinucleotide phosphate (NADPH), the reduced form of NADP$^+$ that catalyzes the reductive biosynthesis of lipids, including myelin, and the reduction/recycling of the antioxidant protein, glutathione.[7,54] Reducing equivalents from NADPH absorb reactive oxygen species generated by biosynthetic and oxidation-reduction (red-ox) reactions, thereby protecting cells from injury. Inhibition of the HMS pathway leads to the generation of harmful reactive oxygen species, and the induction of apoptosis, leading to cell death. Other intermediates of the HMS pathway include the base structure for purine and pyrimidine nucleotides, ribose-5-phosphate, as well as erythrose-4-phosphate, a component of aromatic amino acids.[7,54] The bioavailability of TPP$^+$ impacts the activation and function of thiamine-dependent enzymes, and chronic thiamine deficiency leads to downregulation of enzyme synthesis and function.[8,55,56] Prolonged downregulation or dysfunction of transketolase activity impacts the availability of lipids, cell division and protein synthesis. Cerebral

TABLE 7.1 Classification and Function of Human Thiamine-Dependent Enzymes

Enzyme	Biochemical pathway	Molecular function	Physiologic effect of thiamine deficiency
DECARBOXYLASE SUPERFAMILY			
Pyruvate dehydrogenase complex	Tricarboxylic acid cycle	Oxidative metabolism, production of ATP	↑oxidative stress, lactic acidosis, cell death
TRANSKETOLASE SUPERFAMILY			
Transketolase	Hexose monophosphate shunt	Nonoxidative cellular metabolism and biosynthesis of pentose sugar moiety	Impaired biosynthesis of lipids, aromatic amino acids and nucleotides
2-KETOACID DEHYDROGENASE SUPERFAMILY			
α-Ketoglutarate dehydrogenase complex	Tricarboxylic acid cycle	Oxidative metabolism, production of ATP	↑oxidative stress, lactic acidosis, cell death
Branched-chain α-ketoacid dehydrogenase	Lipid oxidation	Biosynthesis of branched-chain amino acids (e.g., leucine, isoleucine and valine)	Impaired protein synthesis, maple syrup urine disease

demyelination syndromes associated with severe thiamine deficiency, for example, Wernicke–Korsakoff syndrome (cerebral beriberi), central pontine myelinolysis, and Marchiafava–Bignami disease, arise in part from the disruption of myelination associated with inhibition of the HMS pathway, likely in conjunction with cytokine involvement and/or cytotoxic effects of alcohol.[57,58]

SECONDARY EFFECTS OF THIAMINE DEFICIENCY

Chronic exposure to thiamine deficiency and oxidative stress are associated with secondary disruptions and adaptive effects, including the activation of innate immune responses.[11,13,59] Clinical and preclinical evidence suggests that an important underlying factor in these processes is the induction of proinflammatory mediators, including monocytes, astrocytes, and microglia.[13,19] These cellular mediators abnormally activate proinflammatory cytokines (e.g., IL-1β, IL-6, TNF-α) and chemokines (e.g., MCP-1, MIP-1α, MIP-1β, GRO-1), and enhance DNA binding and transcription of nuclear factor kappa-light-chain-enhancer of activated B cells (NF-κB) highly expressed in microglia. Abnormal activation of these inflammatory mediators are associated with weakening of the blood-brain barrier, and the induction of a neurodegenerative cascade that targets selectively the periventricular regions of the brainstem, cerebellar vermis, as well as the thalamic nuclei and mammillary bodies that are more commonly affected in alcoholism.[11,21,27,28,57,60,61] Cytotoxic effects of monocyte chemoattractant protein-1 (MCP-1) activation within the thalamic nucleus, and other brain areas, are proposed to play a role in several neurodegenerative diseases, including alcoholism.[59] Chronic alcohol use in the presence or absence of thiamine deficiency is associated with similar activation of NF-κB, and MCP-1 and related pro-inflammatory cascade, believed to contribute to behavioral disturbances, and continued alcohol use.[13,61] Collective effects of alcohol toxicity and thiamine deficiency on these processes in vulnerable brain regions amplify neurological and functional impairment impacting mood, cognition, and traits related to alcoholism, including impulsiveness, and may synergistically impact the ability to regulate drinking behaviors.

Oxidative stress and inflammation associated with thiamine deficiency also produce regional disturbances in nitric oxide synthase activity, resulting in the overproduction of neuronal nitric oxide (NO).[59,60] Chronic NO elevation inhibits mitochondrial enzymes, exacerbating mitochondrial dysfunction, and may precipitate uncoupling of the electron transport chain.[59,60] Further accumulation of reactive oxygen species increases lipid peroxidation, and induction of apoptosis. Neuronal loss in thiamine deficiency is associated with increased glutamate levels, and NMDA receptor activation, believed to reflect a spreading glutamate-mediated excitotoxic mechanism late in the disease course.[59,60] Additional neuroadaptation of γ-amino butyric acid (GABA)/glutamate signaling, and downregulation/peroxidation of glutamate transporter 1 (GLT-1) have been identified in vulnerable brain regions, and may play a pathogenic role in thiamine deficiency.[59,60] These secondary effects of thiamine deficiency, combined with the effects of chronic alcohol toxicity, may contribute to addiction-related adaptive effects observed in these signaling pathways.

THIAMINE DEFICIENCY, MOOD AND BEHAVIOR

Thiamine deficiency can influence psychological functioning, including personality characteristics, mood, and the expression of psychological symptoms that may impact the ability to regulate alcohol use.[30–33,35,39] Addiction to substances involves allostatic changes in stress and reward pathways, arising from dysregulation of neurochemical components of the ventral striatum, frontal cortex, and extended amygdala.[62] Cognitive decline in Wernicke–Korsakoff syndrome is correlated with a reduction in cholinergic neurons of the basal forebrain, and alterations in hippocampal and cortical structure and function, impacting the extended amygdala that are associated with behavioral and mood disturbances, and alcoholism pathophysiology.[36,37,63–65] Acute thiamine depletion in humans is characterized by a rapid deterioration of mood, with increased irritability and depression symptoms that are readily reversible with thiamine supplementation.[33] Enhanced mood, increased feelings of wellbeing, and decreased fatigue have been reported with thiamine supplementation, with and without active thiamine deficiency.[30–32,35,39,66–68] Depression symptoms were inversely correlated with blood thiamine levels in older adults (50–70 years) from the general population.[69] Overlapping effects of chronic subclinical thiamine deficiency, and alcoholism-induced allostatic changes in stress responsivity have the potential to amplify negative emotional states associated with alcohol withdrawal, precipitating relapse.[38,62]

Disturbances in oxidative metabolism, and increases in inflammatory mediators, such as MCP-1, also undermine neuronal integrity, and impair cortical communication and cognitive efficiency impacting on brain executive function.[14–18] The level of connectivity between frontal and striatal brain structures hinges upon the integrity of cerebral white matter tracts that are compromised in alcoholism and thiamine deficiency, and are negatively correlated with impulsiveness in addiction.[14–18] Loss of frontal-cortical brain connectivity, associated with cerebral demyelination, could reduce inhibitory control over impulse-driven behaviors, leading to an escalation of alcohol and other drug use in these disorders.

THIAMINE ANALOGS

The corrosive effects of alcohol on the gastrointestinal tract impair thiamine absorption and activation, and reduce the bioavailability of traditional, water-soluble, oral thiamine supplements in alcoholism.[24,70] Thus, there is a need to develop alternative adjuvant therapies that are capable of producing a sustained elevation in blood thiamine levels. This is especially important in the outpatient setting, where treatment compliance may be reduced and coupled with ongoing alcohol abuse by the patient.

Allithiamine is a lipid-soluble thiamine analogue found in plants from the *Allium* genus, in the garlic family, that can produce large, sustained elevations in TPP^+ bioavailability in blood and cerebral spinal fluid, and effectively alleviate thiamine deficiency in alcoholism, with a high safety profile.[71] As a natural product, allithiamine is available widely for purchase without prescription at health food stores, or from online retailers. A class of synthetic disulfide and S-acyl thiamine derivatives, for example, benfotiamine (S-benzoylthiamine O-monophosphate), thiamine tetrahydrofurfuryl disulfide (TTFD), and sulbutiamine (isobutyryl thiamine disulfide), with varying TPP^+ pharmacokinetic and safety profiles, have also been developed based upon the molecular structure of allithiamine.[72,73] The most widely accepted and studied in this class is benfotiamine (BF) that has been available commercially in Europe and Asia for decades, and approved for use in Germany for the treatment of sciatica nerve pain.[74]

BF is a lipid-soluble provitamin with a high oral absorption, bioavailability, and safety profile. It is initially dephosphorylated by intestinal alkaline phosphatases in the gut, and rapidly absorbed into the bloodstream, where it is transformed into thiamine monophosphate and diphosphate derivatives by hepatic enzymes (Figure 7.1).[75] BF is the least lipophilic of the currently available thiamine analogs, and has more limited direct penetrance of the blood-brain barrier. BF increases levels of free thiamine

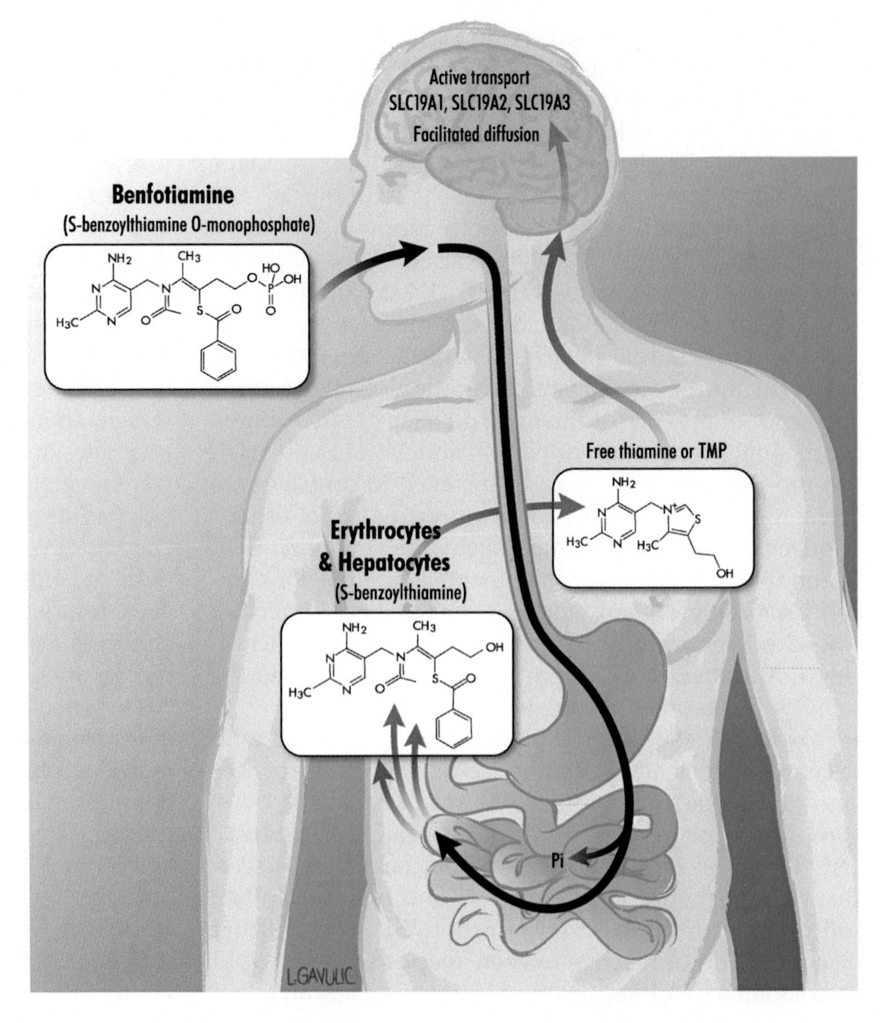

FIGURE 7.1 Pharmacokinetic profile of benfotiamine absorption, molecular activation, and redistribution to the brain. Benfotiamine is dephosphorylated to S-benzoylthiamine by intestinal alkaline phosphatases prior to absorption. Molecular activation of free thiamine to the mono- or diphosphate derivatives is carried out by hepatic enzymes and enter the brain through active transport and/or facilitated diffusion.

and thiamine monophosphate in the bloodstream that enters the brain, as through a class of saturable high-affinity thiamine transporters (SLC19A1, SLC19A2, SLC19A3), as well as facilitated diffusion to maximize brain thiamine exposure (Figure 7.1).[75] Pharmacokinetic profiling of plasma TPP^+ levels in clinical trials of BF in alcoholism has shown rapid and sustained elevation of peripheral TPP^+ after BF treatment.[70,73,75–77] Sustained BF treatment in alcoholism is associated with a dramatic (4x) increase in the activity of thiamine-dependent enzymes,[70,73,75–77] and significant reductions in symptoms of alcoholic and diabetic neuropathy in clinical trials.[74,78–83]

BENFOTIAMINE TRIAL

We carried out a double-blind randomized placebo-controlled clinical trial of BF in chronic severe alcoholism, in order to assess dose tolerability, and the effect of BF on alcohol consumption and alcoholism-related psychometric measures.[50] The study included 120 adult men and women meeting DSM-IV-TR criteria for current alcohol dependence, recruited from the Kansas City Metropolitan area, and examined 6 month outcomes of BF treatment (600 mg p.o. daily), compared with placebo. Participants were recruited as a community sample of active drinkers, not from a treatment center. They were not required to seek treatment or abstain from alcohol, in order to participate in the study, and no formal alcoholism treatment was included in the study design. The study randomization was stratified to control for the effects of family history of alcoholism on severity of illness and course.

A precoded, structured psychosocial interview was administered to all participants that included a 33-item Alcoholism Severity Scale (ASS), the Sobel Drinking Calendar, and the interviewer's rating of the Global Assessment of Functioning (GAF) scale.[84–87] ASS is a structured, multidimensional criterion-referenced psychiatric diagnostic interview that reviews all of the major clinical characteristics of alcoholism, and assesses the lifetime prevalence of 18 syndromes, including alcohol abuse and dependence.[85,88,89] The Sobel Drinking Calendar is a widely used assessment tool for measuring alcohol consumption.[88–90] Psychiatric symptomology and impulsiveness were also assessed, using the Symptom Checklist 90-R (SCL-90-R), and Barratt Impulsivity Scale (BIS).[91]

The study sample was comprised predominantly of African-American (72%), males (71%) with a self-reported family history of alcoholism (85%).[50] The mean age of participants at baseline was 47(7.9) years and they reported ~32(9) years of prior alcohol abuse. Participants displayed a high level of baseline psychiatric distress, as measured by SCL-90-R, and BIS scores four times higher than nonclinical normative ranges. Seventy of the 120 participants completed the entire 6-month study. No serious adverse events were identified, or significant differences in the frequency of side effects associated with BF use.[50] The evaluation of BF treatment response was obscured by a strong participation effect commonly observed in alcoholism studies that was related to the level of lifetime severity of alcoholism illness, or ASS score.

A linear mixed model was used for the analysis of mean daily alcohol consumption, and logistic mixed regression modeling was used for heavy episodic drinking days. Mixed regression models were selected in order to better accommodate missing monthly data. STATA procedures Xtmixed and Xtmelogistic for multilevel panel data were used for analyses. All regression analyses were stratified by family history, and controlled for the effects of gender. The difference between placebo and benfotiamine were examined by testing the interaction of time (month) and treatment. Differences between the placebo and benfotiamine groups were also analyzed with the chi-square statistic, and student's t-test, as appropriate. Two-sided p-values of < 0.05 were considered statistically significant.

Average daily alcohol consumption reported in standardized units of ounces of pure alcohol per day (SDUs/day) decreased as a function of study participation, for both treatment arms, over the course of the 6 month study (Figure 7.2). The quantity of alcohol consumed decreased from an average of 6.1 ± 4.3 SD/day to 2.8 ± 3.4 SD/day. The mean rate of alcohol consumption decreased by 0.63 SDUs/day each month (95% CI = −0.75 to −0.50, $p < 0.001$) in the placebo group, and by 0.73 SDUs/day per month in the benfotiamine group, over the course of the 6 month study. The decreasing rate among benfotiamine treated subjects was not significantly different from the rate in the placebo group (difference = −0.10 SDUs/month, 95% CI = −0.07 to 0.28, $p = 0.25$; see Table 7.2). The difference between the decreasing rate in the benfotiamine group and the decreasing rate in the placebo group is larger within three months, but this difference did not achieve statistical significance (difference = −0.27 SDUs/month, 95% CI = −0.67 to 0.23, $p = 0.19$), possibly due to a high level of observed variance, and low statistical power.

Examination of drinking occasions and patterns found a similar decrease in the likelihood of heavy episodic drinking (consumption of >5 SDUs/day) over time (Figure 7.3). The odds of heavy drinking decreased by 23% per month (95% CI: 21% to 29%, $p < 0.001$) among participants (Table 7.3), and the probability of heavy episodic drinking decreased 10% faster among benfotiamine (95% CI = 6% to 15%, $p < 0.001$) treated subjects, relative to placebo, indicating a selective reduction in heavy

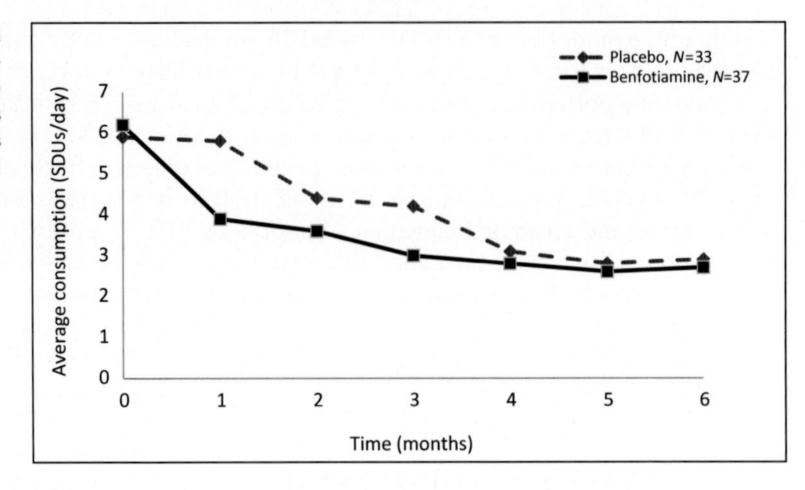

FIGURE 7.2 Average daily alcohol consumption based upon timeline followback assessment. Mean of daily alcohol consumption measured in standard drink units (SDUs) of pure alcohol by volume was calculated in 4-week blocks for completed subjects. A significant effect of time was found as assessed by linear mixed model analysis, $p < 0.001$ (see Table 7.2).

TABLE 7.2 Linear Mixed Model Analyses

Consumption	Correlation coefficient (SE)	Z	p-Value	95% CI
Time	−0.63 (0.66)	−9.54	0.001	−0.75 to −0.50
Treatment	−0.81 (0.79)	−1.02	0.31	−2.36 to 0.74
Gender	0.03 (0.80)	0.04	0.99	−1.53 to 1.59
Time × treatment	−0.10 (0.09)	1.15	0.25	−0.07 to 0.28
Constant	6.44 (0.79)	8.13	0.001	4.88 to 7.99

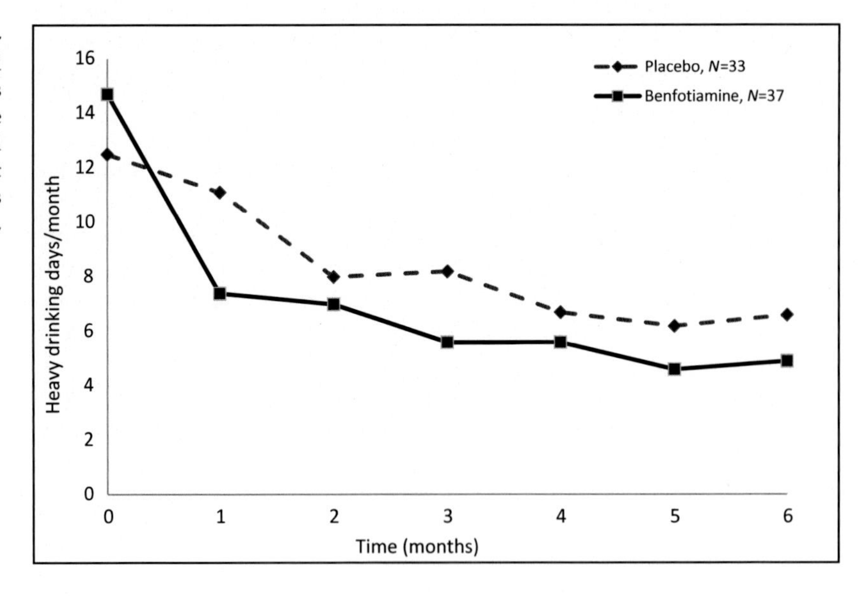

FIGURE 7.3 Heavy alcohol drinking as a function of treatment. Heavy drinking days correspond with alcohol consumption >5 standard drink units (SDUs) of pure alcohol by volume in a single day. The mean number of heavy drinking days was calculated in 4-week blocks for study completers. Significant effect of time, and time by benfotiamine treatment, was found as assessed by logistic mixed model analysis, $p < 0.001$ (see Table 7.3).

TABLE 7.3 Logistic Mixed Model Analysis

Heavy drinking days	Odds ratio	Z	p-Value	95% CI
Time	0.76	−16.0	0.001	0.74 – 0.79
Treatment	0.91	−0.25	0.80	0.44 – 1.89
Gender	1.76	1.54	0.12	0.86 – 3.63
Time × treatment	0.90	−4.45	0.001	0.85 – 0.94

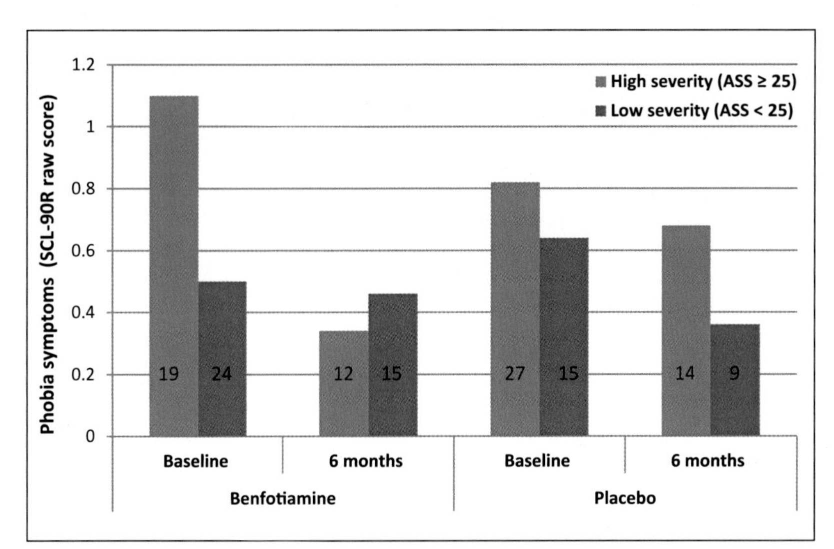

FIGURE 7.4 **Treatment response in alcohol dependent men differ by alcoholism severity level.**

alcohol use with benfotiamine treatment. The number of heavy drinking days decreased from 15 days/month to 5, among benfotiamine treated subjects, compared to a decrease from 12–7 days/month, among placebo. This result suggests that benfotiamine treated subjects may be less likely to engage in binge alcohol drinking, than subjects assigned to the placebo arm.

Gender differences in response to BF treatment were observed in the trajectory of alcohol consumption that decreased significantly more among BF, than placebo treated female participants.[50] Women randomized to BF showed a 45% decrease in alcohol consumption after 1 month, and a 60% decrease over the first 3 months of study participation, while alcohol consumption in women randomized to placebo initially increased by 20%, and then decreased by 13% over the first 3 months.[50] Nine of ten women assigned to the BF treatment arm showed a decrease in alcohol consumption after one month of treatment, compared to just two of eleven women on placebo. Total alcohol consumption from baseline to 6 months also decreased significantly more among BF than placebo treated female participants.[50] The results suggested that pathological drinking among women may be motivated, in part, by subclinical thiamine deficiency, and that treatment with thiamine analogues may reduce alcohol consumption in female alcoholism.

Male response to BF treatment was related to baseline severity of alcoholism illness, with the greatest clinical impact observed in the highest severity group. Males with high alcoholism severity scores (ASS≥25) showed a selective reduction in psychiatric symptoms of anxiety, depression, and phobia (see Figure 7.4). Lifetime severity of alcoholism is likely to correlate with the level of thiamine deficiency, fact that may explain the disparity in response to BF treatment. These data provide further support for the theory that thiamine deficiency impacts the levels of psychiatric distress in severe alcoholism, and supports a possible benefit of thiamine supplementation as an adjuvant therapy in treatment settings. Our studies, as well as previous studies in the literature regarding the use of thiamine in the treatment of alcoholism, will require more research to confirm and characterize the effects of BF, as a thiamine analogue, on outcomes with respect to the severity of alcoholism illness, and to gender.

CONCLUSIONS

Despite efforts to mitigate the impact of malnutrition in alcoholism, high rates of thiamine deficiency are still identified in alcoholic inpatients.[23,24,26–29] High-dose, intravenous vitamin treatments improve acute symptoms of deficiency, but may be insufficient to restore chronic thiamine deficits,[24,25] and traditional, water-soluble, oral supplementation of outpatients has limited efficacy due to reductions in bioavailability, associated with alcoholism.[24,70] The unique properties of BF, an orally administered agent, may represent an effective adjuvant therapy capable of producing a sustained elevation of blood thiamine, related to fat soluble properties, in severely alcohol dependent individuals, in outpatient settings, with minimal supervision and monitoring.[70,73,75–77] Alleviation of thiamine deficiency in alcoholism has been shown to reduce alcohol consumption, and positively impact psychiatric outcomes that may facilitate recovery, particularly in severe cases, and holds promise to reduce suffering associated with this menacing global public health problem.

Summary Points

- Heavy alcohol use impairs nutrient absorption, metabolism, and utilization that may lead to severe deficiency.
- Neurodegenerative effects of chronic vitamin B_1 (thiamine) deficiency on mood and cognitive function may propagate alcohol consumption and abuse.
- High potency thiamine analogues, including benfotiamine, have been developed to improve thiamine status in alcohol dependent individuals.
- The results of a clinical trial of benfotiamine to facilitate recovery from alcohol use disorders are discussed.

Acknowledgment

Dr Manzardo was supported by grants from the Hanlon Charitable Trust.

References

1. Green PH. Alcohol, nutrition and malabsorption. *Clin Gastroenterol* 1983;**12**(2):563–74.
2. Hoyumpa Jr AM. Mechanisms of thiamin deficiency in chronic alcoholism. *Am J Clin Nutr* 1980;**33**(12):2750–61.
3. Lieber CS. Relationships between nutrition, alcohol use, and liver disease. *Alcohol Res Health* 2003;**27**(3):220–31.
4. Said HM. Intestinal absorption of water-soluble vitamins in health and disease. *Biochem J* 2011;**437**(3):357–72.
5. Subramanya SB, Subramanian VS, Said HM. Chronic alcohol consumption and intestinal thiamin absorption: effects on physiological and molecular parameters of the uptake process. *Am J Physiol Gastrointest Liver Physiol* 2010;**299**(1):G23–31.
6. World MJ, Ryle PR, Thomson AD. Alcoholic malnutrition and the small intestine. *Alcohol Alcohol* 1985;**20**(2):89–124.
7. Martin PR, McCool BA, Singleton CK. Molecular genetics of transketolase in the pathogenesis of the Wernicke-Korsakoff syndrome. *Metab Brain Dis* 1995;**10**(1):45–55.
8. Martin PR, Singleton CK, Hiller-Sturmhofel S. The role of thiamine deficiency in alcoholic brain disease. *Alcohol Res Health* 2003;**27**(2): 134–42.
9. Singleton CK, Martin PR. Molecular mechanisms of thiamine utilization. *Curr Mol Med* 2001;**1**(2):197–207.
10. Vogel C, Pleiss J. The modular structure of ThDP-dependent enzymes. *Proteins* 2014;**82**(10):2523–37.
11. de Andrade JA, Gayer CR, Nogueira NP, Paes MC, Bastos VL, Neto Jda C, Alves Jr SC, Coelho RM, da Cunha MG, Gomes RN, Aguila MB, Mandarim-de-Lacerda CA, Bozza PT, da Cunha S. The effect of thiamine deficiency on inflammation, oxidative stress and cellular migration in an experimental model of sepsis. *J Inflamm* 2014;**11**:11.
12. Manzetti S, Zhang J, van der Spoel D. Thiamin function, metabolism, uptake, and transport. *Biochem* 2014;**53**(5):821–35.
13. Yang JY, Xue X, Tian H, Wang XX, Dong YX, Wang F, Zhao YN, Yao XC, Cui W, Wu CF. Role of microglia in ethanol-induced neurodegenerative disease: Pathological and behavioral dysfunction at different developmental stages. *Pharmacol Ther* 2014;**144**(3): 321–7.
14. Liston C, Watts R, Tottenham N, Davidson MC, Niogi S, Ulug AM, Casey BJ. Frontostriatal Microstructure Modulates Efficient Recruitment of Cognitive Control. *Cereb Cortex* 2006;**16**(4):553–60.
15. Moeller FG, Hasan KM, Steinberg JL, Kramer LA, Dougherty DM, Santos RM, Valdes I, Swann AC, Barratt ES, Narayana PA. Reduced anterior corpus callosum white matter integrity is related to increased impulsivity and reduced discriminability in cocaine-dependent subjects: diffusion tensor imaging. *Neuropsychopharmacol* 2005;**30**(3):610–7.
16. Pfefferbaum A, Sullivan EV. Microstructural but not macrostructural disruption of white matter in women with chronic alcoholism. *Neuroimage* 2002;**15**(3):708–18.
17. Pfefferbaum A, Sullivan EV, Hedehus M, Adalsteinsson E, Lim KO, Moseley M. *In vivo* detection and functional correlates of white matter microstructural disruption in chronic alcoholism. *Alcohol Clin Exp Res* 2000;**24**(8):1214–21.
18. Schulte T, Muller-Oehring EM, Pfefferbaum A, Sullivan EV. Neurocircuitry of emotion and cognition in alcoholism: contributions from white matter fiber tractography. *Dialogues Clin Neurosci* 2010;**12**(4):554–60.
19. Yang G, Meng Y, Li W, Yong Y, Fan Z, Ding H, Wei Y, Luo J, Ke ZJ. Neuronal MCP-1 mediates microglia recruitment and neurodegeneration induced by the mild impairment of oxidative metabolism. *Brain Pathol* 2011;**21**(3):279–97.
20. Corrêa Filho JM, Baltieri DA. Psychosocial and clinical predictors of retention in outpatient alcoholism treatment. *Rev Bras Psiquiatr* 2012;**34**(4):413–21.
21. Laureno R. Nutritional cerebellar degeneration, with comments on its relationship to Wernicke disease and alcoholism. *Handb Clin Neurol* 2012;**103**:175–87.
22. Thomson AD. Mechanisms of vitamin deficiency in chronic alcohol misusers and the development of the Wernicke-Korsakoff syndrome. *Alcohol Alcohol Suppl* 2000;**35**(1):2–7.
23. Thomson AD, Jeyasingham MD, Pratt OE, Shaw GK. Nutrition and alcoholic encephalopathies. *Acta Med Scand Suppl* 1987;**717**:55–65.
24. Thomson AD, Guerrini I, Marshall EJ. The evolution and treatment of Korsakoff's syndrome: out of sight, out of mind? *Neuropsychol Rev* 2012;**22**(2):81–92.
25. Markowitz JS, McRae AL, Sonne SC. Oral nutritional supplementation for the alcoholic patient: a brief overview. *Ann Clin Psych* 2000;**12**(3):153–8.
26. de la Monte SM, Kril JJ. Human alcohol-related neuropathology. *Acta Neuropathol* 2014;**127**(1):71–90.
27. Harper C. Thiamine (vitamin B1) deficiency and associated brain damage is still common throughout the world and prevention is simple and safe! *Eur J Neurol* 2006;**13**(10):1078–82.
28. Harper C, Dixon G, Sheedy D, Garrick T. Neuropathological alterations in alcoholic brains. Studies arising from the New South Wales Tissue Resource Centre. *Prog Neuropsychopharmacol Biol Psych* 2003; **27**(6):951–61.
29. Mancinelli R, Ceccanti M, Guiducci MS, Sasso GF, Sebastiani G, Attilia ML, Allen JP. Simultaneous liquid chromatographic assessment of thiamine, thiamine monophosphate and thiamine diphosphate in human erythrocytes: a study on alcoholics. *J Chromatogr B Analyt Technol Biomed Life Sci* 2003;**789**(2):355–63.
30. Benton D, Fordy J, Haller J. The impact of long-term vitamin supplementation on cognitive functioning. *Psychopharmacol* 1995;**117**(3): 298–305.
31. Benton D, Haller J, Fordy J. Vitamin supplementation for 1 year improves mood. *Neuropsychobiol* 1995;**32**(2):98–105.
32. Benton D, Griffiths R, Haller J. Thiamine supplementation mood and cognitive functioning. *Psychopharmacol* 1997;**129**(1):66–71.
33. Brozek J, Caster WO. Psychologic effects of thiamine restriction and deprivation in normal young men. *Am J Clin Nutr* 1957;**5**(2): 109–20.
34. Crews FT, Vetreno RP. Addiction, adolescence, and innate immune gene induction. *Front Psych* 2011;**2**:19.
35. Heseker H, Kübler W, Pudel V, Westenhöfer J. Interaction of vitamins with mental performance. *Bibl Nutr Dieta* 1995;**52**:43–55.

36. Nardone R, Höller Y, Storti M, Christova M, Tezzon F, Golaszewski S, Trinka E, Brigo F. Thiamine deficiency induced neurochemical, neuroanatomical, and neuropsychological alterations: a reappraisal. *Scientific World J* 2013;**2013**:309143.
37. Pitel AL, Zahr NM, Jackson K, Sassoon SA, Rosenbloom MJ, Pfefferbaum A, Sullivan EV. Signs of preclinical Wernicke's encephalopathy and thiamine levels as predictors of neuropsychological deficits in alcoholism without Korsakoff's syndrome. *Neuropsychopharmacol* 2011;**36**(3):580–8.
38. Prisciandaro JJ, Rembold J, Brown DG, Brady KT, Tolliver BK. Predictors of clinical trial dropout in individuals with co-occurring bipolar disorder and alcohol dependence. *Drug Alcohol Depend* 2011;**118**(2–3):493–6.
39. Smidt LJ, Cremin FM, Grivetti LE, Clifford AJ. Influence of thiamin supplementation on the health and general well-being of an elderly Irish population with marginal thiamin deficiency. *J Gerontol* 1991;**46**(1):M16–22.
40. Brady RA, Westerfeld WW. The effect of B-complex vitamins on the voluntary consumption of alcohol by rats. *Q J Stud Alcohol* 1947;**7**: 499–505.
41. Eriksson K, Pekkanen L, Rusi M. The effects of dietary thiamin on voluntary ethanol drinking and ethanol metabolism in the rat. *Br J Nutr* 1980;**43**(1):1–13.
42. Impeduglia G, Martin PR, Kwast M, Hohlstein LA, Roehrich L, Majchrowicz E. Influence of thiamine deficiency on the response to ethanol in two inbred rat strains. *J Pharmacol Exp Ther* 1987;**240**(3): 754–63.
43. Mardones J. On the relationship between deficiency of B vitamins and alcohol intake in rats. *Q J Stud Alcohol* 1951;**12**(4):563–75.
44. Pekkanen L. Pyrithiamin shortens ethanol-induced narcosis and increases voluntary ethanol drinking in rats. *Int J Vitam Nutr Res* 1979;**49**(4):386–90.
45. Pekkanen L. Effects of thiamin deprivation and antagonism on voluntary ethanol intake in rats. *J Nutr* 1980;**110**(5):937–44.
46. Pekkanen L, Eriksson K, Sihvonen ML. Dietarily-induced changes in voluntary ethanol consumption and ethanol metabolism in the rat. *Br J Nutr* 1978;**40**(1):103–13.
47. Pekkanen L, Rusi M. The effects of dietary niacin and riboflavin on voluntary intake and metabolism of ethanol in rats. *Pharmacol Biochem Behav* 1979;**11**(5):575–9.
48. Manzardo AM, Penick EC. A theoretical argument for inherited thiamine insensitivity as one possible biological cause of familial alcoholism. *Alcohol Clin Exp Res* 2006;**30**(9):1545–50.
49. Manzardo AM, Penick EC. Is thiamine deficiency one cause of familial alcoholism? In: Sher L, editor. *Research on the neurobiology of alcohol use disorders. Chapter 5*. New York: Nova Science Publishers, Inc; 2008. p. 65–77.
50. Manzardo AM, He J, Poje A, Penick EC, Campbell J, Butler MG. Double-blind, randomized placebo-controlled clinical trial of benfotiamine for severe alcohol dependence. *Drug Alcohol Depend* 2013;**133**(2):562–70.
51. Schoffeniels E. Thiamine phosphorylated derivatives and bioelectrogenesis. *Arch Int Physiol Biochim* 1983;**91**(3):233–42.
52. Enomoto K, Edwards C. Thiamine blockade of neuromuscular transmission. *Brain Res* 1985;**358**(1–2):316–23.
53. Widmann M, Radloff R, Pleiss J. The Thiamine diphosphate dependent Enzyme Engineering Database: a tool for the systematic analysis of sequence and structure relations. *BMC Biochem* 2010;**11**:9.
54. Schenk G, Duggleby RG, Nixon PF. Properties and functions of the thiamin diphosphate dependent enzyme transketolase. *Int J Biochem Cell Biol* 1998;**30**(12):1297–318.
55. Gibson GE, Ksiezak-Reding H, Sheu KF, Mykytyn V, Blass JP. Correlation of enzymatic, metabolic, and behavioral deficits in thiamin deficiency and its reversal. *Neurochem Res* 1984;**9**(6): 803–14.
56. Lavoie J, Butterworth RF. Reduced activities of thiamine-dependent enzymes in brains of alcoholics in the absence of Wernicke's encephalopathy. *Alcohol Clin Exp Res* 1995;**19**(4):1073–7.
57. Charness ME. Brain lesions in alcoholics. *Alcohol Clin Exp Res* 1993;**17**(1):2–11.
58. Feuerlein W. Neuropsychiatric disorders of alcoholism. *Nutr Metab* 1977;**21**(1–3):163–74.
59. Hazell AS1, Butterworth RF. Update of cell damage mechanisms in thiamine deficiency: focus on oxidative stress, excitotoxicity and inflammation. *Alcohol Alcohol* 2009;**44**(2):141–7.
60. Todd K, Butterworth RF. Mechanisms of selective neuronal cell death due to thiamine deficiency. *Ann NY Acad Sci* 1999;**893**: 404–11.
61. Zahr NM, Alt C, Mayer D, Rohlfing T, Manning-Bog A, Luong R, Sullivan EV, Pfefferbaum A. Associations between in vivo neuroimaging and postmortem brain cytokine markers in a rodent model of Wernicke's encephalopathy. *Exp Neurol* 2014;**261C**:109–19.
62. Koob GF. Theoretical frameworks and mechanistic aspects of alcohol addiction: alcohol addiction as a reward deficit disorder. *Curr Top Behav Neurosci* 2013;**13**:3–30.
63. Arendt T, Bruckner MK, Bigl V, Marcova L. Dendritic reorganisation in the basal forebrain under degenerative conditions and its defects in Alzheimer's disease. II. Ageing, Korsakoff's disease, Parkinson's disease, and Alzheimer's disease. *J Comp Neurol* 1995;**351**(2):189–222.
64. Dirksen CL, Howard JA, Cronin-Golomb A, Oscar-Berman M. Patterns of prefrontal dysfunction in alcoholics with and without Korsakoff's syndrome, patients with Parkinson's disease, and patients with rupture and repair of the anterior communicating artery. *Neuropsychiatr Dis Treat* 2006;**2**(3):327–39.
65. Oscar-Berman M1, Kirkley SM, Gansler DA, Couture A. Comparisons of Korsakoff and non-Korsakoff alcoholics on neuropsychological tests of prefrontal brain functioning. *Alcohol Clin Exp Res* 2004;**28**(4):667–75.
66. Bell IR, Edman JS, Morrow FD, Marby DW, Perrone G, Kayne HL, Greenwald M, Cole JO. Vitamin B 1, B2, and B6. Augmentation of tricyclic antidepressant treatment in geriatric depression with cognitive dysfunction. *J Am Coll Nutr* 1992;**11**:159–63.
67. Benton D, Donohoe RT. The effects of nutrients on mood. *Public Health Nutr* 1999;**2**(3A):403–9.
68. Linton CR, Reynolds MT, Warner NJ. Using thiamine to reduce post-ECT confusion. *Int J Geriatr Psychiatry* 2002;**17**(2):189–92.
69. Zhang G, Ding H, Chen H, Ye X, Li H, Lin X, Ke Z. Thiamine nutritional status and depressive symptoms are inversely associated among older Chinese adults. *J Nutr* 2013;**143**(1):53–8.
70. Baker H, Frank O. Absorption, utilization and clinical effectiveness of allithiamines compared to water-soluble thiamines. *J Nutr Sci Vitaminol (Tokyo)* 1976;(**22 Suppl**):63–8.
71. Fujiwara M. Allithiamine and its properties. *J Nutr Sci Vitaminol (Tokyo)* 1976;(**22 Suppl**):57–62.
72. Lonsdale D. Thiamine tetrahydrofurfuryl disulfide: a little known therapeutic agent. *Med Sci Monit* 2004;**10**(9):RA199–203.
73. Loew D. Pharmacokinetics of thiamine derivatives especially of benfotiamine. *Int J Clin Pharmacol Ther* 1996;**34**(2):47–50.
74. Anonymous. Benfotiamine. *Altern Med Rev* 2006;**11**(3):238–42.
75. Bitsch R, Wolf M, Moller J, Heuzeroth L, Gruneklee D. Bioavailability assessment of the lipophilic benfotiamine as compared to a water-soluble thiamin derivative. *Ann Nutr Metab* 1991;**35**(5): 292–6.
76. Greb A, Bitsch R. Comparative bioavailability of various thiamine derivatives after oral administration. *Int J Clin Pharmacol Ther* 1998;**36**(4):216–21.
77. Schreeb KH, Freudenthaler S, Vormfelde SV, Gundert-Remy U, Gleiter CH. Comparative bioavailability of two vitamin B1 preparations: benfotiamine and thiamine mononitrate. *Eur J Clin Pharmacol* 1997;**52**(4):319–20.

78. Ayazpoor U. Chronic alcohol abuse. Benfotiamine in alcohol damage is a must. *MMW Fortschr Med* 2001;**143**(16):53.
79. Babaei-Jadidi R, Karachalias N, Ahmed N, Battah S, Thornalley PJ. Prevention of incipient diabetic nephropathy by high-dose thiamine and benfotiamine. *Diabetes* 2003;**52**(8):2110–20.
80. Haupt E, Ledermann H, Kopcke W. Benfotiamine in the treatment of diabetic polyneuropathy--a three-week randomized, controlled pilot study (BEDIP study). *Int J Clin Pharmacol Ther* 2005;**43**(2):71–7.
81. Simeonov S, Pavlova M, Mitkov M, Mincheva L, Troev D. Therapeutic efficacy of "Milgamma" in patients with painful diabetic neuropathy. *Folia Med (Plovdiv)* 1997;**39**(4):5–10.
82. Stracke H, Lindemann A, Federlin K. A benfotiamine-vitamin B combination in treatment of diabetic polyneuropathy. *Exp Clin Endocrinol Diab* 1996;**104**(4):311–6.
83. Woelk H, Lehrl S, Bitsch R, Kopcke W. Benfotiamine in treatment of alcoholic polyneuropathy: an 8-week randomized controlled study (BAP I Study). *Alcohol Alcohol* 1998;**33**(6):631–8.
84. American Psychiatric Association. *Diagnostic and statistical manual of mental disorders*. 4th ed. Text revision ed. Washington DC: American Psychiatric Association; 2000. p. 1–943.
85. Knop J, Penick EC, Nickel EJ, Mortensen EL, Sullivan MA, Murtaza S, Jensen P, Manzardo AM, Gabrielli Jr WF. Childhood ADHD and conduct disorder as independent predictors of male alcohol dependence at age 40. *J Stud Alcohol Drugs* 2009;**70**(2):169–77.
86. Sobell LC, Brown J, Leo GI, Sobell MB. The reliability of the Alcohol Timeline Followback when administered by telephone and by computer. *Drug Alcohol Depend* 1996;**42**(1):49–54.
87. Sobell L, Sobell M. *Alcohol timeline followback (TFLB)*. In: Rush AJ, editor. *Handbook of psychiatric measures*. Washington DC: American Psychiatric Association; 2000. p. 477–479.
88. Othmer E, Penick EC, Powell BJ, Read MR, Othmer SC. *Psychiatric diagnostic interview-revised*. Los Angeles, CA: Western Psychological Services; 1989.
89. Othmer E, Penick EC, Powell BJ, Read MR, Othmer SC. *Psychiatric diagnostic interview IV*. Los Angeles, CA: Western Psychiatric Services; 2000.
90. Pedersen ER, LaBrie JW. A within-subjects validation of a group-administered timeline followback for alcohol use. *J Stud Alcohol* 2006;**67**(2):332–5.
91. Derogatis L, Savitz K. The SCL-90-R and the brief symptom inventory (BSI) in primary care. Maruish M, editor. *Handbook of psychological assessment in primary care settings*, vol. 236. Mahwah, New Jersey: Lawrence Erlbaum Associates; 2000 p. 297–334.

CHAPTER

8

Vitamin B Regulation of Alcoholic Liver Disease

Charles H. Halsted, MD, Valentina Medici, MD***

*University of California Davis, The Genome and Biomedical Sciences Facility, Davis, California, USA
**Division of Gastroenterology and Hepatology, University of California Davis, Sacramento, California, USA

EFFECTS OF CHRONIC ALCOHOLISM ON THE AVAILABILITY OF FOLATE, VITAMIN B12, AND VITAMIN B6

Whereas alcohol is used in varied amounts by more than two thirds of the US population, chronic alcoholism, or the habitual use of excessive amounts of alcohol, is estimated to affect about 9% of adults.[1] Although alcohol can be considered a dietary constituent that provides 8 kcal per gram, its metabolism is rapid, and its chronic use carries substantial risks for liver disease, anemia, pancreatitis and pancreatic insufficiency, cardiovascular disease, progressive neurological disorders, and for specific vitamin deficiencies, including those of folate, thiamine, pyridoxine vitamin B6, and vitamin B12.[2] Mechanisms for nutrient deficiencies include anorexia or loss of appetite, particularly in the presence of alcoholic liver disease,[3] intestinal malabsorption,[4,5] and enhanced renal excretion.[6] This chapter will focus on folate, vitamin B12, and vitamin B6, since each of these vitamins play essential roles in hepatic methionine metabolism in regulation of the eventual gene methylation, and expression in the pathogenesis of alcoholic liver disease.

Folate

Folate exists in the diet as pteroylpolyglutamates ($PteGlu_n$) that consists of a pteroyl ring in gamma peptide linkage with up to seven glutamate residues[7] (Figure 8.1). Henceforth, the term folate will refer to dietary, or all endogenous forms, whereas folic acid will be referred to as Pteroylmonoglutamate (PteGlu), the pharmaceutical or endogenous form with a single glutamate residue. During the process of assimilation, the dietary glutamate residues are progressively removed from $PteGlu_n$ by the proximal intestinal brush border enzyme glutamyl carboxypeptidase (GCPII),[8] prior to active transport of PteGlu by the proton coupled folate transporter (PCFT)[9] (Figure 8.2). An additional transporter, the reduced folate carrier (RFC), is also present at the intestinal brush border surface, and may play a supplemental role in the intestinal absorption of dietary methylated and reduced PteGlu.[10] Once in the portal venous system, dietary PteGlu is transported across hepatocyte membranes by folate binding protein (FBP),[11] and $PteGlu_n$ are reconstituted, methylated, and stored in the liver, then hydrolyzed, and to the monoglutamyl form (me-PteGlu) by gamma glutamyl hydrolase, for both recirculation though the biliary system, and release to the systemic circulation, for transport to all organ systems in the body. Folate homeostasis is maintained through renal glomerular filtration and tubular reabsorption by way of the RFC[12] (Figure 8.2).

Prior to the policy of folic acid fortification of the US diet, in 1997, folate deficiency was very common in the chronic alcoholic population,[13] and was found in up to 80% of alcoholics admitted to Boston City Hospital in 1963.[14] The principal clinical sign of folate deficiency is megaloblastic anemia that was recognized in about one third of anemic chronic alcoholic patients admitted to another large municipal US hospital.[15] Although it was presumed that dietary insufficiency was the most common cause of folate deficiency in chronic alcoholics, other studies found rapid reduction in serum folate levels in patients with alcoholic liver disease, placed on low folate diets, presumably due to low liver folate stores,[16] and rapid lowering of serum folate levels after acute alcohol ingestion in normal subjects.[17] In addition, a careful study of folate deficient alcoholic patients found

Molecular Aspects of Alcohol and Nutrition. http://dx.doi.org/10.1016/B978-0-12-800773-0.00008-2

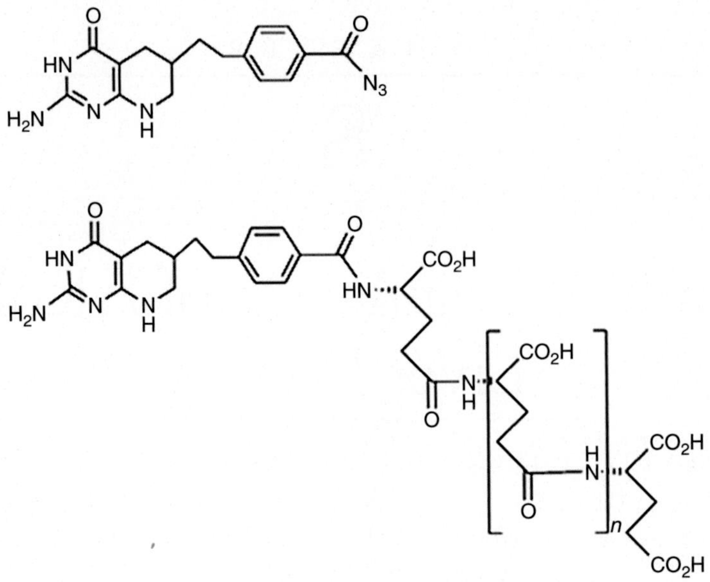

FIGURE 8.1 The folate molecule is represented in its monoglutamyl (PteGlu) metabolic form, and the polyglutamyl ($PteGlu_n$) forms, that predominate in the diet, and as liver storage with up to 6 glutamate residues in gamma linkage.

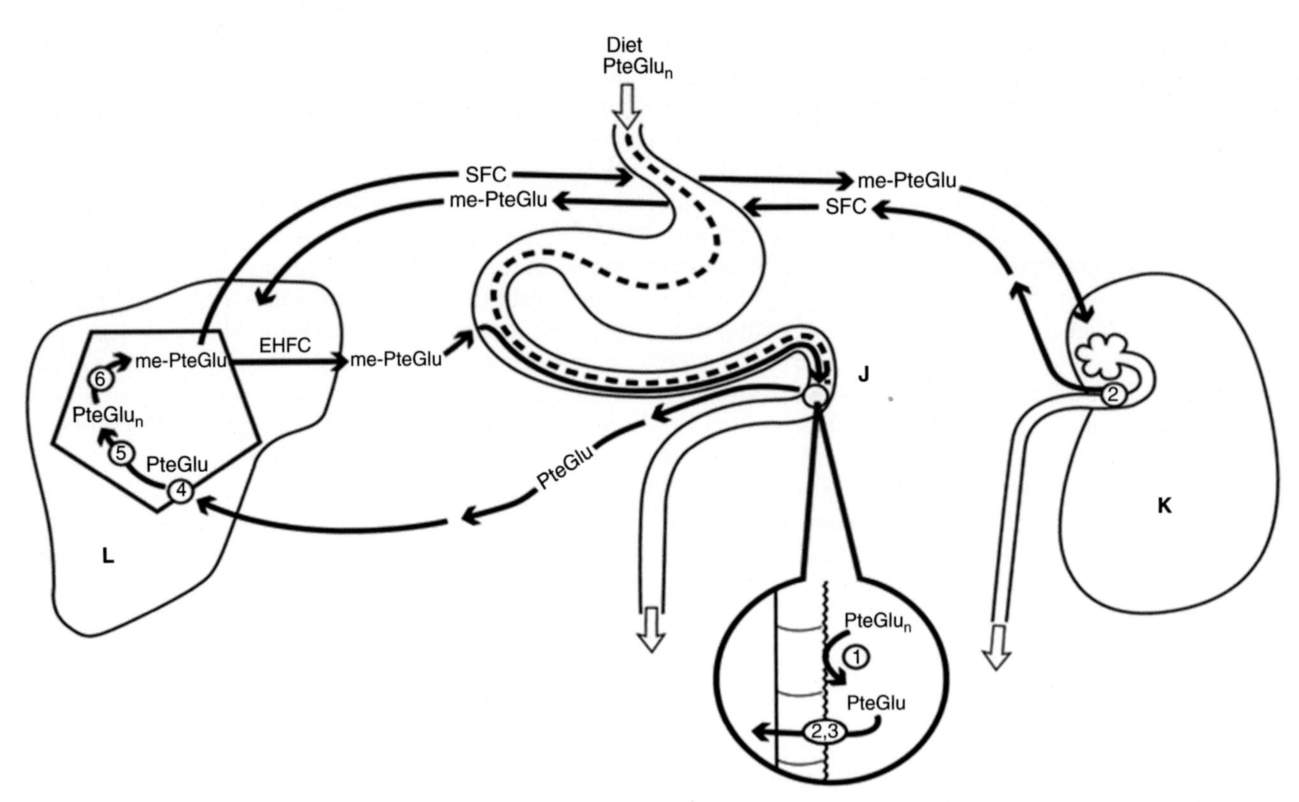

FIGURE 8.2 **Folate homeostasis.** The metabolic pathways of folate metabolism include (1) the cleavage of $PteGlu_n$ residuals by intestinal brush border glutamate carboxipeptidase (GCPII), (2) the intestinal transport or renal tubular transport of PteGlu forms by reduced folate carrier (RFC), (3) the proton coupled folate transporter (PCFT) for intestinal transport of PteGlu, (4) the uptake of PteGlu by the liver plasma membrane (LPM), (5) reconstitution of $PteGlu_n$ by pteroylpolyglutamate synthetase for methylation and hepatic folate storage, (6) gamma glutamyl hydrolase for intracellular hydrolysis, and subsequent methylation of liver me-PteGlu. SFC, systemic folate circulation; EHFC, enterohepatic folate circulation; J, jejunum; L, liver; K, kidney.

abrupt interruption of bone marrow recovery when heavy doses of ethanol ingestion were combined with folic acid treatment.[18] Subsequently, a series of both clinical and experimental studies have described a multitude of causes of ethanol induced altered folate metabolism, including decreased intestinal absorption, reduced liver uptake and storage, and increased renal excretion.

Intestinal folate malabsorption: A series of experiments with chronic alcoholic patients demonstrated decreased intestinal absorption of tritium labeled PteGu that was administered orally, or perfused through a tube placed in the jejunum.[19–21] Intestinal malabsorption of tritium labeled PteGlu was later confirmed in macaque monkeys that had been fed ethanol containing liquid diets, over a two year period.[22] Subsequent studies in a chronic ethanol fed pig model demonstrated decreased activity of GCPII, and expression of RFC in the jejunal mucosa, in association with decreased transport in intestinal brush border vesicles.[23,24] Additional studies in chronic ethanol fed rats demonstrated reduced expression of RFC, and transport of tritium labeled PteGlu across intestinal brush border membranes.[25]

Reduced hepatic uptake and storage: A clinical study of programmed dietary folate deficiency demonstrated that megaloblastic anemia could be induced in patients with established alcoholic liver disease in about half the time described for a normal human subject.[16,26] Since the liver is the major storage organ for folate,[27] these comparative data imply that liver folate stores are greatly reduced in alcoholic liver disease, either because of decreased uptake, increased biliary excretion,[28] or most likely, due to decreased numbers of healthy functional hepatocytes. A subsequent study in the chronic ethanol fed macaque monkey demonstrated reduced hepatic retention of parenterally administered tritium-labeled PteGlu, and reduced liver folate levels with normal pattern of labeled intrahepatic storage $PteGlu_n$ folates,[29] consistent with reduced PteGlu uptake at the liver plasma membrane.

Enhanced renal excretion: Studies in chronic alcoholic subjects, and ethanol fed rats and macaque monkeys, have all documented enhanced renal excretion of folates, compared to normal experimental subjects.[30–32] Subsequent studies in ethanol fed rats showed that increased urinary folate was caused by reduced renal tubular epithelial cell uptake and transport.[33] Although acute ethanol exposure did not affect folic acid binding by brush border membranes,[34] a later study demonstrated that chronic ethanol feeding of rats decreased folic acid transport across basolateral membranes, in association with decreased expressions of both RFC and FBP.[35]

Vitamin B12

Dietary vitamin B12, or cobalamin (Cbl), occurs bound to animal protein, and is required for hematopoiesis and maintenance of neurological integrity. While strict vegetarianism is a potential cause of vitamin B12 deficiency, multiple other causes of deficiency are related to defects in the complex mechanisms of intestinal absorption and circulatory transport of Cbl. A complex series of stages of intestinal absorption of dietary Cbl is initiated by its gastric acid induced release from food binding, and subsequent binding to salivary haptocorrin. Once within the proximal duodenum, Cbl is released from haptocorrin by pancreatic chymotrypsin and trypsin, then bound to intrinsic factor (IF), a protein that is secreted by gastric parietal cells and escorts the vitamin through the intestine to the ileum for its absorptive transport into the portal vein. Here, the Cbl-IF complex is bound to the ileal membrane receptor cubulin for cellular transport of Cbl to portal venous holotranscobalamin (holoTC) that carries it to the liver, and subsequently to all other tissues. Haptocorrin is also involved in the enterohepatic circulation of Cbl, but its functional significance in tissue transport is unclear. A membrane Cbl binding protein also plays a role in transfer of vitamin B12 into tissues.

In addition to rare dietary lack, vitamin B12 deficiency can be traced to hypochlorhydia of gastric contents that prevents its liberation from food binding, congenital or acquired IF deficiency, as in pernicious anemia, intestinal bacterial stasis that may cleave the Cbl-IF complex and prevent binding to ileal receptors, terminal ileal disease such as Crohn disease or surgical resection, and rare congenital deficiency of cubulin production. However, to date, there are no specific data on potential effects of chronic alcoholism on any of these processes, with the exception of pancreatic insufficiency that can impair the release of Cbl from its initial haptocorrin binder, prior to attachment to IF in the duodenum.[36]

Although measurements of serum vitamin B12 levels are typically normal, or even elevated in chronic alcoholics,[37,38] levels of hepatic Cbl were found to be low in liver biopsies from other patients, possibly related to impaired uptake or storage of the vitamin.[39] A recent Polish study of 80 chronic alcoholics demonstrated increased serum vitamin B12 levels that correlated with serum markers of alcoholic liver injury, consistent with inability of damaged hepatocytes to retain the vitamin.[40] Although one clinical study has demonstrated diminished overall intestinal absorption of labeled vitamin B12 in chronic alcoholic subjects, the mechanism is unknown.[41] Summarizing, vitamin B12 deficiency is likely in patients with alcoholic liver disease, but serum levels are misleading, and the deficiency can only be proven by measurements in liver tissue.

Vitamin B6

Vitamin B6, or pyridoxine, is required for both hematopoiesis and neurological function. Serum levels of vitamin B6 were found decreased in association with elevated serum homocysteine levels, in a large Spanish survey of chronic alcoholic patients,[42] and in about two thirds of cirrhotic patients,[43] as well as a cause of sideroblastic anemia in about one quarter of anemic chronic alcoholics admitted to a large urban hospital in New York City.[15] Due to its role in alanine aminotransferase (ALT) synthesis, pyridoxine deficiency may account for the relatively low ratio of serum level of ALT to elevated aspartate aminotransferase (AST) in patients with acute alcoholic hepatitis.[44] Whereas pyridoxine is absorbed throughout the intestine, there are no specific known effects of chronic alcoholism on its uptake. Rather, longstanding evidence suggests that pyridoxine deficiency may be caused in part by increased hepatic degradation,[43] potentially caused by displacement of the vitamin from its protein binding in the liver, by the ethanol metabolite acetaldehyde.[45]

INTERACTIONS OF SELECTED B VITAMINS IN HEPATIC METHIONINE METABOLISM WITH IMPLICATIONS FOR ALCOHOLIC LIVER INJURY

Many injury pathways are involved in alcoholic liver disease, including those of translocation of enteric lipopolysaccharide (LPS) from the intestine to liver, and subsequent activation of inflammatory mediators, hepatic steatosis, necrosis and apoptosis of hepatocytes, and collagen synthesis, in association with the production of reactive oxygen and nitrogen species, induced by ethanol metabolism.[46] For the past decade, we and others have focused our research on the effect of chronic alcoholism and ALD on the hepatic methionine cycle that regulates the production of S-adenosylmethionine (SAM), the principal methyl donor in transmethylation reactions that affect gene methylation and, hence, epigenetic regulation.[47]

As depicted in Figure 8.3, folate, vitamin B12, and vitamin B6 each play integral roles in the methionine metabolic cycle, and folate plays an additional essential role in regulation of DNA stability. Briefly, dietary folates are a mixture of $PteGlu_n$ forms that, after digestion and absorption, and liver transport of PteGlu, are methylated and converted to 5,10-methyltetrahydrofolate (5,10-MTHF) that is substrate for both thymidine synthetase (TS), and methyltetrahydrofolate reductase (MTHFR) reactions for the production of 5-methyltetrahydrofolate (5-MTHF), the original dietary methyl donor. DNA stability is dependent on TS activity for the nucleotide balance of its other substrate deoxyuridine monophosphate (dUMP) and product, deoxythymidine monophosphate (dTMP), and DNA instability is associated with increased DNA strand breaks and oxidation. The activity of TS is regulated by the availability of its substrate 5,10-MTHF; this is dependent upon both the availability of dietary folate, as well as the activity of its alternate MTHFR pathway, for synthesis of 5-MTHF.[47]

In the transmethylation pathway, 5-MTHF is the principal substrate for vitamin B12-dependent methionine synthase (MS) that converts homocysteine to methionine. An alternate enzyme, betaine homocysteine methyltransferase (BHMT), utilizes betaine that is present in the diet, or as a metabolite of choline, and homocysteine, as substrate to produce methionine and dimethylglycine (DMG). Methionine in turn is substrate for methionine

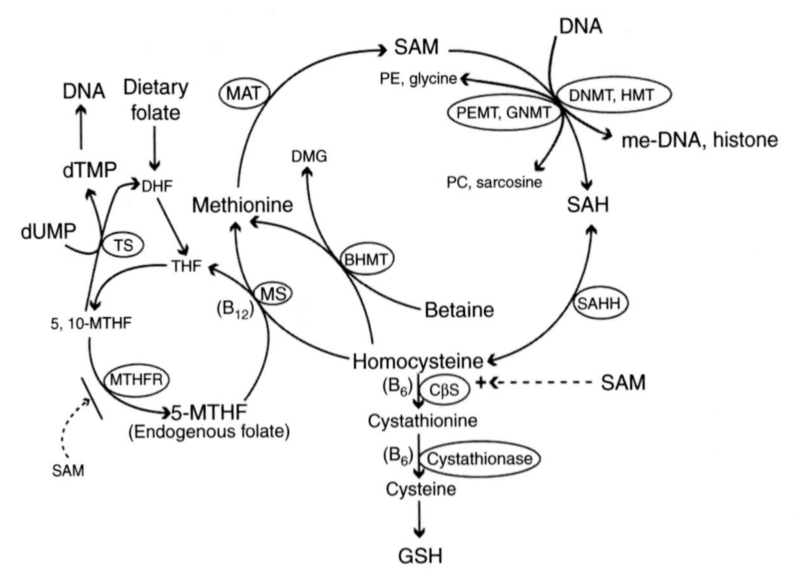

FIGURE 8.3 Hepatic methionine metabolism. B6, vitamin B6; B12, vitamin B12; BHMT, betaine homocysteine methyltransferase; CβS, cystathionine beta synthase; DHF, dihydrofolate; DMG, dimethylglycine; dTMP, deoxythymidine monophosphate; dUMP, deoxyuridine monophosphate; DNMT, DNA methyltransferase; GNMT, glycine N-methyltransferase; GSH, glutathione; HMT, histone methyltransferase; MAT, methionine adenosyltransferase; me, methyl; MS, methionine synthase; 5-MTHF, 5- methyltetrahydrofolate; 5,10-MTHF, 5,10-methylenetetrahydofolate; MTHFR, methylenetetrahydrofolate reductase; PE, phosphatidylethanolamine; PEMT, phosphatidylethanolamine transferase; PC, phosphatidylcholine; SAM, S-adenosylmethionine; SAH, S-adenosylhomocysteine; SAHH, S-adenosylhomocysteine hydrolase; THF, tetrahydrofolate; TS, thymidylate synthase.

adenosyl transferase (MAT) in the production of SAM, the principal methyl donor, for DNA methyltransferases (DNMTs) and histone methyltransferases (MTs). MAT exists in two forms, MAT1A, the principal form in the liver, and MAT2A that exists in fetal liver, and in hepatocellular carcinoma (HCC).[48] In a bi-directional reaction, homocysteine is also both substrate and product of S-adenosylhomocysteine (SAH), that is the principal product of transmethylation reactions with SAM. Since SAH is a potent inhibitor of both DNA and histone transmethylation, the ratio of SAM to SAH (SAM:SAH) is a convenient index of methylation potential.[49] Two additional transmethylation reactions include phosphatidylethanolamine methyltransferase (PEMT) that converts phosphatidylethanolamine (PE) to phosphatidylcholine (PC). This is essential to very low density lipoprotein (VLDL) for export of triglyceride from the liver, and glycine-n-methyltransferase (GNMT) that converts glycine to sarcosine, and constitutes about 1% of soluble rat liver protein. GNMT is the major consumer of SAM, and hence regulator of the SAM:SAH ratio.[50] In addition, SAM plays a regulatory role in the MTHFR reaction, such that in abundance, SAM reduces the activity of MTHFR and production of MTHF, whereas reduced SAM levels promote the MTHFR pathway for preservation of MTHF levels. As a consequence, the enhanced utilization of 5,10 MTHF for 5-MTHF production that occurs with low regulatory SAM levels reduces the availability of 5,10 MTHF for the TS pathway, thereby promoting imbalance of DNA nucleotides as described earlier.[47]

In its third reaction sequence, the trans-sulfuration pathway, homocysteine is substrate for the vitamin B6 regulated cystathionine beta synthase (CβS) that is stabilized and activated by SAM, in production of cystathionine, the substrate for the vitamin B6 dependent cystathioninase reaction for production of cysteine and, eventually, the antioxidant glutathione (GSH). Therefore, consequences of vitamin B6 deficiency include enhancement of homocysteine, due to reduced activities of CβS and cystathionase, and reduced levels of glutathione that promotes an oxidative state.[47]

EFFECTS OF CHRONIC ALCOHOL EXPOSURE ON METHIONINE METABOLIC PATHWAYS

Many studies in animal models have established that chronic alcohol exposure impairs multiple regulatory mechanisms in the methionine metabolic pathway. Several clinical studies demonstrated associations of chronic alcoholism with elevated serum homocysteine levels.[42,51,52] An early study using the chronic ethanol fed rat demonstrated reduced SAM production that was associated with enhanced utilization of the BHMT pathway.[53] Others showed that hepatocytes from ethanol fed rats exhibit increased levels of homocysteine and SAH that are corrected by exposure to the methyl donor betaine, or to SAM.[54–56]

Studies of liver samples from patients with cirrhosis or hepatocellular carcinoma demonstrated reduced transcript levels of MAT1A, BHMT, CBS, and MS, all associated with reduced SAM production.[57] A subsequent clinical study by the same group of liver biopsies from patients with alcoholic steatohepatitis confirmed these findings, together with elevated liver levels of SAH, and reduced levels of the antioxidant GSH.[58] These clinical studies recapitulated findings from the ethanol fed baboon model of ALD that included reductions of both liver SAM and GSH.[59] These findings were confirmed in liver from the ethanol-fed micropig model of alcoholic steatohepatitis, with additional reductions in transcripts of SAHH and MTHFR, and reduced activities of MS, but enhanced activity of the SAM consuming enzyme GNMT.[52] Summarizing, chronic alcoholism and the development of alcoholic liver disease impairs methionine metabolism, primarily through effects upon the transcription and activities of many of its regulatory genes, with resultant elevated serum homocysteine levels, and reduction in the antioxidant GSH that may contribute to the enhanced oxidative stress to the liver observed in the pathogenesis of this disease.

EPIDEMIOLOGY AND PATHOGENESIS OF ALCOHOLIC LIVER DISEASE

Given the high prevalence of alcoholism in the US population, excessive alcohol consumption is among the most frequent etiologies of liver disease in the US, and is one of the major causes of illness and death worldwide. The global burden of mortality related to alcohol drinking increased from 1990 to 2010,[60] and it was estimated that, worldwide, in 2010 almost 500,000 deaths were related to alcoholic liver cirrhosis, whereas liver cancer associated with alcoholic liver disease was related to 80,600 deaths.[61] Furthermore, alcoholic liver disease is the second most frequent indication for liver transplant in the US and in Europe.[62]

Alcoholic liver disease is a spectrum of conditions that range from alcoholic steatosis, steatohepatitis with various stages of fibrosis, and ultimately cirrhosis that is associated with increased risk of hepatocellular carcinoma. Alcoholic steatosis, or fat accumulation in the liver, is very common in alcohol drinkers, affecting at least 80% of chronic alcoholics.[63] Alcoholic steatosis is typically characterized by macrovesicular steatosis with dislocation of hepatocyte nuclei peripherally in the cell, by the deposition of large lipid droplets that are characterized by the accumulation of triglyceride and fatty acid.[64]

A typical feature of alcoholic steatosis is its reversibility, as it can improve significantly, even after a short duration of abstinence. More advanced and less reversible stages of liver disease include steatohepatitis with inflammatory infiltrate that has a predominant neutrophils component, Mallory-Denk bodies, characterized by intracellular cytoskeleton components aggregated in clumped proteins, as well as hepatocyte necrosis and ballooning degeneration. Cirrhosis represents the final and irreversible stage of liver damage, and is characterized by the deposition of bands of fibrosis, alternating with nodules of regenerative hepatocytes that can degenerate into hepatocellular carcinoma. Whereas alcoholic fatty liver is usually asymptomatic, alcoholic hepatitis is characterized by jaundice, fever, ascites, hepatic encephalopathy, and overall rapid worsening of clinical conditions, with a risk of mortality of 30–50% within 6 months.[65] Alcoholic cirrhosis may be asymptomatic as well, but ultimately patients with cirrhosis who continue drinking will develop signs and symptoms of portal hypertension with ascites, peripheral edema, hepatic encephalopathy, and gastrointestinal bleeding associated with rupture of esophageal and gastric varices.

Multiple molecular pathways are involved in the pathogenesis of alcoholic liver disease. Alcohol and its metabolites, in particular acetaldehyde, is associated with increased intestinal permeability, and translocation of bacterial products, such as LPS, through the paracellular intestinal spaces into the portal vein, and to the liver.[66] LPS recognition receptors are present on hepatic Kupffer cells, and include Toll-like receptor 4 (TLR4) that activates downstream signaling pathways leading to nuclear factor κB (NFκB) activation, and increased production of tumor necrosis factor alpha (TNF-α),[67,68] with consequent activation of a cascade of inflammatory factors.[69] Alcoholic fatty liver is the result of dysregulation of several transcription factors involved in lipid metabolism. Sterol regulatory element-binding protein (SREBP)-1, localized in the endoplasmic reticulum (ER) of hepatocytes, is induced in part by TNF-α, and triggers a cascade of genes that are involved in increased lipogenesis.[70] Furthermore, alcohol consumption inhibits fatty acid oxidation by suppressing the expression of peroxisome proliferator-activated receptor alpha (PPAR-α).[71] Oxidative stress in conjunction with reduced defense mechanisms, including reduced glutathione and phosphatidylcholine (PC) synthesis, favors the progression of steatohepatitis, necrosis, and fibrosis, in which hepatic stellate cells (HSCs) are activated from quiescent cells to proliferative and fibrogenic myofibroblasts. Many factors are profibrogenic, and contribute to persistent activation of HSCs in alcoholic liver disease, including acetaldehyde,[72] lipid peroxidation products such as 4-hydroxynonenal,[73] hepatocyte apoptosis,[74] osteopontin accumulation that upregulates type I collagen production,[75] and interleukin-1 (Il-1) accumulation, in association with increased profibrotic markers.[76] In addition to these mechanisms, hepatocyte mitochondrial dysfunction is recognized as a major contributing factor in the progression of alcoholic liver disease damage. Nitroxidative stress is probably responsible for damaging mitochondrial DNA, proteins, and lipids, as a result of increased reactive oxygen and reactive nitrogen species, such as nitric oxide, and increased production of peroxynitrite that can covalently modify mitochondrial components.[77]

REGULATORY EFFECTS OF ALTERED REACTIONS IN METHIONINE METABOLISM AND ITS METABOLITES ON MECHANISMS OF ALCOHOLIC LIVER DISEASE

SAM has been shown to preserve mitochondrial organelle function in ethanol fed rats, whereas its deficiency promotes mitochondrial superoxide production, and the induction of inducible nitric oxide synthase (iNOS)[78] that plays a significant role in the production of nitric oxide and peroxynitrate that, together with reactive oxygen species, induce liver injury.[77] Other groups found that SAM downregulated LPS induced TNF-α in isolated rat Kupffer (macrophage) cells, by a mechanism that was unrelated to the regulatory effect of NFκB on its transcription.[79] Ethanol lowering of the SAM:SAH ratio and methylation capacity also enhanced, and SAM administration lowered LPS-induced phosphodiesterase (PDE4B2), involved in the production of cyclic AMP (cAMP), a principal mediator of TNF-α by way of its activation of inhibitory protein kinase A.[80,81] Another group showed that increased SAH with decreased SAM:SAH ratio had the effect of enhancing liver TNF-α levels, and its cytotoxicity effects in ethanol fed mice,[82] due in part to decreased transport of SAM across mitochondrial membranes.[83]

The SAM:SAH methylation ratio also plays a regulatory role in the activity of proteasome, a multicatalytic enzyme that degrades oxidatively modified proteins and peptides.[84] The proteasome is suppressed by ethanol induction of pro-oxidant cytochrome P450 (CYP2E1),[85] in particular in the presence of lowered SAM:SAH ratio, and decreased methylation of the lysine residue of proteasome 20S.[86] Proteosomal dysfunction in the ethanol-fed rat results in the accumulation of oxidized and ubiquitinated proteins that produce histologically evident Mallory-Denk bodies, in a process that was prevented by supplemental SAM that salvaged proteasome 26S, while decreasing toll-like receptor (TLR-2/4)-dependent proinflammatory cytokines.[87] In addition to these intrinsic interactions of SAM with proteasome induction and inhibition, additional evidence indicates that inhibition of proteasomal activity, in response to ethanol feeding, is associated with downregulation of gene expressions of BHMT, MAT1A, and GNMT, together with reduced DNA and selective histone methylation reactions.[88]

Ethanol-induced steatosis in the liver results from combinations of enhanced lipogenesis, reduced fatty acid oxidation, and reduced export of triglyceride from the liver. Studies in ethanol-fed rats linked steatosis to reduced production of very low density lipoprotein (VLDL), secondary to reduced SAM substrate for methylation of phosphatidylcholine (PC) by PEMT (Figure 8.2) that was prevented by exposure to betaine that modulates SAM production through the BHMT pathway.[89,90] Similar findings were observed in the MAT1A knockout mouse that demonstrated reduced PC and VLDL synthesis corrected by SAM administration.[91] Additional recent studies from the same group focused on the unique effect of ethanol with reduced SAM production on reducing the activity of guanidinoacetate methyltransferase that is essential for muscle creatine synthesis,[92] whereas feeding excessive amounts of the substrate methyl consumer guanidineoacetate increased steatosis in ethanol fed rats, by increased consumption of SAM, and thereby reducing its availability for other methylation reactions.[93]

Emerging data indicate a significant role for altered methionine metabolism in collagen production from hepatic stellate cells (HSC), in the development of fibrosis and cirrhosis. Collagen production from HSC involves the oxidation of retinol (vitamin A) to retinoic acid (RA), by alcohol dehydrogenase III, and subsequently the RA-induced activation of latent transforming growth factor beta (TGF-β).[94] Studies using primary HSC from a mouse model of carbon tetrachloride induced hepatic fibrosis demonstrated the blocking effect on SAM on the promoter region of collagen I.[95]

MAT1A deletion associates with reduced SAM and increased risk of hepatocellular carcinoma (HCC),[96] a lethal complication of alcoholic cirrhosis, and its overexpression prevented tumor growth in a cellular model,[97] while administration of its product SAM prevented HCC in a rat model of this disease.[98] On the other hand, marked elevation of SAM levels in the GNMT knockout model was associated with altered cytokine signaling, and increased incidence of HCC,[99] as well as the accumulation of triglyceride in the liver.[100]

THE ETHANOL-FED MICROPIG MODEL FOR THE INTERACTION OF ETHANOL WITH METHIONINE METABOLISM, AND INDUCTION OF ALCOHOLIC LIVER DISEASE

Initial studies from our laboratory demonstrated the suitability of ethanol feeding of the micropig for induction of histologically evident hepatic steatohepatitis within one year, and collagen production with fibrosis at 2 years.[101] A subsequent study linked these observations to abnormal hepatic methionine metabolism, by showing ethanol induction of elevated serum homocysteine levels after 3 months of feeding, and subsequent findings at one year, of increased hepatocellular apoptosis that associated with imbalance in the dUMP/dTMP nucleotide ratio, reduced MS activity, increased liver SAH, and reduction in the SAM:SAH ratio.[102] In order to recapitulate the model of the folate deficient alcoholic, and in view of the role of folate as MTHF as the precursor of SAM (Figure 8.3), subsequent studies tested the additional effect of low folate diet, with ethanol feeding on the induction of alcoholic liver disease that, in contrast to the prior studies,[101] was achieved after only 3 months of feeding.[103] The principal findings in this model, compared to micropigs fed control diet, folate deficient alone diet, or ethanol alone diet, included all the histological features of alcoholic steatohepatitis, elevated serum homocysteine, reduced liver SAM and the SAM:SAH methylation ratio, increased DNA oxidation and strand breaks, and increased immunohistochemical levels of the oxidation mediator CYP2E1. Consistent with known effects of SAM in regulating CβS activity (Figure 8.3), lowered liver SAM levels correlated well with production of the main antioxidant glutathione (GSH), thereby linking the effects of ethanol induced altered methionine metabolism to oxidative liver injury,[103] and confirming the original evidence for this relationship in the ethanol fed baboon model.[104] The next follow-up study of the folate deficient and ethanol fed micropig recapitulated the same findings on altered hepatic methionine metabolism with reduced SAM and the SAM:SAH ratio, and histopathological evidence of alcoholic steatohepatitis, while demonstrating enhanced transcripts of selected genes related to enhanced oxidative liver injury (CYP2E1) and ER stress pathways for lipid synthesis that included SREBP1-c, stearoyl-coA desaturase (SCD), acetyl-CoA carboxylase (ACC), and fatty acid synthase (FAS).[105] Following on these observations, the essential linkage of altered hepatic methionine metabolism to alcoholic liver disease was proven by demonstrating that all of these findings, including histopathology, changes in SAM and SAH levels, and specific gene activations could be prevented by the inclusion of SAM in the ethanol and folate deficient micropig diets.[106,107] These studies complemented the observation of others that the induction of SREBP1-c and other ER stress genes, related to apoptosis in the intragastric ethanol fed mouse, could be prevented by inclusion of betaine in the diet.[108]

EFFECTS OF SAM IN CLINICAL TRIALS OF TREATMENT OF ALD

The preventive effects of SAM or betaine on the induction of ALD in animal models[106–108] suggested that SAM might be an effective compound for clinical therapy in

ALD patients. An initial 6-month Italian trial evaluated the efficacy of 6 months of oral SAM at 1.2 g/day in treatment of 9 ALD patients, with limited findings of increase in hepatic GSH levels, compared to untreated patients.[109] A later multicenter European study found reduced mortality or liver transplant incidence in 123 ALD patients receiving similar SAM doses over 2 years,[110] but there were no histopathological measurements in either study. More recently, we conducted a randomized control 6-month clinical trial of SAM versus placebo treatment in 37 ALD patients, who were required to maintain abstinence throughout the study. In contrast to the prior two studies, histopathology was scored in all subjects at baseline, and in 14 subjects after treatment. Whereas SAM serum levels were increased in the treatment subjects, there were no differences between the treatment and placebo groups in posttrial biopsy injury scores, or in serum biochemical evidence of liver injury, and we concluded that, while preventive in experimental model, SAM is not a useful therapeutic option in patients whose liver function has already been compromised by significant alcoholic liver injury.[111] A subsequent pathology analysis showed a significant decrease in smooth muscle actin, an index of hepatic stellate cell activation, in biopsies in all abstinent subjects evaluated after treatment but again with no differences between the SAM and placebo groups.[112] Further analysis of sera from alcoholic liver disease patients, compared to a similar number of active drinkers without liver disease, found decreased levels of vitamin B6 in association with elevated cystathionine levels, suggesting that reduced serum vitamin B6, and its dependent enzyme cystathionine, may be relevant to the onset of alcoholic liver disease.[38]

RELATIONSHIPS OF ALTERED METHIONINE METABOLISM TO EPIGENETIC REGULATION OF GENE EXPRESSION IN ALCOHOLIC LIVER INJURY

The term epigenetics refers to the heritable or environmental alterations of DNA structure that influence gene expression, without involving changes in its sequence. More specifically, epigenetic regulation involves remodeling of chromatin by the addition of methyl groups to DNA, or modification of structural histone residues by methylation, acetylation, or phosphorylation, each modifying specific gene transcriptions. SAM is the principal methyl donor and epigenetic regulator of DNA, and the major modifier of histone amino acids that regulate gene expression. DNA methylation occurs on cytosine bases, in particular on CpG dinucleotide clusters or islands, where it primarily serves to silence gene expression.[113,114]

Histone modification can occur on one of four clusters, typically on H3 or H4, with varied effects. Studies of isolated hepatocytes from ethanol fed rats found that H3K4 methylation was associated with up-regulation of several genes, including alcohol dehydrogenase 1, while H3K9 methylation was associated with downregulation of other genes, including CYP2E1.[115] Many of the effects of SAM on gene regulation, during the induction of alcoholic liver injury, may be mediated by histone modifications. For example, the addition of SAM to cultured Kupffer cells inhibited the production of TNF-α by blocking the H3K4 binding site for its LPS inducer,[116] and inhibition of proteasome in ethanol fed rats was associated with reduced H3K9 methylation.[88] Trimethylation of H3K27 that correlates with gene repression was enhanced by SAM, in ethanol fed rats,[117] whereas LPS induction of TNF-α was mediated by H3K4, and prevented by SAM in Kupffer cells.[116] Global DNA hypomethylation and increased DNA strand breaks were associated with decreased SAM in ethanol fed rats,[118] and micropig[103] models, and may be a precursor of HCC.[119]

Our laboratory has conducted two separate studies on the relationship of altered hepatic methionine metabolism to the epigenetic regulation of alcoholic liver injury genes, using the intragastric ethanol fed mouse model.[120,121] In each study, we used CβS heterozygous and wild type mice fed diets, with and without ethanol, since, consistent with the regulatory role of CβS (Figure 8.3), these mice are known to be primed for abnormal methionine metabolism, including significant homocysteinemia.[122] Our initial study found an association of histopathological alcoholic liver injury, with reduction of the SAM:SAH ratio that was negatively correlated with increased gene expressions of ER stress pathway genes SREBP-1c, and growth arrest and apoptotic DNA damage inducible gene (GADD153). Further epigenetic analysis of these findings showed their association with decreased protein levels of H3K9, in promoter regions of the same genes, as well as reduced expression of the H3K9 methyltransferase EHMT that correlated with the SAM:SAH methylation ratio. Summarizing, the effect of ethanol feeding in this mouse model was to reduce the liver SAM:SAH methylation ratio, in association with reduced transcription of a specific histone methyltransferase, with the net effects of decreasing methylation of the specific histone H3K9, and enhancing the expression of the selected ER stress genes.[120]

Our subsequent experiment tested the effect of provision of the alternate methyl donor betaine on prevention of ethanol induced DNA, global and gene-specific hypomethylation, and its effects on relevant genes. Using the same mouse model, the ethanol induced features of alcoholic liver injury were associated with reduced SAM:SAH ratios, together with reduced DNA methylation in gene bodies, but not gene promoter regions in all chromosomes. Ethanol feeding significantly enhanced, and betaine supplementation, prevented the expression of inducible nitric oxide synthase (Nos2), while reducing the expression of perosisome proliferator receptor-alpha (Ppar-α), known to promote fatty acid oxidation. Subsequent pyrosequencing studies showed specific CpG hypomethylation in a gene body region of Nos2 that correlated negatively with its gene expression, and positively with liver SAM and SAM:SAH ratio. iNOS is a primary regulator of the production of nitric oxide that influences lipid peroxidation, mitochondrial disruption, and steatosis, and is associated with the activation of NF-κB and induction of TNF-α[123] that are prevented in Nos-2 knockout mice.[124] Further linking this gene to methionine metabolism, and reduced SAM production, others showed that the excessive production of nitric oxide in the liver was associated with reduced translation and activity of MAT1A, by modification of a specific cysteine residue.[125]

Summarizing, chronic alcoholism and alcoholic liver disease are associated with deficiencies of several B vitamins, including folate, vitamin B12, and vitamin B6. The causes of these deficiencies are both dietary and metabolic, including reduced absorption and liver uptake, with enhanced renal excretion of folates, reduced hepatic retention of vitamin B12, and decreased circulatory transport of vitamin B6. The principal consequences of these deficiencies relate to their impact on hepatic methionine metabolism, in particular reduced production of SAM, and enhanced levels of SAH, with net effects of influencing many pathways of metabolism, and the epigenetic regulation of several gene expressions relevant to liver injury.

CONCLUSION

Key Facts

- Folate, vitamin B12, and vitamin B6 are all involved in the pathogenesis of alcoholic liver disease through their regulation of hepatic methionine metabolism.
- Hepatic methionine metabolism is required for the epigenetic regulation of gene methylation and expression, and is altered in chronic alcoholic liver disease.
- Etiologies of folate deficiency in chronic alcoholism are complex, and include altered intestinal absorption, hepatic cellular uptake, and renal conservation.
- Although serum levels of vitamin B12 in alcoholic liver disease are often normal, deficiency can be demonstrated by decreased concentrations of this vitamin in the liver.
- Altered epigenetic gene regulation in alcoholic liver disease includes abnormal methylation of both selected histones, and gene-specific DNA.

Summary Points

- Chronic alcohol abuse often results in deficiencies and/or abnormal metabolism of folate, vitamin B12, and vitamin B6, all of which are involved in hepatic methionine metabolism.
- Abnormal hepatic methionine metabolism results in abnormal methylation of histones and DNA in the regulation of genes involved in alcoholic liver disease.

References

1. Friedmann PD. Alcohol use in adults. *N Engl J Med* 2013;368(17):1655–6.
2. Leevy CM, Baker H, Tenhove W, Frank O, Cherrick GR. B-Complex Vitamins in Liver Disease of the Alcoholic. *Am J Clin Nutr* 1965;16:339–46.
3. Mendenhall CL, Anderson S, Weesner RE, Goldberg SJ, Crolic KA. Protein-calorie malnutrition associated with alcoholic hepatitis. Veterans Administration Cooperative Study Group on Alcoholic Hepatitis. *Am J Med* 1984;76(2):211–22.
4. Roggin GM, Iber FL, Linscheer WG. Intraluminal fat digestion in the chronic alcoholic. *Gut* 1972;13(2):107–11.
5. Soberon S, Pauley MP, Duplantier R, Fan A, Halsted CH. Metabolic effects of enteral formula feeding in alcoholic hepatitis. *Hepatology* 1987;7(6):1204–9.
6. McMartin KE, Collins TD, Shiao CQ, Vidrine L, Redetzki HM. Study of dose-dependence and urinary folate excretion produced by ethanol in humans and rats. *Alcohol Clin Exp Res* 1986;10(4):419–24.
7. Shane B. Folate metabolism. In: Picciano M, Stokstad ELR, Gregory JF, editors. *Folic acid metabolism in health and disease*. New York, NY: Wiley, Liss; 1990 p. 65–68.
8. Halsted CH, Ling EH, Luthi-Carter R, Villanueva JA, Gardner JM, Coyle JT. Folylpoly-gamma-glutamate carboxypeptidase from pig jejunum. Molecular characterization and relation to glutamate carboxypeptidase II. *J Biol Chem* 1998;273(32):20417–24.
9. Qiu A, Jansen M, Sakaris A, Min SH, Chattopadhyay S, Tsai E, et al. Identification of an intestinal folate transporter and the molecular basis for hereditary folate malabsorption. *Cell* 2006;127(5):917–28.
10. Said HM. Intestinal absorption of water-soluble vitamins in health and disease. *Biochem J* 2011;437(3):357–72.
11. Van Hoozen CM, Ling EH, Halsted CH. Folate binding protein: molecular characterization and transcript distribution in pig liver, kidney and jejunum. *Biochem J* 1996;319(Pt. 3):725–9.
12. Sikka PK, McMartin KE. Determination of folate transport pathways in cultured rat proximal tubule cells. *Chem Biol Interact* 1998;114(1–2):15–31.
13. Leevy CM, Cardi L, Frank O, Gellene R, Baker H. Incidence and significance of hypovitaminemia in a randomly selected municipal hospital population. *Am J Clin Nutr* 1965;17(4):259–71.
14. Herbert V, Zalusky R, Davidson CS. Correlation of folate deficiency with alcoholism and associated macrocytosis, anemia, and liver disease. *Ann Intern Med* 1963;58:977–88.
15. Savage D, Lindenbaum J. Anemia in alcoholics. *Medicine (Baltimore)* 1986;65(5):322–38.
16. Eichner ER, Hillman RS. The evolution of anemia in alcoholic patients. *Am J Med* 1971;50(2):218–32.
17. Eichner ER, Hillman RS. Effect of alcohol on serum folate level. *J Clin Invest* 1973;52(3):584–91.
18. Sullivan LW, Herbert V. Suppression Hematopoiesis by Ethanol. *J Clin Invest* 1964;43:2048–62.

19. Halsted CH, Griggs RC, Harris JW. The effect of alcoholism on the absorption of folic acid (H3-PGA) evaluated by plasma levels and urine excretion. *J Lab Clin Med* 1967;69(1):116–31.
20. Halsted CH, Robles EA, Mezey E. Decreased jejunal uptake of labeled folic acid (3 H-PGA) in alcoholic patients: roles of alcohol and nutrition. *N Engl J Med* 1971;285(13):701–6.
21. Halsted CH, Robles EA, Mezey E. Intestinal malabsorption in folate-deficient alcoholics. *Gastroenterology* 1973;64(4):526–32.
22. Romero JJ, Tamura T, Halsted CH. Intestinal absorption of [3H] folic acid in the chronic alcoholic monkey. *Gastroenterology* 1981; 80(1):99–102.
23. Naughton CA, Chandler CJ, Duplantier RB, Halsted CH. Folate absorption in alcoholic pigs: *in vitro* hydrolysis and transport at the intestinal brush border membrane. *Am J Clin Nutr* 1989;50(6): 1436–41.
24. Villanueva JA, Devlin AM, Halsted CH. Reduced folate carrier: tissue distribution and effects of chronic ethanol intake in the micropig. *Alcohol Clin Exp Res* 2001;25(3):415–20.
25. Hamid A, Wani NA, Rana S, Vaiphei K, Mahmood A, Kaur J. Down-regulation of reduced folate carrier may result in folate malabsorption across intestinal brush border membrane during experimental alcoholism. *FEBS J* 2007;274(24):6317–28.
26. Herbert V. Experimental nutritional folate deficiency in man. *Trans Assoc Am Physicians* 1962;75:307–20.
27. Herbert V. Predicting nutrient deficiency by formula. *N Engl J Med* 1971;284(17):976–7.
28. Steinberg SE, Campbell CL, Hillman RS. The toxic effects of alcohol on folate metabolism. *Clin Toxicol* 1980;17(3):407–11.
29. Tamura T, Romero JJ, Watson JE, Gong EJ, Halsted CH. Hepatic folate metabolism in the chronic alcoholic monkey. *J Lab Clin Med* 1981;97(5):654–61.
30. McMartin KE, Collins TD, Eisenga BH, Fortney T, Bates WR, Bairnsfather L. Effects of chronic ethanol and diet treatment on urinary folate excretion and development of folate deficiency in the rat. *J Nutr* 1989;119(10):1490–7.
31. Russell RM, Rosenberg IH, Wilson PD, Iber FL, Oaks EB, Giovetti AC, et al. Increased urinary excretion and prolonged turnover time of folic acid during ethanol ingestion. *Am J Clin Nutr* 1983;38(1): 64–70.
32. Tamura T, Halsted CH. Folate turnover in chronically alcoholic monkeys. *J Lab Clin Med* 1983;101(4):623–8.
33. Romanoff RL, Ross DM, McMartin KE. Acute ethanol exposure inhibits renal folate transport, but repeated exposure upregulates folate transport proteins in rats and human cells. *J Nutr* 2007;137(5): 1260–5.
34. Ross DM, McMartin KE. Effect of ethanol on folate binding by isolated rat renal brush border membranes. *Alcohol* 1996;13(5): 449–54.
35. Hamid A, Kaur J. Decreased expression of transporters reduces folate uptake across renal absorptive surfaces in experimental alcoholism. *J Membr Biol* 2007;220(1–3):69–77.
36. Marcoullis G, Parmentier Y, Nicolas JP, Jimenez M, Gerard P. Cobalamin malabsorption due to nondegradation of R proteins in the human intestine. Inhibited cobalamin absorption in exocrine pancreatic dysfunction. *J Clin Invest* 1980;66(3):430–40.
37. Cravo ML, Camilo ME. Hyperhomocysteinemia in chronic alcoholism: relations to folic acid and vitamins B(6) and B(12) status. *Nutrition* 2000;16(4):296–302.
38. Medici V, Peerson JM, Stabler SP, French SW, Gregory III JF, Virata MC, et al. Impaired homocysteine transsulfuration is an indicator of alcoholic liver disease. *J Hepatol* 2010;53(3):551–7.
39. Kanazawa S, Herbert V. Total corrinoid, cobalamin (vitamin B12), and cobalamin analogue levels may be normal in serum despite cobalamin in liver depletion in patients with alcoholism. *Lab Invest* 1985;53(1):108–10.
40. Cylwik B, Czygier M, Daniluk M, Chrostek L, Szmitkowski M. Vitamin B12 concentration in the blood of alcoholics. *Pol Merkur Lekarski* 2010;28(164):122–5.
41. Lindenbaum J, Lieber CS. Alcohol-induced malabsorption of vitamin B12 in man. *Nature* 1969;224(5221):806.
42. Cravo ML, Gloria LM, Selhub J, Nadeau MR, Camilo ME, Resende MP, et al. Hyperhomocysteinemia in chronic alcoholism: correlation with folate, vitamin B-12, and vitamin B-6 status. *Am J Clin Nutr* 1996;63(2):220–4.
43. Labadarios D, Rossouw JE, McConnell JB, Davis M, Williams R. Vitamin B6 deficiency in chronic liver disease--evidence for increased degradation of pyridoxal-5′-phosphate. *Gut* 1977;18(1):23–7.
44. Diehl AM, Potter J, Boitnott J, Van Duyn MA, Herlong HF, Mezey E. Relationship between pyridoxal 5′-phosphate deficiency and aminotransferase levels in alcoholic hepatitis. *Gastroenterology* 1984;86(4):632–6.
45. Lumeng L. The role of acetaldehyde in mediating the deleterious effect of ethanol on pyridoxal 5′-phosphate metabolism. *J Clin Invest* 1978;62(2):286–93.
46. Orman ES, Odena G, Bataller R. Alcoholic liver disease: Pathogenesis, management, and novel targets for therapy. *J Gastroenterol Hepatol* 2013;28 (Suppl. 1):77–84.
47. Halsted CH. Nutrition and alcoholic liver disease. *Semin Liver Dis* 2004;24(3):289–304.
48. Ramani K, Yang H, Kuhlenkamp J, Tomasi L, Tsukamoto H, Mato JM, et al. Changes in the expression of methionine adenosyltransferase genes and S-adenosylmethionine homeostasis during hepatic stellate cell activation. *Hepatology* 2010;51(3):986–95.
49. Clarke S, Banfield K. S-adenosylmethionine-dependent methyltransferases. In: Carmel R, Jacobsen D, editors. *Homocysteine in health and disease*. Cambridge: Cambridge University Press; 2001. p. 63–78.
50. Mato JM, Lu SC. Role of S-adenosyl-L-methionine in liver health and injury. *Hepatology* 2007;45(5):1306–12.
51. Hultberg B, Berglund M, Andersson A, Frank A. Elevated plasma homocysteine in alcoholics. *Alcohol Clin Exp Res* 1993;17(3):687–9.
52. Villanueva JA, Halsted CH. Hepatic transmethylation reactions in micropigs with alcoholic liver disease. *Hepatology* 2004;39(5): 1303–10.
53. Trimble KC, Molloy AM, Scott JM, Weir DG. The effect of ethanol on one-carbon metabolism: increased methionine catabolism and lipotrope methyl-group wastage. *Hepatology* 1993;18(4):984–9.
54. Barak AJ, Beckenhauer HC, Mailliard ME, Kharbanda KK, Tuma DJ. Betaine lowers elevated s-adenosylhomocysteine levels in hepatocytes from ethanol-fed rats. *J Nutr* 2003;133(9):2845–8.
55. Kharbanda KK, Rogers DD, 2nd, Mailliard ME, Siford GL, Barak AJ, Beckenhauer HC, et al. A comparison of the effects of betaine and S-adenosylmethionine on ethanol-induced changes in methionine metabolism and steatosis in rat hepatocytes. *J Nutr* 2005;135(3): 519–24.
56. Kharbanda KK, Rogers II DD, Mailliard ME, Siford GL, Barak AJ, Beckenhauer HC, et al. Role of elevated S-adenosylhomocysteine in rat hepatocyte apoptosis: protection by betaine. *Biochem Pharmacol* 2005;70(12):1883–90.
57. Avila MA, Berasain C, Torres L, Martin-Duce A, Corrales FJ, Yang H, et al. Reduced mRNA abundance of the main enzymes involved in methionine metabolism in human liver cirrhosis and hepatocellular carcinoma. *J Hepatol* 2000;33(6):907–14.
58. Lee TD, Sadda MR, Mendler MH, Bottiglieri T, Kanel G, Mato JM, et al. Abnormal hepatic methionine and glutathione metabolism in patients with alcoholic hepatitis. *Alcohol Clin Exp Res* 2004;28(1): 173–81.
59. Lieber CS, Casini A, DeCarli LM, Kim CI, Lowe N, Sasaki R, et al. S-adenosyl-L-methionine attenuates alcohol-induced liver injury in the baboon. *Hepatology* 1990;11(2):165–72.

60. Rehm J, Shield KD. Global alcohol-attributable deaths from cancer, liver cirrhosis, and injury in 2010. *Alcohol Res* 2013;35(2):174–83.
61. Rehm J, Samokhvalov AV, Shield KD. Global burden of alcoholic liver diseases. *J Hepatol* 2013;59(1):160–8.
62. Testino G, Burra P, Bonino F, Piani F, Sumberaz A, Peressutti R, et al. Acute alcoholic hepatitis, end stage alcoholic liver disease and liver transplantation: An Italian position statement. *World J Gastroenterol* 2014;20(40):14642–51.
63. Becker U, Deis A, Sorensen TI, Gronbaek M, Borch-Johnsen K, Muller CF, et al. Prediction of risk of liver disease by alcohol intake, sex, and age: a prospective population study. *Hepatology* 1996;23(5):1025–9.
64. Lefkowitch JH. Morphology of alcoholic liver disease. *Clin Liver Dis* 2005;9(1):37–53.
65. O'Shea RS, Dasarathy S, McCullough AJ. Alcoholic liver disease. *Hepatology* 2010;51(1):307–28.
66. Rao RK. Acetaldehyde-induced barrier disruption and paracellular permeability in Caco-2 cell monolayer. *Methods Mol Biol* 2008;447:171–83.
67. Hill DB, Barve S, Joshi-Barve S, McClain C. Increased monocyte nuclear factor-kappaB activation and tumor necrosis factor production in alcoholic hepatitis. *J Lab Clin Med* 2000;135(5):387–95.
68. Kawai T, Akira S. TLR signaling. *Semin Immunol* 2007;19(1):24–32.
69. Hritz I, Mandrekar P, Velayudham A, Catalano D, Dolganiuc A, Kodys K, et al. The critical role of toll-like receptor (TLR) 4 in alcoholic liver disease is independent of the common TLR adapter MyD88. *Hepatology* 2008;48(4):1224–31.
70. Endo M, Masaki T, Seike M, Yoshimatsu H. TNF-alpha induces hepatic steatosis in mice by enhancing gene expression of sterol regulatory element binding protein-1c (SREBP-1c). *Exp Biol Med (Maywood)* 2007;232(5):614–21.
71. Wagner M, Zollner G, Trauner M. Nuclear receptors in liver disease. *Hepatology* 2011;53(3):1023–34.
72. Liu Y, Brymora J, Zhang H, Smith B, Ramezani-Moghadam M, George J, et al. Leptin and acetaldehyde synergistically promotes alphaSMA expression in hepatic stellate cells by an interleukin 6-dependent mechanism. *Alcohol Clin Exp Res* 2011;35(5):921–8.
73. Smathers RL, Galligan JJ, Stewart BJ, Petersen DR. Overview of lipid peroxidation products and hepatic protein modification in alcoholic liver disease. *Chem Biol Interact* 2011;192(1–2):107–12.
74. Suh YG, Jeong WI. Hepatic stellate cells and innate immunity in alcoholic liver disease. *World J Gastroenterol* 2011;17(20):2543–51.
75. Urtasun R, Lopategi A, George J, Leung TM, Lu Y, Wang X, et al. Osteopontin, an oxidant stress sensitive cytokine, up-regulates collagen-I via integrin alpha(V)beta(3) engagement and PI3K/pAkt/NFkappaB signaling. *Hepatology* 2012;55(2):594–608.
76. Mathews S, Gao B. Therapeutic potential of interleukin 1 inhibitors in the treatment of alcoholic liver disease. *Hepatology* 2013;57(5):2078–80.
77. Song BJ, Abdelmegeed MA, Henderson LE, Yoo SH, Wan J, Purohit V, et al. Increased nitroxidative stress promotes mitochondrial dysfunction in alcoholic and nonalcoholic fatty liver disease. *Oxid Med Cell Longev* 2013;2013:781050.
78. Bailey SM, Robinson G, Pinner A, Chamlee L, Ulasova E, Pompilius M, et al. S-adenosylmethionine prevents chronic alcohol-induced mitochondrial dysfunction in the rat liver. *Am J Physiol Gastrointest Liver Physiol* 2006;291(5):G857–67.
79. Veal N, Hsieh CL, Xiong S, Mato JM, Lu S, Tsukamoto H. Inhibition of lipopolysaccharide-stimulated TNF-alpha promoter activity by S-adenosylmethionine and 5′-methylthioadenosine. *Am J Physiol Gastrointest Liver Physiol* 2004;287(2):G352–62.
80. Gobejishvili L, Avila DV, Barker DF, Ghare S, Henderson D, Brock GN, et al. S-adenosylmethionine decreases lipopolysaccharide-induced phosphodiesterase 4B2 and attenuates tumor necrosis factor expression via cAMP/protein kinase A pathway. *J Pharmacol Exp Ther* 2011;337(2):433–43.
81. Gobejishvili L, Barve S, Joshi-Barve S, McClain C. Enhanced PDE4B expression augments LPS-inducible TNF expression in ethanol-primed monocytes: relevance to alcoholic liver disease. *Am J Physiol Gastrointest Liver Physiol* 2008;295(4):G718–24.
82. Song Z, Zhou Z, Uriarte S, Wang L, Kang YJ, Chen T, et al. S-adenosylhomocysteine sensitizes to TNF-alpha hepatotoxicity in mice and liver cells: a possible etiological factor in alcoholic liver disease. *Hepatology* 2004;40(4):989–97.
83. Song Z, Zhou Z, Song M, Uriarte S, Chen T, Deaciuc I, et al. Alcohol-induced S-adenosylhomocysteine accumulation in the liver sensitizes to TNF hepatotoxicity: possible involvement of mitochondrial S-adenosylmethionine transport. *Biochem Pharmacol* 2007;74(3):521–31.
84. Kharbanda KK, Bardag-Gorce F, Barve S, Molina PE, Osna NA. Impact of altered methylation in cytokine signaling and proteasome function in alcohol and viral-mediated diseases. *Alcohol Clin Exp Res* 2013;37(1):1–7.
85. Bardag-Gorce F, French BA, Nan L, Song H, Nguyen SK, Yong H, et al. CYP2E1 induced by ethanol causes oxidative stress, proteasome inhibition and cytokeratin aggresome (Mallory body-like) formation. *Exp Mol Pathol* 2006;81(3):191–201.
86. Osna NA, White RL, Donohue Jr TM, Beard MR, Tuma DJ, Kharbanda KK. Impaired methylation as a novel mechanism for proteasome suppression in liver cells. *Biochem Biophys Res Commun* 2010;391(2):1291–6.
87. Bardag-Gorce F, Oliva J, Li J, French BA, French SW. SAMe prevents the induction of the immunoproteasome and preserves the 26S proteasome in the DDC-induced MDB mouse model. *Exp Mol Pathol* 2010;88(3):353–62.
88. Oliva J, Dedes J, Li J, French SW, Bardag-Gorce F. Epigenetics of proteasome inhibition in the liver of rats fed ethanol chronically. *World J Gastroenterol* 2009;15(6):705–12.
89. Kharbanda KK, Mailliard ME, Baldwin CR, Beckenhauer HC, Sorrell MF, Tuma DJ. Betaine attenuates alcoholic steatosis by restoring phosphatidylcholine generation via the phosphatidylethanolamine methyltransferase pathway. *J Hepatol* 2007;46(2):314–21.
90. Kharbanda KK, Todero SL, Ward BW, Cannella III JJ, Tuma DJ. Betaine administration corrects ethanol-induced defective VLDL secretion. *Mol Cell Biochem* 2009;327(1–2):75–8.
91. Cano A, Buque X, Martinez-Una M, Aurrekoetxea I, Menor A, Garcia-Rodriguez JL, et al. Methionine adenosyltransferase 1A gene deletion disrupts hepatic very low-density lipoprotein assembly in mice. *Hepatology* 2011;54(6):1975–86.
92. Kharbanda KK, Todero SL, Moats JC, Harris RM, Osna NA, Thomes PG, et al. Alcohol consumption decreases rat hepatic creatine biosynthesis via altered guanidinoacetate methyltransferase activity. *Alcohol Clin Exp Res* 2014;38(3):641–8.
93. Kharbanda KK, Todero SL, Thomes PG, Orlicky DJ, Osna NA, French SW, et al. Increased methylation demand exacerbates ethanol-induced liver injury. *Exp Mol Pathol* 2014;97(1):49–56.
94. Yi HS, Lee YS, Byun JS, Seo W, Jeong JM, Park O, et al. Alcohol dehydrogenase III exacerbates liver fibrosis by enhancing stellate cell activation and suppressing natural killer cells in mice. *Hepatology* 2014;60(3):1044–53.
95. Nieto N, Cederbaum AI. S-adenosylmethionine blocks collagen I production by preventing transforming growth factor-beta induction of the COL1A2 promoter. *J Biol Chem* 2005;280(35):30963–74.
96. Martinez-Lopez N, Varela-Rey M, Ariz U, Embade N, Vazquez-Chantada M, Fernandez-Ramos D, et al. S-adenosylmethionine and proliferation: new pathways, new targets. *Biochem Soc Trans* 2008;36(Pt. 5):848–52.

97. Li J, Ramani K, Sun Z, Zee C, Grant EG, Yang H, et al. Forced expression of methionine adenosyltransferase 1A in human hepatoma cells suppresses *in vivo* tumorigenicity in mice. *Am J Pathol* 2010;176(5):2456–66.
98. Lu SC, Ramani K, Ou X, Lin M, Yu V, Ko K, et al. S-adenosylmethionine in the chemoprevention and treatment of hepatocellular carcinoma in a rat model. *Hepatology* 2009;**50**(2):462–71.
99. Martinez-Chantar ML, Vazquez-Chantada M, Ariz U, Martinez N, Varela M, Luka Z, et al. Loss of the glycine N-methyltransferase gene leads to steatosis and hepatocellular carcinoma in mice. *Hepatology* 2008;**47**(4):1191–9.
100. Martinez-Una M, Varela-Rey M, Cano A, Fernandez-Ares L, Beraza N, Aurrekoetxea I, et al. Excess S-adenosylmethionine reroutes phosphatidylethanolamine towards phosphatidylcholine and triglyceride synthesis. *Hepatology* 2013;**58**(4):1296–305.
101. Halsted CH, Villanueva J, Chandler CJ, Ruebner B, Munn RJ, Parkkila S, et al. Centrilobular distribution of acetaldehyde and collagen in the ethanol- fed micropig. *Hepatology* 1993;**18**(4):954–60.
102. Halsted CH, Villanueva J, Chandler CJ, Stabler SP, Allen RH, Muskhelishvili L, et al. Ethanol feeding of micropigs alters methionine metabolism and increases hepatocellular apoptosis and proliferation. *Hepatology* 1996;**23**(3):497–505.
103. Halsted CH, Villanueva JA, Devlin AM, Niemela O, Parkkila S, Garrow TA, et al. Folate deficiency disturbs hepatic methionine metabolism and promotes liver injury in the ethanol-fed micropig. *Proc Natl Acad Sci USA* 2002;**99**(15):10072–7.
104. Lieber CS, DeCarli LM. An experimental model of alcohol feeding and liver injury in the baboon. *J Med Primatol* 1974;**3**(3):153–63.
105. Esfandiari F, Villanueva JA, Wong DH, French SW, Halsted CH. Chronic ethanol feeding and folate deficiency activate hepatic endoplasmic reticulum stress pathway in micropigs. *Am J Physiol Gastrointest Liver Physiol* 2005;**289**(1):G54–63.
106. Esfandiari F, You M, Villanueva JA, Wong DH, French SW, Halsted CH. S-adenosylmethionine attenuates hepatic lipid synthesis in micropigs fed ethanol with a folate-deficient diet. *Alcohol Clin Exp Res* 2007;**31**(7):1231–9.
107. Villanueva JA, Esfandiari F, White ME, Devaraj S, French SW, Halsted CH. S-Adenosylmethionine Attenuates Oxidative Liver Injury in Micropigs Fed Ethanol With a Folate-Deficient Diet. *Alcohol Clin Exp Res* 2007;**31**:1934–43.
108. Ji C, Kaplowitz N. Betaine decreases hyperhomocysteinemia, endoplasmic reticulum stress, and liver injury in alcohol-fed mice. *Gastroenterology* 2003;**124**(5):1488–99.
109. Vendemiale G, Altomare E, Trizio T, Le Grazie C, Di Padova C, Salerno MT, et al. Effects of oral S-adenosyl-L-methionine on hepatic glutathione in patients with liver disease. *Scand J Gastroenterol* 1989;**24**(4):407–15.
110. Mato JM, Camara J, Fernandez de Paz J, Caballeria L, Coll S, Caballero A, et al. S-adenosylmethionine in alcoholic liver cirrhosis: a randomized, placebo-controlled, double-blind, multicenter clinical trial. *J Hepatol* 1999;**30**(6):1081–9.
111. Medici V, Virata MC, Peerson JM, Stabler SP, French SW, Gregory III JF, et al. S-adenosyl-L-methionine treatment for alcoholic liver disease: a double-blinded, randomized, placebo-controlled trial. *Alcohol Clin Exp Res* 2011;**35**(11):1960–5.
112. Le MD, Enbom E, Traum PK, Medici V, Halsted CH, French SW. Alcoholic liver disease patients treated with S-adenosyl-L-methionine: an in-depth look at liver morphologic data comparing pre and post treatment liver biopsies. *Exp Mol Pathol* 2013;**95**(2):187–91.
113. Mann DA. Epigenetics in liver disease. *Hepatology* 2014;**60**(4):1418–25.
114. Mandrekar P. Epigenetic regulation in alcoholic liver disease. *World J Gastroenterol* 2011;**17**(20):2456–64.
115. Pal-Bhadra M, Bhadra U, Jackson DE, Mamatha L, Park PH, Shukla SD. Distinct methylation patterns in histone H3 at Lys-4 and Lys-9 correlate with up- & down-regulation of genes by ethanol in hepatocytes. *Life Sci* 2007;**81**(12):979–87.
116. Ara AI, Xia M, Ramani K, Mato JM, Lu SC. S-adenosylmethionine inhibits lipopolysaccharide-induced gene expression via modulation of histone methylation. *Hepatology* 2008;**47**(5):1655–66.
117. Bardag-Gorce F, Li J, Oliva J, Lu SC, French BA, French SW. The cyclic pattern of blood alcohol levels during continuous ethanol feeding in rats: the effect of feeding S-adenosylmethionine. *Exp Mol Pathol* 2010;**88**(3):380–7.
118. Lu SC, Huang ZZ, Yang H, Mato JM, Avila MA, Tsukamoto H. Changes in methionine adenosyltransferase and S-adenosylmethionine homeostasis in alcoholic rat liver. *Am J Physiol Gastrointest Liver Physiol* 2000;**279**(1):G178–85.
119. Martínez-Chantar ML, Corrales FJ, Martínez-Cruz LA, García-Trevijano ER, Huang ZZ, Chen L, Kanel G, Avila MA, Mato JM, Lu SC. Spontaneous oxidative stress and liver tumors in mice lacking methionine adenosyltransferase 1A. *FASEB J* 2002;**16**(10):1292–4.
120. Esfandiari F, Medici V, Wong DH, Jose S, Dolatshahi M, Quinlivan E, et al. Epigenetic regulation of hepatic endoplasmic reticulum stress pathways in the ethanol-fed cystathionine beta synthase-deficient mouse. *Hepatology* 2010;**51**(3):932–41.
121. Medici V, Schroeder DI, Woods R, LaSalle JM, Geng Y, Shibata NM, et al. Methylation and gene expression responses to ethanol feeding and betaine supplementation in the cystathionine beta synthase-deficient mouse. *Alcohol Clin Exp Res* 2014;**38**(6):1540–9.
122. Watanabe M, Osada J, Aratani Y, Kluckman K, Reddick R, Malinow MR, et al. Mice deficient in cystathionine beta-synthase: animal models for mild and severe homocyst(e)inemia. *Proc Natl Acad Sci USA* 1995;**92**(5):1585–9.
123. Yuan GJ, Zhou XR, Gong ZJ, Zhang P, Sun XM, Zheng SH. Expression and activity of inducible nitric oxide synthase and endothelial nitric oxide synthase correlate with ethanol-induced liver injury. *World J Gastroenterol* 2006;**12**(15):2375–81.
124. Venkatraman A, Shiva S, Wigley A, Ulasova E, Chhieng D, Bailey SM, et al. The role of iNOS in alcohol-dependent hepatotoxicity and mitochondrial dysfunction in mice. *Hepatology* 2004;**40**(3):565–73.
125. Avila MA, Mingorance J, Martinez-Chantar ML, Casado M, Martin-Sanz P, Bosca L, et al. Regulation of rat liver S-adenosylmethionine synthetase during septic shock: role of nitric oxide. *Hepatology* 1997;**25**(2):391–6.

CHAPTER

9

Interactions Vitamin D – Bone Changes in Alcoholics

Emilio González-Reimers, PhD, Francisco Santolaria-Fernández, MD, Geraldine Quintero-Platt, MD, Antonio Martínez-Riera, MD

Servicio de Medicina Interna, Hospital Universitario de Canarias, Universidad de La Laguna, Tenerife, Canary Islands, Spain

INTRODUCTION

Bone alterations among alcoholics have been described for many years now. Saville[1] stated that the skeleton of alcoholics showed changes similar to aged individuals, thus describing a reduced bone mass in relation to age. In the forthcoming decades, it became progressively clear that ethanol itself may disrupt bone metabolism,[2,3] but that there are many other alterations in the alcoholic patient that also contribute to changes in bone structure, and the development of bone disease.

Decreased bone mass results from decreased bone synthesis, and/or increased bone resorption. Both mechanisms are involved in heavy alcoholism. Many excellent original works and reviews have been published on this subject,[4–7] so we will review in this work only those aspects of bone disease in alcoholics related with vitamin D deficiency. Also, in chronic alcoholics, several associated endocrine alterations, such as deranged metabolism of the gonadal axis, corticosteroids, IGF-1, and estrogens, as well as malnutrition, altered nutrient intake, and increased cytokine secretion, are all involved in skeletal alterations.[8]

One of the important hormones involved in bone homeostasis is vitamin D. Low levels of vitamin D have been reported for alcoholics for many years now. Moreover, it has become clear in recent times that the consequences of vitamin D deficiency lay far beyond the direct actions on bone (Figure 9.1). In this chapter we will review some of the mechanism – both skeletal and extraskeletal – by which vitamin D deficiency may contribute to bone alterations among alcoholics.

VITAMIN D METABOLISM

Vitamin D, in its active form, is one of the most important factors involved in bone homeostasis, mainly through its action on serum calcium levels. This hormone results from the synthesis, by liver and kidney, of an active compound: 1,25 $(OH)_2$ vitamin D, formed after liver 25-hydroxylation of vitamin D and posterior 1-α hydroxylation of 25 hydroxivitamin D by the kidneys (and many other tissues), a metabolic pathway disentangled many years ago.

Vitamin D is a liposoluble vitamin. Dietary sources are, especially, fish oils, and, to a lesser extent, milk and some cereals. However, for many years now, as recorded by classic reviews,[9] it is well known that most vitamin D is synthesized in the active part of the epidermis, from 7 dehydrocholesterol, whose double bonds between C5 and C6, and another between C7 and C8, absorb 290–315 ultraviolet light. This opens the B ring of the cyclopentane perhydrophenanthrene moiety, forming 9–10 secosterol (previtamin D3) that spontaneously (thermically) isomerizes into 9–10 secosterol (vitamin D3), a process that takes place at body temperature, within 3 days.[10] Further exposition to ultraviolet light induces transformation of this compound into biologically inactive products – mainly lumisterol – thus preventing excessive skin synthesis of vitamin D in sun-exposed skin. Melanin competes with 7-dehydrocholesterol for ultraviolet light, so vitamin D synthesis is less intense in dark-colored, rather than in light-colored skin.[11] Latitude limits the effect, converting food, a unique source of vitamin D in winter months[12] in high latitude areas,

Molecular Aspects of Alcohol and Nutrition. http://dx.doi.org/10.1016/B978-0-12-800773-0.00009-4

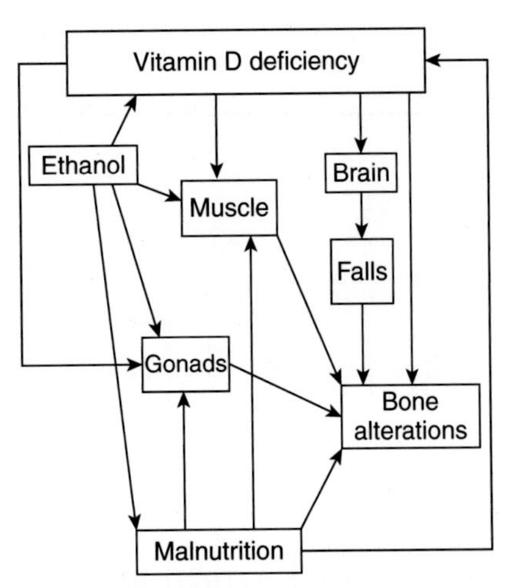

FIGURE 9.1 Vitamin D deficiency exerts direct effects on bone, but also indirect ones. The main extraskeletal alterations, which may influence on bone changes include muscle disease, brain alterations, and hypogonadism. Both ethanol and malnutrition also contribute to all these changes.

since ozone absorbs a great quantity of UV radiation. In temperate zones, no vitamin D synthesis in the skin takes place from November to February, due to the fact that the more oblique sunrays are obliged to traverse a thicker UV absorbing ozone band.[13] Over 70 years age, skin synthesis of vitamin D decreases. McLaughlin and Holick,[14] in skin pieces of 6.2 cm^2, observed that skin weight decreased with age (skin weight of individuals 77–88 years old is about 70–75% of that of younger age), as well as vitamin D synthesis (only 35–45% of that synthesized by young individuals). Later studies have also shown a direct correlation between skinfold thickness (measured on the back of the hand), and serum 25 OH D among 433 women, also finding a decline in skinfold thickness among the oldest ones.[15] There is no apparent effect of age on vitamin D intestinal absorption,[16] although others do report an inhibition of calcium absorption with age.[17]

Either absorbed or skin-synthesized, vitamin D is transported to the liver by a specific carrier (vitamin D-binding globulin, an α-1 globulin). In the liver, it is transformed into 25-OH D3, where it exerts a nonperfect feedback loop effect,[18] so that increased dietary absorption or synthesis leads to raised vitamin D levels.[19]

The active form, 1,25 dihydroxyvitamin D, is synthesized in the kidney, under the action of 1-α hydroxylase.[20] Hypocalcemia, possibly through PTH, and/or phosphate, activate 1-α hydroxylase, whereas higher levels of 1,25 dihydroxyvitamin D inhibit it. In the kidney, there is also a 24-hydroxylase, so that it is possible that some amounts of 24,25-dihydroxyvitamin D are also synthesized. Fibroblast growth factor (FGF) 23 and α-Klotho suppress the expression of extrarenal[21] and renal 1-α hydroxylase, and enhances 24-α hydroxylase.[22] This hormone is less active than the 1,25 isoform.[23] Age decreases the rate of synthesis of 1,25,[24] increasing the ratio 24,25 to 1,25 dihydroxyvitamin D. Expression of 1-α hydroxylase is widespread,[25] including tumor tissue.[26]

The active form of vitamin D promotes intestinal calcium absorption. In addition to passive absorption of calcium via paracellular pathway (tight junctions), active transcellular absorption is enhanced by vitamin D. This process involves three steps: calcium entry from the intestinal lumen to the apical pole of the enterocyte, cytoplasmic traslocation, bound to calbindin, to the baso-lateral membrane, and extrusion by an active pump system. Active uptake is mediated by the receptors transient receptor potential vanilloid (TRPV) type 5 and type 6, although there is some controversy regarding this[27]; diffusion thorough cytoplasm is facilitated by calbindin, as is extrusion, by two mechanisms: an ATP-dependent Ca^{2+}-ATPase (PMCA1b), and a Na^+/Ca^{2+} exchanger (NCX1). Vitamin D increases the expression of all these proteins involved in active calcium absorption.[28] This induction allows a more efficient calcium absorption (from 10% to 15% of dietary intake to 30–40%.[29] Interestingly, prolactin and estrogens also activate TRPV6 receptor[30]; glucocorticoids inhibit it. Prolactin also enhances expression of 1-α hydroxylase.[31]

VITAMIN D AMONG ALCOHOLICS

Several reports have described vitamin D deficiency among alcoholics. Naude et al. among 81 South African children aged 12–16 with a binge drinking consumption pattern, found that 48.8% of them showed vitamin D deficiency (25 hydroxyvitamin D levels below 20 ng/mL), and 88.8% showed vitamin D insufficiency (below 30 ng/mL), significantly more than among nonalcohol abusers.[32] In Nepal, vitamin deficiency (25 hydroxyvitamin D <50 nmol/L) reached figures of 64% among 174 alcoholic patients.[33] Sobral-Oliveira et al. found low mean values (23.3 ± 7.9 ng/mL) among 60 alcoholic patients without pancreatitis, vitamin D levels showing an inverse correlation with ethanol intake.[34] Among 80 adult Danish patients, Wilkens-Knudsen et al. also report 64% of alcoholic patients with vitamin D insufficiency (25-OH-vit D <50 nmol/L).[35] Median values of 24 ng/mL were also reported for 90 alcoholic patients, significantly lower than in controls. In that study, alcoholic patients with fracture showed even lower values.[36] Diamond et al.[3] and Malik et al.[37] also found lower 25 OH vitamin D levels in alcoholics, than in controls. In contrast, Santori et al.[8] and Kim et al.[38] did not find differences between alcoholics and controls: serum values were lower, but not significantly different; in 1985, Wilkinson et al. failed to find differences in serum vitamin D between

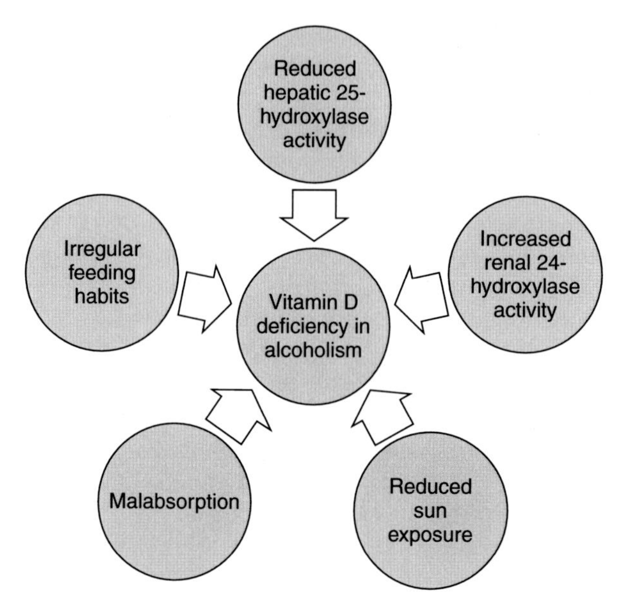

FIGURE 9.2 **Main mechanisms involved in the pathogenesis of vitamin D deficiency in alcoholics.**

alcoholic patients with, and without, fractures.[39] Bikle et al. also reported, in 1985, normal vitamin D values among alcoholics.[40]

There are several factors that may lead to low 25 vitamin D levels among alcoholics, including age, ethnicity, gender, smoking, dietary habits, and latitude and seasonal effect, in addition to some specific alterations (Figure 9.2). Given the malabsorption associated with portal hypertension, it is logical that cirrhotic patients show even lower vitamin D levels. Altered biliary secretion and coexisting pancreatic insufficiency may cause impaired absorption of the liposoluble vitamin D precursor, and contribute to low levels in heavy alcoholics. In a classic study performed in 1986,[41] seven out of 12 noncirrhotic alcoholics who consumed a similar amount of vitamin D than controls showed serum 25 hydroxyvitamin D values below 20 mg/mL. A similar result was observed by Laitinen et al.: in a study on 38 alcoholic patients aged 30–55 years, despite similar vitamin D consumption, patients showed significantly lower vitamin D 1,25 and 24,25 (dihydroxyvitamin D) than controls.[42] Both studies suggest deranged absorption and/or metabolism of vitamin D in alcoholics. Poor intake or reduced sun exposure, in relation with bizarre lifestyle and behavior of these patients may aggravate the scenario. In 1990, Pun et al. in a study on 69 (nonalcoholic) patients with femoral neck fracture report lower levels of vitamin D among them, when compared with nonfractured individuals.[43] The coexistence of low vitamin D and hip fracture was associated with lifestyle characteristics, and reduced sun exposure. However, in a study on 120 alcoholic patients we found that there is a trend to lower vitamin D levels among those with impaired social environment (separated, without a job), but without statistical significance.

Some data also support the hypothesis of a direct effect of ethanol on the metabolism of vitamin D precursors: it was shown that chronic ethanol consumption alters renal metabolism of 25 OH D3, inducing the synthesis of 24 hydroxylase, leading to the formation of the inactive metabolite 24-25 (OH)2 D3.[44]

VITAMIN D DEFICIENCY AND BONE METABOLISM

The decreased vitamin D levels observed in alcoholics may have several effects on bone health, both by direct effects on the skeleton, and by indirect ones.

Direct Effects on Bone

Bone remodeling is a process in which bone resorption ensues, providing calcium, phosphorus, or carbonate and bicarbonate salts to the extracellular medium, a pathway that is coupled to bone synthesis, so that total bone mass is preserved. Bone resorption is carried out by osteoclasts, whereas bone synthesis depends on osteoblast activity. Bone formation includes synthesis of a collagen matrix, an osteoid, and further deposition of hydroxyapatite crystals on this osteoid. An inbalance between bone synthesis and bone resorption leads to osteoporosis, characterized by decreased bone mass. A situation in which osteoid forms but is not mineralized is called osteomalacia (or rachitis in infancy), histologically defined by an increase in unmineralized osteoid.

In situations characterized by vitamin D deficiency, calcium absorption is impaired, and hypocalcemia ensues. This leads to activation of the calcium sensing receptor of chief parathyroideal cells that induce increased secretion of parathyroid hormone (PTH) that, in turn, increases bone resorption by osteoclasts, leading to decreased bone mass. Therefore, osteoporosis is a major consequence of vitamin D deficiency.

Although the classical paradigmatic disease associated with vitamin D deficiency is osteomalacia, it has been shown that osteomalacia is rarely seen today. It only occurs only when both dietary calcium and vitamin D levels are severely reduced; in fact, if dietary calcium is adequate osteomalacia does not ensue even if vitamin D levels are low.[45] This observation led to put into question whether vitamin D exerts or not a direct effect on osteoblasts, a matter which has been largely debated. Adding confusion to this item, and making research more difficult, human beings and rats respond differently than mice in relation to the effects of 1,25 dihydroxyvitamin D on osteoblasts.[46] However, all the major bone cells show vitamin D receptors and, moreover, 1-α hydroxylase activity.[45]

It is out of doubt that vitamin D receptor is present in osteoblasts, and its activity is modulated by vitamin D, PTH, glucocorticoids, epidermal growth factor, and transforming growth factor β. Vitamin D promotes osteoblast growth and differentiation, and exerts a positive effect on bone mineralization, not only supplying calcium, but also preparing the osteoblast for mineralization before this process begins. It stimulates osteocalcin (in human), and alkaline phosphatase synthesis.[47] It also seems that vitamin D exerts a protective effect against excessive mineralization. Activin A is a member of the superfamily of transforming growth factor β, with inhibitory effects on bone mineralization.[48] Its expression in human osteoblasts is enhanced by 1,25 dihydroxyvitamin D.[49]

Vitamin D also induces bone resorption, an effect observed more than 60 years ago and confirmed later,[50] although, paradoxically, the beneficial (therapeutic) effects of vitamin D on bone are due to an inhibition of osteoclastogenesis.[51] *In vitro*, it was shown that vitamin D is one of the most potent factors that enhance the expression of RANKL on the surface of osteoblasts. This RANKL (formerly called osteoclast differentiation factor) plays a major role in the generation and differentiation of osteoclasts from splenocytes. In this sense, vitamin D acts synergistically with other compounds, such as PTH, prostaglandin E2, and IL-11, promoting osteoclast activation,[51] and secretion of cytokines involved in the crosstalk with osteoblasts and bone marrow stromal cells. However, this effect was not reproduced in monocytes.[52]

This bone-resorbing effect of vitamin D is in sharp contrast with the usefulness of vitamin D/vitamin D analogs in treating patients with osteoporosis, a therapeutic paradox that underlying mechanisms have not been elucidated.[51] From a finalist point of view, it has been interpreted as a result of the primordial effect of vitamin D on serum calcium: if calcium intestinal absorption is not sufficient to maintain normal calcium levels, vitamin D, acting synergistically with PTH, increases renal tubular calcium reabsorption and activates osteoclast.[30]

Impaired calcium absorption in alcoholics may also occur despite adequate vitamin D levels.[53] Either related or not to vitamin D deficiency, decreased calcium levels activate parathyroid glands, leading to increased parathyroid hormone (PTH) levels that, in turn, activate osteoclasts, leading to reduced bone mass. Consequently with this, the main lesion observed among alcoholics is osteoporosis.[54] Osteomalacia, in fact, has been only rarely described in some predominantly cholestatic forms of liver injury.[55]

Ethanol triggers an increased gut permeability to both Gram+ and Gram– bacteria[56] that reach the liver and activate Kupffer cells, after binding to toll-like receptors (TLR), mainly TLR-4.[57,58] Activated Kupffer cells secrete proinflammatory cytokines, especially TNF-α, and also induce the formation of reactive oxygen species (ROS), also derived from ethanol metabolism by other pathways.[59] Therefore, alcoholics show increased TNF-α and increased ROS production. Both compounds may be important in direct effects of ethanol on osteoblast differentiation.

As commented, osteoporosis is defined by a decrease in bone mass, due either to a decreased synthesis or enhanced bone resorption. Ethanol directly affects bone synthesis,[3] logically reflected in decreased serum levels of osteocalcin.[7] Giuliani et al. described a direct toxic effect of ethanol and, especially, acetaldehyde, on osteoblastogenesis.[61] More recent reports suggest that ethanol activates the TNF-α signaling pathway shifting bone marrow stem cells, a common precursor of both adipocytes and osteoblast, away from the osteogenic differentiation.[62] In addition, ethanol-mediated increased production of reactive oxygen species (ROS) may be involved in suppression of the Wnt1 β-catenin, also reducing osteogenesis.[63] Therefore, in alcoholics, negative effects of vitamin D on calcium homeostasis and, possibly, on osteoblast maturation and differentiation (possibly, also at the stage of mesenchymal precursors,[46] become aggravated by a direct effect of ethanol on osteoblast differentiation (and indirectly by increased circulating TNF and ROS generation).

Effects on Muscle

Vitamin D receptor in human muscle cells were definitively isolated about 15 years ago,[64] confirming similar previous findings in rats.[65] This receptor is a nuclear hormone receptor that acts in conjunction with coactivators, such as retinoid X receptor and steroid receptor coactivator 3, and other interacting proteins,[66] controlling the transcription of target genes.[67] Binding of this complex to 1,25 dihydroxyvitamin D modulates gene expression coding for a number of different proteins by binding to specific gene promoter regions, known as vitamin D response elements. In muscle, these genomic actions promote transcription of genes involved in muscle cell differentiation and proliferation, and also proteins related with calcium metabolism, and others, such as insulin-like growth factor binding protein 3 (IGFBP3), involved in skeletal growth.[67,68] In addition, there are non genomic effects related with the activation of membrane calcium channels.[66] Moreover, VDR polymorphisms are associated with differences in muscle strength.[69] There are also data that support the existence of 1 alpha hydroxylase in muscle cells, encoded by the gene *Cyp27b1*.[70]

Severe vitamin D deficiency causes muscle pain and proximal weakness, causing marked disability,[71] and is also accompanied by atrophy of muscle fibers, a phenomenon described already nearly 40 years ago.[72] This result is in accordance with an experimental study in which we documented a relation between vitamin D and fiber muscle area among alcohol treated rats, suggesting that, in alcohol experimental models, vitamin D deficiency may also alter muscle function and structure.[73]

Muscle activity is fully necessary for bone maintenance, probably acting via WNt β-catenin pathway, an activation that leads to osteoblast proliferation and increased bone formation.[74] This system is antagonized by sclerostin that binds to low-density lipoprotein receptor-related proteins 5 and 6 (LRP5 and LRP6), and inhibits osteoblast function, differentiation, and survival,[75] and, possibly, by connexin that promotes osteolysis triggered by unloading.[76]

In addition to reduced bone mass and increased prevalence of fracture, alcoholics usually show a chronic myopathy, defined by muscle fiber atrophy, especially type IIb fibers,[77] and, as mentioned before, vitamin D deficiency is highly prevalent among them. It is tempting to speculate that, perhaps, vitamin D deficiency does play a role in muscle atrophy observed in these individuals, and, indirectly, on reduced bone mass. Hickish et al. showed low vitamin D levels in alcoholics (both cirrhotics and noncirrhotics), and decreased muscle strength, but concluded that no relation existed between both phenomena.[78] In a preliminary analysis on 62 male heavy alcoholic patients, we have found a relation between vitamin D and handgrip strength (Figure 9.3), as well as a relation with reduced lean mass (Figure 9.4). Sorensen et al. in a classic study of 1979 treated 11 patients with age-related bone loss with 1-alpha-hydroxycholecalciferol and calcium, for 3–6 months. These patients also underwent muscle biopsy (*vastus lateralis*) before and after treatment. They found atrophy and reduction in the proportion of Type II-a fibers that recovered after treatment.[79]

To our knowledge, the possibility that alcoholic myopathy, triggered or aggravated by vitamin D deficiency, leads to osteopenia, remains speculative. Due to the fact that vitamin D affects magnesium and phosphorus metabolism, and that alterations in these elements may affect muscle function, it is still not clear that impaired muscle performance is entirely due to vitamin D deficiency *per se*. Hypogonadism may be another consequence of low vitamin D levels.[80] In addition to the possible influence of low vitamin D levels, a double defect involving gonads and hypothalamic-hypophyseal axis has been described in alcoholics.[81] Low testosterone may lead to muscle atrophy and decreased bone mass, and, theoretically, this mechanism could also be involved in vitamin D related osteopenia of the alcoholic, although it remains speculative.

Vitamin D Deficiency, Falls and Fractures

While it is a likely mechanistic possibility that the negative effect of vitamin D deficiency on muscle function is that it may also lead to decreased bone mass, it is more intuitive that altered muscle performance may also cause an increased rate of bone fractures due to falls.[82] In any case, vitamin D deficiency and bone fractures are related to each other: among heavy male alcoholics, vitamin D levels were lower among those with rib fractures (Figure 9.5). The relation of reduced vitamin D levels to

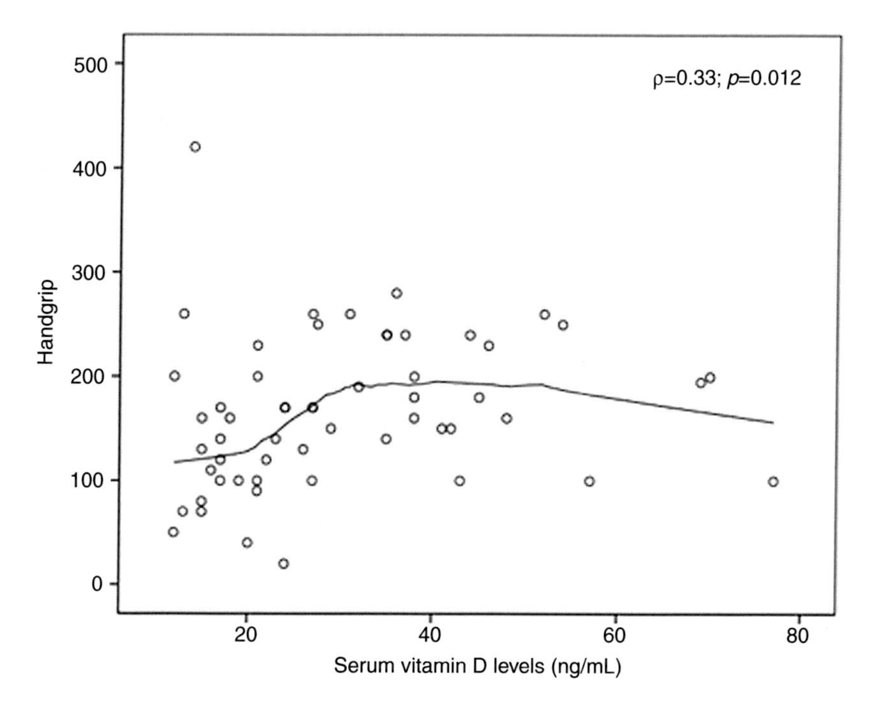

FIGURE 9.3 **Relationship between vitamin D levels and handgrip strength (in lbs) among heavy alcoholic men.**

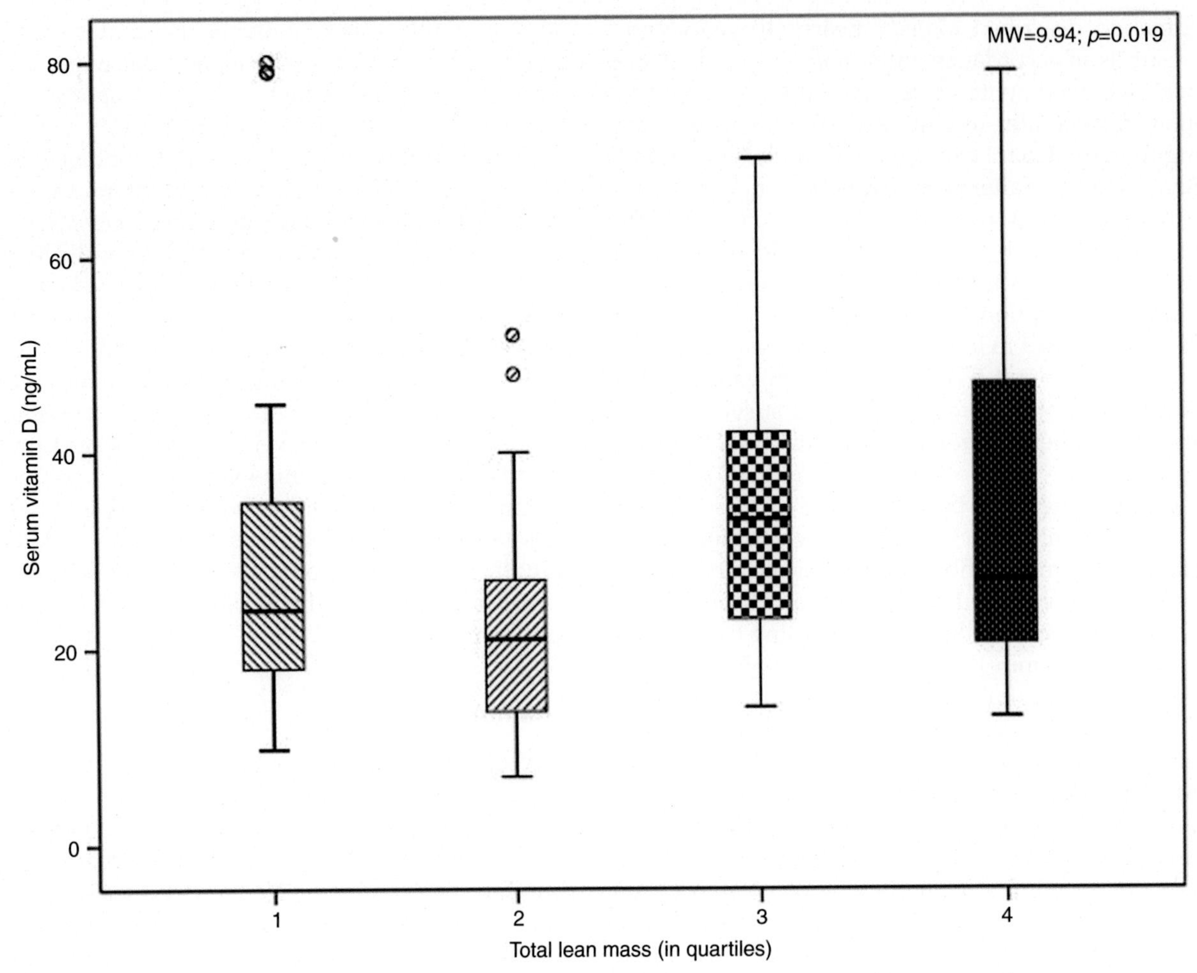

FIGURE 9.4 **Lower vitamin D levels are observed among alcoholic individuals with lowest total lean mass values.**

an increased falls rate has been analyzed in several studies. Dhesi et al. found vitamin D deficiency (<20 μg/L) in 72.5% that was severe (<12 μg/L) in 31.8% of 400 consecutive patients, aged 65–97 years, mostly women (73.3%), attended in a falls clinic.[83] Flicker et al. observed that serum vitamin D levels below 25 nmol/L were present in 22% in low-level care, and 45% in high-level care, out of 667 and 952 women, respectively, and that there was a relationship between vitamin D levels and the time to the first fall, even after adjusting for cognitive status, weight, previous Colles fracture, open-air activity, and psychotropic drug use.[84] Hazard ratio yields a result that implies a 20% reduction in the risk of falling with a doubling of serum vitamin D levels. Among 102 elderly patients, the prevalence of vitamin D insufficiency was 72.6%, and 27.5% showed frank deficiency. Deficiency was more common among those attending a falls clinic.[85] Reduced sunlight exposure may play a contributory role among elderly people (and among a certain subset of alcoholics), together with impaired intestinal vitamin D absorption, and impaired hydroxylation in the liver and kidneys.[86] Indeed, Prince et al. in a study performed in Perth (Australia), with enough available sun, found that treatment with ergocalciferol 1000 U/day was associated to a 19% reduction of the relative risk of falling, among elderly (70–90 years-old) women with a history of previous falling during the year prior to inclusion in the study.[87] This reduction was mainly accounted for falls taking place in winter and spring. In contrast, some studies failed to find a reduction in falls with vitamin supplementation, such as that of Burleigh et al. in which 255 geriatric patients did not benefit from 800 U vitamin D + 1200 mg calcium supplements, compared with those who received calcium alone.[88]

Although most of the studies have been performed on an elderly population, it is probable that falls also play a role in alcoholics, although results are conflicting. In a cross-sectional study involving nearly 6000 individuals, aged 65 or more, it was found that those who drank more than two drinks per day (self reported) had a higher risk of two or more falls, compared with abstainers, although they had higher femoral neck and total spine bone mineral density.[89] While populational studies are subjected to the effect of many confusion factors, alcohol-related poly-

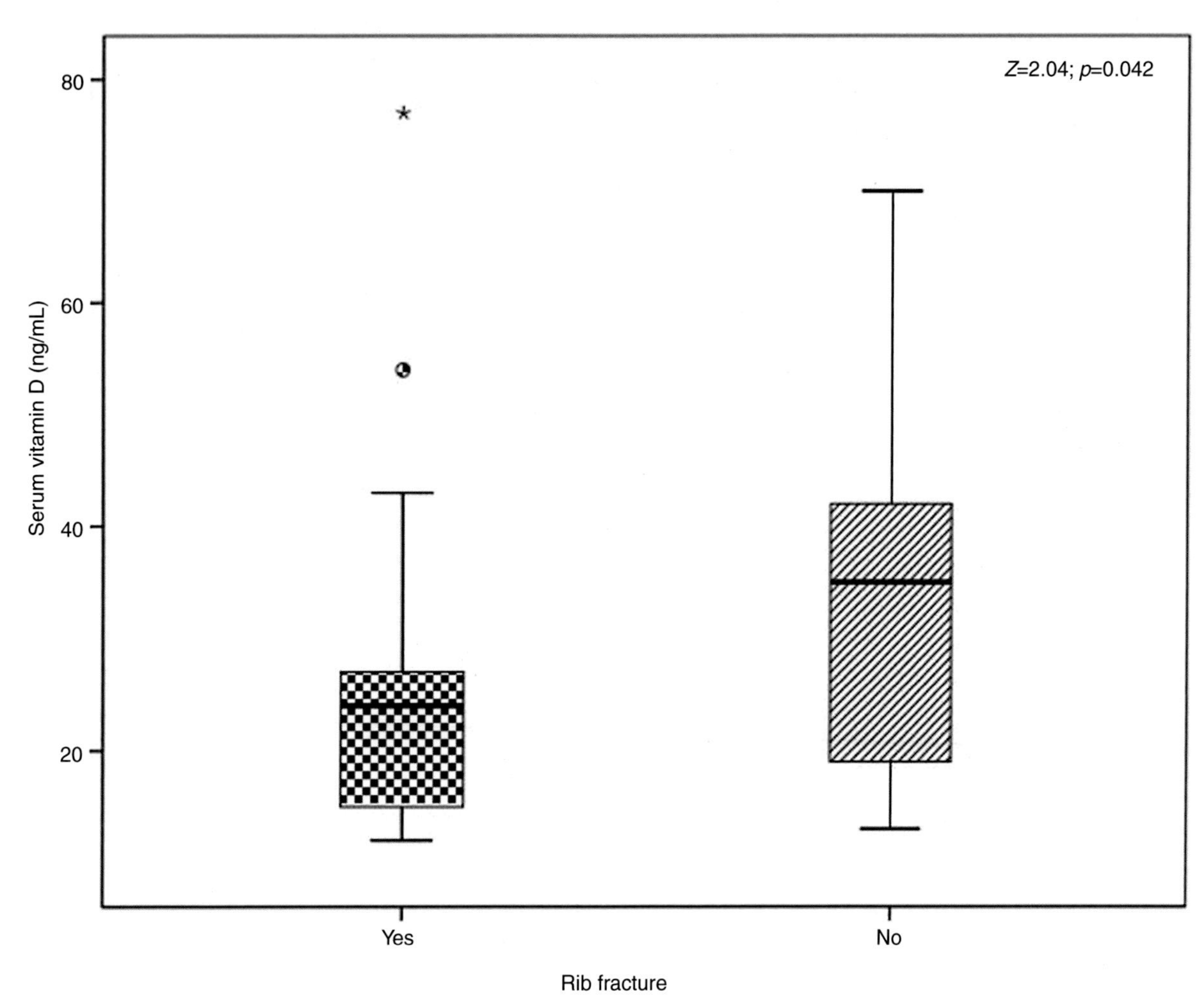

FIGURE 9.5 **Rib fractures are associated to lower vitamin D levels among 25 alcoholics with fractures, compared with 25 alcoholics without fractures.**

neuropathy surely contributes to an increased risk of falls, as it seems to happen among a geriatric population,[90] as well as cerebral and cerebellar alterations, although all this is speculative. As with muscle function impairment, both ethanol and vitamin D deficiency may exert synergistic, deleterious effects on brain and cerebellum.

Vitamin D Deficiency and Brain Alteration

Brain atrophy is related to low vitamin D levels,[91] although it is unclear if there is or not a causal mechanism. In any case, vitamin D exerts positive effects on neuronal differentiation, migration and proliferation, and apoptosis; low vitamin D levels have been described in Alzheimer's, depression, and cognitive impairment. Brain atrophy, a common finding in heavy alcoholics, may be also related to gait instability that makes the patient more prone to falls. Sato et al. in a study on patients affected with Alzheimers's disease, found that 54% of them showed very low vitamin D levels in the so called osteomalatic range (<5 ng/mL), whereas a further 26% showed values below 10 ng/mL. Among them, vitamin D consumption was low (<100 U/day), and most were sun-deprived. The same group analyzed the results of controlled trials that show that vitamin D supplementation and/or sunlight, together with calcium and vitamin K2 (in one trial), decreased the rate of hip fracture.[92] Indeed, decreased dietary intake and diminished sunlight exposure are main factors involved in vitamin D deficiency in the elderly,[86] reduced skin thickness also being an important determinant of reduced synthesis.[15] Vitamin D receptor is present in the brain, and vitamin D deficiency affects not only brain development, with lasting effects until adulthood,[91] but also exerts antioxidant and immunomodulatory effects on the adult brain.[93] The extent of these effects of vitamin D in alcoholics, in whom brain atrophy seems to be heavily dependent upon proinflammatory cytokines and oxidative damage,[60] is not fully known.

Another striking effect of vitamin D that may indirectly affect brain performance is the relation between low vitamin D levels and blood pressure. There is an inverse relation between vitamin D levels, and plasma renin activity. Vascular smooth muscle cells from vitamin D receptor knockout mice produced more angiotensin II and ROS than wild type mice,[94] and angiotensin production increased several fold in VDR knockout mice, leading to hypertension and cardiac hypertrophy. In addition, vitamin D also exerts antihypertensive effects by angiotensin-independent pathways, promoting local endothelial relaxation. There is also an inverse relation between insulin resistance and vitamin D. Interestingly, as with other alterations described in this review, the renin-angiotensin system may be also involved in the genesis of hypertension among alcoholics.[95] Brain hypertensive damage includes multiple lacunar infarcts, and is a well-recognized effect of hypertension in alcoholics, although its relation with falls and fractures is uncertain.

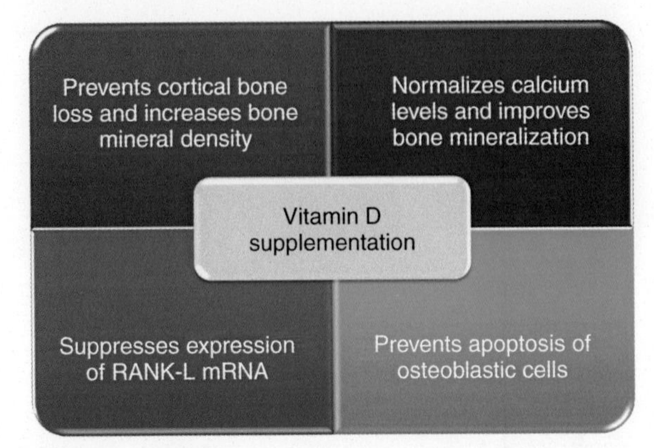

FIGURE 9.6 **Effects of vitamin D supplementation in animal models of alcoholism.**

EFFECTS OF THERAPY

Effects of Abstinence

In a study on 138 patients, multiple stepwise regression analysis identified age, poor activities of daily living (ADL), ongoing drinking, absence of cirrhosis, depression, and dementia as determinants of low bone density.[96] Abstinence may improve bone mass, but smoking cessation, and the adoption of a healthy lifestyle, are also essential measures. In a study on 77 alcoholics followed 6 months, we showed that abstainers gained, or did not lose, bone mineral density, in contrast with ongoing drinkers,[97] confirming a previous observation by Peris et al.[5] who followed their patients during a longer period. Lindholm et al. in 1991,[6] also observed that bone mass of alcoholics who had stop drinking two years prior to the study was normal, in contrast to that of active drinkers. Therefore, it seems that ethanol abstinence is the first measure to be implemented in these patients.

Preclinical Studies and Clinical Trials

There is little evidence regarding the efficacy of vitamin D supplementation on the risk of fractures or osteoporosis in men. While, in some studies that included men, supplementation with vitamin D did not improve bone mass, other studies showed only small improvements in bone mineral density, but a marked decrease in the rate of fractures.[98] However, to our knowledge, there are very few clinical trials that study the use of vitamin D in alcoholic patients. The recommendation of vitamin D supplementation in these patients is largely based on experimental evidence obtained from animal models of alcoholic bone disease, and on extrapolations from studies performed on postmenopausal women (Figure 9.6).

Mercer[99] studied mice exposed to ethanol, and found that vitamin D supplementation prevented cortical bone loss associated with ethanol exposure, normalized serum calcium levels, and suppressed expression of RANK-L mRNA. Vitamin D also prevented apoptosis of osteoblastic cells. While this outlines the role of vitamin D supplementation in chronic alcoholism, Wezeman[100] studied rats exposed to acute excesses of alcohol, in order to reproduce a situation of binge drinking. They found that vitamin D supplementation increased cancellous bone mineral density values.

On the other hand, one of the clinical trials available is the one performed by Mobarhan et al. on 56 patients with alcoholic cirrhosis that were randomized to receive placebo, or vitamin D in the form of D2 or 25 OH D3. They showed that supplementation with both forms of vitamin D increased bone mineral density, and when D2 was administered, it led to an improvement in static measures of bone remodeling.[7] This study was published three decades ago, and larger studies are necessary to study the effect of vitamin D supplementation on alcoholic bone disease.

The drugs currently used to treat osteoporosis are calcitriol (vitamin D3), alfacalcidol, and eldecalcitol. These drugs inhibit bone resorption and, therefore, increase bone mineral density. Eldecalcitol is an analog of 1α,25 -dihydroxyvitamin D3 that bears a hydroxypropoxy residue at the 2B position. It has a longer half-life than calcitriol, and it increases bone mineral density more than the other commercially available forms of vitamin D, but it suppresses PTH less than the two other forms. Further trials comparing the efficacy of these forms of vitamin D are needed, in both alcoholic and nonalcoholic patients.

CONCLUSION

Vitamin D deficiency exerts direct effects on bone, but also indirect ones. The main extraskeletal alterations, which may influence on bone changes include muscle disease, brain alterations, and hypogonadism. Both ethanol and malnutrition also contribute to all these changes. Increased activity of renal 24-hydroxylase leads to the synthesis of the inactive form of vitamin D: 24,25-dihydroxyvitamin D. On the other hand, there is decreased activity of hepatic 25-hydroxylase, the first step in the synthesis of the active form of vitamin D. Although clinical studies are still needed, animal models of both acute and chronic exposure to ethanol have shown beneficial effects of vitamin D supplementation on bone metabolism.

Key Facts

- Vitamin D deficiency is frequent among alcoholics, and is mainly due to malabsorption, abnormal biliary secretion, pancreatic insufficiency, inadequate diet, and reduced sun exposure. Ethanol also directly affects vitamin D metabolism, leading to the synthesis of inactive metabolites.
- Vitamin D deficiency leads to osteoporosis that is the result of an imbalance between bone synthesis (carried out by osteoblasts), and bone resorption (carried out by osteoclasts).
- In alcoholics, vitamin D deficiency has negative effects on calcium homeostasis, and hinders osteoblast maturation and differentiation. Alcohol also directly shifts osteoblasts away from osteogenic differentiation.
- Vitamin D receptors are found both in bone, and muscle cells. Bone and muscle function are intertwined; in fact, muscle activity promotes osteoblast proliferation through the Wnt β-catenin pathway.
- Falls and bone fractures are common among alcoholics. Vitamin D promotes muscle function and neuronal differentiation. Altered muscle function and brain atrophy with gait instability favor falls.
- Bone mass may be increased by alcohol abstinence, and by vitamin D supplementation. However, clinical trials that study the use of vitamin D supplementation in alcoholics are needed.

Summary Points

- Vitamin D deficiency is frequent among alcoholics and has negative effects on the musculoskeletal and nervous systems.
- Osteoporosis is the main metabolic bone disease seen in alcoholics.
- Alcoholic patients are prone to falls and therefore to bone fractures.
- Excessive alcohol intake increases bone resorption while hindering bone synthesis.
- Vitamin D directly promotes increased bone mass. Indirectly, it favors muscle function which is needed for bone maintenance.
- Alcohol abstinence may improve bone mass and vitamin D supplementation may be useful in alcoholics.

Abbreviations

ADL	Activities of daily living
ATP	Adenosine triphosphate
FGF	Fibroblast growth factor
IL	Interleukin
IGF-1	Insulin-like growth factor 1
IGFBP 3	Insulin-like growth factor binding protein 3
LRP5 and LRP6	Lipoprotein receptor-related proteins 5 and 6
mRNA	Messenger ribonucleic acid
PTH	Parathyroid hormone
RANK	Receptor Activator for Nuclear Factor κ B
RANK-L	Receptor Activator for Nuclear Factor κ B Ligand
ROS	Reactive oxygen species
TLR	Toll-like receptors
TNF-α	Tumor necrosis factor alpha
TRPV	Transient receptor potential vanilloid
UV	Ultraviolet
VDR	Vitamin D receptor
25 OH D	25-hydroxy vitamin *D*
24-25 (OH)2 D3	24,25-Dihydroxycholecalciferol or 24,25-dihydroxyvitamin D3
1-25 (OH)2 D3	Calcitriol or 1,25-dihydroxycholecalciferol or 1,25-dihydroxyvitamin D_3

References

1. Saville PD. Changes in Bone Mass with Age and Alcoholism. *J Bone Joint Surg Am* 1965;**47**:492–9.
2. Oppenheim WL. The battered alcoholic syndrome. *J Trauma* 1977;**17**:850–6.
3. Diamond T, Stiel D, Lunzer M, Wilkinson M, Posen S. Ethanol reduces bone formation and may cause osteoporosis. *Am J Med* 1989;**86**:282–8.
4. López-Larramona G, Lucendo AJ, González-Delgado L. Alcoholic liver disease and changes in bone mineral density. *Rev Esp Enferm Dig* 2013;**105**:609–21.
5. Peris P, Parés A, Guañabens N, Del Río L, Pons F, Martínez de Osaba MJ, Monegal A, Caballería J, Rodés J, Muñoz-Gómez J. Bone mass improves in alcoholics after two years of abstinence. *J Bone Miner Res* 1994;**10**:1607–12.
6. Lindholm J, Steiniche T, Rasmussen E, Thamsborg G, Nielsen IO, Brockestedt-Rasmussen H, Storm T, Hyldstrup L, Schou C. Bone disorders in men with chronic alcoholism: a reversible disease? *J Clin Endocrinol Metab* 1991;**73**:118–24.
7. Mobarhan SA, Russell RM, Recker RR, Posner DB, Iber FL, Miller P. Metabolic bone disease in alcoholic cirrhosis: a comparison of the effect of vitamin D2, 25-hydroxyvitamin D, or supportive treatment. *Hepatology* 1984;**4**:266–73.
8. Santori C, Ceccanti M, Diacinti D, Attilia ML, Toppo L, D'Erasmo E, Romagnoli E, Mascia ML, Cipriani C, Prastaro A, Carnevale V, Minisola S. Skeletal turnover, bone mineral density, and fractures in male chronic abusers of alcohol. *J Endocrinol Invest* 2008;**31**: 321–6.
9. Stamp TC. Vitamin D metabolism. Recent advances. *Arch Dis Child* 1973;**48**:2–7.
10. Holick MF, MacLaughlin JA, Clark MB, Holick SA, Potts Jr JT, Anderson RR, Blank IH, Parrish JA, Elias P. Photosynthesis of previtamin D3 in human skin and the physiologic consequences. *Science* 1980;**210**:203–5.

11. Holick MF. Photosynthesis of vitamin D in the skin: effect of environmental and life-style variables. *Fed Proc* 1987;**46**:1876–82.
12. Slater J, Larcombe L, Green C, Slivinski C, Singer M, Denechezhe L, Whaley C, Nickerson P, Orr P. Dietary intake of vitamin D in a northern Canadian Dené First Nation community. *Int J Circumpolar Health* 2013;**72**, 10.3402/ijch.v72iO.20723.
13. Webb AR, Kline L, Holick MF. Influence of season and latitude on the cutaneous synthesis of vitamin D3: exposure to winter sunlight in Boston and Edmonton will not promote vitamin D3 synthesis in human skin. *J Clin Endocrinol Metab* 1988;**67**: 373–8.
14. MacLaughlin J, Holick MF. Aging decreases the capacity of human skin to produce vitamin D3. *J Clin Invest* 1985;**76**:1536–8.
15. Need AG, Morris HA, Horowitz M, Nordin C. Effects of skin thickness, age, body fat, and sunlight on serum 25-hydroxyvitamin D. *Age Ageing* 2002;**31**:267–71.
16. Borel P, Caillaud D, Cano NJ. Vitamin D bioavailability: State of the art. *Crit Rev Food Sci Nutr* 2015;**55**(9):1193–205.
17. Pattanaungkul S, Riggs BL, Yergey AL, Vieira NE, O'Fallon WM, Khosla S. Relationship of intestinal calcium absorption to 1,25-dihydroxyvitamin D [1,25(OH)2D] levels in young versus elderly women: evidence for age-related intestinal resistance to 1,25(OH)2D action. *J Clin Endocrinol Metab* 2000;**85**:4023–7.
18. O'Mahony L, Stepien M, Gibney MJ, Nugent AP, Brennan L. The potential role of vitamin D enhanced foods in improving vitamin D status. *Nutrients* 2011;**3**:1023–41.
19. Conti G, Chirico V, Lacquaniti A, Silipigni L, Fede C, Vitale A, Fede C. Vitamin D intoxication in two brothers: be careful to dietary supplements. *J Pediatr Endocrinol Metab* 2014;**27**(7–8): 763–7.
20. Holick MF, Garabedian M, Schnoes HK, DeLuca HF. Relationship of 25-hydroxyvitamin D3 side chain structure to biological activity. *J Biol Chem* 1975;**250**:226–30.
21. Bacchetta J, Sea JL, Chun RF, Lisse TS, Wesseling-Perry K, Gales B, Adams JS, Salusky IB, Hewison M. Fibroblast growth factor 23 inhibits extrarenal synthesis of 1,25-dihydroxyvitamin D in human monocytes. *J Bone Miner Res* 2013;**28**:46–55.
22. Shimada T, Hasegawa H, Yamazaki Y, Muto T, Hino R, Takeuchi Y, Fujita T, Nakahara K, Fukumoto S, Yamashita T. FGF-23 is a potent regulator of vitamin D metabolism and phosphate homeostasis. *J Bone Miner Res* 2004;**19**:429–35.
23. Tanaka Y, DeLuca HF, Kobayashi Y, Taguchi T, Ikekawa N, Morisaki M. Biological activity of 24,24-difluoro-25-hydroxyvitamin D3. Effect of blocking of 24-hydroxylation on the functions of vitamin D. *J Biol Chem* 1979;**254**:7163–7.
24. Armbrecht HJ, Zenser TV, Davis BB. Effect of age on the conversion of 25-hydroxyvitamin D3 to 1,25-dihydroxyvitamin D3 by kidney of rat. *J Clin Invest* 1980;**66**:1118–23.
25. Ryan JW, Anderson PH, Turner AG, Morris HA. Vitamin D activities and metabolic bone disease. *Clin Chim Acta* 2013;**425**: 148–52.
26. McCarthy K, Laban C, Bustin SA, Ogunkolade W, Khalaf S, Carpenter R, Jenkins PJ. Expression of 25-hydroxyvitamin D-1-alpha-hydroxylase, and vitamin D receptor mRNA in normal and malignant breast tissue. *Anticancer Res* 2009;**29**:155–7.
27. Kutuzova GD, Sundersingh F, Vaughan J, Tadi BP, Ansay SE, Christakos S, Deluca HF. TRPV6 is not required for 1alpha, 25-dihydroxyvitamin D3-induced intestinal calcium absorption *in vivo*. *Proc Natl Acad Sci USA* 2008;**105**:19655–9.
28. Hoenderop JG, Nilius B, Bindels RJ. Calcium absorption across epithelia. *Physiol Rev* 2005;**85**:373–422.
29. Holick MF. Vitamin D deficiency. *N Eng J Med* 2007;**357**:266–81.
30. Christakos S, Dhawan P, Porta A, Mady LJ, Seth T. Vitamin D and intestinal calcium absorption. *Mol Cell Endocrinol* 2011;**347**:25–9.
31. Ajibade DV, Dhawan P, Fechner AJ, Meyer MB, Pike JW, Christakos S. Evidence for a role of prolactin in calcium homeostasis: regulation of intestinal transient receptor potential vanilloid type 6, intestinal calcium absorption, and the 25-hydroxyvitamin D(3) 1alpha hydroxylase gene by prolactin. *Endocrinology* 2010;**151**: 2974–84.
32. Naude CE, Carey PD, Laubscher R, Fein G, Senekal M. Vitamin D and calcium status in South African adolescents with alcohol use disorders. *Nutrients* 2012;**4**:1076–94.
33. Neupane SP, Lien L, Hilberg T, Bramness JG. Vitamin D deficiency in alcohol-use disorders and its relationship to comorbid major depression: a cross-sectional study of inpatients in Nepal. *Drug Alcohol Depend* 2013;**133**:480–5.
34. Sobral-Oliveira MB, Faintuch J, Guarita DR, Oliveira CP, Carrilho FJ. Nutritional profile of asymptomatic alcoholic patients. *Arq Gastroenterol* 2011;**48**:112–8.
35. Wilkens-Knudsen A, Jensen JE, Nordgaard-Lassen I, Almdal T, Kondrup J, Becker U. Nutritional intake and status in persons with alcohol dependency: data from an outpatient treatment programme. *Eur J Nutr* 2014;**53**(7):1483–92.
36. González-Reimers E, Alvisa-Negrín J, Santolaria-Fernández F, Candelaria, Martín-González M, Hernández-Betancor I, Fernández-Rodríguez CM, Viña-Rodríguez J, González-Díaz A. Vitamin D and nutritional status are related to bone fractures in alcoholics. *Alcohol Alcohol* 2011;**46**:148–55.
37. Malik P, Gasser RW, Kemmler G, Moncayo R, Finkenstedt G, Hurz M, Fleischhacker WW. Low bone mineral density and impaired bone metabolism in young alcoholic patients without liver cirrhosis: a cross-sectional study. *Alcohol Clin Exp Res* 2009;**33**:375–81.
38. Kim MJ, Shim MS, Kim MK, Lee Y, Shin YG, Chung CH, Kwon SO. Effect of chronic alcohol ingestion on bone mineral density in males without liver cirrhosis. *Korean J Intern Med* 2003;**18**:174–80.
39. Wilkinson G, Cundy T, Parsons V, Lawson-Matthew P. Metabolic bone disease and fractures in male alcoholics: a pilot study. *Br J Addict* 1985;**80**:65–8.
40. Bikle DD, Genant HK, Cann C, Recker RR, Halloran BP, Strewler GJ. Bone disease in alcohol abuse. *Ann Intern Med* 1985;**103**:42–8.
41. Bjørneboe GE, Johnsen J, Bjørneboe A, Rousseau B, Pedersen JI, Norum KR, Mørland J, Drevon CA. Effect of alcohol consumption on serum concentration of 25-hydroxyvitamin D3, retinol, and retinol-binding protein. *Am J Clin Nutr* 1986;**44**:678–82.
42. Laitinen K, Välimäki M, Lamberg-Allardt C, Kivisaari L, Lalla M, Kärkkäinen M, Ylikahri R. Deranged vitamin D metabolism but normal bone mineral density in Finnish noncirrhotic male alcoholics. *Alcohol Clin Exp Res* 1990;**14**:551–6.
43. Pun KK, Wong FH, Wang C, Lau P, Ho PW, Pun WK, Chow SP, Cheng CL, Leong JC, Young RT. Vitamin D status among patients with fractured neck of femur in Hong Kong. *Bone* 1990;**11**:365–8.
44. Shankar K, Liu X, Singhal R, Chen JR, Nagarajan S, Badger TM, Ronis MJ. Chronic ethanol consumption leads to disruption of vitamin D homeostasis associated with induction of renal 1,25 dihydroxyvitamin D3 24 hydroxylase (CYP24A1). *Endocrinology* 2008;**149**:1748–56.
45. Morris HA, Anderson PH. Autrocrine and paracrine actions of vitamin D. *Clin Biochem Rev* 2010;**31**:129–38.
46. van de Peppel J, van Leeuwen JP. Vitamin D and gene networks in human osteoblasts. *Front Physiol* 2014;**5**:137.
47. Woeckel VJ, Alves RD, Swagemakers SM, Eijken M, Chiba H, van der Eerden BC, van Leeuwen JP. 1Alpha, 25-(OH)2D3 acts in the early phase of osteoblast differentiation to enhance mineralization via accelerated production of mature matrix vesicles. *J Cell Physiol* 2010;**225**:593–600.
48. Alves RD, Eijken M, Bezstarosti K, Demmers JA, van Leeuwen JP. Activin A suppresses osteoblast mineralization capacity by altering extracellular matrix (ECM) composition and impairing matrix vesicle (MV) production. *Mol Cell Proteomics* 2013;**12**:2890–900.
49. Woeckel VJ, van der Eerden BC, Schreuders-Koedam M, Eijken M, Van Leeuwen JP. 1α,25-dihydroxyvitamin D3 stimulates activin A

production to fine-tune osteoblast-induced mineralization. *J Cell Physiol* 2013;**228**:2167–74.
50. Raisz LG, Trummel CL, Holick MF, DeLuca HF. 1,25-dihydroxycholecalciferol: a potent stimulator of bone resorption in tissue culture. *Science* 1972;**175**(4023):768–9.
51. Takahashi N, Udagawa N, Suda T. Vitamin D endocrine system and osteoclasts. *Bonekey Rep* 2014;**3**:495.
52. Luo J, Wen H, Guo H, Cai Q, Li S, Li X. 1,25-dihydroxyvitamin D3 inhibits the RANKL pathway and impacts on the production of pathway-associated cytokines in early rheumatoid arthritis. *Biomed Res Int* 2013;**2013**:101805.
53. Krawitt EL. Effect of ethanol ingestion on duodenal calcium transport. *J Lab Clin Med* 1975;**85**:665–71.
54. Jorge-Hernández JA, González-Reimers E, Torres-Ramírez A, Santolaria-Fernandez F, Gonzalez-Garcia C, Batista-Lopez JN, Pestana-Pestana M, Hernandez-Nieto L. Bone changes in alcoholic liver cirrhosis: a histomorphometrical analysis of 52 cases. *Dig Dis Sci* 1988;**33**:1089–95.
55. Lalor BC, France MW, Powell D, Adams PH, Counihan TB. Bone and mineral metabolism and chronic alcohol abuse. *Quart J Med* 1986;**59**:497–511.
56. Wang HJ, Zakhari S, Jung MK. Alcohol, inflammation, and gut-liver-brain interactions in tissue damage and disease development. *World J Gastroenterol* 2010;**16**:1304–11.
57. Gustot T, Lemmers A, Moreno C, Nagy N, Quertinmont E, Nicaise C, Franchimont D, Louis H, Devière J, Le Moine O. Differential liver sensitization to toll-like receptor pathways in mice with alcoholic fatty liver. *Hepatology* 2006;**43**:989–1000.
58. Hritz I, Mandrekar P, Velayudham A, Catalano D, Dolganiuc A, Kodys K, Kurt-Jones E, Szabo G. The critical role of toll-like receptor (TLR) 4 in alcoholic liver disease is independent of the common TLR adapter MyD88. *Hepatology* 2008;**48**:1224–31.
59. Tuma DJ. Role of malondialdehyde-acetaldehyde adducts in liver injury. *Free Radic Biol Med* 2002;**32**:303–8.
60. Crews FT, Bechara R, Brown LA, Guidot DM, Mandrekar P, Oak S, Qin L, Szabo G, Wheeler M, Zou J. Cytokines and alcohol. *Alcohol Clin Exp Res* 2006;**30**:720–30.
61. Giuliani N, Girasole G, Vescovi PP, Passeri G, Pedrazzoni M. Ethanol and acetaldehyde inhibit the formation of early osteoblast progenitors in murine and human bone marrow cultures. *Alcohol Clin Exp Res* 1999;**23**:381–5.
62. Chen Y, Chen L, Yin Q, Gao H, Dong P, Zhang X, Kang J. Reciprocal interferences of TNF-α and Wnt1/β-catenin signaling axes shift bone marrow-derived stem cells towards osteoblast lineage after ethanol exposure. *Cell Physiol Biochem* 2013;**32**:755–65.
63. Chen Y, Gao H, Yin Q, Chen L, Dong P, Zhang X, Kang J. ER stress activating ATF4/CHOP-TNF-α signaling pathway contributes to alcohol-induced disruption of osteogenic lineage of multipotential mesenchymal stem cell. *Cell Physiol Biochem* 2013;**32**:743–54.
64. Bischoff HA, Borchers M, Gudat F, Duermueller U, Theiler R, Stähelin HB, Dick W. *In situ* detection of 1,25-dihydroxyvitamin D3 receptor in human skeletal muscle tissue. *Histochem J* 2001;**33**: 19–24.
65. Simpson RU, Thomas GA, Arnold AJ. Identification of 1,25-dihydroxyvitamin D3 receptors and activities in muscle. *J Biol Chem* 1985;**260**:8882–91.
66. Marshall PA, Hernández Z, Kaneko I, Widener T, Tabacaru C, Aguayo I, Jurutka PW. Discovery of novel vitamin D receptor interacting proteins that modulate 1,25-dihydroxyvitamin D_3 signaling. *J Steroid Biochem Mol Biol* 2012;**132**:147–59.
67. Hamilton B. Vitamin D and Human Skeletal Muscle. *Scand J Med Sci Sports* 2010;**20**:182–90.
68. Ceglia L, Harris SS. Vitamin D and its role in skeletal muscle. *Calcif Tissue Int* 2013;**92**:151–62.
69. Geusens P, Vandevyver C, Vanhoof J, Cassiman JJ, Boonen S, Raus J. Quadriceps and grip strength are related to vitamin D receptor genotype in elderly nonobese women. *J Bone Miner Res* 1997;**12**:2082–8.
70. Srikuea R, Zhang X, Park-Sarge OK, Esser KA. VDR and CYP27B1 are expressed in C2C12 cells and regenerating skeletal muscle: potential role in suppression of myoblast proliferation. *Am J Physiol Cell Physiol* 2012;**303**:C396–405.
71. Prabhala A, Garg R, Dandona P. Severe myopathy associated with vitamin D deficiency in western New York. *Arch Intern Med* 2000;**160**:1199–203.
72. Irani PF. Electromyography in nutritional osteomalacic myopathy. *J Neurol Neurosurg Psychiatry* 1976;**39**:686–93.
73. González-Reimers E, Durán-Castellón MC, López-Lirola A, Santolaria-Fernández F, Abreu-González P, Alvisa-Negrín J, Sánchez-Pérez MJ. Alcoholic myopathy: vitamin D deficiency is related to muscle fibre atrophy in a murine model. *Alcohol Alcohol* 2010;**45**:223–30.
74. Burgers TA, Williams BO. Regulation of Wnt/β-catenin signaling within and from osteocytes. *Bone* 2013;**54**:244–9.
75. Lewiecki EM. Role of sclerostin in bone and cartilage and its potential as a therapeutic target in bone diseases. *Ther Adv Musculoskelet Dis* 2014;**6**:48–57.
76. Lloyd SA, Loiselle AE, Zhang Y, Donahue HJ. Evidence for the role of connexin 43-mediated intercellular communication in the process of intracortical bone resorption via osteocytic osteolysis. *BMC Musculoskelet Disord* 2014;**15**:122.
77. Preedy VR, Adachi J, Ueno Y, Ahmed S, Mantle D, Mullatti N, Rajendram R, Peters TJ. Alcoholic skeletal muscle myopathy: definitions, features, contribution of neuropathy, impact and diagnosis. *Eur J Neurol* 2001;**8**:677–87.
78. Hickish T, Colston KW, Bland JM, Maxwell JD. Vitamin D deficiency and muscle strength in male alcoholics. *Clin Sci (Lond)* 1989;**77**:171–6.
79. Sørensen OH, Lund B, Saltin B, Lund B, Andersen RB, Hjorth L, Melsen F, Mosekilde L. Myopathy in bone loss of ageing: improvement by treatment with 1 alpha-hydroxycholecalciferol and calcium. *Clin Sci (Lond)* 1979;**56**:157–61.
80. Lee DM, Tajar A, Pye SR, Boonen S, Vanderschueren D, Bouillon R, O'Neill TW, Bartfai G, Casanueva FF, Finn JD, Forti G, Giwercman A, Han TS, Huhtaniemi IT, Kula K, Lean ME, Pendleton N, Punab M, Wu FC. EMAS study group. Association of hypogonadism with vitamin D status: the European Male Ageing Study. *Eur J Endocrinol* 2012;**166**:77–85.
81. Van Thiel DH, Lester R. Alcoholism: its effect on hypothalamic pituitary gonadal function. *Gastroenterology* 1976;**71**:318–27.
82. Chapuy MC, Arlot ME, Duboeuf F, Brun J, Crouzet B, Arnaud S, Delmas PD, Meunier PJ. Vitamin D3 and calcium to prevent hip fractures in the elderly women. *N Engl J Med* 1992;**327**:1637–42.
83. Dhesi JK, Moniz C, Close JC, Jackson SH, Allain TJ. A rationale for vitamin D prescribing in a falls clinic population. *Am J Clin Nutr* 1993;**58**:882–5.
84. Flicker L, Mead K, MacInnis RJ, Nowson C, Scherer S, Stein MS, Thomas J, Hopper JL, Wark JD. Serum vitamin D and falls in older women in residential care in Australia. *J Am Geriatr Soc* 2003;**51**:1533–8.
85. Burleigh E, Potter J. Vitamin D deficiency in outpatients -a Scottish perspective. *Scott Med J* 2006;**51**:27–31.
86. Janssen HC, Samson MM, Verhaar HJ. Vitamin D deficiency, muscle function, and falls in elderly people. *Am J Clin Nutr* 2002;**75**:611–5.
87. Prince RL, Austin N, Devine A, Dick IM, Bruce D, Zhu K. Effects of ergocalciferol added to calcium on the risk of falls in elderly high-risk women. *Arch Intern Med* 2008;**168**:103–8.
88. Burleigh E, McColl J, Potter J. Does vitamin D stop inpatients falling? A randomised controlled trial. *Age Ageing* 2007;**36**:507–13.
89. Cawthon PM, Harrison SL, Barrett-Connor E, Fink HA, Cauley JA, Lewis CE, Orwoll ES, Cummings SR. Alcohol intake and its

relationship with bone mineral density, falls, and fracture risk in older men. *J Am Geriatr Soc* 2006;**54**:1649–57.
90. Leblhuber F, Schroecksnadel K, Beran-Praher M, Haller H, Steiner K, Fuchs D. Polyneuropathy and dementia in old age: common inflammatory and vascular parameters. *J Neural Transm* 2011;**118**: 721–5.
91. Féron F, Burne TH, Brown J, Smith E, McGrath JJ, Mackay-Sim A, Eyles DW. Developmental Vitamin D3 deficiency alters the adult rat brain. *Brain Res Bull* 2005;**65**:141–148.
92. Iwamoto J, Sato Y, Tanaka K, Takeda T, Matsumoto H. Prevention of hip fractures by exposure to sunlight and pharmacotherapy in patients with Alzheimer's disease. *Aging Clin Exp Res* 2009;**21**:277–81.
93. Briones TL, Darwish H. Decrease in age-related tau hyperphosphorylation and cognitive improvement following vitamin D supplementation are associated with modulation of brainenergy metabolism and redox state. *Neuroscience* 2014;**262**: 143–55.
94. Valcheva P, Cardus A, Panizo S, Parisi E, Bozic M, Lopez Novoa JM, Dusso A, Fernández E, Valdivielso JM. Lack of vitamin D receptor causes stress-induced premature senescence in vascular smooth muscle cells through enhanced local angiotensin-II signals. *Atherosclerosis* 2014;**235**:247–55.
95. Husain K, Ansari RA, Ferder L. Alcohol-induced hypertension: mechanism and prevention. *World J Cardiol* 2014;**6**(5):245–52.
96. Matsui T, Yokoyama A, Matsushita S, Ogawa R, Mori S, Hayashi E, Roh S, Higuchi S, Arai H, Maruyama K. Effect of a comprehensive lifestyle modification program on the bone density of male heavy drinkers. *Alcohol Clin Exp Res* 2010;**34**:869–75.
97. Alvisa-Negrín J, González-Reimers E, Santolaria-Fernández F, García-Valdecasas-Campelo E, Valls MR, Pelazas-González R, Durán-Castellón MC, de Los Angeles Gómez-Rodríguez M. Osteopenia in alcoholics: effect of alcohol abstinence. *Alcohol Alcohol* 2009;**44**:468–75.
98. Dawson-Hughes B, Harris SS, Krall EA, Dallal GE. Effect of calcium and vitamin D supplementation on bone density in men and women 65 years of age or older. *N Engl J Med* 1997;**337**:670–6.
99. Mercer KE, Wynne RA, Lazarenko OP, Lumpkin CK, Hogue WR, Suva LJ, Chen JR, Mason AZ, Badger TM, Ronis MJ. Vitamin D supplementation protects against bone loss associated with chronic alcohol administration in female mice. *J Pharmacol Exp Ther* 2012;**343**:401–12.
100. Wezeman FH, Juknelis D, Himes R, Callaci JJ. Vitamin D and ibandronate prevent cancellous bone loss associated with binge alcohol treatment in male rats. *Bone* 2007;**41**:639–645s.

CHAPTER

10

Antioxidant Treatment and Alcoholism

*Camila S. Silva, PhD**, Guilherme V. Portari, PhD**, Helio Vannucchi, MD, PhD*†

*Division of Biochemical Toxicology, National Center for Toxicological Research, US Food and Drug Administration, Jefferson, AR, USA

**Department of Nutrition, Federal University of Triângulo Mineiro, Uberaba, MG, Brazil

†Department of Internal Medicine, Division of Nutrition, University of São Paulo, Ribeirão Preto, Brazil

IMPORTANCE OF THE ANTIOXIDANT SYSTEMS TO THE ORGANISM

Cells have endogenous antioxidant defense mechanisms composed of enzymes (such as superoxide dismutase, catalase, glutathione peroxidase, and heme oxygenase), peptides (glutathione), and proteins (thioredoxin), all of which operate along with nutritional compounds (carotenoids, flavonoids, vitamin C, vitamin E) provided by dietary sources.[1] Together, these systems protect cellular structures against damage induced by free-radicals. Nevertheless, an elevation of reactive oxygen species (ROS) synthesis may compromise the antioxidant status, resulting in cellular oxidative stress. Consequently, the ability of ROS to modify the structure of cellular components, change DNA polymerase activity, and modulate gene expression and protein synthesis, results in a variety of pathological conditions, including alcohol-induced liver injury.[2] The ability of acute and chronic alcohol intake to increase ROS production, and potentiate cellular damage, has been demonstrated in a number of cells, systems, and species, including humans.[1,3,4]

One of the main enzymatic antioxidants is superoxide dismutase (SOD), the expression of which is affected by the status of zinc (Zn), copper (Cu), and manganese (Mn).[5] SOD catalyzes the dismutation of the superoxide radical into hydrogen peroxide and oxygen, with the hydrogen peroxide being subsequently detoxified by catalase and/or glutathione peroxidase (GPx).[6] SOD exists in several isoforms, differing in the nature of the active metal center and amino acid constituents. Cu/Zn-SOD is found in the cytosol and mitochondrial membranes, whereas Mn-SOD is located in the mitochondrial matrix; both isoforms are essential for preventing oxidative damage in cells.[7]

Catalase is present in peroxisomes, and promotes the conversion of hydrogen peroxide to water and molecular oxygen.[6] Glutathione peroxidase, a selenium-dependent enzyme, catalyzes the hydrogen atom transfer from reduced glutathione (GSH) to hydrogen peroxide that results in the production of water and oxidized glutathione (GSSG). This pathway prevents the lipid peroxidation. GSSG is recycled back to GSH through the action of NADPH-dependent glutathione reductase.

GSH is an endogenous antioxidant, indispensable in the metabolism of superoxide, hydrogen peroxide, and hydroxyl radicals generated during the oxidation of ethanol via cytochrome P450 (CYP) 2E1. Additionally, GSH is a cofactor of glutathione transferases in the detoxification of products derived from xenobiotic metabolism and other reactive molecules produced in cells. Therefore, GSH is considered one of the most important antioxidant in cells.[1]

Nonenzymatic antioxidants, in particular α-tocopherol (vitamin E) and ascorbate (vitamin C), are also found in cells. α-Tocopherol represents the most active form of vitamin E in humans, and the most abundant antioxidant in lipid phase of membranes.[1,8] α-Tocopherol is well known for its ability to act as a lipid soluble radical scavenger, and protect cellular membranes against radical-induced damage.[9] Specifically, α-tocopherol interacts with peroxyl and hydroxyl radicals, and prevents the propagation of chain reactions induced by free radicals in biological membranes.[10] The metabolite produced in this reaction is recycled back to vitamin E by GSH and ascorbate[1]; thus, vitamin E effectively decreases the susceptibility of membranes to lipid peroxidation.[11,12] In animals, vitamin E deficiency has been reported to cause lipid peroxidation.[3,13]

Molecular Aspects of Alcohol and Nutrition. http://dx.doi.org/10.1016/B978-0-12-800773-0.00010-0

Low concentrations of endogenous antioxidants (such as SOD, catalase, GPx, and GSH) result in increased susceptibility to cell oxidative damage induced by free-radicals.[13]

ANTIOXIDANT STATUS IN ALCOHOLISM

The metabolism of alcohol leads to elevated synthesis of ROS, and simultaneously causes changes in the antioxidant status of cells, such as decreases in the levels of vitamins A, C, and E, selenium, zinc, and GSH, and lower activity of SOD and GPx.[14–17]

Continuous ethanol intake induces free-radical synthesis, resulting in an imbalance in antioxidant status. Microsomal ethanol oxidation leads to high levels of acetaldehyde and ROS that are responsible for oxidative stress, and membrane damage. Oxidative stress caused by chronic ethanol exposure is believed to promote necrosis and/or apoptosis, through a change in mitochondrial permeability, as hepatocytes become sensitized to the action of proapoptotic tumor necrosis factor-α (TNF-α). Furthermore, oxidative processes can contribute to hepatic fibrosis through the release of profibrotic cytokines, and increased collagen gene expression in the liver.[18] Finally, antigens produced in reactions between lipid peroxidation products and hepatic proteins activate an immune response that might represent the mechanism by which ethanol-induced oxidative stress facilitates the formation of chronic hepatic inflammation.[18]

Hepatotoxicity and oxidative stress play critical roles in liver injury and hepatocarcinogenesis.[19,20] ROS have the potential to oxidize vital structures and macromolecules, leading to DNA damage, lipid peroxidation, and protein modifications that can compromise cellular integrity.[21] These events are illustrated in Figure 10.1.

Peroxyl radicals produced during the lipid peroxidation react with membrane proteins, modify enzymes, receptors, and signal transduction systems, and also oxidize cholesterol.[22,23] Antioxidants protect against lipid peroxidation of cellular membranes by inhibiting chain reactions induced by free radicals.[24]

Animal studies, utilizing acute and chronic alcohol intake, can cause reduced levels of α-tocopherol in the liver.[3,24–26] Giavarotti et al.[27] reported that rats fed alcohol had decreased levels of hepatic α-tocopherol and β-carotene. Jordão Jr. et al.[3] showed that vitamin E deficiency itself induces lipid peroxidation in the liver of rats exposed to acute doses of ethanol. Alcohol also leads to decreased levels of metal ions in liver, including selenium, copper, and zinc, all of which are essential cofactors of antioxidant enzymes.[17,24] The serum selenium/malondialdehyde ratio has been proposed as an indicator of hepatic damage caused by alcohol consumption, and selenium has been suggested as a possible antioxidant treatment for liver disease in alcoholic patients.[17]

In oxidative stress, there is high GSH uptake, in addition to its lowered synthesis, resulting in its depletion in experimental models of alcoholism.[28,29] Under certain circumstances, ethanol exposure may lead to increased hepatic GSH levels,[25,30,31] as a result of a recovery of its synthesis during oxidative stress. Rats chronically administered ethanol show elevated GSH levels, due to efficient glutathione recycling.[32,33]

There are divergent reports in the literature regarding the antioxidant status in experimental models for chronic alcoholism. For example, human and animal studies have shown that α-tocopherol levels can be both depleted,[34,35] or unaltered,[36–39] following ethanol exposure. In patients with cirrhosis, diminished hepatic vitamin E levels have been observed.[40] Reports concerning the effect of ethanol on hepatic GSH levels have indicated

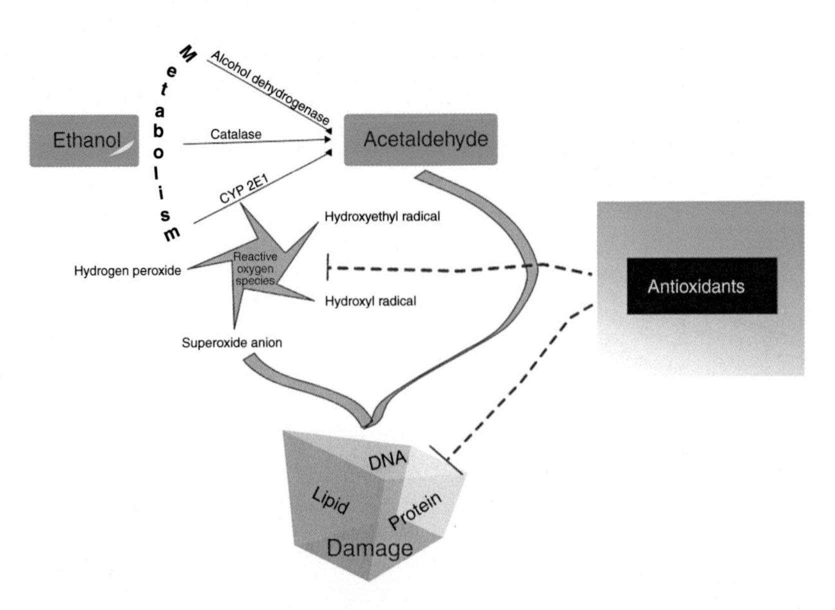

FIGURE 10.1 **Relation between ethanol metabolism, free-radical synthesis, and antioxidants.**

depletion,[36,41] increases,[22,31,33,42] or no change,[43] fact that seems to depend on the dosage, period of exposure, and route of ethanol administration.

In experimental models, the chronic exposure to ethanol through either a Lieber–DeCarli liquid diet or intragastric infusion resulted in a decreased GSH content in the liver.[28] On the other hand, Bailey et al.[14] found elevated levels of GSH, following the chronic intake of ethanol (a Lieber–DeCarli diet for 31 days), while Deaciuc et al.[43] reported unaltered GSH levels in animals fed a diet containing ethanol for seven weeks.

Accordingly, because of the effect of alcohol metabolism on the antioxidant status, several antioxidant treatments have been investigated throughout the last decades, in an attempt to alleviate the ethanol-induced effects.

ANTIOXIDANT TREATMENT FOR ALCOHOLISM

Experimental and clinical studies have investigated the use of antioxidants, including GSH precursors, carotenoids, vitamin C, vitamin E, selenium, curcumin, epigallocatechin gallate, ginsenosides, puerarin, quercetin, resveratrol, and silymarin, in the prevention and/or treatment of clinical conditions related to the alcoholism. In this chapter, we review the main outcomes from a number of these studies.

α-Tocopherol and Vitamin C

α-Tocopherol is well known for its ability to act as a hydroperoxyl radical scavenger, therefore playing an important role in protecting the organism against oxidative damage.[9]

There have been a number of studies concerning the effect of vitamin E supplementation in alcohol abuse. The results show either protection, or no benefit, and this seems to depend on the specific organism, the time and dose of treatment, and the diet characteristics.

Besides its well-established antioxidant function, vitamin E plays important roles in cell signaling, by acting as a regulator of gene expression. Interestingly, vitamin E supplements have been reported to modulate either the immune response, or interaction of inflammatory cells with target tissues.[44,45] In a rat model of alcoholic chronic pancreatitis, α-tocopherol supplementation was shown to lead to a gene expression profile characterized by an anti-inflammatory response in the pancreas,[38] namely, downregulation of genes with a proinflammatory profile (COX-2, α-SMA, TNF-α, IL-6, and MIP-3α), and upregulation of the gene encoding Pap (Reg 3b), a protein with a protective role against pancreatic injury that operates in tissue regeneration.[38]

In another study where rats were exposed to ethanol for six weeks, α-tocopherol supplementation prevented Mallory bodies, and inflammatory infiltration in liver.[39] Mallory bodies consist of aggregates of modified proteins, or adducts of protein, formed as a result of lipid peroxidation.[46] However, the antioxidant treatment did not prevent apoptosis. Despite the fact that normal hepatic α-tocopherol levels were found in the alcohol-fed animals that did not receive antioxidant supplementation, ethanol is likely to have caused a temporary antioxidant imbalance, since Mallory bodies were formed.[39] This formation is due to the inhibition of the ubiquitin-proteasome proteolytic pathway, as a result of oxidative stress.[46] Moreover, differentially expressed genes showed an alteration of cellular homeostasis, resulting from chronic ethanol intake.[39]

Many studies have implicated alcohol-induced oxidative stress in the pathogenesis of reduced protein synthesis.[47,48] Some studies showed that supplemental α-tocopherol can prevent the inhibition of protein synthesis,[49,50] but Reilly et al.[51] reported that α-tocopherol supplementation was not able to restore the rate of protein synthesis in the skeletal muscle of rats acutely exposed to ethanol. The same effect was also found later in a study investigating α-tocopherol supplementation during both conditions, acute and chronic exposure.[52]

Tran et al.[53] reported that, unlike previous studies showing a neuroprotective potential of antioxidants on alcohol-mediated cerebellar damage, vitamin E supplementation failed to protect rat pups against structural and functional damage in cerebellum, in a model of binge ethanol exposure on postnatal day 4.

On the other hand, the administration of vitamin E to pregnant rats exposed to alcohol, from gestation day 7 throughout lactation, was reported to alleviate the brain atrophy and DNA damage, and to reduce alcohol-induced homocysteine levels.[54]

Kaur et al.[55] suggested vitamin E is a potential therapeutic agent for ethanol-induced oxidative liver injury, because it exhibited beneficial effects on mice exposed to ethanol for 15 days. They reported that vitamin E restored redox status, decreased apoptosis, and opposed oxidative stress.

Lee et al.[56] reported that rats fed a low-fat ethanol diet for five weeks exhibited oxidative stress, and consequent liver toxicity, and that vitamin E supplementation restored plasma levels of activities of aspartate aminotransferase (AST), total radical-trapping antioxidant potential (TRAP), and conjugated dienes to control levels. In contrast, although vitamin C supplementation increased plasma GSH and hepatic S-adenosylmethionine concentrations in ethanol-fed rats, it did not protect against lipid peroxidation, unlike the group supplemented with vitamin E. They suggested that a high intake of vitamin C may induce hepatic CYP2E1-linked monooxygenases that cause adverse pro-oxidant outcomes, depending on the dose.

Prathibha et al.[57] reported that coadministration of α-tocopherol and ascorbic acid to rats fed alcohol for 90 days resulted in reduced liver fibrosis (20% reduction), compared to rats given individual antioxidant therapies. The combined treatment attenuated several parameters that had been increased by alcohol exposure, including markers of hepatic function, lipid peroxidation, protein carbonyls, and the expression of nuclear factor kappa B (NFκB), tumor necrosis factor α, transforming growth factor β 1, CYP2E1, and collagen Type I. In addition, the combined treatment increased the activity of antioxidant enzymes that had been inhibited by ethanol.

Carotenoids

Carotenoids constitute a complex class of over 600 organic pigments, naturally produced by photosynthetic organisms. About 50 carotenoids are present in human diets, and approximately 20 can be found in human blood and tissues.[58] Because of their antioxidant properties, some carotenoids, such as β-carotene (pro-vitamin A), have been studied as potential agents against oxidative stress induced by alcohol abuse; however, reported outcomes are conflicting and inconclusive.

In a rat chronic model for alcoholism, β-carotene beadlet (5 mg β-carotene/kg body weight) supplementation for 10 weeks resulted in maintenance of enzymatic AST and alanine aminotransferase (ALT), normal hepatic histology, and increased hepatic GSH levels.[36] Hepatocytes obtained from rats fed a Lieber–DeCarli diet for 4 weeks, and then supplemented with β-carotene *in vitro* for 24 h, displayed increased GSH levels, possibly due to a decreased activity of GPx.[59] In contrast, some studies have reported β-carotene toxicity associated with ethanol intake. For example, the pioneering ones were described by Lieber and coworkers,[60] who showed hepatoxicity in baboons, namely inflammatory response with autophagic vacuoles, and alterations of the endoplasmic reticulum and mitochondria. In another study, rats fed a Lieber–DeCarli diet and supplemented with crystalline β-carotene, or β-carotene beadlets, for 8 weeks, had increased levels of the lipid peroxidation markers F2-isoprostane and 4-hydroxynonenal, and mitochondrial damage, as demonstrated by increased levels of the mitochondrial enzyme glutamate dehydrogenase in the bloodstream.[61] These authors further found that β-carotene potentiated the induction of CYP2E1 by ethanol, and also upregulated CYP4A1.[62]

Our group has shown a pro-oxidant effect of β-carotene supplementation in rats administered ethanol in drinking water for 28 days. Changes were observed in hepatic GSH profile, and increased levels of thiobarbituric acid reactive species (TBARS) were found.[34] This paradoxical effect is believed to occur due to the formation, under some circumstances, of cleavage products with pro-oxidant activity (apo-carotenoids), when β-carotene is exposed to free-radicals, such as occurs in alcoholism.[63] Furthermore, in rodents, the chronic alcohol exposure upregulates hepatic expression of the enzymes responsible for β-carotene cleavage: carotenoid 15,15′-monooxygenase (vitamin A production), and carotenoid 9′10′-monooxygenase (apo-carotenoid production).[64]

Curcumin

Curcumin (diferuloylmethane) is a polyphenolic compound isolated from rhizomes of the plant *Curcuma longa*. Curcumin has been used in ancient Asian medicine, and has been described to possess antioxidant and anti-inflammatory properties.[65–67] Many studies have reported a protective effect of curcumin against alcohol-induced oxidative stress.[68–73]

The treatment of ethanol-fed rats with curcumin, via intragastric tube, attenuated ethanol-induced liver injury through the inhibition of NF-κB activation and oxidative stress.[71] A murine chronic model for alcoholic liver disease, displaying increased ROS and malondialdehyde (MDA) levels, GSH depletion, and the impairment of antioxidant responses, had each of these parameters values reversed by curcumin treatment.[70] In mice acutely exposed to ethanol, curcumin increased the activity of SOD, GPx, and the total antioxidant capacity (T-AOC) in liver, and decreased the serum levels of aspartate aminotransferase (AST), and alanine aminotransferase (ALT).[72] In a rat model of acute alcoholism, curcumin, administered in a chelated form (Zn(II)-curcumin), was more effective than curcumin itself at the same dose. It increased SOD activity, prevented the alcohol-induced elevation of serum levels of MDA, decrease GSH levels, and inhibited liver injury.[73]

Epigallocatechin Gallate

Epigallocatechin gallate (EGCG), or epigallocatechin-3-gallate, is the major polyphenol in green tea (*Camellia sinensis*), and is known for its antioxidant character.[74] In Japan, tea has been used for decades to alleviate the discomfort caused by excessive alcohol intake. Kakuda et al.[75] investigated the effects of green tea extract on alcohol metabolism in ICR mice (high resistance to ethanol), and found that the treatment resulted in lower levels of alcohol and acetaldehyde in blood and liver. In addition, the administration of the EGCG lowered the acetate and acetone content in blood and liver, suggesting a detoxification effect.

Green tea consumption has exhibited protective effects on experimental animals and in humans under pathological conditions where oxidative stress is involved.[76,77]

Arteel et al.[78] suggested that antioxidants found in green tea are able to prevent early alcohol-induced liver injury, most likely by preventing oxidative stress. According to their study, green tea extract almost completely blocked lipid peroxidation in alcohol-fed rats, and also reduced the serum levels of ALT and TNF-α. Similar protective effects of EGCG were found subsequently in rats fed a liquid diet containing ethanol. EGCG was shown to reduce the ethanol-induced liver injury, and the serum levels of AST and ALT.[79]

EGCG has been described to suppress ethanol-induced expression of Mn-SOD and Cu/Zn-SOD, and abolish the lipid peroxidation of cellular membrane in pancreatic stellate cells cultured *in vitro*, suggesting a preventive role in pancreatic fibrosis through an antioxidant effect.[80] Kaviarasan et al.[81] also demonstrated a decrease in ethanol-induced lipid peroxidation in rats treated with EGCG.

Additional studies have suggested a potential for EGCG to protect against alcoholic neuropathy through the modulation of oxide-inflammatory cascade.[82,83] Interestingly, a combination of three natural compounds (taurine, EGCG, and genistein) exhibited benefits in the treatment of alcohol-induced liver fibrosis in rats, possibly through diminishing the synthesis and release of inflammatory mediators.[84] The results showed increased activity of SOD and GPx in liver, decreased hepatic MDA content, and reduced levels of serum ALT and AST. Additionally, the treatment inhibited the pro-inflammatory serum cytokines IL-6, TNF-α, and MPO that had been upregulated by the ethanol. Together, these findings supported that this combination could represent a new protective approach, and could be useful for the prevention and treatment of liver fibrosis, and even cirrhosis.[84]

Ginsenosides

Ginsenosides are a group of glycosylated triterpenes, also known as saponins, that represent the active compounds found in the root *Panax*.[85] These compounds have been reported to display antioxidant potential among other properties. *Panax ginseng* (ginseng) has been used in Asian medicine for over 2000 years, and has been shown to exert pharmacological and physiological effects in humans, as well as animals.[86]

Panax notoginseng has been shown to inhibit hepatic lipid peroxidation, and suggested as a hepatoprotective agent against chronic ethanol-induced toxicity in mice.[87] Subsequently, Lee et al.[88] reported that black ginseng (steamed red ginseng) was able to inhibit ethanol-induced teratogenesis in mouse embryos, *in vitro*. This result was attributed to the antioxidant activity of ginseng that restored the gene expression of antioxidant enzymes (cytosolic GPx, phospholipid hydroperoxide GPx, and selenoprotein P) that had been depressed by the ethanol exposure.

Li et al.[89] reported that saponins from *Panax japonicus* exerted protective effect against alcohol-induced hepatic damage in mice that was believed to have occurred through antioxidant action. The saponins served as hydroxyl radical scavengers in normal human liver cells (L-02) exposed to alcohol *in vitro*, whereas in mice they decreased the serum levels of AST, ALT, and MDA, and restored the GPx and SOD activities, parameters altered due to ethanol exposure. Moreover, the treatment of the mice with saponins also restored the mRNA levels of GPx3, SOD1, and SOD3, while the transcript levels for CAT, GPX1, and SOD2 remained low.

Park et al.[90] proposed that the Korean red ginseng and its primary ginsenosides (Rg3 and Rh2) are able to inhibit oxidative injury induced by alcohol, through the suppression of MAPK pathway in mouse hepatocytes (TIB-73 cells) *in vitro*. In a rat model of alcohol-induced liver fibrosis, ginsenoside Rg1 exhibited antioxidant and anti-inflammatory effects, and protective action against liver fibrosis, probably through promoting the nuclear translocation of Nrf2 and expression of antioxidant enzymes. Ginsenoside Rg1 repressed the serum levels of enzymes induced by ethanol (ALT, AST, lactate dehydrogenase, and alkaline phosphatase), lowered the hepatic MDA content, and increased the activity of antioxidant enzymes (SOD, GPx, and catalase) in the liver. In addition, this ginsenoside upregulated the expression of Nrf2, a transcription factor responsible for regulating the response to oxidative stress.[91]

Ginsenoside-free molecules (GFM) from steam-dried ginseng berries were recently proposed as a new and promising treatment against ethanol-induced hangover, because of its effects on alcohol metabolism. The most active ingredient appeared to be linoleic acid. GFM was shown to scavenge free-radicals, increase the expression of primary enzymes in HepG2 (human hepatocellular carcinoma) cells, and protect them from ethanol action. Furthermore, in mice, GFM lowered the blood levels of ethanol and acetaldehyde.[92]

GSH and Precursors

In animals, dietary GSH supplementation is effective in maintaining GSH homeostasis, but it requires chronic GSH ingestion to prevent oxidative damage.[93–95] However, dietary supplementation with GSH or GSH precursors has also been shown to restore GSH levels, and reduce the risk of alcoholic lung injury,[93,96] and has been suggested as a better therapeutic approach.[96,97]

Chronic alcohol abuse has been shown to deplete GSH levels in bronchoalveolar lavage fluid of human subjects,[98] and in alveolar macrophages of rats,[99] and this is associated with lung oxidative stress and phagocytosis dysfunction,[95,99] through the positive regulation of NADPH oxidases.[96] Since this clinical condition increases the susceptibility to respiratory infections, antioxidant treatment with oral GSH has been tested. In rats, it restored the GSH pool and alveolar macrophage phagocytic function; furthermore, it attenuated the alcohol-induced oxidative stress in lungs by down regulating NADPH oxidases.[96] However, additional studies are still required to assure the effectiveness of this treatment as a novel therapeutic approach.

N-Acetylcysteine is a precursor of cysteine needed for GSH synthesis, and has been tested as a therapy to alleviate adverse effects in alcohol abuse.[97,100] *N*-Acetylcysteine has the potential for treating alcoholic lung disease due to its property of restoring GSH homeostasis in the lung, by either GSH replenishment, or increasing its endogenous synthesis. Additionally, *N*-acetylcysteine can also restore antioxidant enzymes (glutathione reductase and SOD).[97] Furthermore, rats in alcohol withdrawal supplemented with *N*-acetylcysteine seem to show a better serum lipids profile and antioxidant response in liver.[100]

N-acetylcysteine is effective in treating alcohol-fed animals against liver fat accumulation, inflammation, and injury.[101–103] The use of another glutathione precursor, S-adenosylmethionine (SAMe), was shown to attenuate liver injury in alcohol-fed baboons;[104] however, it presented only a limited improvement of survival in alcoholic hepatitis patients.[105]

In an experimental model of alcohol-related steatohepatitis, Setshedi et al.[102] showed that the *N*-acetylcysteine treatment diminished the severity of oxidative stress/inflammation, but not the persistence of insulin/IGF signaling impairment in liver, fact that indicates the need for additional therapeutic agents, such as insulin sensitizers.

Polyenylphosphatidylcholine (PPC)

Polyenylphosphatidylcholine (PPC) is an antioxidant that restores the phospholipids in damaged membranes, and reactivates the enzymes required for the phospholipid regeneration. PPC supplementation of alcohol-fed animals has prevented depletion of SAMe in rats and baboons;[104,106] moreover, it has protected the baboons against liver fibrosis and cirrhosis by stimulating collagenase,[107] and also against oxidative stress by opposing lipid peroxidation.[16]

Puerarin

A *Puerariae radix* extract has been shown to mitigate liver injury and lipid accumulation induced by chronic alcohol intake,[108] that has been attributed to three major isoflavonoids present in this plant – puerarin, daidzin, and daidzein. *Puerariae radix* extract is obtained from the root of *Pueraria lobata*, also known as kudzu, a plant used in traditional Chinese medicine to treat alcoholism. This treatment increased the activity of the antioxidant enzymes Cu/Zn SOD and catalase, and attenuated the hepatic oxidative damage in rats exposed to ethanol.[109] Daidzin has been identified as a selective inhibitor of aldehyde dehydrogenase (ALDH-2),[110–112] and puerarin has been found to protect against acute alcoholic liver injury and alcoholism-related disorders by inhibiting the oxidative stress.[113,114]

Quercetin

Quercetin, a flavonoid antioxidant widely found in fruits, vegetables, and grains, has been reported to attenuate the oxidative stress during chronical alcoholism. For examples, Molina et al.[115] showed that quercetin treatment reduced the hepatic lipid peroxidation, and increased the activity of antioxidant enzymes in mice exposed to ethanol. In rats with alcohol-induced neuropathy, quercetin was reported to alleviate allodynia, hyperalgesia, and impaired motor coordination and nerve conduction. It also decreased the levels of membrane-bound $Na(^{+})$-$K(^{+})$-ATPase, MDA, myeloperoxidase, and nitric oxide, suggesting a neuroprotective effect.[116]

Resveratrol

Resveratrol is an antioxidant and anti-inflammatory polyphenol present in purple grape juice, peanuts, and red wine.[117] Resveratrol has been reported to display a protective effect against ethanol-induced hepatotoxicity.[118,119] For example, Bujanda et al.[120] indicated that resveratrol reduced the mortality and liver damage in mice submitted to chronic intoxication with up to 40% v/v alcohol in drinking water. The mechanisms of action were not clear, but the authors suggest these effects might be due to the antioxidant and anti-inflammatory properties of resveratrol.

Resveratrol has been reported to upregulate sirtuin 1 (SIRT1), the expression of which is suppressed by ethanol intake, an effect that may be linked to liver steatosis in mice chronically exposed to alcohol feeding.[73] SIRT1 has been proposed as a therapeutic target for treatment of human alcoholic fatty liver disease;[73,121] however, Oliva et al.[122] determined that ethanol affects SIRT1/PGC1α pathway, and resveratrol prevents this activation by ethanol, but not the liver pathology in rats fed ethanol.

Because oxidative stress is known to play an important role in alcohol-induced neurotoxicity, resveratrol was tested in ethanol-treated embryonic dorsal root ganglion (DRG) neurons *in vitro*.[123] In this study, resveratrol was shown to protect the DRG neurons from neurotoxicity, and attenuate some effects increased by ethanol, such as the levels of ROS, MDA, nitrite, glutathione, and SOD.

Selenium

Due to its antioxidant properties, selenium has been proposed for the prevention and treatment of liver cancer in alcoholics.[124] Selenium is an essential and powerful antioxidant as it is the cofactor of the GPx enzyme that operates closely with the vitamins E and C to neutralize the free radicals.

In a study involving rats and a Lieber–DeCarli experimental design, González-Reimers et al.[125] showed that the administration of seleniomethionine prevented the appearance of early signs of ethanol-mediated liver injury, by decreasing liver fat amount and hepatocyte ballooning. Moreover, the treatment increased GPx activity, and the serum and liver content of selenium.

Silymarin

Silymarin is a flavonoid derived from *Silybum marianum*, a plant commonly known as milk thistle. Silymarin exhibits hepatoprotective and regenerative properties,[126,127] and is one of the most widely used natural compounds for the treatment of hepatic diseases worldwide, due to its antioxidant, anti-inflammatory, and antifibrotic activities. Silymarin also increases hepatic levels of glutathione by raising cysteine availability, while inhibiting its catabolism to taurine that may enhance the antioxidant defense in liver.[128]

Ferenci et al.[129] reported a slight increase in survival of patients with alcoholic cirrhosis, after silymarin treatment; however, in another study, no effect on survival or on the course of the disease was observed in alcoholic patients with liver cirrhosis.[130] Silymarin was found to oppose oxidative stress, and to slow the progression of hepatic fibrosis in alcohol-fed baboons.[131]

The potential benefit of silymarin in liver diseases therapy remains a controversial issue.[127] In spite of substantial evidence suggesting that this treatment represents a promising approach to attenuate liver diseases, there are some contradictory data. It is known, though, that silymarin is innocuous. Additional molecular and clinical studies investigating this compound are needed in order to find out an effective intervention.

Sulfur-Containing Compounds

Although sulfur-containing compounds are chemically distinct and cannot be grouped into a single class, some have shown potential for the treatment of oxidative stress induced by ethanol metabolism. α-Lipoic acid is a natural eight carbon disulfide compound that plays a metabolic role as cofactor of pyruvate dehydrogenase complex.[132] In addition, α-lipoic acid and its reduced form, dihydrolipoic acid, are potent antioxidants, in both aqueous or lipid phase, and act synergistically with other antioxidants (glutathione, and vitamins C and E), and are able to recycle ascorbic acid and vitamin E.[133,134] In cultured mouse hippocampal cells, incubated with ethanol for 24 h, α-lipoic acid administration reduced the protein carbonyl content, maintained glutathione levels, and decreased the ethanol-induced neurotoxicity.[133] Likewise, Shirpoor et al.[134] also showed the protective effect of α-lipoic acid upon ethanol-induced DNA damage, lipid peroxidation, and protein oxidation, in the developing rat hippocampus and cerebellum. Nonetheless, in a randomized double-blind trial with precirrhotic patients, the supplementation with 300 mg α-lipoic acid per day, for 6 months, did not restore serum gamma glutamyl transpeptidase and AST levels, or improve histological changes in liver, in either drinking, or abstinent patients.[132]

Thiamine (vitamin B1) is a water soluble sulfur-containing vitamin that plays an essential role in energy metabolism, from the tricarboxylic acid cycle.[135] Alcoholics are usually deficient in thiamine due to inadequate intake, absorption, and metabolism, and often require supplementation with physiological doses for treatment or prevention of beriberi and Wernick-Korsakoff Syndrome.[135,136] Although there are no clinical trials, some animal studies have shown an antioxidant effect of thiamine at supraphysiological doses. In an acute alcohol intoxication model, thiamine supplementation prevented changes in parameters related to lipid peroxidation, possibly due to the interaction and regeneration of glutathione.[4] Moreover, thiamine supplementation was effective in treating hepatic- and neuro-toxicity in two studies, in which toxicity was induced by 90 days of alcohol administration, followed by 30 days of abstinence and antioxidant treatment.[136,137]

Table 10.1 summarizes the antioxidants reviewed in this chapter, and the main outcomes described by studies about their use, in different organisms and experimental models for alcohol exposure.

CONCLUSIONS

A large number of animal and cell culture studies have shown the benefits of natural antioxidant treatments, not only in alcoholism, but also in several other conditions where oxidative damage ensues. Although there are promising antioxidant candidates that have good bioavailability, safety, and low cost for translation to the clinic, most has not achieved satisfactory outcomes when assessed in controlled clinical trials. Part of this failure may be due to the fact that when an antioxidant imbalance occurs, it may be the direct causation of the observed pathologies. In such instances, antioxidant therapy will unlikely be successful. An additional problem is that the methods available to measure the antioxidant efficacy may not be sufficiently accurate, fact

TABLE 10.1 Summary of the Main Antioxidants Used as Treatment in Studies Related to Alcoholism and Respective Outcomes

Antioxidant	Organism	Outcomes
α-Tocopherol	• Ethanol-fed rats	• Prevented Mallory bodies and inflammatory infiltration in liver.[39] • Restored plasma levels of activities of aspartate aminotransferase, total radical-trapping antioxidant potential, and conjugated dienes.[56]
	• Ethanol-fed pregnant rats	• Alleviated the brain atrophy and DNA damage, and reduced homocysteine levels of offspring rats.[54]
	• Ethanol-fed mice	• Recovery of redox status.[55]
Ascorbic acid	• Ethanol-fed rats	• Increased plasma GSH and hepatic S-adenosylmethionine concentrations.[56]
α-Tocopherol + ascorbic acid	• Ethanol-fed rats	• Coadministration of α-tocopherol and ascorbic acid attenuated lipid peroxidation, protein carbonyls, and liver fibrosis; increased the activity of antioxidant enzymes.[57]
β-Carotene	• Ethanol-fed rats	• Pro-oxidant effect.[34] • Maintenance of enzymatic aspartate and alanine aminotransferases, normal hepatic histology; increased hepatic GSH levels.[36] • Increased levels of F2-isoprostane and 4-hydroxynonenal; increased levels of the mitochondrial enzyme glutamate dehydrogenase in the bloodstream.[61] • Potentiated induction of CYP2E1.[62]
	• Hepatocytes obtained from rats fed a Lieber–DeCarli diet for 4 weeks	• Increased GSH levels.[59]
	• Ethanol-fed baboons	• Inflammatory response with autophagic vacuoles, and alterations of the endoplasmic reticulum and mitochondria.[60]
Curcumin	• Ethanol-fed rats	• Inhibition of NF-κB activation and oxidative stress.[71] • Increased superoxide dismutase activity, prevented increase of serum levels of malondialdehyde and decrease of GSH levels; inhibited liver injury.[73]
	• Ethanol-fed mice	• Decreased ROS and malondialdehyde levels; restored GSH; improved antioxidant responses.[70] • Increased the activity of superoxide dismutase, glutathione peroxidase, and the total antioxidant capacity (T-AOC) in liver; decreased the serum levels of aspartate and alanine aminotransferase.[72]
Epigallocatechin gallate	• Ethanol-fed mice	• Lowered levels of alcohol, acetaldehyde, acetate, and acetone in blood and liver, suggesting a detoxification effect.[75]
	• Ethanol-fed rats	• Decreased lipid peroxidation[78,81]; reduced the serum levels of alanine aminotransferase and TNF-α.[78] • Reduced the ethanol-induced liver injury, and the serum levels of aspartate and alanine aminotransferases.[79]
	• Pancreatic stellate cells (*in vitro*).	• Suppressed ethanol-induced expression of Mn-SOD and Cu/Zn-SOD and interrupted the lipid peroxidation of cellular membrane.[80]
Epigallocatechin gallate in association with taurine and genistein	• Rats with alcohol-induced liver fibrosis	• Increased the hepatic activity of superoxide dismutase and glutathione peroxidase; decreased hepatic malondialdehyde content; reduced levels of serum aspartate and alanine aminotransferases; inhibited the proinflammatory serum cytokines IL-6, TNF-α, and myeloperoxidases.[84]
Ginsenosides	• Ethanol-fed mice	• Inhibited hepatic lipid peroxidation.[87] • Decreased the serum levels of aspartate aminotransferase, alanine aminotransferases and malondialdehyde; restored the superoxide dismutase and glutathione peroxidase activities.[89]
	• Mouse embryos (*in vitro*)	• Inhibited ethanol-induced teratogenesis through antioxidant activity (restored the gene expression of antioxidant enzymes glutathione peroxidases and selenoprotein P).[88]

TABLE 10.1 Summary of the Main Antioxidants Used as Treatment in Studies Related to Alcoholism and Respective Outcomes *(cont.)*

Antioxidant	Organism	Outcomes
	• Mouse hepatocytes (TIB-73 cells) *in vitro*	• Inhibited the oxidative injury induced by alcohol through suppression of MAPK pathway.[90]
	• Rats with alcohol-induced liver fibrosis	• Upregulated the expression of the transcription factor Nrf2. Repressed the serum levels of enzymes induced by ethanol (aspartate aminotransferase, alanine aminotransferase, lactate dehydrogenase, and alkaline phosphatase); lowered the hepatic malondialdehyde content; increased the activity of antioxidant enzymes (superoxide dismutase, glutathione peroxidase, and catalase) in the liver.[91]
GSH and precursors		
• GSH	• Ethanol-fed rats	• Restored GSH levels.[93,96]
• *N*-Acetylcysteine	• Ethanol-fed rats	• Restored GSH levels.[93] • Rats in alcohol withdrawal supplemented with *N*-acetylcysteine seem to show better serum lipids profile, and antioxidant response in liver.[100]
	• Rats with alcohol-related steatohepatitis	• Reduced severity of oxidative stress/inflammation, but not the persistence of insulin/IGF signaling impairment in liver.[102]
• S-Adenosylmethionine (SAMe)	• Ethanol-fed baboons • Alcoholic hepatitis patients	• Attenuated liver injury.[104] • Limited improvement of survival.[105]
Polyenylphosphatidylcholine (PPC)	• Ethanol-fed rats	• Prevented depletion of SAMe.[104]
	• Ethanol-fed baboons	• Decreased oxidative stress by opposing lipid peroxidation.[16] • Prevented depletion of SAMe.[106] • Protected against liver fibrosis and cirrhosis by stimulating collagenase.[107]
Puerarin	• Ethanol-fed rats	• Increased the activity of Cu/Zn superoxide dismutase and catalase; attenuated the hepatic oxidative damage.[109]
Quercetin	• Ethanol-fed mice	• Reduced the hepatic lipid peroxidation and increased the activity of antioxidant enzymes.[115]
	• Rats with alcohol-induced neuropathy	• Alleviated allodynia, hyperalgesia, and impaired motor coordination and nerve conduction. It also decreased the levels of membrane-bound $Na(^{+})$-$K(^{+})$-ATPase, malondialdehyde, myeloperoxidase, and nitric oxide, suggesting a neuroprotective effect.[116]
Resveratrol	• Ethanol-fed rats	• Protective effect against ethanol-induced hepatotoxicity.[118,119]
	• Ethanol-fed mice	• Antioxidant and anti-inflammatory properties.[120]
	• Embryonic dorsal root ganglion neurons (*in vitro*)	• Protected the cells from neurotoxicity; attenuated the levels of reactive oxygen species and malondialdehyde.[123]
Selenium	• Ethanol-fed rats	• Decreased liver fat content and hepatocyte ballooning; increased glutathione peroxidase activity and the serum and liver content of selenium.[125]
Silymarin	• Ethanol-fed mice	• Increased hepatic levels of glutathione.[128]
	• Ethanol-fed baboons	• Opposed the oxidative stress and slowed the progression of hepatic fibrosis.[131]
Sulfur-containing compounds		
• α-Lipoic acid/ dihydrolipoic acid	• Mouse hippocampal cells (*in vitro*)	• Act synergistically with other antioxidants (glutathione and vitamins C and E).[133,134]
	• Precirrhotic patients	• Did not restore serum gamma glutamyl transpeptidase and aspartate aminotransferase levels, or improve histological changes in liver in either drinking or abstinent patients.[132]
• Thiamine	• Ethanol-fed rats	• Prevented changes in parameters related to lipid peroxidation.[4]

that can confound the interpretation of results; thus, an unsuccessful outcome in a clinical trial could be due to insufficient antioxidant dosage, or lack of tools capable to detect changes in a particular marker being examined. Additionally, it is still controversial if antioxidant therapy may compromise the function of ROS in innate immune responses through neutrophils, or whether it would deregulate important signaling mechanisms. Therefore, the development of more precise methods to evaluate the antioxidant efficacy is essential for the clinical trials, and the design of a safe antioxidant therapy in alcoholism. More studies are needed, in particular well conducted clinical trials and assessment of wide endpoints, to fully understand the potential and limitations of natural antioxidant compounds.

Key Facts

- Reactive oxygen species are highly reactive molecules with the ability to oxidize cell structures and biomolecules, and consequently cause DNA damage, lipid peroxidation, and protein modification that can harm the cell integrity.
- Ethanol abuse leads to increased reactive oxygen species synthesis.
- Oxidative stress is defined as the imbalance caused by an increased free-radicals production, and deficient antioxidant status.
- Antioxidants are molecules that exhibit the role to inhibit the oxidation of other molecules.
- Cellular antioxidant defense mechanisms, composed by enzymes, peptides, proteins, and compounds derived from diet, protect cell structures and biomolecules against damage caused by reactive oxygen species.

Summary Points

- Continuous ethanol intake impairs cellular antioxidant status. As such, oxidative stress induced by alcoholism plays a critical role in the pathogenesis of disorders in some vital organs, especially the liver.
- Many studies in alcoholism have investigated antioxidant compounds with the aim of finding an effective and safe alternative to restore the antioxidant balance, and attenuate cellular damage induced by free-radical molecules.
- This chapter reviews studies on several antioxidants and their use in clinical settings related to alcohol abuse.
- The antioxidants described in this review are: carotenoids, curcumin, epigallocatechin gallate, ginsenosides, puerarin, quercetin, resveratrol, selenium, silymarin, vitamin C, vitamin E, and sulfur-containing compounds.
- There are promising antioxidant candidates; however, the identification of specific biomarkers of oxidative damage to assess their efficacy is indispensable for the design of a safe antioxidant in the treatment of clinical conditions related to alcoholism.

Disclaimer

The views expressed in this manuscript do not necessarily reflect those of the US Food and Drug Administration.

References

1. Cederbaum AI, Lu Y, Wu D. Role of oxidative stress in alcohol-induced liver injury. *Arch Toxicol* 2009;**83**:519–48.
2. Cooke MS, Evans MD, Herbert KE, Lunec J. Urinary 8-oxo-2'-deoxyguanosine--source, significance and supplements. *Free Radic Res* 2000;**32**:381–97.
3. Jordão Jr AA, Chiarello PG, Arantes MR, Meirelles MS, Vannucchi H. Effect of an acute dose of ethanol on lipid peroxidation in rats: action of vitamin E. *Food Chem Toxicol* 2004;**42**:459–64.
4. Portari GV, Marchini JS, Vannucchi H, Jordão AA. Antioxidant effect of thiamine on acutely alcoholized rats and lack of efficacy using thiamine or glucose to reduce blood alcohol content. *Basic Clin Pharmacol Toxicol* 2008;**103**:482–6.
5. Inoue M, Sato EF, Nishikawa M, et al. Mitochondrial generation of reactive oxygen species and its role in aerobic life. *Curr Med Chem* 2003;**10**:2495–505.
6. Valko M, Rhodes CJ, Moncol J, Izakovic M, Mazur M. Free radicals, metals and antioxidants in oxidative stress-induced cancer. *Chem Biol Interact* 2006;**160**:1–40.
7. Michiels C, Raes M, Toussaint O, Remacle J. Importance of Se-glutathione peroxidase, catalase, and Cu/Zn-SOD for cell survival against oxidative stress. *Free Radic Biol Med* 1994;**17**:235–48.
8. Zingg JM. Vitamin E: an overview of major research directions. *Mol Aspects Med* 2007;**28**:400–22.
9. Lebold KM, Traber MG. Interactions between alpha-tocopherol, polyunsaturated fatty acids, and lipoxygenases during embryogenesis. *Free Radic Biol Med* 2014;**66**:13–9.
10. Traber MG. Cellular and molecular mechanisms of oxidants and antioxidants. *Miner Electrolyte Metab* 1997;**23**:135–9.
11. Brockes C, Buchli C, Locher R, Koch J, Vetter W. Vitamin E prevents extensive lipid peroxidation in patients with hypertension. *Br J Biomed Sci* 2003;**60**:5–8.
12. Pirozhkov SV, Eskelson CD, Watson RR, Hunter GC, Piotrowski JJ, Bernhard V. Effect of chronic consumption of ethanol and vitamin E on fatty acid composition and lipid peroxidation in rat heart tissue. *Alcohol* 1992;**9**:329–34.
13. McDonough KH. Antioxidant nutrients and alcohol. *Toxicology* 2003;**189**:89–97.
14. Bailey SM, Patel VB, Young TA, Asayama K, Cunningham CC. Chronic ethanol consumption alters the glutathione/glutathione peroxidase-1 system and protein oxidation status in rat liver. *Alcohol Clin Exp Res* 2001;**25**:726–33.
15. Lecomte E, Herbeth B, Pirollet P, et al. Effect of alcohol consumption on blood antioxidant nutrients and oxidative stress indicators. *Am J Clin Nutr* 1994;**60**:255–61.
16. Lieber CS, Leo MA, Aleynik SI, Aleynik MK, DeCarli LM. Polyenylphosphatidylcholine decreases alcohol-induced oxidative stress in the baboon. *Alcohol Clin Exp Res* 1997;**21**:375–9.

17. Rua RM, Ojeda ML, Nogales F, et al. Serum selenium levels and oxidative balance as differential markers in hepatic damage caused by alcohol. *Life Sci* 2014;**94**:158–63.
18. Albano E. Oxidative mechanisms in the pathogenesis of alcoholic liver disease. *Mol Aspects Med* 2008;**29**:9–16.
19. Bradford BU, Kono H, Isayama F, et al. Cytochrome P450 CYP2E1, but not nicotinamide adenine dinucleotide phosphate oxidase, is required for ethanol-induced oxidative DNA damage in rodent liver. *Hepatology* 2005;**41**:336–44.
20. Liu LG, Yan H, Yao P, et al. CYP2E1-dependent hepatotoxicity and oxidative damage after ethanol administration in human primary hepatocytes. *World J Gastroenterol* 2005;**11**:4530–5.
21. Das SK, Vasudevan DM. Alcohol-induced oxidative stress. *Life Sci* 2007;**81**:177–87.
22. Abbey M, Nestel PJ, Baghurst PA. Antioxidant vitamins and low-density-lipoprotein oxidation. *Am J Clin Nutr* 1993;**58**:525–32.
23. Bjorneboe A, Bjorneboe GE, Drevon CA. Absorption, transport and distribution of vitamin E. *J Nutr* 1990;**120**:233–42.
24. Koch OR, Pani G, Borrello S, et al. Oxidative stress and antioxidant defenses in ethanol-induced cell injury. *Mol Aspects Med* 2004;**25**: 191–8.
25. Kawase T, Kato S, Lieber CS. Lipid peroxidation and antioxidant defense systems in rat liver after chronic ethanol feeding. *Hepatol* 1989;**10**:815–21.
26. Tyopponen JT, Lindros KO. Combined vitamin E deficiency and ethanol pretreatment: liver glutathione and enzyme changes. *Int J Vitam Nutr Res* 1986;**56**:241–5.
27. Giavarotti L, D'Almeida V, Giavarotti KA, et al. Liver necrosis induced by acute intraperitoneal ethanol administration in aged rats. *Free Radic Res* 2002;**36**:269–75.
28. Fernandez-Checa JC, Ookhtens M, Kaplowitz N. Effect of chronic ethanol feeding on rat hepatocytic glutathione. Compartmentation, efflux, and response to incubation with ethanol. *J Clin Invest* 1987;**80**:57–62.
29. Hirano T, Kaplowitz N, Tsukamoto H, Kamimura S, Fernandez-Checa JC. Hepatic mitochondrial glutathione depletion and progression of experimental alcoholic liver disease in rats. *Hepatology* 1992;**16**:1423–7.
30. Kerai MD, Waterfield CJ, Kenyon SH, Asker DS, Timbrell JA. Taurine: protective properties against ethanol-induced hepatic steatosis and lipid peroxidation during chronic ethanol consumption in rats. *Amino Acids* 1998;**15**:53–76.
31. Kerai MD, Waterfield CJ, Kenyon SH, Asker DS, Timbrell JA. Reversal of ethanol-induced hepatic steatosis and lipid peroxidation by taurine: a study in rats. *Alcohol Alcohol* 1999;**34**:529–41.
32. Morton S, Mitchell MC. Effects of chronic ethanol feeding on glutathione turnover in the rat. *Biochem Pharmacol* 1985;**34**:1559–63.
33. Oh SI, Kim CI, Chun HJ, Park SC. Chronic ethanol consumption affects glutathione status in rat liver. *J Nutr* 1998;**128**:758–63.
34. Portari GV, Jordão Jr AA, Meirelles MS, Marchini JS, Vannucchi H. Effect of beta-carotene supplementation on rats submitted to chronic ethanol ingestion. *Drug Chem Toxicol* 2003;**26**:191–8.
35. Sadrzadeh SM, Nanji AA, Meydani M. Effect of chronic ethanol feeding on plasma and liver alpha- and gamma-tocopherol levels in normal and vitamin E-deficient rats. Relationship to lipid peroxidation. *Biochem Pharmacol* 1994;**47**:2005–10.
36. Lin WT, Huang CC, Lin TJ, et al. Effects of beta-carotene on antioxidant status in rats with chronic alcohol consumption. *Cell Biochem Funct* 2009;**27**:344–50.
37. Meydani M, Seitz HK, Blumberg JB, Russell RM. Effect of chronic ethanol feeding on hepatic and extrahepatic distribution of vitamin E in rats. *Alcohol Clin Exp Res* 1991;**15**:771–4.
38. Monteiro TH, Silva CS, Cordeiro Simoes Ambrosio LM, Zucoloto S, Vannucchi H. Vitamin E alters inflammatory gene expression in alcoholic chronic pancreatitis. *J Nutrigenet Nutrigenomics* 2012;**5**: 94–105.
39. Silva CS, Monteiro TH, Simoes-Ambrosio LM, et al. Effects of alpha-tocopherol supplementation on liver of rats chronically exposed to ethanol. *J Nutrigenet Nutrigenom* 2013;**6**:125–36.
40. Leo MA, Rosman AS, Lieber CS. Differential depletion of carotenoids and tocopherol in liver disease. *Hepatol* 1993;**17**:977–86.
41. Kim YC, Kim SY, Sohn YR. Effect of age increase on metabolism and toxicity of ethanol in female rats. *Life Sci* 2003;**74**:509–19.
42. Chen LH, Xi S, Cohen DA. Liver antioxidant defenses in mice fed ethanol and the AIN-76A diet. *Alcohol* 1995;**12**:453–7.
43. Deaciuc IV, Fortunato F, D'Souza NB, et al. Modulation of caspase-3 activity and Fas ligand mRNA expression in rat liver cells in vivo by alcohol and lipopolysaccharide. *Alcohol Clin Exp Res* 1999;**23**:349–56.
44. Ozer NK, Boscoboinik D, Azzi A. New roles of low density lipoproteins and vitamin E in the pathogenesis of atherosclerosis. *Biochem Mol Biol Int* 1995;**35**:117–24.
45. Steiner M. Influence of vitamin E on platelet function in humans. *J Am Coll Nutr* 1991;**10**:466–73.
46. Bardag-Gorce F, French BA, Nan L, et al. CYP2E1 induced by ethanol causes oxidative stress, proteasome inhibition and cytokeratin aggresome (Mallory body-like) formation. *Exp Mol Pathol* 2006;**81**:191–201.
47. Adachi J, Asano M, Ueno Y, et al. 7alpha- and 7beta-hydroperoxycholest-5-en-3beta-ol in muscle as indices of oxidative stress: response to ethanol dosage in rats. *Alcohol Clin Exp Res* 2000;**24**:675–81.
48. Preedy VR, Patel VB, Reilly ME, Richardson PJ, Falkous G, Mantle D. Oxidants, antioxidants and alcohol: implications for skeletal and cardiac muscle. *Front Biosci* 1999;**4**:e58–66.
49. Fraga CG, Zamora R, Tappel AL. Damage to protein synthesis concurrent with lipid peroxidation in rat liver slices: effect of halogenated compounds, peroxides, and vitamin E1. *Arch Biochem Biophys* 1989;**270**:84–91.
50. Uto A, Dux E, Kusumoto M, Hossmann KA. Delayed neuronal death after brief histotoxic hypoxia in vitro. *J Neurochem* 1995;**64**: 2185–92.
51. Reilly ME, Patel VB, Peters TJ, Preedy VR. *In vivo* rates of skeletal muscle protein synthesis in rats are decreased by acute ethanol treatment but are not ameliorated by supplemental alpha-tocopherol. *J Nutr* 2000;**130**:3045–9.
52. Koll M, Beeso JA, Kelly FJ, et al. Chronic alpha-tocopherol supplementation in rats does not ameliorate either chronic or acute alcohol-induced changes in muscle protein metabolism. *Clin Sci (Lond)* 2003;**104**:287–94.
53. Tran TD, Jackson HD, Horn KH, Goodlett CR. Vitamin E does not protect against neonatal ethanol-induced cerebellar damage or deficits in eyeblink classical conditioning in rats. *Alcohol Clin Exp Res* 2005;**29**:117–29.
54. Shirpoor A, Salami S, Khadem-Ansari MH, Minassian S, Yegiazarian M. Protective effect of vitamin E against ethanol-induced hyperhomocysteinemia, DNA damage, and atrophy in the developing male rat brain. *Alcohol Clin Exp Res* 2009;**33**:1181–6.
55. Kaur J, Shalini S, Bansal MP. Influence of vitamin E on alcohol-induced changes in antioxidant defenses in mice liver. *Toxicol Mech Methods* 2010;**20**:82–9.
56. Lee SJ, Kim SY, Min H. Effects of vitamin C and E supplementation on oxidative stress and liver toxicity in rats fed a low-fat ethanol diet. *Nutr Res Pract* 2013;**7**:109–14.
57. Prathibha P, Rejitha S, Harikrishnan R, Das SS, Abhilash PA, Indira M. Additive effect of alpha-tocopherol and ascorbic acid in combating ethanol-induced hepatic fibrosis. *Redox Rep* 2013;**18**:36–46.
58. Fiedor J, Burda K. Potential role of carotenoids as antioxidants in human health and disease. *Nutrients* 2014;**6**:466–88.
59. Yang SC, Huang CC, Chu JS, Chen JR. Effects of beta-carotene on cell viability and antioxidant status of hepatocytes from chronically ethanol-fed rats. *Br J Nutr* 2004;**92**:209–15.

60. Leo MA, Aleynik SI, Aleynik MK, Lieber CS. beta-Carotene beadlets potentiate hepatotoxicity of alcohol. *Am J Clin Nutr* 1997;**66**: 1461–9.
61. Leo MA, Kim C, Lowe N, Lieber CS. Interaction of ethanol with beta-carotene: delayed blood clearance and enhanced hepatotoxicity. *Hepatology* 1992;**15**:883–91.
62. Kessova IG, Leo MA, Lieber CS. Effect of beta-carotene on hepatic cytochrome P-450 in ethanol-fed rats. *Alcohol Clin Exp Res* 2001;**25**:1368–72.
63. Siems W, Sommerburg O, Schild L, Augustin W, Langhans CD, Wiswedel I. Beta-carotene cleavage products induce oxidative stress in vitro by impairing mitochondrial respiration. *FASEB J* 2002;**16**:1289–91.
64. Luvizotto RA, Nascimento AF, Veeramachaneni S, Liu C, Wang XD. Chronic alcohol intake upregulates hepatic expression of carotenoid cleavage enzymes and PPAR in rats. *J Nutr* 2010;**140**: 1808–14.
65. Sharma OP. Antioxidant activity of curcumin and related compounds. *Biochem Pharmacol* 1976;**25**:1811–2.
66. Srimal RC, Dhawan BN. Pharmacology of diferuloyl methane (curcumin), a non-steroidal anti-inflammatory agent. *J Pharm Pharmacol* 1973;**25**:447–52.
67. Strimpakos AS, Sharma RA. Curcumin: preventive and therapeutic properties in laboratory studies and clinical trials. *Antioxid Redox Sign* 2008;**10**:511–45.
68. Bao W, Li K, Rong S, et al. Curcumin alleviates ethanol-induced hepatocytes oxidative damage involving heme oxygenase-1 induction. *J Ethnopharmacol* 2010;**128**:549–53.
69. Pyun CW, Kim JH, Han KH, Hong GE, Lee CH. *In vivo* protective effects of dietary curcumin and capsaicin against alcohol-induced oxidative stress. *Biofactors* 2014;**40**(5):494–500.
70. Rong S, Zhao Y, Bao W, et al. Curcumin prevents chronic alcohol-induced liver disease involving decreasing ROS generation and enhancing antioxidative capacity. *Phytomedicine* 2012;**19**:545–50.
71. Samuhasaneeto S, Thong-Ngam D, Kulaputana O, Suyasunanont D, Klaikeaw N. Curcumin decreased oxidative stress, inhibited NF-kappaB activation, and improved liver pathology in ethanol-induced liver injury in rats. *J Biomed Biotechnol* 2009;**2009**:981963.
72. Zeng Y, Liu J, Huang Z, Pan X, Zhang L. Effect of curcumin on antioxidant function in the mice with acute alcoholic liver injury. *Wei Sheng Yan Jiu* 2014;**43**:282–5.
73. You M, Liang X, Ajmo JM, Ness GC. Involvement of mammalian sirtuin 1 in the action of ethanol in the liver. *Am J Physiol Gastrointest Liver Physiol* 2008;**294**:G892–8.
74. Kim HS, Quon MJ, Kim JA. New insights into the mechanisms of polyphenols beyond antioxidant properties; lessons from the green tea polyphenol, epigallocatechin 3-gallate. *Redox Biol* 2014;**2**: 187–95.
75. Kakuda T, Sakane I, Takihara T, Tsukamoto S, Kanegae T, Nagoya T. Effects of tea (Camellia sinensis) chemical compounds on ethanol metabolism in ICR mice. *Biosci Biotechnol Biochem* 1996;**60**:1450–4.
76. Sai K, Kai S, Umemura T, et al. Protective effects of green tea on hepatotoxicity, oxidative DNA damage and cell proliferation in the rat liver induced by repeated oral administration of 2-nitropropane. *Food Chem Toxicol* 1998;**36**:1043–51.
77. Surh Y. Molecular mechanisms of chemopreventive effects of selected dietary and medicinal phenolic substances. *Mutat Res* 1999;**428**:305–27.
78. Arteel GE, Uesugi T, Bevan LN, et al. Green tea extract protects against early alcohol-induced liver injury in rats. *Biol Chem* 2002; **383**:663–70.
79. Yun JW, Kim YK, Lee BS, et al. Effect of dietary epigallocatechin-3-gallate on cytochrome P450 2E1-dependent alcoholic liver damage: enhancement of fatty acid oxidation. *Biosci Biotechnol Biochem* 2007;**71**:2999–3006.
80. Asaumi H, Watanabe S, Taguchi M, et al. Green tea polyphenol (-)-epigallocatechin-3-gallate inhibits ethanol-induced activation of pancreatic stellate cells. *Eur J Clin Invest* 2006;**36**:113–22.
81. Kaviarasan S, Sundarapandiyan R, Anuradha CV. Epigallocatechin gallate, a green tea phytochemical, attenuates alcohol-induced hepatic protein and lipid damage. *Toxicol Mech Methods* 2008;**18**: 645–52.
82. Tiwari V, Kuhad A, Chopra K. Epigallocatechin-3-gallate ameliorates alcohol-induced cognitive dysfunctions and apoptotic neurodegeneration in the developing rat brain. *Int J Neuropsychopharmacol* 2010;**13**:1053–66.
83. Tiwari V, Kuhad A, Chopra K. Amelioration of functional, biochemical and molecular deficits by epigallocatechin gallate in experimental model of alcoholic neuropathy. *Eur J Pain* 2011;**15**: 286–92.
84. Zhuo L, Liao M, Zheng L, et al. Combination therapy with taurine, epigallocatechin gallate and genistein for protection against hepatic fibrosis induced by alcohol in rats. *Biol Pharm Bull* 2012;**35**: 1802–10.
85. Murthy HN, Georgiev MI, Kim YS, et al. Ginsenosides: prospective for sustainable biotechnological production. *Appl Microbiol Biotechnol* 2014;**98**:6243–54.
86. Attele AS, Wu JA, Yuan CS. Ginseng pharmacology: multiple constituents and multiple actions. *Biochem Pharmacol* 1999;**58**:1685–93.
87. Lin CF, Wong KL, Wu RS, Huang TC, Liu CF. Protection by hot water extract of Panax notoginseng on chronic ethanol-induced hepatotoxicity. *Phytother Res* 2003;**17**:1119–22.
88. Lee SR, Kim MR, Yon JM, et al. Black ginseng inhibits ethanol-induced teratogenesis in cultured mouse embryos through its effects on antioxidant activity. *Toxicol In Vitro* 2009;**23**:47–52.
89. Li YG, Ji DF, Zhong S, Shi LG, Hu GY, Chen S. Saponins from Panax japonicus protect against alcohol-induced hepatic injury in mice by up-regulating the expression of GPX3, SOD1 and SOD3. *Alcohol Alcohol* 2010;**45**:320–31.
90. Park HM, Kim SJ, Mun AR, et al. Korean red ginseng and its primary ginsenosides inhibit ethanol-induced oxidative injury by suppression of the MAPK pathway in TIB-73 cells. *J Ethnopharmacol* 2012;**141**:1071–6.
91. Li JP, Gao Y, Chu SF, et al. Nrf2 pathway activation contributes to anti-fibrosis effects of ginsenoside Rg1 in a rat model of alcohol- and CCl4-induced hepatic fibrosis. *Acta Pharmacol Sin* 2014;**35**: 1031–44.
92. Ik Lee D, Tae Kim S, Hoon Lee D, Min Yu J, Kil Jang S, Soo Joo S. Ginsenoside-free molecules from steam-dried ginseng berry promote ethanol metabolism: an alternative choice for an alcohol hangover. *J Food Sci* 2014;**79**:C1323–30.
93. Guidot DM, Brown LA. Mitochondrial glutathione replacement restores surfactant synthesis and secretion in alveolar epithelial cells of ethanol-fed rats. *Alcohol Clin Exp Res* 2000;**24**:1070–6.
94. Guidot DM, Modelska K, Lois M, et al. Ethanol ingestion via glutathione depletion impairs alveolar epithelial barrier function in rats. *Am J Physiol Lung Cell Mol Physiol* 2000;**279**:L127–35.
95. Holguin F, Moss I, Brown LA, Guidot DM. Chronic ethanol ingestion impairs alveolar type II cell glutathione homeostasis and function and predisposes to endotoxin-mediated acute edematous lung injury in rats. *J Clin Invest* 1998;**101**:761–8.
96. Yeligar SM, Harris FL, Hart CM, Brown LA. Glutathione attenuates ethanol-induced alveolar macrophage oxidative stress and dysfunction by downregulating NADPH oxidases. *Am J Physiol Lung Cell Mol Physiol* 2014;**306**:L429–41.
97. Kaphalia L, Calhoun WJ. Alcoholic lung injury: metabolic, biochemical and immunological aspects. *Toxicol Lett* 2013;**222**:171–9.
98. Moss M, Guidot DM, Wong-Lambertina M, Ten Hoor T, Perez RL, Brown LA. The effects of chronic alcohol abuse on pulmonary glutathione homeostasis. *Am J Respir Crit Care Med* 2000;**161**:414–9.

99. Brown LA, Ping XD, Harris FL, Gauthier TW. Glutathione availability modulates alveolar macrophage function in the chronic ethanol-fed rat. *Am J Physiol Lung Cell Mol Physiol* 2007;**292**:L824–32.
100. Ferreira Seiva FR, Amauchi JF, Ribeiro Rocha KK, et al. Effects of N-acetylcysteine on alcohol abstinence and alcohol-induced adverse effects in rats. *Alcohol* 2009;**43**:127–35.
101. Day CP. Treatment of alcoholic liver disease. *Liver Transpl* 2007;**13**: S69–75.
102. Setshedi M, Longato L, Petersen DR, et al. Limited therapeutic effect of N-acetylcysteine on hepatic insulin resistance in an experimental model of alcohol-induced steatohepatitis. *Alcohol Clin Exp Res* 2011;**35**:2139–51.
103. Wang HJ, Gao B, Zakhari S, Nagy LE. Inflammation in alcoholic liver disease. *Annu Rev Nutr* 2012;**32**:343–68.
104. Lieber CS, Casini A, DeCarli LM, et al. S-adenosyl-L-methionine attenuates alcohol-induced liver injury in the baboon. *Hepatol* 1990;**11**:165–72.
105. Mato JM, Camara J, Fernandez de Paz J, et al. S-adenosylmethionine in alcoholic liver cirrhosis: a randomized, placebo-controlled, double-blind, multicenter clinical trial. *J Hepatol* 1999;**30**:1081–9.
106. Aleynik SI, Lieber CS. Polyenylphosphatidylcholine corrects the alcohol-induced hepatic oxidative stress by restoring s-adenosylmethionine. *Alcohol Alcohol* 2003;**38**:208–12.
107. Lieber CS, Robins SJ, Li J, et al. Phosphatidylcholine protects against fibrosis and cirrhosis in the baboon. *Gastroenterol* 1994;**106**: 152–9.
108. Peng JH, Cui T, Sun ZL, et al. Effects of Puerariae Radix Extract on Endotoxin Receptors and TNF-alpha Expression Induced by Gut-Derived Endotoxin in Chronic Alcoholic Liver Injury. *Evid Based Complement Alternat Med* 2012;**2012**:234987.
109. Lee MK, Cho SY, Jang JY, et al. Effects of Puerariae Flos and Puerariae Radix extracts on antioxidant enzymes in ethanol-treated rats. *Am J Chin Med* 2001;**29**:343–54.
110. Keung WM, Klyosov AA, Vallee BL. Daidzin inhibits mitochondrial aldehyde dehydrogenase and suppresses ethanol intake of Syrian golden hamsters. *Proc Natl Acad Sci USA* 1997;**94**:1675–9.
111. Keung WM, Vallee BL. Daidzin and its antidipsotropic analogs inhibit serotonin and dopamine metabolism in isolated mitochondria. *Proc Natl Acad Sci USA* 1998;**95**:2198–203.
112. Rooke N, Li DJ, Li J, Keung WM. The mitochondrial monoamine oxidase-aldehyde dehydrogenase pathway: a potential site of action of daidzin. *J Med Chem* 2000;**43**:4169–79.
113. Zhang Z, Li S, Jiang J, Yu P, Liang J, Wang Y. Preventive effects of Flos Perariae (Gehua) water extract and its active ingredient puerarin in rodent alcoholism models. *Chin Med* 2010;**5**:36.
114. Zhao M, Du YQ, Yuan L, Wang NN. Protective effect of puerarin on acute alcoholic liver injury. *Am J Chin Med* 2010;**38**:241–9.
115. Molina MF, Sanchez-Reus I, Iglesias I, Benedi J. Quercetin, a flavonoid antioxidant, prevents and protects against ethanol-induced oxidative stress in mouse liver. *Biol Pharm Bull* 2003;**26**:1398–402.
116. Raygude KS, Kandhare AD, Ghosh P, Ghule AE, Bodhankar SL. Evaluation of ameliorative effect of quercetin in experimental model of alcoholic neuropathy in rats. *Inflammopharmacol* 2012;**20**:331–41.
117. Manzo-Avalos S, Saavedra-Molina A. Cellular and mitochondrial effects of alcohol consumption. *Int J Environ Res Public Health* 2010;**7**:4281–304.
118. Kasdallah-Grissa A, Mornagui B, Aouani E, et al. Resveratrol, a red wine polyphenol, attenuates ethanol-induced oxidative stress in rat liver. *Life Sci* 2007;**80**:1033–9.
119. Kasdallah-Grissa A, Mornagui B, Aouani E, et al. Protective effect of resveratrol on ethanol-induced lipid peroxidation in rats. *Alcohol Alcohol* 2006;**41**:236–9.
120. Bujanda L, Garcia-Barcina M, Gutierrez-de Juan V, et al. Effect of resveratrol on alcohol-induced mortality and liver lesions in mice. *BMC Gastroenterol* 2006;**6**:35.
121. Ajmo JM, Liang X, Rogers CQ, Pennock B, You M. Resveratrol alleviates alcoholic fatty liver in mice. *Am J Physiol Gastrointest Liver Physiol* 2008;**295**:G833–42.
122. Oliva J, French BA, Li J, Bardag-Gorce F, Fu P, French SW. Sirt1 is involved in energy metabolism: the role of chronic ethanol feeding and resveratrol. *Exp Mol Pathol* 2008;**85**:155–9.
123. Yuan H, Zhang W, Li H, Chen C, Liu H, Li Z. Neuroprotective effects of resveratrol on embryonic dorsal root ganglion neurons with neurotoxicity induced by ethanol. *Food Chem Toxicol* 2013;**55**:192–201.
124. Darvesh AS, Bishayee A. Selenium in the prevention and treatment of hepatocellular carcinoma. *Anticancer Agents Med Chem* 2010;**10**:338–45.
125. González-Reimers E, Monedero-Prieto MJ, González-Perez JM, et al. Relative and combined effects of selenium, protein deficiency and ethanol on hepatocyte ballooning and liver steatosis. *Biol Trace Elem Res* 2013;**154**:281–7.
126. Bergheim I, McClain CJ, Arteel GE. Treatment of alcoholic liver disease. *Dig Dis* 2005;**23**:275–84.
127. Vargas-Mendoza N, Madrigal-Santillan E, Morales-Gonzalez A, et al. Hepatoprotective effect of silymarin. *World J Hepatol* 2014;**6**:144–9.
128. Kwon do Y, Jung YS, Kim SJ, Kim YS, Choi DW, Kim YC. Alterations in sulfur amino acid metabolism in mice treated with silymarin: a novel mechanism of its action involved in enhancement of the antioxidant defense in liver. *Planta Med* 2013;**79**:997–1002.
129. Ferenci P, Dragosics B, Dittrich H, et al. Randomized controlled trial of silymarin treatment in patients with cirrhosis of the liver. *J Hepatol* 1989;**9**:105–13.
130. Pares A, Planas R, Torres M, et al. Effects of silymarin in alcoholic patients with cirrhosis of the liver: results of a controlled, double-blind, randomized and multicenter trial. *J Hepatol* 1998;**28**:615–21.
131. Lieber CS, Leo MA, Cao Q, Ren C, DeCarli LM. Silymarin retards the progression of alcohol-induced hepatic fibrosis in baboons. *J Clin Gastroenterol* 2003;**37**:336–9.
132. Marshall AW, Graul RS, Morgan MY, Sherlock S. Treatment of alcohol-related liver disease with thioctic acid: a six month randomised double-blind trial. *Gut* 1982;**23**:1088–93.
133. Pirlich M, Kiok K, Sandig G, Lochs H, Grune T. Alpha-lipoic acid prevents ethanol-induced protein oxidation in mouse hippocampal HT22 cells. *Neurosci Lett* 2002;**328**:93–6.
134. Shirpoor A, Minassian S, Salami S, Khadem-Ansari MH, Yeghiazaryan M. Alpha--lipoic acid decreases DNA damage and oxidative stress induced by alcohol in the developing hippocampus and cerebellum of rat. *Cell Physiol Biochem* 2008;**22**:769–76.
135. Portari GV, Vannucchi H, Jordão Jr AA. Liver, plasma and erythrocyte levels of thiamine and its phosphate esters in rats with acute ethanol intoxication: a comparison of thiamine and benfotiamine administration. *Eur J Pharm Sci* 2013;**48**:799–802.
136. Vidhya A, Renjugopal V, Indira M. Impact of thiamine supplementation in the reversal of ethanol induced toxicity in rats. *Indian J Physiol Pharmacol* 2013;**57**:406–17.
137. Ambadath V, Venu RG, Madambath I. Comparative study of the efficacy of ascorbic acid, quercetin, and thiamine for reversing ethanol-induced toxicity. *J Med Food* 2010;**13**:1485–9.

CHAPTER

11

Selenium Dietary Supplementation and Oxidative Balance in Alcoholism

Olimpia Carreras, PhD, María Luisa Ojeda, PhD, Fátima Nogales, MD

Department of Physiology, Faculty of Pharmacy, Seville University, Seville, Spain

INTRODUCTION

Chronic ethanol consumption produces a wide range of gastrointestinal effects, including alteration in the intake, absorption, and utilization of various nutrients.[1–4] Furthermore, alcohol consumption can cause major health and addiction problems, the consequence of toxic compounds produced by ethanol metabolism. It is important to remember that the liver is the main tissue involved in the oxidative metabolism of this compound. Therefore, this tissue is exposed to greater oxidation levels, alcohol being one of the major causes of alcoholic liver diseases (ALD) such as steatosis, steatohepatitis, fibrosis and cirrhosis.[5]

The current understanding of alcohol toxicity to organs suggests that alcohol initiates injury by generating oxidative and nonoxidative ethanol metabolites, and by the translocation of gut-derived endotoxins into the bloodstream.[6–7] Alcohol-induced oxidative stress is related to ethanol metabolism. Three pathways are regarded as the classic methods of ethanol metabolism, involving enzymes such as alcohol dehydrogenase, the microsomal ethanol-oxidation system, and catalase. Each of the three pathways leads to the formation of free radicals that are counteracted by the activity of the antioxidant defense system.[8] Free radicals can damage all types of biological molecules, including lipids, proteins, carbohydrates, and DNA. In fact, alcohol-dependent individuals are vulnerable to excessive free radical production; notably, oxidative DNA damage persisted after a 1-week detoxification period.[9]

There are four endogenous antioxidant defense enzymes in place to protect the organism from the adverse effects of free radicals, called glutathione peroxidase (GPx), glutathione reductase (GR), superoxide dismutase (SOD), and catalase (CAT). One of them, GPx, has selenium in its key active site, forming part of the protein family called selenoproteins that have several essential biological functions.[10]

Selenium (Se), an essential trace mineral, is present in many foods, such as bread and cereals, meat, fish, eggs and milk/dairy products. It is also added to other foods, and is available as a dietary supplement. Despite the existence of different selenium compounds in food, many people rarely eat foods that are good sources of selenium.[11] Intake recommendations for selenium in the dietary reference intake developed by the Recommended Dietary Allowance (RDA) is, in male and female, from 40 to 70 and 55 μg/day respectively; infants: 15 μg/day; children: 20–30 μg/day; pregnant women: 65 μg/day; and lactating mothers: 75 μg/day, changing with the metabolic stage. By 1957, Se was considered a toxic element, as both high and low intakes of organic/inorganic selenium cause several alterations. Se deficiency has been linked to an increased risk of infection, male infertility, low mood, a higher incidence of cancer, and Keshan disease, a miocardiopathy located in China.[12–13] However, high-dosage Se supplements (95–120 μg/L) are highly toxic.[14] Thus, the unregulated intake of dietary or pharmacological Se supplements has negative consequences on DNA integrity,[15] and, as such, their intake is considered to be a double-edged sword.[16]

Current opinion is that Se is of fundamental importance to human health. As previously mentioned, the biological effects of Se are mediated by selenium-containing proteins (selenoproteins). Although selenoproteins present diverse molecular pathways and biological functions, all of these proteins contain at least one selenocysteine (Se-Cys), a selenium-containing amino acid, and most have oxidoreductase functions. Se is associated with decreased oxidative stress, and

Molecular Aspects of Alcohol and Nutrition. http://dx.doi.org/10.1016/B978-0-12-800773-0.00011-2

it is involved in the antioxidant defense system. Se is, therefore, required as a micronutrient for the function of selenoproteins.[17] 25 selenoproteins have been identified in the human proteome.[18] Most selenoproteins have antioxidant activities, although there are other selenoproteins that participate in different processes, including the regulation of thyroid hormones (thioredoxin reductase (TrxRs), and iodothyronine deiodinase (DIOs) families); the reduction of oxidized proteins and membranes (GPx family); redox regulation processes (SelW, SelH, SelT, SelV); the regulation of apoptosis (SeP15), and selenium transport, storage, and immunomodulation (SelP: is the only selenoprotein that contains 10 Se-Cys residues). However, the biochemical role for most selenoproteins is still partially unknown.[19]

In recent years, several families of human selenoproteins have been partially characterized with respect to their function.[20–23] The GPx family that comprises at least eight members (GPx1-GPx8) in mammals has an antioxidant activity.[24] Among them, GPx1 can catalyze the reduction of hydrogen peroxide (H_2O_2) and lipid hydroperoxides, using GSH as a reduced cofactor.[25] GPx1 is one of the selenoproteins that is most sensitive to changes in both Se status, and oxidative stress conditions.[26] GPx2 is mainly expressed in the gastrointestinal tract, and may serve as a first line of defense against oxidative stress induced by ingested oxidants. GPx3 is found in extracellular space and plasma with a lower activity (10-fold) than GPx1. GPx4 is present in cell membrane, since it is the only GPx member that detoxifies lipid peroxides inside the membrane of the cell, and, because of this, it is also known as phospholipid hydroperoxide GPx. Its function is also related to the transcriptional factor NFκB and apoptosis. GPx5, containing a cysteine (Cys) instead of Se-Cys in the active center is characterized as a secreted protein in the epididymis. GPx6 is a selenoprotein found in humans, but not in rats or mice, and is expressed in the olfactory epithelium and embryonic tissues, with an as yet unknown function. Also, GPx7 and GPx8 are Cys-GPxs with low GPx activity.[24,27–28]

Therefore, it is postulated that Se modulates oxidative stress mainly through the GPxs.[29] This property, together with others, means that Se deficiency causes several biological alterations. From the research by Rua et al.,[7] it could be concluded that Se levels are drastically reduced in alcoholic liver disease patients, showing that this element has a direct correlation with GPx activity and lipid oxidation. This suggests that the Se/MDA ratio could be an indicator of hepatic damage caused by alcohol consumption. These data point to Se as a possible dietary antioxidant therapy.

Taking into account that alcohol is a prooxidant, and knowing that Se levels and GPx activity are depleted in humans, data in animals should also be analyzed. Accordingly, it has been demonstrated in Wistar rats that ethanol affects Se absorption, retention, body distribution, selenoprotein expression, and antioxidative action.[30–36] However, dietary Se supplementation to the dams mitigated the adverse effects of alcohol in offspring exposed to ethanol,[32,34,36] increasing the expression of GPx1, GPx4, and SelP,[31] thus confirming Se as a mineral with antioxidant capacity.

ALCOHOL AND MALNUTRITION: SELENIUM SUPPLEMENTATION

As mentioned above, extensive evidence indicates that ethanol consumption affects human health because, among other factors, it compromises nutritional status,[1–3,37] causing primary and secondary malnutrition. Primary malnutrition occurs when alcohol replaces other nutrients in the diet, resulting in an overall reduced nutrient intake. Secondary malnutrition occurs when the drinker consumes adequate nutrients, but alcohol interferes with the absorption of those nutrients from the intestine, so they are not available to the body.[4] Therefore, excessive alcohol consumption may lead to the deficiency of micronutrients, including Se.[38]

Since the beginning of the 1980s, several studies stated that both chronic and acute alcoholism decreased the serum selenium levels.[39–40] These levels were especially lower in alcoholic patients with liver disease.[7,41] Therefore, lower Se levels are related to overall nutritional status, and the hepatic dysfunction produced.[41–43] Similar results were verified in rats that chronically consume alcohol.[44]

Parallel with these results, it has been observed that ethanol consumption influences nutritional status during gestation and lactation, because it alters the maternal endocrine function, disrupts hormonal relationships between mother and fetus, and affects their progeny's future health.[45] In experiments with mother rats exposed to chronic ethanol consumption and pair-feed dams, a growth deficit was found in their progeny, a deficit that was more exacerbated in the offspring of mothers exposed to ethanol.[33] This is provoked in part by a lower Se intake during the lactation period, itself caused by a decrease in milk Se content, and a decrease in the amount of milk consumed in offspring, coinciding with primary malnutrition. Administration of a Se-supplemented diet to ethanol-exposed rats prevents Se intake deficiency in the mothers and their offspring, improving Se status.

Ethanol also causes secondary malnutrition because it alters intestinal absorption, inducing morphological and biochemical alterations in the intestinal mucosa, and compromising the absorption of several nutrients, such as Se.[46–47] The intestinal absorption of Se depends on the forms that this compound presents in the diet, as well as the presence of other substrates. Organic species such as selenomethionine (Se-Met), that is predominant in foods, or selenocysteine (Se-Cys), show greater absorption than inorganic species, and are transported by Na^+-dependent neutral amino acid transporters.[48–49] These transporters are preferably located in the apical membrane of enterocytes, and are partially inhibited by their respective S-analog amino acids. Additionally, sodium selenite (inorganic form) is absorbed preferentially by simple diffusion, utilized almost exclusively by the paracellular pathway.[50] The uptake of these forms of Se from apical to basolateral membrane, is time-dependent, and they showed species-dependent transport efficiencies, ranking from highest to lowest: Se-Met > Se-Cys > sodium selenite.[50–51]

Although no specific studies linking Se absorption with alcohol intake have been developed, it has been established that chronic or acute alcohol exposure increases the permeability of the gut to macromolecules,[52] so a greater absorption of Se in any of its chemical forms might be expected. In this line, certain studies have examined the effect of ethanol on intestinal absorption during pregnancy and lactation, since besides altering the maternal nutrient transport, ethanol provokes anomalies in fetal growth, and can cause damage in the intestinal development of the offspring.[46,53] Thus, it has been demonstrated that ethanol treatment reduces the maximum transport of methionine, and inhibits Na^+-dependent methionine transporters in pregnant rats.[46,53] These transporters are also utilized by Se-Met. Therefore, ethanol could also inhibit the transport of Se-Met through intestinal cells during this physiological stage.

Moreover, during postnatal development, maternal exposure to ethanol decreases the pups' intestinal perimeter, as well as its length and weight.[36,53] These data are consistent with the research by Delgado et al.[54] who compared the intestinal development in offspring born to Se-deficient or Se-supplemented rats. Se-deficient offspring presented a decrease in all of the intestinal parameters, while the intestinal mucosa weight was greater in the Se-supplemented pups. This study suggests that since Se is the only deficient nutrient in the overall diet provided to Se-deficient mothers, this micronutrient is essential to the growth and maturity of the duodenum. Se appears to be crucial for the development of the intestinal mucosa, since it forms part of GPx2 that is expressed in the gastrointestinal tract. GPx2 induces cell proliferation and protects the epithelium from oxidative stress.[55]

The kinetic parameters of Se intestinal absorption showed that ethanol-exposed pups presented increased levels of Se-Met absorption (the main form of selenium found in milk), presenting the highest value of maximal velocity (V_{max}), compared with the control group.[36] When analyzing the role of Se supplementation on Se-Met absorption of ethanol-exposed pups, a significant increase with respect to control groups, supplemented with Se or not, was found, as well as a V_{max} value similar to non-Se-supplemented ethanol-exposed pups. This is possible because ethanol destabilizes the intercellular junctions of the epithelium, and promotes the paracellular pathway,[52] and thereby increases Se absorption. Additionally, in order to know the transporter's affinity to Se-Met in the progeny, an affinity constant (*Km*) was determined in ethanol-exposed groups, supplemented or not. Affinity was lower in the nonsupplemented ethanol group, followed by the Se-supplemented ethanol group.[36] The results suggested that Se-supplemented diets can enhance the transporter's affinity through Se-Met, probably because this element increases GPx2 activity that protects against oxidation, improving the structure of lipid membrane and the transporter.

However, it is difficult to determine the true intestinal absorption of nutrients in the body without using invasive methods; therefore, sometimes the apparent absorption of the nutrients is used. This parameter was determined from the quantity of nutrients ingested in diet, and excreted by feces and urine. In general terms, apparent Se absorption is unchanged after exposure to alcohol. In this context, Cho et al.[56] found no change in Seabsorption or retention, even in rats consuming an Se-deficient diet that were exposed to chronic ethanol consumption. In addition, it has been demonstrated that, during gestation and lactation, two special physiological periods, apparent Se absorption did not change with alcohol exposure, but the Se balance (or retention) decreased, since the maternal Se excretion pattern changed.[33] Thus, although ethanol-exposed rats ingested a lower amount of food and, therefore, of Se, they presented serum Se levels similar to control rats. However, they eliminated more Se in urine and less in feces. On the contrary, Se-supplemented diets enhanced its absorption and retention in organism, since Se intake increased, and its excretion via feces and urine diminished. It appears that the Se-supplemented diets employed increased the concentration of this antioxidant element in different tissues, and improved the nutritional status.[30] This circumstance may also occur in the progeny.

ALCOHOL AND SELENIUM TISSUE DISTRIBUTION: SELENIUM SUPPLEMENTATION

The distribution of Se in the organism, and its bioavailability, depend to a large extent on its chemical form in the diet;[57] also on sex and age;[26,58–59] the general oxidative corporal balance,[32,35] and, of course, on special physiological stages, such as gestation and lactation.[30]

Moreover, Se tissue distribution is not homogeneous; some organs have higher deposits of this element than others. Liver and kidney are rich in Se deposits; even in a deficient Se status, these tissues accumulate the nutrient in a nonmetabolized pool, when high doses are administered.[60] In mice and weanling rats, Se concentration in kidney was higher than in liver, and their Se deposits clearly increase after Se dietary supplementation.[61–62] Another tissue especially related to corporal Se distribution is the skeletal muscle that retains great amounts of Se when the diet is rich in this nutrient, and loses it when body Se levels decrease. This indicates that muscle redistributes Se to other tissues that have higher metabolic priority for it, such as testes.[63] In the few studies where Se tissue distribution has been measured, it was found that the Se deposited in the organs of beef cattle were, in the order of decreasing concentrations: kidney > liver > skeletal muscle > spleen;[64] and in male rats: liver > red blood cells > heart, hair and nails > skeletal muscle.[57] However, a complete map of organ Se distribution in control adult rats is still not well-established.

Interestingly, Se deposits determine in part the function of the proteins, or selenoproteins, in which it is incorporated. The members of this family exhibit diverse patterns of tissue distribution, ranging from ubiquitous, such as GPx1, to tissue-specific expression.[28] In mice, Hoffmann et al.[65] have evaluated mRNA levels for the entire murine selenoproteome in brain, testes, liver, kidney, lung, heart, intestine, and spleen, finding that the specific pattern of mRNA abundance differs between tissues. However, there were some common features of selenoproteome mRNA between tissues. For example, Gpx4 was the mostly highly-detected, present in nearly all of the tissues examined. SelP is another, highly abundant mRNA, especially in liver. Gpx3 was found in abundance in kidney, but also in heart. However, there is also a significant difference among tissues, in selenoprotein synthesis.

This different pattern of selenoproteins expression depends not only on the tissue, but also on Se deposits. Therefore, degradation of selenoprotein mRNAs under conditions of low Se availability is not uniform, with some transcripts clearly more sensitive to nonsense-mediated decay than others.[62] In this context, and in order to understand the hierarchy of selenoprotein expression at the transcriptome levels, from Se deficiency to toxicity in different tissues, Sunde and Raines[66] have demonstrated that Se deficiency in weanling rats downregulated only seven selenoproteins genes in rat liver, with three highly regulated Selh, Gpx1, and Selt. It was also found that there were no gene expression changes to supernutritional Se intake, and that 1,193 transcripts were altered after toxic Se status. These altered genes were specially related to an increase in reactive oxygen species, development, glucose metabolism, and angiogenesis, among which four selenoproteins: Selm, Selpw1, Txnrd1, and Gpx3, were upregulated. Barnes et al.[62] also found that a Se deficient status causes a decrease of selenoprotein mRNA of only Gpx1, Sepw1, and Selh in different tissues, such as kidney, liver, muscle, and testes of weanling rats, as well as a decrease of Gpx3 in serum. From this information, the selenoproteins hierarchy has been formulated. This term describes that selenoproteins are not uniformly supplied with Se, especially when availability is limited. Those that disappear rapidly under such conditions, such as GPx1, rank low in this hierarchy.[24]

As ethanol consumption affects oxidative balance, Se intake, absorption, and retention,[33] it could be expected that this drug affects Se tissue distribution, and tissue selenoprotein expression. However, there is little in the bibliography concerning this issue. In this context, Cho et al.[56] suggest that ethanol intake in weanling rats shunted Se to liver and to blood, and away from certain other tissues, such as kidney. In contrast to this observation, reduced liver Se concentrations were reported in male rats that had been pair-fed-ethanol-containing (20–35%) liquid diets.[67]

Using three experimental groups of Wistar rats – control, chronic ethanol, and pair-fed – Ojeda et al.[33] demonstrated that ethanol consumption by dams affects Se deposits and antioxidant balance in mothers, and in their progeny. It was also demonstrated that these effects are caused by a reduction in Se intake, and a direct alcohol-generated oxidation action. This study is especially interesting because all of the essential trace elements required for offspring development are transferred from the dam, via either the placenta or milk. Stress factors, such as alcoholism, or malnutrition during gestation and lactation in the dam, can affect the fetus' postnatal development.[68] In this context, the hypothesis of "fetal programming" was described, asserting that tissues can be programmed *in utero* during critical periods of development, with adverse consequences for their function in later life, especially if the fetus is undernourished.[69] Moreover, it has been well established that ethanol provokes teratogenia by promoting oxidative damage,[70] disturbing embryogenesis,[71] and causing fetal alcohol spectral disorders (FASD).[72] These effects could be even worse if the mother consumed ethanol while breastfeeding.[73]

Continuing this line of research, a specific study in chronic Se-supplemented ethanol dams demonstrated that Se tissue distribution in dams is altered by ethanol, reducing Se levels in the tissues with lower requirements: cortex, skeletal muscle, mammary gland, and salivary gland, but not affecting deposits in kidney, the tissue with the greatest levels of Se.[30] However, ethanol increased Se levels in spleen, heart, and liver. It appears that ethanol tends to retain more Se in spleen, heart, and liver, at the expense of other tissue (mainly skeletal muscle), in order to improve the redox balance and immunity, both of which are especially altered after ethanol consumption. Dietary Se supplementation to ethanol dams restored Se deposits to basal levels in all tissues, except for cortex. It also sequestered Se to the spleen, heart, and liver, these levels being even higher than in the control animals. Furthermore, this treatment also increased Se levels in the kidney.

As was reported,[33] Se deposits in several organs differ in order of decreasing concentrations in dams and in offspring. Moreover, alternative Se elimination routes (nails, hair and lung) participate more actively in pups, than during adulthood. Therefore, the changes provoked by ethanol consumption in offspring should be evaluated. To gain a more in-depth insight into the matter, Ojeda et al.[34] examined body distribution of Se in offspring exposed to ethanol during gestation and lactation, in order to detect the organs that were the most compromised by this drug and, therefore, identify which could suffer future oxidative damage. A Se-supplemented diet that could restore these values was also administered to dams. In control suckling pups, the Se deposits levels were in order of decreasing concentrations: kidney > liver > lung-testes > spleen-heart > pancreas-brain. It was found that ethanol-exposed pups maintained the tissue order in decreasing Se concentration, with the exception of testes and pancreas, but they had lower deposits in heart, liver, kidney, and testes, and higher in pancreas and in serum. The Se-supplemented diet used increased all of these impaired levels and restored Se pancreas concentration to a control status. This oral therapy mainly displaces Se to serum, kidney, and spleen. It is reported that kidney represents the principal site of GPx3 production and secretion to blood.[74] Perhaps this supplementation in the form of selenite increases GPx3 activity, and improves serum antioxidant activity, in order to maintain the first line of defense against ethanol-provoked oxidative stress. This hypothesis agrees with Payne and Southern,[75] who reported that Se stored in tissues could be utilized to maintain plasma GPx activity during periods of low Se intake. The higher levels of Se found in the spleen point to this tissue as a target tissue for Se storage, when there is a supra-normal intake of Se in pups, as is already well-known for liver, kidney, and skeletal muscle in adults. Nevertheless, a critical question is to elucidate the mechanisms by which the substance is selectively accumulated by key target organs. It must be pointed out that different LDL receptors are required for Se uptake into the organs, such as apolipoprotein E receptor-2 (ApoER-2) or megalin,[76] and that they are expressed differently, in function of the tissue involved.[77]

Recently, Jotty et al.[31] have studied the expression of the three main hepatic selenoproteins, Gpx1, Gpx4, and SelP,[65] in ethanol-exposed pups, in order to increase the knowledge about Se tissue deposits in offspring, and its biological implication, especially in the liver, as this is one of the most important Se reservoir tissues in pups. The liver is also the main ethanol metabolizing tissue that receives the oxidative products generated by this drug.[78] A Se-supplemented group was also evaluated. It was concluded that the depletion of hepatic Se levels after ethanol exposure in pups affects the expression of the three main hepatic selenoproteins in different manners. GPx1 synthesis decreased proportionally, according to the hepatic Se depletion caused by this drug, increasing after Se supplementation. In part, this was to be expected as this protein ranks low in the hierarchy of selenoproteins.[24] However, GPx1 activity was also affected by the direct action of ethanol, since it consumes some of its cofactors, GSH and NADPH.[79] Despite Se liver deposits being down-regulated after ethanol exposure, SelP expression increased in ethanol pups, reaching a plateau value similar to the supplemented pups. The difference between these pups was that, in ethanol pups, the SelP was delivered to the blood in order to displace broadly Se to other tissues,[80] while, in supplemented pups, SelP remained in the liver, acting as a reservoir. With regard to GPx4, its synthesis was greatly increased after ethanol exposure, and it increased even more after supplementation. It seems that this selenoprotein expression has an important role during breastfeeding, and after alcohol consumption. It might occur in order to prevent the effects of ethanol in membrane phospholipids oxidation (during lactation, the amount of phospholipids are greater than in adulthood),[81] to increase its anti-inflamatory properties,[82] or due to its antiapoptotic effects.[83] It could also be acting as a new reservoir target during this period, or it could have a pivotal role in the embryogenesis, midgestation, or breastfeeding periods, as suggested by different authors.[24,55,84–85]

ALCOHOL AND OXIDATIVE BALANCE: SELENIUM SUPPLEMENTATION

Ethanol causes its main body damage primarily by oxidative action. The oxidation provoked by ethanol depends on several factors, such as drinking patterns (chronic or acute);[86–87] the amount of alcohol consumed (moderate or excessive); the body organ or system affected;[32,35] the sex and age of the user, and their physiological status.[88] In rats, chronic ethanol consumption significantly decreased cytosolic and mitochondrial GPx activities, and caused a parallel increase in the oxidation of proteins in hepatocytes.[89–90] Furthermore, long-term ethanol abuse can also decrease intracellular antioxidant capacity by altering the activities of the endogenous antioxidant enzymes, such as SOD, CAT, GPx, and GR, as well as the intracellular levels of reduced glutathione (GSH) in numerous organs, such as liver, brain, and heart.[86,91–92] This intracellular redox imbalance results in the oxidation of proteins, lipids and DNA.

According to these results, several interventions have been proposed, in order to attenuate this oxidative damage. Such interventions include supplementation with antioxidants, molecules that possess the ability to stabilize free oxygen radicals.[93] As mentioned above, Se is a trace element, with an antioxidant property, since it forms part of selenoproteins, as GPx, SelP, and TrxR – that could decrease as a consequence of alcohol abuse.[13] Therefore, there are many studies that propose Se supplementation as a mechanism for enhancing antioxidant defense, and as a protection from alcohol-induced injuries. Thus, it has been found that the addition of Se to the diet, alone or with other antioxidants, reduces alcohol-induced hepatic damage.[43,47,94–96] Similarly, Kim et al.[97] confirmed that selenium inhibits ethanol-induced gastric mucosa lesions, as it prevents lipid peroxidation, and activates the radical-scavenging enzymes in a dose-dependent manner. This is possible because dietary Se enhances antioxidant enzymes' activity, particularly that of the selenoprotein GPx, as well as SOD, GR, and CAT enzymes.[43,96–97] Moreover, different authors have demonstrated that, after ethanol exposure, Se has a protective function in brain, testes, and serum.[98–100]

The deleterious effects of alcohol become even more harmful during pregnancy and lactation; not only does it affect pregnant or nursing mothers, but it can also cause a wide range of adverse effects in the developing fetus, and their offspring.[101] In addition, fetuses are more susceptible to ethanol than their mothers. Therefore, it has been demonstrated that the offspring of rats fed nontoxic levels of ethanol presented a decrease in liver growth and antioxidant activity, while their mothers were not affected.[102] In line with the above, breastfeeding pups exposed to ethanol during gestation and lactation showed important differences in their antioxidant body profile. These pups presented a decrease in liver Se levels and GPx activity, together with an increase in protein oxidation (Figure 11.1).[32] The reduction in hepatic Se deposits was directly related to GPx1 expression, since as indicated previously, the synthesis of this selenoprotein is highly regulated by Se deficiency.[31] Hepatic GPx activity is related to GPx1 expression; however, in this study, GPx1 activity also decreased due to the direct action of ethanol exposure. Furthermore, this drug produced an increase in serum Se levels and GPx activity in offspring.[32] This may occur because alcohol helps to maintain a greater amount of serum Se, at the expense of other tissues,[34] in order to increase extracellular GPx activity (in serum). It was also determined that, in pups, ethanol does not affect renal GPx activity, but it decreases Se deposits in their kidney tissue (Figure 11.2).[35] In kidney, there was no correlation between Se levels and GPx activity. The explanation for this is that the kidney is the main tissue for synthesizing extracellular GPx (GPx3), so that once synthesized, it is released into the plasma.

In line with this investigation, it has also been observed that alcohol exposure in the offspring, during pregnancy and lactation, caused an imbalance in the activity of GR, CAT, and SOD in the liver and kidney. This imbalance was due to oxidative stress and, as a consequence, of protein, and lipid oxidation was produced, being different in each tissue (Figures 11.1 and 11.2).[32,35] These results are in agreement with other studies involving chronic alcohol exposure in adult rats. Whitin et al.[74] conclude that the mechanism of antioxidant protection against long-term alcohol consumption presents peculiarities linked to the organ type.

Finally, since Se supplementation in adults had a protective role against alcohol-induced oxidation, several studies have researched into the effect of adding this micronutrient, via the maternal diet, on their progeny. Thus, coadministration of Se to alcohol-exposed

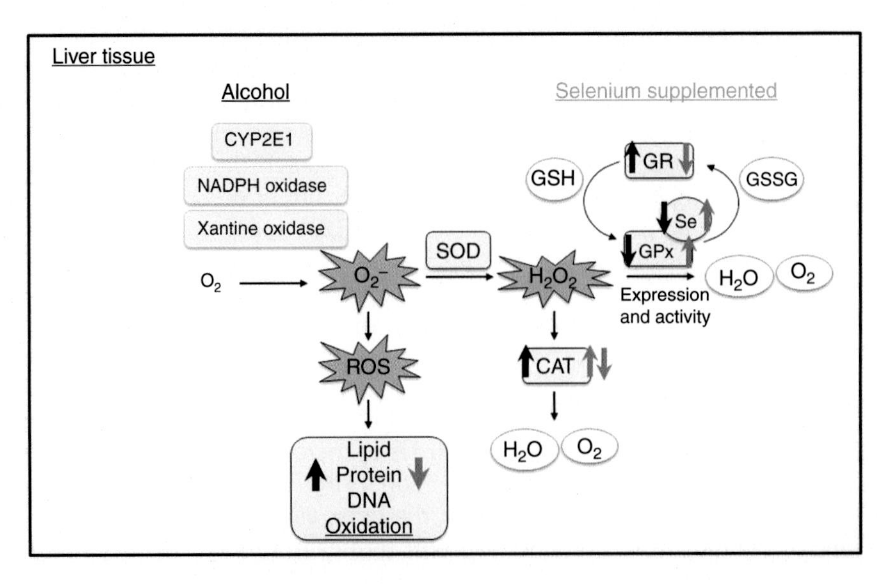

FIGURE 11.1 Oxidation and antioxidant balance in the liver of pup born from mother exposed to alcohol during gestation and lactation, and supplemented, or no, with Se. Black arrows show the effect of alcohol in offspring exposed to this drug, with respect to the control group. Grey arrows show the effect of Se in alcohol exposed pups, and Se supplemented, compared to non-supplemented alcohol exposed pups. ↑, Increases; ↓, Decreases; ↑↓, No changes.

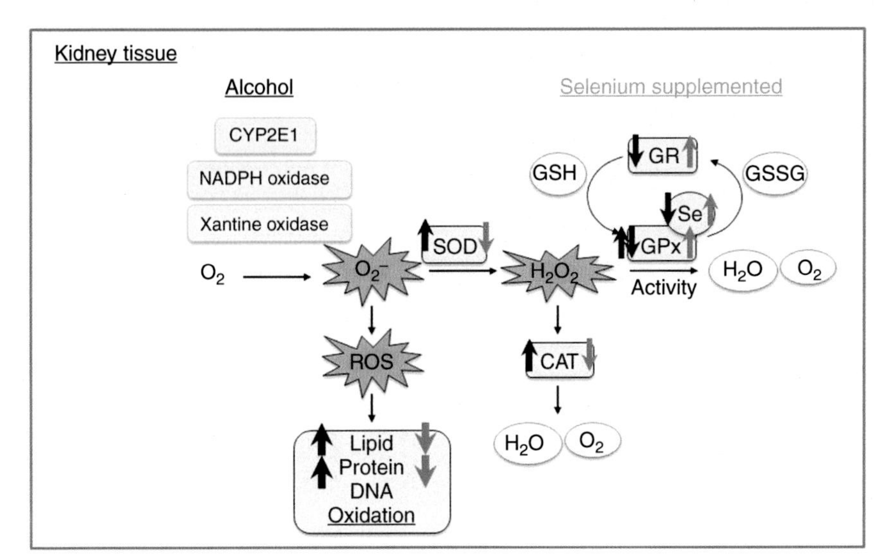

FIGURE 11.2 Oxidation and antioxidant balance in the kidney of pup born from mother exposed to alcohol during gestation and lactation, and supplemented, or no, with Se. Black arrows show the effect of alcohol in offspring exposed to this drug, with respect to the control group. Grey arrows show the effect of Se in alcohol exposed pups, and Se supplemented, compared to non-supplemented alcohol exposed pups. ↑, Increases; ↓, Decreases; ↑↓, No changes.

pups restored the GPx imbalance in liver provoked by ethanol, and decreased the activity of GR, CAT, and peroxidation protein products (Figure 11.1).[32] These results agree with the reports that Se-enriched diets were able to reduce the consequences of oxidative stress in the liver.[103] Furthermore, it was demonstrated that Se supplementation protects the kidney tissue of offspring against oxidative damage provoked by ethanol exposure during gestation and lactation (Figure 11.2). Therefore, although the supplemented diet does not improve the glomerular filtration function altered by alcohol, this treatment improved renal development and protein content, and modified antioxidant enzymes' activity, decreasing lipid and protein oxidation.[35] Thus, all results exposed suggest that Se could be part of an effective therapy in neutralizing the damage of ethanol consumption in pups.

CONCLUSION

Summary Points

- Alcohol consumption causes major health problems by its oxidative metabolism at a hepatic level, leading to an imbalance of antioxidant enzymes' activities and increasing reactive oxygen species.
- Selenium is essential to health because it presents antioxidant properties, and forms part of selenoproteins, such as the endogenous antioxidant enzyme glutathione peroxidase.
- Alcohol intake, acute or chronic, decreases Se deposits in the organism, as well as selenoproteins' expressions and activity, exacerbating the oxidative problems induced by alcoholism.
- The role of dietary Se supplementation as a means of acting against the oxidative adverse effects that alcohol provokes is currently under deep discussion. In chronic alcoholic patients, this element has a direct correlation with GPx activity and lipid oxidation, suggesting that the Se/MDA ratio could be an indicator of hepatic damage caused by alcohol consumption. These data point to Se as a possible dietary antioxidant therapy.

Abbreviations

ALD Alcoholic liver disease
APOER-2 Apolipoprotein E receptor-2
CAT Catalase
DIOs Iodothyronine deiodinase
FASD Fetal alcohol spectral disorders
GPx Glutathione peroxidase
GR Glutathione reductase
Km Affinity constant
RDA Recommended dietary allowance
Se Selenium
Sel P Selenoprotein P
Se-Cys Selenocysteine
Se-Met Selenomethionine
SOD Superoxide dismutase
TrxRs Thioredoxin reductase
V_{max} Maximal velocity

References

1. Carreras O, Vazquez AL, Rubio JM, Delgado MJ, Murillo ML. Comparative effects of intestinal absorption of folic acid and methyltetrahydrofolic acid in chronic ethanol-fed rats. *Ann Nutr Metab* 1994;**38**:221–5.
2. Carreras O, Vazquez AL, Rubio JM, Delgado MJ, Murillo ML. The effect of ethanol on intestinal L-leucine absorption in rats. *Arch Int Physiol Biochim Biophys* 1993;**101**:13–6.
3. Carreras O, Vazquez AL, Rubio JM, Delgado MJ, Murillo ML. Effect of chronic ethanol on D-galactose absorption by the rat whole intestinal surface. *Alcohol* 1992;**9**:83–6.
4. Lieber CS. Relationships between nutrition, alcohol use, and liver disease. *Alcohol Res Health* 2003;**27**:220–31.

5. Mathews S, Xu M, Wang H, Bertola A, Gao B. Animals models of gastrointestinal and liver diseases. Animal models of alcohol-induced liver disease: pathophysiology, translational relevance, and challenges. *Am J Physiol Gastrointest Liver Physiol* 2014;**306**:G819–23.
6. Seth D, D'Souza El-Guindy NB, Apte M, et al. Alcohol, signaling, and ECM turnover. *Alcohol Clin Exp Res* 2010;**34**:4–18.
7. Rua RM, Ojeda ML, Nogales F, et al. Serum selenium levels and oxidative balance as differential markers in hepatic damage caused by alcohol. *Life Sci* 2014;**94**:158–63.
8. Zima T, Fialová L, Mestek O, et al. Oxidative stress, metabolism of ethanol and alcohol-related diseases. *J Biomed Sci* 2001;**8**:59–70.
9. Chen CH, Pan CH, Chen CC, Huang MC. Increased oxidative DNA damage in patients with alcohol dependence and its correlation with alcohol withdrawal severity. *Alcohol Clin Exp Res* 2011;**35**:338–44.
10. Steinbrenner H, Sies H. Protection against reactive oxygen species by selenoproteins. *Biochim Biophys Acta* 2009;**1790**:1478–85.
11. Roman M, Jitaru P, Barbante C. Selenium biochemistry and its role for human health. *Metallomics* 2014;**6**:25–54.
12. Moghadaszadeh B, Beggs AH. Selenoproteins and their impact on human health through diverse physiological pathways physiology. *Physiology (Bethesda)* 2006;**21**:307–15.
13. Pappas AC, Zoidis E, Surai PF, Zervas G. Selenoproteins and maternal nutrition. *Comp Biochem Physiol B Biochem Mol Biol* 2008;**151**:361–72.
14. Kristal AR, Darke AK, Morris JS, et al. Baseline selenium status and effects of selenium and vitamin E supplementation on prostate cancer risk. *J Natl Cancer Inst* 2014;**106**:djt456.
15. Algotar AM, Stratton MS, Stratton SP, Hsu CH, Ahmann FR. No effect of selenium supplementation on serum glucose levels in men with prostate cancer. *Am J Med* 2010;**123**:765–8.
16. Brozmanová J, Mániková D, Vlčková V, Chovanec M. Selenium: a double-edged sword for defense and offence in cancer. *Arch Toxicol* 2010;**84**:919–38.
17. Joseph J, Loscalzo J. Selenistasis: epistatic effects of selenium on cardiovascular phenotype. *Nutrients* 2013;**5**:340–58.
18. Papp LV, Lu J, Holmgren A, Khanna KK. From selenium to selenoproteins: synthesis, identity, and their role in human health. *Antioxid Redox Signal* 2007;**9**:706–75.
19. Reilly C. *Selenium in Food and Health*. 2nd ed. New York: Springer-Verlag; 2006.
20. Köhrle J, Brigelius-Flohe R, Böck A, Gartner R, Meyer O, Flohe L. Selenium in biology: facts and medical perspectives. *Biol Chem* 2000;**381**:849–64.
21. Birringer M, Pilawa S, Flohe L. Trends in selenium biochemistry. *Nat Prod Rep* 2002;**19**:693–718.
22. Kryukov GV, Castellano S, Novoselov SV, et al. Characterization of mammalian selenoproteomes. *Science* 2003;**300**:1439–43.
23. Gladyshev VN. Characterization of mammalian selenoproteomes. *Science* 2003;**300**:1439–43.
24. Briguelius-Flohé R, Maiorino M. Glutathione peroxidases. *Biochim Biophys Acta* 2013;**1830**:3289–303.
25. Gromer S, Eubel J, Lee B, Jacob J. Human selenoproteins at a glance. *Cell Moll Life Science* 2005;**62**:2414–37.
26. Sunde RA, Thompson KM. Dietary selenium requirements based on tissue selenium concentration and glutathione peroxidase activities in old female rats. *J Trace Elem Med Biol* 2009;**23**:132–7.
27. Papp LV, Holmgren A, Khanna KK. Selenium and selenoproteins in health and disease. *Antioxid Redox Signal* 2010;**12**:735–93.
28. Reeves MA, Hoffmann PR. The human selenoproteome: recent insights into functions and regulation. *Cell Mol Life Sci* 2009;**66**:2457–78.
29. Rayman MP. The importance of selenium to human health. *Lancet* 2000;**356**:233–41.
30. Jotty K, Ojeda ML, Nogales F, Rubio JM, Murillo ML, Carreras O. Selenium tissue distribution changes after ethanol exposure during gestation and lactation: Selenite as a therapy. *Food Chem Toxicol* 2009;**47**:2484–9.
31. Jotty K, Ojeda ML, Nogales F, Murillo ML, Carreras O. Selenium dietary supplementation as a mechanism to restore hepatic selenoprotein regulation in rat pups exposed to alcohol. *Alcohol* 2013;**47**:545–52.
32. Ojeda ML, Nogales F, Vázquez B, Delgado MJ, Murillo ML, Carreras O. Alcohol, gestation and breastfeeding: selenium as an antioxidant therapy. *Alcohol Alcohol* 2009;**44**:272–7.
33. Ojeda ML, Vázquez B, Nogales F, Murillo ML, Carreras O. Ethanol consumption by Wistar rat dams affects selenium bioavailability and antioxidant balance in their progeny. *Int J Environ Res Public Health* 2009;**6**:2139–49.
34. Ojeda ML, Jotty K, Nogales F, Murillo ML, Carreras O. Selenium or selenium plus folic acid intake improves the detrimental effects of ethanol on pups' selenium balance. *Food Chem Toxicol* 2010;**48**:3486–91.
35. Ojeda ML, Nogales F, Murillo ML, Carreras O. Selenium or selenium plus folic acid-supplemented diets ameliorate renal oxidation in ethanol-exposed pups. *Alcohol Clin Exp Res* 2012;**36**:1863–72.
36. Nogales F, Ojeda ML, Delgado MJ, et al. Effects of antioxidant supplementation on duodenal Se-Met absorption in ethanol-exposed rat offspring *in vivo*. *J Reprod Dev* 2011;**57**:708–14.
37. Bondy SC, Pearson KR. Ethanol-induced oxidative stress and nutritional status. *Alcohol Clin Exp Res* 1993;**17**:651–4.
38. Sher L. Depression and suicidal behavior in alcohol abusing adolescents: possible role of selenium deficiency. *Minerva Pediatr* 2008;**60**:201–9.
39. Dutta SK, Miller PA, Greenberg LB, Levander OA. Selenium acute alcoholism. *Am J Clin Nutr* 1983;**38**:713–8.
40. Dworkin BM, Rosenthal WS, Stahl ER, Panesar NK. Decreased hepatic selenium content in alcoholic cirrhosis. *Dig Dis Sci* 2003;**33**:1213–7.
41. Thuluvath PJ, Triger DR. Selenium in chronic liver disease. *J Hepatol* 1992;**14**:176–82.
42. Loguercio C, De Girolamo V, Federico A, et al. Relationship of blood trace elements to liver damage, nutritional status, and oxidative stress in chronic nonalcoholic liver disease. *Biol Trace Elem Res* 2001;**81**:245–54.
43. González-Reimers E, Galindo-Martín L, Santolaria-Fernández F, et al. Prognostic value of serum selenium levels in alcoholics. *Biol Trace Elem Res* 2008;**125**:22–9.
44. Park T, Cho K, Park SH, Lee DH, Kim HW. Taurine normalizes blood levels and urinary loss of selenium, chromium, and manganese in rats chronically consuming alcohol. *Adv Exp Med Biol* 2009;**643**:407–14.
45. Zhang X, Sliwowska JH, Weinberg J. Prenatal alcohol exposure and fetal programming: effects on neuroendocrine and immune function. *Exp Biol Med (Maywood)* 2005;**230**:376–88.
46. Polache A, Martin-Algarra RV, Guerri C. Effects of chronic alcohol consumption on enzyme activities and active methionine absorption in the small intestine of pregnant rats. *Alcohol Clin Exp Res* 1996;**20**:1237–42.
47. Koyuturk M, Bolkent S, Ozdil S, Arbak S, Yanardag R. The protective effect of vitamin C, vitamin E and selenium combination therapy on ethanol-induced duodenal mucosal injury. *Hum Exp Toxicol* 2004;**23**:391–8.
48. Wolffram S, Berger B, Grenacher B, Scharrer E. Transport of selenoamino acids and their sulfur analogues across the intestinal brush border membrane of pigs. *J Nutr* 1989;**119**:706–12.
49. Nickel A, Kottra G, Schmidt G, Danier J, Hofmann T, Daniel H. Characteristics of transport of selenoamino acids by epithelial amino acid transporters. *Chem Biol Interact* 2009;**177**:234–41.

50. Thiry C, Ruttens A, Pussemier L, Schneider YJ. An *in vitro* investigation of species-dependent intestinal transport of selenium and the impact of this process on selenium bioavailability. *Br J Nutr* 2013;**109**:2126–34.
51. Wang Y, Fu L. Forms of selenium affect its transport, uptake and glutathione peroxidase activity in the Caco-2 cell model. *Biol Trace Elem Res* 2012;**149**:110–6.
52. Bode C, Bode JC. Effect of alcohol consumption on the gut. *Best Pract Res Clin Gastroenterol* 2003;**17**:575–92.
53. Bhalla S, Mahmood S, Mahmood A. Effect of prenatal exposure to ethanol on postnatal development of intestinal transport functions in rats. *Eur J Nutr* 2004;**43**:109–15.
54. Delgado MJ, Nogales F, Ojeda ML, Murillo ML, Carreras O. Effect of dietary selenite on development and intestinal absorption in offspring rats. *Life Sci* 2011;**88**:150–5.
55. Reeves MA, Hoffmann PR. The human selenoproteome: recent insights into functions and regulation. *Cell Mol Life Sci* 2009;**66**:2457–78.
56. Cho HK, Yang FL, Snook JT. Effect of chronic ethanol consumption on selenium status and utilization in rats. *Alcohol* 1991;**8**:91–6.
57. Behne D, Gessner H, Kyriakopoulos A. Information on the selenium status of several body compartments of rats from the selenium concentrations in blood fractions, hair and nails. *J Trace Elem Med Biol* 1996;**10**:174–9.
58. Debski B, Zarski TP, Milner JA. The influence of age and sex on selenium distribution and glutathione peroxidase activity in plasma and erythrocytes of selenium-adequate and supplemented rats. *J Physiol Pharmacol* 1992;**43**:299–306.
59. Weiss SL, Evenson JK, Thompson KM, Sunde RA. The selenium requirement for glutathione peroxidase mRNA level is half of the selenium requirement for glutathione peroxidase activity in female rats. *J Nutr* 1996;**126**:2260–7.
60. Loeschner K, Hadrup N, Hansen M, et al. Absorption, distribution, metabolism and excretion of selenium following oral administration of elemental selenium nanoparticles or selenite in rats. *Metallomics* 2014;**6**:330–7.
61. Sunde RA, Raines AM, Barnes KM, Evenson JK. Selenium status highly regulates selenoprotein mRNA levels for only a subset of the selenoproteins in the selenoproteome. *Biosci Rep* 2009;**29**:329–38.
62. Barnes KM, Evenson JK, Raines AM, Sunde RA. Transcript analysis of the selenoproteome indicates that dietary selenium requirements of rats based on selenium-regulated selenoprotein mRNA levels are uniformly less than those based on glutathione peroxidase activity. *J Nutr* 2009;**139**:106–99.
63. Hawkes WC, Alkan Z, Wong K. Selenium supplementation does not affect testicular selenium status or semen quality in North American men. *J Androl* 2009;**30**:525–33.
64. Lawler TL, Taylor JB, Finley JW, Caton JS. Effect of supranutritional and organically bound selenium on performance, carcass characteristics, and selenium distribution in finishing beef steers. *J Anim Sci* 2004;**82**:1488–93.
65. Hoffmann PT, Höge SC, Li P, Hoffmann FW, Hashimoto AC, Berry MJ. The selenoproteome exhibits widely varying, tissue-specific dependence on selenoprotein P for selenium supply. *Nucleic Acids Res* 2007;**35**:3963–73.
66. Sunde RA, Raines AM. Selenium regulation of the selenoprotein and nonselenoprotein transcriptomes in rodents. *Adv Nutr* 2011;**2**:138–50.
67. Smith-Kielland A, Aaseth J, Thomassen Y. Effect of long-term ethanol intake on the content of selenium in rat liver: relation to the rate of hepatic protein synthesis. *Acta Pharmacol Toxicol (Copenh)* 1986;**58**:237–9.
68. Merlot E, Couret D, Otten W. Prenatal stress, fetal imprinting and immunity. *Brain Behav Immun* 2008;**22**:42–51.
69. Fowden AL, Forhead AJ. Endocrine mechanisms of intrauterine programming. *Reproduction* 2004;**127**:515–26.
70. Ostrowska J, Luczaj W, Kasacka I, Rozanski A, Skrzydlewska E. Green tea protects against ethanol-induced lipid peroxidation in rat organs. *Alcohol* 2004;**32**:25–32.
71. Wentzel P, Eriksson UJ. Ethanol-induced fetal dysmorphogenesis in the mouse is diminished by high antioxidative capacity of the mother. *Toxicol Sci* 2006;**92**:416–22.
72. Miller L, Shapiro AM, Wells PG. Embryonic catalase protects against ethanol-initiated DNA oxidation and teratogenesis in acatalasemic and transgenic human catalase-expressing mice. *Toxicol Sci* 2013;**134**:400–11.
73. Murillo-Fuentes ML, Artillo R, Ojeda ML, Delgado MJ, Murillo ML, Carreras O. Effects of prenatal or postnatal ethanol consumption on Zn intestinal absorption and excretion in rats. *Alcohol* 2007;**42**:3–10.
74. Whitin JC, Vaharme S, Tham DM, Cohen HJ. Extracellular glutathione peroxidase is secreted basolaterally by human renal proximal tubule cells. *Am J Physiol Renal Physiol* 2002;**283**:F20–8.
75. Payne RL, Southern LL. Changes in glutathione peroxidase and tissue selenium concentrations of broilers after consuming a diet adequate in selenium. *Poult Sci* 2005;**84**:1268–76.
76. Olson GE, Winfrey VP, Hill KE, Burk RF. Megalin mediates selenoprotein P uptake by kidney proximal tubule epithelial cells. *J Biol Chem* 2008;**283**:6854–60.
77. Herz J, Bock HH. Lipoprotein receptors in the nervous system. *Annu Rev Biochem* 2002;**71**:405–34.
78. Lieber CS. Pathogenesis and treatment of alcoholic liver disease: progress over the last 50 years. *Rocz Akad Med Bialymst* 2005;**50**:7–20.
79. Ting JW, Lautt WW. The effect of acute, chronic, and prenatal ethanol exposure on insulin sensitivity. *Pharmacol Ther* 2006;**111**:346–73.
80. Steinbrenner H, Speckmann B, Pinto A, Sies H. High selenium intake and increased diabetes risk: experimental evidence for interplay between selenium and carbohydrate metabolism. *J Clin Biochem Nutr* 2010;**48**:40–5.
81. Ojeda ML, Delgado-Villa MJ, Llopis R, Murillo ML, Carreras O. Lipid metabolism in ethanol-treated rat pups and adults: effects of folic acid. *Alcohol* 2008;**43**:544–50.
82. Brigelius-Flohe R. Glutathione peroxidases and redox-regulated transcription factors. *Biol Chem* 2006;**387**:1329–35.
83. Liang H, Ran Q, Jang YC, et al. Glutathione peroxidase 4 differentially regulates the release of apoptogenic proteins from mitochondria. *Free Radic Biol Med* 2009;**47**:312–20.
84. Yant LJ, Ran Q, Rao L, et al. The selenoprotein GPX4 is essential for mouse development and protects from radiation and oxidative damage insults. *Free Radic Biol Med* 2003;**34**:402–96.
85. Ufer C, Wang CC. The roles of glutathione peroxidases during embryo development. *Front Mol Neurosci* 2011;**4**:12.
86. Kalaz EB, Evran B, Develi S, Erata GÖ, Uysal M, Koçak-Toker N. Effect of binge ethanol treatment on prooxidant-antioxidant balance in rat heart tissue. *Pathophysiol* 2012;**19**:49–53.
87. Rendón-Ramírez A, Cortés-Couto M, Martínez-Rizo AB, Muñiz-Hernández S, Velázquez-Fernández JB. Oxidative damage in young alcohol drinkers: a preliminary study. *Alcohol* 2013;**47**:501–4.
88. Mallikarjuna K, Shanmugam KR, Nishanth K, et al. Alcohol-induced deterioration in primary antioxidant and glutathione family enzymes reversed by exercise training in the liver of old rats. *Alcohol* 2010;**44**:523–9.
89. Oh SI, Kim CI, Chun HJ, Park SC. Chronic ethanol consumption affects glutathione status in rat liver. *J Nutr* 1998;**128**:758–63.
90. Bailey SM, Patel VB, Young TA, Asayama K, Cunningham CC. Chronic ethanol consumption alters the glutathione/glutathione peroxidase-1 system and protein oxidation status in rat liver. *Alcohol Clin Exp Res* 2001;**25**:726–33.
91. Cederbaum AI, Lu Y, Wu D. Role of oxidative stress in alcohol-induced liver injury. *Arch Toxicol* 2009;**83**:519–48.

92. Brocardo PS, Gil-Mohapel J, Christie BR. The role of oxidative stress in fetal alcohol spectrum disorders. *Brain Res Rev* 2011;**67**:209–25.
93. Cohen-Kerem R, Koren G. Antioxidants and fetal protection against ethanol teratogenicity. I. Review of the experimental data and implications to humans. *Neurotoxicol Teratol* 2003;**25**:1–9.
94. Yu L, Yang S, Sun L, Jiang YF, Zhu LY. Effects of selenium-enriched agaricus blazei murill on liver metabolic dysfunction in mice, a comparison with selenium-deficient agaricus blazei murill and sodium selenite. *Biol Trace Elem Res* 2014;**160**:79–84.
95. Markiewicz-Górka I, Zawadzki M, Januszewska L, Hombek-Urban K, Pawlas K. Influence of selenium and/or magnesium on alleviation alcohol induced oxidative stress in rats, normalization function of liver and changes in serum lipid parameters. *Hum Exp Toxicol* 2011;**30**:1811–27.
96. Yanardag R, Ozsoy-Sacan O, Ozdil S, Bolkent S. Combined effects of vitamin C, vitamin E, and sodium selenate supplementation on absolute ethanol-induced injury in various organs of rats. *Int J Toxicol* 2007;**26**:513–23.
97. Kim JH, Park SH, Nam SW, Choi YH. Gastroprotective effect of selenium on ethanol-induced gastric damage in rats. *Int J Mol Sci* 2012;**13**:5740–50.
98. Lamarche F, Signorini-Allibe N, Gonthier B, Barret L. Influence of vitamin E, sodium selenite, and astrocyte-conditioned medium on neuronal survival after chronic exposure to ethanol. *Alcohol* 2004;**33**:127–38.
99. Swathy SS, Panicker S, Indira M. Effect of exogenous selenium on the testicular toxicity induced by ethanol in rats. *Indian J Physiol Pharmacol* 2006;**50**:215–24.
100. Dey Sarkar P, Ramprasad N, Dey Sarkar I, Shivaprakash TM. Study of oxidative stress and trace element levels in patients with alcoholic and non-alcoholic coronary artery disease. *Indian J Physiol Pharmacol* 2007;**51**:141–6.
101. Gil-Mohapel J, Boehme F, Patten A, et al. Altered adult hippocampal neuronal maturation in a rat model of fetal alcohol syndrome. *Brain Res* 2011;**1384**:29–41.
102. Addolorato G, Gasbarrini A, Marcoccia S, et al. Prenatal exposure to ethanol in rats: effects on liver energy level and antioxidant status in mothers, fetuses, and newborns. *Alcohol* 1997;**14**:569–73.
103. Bordoni A, Danesi F, Malaguti M, et al. Dietary Selenium for the counteraction of oxidative damage: fortified foods or supplements? *Br J Nutr* 2008;**99**:191–7.

CHAPTER

12

Role of Zinc in Alcoholic Liver Disease

Wei Zhong, DVM, PhD, Qian Sun, PhD, Zhanxiang Zhou, PhD

Center for Translational Biomedical Research, University of North Carolina at Greensboro, Kannapolis, NC, USA

INTRODUCTION

Zinc deficiency has been well-documented in patients with alcoholic liver disease (ALD). The zinc levels in the blood and liver of patients with ALD have been reported to be significantly reduced, compared to that of the normal subjects. The severity of reduction of the hepatic zinc level tends to increase along with the disease progression from alcoholic steatosis to cirrhosis.[1] Alcohol-induced disorder in zinc metabolism has also been found in animal models, and the reduction of zinc level in the liver could be detected as early as 2 weeks after alcohol feeding in rodents.[2,3] Dietary zinc supplementation has been shown to prevent/reverse alcohol-induced liver injury.[4] Zinc supplementation improves hepatic zinc homeostasis, and restores the function of zinc proteins, particularly the zinc finger transcription factors, hepatocyte nuclear factor 4α (HNF-4α) and peroxisome proliferation activator α (PPAR-α).[4] This chapter describes the mechanisms of alcohol consumption-induced zinc dyshomeostasis, the role of zinc deficiency in the pathogenesis of ALD, and the benefits of dietary zinc supplementation in preventing/reversing the progression of ALD. In addition, information from clinical trials on the efficiency of zinc supplementation, as a dietary intervention for treating human ALD, are also provided.

ZINC METABOLISM AND FUNCTION

Zinc is the second most abundant trace element in the body.[5] The total body zinc content has been estimated to be 2–3 g. Zinc is found in all organs and tissues, with higher amounts in the liver, kidney, bone, muscle, and skin.[6] In the United States, the Recommended Dietary Allowance (RDA) for zinc is 11 mg/day for men, and 8 mg/day for women older than 19 years old.[7]

Zinc plays an important role in maintaining normal physiological processes, such as metabolism, signaling transduction, cell growth, and differentiation.[6] Zinc participates in these cellular functions through coordination to zinc proteins, thereby regulating protein functions. Zinc serves as a catalytic cofactor for more than three hundreds of metalloenzymes, and a structural component of zinc finger motif for a large number of zinc proteins, including zinc finger transcription factors and hormones. While loss of zinc from metalloenzymes leads to enzyme inactivation, loss of zinc from zinc proteins prevents interaction of the proteins with their targets, including DNA, RNA, protein, and lipid. A classic example showing the importance of zinc in maintaining the enzymatic activity is that removing zinc ions from alcohol dehydrogenase (ADH) leads to a complete loss of its catalytic activity.[8] The vital role of zinc in maintaining gene regulation function of zinc finger transcription factors has also been defined by loss of DNA binding ability, after removal of zinc from the zinc finger motif.[4,9]

Cellular zinc homeostasis is achieved by multiple processes, including zinc influx into the cells, zinc efflux out of the cells, and compartmentalization and storage within the cells.[6,10] These processes are accomplished by a diverse family of proteins, including zinc transporters and reservoirs. Zinc transporters distribute on the plasma membrane and organelle membrane, and consist of two families, zinc importers (SLC39/ZIPs),[11] and zinc exporters (SLC30/ZnTs).[12] The ZIP family has 14 members, from ZIP1 to ZIP14, and they move zinc from extracellular space or intracellular organelles into the cytosol. The ZnT family contains 10 members, from ZnT1 to ZnT10, and they are responsible for exporting zinc from cytosol to extracellular space or intracellular organelles. Generally speaking, the cytosolic zinc levels are regulated positively by ZIPs, and negatively by ZnTs, while the zinc levels of organelles are regulated positively by ZnTs, and negatively by ZIPs.

Molecular Aspects of Alcohol and Nutrition. http://dx.doi.org/10.1016/B978-0-12-800773-0.00012-4

Metallothioneins (MTs) and glutathione (GSH), particularly the former, function as the reservoirs of intracellular zinc. There are four MT isoforms in mammals, MT-I to MT-IV, and MT-I and MT-II distribute in the liver. The structure of MT is composed of two zinc/thiolate clusters; one cluster binds to three zinc ions and the other cluster binds to four zinc ions.[13] One MT molecule can bind up to seven zinc ions under physiological conditions, and thus functions as the major zinc reservoir.[5,14] There is virtually no free zinc in the cell,[15] and the dynamic activities of a cell require constant zinc transfer from one location to another, or from one zinc binding site to another. In cell-free systems, oxidatively-released zinc from MT has been shown to be accepted by zinc-binding proteins, including carbonic anhydrase,[16] alkaline phosphatase,[17] thermolysin,[17] pyridoxal kinase,[18] estrogen receptor,[19] and zinc finger peptides.[20] These proteins and peptides are involved in a broad spectrum of cellular functions, including signal transduction, second messenger metabolism, gene expression, and protein phosphorylation. It has been reported that the zinc level in the liver correlates well with the MT level. MT overexpression in the liver of the MT transgenic (MT-TG) mice is associated with a two-fold increase in hepatic zinc concentration, while MT-knockout (MT-KO) mice showed a decrease in hepatic zinc level.[21–23] In addition, MTs also participate in intracellular zinc trafficking. Previous studies have shown that MTs translocate from cytosol to nuclei, or the mitochondria intermembrane spaces, and release zinc.[24,25] Hepatic MTs and GSH are sensitive to dietary zinc level, and zinc treatment can significantly increase the concentrations of MTs and GSH in the liver.[26] One of the most important functions of both MTs and GSH is to transfer zinc to apoproteins to form functional zinc proteins, thereby regulating a variety of cellular processes.

OCCURRENCE OF ZINC DEFICIENCY IN ALD

The liver is the major organ responsible for alcohol metabolism.[27] Alcohol metabolism in the liver generates reactive oxygen species (ROS) and toxic aldehydes, including acetaldehydes, the first metabolite generated from alcohol metabolism, and lipid aldehydes, such as 4-hydroxynonenol (4-HNE) and malondialdehyde (MDA).[28] A profound oxidative stress condition is a fundamental cellular disorder caused by alcohol abuse. Zinc binds to proteins at the thiol group that is susceptible to the attack caused by pro-oxidants. Therefore, disturbance of zinc homeostasis and inactivation of zinc proteins in the liver is highly expected in ALD.

ALD has been reported to be associated with hypozincemia (low serum zinc level), and reduction of zinc level in the liver, for over half a century.[29] In a representative study, the average serum zinc concentration in alcoholic patients was 7.52 μmol/L, which was significantly lower than an average of 12.69 μmol/L in healthy subjects.[30] Moreover, the decrease in serum zinc level correlates positively with the severity of liver damage.[30,31] Clinical studies also demonstrated that the severity of zinc deficiency correlates with the severity of ALD. In comparison to normal subjects, patients with alcoholic cirrhosis showed a 37% decrease in hepatic zinc level, while a 24% decrease was found in noncirrhotic patients with ALD.[31] Patients with hepatic encephalopathy have more severe serum zinc depletion than hepatic cirrhosis only patients.[32] Zinc in the serum is bound mainly to albumin,[33] and a decrease in serum albumin level in association with zinc reduction was found in alcoholic patients.[34] It is well known that heavy drinkers have poor nutrition status, due to decreased food intake to compensate calories from alcohol intake. Besides, the liver is the organ responsible for albumin synthesis, and impaired liver function would further exacerbate zinc deficiency in advanced ALD patients. Manifestations of zinc deficiency that are relevant to ALD include skin lesions, anorexia, depressed wound healing, hypogonadism, altered immune function, impaired night vision, and depressed mental function, with possible encephalopathy.[1,35]

Acquired zinc deficiency has also been observed in animal models of ALD. Alcohol exposure reduces hepatic zinc level in both rats and mice,[4,36] and this reduction could happen as early as 2 weeks of alcohol exposure.[2,3] In addition, liver mitochondrion contains the highest zinc level, and is the most rapidly and remarkably depleted subcellular compartment upon alcohol intoxication. Plasma and muscle zinc levels decline following hepatic zinc loss.[3] Moreover, marginal dietary zinc deficiency synergistically decreased hepatic zinc level in the alcohol-fed mice.[37] It is noteworthy that the serum zinc level could be elevated at the early stage of ALD. It is assumed that liver zinc pool is the first site attacked by toxic alcohol metabolites, which decreases the ability of the liver to maintain zinc coordination in metalloproteins, thereby increasing hepatic zinc release to the blood. As a result, a transient elevation of blood zinc level was observed in alcoholics with normal or fatty liver,[38] and a mouse model of alcoholic fatty liver disease exhibited the same phenomenon.[2] However, sustained decrease in hepatic zinc level worsens liver function, and decreases blood albumin level that ultimately causes decreased serum zinc level.[38] Therefore, serum zinc level may not be a good indicator for the assessment of organ zinc status at an early stage of ALD.

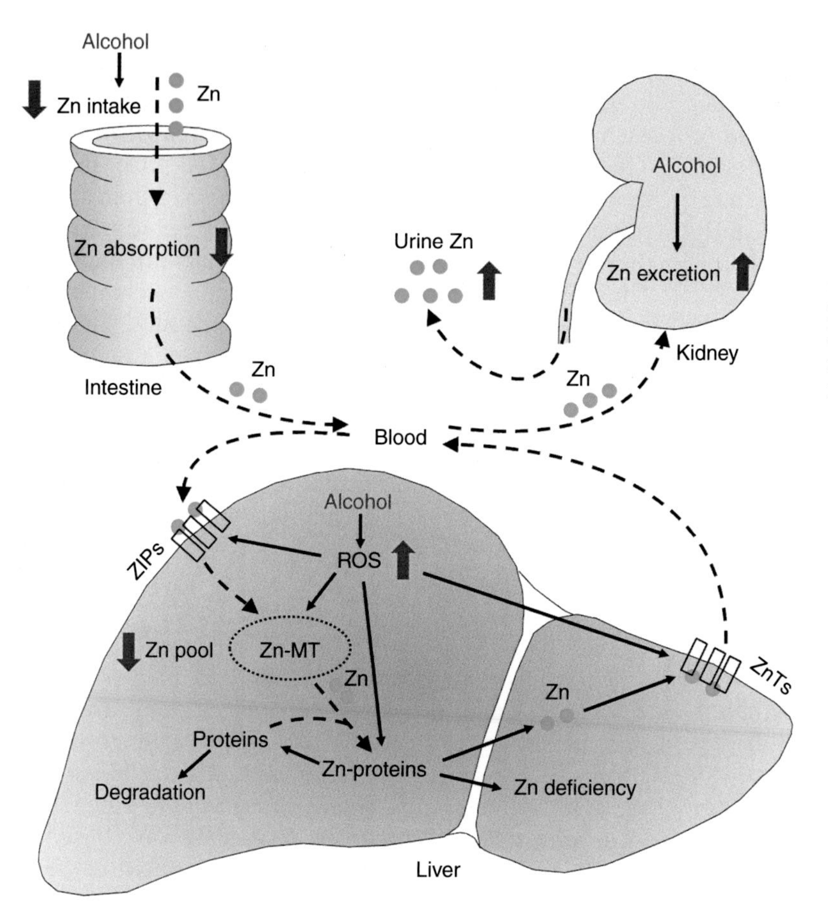

FIGURE 12.1 Mechanisms of zinc dyshomeostasis in ALD. Alcohol abuse causes inadequate zinc intake, reduced intestinal zinc absorption, disturbed hepatic zinc metabolism, and increased urine zinc excretion. Generation of oxidative stress, in association with alcohol metabolism, affects zinc transporters and zinc reservoir proteins, leading to a reduction of the cellular zinc pool. Oxidative stress may also cause zinc release from zinc proteins, and consequently, an excess zinc export from the liver. Dash lines indicate zinc transportation under physiological conditions, while solid lines represent alcohol-induced pathological changes in zinc metabolism. Zn, zinc; ROS, oxidative stress; MT, metallothionein.

MECHANISMS OF ZINC DYSHOMEOSTASIS IN ALD

Considerable efforts have been made during the past 50 years toward the understanding of the mechanism of how alcohol consumption disturbs zinc homeostasis at systemic, cellular, and molecular levels. As summarized in Figure 12.1, several mechanisms have been suggested to account for alcohol-induced zinc deficiency, including inadequate dietary zinc intake, reduced intestinal absorption, disturbed hepatic zinc metabolism, and increased urine loss. Generation of oxidative stress, in association with alcohol metabolism, is likely a key player in mediating alcohol-induced zinc dyshomeostasis at the molecular level. Zinc transporters and zinc reservoir proteins are the major players involved in zinc absorption from intestine, zinc excretion from the kidney, and cellular zinc homeostasis, while both of them are affected by elevated oxidative stress.

Inadequate Dietary Zinc Intake

Zinc is present in a wide variety of foods. Seafood, especially mollusks and oysters, and red meats, especially organ meats, are very good sources of zinc. Pork, poultry, and dairy products are other good sources of zinc. Plants are poor zinc sources.[39,40] Zinc in food is combined with amino acids and nucleic acid. Hydrochloric acid, protease, and nuclease are needed to release zinc from food before absorption.[41] Because alcohol contributes calories to the body, studies have been conducted to assess the nutrition status of alcoholics by evaluating their dietary intake and measuring their weight, height, and fat mass. It has been found that, with increased alcohol intake, the percentage of energy derived from protein, fat, and carbohydrate was decreased, as well as vitamins and minerals.[42] Along with that is a progressively lowered body mass index (BMI) among drinkers.[43] Patients with ALD usually have poor diet; the more they drink, the less nutritional quality of food they consume. With the progression of ALD, impaired gastrointestinal and liver function would further reduce food intake, thereby reducing zinc intake.[44]

Reduced Zinc Absorption From the Intestine

Absorption of zinc occurs throughout the small intestine, with jejunum being the most important site. Both human and animal studies have shown that alcohol consumption impairs the absorption of zinc from the intestine. Using a dual isotope absorption technique, the intestinal absorption rate of $zinc^{65}$ was estimated to be 56% in normal subjects, while the rate was reduced to 37% (a 34% decrease) in alcoholic patients.[45,46] In a rat model of ALD, chronic alcohol ingestion significantly impaired zinc absorption in the ileum, as early as after 4 weeks of alcohol feeding.[47]

Perturbed zinc transportation is believed, at least partially, to be responsible for the decreased intestinal absorption. Both endogenous and exogenous zinc in the intestinal lumen is transported into the intestinal epithelial cells by ZIP4 that locates at the apical plasma membrane. After getting into the cells, zinc is used or stored within the enterocyte, while the rest is transported across the basolateral membrane, toward the circulation system, by ZnT1.[48] The expression of ZIP4 has been shown to be down-regulated by alcohol exposure in mice[49] that may account for decreased absorption of zinc. However, the effect of alcohol on the expression of other zinc transporters in the intestine remains unknown, and future studies are needed to fill in this gap. In addition, dietary fibers/phytates can also reduce zinc absorption. The Western world's shift from consumption of meat proteins to cereal proteins, containing high levels of fibers, may worsen alcohol-decreased intestinal zinc absorption.[50,51]

Disturbed Zinc Metabolism in the Liver

Zinc absorbed from the intestine travels in the blood through binding mainly to albumin and, to a lesser extent, to transferrin, α-2 macroglobulin, and immunoglobulin. The loosely-bound zinc is then transported into the liver through the port vein. As mentioned above, zinc transporters (ZIPs and ZnTs) are responsible for transporting zinc in and out of the liver, as well as the subcellular compartments of the hepatocytes. In a mouse model of ALD, all of the 14 ZIPs, and 9 out of 10 of the ZnTs (except ZnT2) were detected at the mRNA level in the liver. Four hepatic zinc transporters were significantly perturbed in alcohol-fed mice, compared to the controls at the protein level; ZIP5 and ZIP14 were decreased, while ZIP7 and ZnT7 were increased.[2] Moreover, *in vitro* treatment of hepatocytes with a lipid peroxidation product (4-HNE) or hydrogen peroxide mimicked the alterations of zinc transporters as observed *in vivo*, which suggests an oxidative stress-dependent mechanism for alcohol-perturbed zinc transport in the liver.

Under oxidative stress, zinc could be released from MTs and GSH due to oxidation of the sulfur donor atoms of the cysteine ligands. Indeed, downregulation of hepatic MT and GSH is accompanied by alcohol-induced oxidative stress.[26] Reduction of hepatic MTs and GSH not only leads to a shrink of the cellular zinc pool, but also impairs zinc trafficking and transferring to zinc proteins. Based on the fact that zinc proteins, such as HNF-4α, PPAR-α, and superoxide dismutase 1 (Cu/Zn-SOD), in the liver of alcohol-fed mice, showed more significant decreases at their activities than at their protein levels,[2] it is assumed that zinc coordination to these zinc proteins is reduced under alcohol exposure. Thus, oxidative stress is likely a common mechanism for mobilizing zinc from zinc proteins in the liver. Since the cellular free zinc pool is tightly controlled,[15] the free zinc mobilized from the zinc proteins would be transported out of the hepatocytes.[52] ZnT1 has been shown to play a major role in efflux of the cytosolic free zinc.[12] Under chronic alcohol abuse condition, sustained zinc release from the hepatocytes to the blood will eventually lead to zinc depletion in the liver.

Increased Zinc Excretion From the Urine

Multiple organ systems contribute to the loss of body zinc, including gastrointestinal tract, kidney, and skin. Endogenous zinc secreted by the intestinal mucosa, pancreas, and liver into the gastrointestinal tract, along with those in sloughed enterocytes, contribute to zinc loss in feces, and the zinc content in feces may vary based on zinc levels in food and the body requirement of zinc.[53] Under physiological condition, the amount of zinc lost from the urine and sweat, as well as from semen and menses, is relatively small and constant, compared to that from the feces. Urine zinc is derived from a minor percentage of plasma zinc binding with amino acid, especially cysteine and histidine.[54]

Under alcohol abuse condition, zinc loss from the urine has been suggested as an important mechanism underlying the pathogenesis of zinc deficiency at systemic level. An increased excretion of urinary zinc (hyperzincuria) has been well documented in alcoholic cirrhosis. In comparison to urinary excretion of 734 μg zinc/day in healthy controls, urinary zinc excretion was increased to 1,777 μg/day in patients with alcoholic cirrhosis.[1] To determine the link between alcohol consumption and zinc imbalance, prolonged alcohol administration was performed in alcoholic subjects who were admitted to a metabolic ward for 34 days.[55] Alcohol was administered orally for 17 days, every 2 h for 18 h each day, to maintain a blood alcohol level of 100 mg/dL, followed by 17 days of abstinence. The urinary zinc excretion during the drinking period was nearly twice as high as during the nondrinking period. The serum zinc level was decreased with alcohol consumption, and normalized after abstinence.

The mechanisms of alcohol-elevated urinary excretion of zinc still remain largely unknown. In a 7-day alcohol intoxication experiment with dogs, elevated rates of both renal filtration and reabsorption of zinc were observed.[56] However, the increased filtration rate was too overwhelming to be compensated by the reabsorption capacity, which resulting in a significant increase in the urinary excretion of zinc. Moreover, it has been found that zinc transporters, such as ZnT1, ZnT2, ZnT4, and ZIP1, exist in the kidney.[57] Renal zinc transporters may also participate in controlling zinc reabsorption and homeostasis in ALD. In addition, increased urine zinc level in cirrhotic patients could be related to tubular cell zinc turnover and leakage.[58] Other factors, such as increased glucagon and concentrations of cysteine and histidine in the urine and plasma, may also contribute to the increase of urinary zinc excretion.[59]

BENEFICIAL EFFECTS OF ZINC ON ALD

Zinc has long been regarded as an essential and necessary nutrient for maintaining liver function. In patients with ALD, hepatic zinc decrease has been found to correlate with liver dysfunction, as indicated by decreased serum albumin, increased serum bilirubin, and decreased galactose elimination capacity and antipyrine clearance.[34,60,61] Mechanistic studies in animal models have shown that zinc functions as a hepatoprotective agent through regulation of antioxidant defense, cell proliferation, and cell death, in the development and progression of ALD. Zinc supplementation attenuates alcohol-induced hepatic zinc depletion, suppresses inflammation and cell death, and stimulates fatty acid oxidation in the liver (Figure 12.2). Extrahepatic actions of zinc, such as regulation of intestinal barrier and adipose adipokine secretion, are also implicated in the beneficial effect of zinc on alcoholic liver injury.

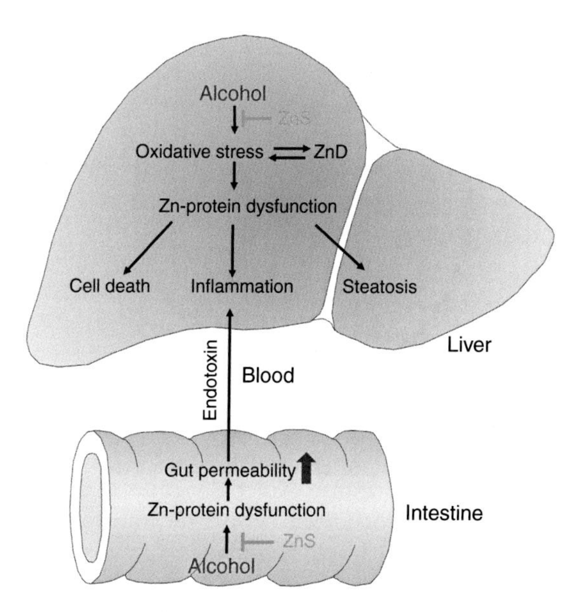

FIGURE 12.2 **Beneficial effects of zinc on ALD.** Zinc supplementation attenuates alcohol-induced oxidative stress, alleviates hepatic lipid accumulation, and suppresses inflammation and cell death. Extrahepatic actions of zinc, such as restoring the intestinal barrier function, and consequently preventing endotoxemia and hepatic inflammation, are also implicated in the beneficial effects of zinc on alcoholic liver injury. ZnS, zinc supplementation; ZnD, zinc deficiency.

Attenuation of Alcoholic Steatosis

Alcoholic steatosis (fatty liver) is one of the earliest pathological changes in ALD. Accumulation of lipids in the hepatocyte makes the liver susceptible to inflammatory mediators or other toxic agents that lead to further progression to hepatitis and eventually fibrosis.[62] Alcoholic steatosis is a reversible stage of liver damage, and alleviating steatosis would halt or slow the progression of ALD. While previous studies have shown that alterations in fatty acid synthesis, fatty acid oxidation, and triglyceride-rich very low density lipoprotein (VLDL) secretion are involved in the pathogenesis of alcoholic steatosis,[63–65] the molecular mechanisms underlying the alcohol effects on hepatic lipid homeostasis have not been fully defined. Increasing evidence suggests that zinc plays a critical role in the regulation of hepatic lipid metabolism. A lower hepatic zinc level was associated with steatosis in leptin receptor deficient rats.[66] Feeding rats or mice with a zinc-deficient diet (a single nutrient deficiency) caused hepatic lipid accumulation, in association with dysregulation of a large number of genes involved in lipid metabolism.[37,67–69]

Given the fact that dietary zinc deficiency facilitates hepatic lipid accumulation, and alcohol causes hepatic zinc deficiency, zinc supplementation would provide beneficial effect on alcoholic steatosis. In a mouse model of ALD, the therapeutic effect of 4 weeks of dietary zinc supplementation on preestablished liver injury by 12 weeks of alcohol exposure has been tested.[4] Chronic alcohol exposure caused liver steatosis, as indicated by the accumulation of lipid droplets and increased concentrations of lipid species. Zinc supplementation remarkably decreased the levels of triglyceride, cholesterol, and free fatty acids, and reduced the number and size of lipid droplets in the liver of alcohol-fed mice. Two major mechanisms may account for the reversal effect of zinc on alcoholic steatosis. First, zinc supplementation speeded up fatty acid β-oxidation. Second, alcohol exposure impaired hepatic lipid export by suppressing VLDL secretion that was normalized by zinc supplementation. Accordingly, zinc supplementation upregulated hepatic expressions of mitochondrial fatty acid β-oxidation enzyme, long chain acyl-CoA dehydrogenase (Acadl), and two critical genes related to VLDL secretion, the microsomal triglyceride transport protein (Mttp), and apolipoprotein B (Apob). These findings suggest that zinc depletion is a causal factor in the development of alcoholic steatosis, and dietary zinc supplementation is able to reverse alcoholic steatosis by enhancing hepatic fatty acid β-oxidation and VLDL secretion.

Suppression of Alcohol-Induced Hepatic Inflammation

It is well established that ALD is associated with imbalanced immune responses and increased production of proinflammatory cytokines/chemokines.[70,71] Inflammatory cytokines/chemokines, such as tumor necrosis factor alpha (TNF-α), interleukin-1β (IL-1β), monocyte chemoattractant protein-1 (MCP-1) and interleukin-8 (IL-8), and hepatic acute-phase cytokines, such as interleukin-6 (IL-6), play a pivotal role in modulating organ injury and metabolic complications in ALD. Zinc deprivation has been shown to mediate alcohol-induced hepatic IL-8 expression through inhibiting histone deacetylases, and activating nuclear factor-κB (NF-κB).[72] Marginal dietary zinc deficiency exaggerated alcohol-induced hepatic neutrophil infiltration, and inflammatory cytokine production in mice.[37] These findings indicate that alcohol-induced zinc deficiency is an important determinant of hepatic inflammation.

Zinc is an immune-modulatory trace element, and a growing body of evidence suggests that zinc may be beneficial against the progression of inflammatory diseases, including ALD. Dietary zinc supplementation inhibited hepatic TNF-α expression and plasma TNF-α level in a mouse model of chronic alcohol consumption.[73] Supplementation with zinc to alcohol-fed rats increased the expression of the master transcription factor PU.1 in alveolar macrophages, reversed alveolar macrophage dysfunction, and enhanced bacterial clearance.[74] Dietary zinc supplementation throughout pregnancy also improved postnatal survival through alleviating inflammatory response, in mice prenatally exposed to alcohol.[75] A major mechanism of zinc in preventing and/or ameliorating inflammation is likely dependent on the modulation of NF-κB and PPAR signaling.[76] The zinc finger protein A20 is known to regulate negatively the NF-κB signaling in TNFR and TLR4 signaling pathways.[77] Whether A20 mediates the anti-inflammatory action of zinc in ALD remains a subject to be investigated.

Abrogation of Alcohol-Induced Hepatic Cell Death

Apoptotic cell death of hepatocytes, which has been repeatedly demonstrated in both clinical and experimental studies, is a feature of alcohol-induced liver damage.[78,79] One of the well-recognized mechanisms underlying alcohol-induced hepatocyte apoptosis is the death receptor-mediated pathway (namely the extrinsic pathway), in particular, the TNF-α/TNF-R1, and FasL/Fas systems are highly involved.[80] Elevated expression of TNF-R1 has been found in alcoholic steatohepatitis patients, and in alcohol-exposed hepatocytes.[81,82] A TNF-R1 knockout mice model showed that abrogating TNF-α signaling led to a reduction of apoptosis after 4 weeks of alcohol exposure.[83] The liver is very sensitive to the FasL/Fas signaling-induced cell death, and administration of Jo2 (a FasL agonist) has been shown to cause massive apoptosis in the liver.[84] In patients with alcoholic hepatitis, the Fas and FasL expression in the liver was significantly increased.[78,81]

Zinc is known to play a vital role in cell proliferation. Zinc deprivation in cell culture studies has been shown to induce apoptosis in diverse cell lines.[85–89] Zinc deprivation also potentiates death receptor-mediated apoptosis.[90] It has been reported that marginal dietary zinc deficiency synergistically elevated chronic alcohol exposure-induced hepatic expression of TNF-R1 and Fas in mice.[37] In accordance, the plasma alanine aminotransferase (ALT) level was significantly higher in the alcohol-fed mice with zinc deficient diet, compared to those with zinc adequate diet.[37] On the other hand, zinc treatment has been shown to attenuate apoptotic cell death in the liver, in a variety of models including D-galactosamine/TNF-α,[80,85] acute alcohol intoxication,[91] and chronic alcohol exposure.[73] Zinc supplementation also attenuated alcoholic hepatitis, and reduced the number of TUNEL (terminal deoxynucleotidyl transferase dUTP nick end labeling)-positive cells, in association with inhibition of caspase activities in mice.[73] Proapoptotic genes, including Tnfα, Tnfr1, Fasl, Fas, Faf-1 (Fas associated factor 1), and caspase-3 in the liver that were upregulated by alcohol exposure, were attenuated by dietary zinc supplementation.[73] These findings suggest that the antiapoptotic action of zinc is likely through suppression and death receptor signaling pathways.

Extrahepatic Actions of Zinc Against Alcoholic Toxicity

Intestinal Permeability and Endotoxemia

Endotoxin is lipopolysaccharide (LPS) derived from the cell wall of Gram-negative bacteria that inhabit the lumen of the distal intestine. Normally, only a trace amount of endotoxin can penetrate from the intestine into the circulation system, due to the barrier function of the intestine.[92] Alcohol consumption is known to increase the intestinal permeability to endotoxin, leading to elevated endotoxin in the blood (namely endotoxemia) and hepatic inflammation.[93] Endotoxemia has been well documented in patients with ALD, and the importance of endotoxemia in the development of alcoholic hepatitis has been well defined.[92] Clinical studies have shown that the blood endotoxin level correlates well with the hepatic TNF-α level, and the severity of ALD.[94–96]

Zinc plays a critical role in the maintenance of intestinal integrity and function. Prolonged zinc deficiency has been shown to induce morphological atrophy of intestinal mucosa.[97] Previous studies have demonstrated a direct link between zinc deficiency and intestinal epithelial barrier dysfunction. Caco-2 cells cultured in a zinc deficient medium exhibited a decrease in epithelial barrier function, as indicated by lower values of transepithelial electrical resistance.[98] Zinc deficiency caused delocalization of junction proteins, including ZO-1, occludin, β-cadherin, and disorganization of cytoskeleton proteins such as F-actin and β-tubulin. Zinc deficiency also accelerated neutrophil migration through the Caco-2 monolayer, and stimulated secretion of IL-8, epithelial neutrophil activating peptide-78, and growth-related oncogene-α.[98] A mouse model of chronic alcohol exposure demonstrated a significant elevation of plasma endotoxin, in association with generation of liver injury.[99] *Ex vivo* assay of the intestinal barrier function showed that an increased permeability of the ileum, but not duodenum or jejunum, to FITC-dextran (MW. 4,000) was induced by alcohol exposure. In accordance, intestinal zinc concentration was reduced in the ileum, rather than in other parts of the small intestine. Moreover, alcohol treatment and zinc deprivation from cell culture medium synergistically decreased the expression of tight junction proteins, including occludin, claudin-1, and ZO-1 in Caco-2 cells, indicating that marginal dietary zinc deficiency may not only directly affect the intestinal barrier, but also sensitize the deleterious effects of alcohol on the intestinal barrier function.

On the other hand, zinc supplementation has been shown to tighten the leaky intestine, and to preserve the intestinal barrier function under a variety of disease conditions. Patients with Crohn's disease showed an increase in intestinal permeability, as indicated by an elevated excretion ratio of lactulose/mannitol that was attenuated by oral zinc sulfate supplementation for 8 weeks.[100] Oral zinc supplementation in patients with acute shigellosis or diarrhea improved the intestinal barrier function, as indicated by reduction of lactulose:mannitol ratio.[101,102] Zinc supplementation also prevented epithelial barrier dysfunction in association with indomethacin-induced small intestine injury in rats.[103] Evaluation of the tight junction ultrastructure by electron microscopy demonstrated that dietary zinc supplementation to rats with colitis reduced the number of opened tight junction complexes in the colon epithelium.[104]

The effects of zinc treatment on alcohol-induced leaky intestine have been tested in both acute and chronic rodent models of ALD. In an acute model of ALD, mice were treated with three intragastric doses of zinc sulfate at 5 mg element zinc/kg, in a 12-h interval, prior to an oral dose of alcohol; zinc pretreatment abrogated alcohol-induced endotoxemia and liver damage.[105] Moreover, alcohol intoxication increased the penetration of intragastrically administrated LPS to the blood that was attenuated by zinc supplementation.[105] Dietary zinc supplementation also attenuated alcohol-increased ileum permeability to macromolecules, in rodents that chronically consumed alcohol (unpublished data from the authors' laboratory). These findings indicate that zinc prevention of alcohol-increased intestinal permeability contributes to the beneficial effects of zinc on alcohol-induced liver damage.

Adipose Tissues and Adipokines

White adipose tissue (WAT) plays a critical role in whole body energy homeostasis. It stores excess energy in form of triglycerides, and releases fatty acids via lipolysis for usage by other organs. However, excessive release of fatty acids from WAT may lead to ectopic lipid storage, under disease conditions including ALD.[106,107] WAT also regulates lipid metabolism in other organs through secreting adipokines, including adiponectin and leptin.[108] In the liver, adiponectin and leptin negatively regulate lipid content by stimulating fatty acid oxidation.[109,110] A recent study reported that alcohol exposure dramatically reduced plasma leptin level, in association with reduction of WAT mass, and normalization of plasma leptin level by administrating exogenous leptin stimulated fatty acid oxidation, and attenuated alcoholic fatty liver in mice.[111] Interestingly, dietary zinc deficiency worsened alcohol-induced decline of plasma leptin level[37] that indicates an extrahepatic role of zinc in regulating WAT function. The mechanism by which zinc impacts on adipose tissue and the release of adipokines is a subject of investigation.

MOLECULAR MECHANISMS OF ZINC PROTECTION AGAINST ALD

As noted previously, zinc supplementation prevents/reverses alcohol-induced liver injury through multiple aspects, involving both hepatic and extrahepatic factors. Mechanistic studies have shown that the beneficial effect of zinc supplementation on ALD involves the suppression of oxidative stress and restoration of zinc finger transcription factors.

Inhibition of Oxidative Stress

Suppression of ROS Generation

ADH is the major enzyme responsible for alcohol metabolism in the liver, at normal physiological condition. However, chronic alcohol consumption induces cytochrome P450 2E1 (CYP2E1), instead of ADH.[112,113]

Alcohol metabolism via CYP2E1 pathway generates ROS, and it has been well defined that hepatic CYP2E1 induction represents a major mechanism for alcohol-induced oxidative stress.[114] Zinc is a cofactor of ADH, and removal of zinc from ADH led to a complete loss of its catalytic activity.[8] In alcoholic patients, a significant reduction of hepatic ADH activity is associated with zinc decrease in the liver.[115] Thus, alcohol-induced hepatic zinc depletion is most likely linked to a shift of alcohol metabolic pathway from ADH to CYP2E1 that favors ROS generation. Indeed, dietary supplementation with zinc sulfate suppressed alcohol-elevated CYP2E1 activity, while it increased ADH activity in the liver of alcohol-fed mice.[26] Meanwhile, the accumulation of ROS and lipid peroxidation products, including 4-HNE and MDA, in the liver was also attenuated by zinc supplementation.

Upregulation of Antioxidant Molecules

Zinc plays an important role in the regulation of cellular GSH level.[116] GSH is the most important molecule involved in cellular antioxidant defense, and selective GSH depletion in mitochondria has been repeatedly reported in rodent models of ALD.[117,118] Dietary supplementation with S-adenosyl-l-methionine has been shown to replenish the mitochondrial GSH pool, thereby attenuating mitochondrial dysfunction, and preventing alcohol-induced liver injury.[119] A positive correlation has been found between zinc and GSH. Zinc depletion with zinc chelator induced a dose-dependent decrease in cellular GSH level, in cultured keratinocytes, hepatocytes, and hepatic stellate cells.[86,120] The levels of GSH in blood and liver were found to be correlated well with dietary zinc status.[121,122] In a mouse model of ALD, zinc supplementation effectively prevented alcohol-induced GSH decrease, both in the cytosol and mitochondria.[26] Furthermore, zinc supplementation increased glutathione reductase activity, indicating that zinc protects the GSH pool at least partially, through enhancing the reduction of oxidized glutathione. Zinc supplementation also prevented the alcohol-induced decrease of glutathione peroxidase activity in the liver.[26] Therefore, GSH-related antioxidant capacity is improved in the alcohol-intoxicated liver by zinc supplementation.

The antioxidative action of zinc is also achieved by induction of MT and Cu/Zn-SOD. Apart from the well-known role in cellular buffering, trafficking, and transferring zinc, MT also has potent antioxidative function. MTs are comprised of 61–68 amino acids with a highly conserved sequence of 20 cysteine residues,[123] and therefore they share a great similarity with GSH. Importantly, the thiol groups in MT are preferential attacking targets for free radicals, compared to the sulfhydryl residues from GSH, or protein fractions.[124] The fact that zinc supplementation induced MT in the liver and intestine suggests a link between MT and zinc action, in ALD.[22,125] Indeed, transgenic overexpression of MT in mice significantly reduced the accumulation of superoxide anion, lipid peroxidation products, and oxidized GSH in the liver after acute alcohol intoxication.[126] These results indicate that MTs are effective agents in the protection against alcohol-induced liver injury, and the hepatic protection by MTs is likely through the inhibition of alcohol-induced oxidative stress. Although the hepatic and extrahepatic effects of zinc could be independent of MT,[22,125] MT is critical to maintain high levels of zinc in the liver, suggesting that the protective actions by MT in the liver are mediated, at least partially, by zinc. Low levels of MT in the liver reduce the endogenous zinc reservoir, and sensitize the organ to alcohol-induced injury.[26] The coordination between zinc and MT is that MT maintains high levels of zinc in the liver, and releases zinc under oxidative stress conditions, leading to hepatoprotective action.

Zinc is a structural component of Cu/Zn-SOD. Measurement of Cu/Zn- and Mn-SOD, in hepatocytes from liver biopsies of patients with ALD, showed that the amount of Cu/Zn-SOD, rather than Mn-SOD, was significantly lower in alcoholics, than that in the healthy subjects.[127] Animal studies showed that alcohol exposure decreased hepatic zinc level, in association with reduced Cu/Zn-SOD activity,[2] and dietary supplementation with zinc normalized hepatic SOD activity that was decreased upon alcohol exposure in mice.[4]

Reactivation of PPAR-α and HNF-4α

PPAR-α and HNF-4α are zinc finger transcription factors that belong to the nuclear hormone receptor superfamily.[128] PPAR-α is a key transcriptional regulator of lipid metabolism and transport, fatty acid oxidation, and glucose homeostasis.[129] It is one of the most well-recognized transcription factors in mechanistic studies on the pathogenesis of alcoholic steatosis.[64,130] Chronic alcohol exposure in mice has been reported to suppress the DNA binding activity of PPAR-α,[131] with its protein level being decreased[2,4] or unaffected.[131] Treatment with the PPAR-α agonist Wy14,643 attenuated alcoholic steatosis,[131] hepatitis,[132] and fibrosis,[133] in association with elevated PPAR-α expression and/or activity. HNF-4α is a master regulator of hepatic gene expression,[134] and it has been shown to bind to the promoter of more than 1200 genes involved in most aspects of hepatocyte function.[135] Previous studies have shown that chronic alcohol exposure decreased the DNA binding ability of HNF-4α in the liver[4] and intestine.[136]

Hepatic PPAR-α and HNF-4α

Zinc is a central component for the zinc-finger motif of PPAR-α and HNF-4α that is required for their DNA binding activity. Elimination of zinc coordination by ROS from the zinc finger motif disassembles the zinc finger structure of these proteins, leading to defective DNA binding, and decreased target genes transcription.[9] Thus, inactivation of zinc transcription factors may account for alcohol-induced cellular disorders. The possible link between zinc and zinc finger transcription factors in ALD has been explored both in mouse models of chronic alcohol exposure, and in alcohol-intoxicated hepatoma cell models. In a mouse model of 3-month alcohol exposure, dietary zinc supplementation for the last 4 weeks reversed alcohol pre-established increase in hepatic triglyceride, cholesterol, and free fatty acids, and accelerated fatty acid oxidation and VLDL secretion in the liver.[4] In accordance, the impaired DNA binding activity of HNF-4α and PPAR-α was restored by zinc supplementation. The direct link between zinc and zinc finger transcription factors was further defined by experimental zinc depletion in HepG2 cells.[4] Zinc deprivation did not significantly affect the expression of HNF-4α and PPAR-α at protein levels; however, the DNA binding activity of HNF-4α and PPAR-α was significantly decreased. Adding back zinc to the medium reactivated both proteins. The expression of HNF-4α- and PPAR-α-regulated proteins, that are related to fatty acids β-oxidation (ACADL) and lipid secretion (MTTP, ApoB), were downregulated by zinc deprivation. As a result, excess lipid droplets were accumulated in alcohol-treated HepG2 cells with zinc deprivation. These results suggest that alcohol exposure reduces the DNA binding activity of HNF-4α and PPAR-α, at least partially, through mobilizing zinc from these zinc finger transcription factors. Reactivation of HNF-4α and PPAR-α is likely the most important molecular mechanism underlying the beneficial effect of zinc on hepatic lipid homeostasis.

Intestinal HNF-4α

HNF-4α expression is also detected in the epithelia of the intestine, pancreas, and kidney, where it exerts functional roles in regulating epithelial junctions, cell proliferation, and inflammation.[137] Intestinal HNF-4α has been detected in all the segments of the gastrointestinal tract, except for the stomach, and its distribution gradually increased along the intestine, from the proximal to the distal.[136] Knockout of HNF-4α in the intestinal epithelial cells, in mice, increased the permeability to experimental inflammatory bowel disease.[138,139] Importantly, the intestinal permeability was significantly elevated in the intestinal epithelial HNF-4α null mice with acute colitis. Chronic alcohol exposure inhibited HNF-4α DNA binding activity in all the intestinal segments, with the most prominent effect in the ileum.[136] In accordance, decreased HNF-4α protein level in the intestine was detected. A highly positive correlation was found between the intestinal HNF-4α mRNA and the intestinal tight junction protein mRNAs. Caco-2 cell culture studies further defined the role of HNF-4α in the regulation of tight junction proteins, and epithelial barrier function.[136] Knockdown of HNF-4α by siRNA transfection resulted in a decreased expression of claudin-1, occludin, and ZO-1, in association with disruption of the epithelial barrier function. Furthermore, experimental zinc deprivation in Caco-2 cells suppressed HNF-4α activity in a dose-dependent manner. Dietary zinc supplementation restored intestinal barrier function in alcohol-fed rats. In accordance, the ileal HNF-4α activity was reactivated, in association with increased expression of tight junction proteins in zinc supplementation rats (unpublished data from the authors' laboratory).

ZINC SUPPLEMENTATION IN HUMAN ALD

There have been multiple studies showing that zinc supplementation reverses manifestations of zinc deficiency in ALD, such as impaired night vision, skin lesions, and, in some cases, encephalopathy and immune dysfunction.[1] Studies have been performed to determine the duration and amounts of zinc necessary to improve serum and hepatic zinc in alcoholic patients.

To assess the duration of zinc intake necessary to normalize serum and hepatic zinc concentrations, zinc supplementation in patients with ALD has been performed in a time-dependent manner. Alcoholic patients without cirrhosis received zinc sulfate at 600 mg/day, for 10 days, and alcoholic patients with cirrhosis received the same dose of zinc sulfate for 10, 30, and 60 days, respectively.[140] Serum zinc concentrations were increased to normal values in all groups, during 10–60 days of zinc supplementation. Zinc concentrations in the liver biopsies were increased significantly in patients with cirrhosis, after zinc supplementation for 10 or 60 days, though some patients remained under normal values. No adverse reactions of zinc supplementation were observed in that study.

A long-term oral zinc supplementation (200 mg three times a day, for 2–3 months) to cirrhotic patients, including alcoholic cirrhosis, produced beneficial effects on both metabolic function, and nutritional parameters of the liver.[141] Quantitative liver function tests, including galactose elimination capacity and antipyrine clearance, demonstrated that oral zinc supplementation significantly improved liver metabolic function. Similarly, the Child-Pugh score, an overall estimation of hepatocellular failure, was improved by an average of more than 1 point, after zinc supplementation. Zinc supplementation also significantly improved nutritional parameters, including urinary excretion of creatinine, serum prealbumin, retinol binding protein, and insulin-like growth factor 1 (IGF-1). In particular, the serum IGF-1 was increased by an average of 30%, after zinc therapy. However, the nutritional parameters remained on average below the lower limit of the normal range. Glucose disappearance was improved by greater than 30%, in response to zinc therapy. There were no changes in pancreatic insulin secretion and systemic delivery, or in the hepatic extraction of insulin. Insulin sensitivity, which was reduced by 80% before treatment, did not change. Glucose effectiveness was nearly halved in cirrhosis before treatment, and significantly increased after zinc therapy. These data suggest that zinc treatment in advanced cirrhosis improves glucose tolerance, via an increasing effect of glucose *per se* on glucose metabolism.

In another study, polaprezinc, a synthetic zinc-containing compound with 34 mg of elemental zinc, was administrated daily, for 24 weeks, to patients with chronic hepatitis or cirrhosis, including ALD.[142] Oral polaprezinc supplementation increased serum zinc concentrations by 16.6% and 31.6%, at 12 and 24 weeks, respectively. Serum type IV collagen levels that reflect liver fibrosis were decreased significantly at 24 weeks, in comparison to the baseline in the patients showing increased serum zinc level after polaprezinc supplementation. The tissue inhibitors of metalloproteinase-1 (TIMP-1) levels were reduced significantly in these patients. The results suggest that zinc may have an inhibitory effect on liver fibrosis through downregulation of TIMP-1.

CONCLUSIONS

Zinc deficiency is a causal factor in the development and progression of ALD. Inactivation of zinc proteins, due to zinc release under oxidative stress condition, at least partially accounts for alcohol-induced metabolic disorders and cell injury. Dietary zinc supplementation in experimental models provides protection against alcohol-induced liver injury, through modulating multiple pathways, including oxidative stress, alcohol metabolism, lipid metabolism, cytokine production, and cell death signaling. Data from human studies demonstrate the reversal of certain zinc deficiency signs and symptoms, with zinc supplementation. Taken as a whole, all the observations from animal and clinical studies suggest an exciting possibility that zinc could be used as a dietary intervention for treating ALD. However, it should be pointed out that even though zinc is relatively harmless, compared to several other metal ions with similar chemical properties, high doses of zinc administration may affect copper homeostasis. Careful monitoring of zinc and copper status during the therapy is therefore highly recommended.

Key Facts of Zinc in Alcoholic Liver Disease

- Zinc is the second most abundant trace mineral in the body. Zinc is essential for people to stay healthy.
- Alcohol abuse reduces zinc level in the body. The severity of hepatic zinc reduction increases along with the disease progression.
- Zinc deficiency may lead to impaired night vision, skin lesions, depressed wound healing, anorexia, hypogonadism, altered immune function, and depressed mental function.
- Alcohol metabolism generates oxidative stress that is likely a key player in mediating alcohol-induced zinc deficiency.
- Dietary zinc supplementation improves hepatic zinc homeostasis, antioxidant defense, and tissue repair, leading to reversal of alcoholic liver disease.
- Zinc supplementation is recommended for people consuming alcohol.

Summary Points

- This chapter focuses on the role of zinc in alcoholic liver disease.
- Zinc deficiency has been well-documented in alcoholic patients, and experimental models with alcoholic liver disease.
- The mechanisms underlying alcohol-induced zinc deficiency include reduced dietary zinc intake, perturbed intestinal absorption, disturbed hepatic zinc metabolism, and increased urinary zinc excretion.
- Dietary zinc supplementation has been shown to prevent/reverse alcohol-induced liver injury, both in animal models and clinical studies of alcoholic liver disease.
- The beneficial effects of zinc are achieved by both hepatic actions and extrahepatic actions.
- Inhibition of oxidative stress and restoration of alcohol-inactivated zinc finger transcription factors represent important molecular mechanisms of zinc actions.

References

1. McClain CJ, Antonow DR, Cohen DA, Shedlofsky SI. Zinc metabolism in alcoholic liver disease. *Alcohol Clin Exp Res* 1986;**10**(6):582–9.
2. Sun Q, Li Q, Zhong W, et al. Dysregulation of hepatic zinc transporters in a mouse model of alcoholic liver disease. *Am J Physiol Gastrointest Liver Physiol* 2014;**307**(3):G313–22.
3. Wang J, Pierson Jr RN. Distribution of zinc in skeletal muscle and liver tissue in normal and dietary controlled alcoholic rats. *J Lab Clin Med* 1975;**85**(1):50–8.
4. Kang X, Zhong W, Liu J, et al. Zinc supplementation reverses alcohol-induced steatosis in mice through reactivating hepatocyte nuclear factor-4alpha and peroxisome proliferator-activated receptor-alpha. *Hepatol* 2009;**50**(4):1241–50.

5. Zhou Z. Zinc and alcoholic liver disease. *Dig Dis* 2010;**28**(6):745–50.
6. Krebs NE, Hambidge KM. Zinc metabolism and homeostasis: the application of tracer techniques to human zinc physiology. *Biometals* 2001;**14**(3–4):397–412.
7. Maret W, Sandstead HH. Zinc requirements and the risks and benefits of zinc supplementation. *J Trace Elem Med Biol* 2006;**20**(1):3–18.
8. Hao Q, Maret W. Aldehydes release zinc from proteins. A pathway from oxidative stress/lipid peroxidation to cellular functions of zinc. *FEBS J* 2006;**273**(18):4300–10.
9. Webster KA, Prentice H, Bishopric NH. Oxidation of zinc finger transcription factors: physiological consequences. *Antioxid Redox Signal* 2001;**3**(4):535–48.
10. Solomons NW. Update on zinc biology. *Ann Nutr Metab* 2013; **62**(Suppl. 1):8–17.
11. Jeong J, Eide DJ. The SLC39 family of zinc transporters. *Mol Aspects Med* 2013;**34**(2–3):612–9.
12. Huang L, Tepaamorndech S. The SLC30 family of zinc transporters - a review of current understanding of their biological and pathophysiological roles. *Mol Aspects Med* 2013;**34**(2–3):548–60.
13. Vasák M. Advances in metallothionein structure and functions. *J Trace Elem Med Biol* 2005;**19**(1):13–7.
14. Kang YJ, Zhou Z. Zinc prevention and treatment of alcoholic liver disease. *Mol Aspects Med* 2005;**26**(4–5):391–404.
15. Outten CE, O'Halloran TV. Femtomolar sensitivity of metalloregulatory proteins controlling zinc homeostasis. *Science* 2001; **292**(5526):2488–92.
16. Li TY, Kraker AJ, Shaw III CF, Petering DH. Ligand substitution reactions of metallothioneins with EDTA and apo-carbonic anhydrase. *Proc Natl Acad Sci USA* 1980;**77**(11):6334–8.
17. Udom AO, Brady FO. Reactivation in vitro of zinc-requiring apoenzymes by rat liver zinc-thionein. *Biochem J* 1980;**187**(2):329–35.
18. Hao R, Pfeiffer RF, Ebadi M. Purification and characterization of metallothionein and its activation of pyridoxal phosphokinase in trout (Salmo gairdneri) brain. *Comp Biochem Physiol B* 1993;**104**(2): 293–8.
19. Cano-Gauci DF, Sarkar B. Reversible zinc exchange between metallothionein and the estrogen receptor zinc finger. *FEBS Lett* 1996;**386**(1):1–4.
20. Hathout Y, Fabris DCF. Stoichiometry in zinc ion transfer from metallothionein to zinc finger peptides. *Int J Mass Spectrom* 2001;**204**(1–3):1–6.
21. Masters BA, Kelly EJ, Quaife CJ, Brinster RL, Palmiter RD. Targeted disruption of metallothionein I and II genes increases sensitivity to cadmium. *Proc Natl Acad Sci USA* 1994;**91**(2):584–8.
22. Zhou Z, Sun X, Lambert JC, Saari JT, Kang YJ. Metallothionein-independent zinc protection from alcoholic liver injury. *Am J Pathol* 2002;**160**(6):2267–74.
23. Iszard MB, Liu J, Liu Y, et al. Characterization of metallothionein-I-transgenic mice. *Toxicol Appl Pharmacol* 1995;**133**(2):305–12.
24. Krezel A, Hao Q, Maret W. The zinc/thiolate redox biochemistry of metallothionein and the control of zinc ion fluctuations in cell signaling. *Arch Biochem Biophys* 2007;**463**(2):188–200.
25. Kang YJ. The antioxidant function of metallothionein in the heart. *Proc Soc Exp Biol Med* 1999;**222**(3):263–73.
26. Zhou Z, Wang L, Song Z, Saari JT, McClain CJ, Kang YJ. Zinc supplementation prevents alcoholic liver injury in mice through attenuation of oxidative stress. *Am J Pathol* 2005;**166**(6):1681–90.
27. Zakhari S. Overview: how is alcohol metabolized by the body? *Alcohol Res Health* 2006;**29**(4):245–54.
28. Smathers RL, Galligan JJ, Stewart BJ, Petersen DR. Overview of lipid peroxidation products and hepatic protein modification in alcoholic liver disease. *Chem Biol Interact* 2011;**192**(1–2):107–12.
29. Bartholomay AF, Robin ED, Vallee RL, Wacker WE. Zinc metabolism in hepatic dysfunction. I. Serum zinc concentrations in Laennec's cirrhosis and their validation by sequential analysis. *N Engl J Med* 1956;**255**(9):403–8.
30. Goode HF, Kelleher J, Walker BE. Relation between zinc status and hepatic functional reserve in patients with liver disease. *Gut* 1990;**31**(6):694–7.
31. Rodríguez-Moreno F, González-Reimers E, Santolaria-Fernández F, et al. Zinc, copper, manganese, and iron in chronic alcoholic liver disease. *Alcohol* 1997;**14**(1):39–44.
32. Rahelić D, Kujundzić M, Romić Z, Brkić K, Petrovecki M. Serum concentration of zinc, copper, manganese and magnesium in patients with liver cirrhosis. *Coll Antropol* 2006;**30**(3):523–8.
33. Foote JW, Delves HT. Albumin bound and alpha 2-macroglobulin bound zinc concentrations in the sera of healthy adults. *J Clin Pathol* 1984;**37**(9):1050–4.
34. Wu CT, Lee JN, Shen WW, Lee SL. Serum zinc, copper, and ceruloplasmin levels in male alcoholics. *Biol Psychiatry* 1984;**19**(9): 1333–8.
35. Mohammad MK, Zhou Z, Cave M, Barve A, McClain CJ. Zinc and liver disease. *Nutr Clin Pract* 2012;**27**(1):8–20.
36. Barak AJ, Beckenhauer HC, Kerrigan FJ. Zinc and manganese levels in serum and liver after alcohol feeding and development of fatty cirrhosis in rats. *Gut* 1967;**8**(5):454–7.
37. Zhong W, Zhao Y, Sun X, Song Z, McClain CJ, Zhou Z. Dietary zinc deficiency exaggerates ethanol-induced liver injury in mice: involvement of intrahepatic and extrahepatic factors. *PLoS One* 2013;**8**(10):e76522.
38. Hartoma TR, Sotaniemi EA, Pelkonen O, Ahlqvist J. Serum zinc and serum copper and indices of drug metabolism in alcoholics. *Eur J Clin Pharmacol* 1977;**12**(2):147–51.
39. Murphy EW, Willis BW, Watt BK. Provisional tables on the zinc content of foods. *J Am Diet Assoc* 1975;**66**(4):345–55.
40. Krebs NF. Dietary zinc and iron sources, physical growth and cognitive development of breastfed infants. *J Nutr* 2000;**130**(2S Suppl): 358S–60S.
41. Wapnir RA. Protein digestion and the absorption of mineral elements. *Adv Exp Med Biol* 1989;**249**:95–115.
42. Lieber CS. Relationships between nutrition, alcohol use, and liver disease. *Alcohol Res Health* 2003;**27**(3):220–31.
43. Gearhardt AN, Corbin WR. Body mass index and alcohol consumption: family history of alcoholism as a moderator. *Psychol Addict Behav* 2009;**23**(2):216–25.
44. Manari AP, Preedy VR, Peters TJ. Nutritional intake of hazardous drinkers and dependent alcoholics in the UK. *Addict Biol* 2003;**8**(2):201–10.
45. Dinsmore WW, Callender ME, McMaster D, Love AH. The absorption of zinc from a standardized meal in alcoholics and in normal volunteers. *Am J Clin Nutr* 1985;**42**(4):688–93.
46. Valberg LS, Flanagan PR, Ghent CN, Chamberlain MJ. Zinc absorption and leukocyte zinc in alcoholic and nonalcoholic cirrhosis. *Dig Dis Sci* 1985;**30**(4):329–33.
47. Antonson DL, Vanderhoof JA. Effect of chronic ethanol ingestion on zinc absorption in rat small intestine. *Dig Dis Sci* 1983;**28**(7): 604–8.
48. Wang X, Zhou B. Dietary zinc absorption: A play of Zips and ZnTs in the gut. *IUBMB Life* 2010;**62**(3):176–82.
49. Dufner-Beattie J, Wang F, Kuo YM, Gitschier J, Eide D, Andrews GK. The acrodermatitis enteropathica gene ZIP4 encodes a tissue-specific, zinc-regulated zinc transporter in mice. *J Biol Chem* 2003; **278**(35):33474–81.
50. Lonnerdal B. Dietary factors influencing zinc absorption. *J Nutr* 2000;**130**(5S Suppl):1378S–83S.
51. Kim J, Paik HY, Joung H, Woodhouse LR, Li S, King JC. Effect of dietary phytate on zinc homeostasis in young and elderly Korean women. *J Am Coll Nutr* 2007;**26**(1):1–9.
52. Vallee BL, Falchuk KH. The biochemical basis of zinc physiology. *Physiol Rev* 1993;**73**(1):79–118.
53. King JC, Shames DM, Woodhouse LR. Zinc homeostasis in humans. *J Nutr* 2000;**130**(5S Suppl):1360S–6S.

54. Zlotkin SH. Nutrient interactions with total parenteral nutrition: effect of histidine and cysteine intake on urinary zinc excretion. *J Pediatr* 1989;**114**(5):859–64.
55. Russell RM. Vitamin A and zinc metabolism in alcoholism. *Am J Clin Nutr* 1980;**33**(12):2741–9.
56. Sargent WQ, Simpson JR, Beard JD. The effects of acute and chronic ethanol administration on divalent cation excretion. *J Pharmacol Exp Ther* 1974;**190**(3):507–14.
57. Ranaldi G, Perozzi G, Truong-Tran A, Zalewski P, Murgia C. Intracellular distribution of labile Zn(II) and zinc transporter expression in kidney and MDCK cells. *Am J Physiol Renal Physiol* 2002; **283**(6):F1365–75.
58. Sullivan JF, Heaney RP. Zinc metabolism in alcoholic liver disease. *Am J Clin Nutr* 1970;**23**(2):170–7.
59. Medici V, Halsted CH. Folate, alcohol, and liver disease. *Mol Nutr Food Res* 2013;**57**(4):596–606.
60. Atukorala TM, Herath CA, Ramachandran S. Zinc and vitamin A status of alcoholics in a medical unit in Sri Lanka. *Alcohol Alcohol* 1986;**21**(3):269–75.
61. Bell MS, Vermeulen LC, Sperling KB. Pharmacotherapy with botulinum toxin: harnessing nature's most potent neurotoxin. *Pharmacother* 2000;**20**(9):1079–91.
62. Purohit V, Russo D, Coates PM. Role of fatty liver, dietary fatty acid supplements, and obesity in the progression of alcoholic liver disease: introduction and summary of the symposium. *Alcohol* 2004;**34**(1):3–8.
63. Lakshman MR. Some novel insights into the pathogenesis of alcoholic steatosis. *Alcohol* 2004;**34**(1):45–8.
64. Nagy LE. Molecular aspects of alcohol metabolism: transcription factors involved in early ethanol-induced liver injury. *Annu Rev Nutr* 2004;**24**:55–78.
65. Crabb DW, Liangpunsakul S. Alcohol and lipid metabolism. *J Gastroenterol Hepatol* 2006;**21**(Suppl. 3):S56–60.
66. Tomita K, Azuma T, Kitamura N, et al. Leptin deficiency enhances sensitivity of rats to alcoholic steatohepatitis through suppression of metallothionein. *Am J Physiol Gastrointest Liver Physiol* 2004; **287**(5):G1078–85.
67. tom Dieck H, Döring F, Roth HP, Daniel H. Changes in rat hepatic gene expression in response to zinc deficiency as assessed by DNA arrays. *J Nutr* 2003;**133**(4):1004–10.
68. tom Dieck H, Döring F, Fuchs D, Roth HP, Daniel H. Transcriptome and proteome analysis identifies the pathways that increase hepatic lipid accumulation in zinc-deficient rats. *J Nutr* 2005; **135**(2):199–205.
69. Yousef MI, El-Hendy HA, El-Demerdash FM, Elagamy EI. Dietary zinc deficiency induced-changes in the activity of enzymes and the levels of free radicals, lipids and protein electrophoretic behavior in growing rats. *Toxicol* 2002;**175**(1–3):223–34.
70. McClain CJ, Song Z, Barve SS, Hill DB, Deaciuc I. Recent advances in alcoholic liver disease. IV. Dysregulated cytokine metabolism in alcoholic liver disease. *Am J Physiol Gastrointest Liver Physiol* 2004;**287**(3):G497–502.
71. Tilg H, Diehl AM. Cytokines in alcoholic and nonalcoholic steatohepatitis. *N Engl J Med* 2000;**343**(20):1467–76.
72. Zhao Y, Zhong W, Sun X, et al. Zinc deprivation mediates alcohol-induced hepatocyte IL-8 analog expression in rodents via an epigenetic mechanism. *Am J Pathol* 2011;**179**(2):693–702.
73. Zhou Z, Liu J, Song Z, McClain CJ, Kang YJ. Zinc supplementation inhibits hepatic apoptosis in mice subjected to a long-term ethanol exposure. *Exp Biol Med (Maywood)* 2008;**233**(5):540–8.
74. Mehta AJ, Joshi PC, Fan X, et al. Zinc supplementation restores PU.1 and Nrf2 nuclear binding in alveolar macrophages and improves redox balance and bacterial clearance in the lungs of alcohol-fed rats. *Alcohol Clin Exp Res* 2011;**35**(8):1519–28.
75. Summers BL, Rofe AM, Coyle P. Dietary zinc supplementation throughout pregnancy protects against fetal dysmorphology and improves postnatal survival after prenatal ethanol exposure in mice. *Alcohol Clin Exp Res* 2009;**33**(4):591–600.
76. Shen H, Oesterling E, Stromberg A, Toborek M, MacDonald R, Hennig B. Zinc deficiency induces vascular pro-inflammatory parameters associated with NF-kappaB and PPAR signaling. *J Am Coll Nutr* 2008;**27**(5):577–87.
77. Shembade N, Ma A, Harhaj EW. Inhibition of NF-kappaB signaling by A20 through disruption of ubiquitin enzyme complexes. *Science* 2010;**327**(5969):1135–9.
78. Natori S, Rust C, Stadheim LM, Srinivasan A, Burgart LJ, Gores GJ. Hepatocyte apoptosis is a pathologic feature of human alcoholic hepatitis. *J Hepatol* 2001;**34**(2):248–53.
79. Ziol M, Tepper M, Lohez M, et al. Clinical and biological relevance of hepatocyte apoptosis in alcoholic hepatitis. *J Hepatol* 2001;**34**(2):254–60.
80. Lambert JC, Zhou Z, Kang YJ. Suppression of Fas-mediated signaling pathway is involved in zinc inhibition of ethanol-induced liver apoptosis. *Exp Biol Med (Maywood)* 2003;**228**(4):406–12.
81. Ribeiro PS, Cortez-Pinto H, Sola S, et al. Hepatocyte apoptosis, expression of death receptors, and activation of NF-kappaB in the liver of nonalcoholic and alcoholic steatohepatitis patients. *Am J Gastroenterol* 2004;**99**(9):1708–17.
82. García-Ruiz C, Morales A, Colell A, et al. Feeding S-adenosyl-L-methionine attenuates both ethanol-induced depletion of mitochondrial glutathione and mitochondrial dysfunction in periportal and perivenous rat hepatocytes. *Hepatol* 1995;**21**(1):207–14.
83. Ji C, Kaplowitz N. Betaine decreases hyperhomocysteinemia, endoplasmic reticulum stress, and liver injury in alcohol-fed mice. *Gastroenterol* 2003;**124**(5):1488–99.
84. Galle PR, Hofmann WJ, Walczak H, et al. Involvement of the CD95 (APO-1/Fas) receptor and ligand in liver damage. *J Exp Med* 1995;**182**(5):1223–30.
85. Meerarani P, Ramadass P, Toborek M, Bauer HC, Bauer H, Hennig B. Zinc protects against apoptosis of endothelial cells induced by linoleic acid and tumor necrosis factor alpha. *Am J Clin Nutr* 2000;**71**(1):81–7.
86. Nakatani T, Tawaramoto M, Opare Kennedy D, Kojima A, Matsui-Yuasa I. Apoptosis induced by chelation of intracellular zinc is associated with depletion of cellular reduced glutathione level in rat hepatocytes. *Chem Biol Interact* 2000;**125**(3):151–63.
87. Cao J, Bobo JA, Liuzzi JP, Cousins RJ. Effects of intracellular zinc depletion on metallothionein and ZIP2 transporter expression and apoptosis. *J Leukoc Biol* 2001;**70**(4):559–66.
88. Pang W, Leng X, Lu H, et al. Depletion of intracellular zinc induces apoptosis of cultured hippocampal neurons through suppression of ERK signaling pathway and activation of caspase-3. *Neurosci Lett* 2013;**552**:140–5.
89. Guo B, Yang M, Liang D, Yang L, Cao J, Zhang L. Cell apoptosis induced by zinc deficiency in osteoblastic MC3T3-E1 cells via a mitochondrial-mediated pathway. *Mol Cell Biochem* 2012;**361**(1–2): 209–16.
90. Bao S, Knoell DL. Zinc modulates airway epithelium susceptibility to death receptor-mediated apoptosis. *Am J Physiol Lung Cell Mol Physiol* 2006;**290**(3):L433–41.
91. Zhou Z, Kang X, Jiang Y, et al. Preservation of hepatocyte nuclear factor-4alpha is associated with zinc protection against TNF-alpha hepatotoxicity in mice. *Exp Biol Med (Maywood)* 2007;**232**(5): 622–8.
92. Rao R. Endotoxemia and gut barrier dysfunction in alcoholic liver disease. *Hepatol* 2009;**50**(2):638–44.
93. Thurman II RG. Alcoholic liver injury involves activation of Kupffer cells by endotoxin. *Am J Physiol* 1998;**275**(4 Pt 1):G605–11.
94. Fujimoto M, Uemura M, Nakatani Y, et al. Plasma endotoxin and serum cytokine levels in patients with alcoholic hepatitis: relation to severity of liver disturbance. *Alcohol Clin Exp Res* 2000; **24**(4 Suppl):48S–54S.

95. Hanck C, Rossol S, Bocker U, Tokus M, Singer MV. Presence of plasma endotoxin is correlated with tumour necrosis factor receptor levels and disease activity in alcoholic cirrhosis. *Alcohol Alcohol* 1998;**33**(6):606–8.
96. Urbaschek R, McCuskey RS, Rudi V, et al. Endotoxin, endotoxin-neutralizing-capacity, sCD14, sICAM-1, and cytokines in patients with various degrees of alcoholic liver disease. *Alcohol Clin Exp Res* 2001;**25**(2):261–8.
97. Ziegler TR, Evans ME, Fernández-Estivaríz C, Jones DP. Trophic and cytoprotective nutrition for intestinal adaptation, mucosal repair, and barrier function. *Annu Rev Nutr* 2003;**23**:229–61.
98. Finamore A, Massimi M, Conti Devirgiliis L, Mengheri E. Zinc deficiency induces membrane barrier damage and increases neutrophil transmigration in Caco-2 cells. *J Nutr* 2008;**138**(9): 1664–70.
99. Zhong W, McClain CJ, Cave M, Kang YJ, Zhou Z. The role of zinc deficiency in alcohol-induced intestinal barrier dysfunction. *Am J Physiol Gastrointest Liver Physiol* 2010;**298**(5):G625–33.
100. Sturniolo GC, Di Leo V, Ferronato A, D'Odorico A, D'Inca R. Zinc supplementation tightens "leaky gut" in Crohn's disease. *Inflamm Bowel Dis* 2001;**7**(2):94–8.
101. Roy SK, Behrens RH, Haider R, et al. Impact of zinc supplementation on intestinal permeability in Bangladeshi children with acute diarrhoea and persistent diarrhoea syndrome. *J Pediatr Gastroenterol Nutr* 1992;**15**(3):289–96.
102. Alam AN, Sarker SA, Wahed MA, Khatun M, Rahaman MM. Enteric protein loss and intestinal permeability changes in children during acute shigellosis and after recovery: effect of zinc supplementation. *Gut* 1994;**35**(12):1707–11.
103. Sivalingam N, Pichandi S, Chapla A, Dinakaran A, Jacob M. Zinc protects against indomethacin-induced damage in the rat small intestine. *Eur J Pharmacol* 2011;**654**(1):106–16.
104. Sturniolo GC, Fries W, Mazzon E, Di Leo V, Barollo M, D'Inca R. Effect of zinc supplementation on intestinal permeability in experimental colitis. *J Lab Clin Med* 2002;**139**(5):311–5.
105. Lambert JC, Zhou Z, Wang L, Song Z, McClain CJ, Kang YJ. Prevention of alterations in intestinal permeability is involved in zinc inhibition of acute ethanol-induced liver damage in mice. *J Pharmacol Exp Ther* 2003;**305**(3):880–6.
106. Wei X, Shi X, Zhong W, et al. Chronic alcohol exposure disturbs lipid homeostasis at the adipose tissue-liver axis in mice: analysis of triacylglycerols using high-resolution mass spectrometry in combination with in vivo metabolite deuterium labeling. *PLoS One* 2013;**8**(2):e55382.
107. Zhong W, Zhao Y, Tang Y, et al. Chronic alcohol exposure stimulates adipose tissue lipolysis in mice: role of reverse triglyceride transport in the pathogenesis of alcoholic steatosis. *Am J Pathol* 2012;**180**(3):998–1007.
108. Havel PJ. Control of energy homeostasis and insulin action by adipocyte hormones: leptin, acylation stimulating protein, and adiponectin. *Curr Opin Lipidol* 2002;**13**(1):51–9.
109. Yamauchi T, Kamon J, Minokoshi Y, et al. Adiponectin stimulates glucose utilization and fatty-acid oxidation by activating AMP-activated protein kinase. *Nat Med* 2002;**8**(11):1288–95.
110. Wein S, Ukropec J, Gasperíková D, Klimes I, Seböková E. Concerted action of leptin in regulation of fatty acid oxidation in skeletal muscle and liver. *Exp Clin Endocrinol Diabetes* 2007;**115**(4): 244–51.
111. Tan X, Sun X, Li Q, et al. Leptin deficiency contributes to the pathogenesis of alcoholic fatty liver disease in mice. *Am J Pathol* 2012;**181**(4):1279–86.
112. Lieber CS. Ethanol metabolism, cirrhosis and alcoholism. *Clin Chim Acta* 1997;**257**(1):59–84.
113. McClearn GE, Bennett EL, Hebert M, Kakihana R, Schlesinger K. Alcohol dehydrogenase activity and previous ethanol consumption in mice. *Nature* 1964;**203**:793–4.
114. Cederbaum AI, Lu Y, Wu D. Role of oxidative stress in alcohol-induced liver injury. *Arch Toxicol* 2009;**83**(6):519–48.
115. Mills PR, Fell GS, Bessent RG, Nelson LM, Russell RI. A study of zinc metabolism in alcoholic cirrhosis. *Clin Sci (Lond)* 1983;**64**(5):527–35.
116. Parat MO, Richard MJ, Beani JC, Favier A. Involvement of zinc in intracellular oxidant/antioxidant balance. *Biol Trace Elem Res* 1997;**60**(3):187–204.
117. Hoek JB, Cahill A, Pastorino JG. Alcohol and mitochondria: a dysfunctional relationship. *Gastroenterol* 2002;**122**(7):2049–63.
118. Nordmann R. Alcohol and antioxidant systems. *Alcohol Alcohol* 1994;**29**(5):513–22.
119. Lieber CS. S-Adenosyl-L-methionine and alcoholic liver disease in animal models: implications for early intervention in human beings. *Alcohol* 2002;**27**(3):173–7.
120. Kojima-Yuasa A, Ohkita T, Yukami K, et al. Involvement of intracellular glutathione in zinc deficiency-induced activation of hepatic stellate cells. *Chem Biol Interact* 2003;**146**(1): 89–99.
121. Ozturk A, Baltaci AK, Mogulkoc R, et al. Effects of zinc deficiency and supplementation on malondialdehyde and glutathione levels in blood and tissues of rats performing swimming exercise. *Biol Trace Elem Res* 2003;**94**(2):157–66.
122. Shaheen AA, el-Fattah AA. Effect of dietary zinc on lipid peroxidation, glutathione, protein thiols levels and superoxide dismutase activity in rat tissues. *Int J Biochem Cell Biol* 1995;**27**(1): 89–95.
123. Kägi JH. Overview of metallothionein. *Methods Enzymol* 1991; **205**:613–26.
124. Thornalley PJ, Vasák M. Possible role for metallothionein in protection against radiation-induced oxidative stress. Kinetics and mechanism of its reaction with superoxide and hydroxyl radicals. *Biochim Biophys Acta* 1985;**827**(1):36–44.
125. Lambert JC, Zhou Z, Wang L, Song Z, McClain CJ, Kang YJ. Preservation of intestinal structural integrity by zinc is independent of metallothionein in alcohol-intoxicated mice. *Am J Pathol* 2004;**164**(6):1959–66.
126. Zhou Z, Sun X, James Kang Y. Metallothionein protection against alcoholic liver injury through inhibition of oxidative stress. *Exp Biol Med (Maywood)* 2002;**227**(3):214–22.
127. Zhao M, Matter K, Laissue JA, Zimmermann A. Copper/zinc and manganese superoxide dismutases in alcoholic liver disease: immunohistochemical quantitation. *Histol Histopathol* 1996;**11**(4): 899–907.
128. Wagner M, Zollner G, Trauner M. Nuclear receptors in liver disease. *Hepatology* 2011;**53**(3):1023–34.
129. Yu S, Rao S, Reddy JK. Peroxisome proliferator-activated receptors, fatty acid oxidation, steatohepatitis and hepatocarcinogenesis. *Curr Mol Med* 2003;**3**(6):561–72.
130. Crabb DW, Galli A, Fischer M, You M. Molecular mechanisms of alcoholic fatty liver: role of peroxisome proliferator-activated receptor alpha. *Alcohol* 2004;**34**(1):35–8.
131. Fischer M, You M, Matsumoto M, Crabb DW. Peroxisome proliferator-activated receptor alpha (PPARalpha) agonist treatment reverses PPARalpha dysfunction and abnormalities in hepatic lipid metabolism in ethanol-fed mice. *J Biol Chem* 2003;**278**(30): 27997–8004.
132. Kong L, Ren W, Li W, et al. Activation of peroxisome proliferator activated receptor alpha ameliorates ethanol induced steatohepatitis in mice. *Lipids Health Dis* 2011;**10**:246.
133. Nan YM, Kong LB, Ren WG, et al. Activation of peroxisome proliferator activated receptor alpha ameliorates ethanol mediated liver fibrosis in mice. *Lipids Health Dis* 2013;**12**:11.
134. Watt AJ, Garrison WD, Duncan SA. HNF4: a central regulator of hepatocyte differentiation and function. *Hepatol* 2003;**37**(6): 1249–53.

135. Odom DT, Zizlsperger N, Gordon DB, et al. Control of pancreas and liver gene expression by HNF transcription factors. *Science* 2004;**303**(5662):1378–81.
136. Zhong W, Zhao Y, McClain CJ, Kang YJ, Zhou Z. Inactivation of hepatocyte nuclear factor-4{alpha} mediates alcohol-induced downregulation of intestinal tight junction proteins. *Am J Physiol Gastrointest Liver Physiol* 2010;**299**(3):G643–51.
137. Babeu JP, Boudreau F. Hepatocyte nuclear factor 4-alpha involvement in liver and intestinal inflammatory networks. *World J Gastroenterol* 2014;**20**(1):22–30.
138. Darsigny M, Babeu JP, Dupuis AA, et al. Loss of hepatocyte-nuclear-factor-4alpha affects colonic ion transport and causes chronic inflammation resembling inflammatory bowel disease in mice. *PLoS One* 2009;**4**(10):e7609.
139. Ahn SH, Shah YM, Inoue J, et al. Hepatocyte nuclear factor 4alpha in the intestinal epithelial cells protects against inflammatory bowel disease. *Inflamm Bowel Dis* 2008;**14**(7):908–20.
140. Zarski JP, Arnaud J, Labadie H, Beaugrand M, Favier A, Rachail M. Serum and tissue concentrations of zinc after oral supplementation in chronic alcoholics with or without cirrhosis. *Gastroenterol Clin Biol* 1987;**11**(12):856–60.
141. Bainchi GP, Marchesini G, Brizi M, et al. Nutritional effects of oral zinc supplementation in cirrhosis. *Nutr Res* 2000;**20**(8): 1079–89.
142. Takahashi M, Saito H, Higashimoto M, Hibi T. Possible inhibitory effect of oral zinc supplementation on hepatic fibrosis through downregulation of TIMP-1: A pilot study. *Hepatol Res* 2007; **37**(6):405–9.

CHAPTER

13

Interactions Between Alcohol and Folate

Bogdan Cylwik, MD, Lech Chrostek, PhD***

*Department of Paediatric Laboratory Diagnostics, Medical University of Bialystok, Bialystok, Poland
**Department of Biochemical Diagnostics, Medical University of Bialystok, Bialystok, Poland

INTRODUCTION

Folates are the common name of folic acid and its derivatives (Figure 13.1). They are water-soluble B vitamins that naturally occur in many foods. Rich sources of this vitamin are: liver, yeast, dark green vegetables, lentils, beans, dairy products, seafood, meat, nuts, and fresh fruits.[1,2] In some countries, food products are fortified with folic acid. Folic acid is synthetically produced as a pteroylmonoglutamate and added to cereal grain products or supplements, and multivitamin preparations.[3] Recommendations for the daily intake of folates for adults show certain differences, depending on the geographical region. Thus, some countries recommend to consume 400 μg (e.g., USA, Canada), 300 μg (e.g., Poland, Germany, France, Netherlands), or 200 μg only (United Kingdom).[4] Daily folic acid supplementation in pregnant women is 400 μg.[5] The total content of folates in the human body is estimated to be 20–70 mg.[6,7] There are four pools of this vitamin: serum, erythrocytes, bile, and liver. Liver contains about one-half of the body folate stores.[8] In healthy subjects, the total content of folates is sufficient for about 150 days.[9] The expected values for serum/plasma folate concentrations are 3.1–20.5 ng/mL.[10] Folate deficiency is typically linked with serum levels less than 3.5 ng/mL, or red blood cells levels less than 150 ng/mL. The serum folate level depends on many factors, such as dietary habits or environmental and geographical conditions. The adults' daily requirement for folates is 3 μg/kg of body weight, with a minimal daily amount of 50 μg.[4] Folates perform a very important role in the human body. Their biological functions are to enhance the body's metabolism, and to promote and restore cell growth. They are necessary for cell division, red blood cell formation and circulation support, proper normal nerve and brain function, growth and development of the fetus, and cancer and stroke prevention. The best known disease associated with folate deficiency is megaloblastic anemia. Folate deficiency can be caused by low dietary intake, malabsorption due to gastrointestinal diseases (e.g., inflammatory bowel disease, celiac disease), inadequate utilization due to enzyme deficiencies, prolonged treatment with some drugs (e.g., sulphonamides, methotrexate, sulfasalazine, oral contraceptives), alcohol abuse, and an excessive folate demand (e.g., pregnancy, neonatal growth).[2,11] Folate metabolism includes conversion of dietary polyglutamates to monoglutamates form in the intestinal lumen, intestinal absorption and modification, uptake and metabolism in the liver, transport to the other tissues, intracellular conversion to polyglutamates, reconversion to monoglutamates, renal tubular reabsorption, and excretion.[6,7,12–15]

Folates Absorption

Folates absorbed in the digestive tract derive from two sources. The main is food, and the second one is normal microflora of the colon.[16] Folates are absorbed in two sites of the digestive tract, mainly in the proximal small intestine (jejunum, duodenum),[17] and smaller amounts in the large intestine (colon).[16] The colon folates absorption rate is estimated to be about 50 times lower than in the small intestine. However, due to its continuous production by bacteria residing in the colon, and prolonged absorption (approximately 15 times longer), it may significantly contribute to total folate absorption.[18] Typically, the different forms of folate occur in food, both active (reduced, e.g., dihydrofolate, tetrahydrofolate) and inactive (oxidized, e.g., folic acid), as well as pteroylpolyglutamates, and pteroylmonoglutamates. Most of dietary folates exist as a mixture of reduced folate pteroylpolyglutamates (preferably, 5-methyl-THF and 10-formyl-THF), and in this form cannot be absorbed in the intestine. They are hydrolyzed to one-short-chain forms by the lysosomal exopeptidase – γ-glutamyl hydrolase (GGH) (E.C.3.4.19.9), located on the intestinal brush-border surface. After hydrolysis,

Molecular Aspects of Alcohol and Nutrition. http://dx.doi.org/10.1016/B978-0-12-800773-0.00013-6

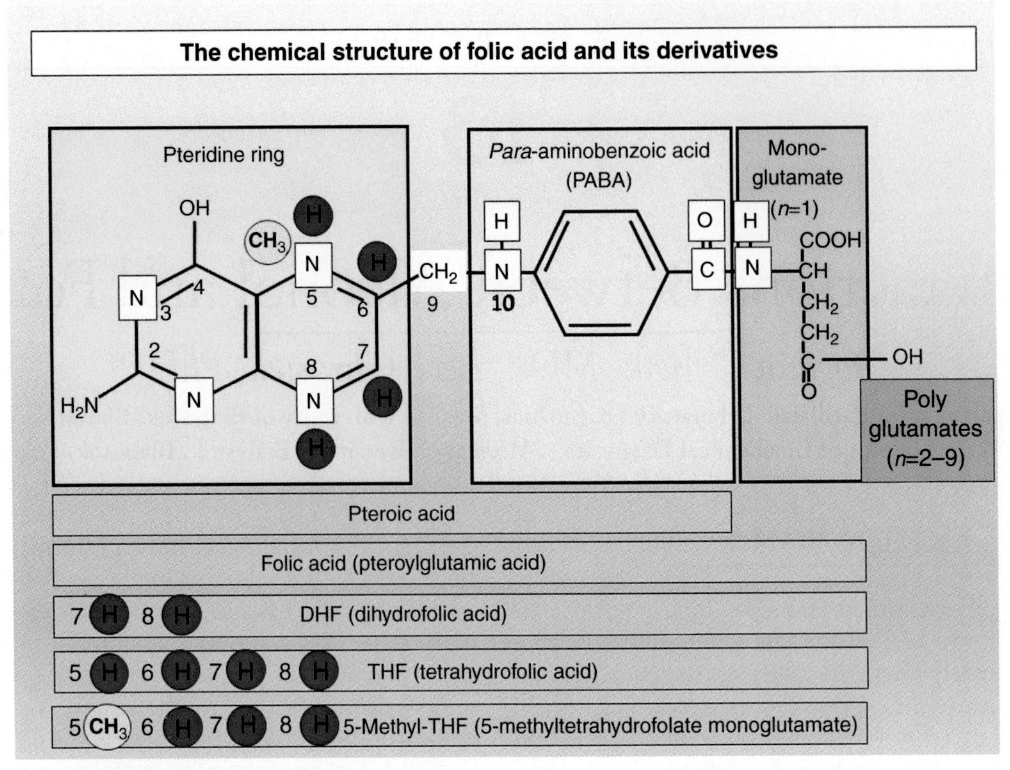

FIGURE 13.1 The chemical structure of folic acid and its derivatives. The chemical structure of folic acid and its derivatives is presented. Folates are a group of heterocyclic compounds that are made up of three units: pterin (composed of two pteridine core rings), *para*-aminobenzoic acid (PABA), then creating pteroic acid, and one glutamic acid, then forming pteroylglutamate (PteGlu). In humans, additional residues of glutamic acid may be attached to PteGlu (up to nine residues) by γ-peptide linkage, then creating pteroylpolyglutamate forms ($PteGlu_n$). Dietary folic acid is not a biologically active compound *per se* until it is converted to its reduced form. Folic acid is partially reduced to dihydrofolic acid (DHF) at the two positions, 7 and 8 of the pyrazine ring, while tetrahydrofolic acid (THF) is reduced fully at four positions: 5, 6, 7, and 8 of the ring. THF in the body mainly occurs as a 5-methyl-THF, 5,10-methylene-THF, and 10-formyl-THF derivatives.

pteroylmonoglutamate derivatives are obtained that are then transported across the luminal membrane into enterocytes.[19] GGH activity is found in many other tissues, such as, the prostate, brain, kidney, placenta, colon, and tumor vasculature.[20] A second enzyme that exhibits an endopeptidase activity is placed intracellularly in the jejunum mucosa.[21] The enzymatic process is not complete; therefore, the bioavailability of natural folates is limited to 50–60%, while the bioavailability of synthetic folic acid is greater, at 85%.[22] Most folate absorption occurs in the apical brush-border membrane (BBM) of the proximal small intestine, and requires the presence of saturable, carrier-mediated, pH and energy-dependent transport mechanisms.[17,23–26] There are three mechanisms of folate transport: active by specific transporters – mainly, the protein-coupled folate transporter (PCFT), and reduced-folate carrier (RFC), as well as passive diffusion.[24–26] Folate transport by PCFT is pH-dependent, saturable, and sodium-independent; it takes place in an acid microclimate (low pH optimum 5.5), and occurs in the small intestine, colon, liver, kidney, placenta, and brain.[27] The PCFT exhibits a high affinity for folates (5-methyl-THF) and folic acid,[13,27] but RFC for reduced folates (preferentially, 5-methyl-THF), and also methotrexate, thiamine pyrophosphate, and a very low affinity for folic acid.[14,24] RFC is expressed in the small intestine, hepatocytes, renal epithelial cells, and brain, and takes place in a pH neutral of 7.4.[14] Another type of folate transport is passive diffusion through cell membranes that plays a minimal role in the intestine because of the hydrophilic nature of the charged folate molecule.[26] This transport may concern deconjugated folates without modification, and unreduced synthetic folic acid consumed in excess.[26] In addition to PCFT and RFC, the BBM of the small intestine contains the folate-binding proteins (FBPs) that bind folic acid with a high affinity, and folate receptors (FR): FRα and FRβ that bind 5-methyl-THF and various antifolate drugs.[28,29] Once the folates are inside the intestinal absorptive cells, they are modified, reduced, and methylated. First, the monoglutamyl folates are elongated using the enzyme tetrahydrofolate synthase (E.C. 6.3.2.17) to polyglutamates that can be retained, stored, and metabolized in the cell.[30] The intracellular folates pool consists of 95% polyglutamates. The folates are reduced by dihydrofolate reductase (DHFR) (E.C. 1.5.1.3), to dihydrofolate (DHF), and tetrahydro folate (THF) that is methylated by methylenetetrahydrofolate reductase (MTHFR) (E.C. 1.5.1.20) to 5-methyltetrahydrofolate (5-methyl-THF).[31] The 5-methyl-THF and the intracellular monoglutamates pass through the basolateral membrane of enterocyte to the plasma, by multidrug resistance-associated protein (MRP, mainly MRP2) family of ABC cassette exporters, and enter the portal circulation.[32] Out of folates, serum mainly transports 5-methyl-THF (about 80%) that is bound unspecifically to proteins – for example, with albumin – or is specifically linked to a soluble folate-binding protein.[33] One third of folates is not bound and remains free. The highest plasma concentration is normally attained after 1–2 h after folate intake.

Liver Uptake

After intestinal absorption, folates are transported via the mesenteric veins to the hepatic portal vein, and are taken up by the liver.[34] Liver can retain 10–20% of absorbed folates (the so-called liver first-pass effect),[35] and this pool can accumulate 50% of the total folate body content.[8,36] Folates that reach the liver may be directed into three metabolic ways. The folate monoglutamates can be converted to polyglutamate forms by the enzyme tetrahydrofolate synthase (E.C. 6.3.2.17). This enzyme can attach up to eight glutamate residues.[37] In the second way, the folate monoglutamates are partially secreted into the bile, and excreted to the duodenum and small intestine. The part of these is then reabsorbed, creating enterohepatic folates circulation.[38] In the third way, polyglutamates stores are hydrolyzed to monoglutamates by the enzyme glutamate carboxypeptidase II (E.C. 3.4.17.21), depending on the body's requirements. Apart from the liver, pancreas is the second richest store of folates and, therefore, may play an important role in their homeostasis.[35]

Excretion of Folates

The excess of folates is excreted in the urine and the feces, as an intact form and/or its metabolites.[7,39] In humans, folates daily excretion is estimated to be 0.3–0.8% of the folate body pool. Most of them are excreted with urine, as metabolites (>300 nmol/day of *para*-aminobenzoylglutamate), and only about 5% of ingested folate at physiological doses, as an unchanged form (3–75 nmol/day of intact folates).[7] Metabolites of folates are produced in tissues by the irreversible oxidative cleavage of the C9–N10 bond, forming various pterins and folate-derived amines, including *para*-aminobenzoylglutamate (*p*ABG), and its acetylated form, *para*-acetamidobenzoylglutamate (ApABG).[40] Plasma folic acid is excreted mainly through the kidneys, as a 5-formyl tetrahydrofolate. In normal conditions, kidneys uptake, filter, and reabsorb folates from primary urine by mechanisms involving the binding of folate to FBP in the BBM of proximal tubule cells, transport by RFC, PCFT, and organic anion transporters (OAT).[41–43] In turn, folates contained in feces originate from bacterial production, lysed enterocytes, and bile.[44] Their content in adults is about 400 nmol/day. It is estimated that up to 20% of fecal folate derive from nonabsorbed food folate, that confirms the incomplete bioavailability of folates.[44,45]

Folate Metabolism

Folic acid cannot be synthesized in the human body, and must be ingested with the diet. Mammals may only produce the pteridine ring, but cannot join it to the other components of folate. Dietary folic acid is not a biologically active compound *per se* until it is converted by the action of the enzyme DHFR, partially into the intermediate DHF, or completely into the THF, and other derivatives. Folic acid is partially reduced to dihydrofolate at the two positions, 7 and 8 of the pyrazine ring, while tetrahydrofolate is fully reduced at four positions: 5, 6, 7, and 8 of the ring (Figure 13.1). The functional folates have one-carbon groups derived from several metabolic precursors. In most organisms, the major source of one-carbon groups is carbon 3 of serine, derived from glycolytic intermediates. It is transferred to THF in a reaction catalyzed by serine hydroxymethyltransferase (E.C. 2.1.2.1), forming 5,10-methylene-THF and glycine. THF in the body mainly occurs as a 5-methyl-THF, 5,10-methylene-THF, and 10-formyl-THF derivatives. In serum, 5-methyl-THF derivatives are present as monoglutamates, but in tissues, as a polyglutamates.[12,37]

The basic role of folate within the cell is to accept or donate one-carbon groups in key metabolic processes. The term one-carbon groups refers to the following methyl-, methylene-, hydroxymethyl-, methenyl-, formyl-, and formimino- groups that are transferred from donor to another molecule, and those pathways are generally named one-carbon metabolism. Those single-carbon groups are attached to N5 of the pteridine ring, N10 of the *para*-aminobenzoate, or both nitrogen of the tetrahydrofolate. The THF-carried one-carbon units may exist in any of three oxidation states, from the most oxidized (5-formyl-THF), through intermediate (5,10-methylene-THF), to most reduced (5-methyl-THF). In the intracellular folate metabolism, all chemical groups have the ability to interconvert at different oxidation level. For example, 5-formyl-THF and 10-formyl-THF derivates can be converted to 5,10-methenyl-THF, 5,10-methylene-THF and, in an irreversible step, to 5-methyl-THF, depending on the needs of the cell. Folate-one-carbon metabolism is compartmentalized in the cytoplasm and mitochondria. In the cytoplasm, the one-carbon forms of folate are required cofactors for the synthesis of three products:

10-formyl-THF for the synthesis of the purine ring, methylene-THF for the conversion of deoxyuridine monophosphate (dUMP) to deoxythymidine monophosphate (dTMP), and 5-methyl-THF for the remethylation of homocysteine to methionine. In some cells, an additional source of one-carbon groups in the cytoplasm is serine, histidine, and purine catabolism. In the mitochondria, one-carbon metabolism is needed for the synthesis of formate, formylated methionyl-tRNA, and glycine. In most cells, the primary source of formate is serine catabolism in the mitochondria, but in the liver, kidney, and astrocytes formate is derived from dimethylglycine, sarcosine, and glycine. The main reactions of folate-mediated one-carbon metabolism are presented in Figure 13.2.[12,37]

Folates primarily participate in the metabolism of nucleic acids (Figure 13.3). They are essential for the support of DNA biosynthesis and repair. THF-polyglutamates are used as substrates in nucleotide synthesis (pyrimidine and purine), and methylation cycle. During the formation of pyrimidine, the 5,10-methylene-THF donates a methyl group to the deoxyuridylate monophosphate (dUMP), forming *de novo* deoxythymidine monophosphate (dTMP) and dihydrofolate (DHF) (thymidylate synthase (E.C. 2.1.1.45) (Figure 13.2). DHF is an inactive product, and must be regenerated to THF by the enzyme DHFR. The DHF reductase plays a key role in the folate homeostasis. It is believed that blocking of its activity may be useful in cancer treatment. Another two enzymes support *de novo* purine synthesis processes, leading to the formation of precursors to adenine and guanine. In the first reaction, 10-formyl-THF provides the one-carbon units to *N*1-(5-P-D-rib)glycinamide, creating THF and *N*2-formyl-*N*1-(5-P-D-rib)glycinamide by the enzyme phosphoribosylglycinamide formyltransferase (E.C. 2.1.2.2) (Figure 13.2). In the second reaction, 10-formyl-THF transfers the single-carbon groups to 5-amino-1-(5-P-D-rib)imidazole-4-carboxamide, generating THF and 5-formamido-1-(5-P-D-rib)imidazole-4-carboxamide by the phosphoribosylaminoimidazolecarboxamide formyltransferase (E.C. 2.1.2.3) (Figure 13.2). The presence of vitamin B_{12} that transfers the methyl group from 5-methyl-THF to homocysteine is necessary to the formation of THF. On the one hand, deficiency of vitamin B_{12} causes a defect in DNA synthesis, and leads to impaired erythropoiesis; on the other, methionine deficiency results in symptoms of neuropathy.

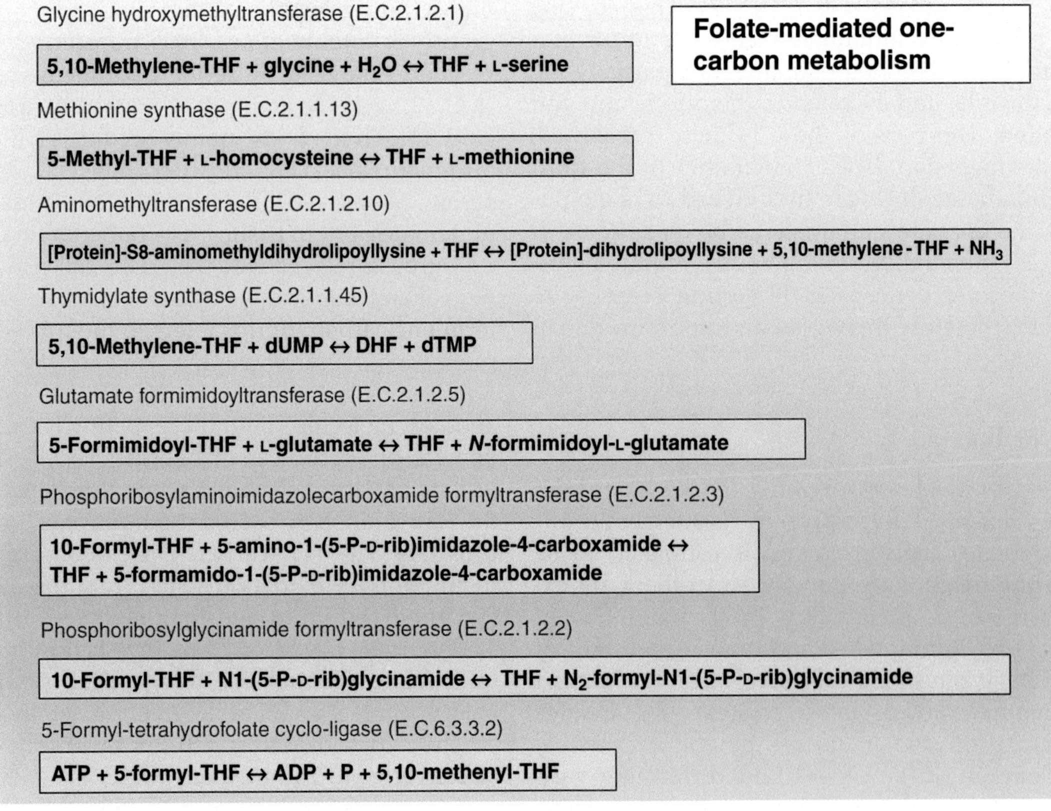

FIGURE 13.2 **Folate-mediated one-carbon metabolism.** The reactions in which tetrahydrofolate (THF) is a cofactor in the transfer and utilization of one-carbon-groups are shown. The term one-carbon groups refers to the methyl-, methylene-, hydroxymethyl-, methenyl-, formyl-, and formimino-groups that are transferred from donor to another molecule, and those pathways are generally named one-carbon metabolism. Folate mediated one-carbon metabolism is compartmentalized in the cytoplasm, mitochondria, and nucleus. DHF, dihydrofolate; dUMP, deoxyuridylate; dTMP, deoxythymidylate.

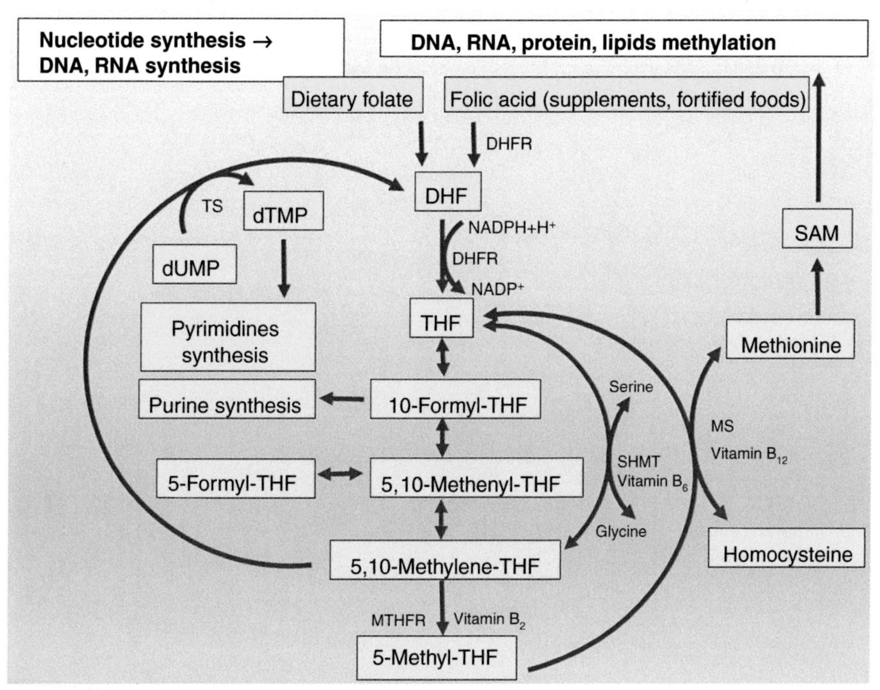

FIGURE 13.3 **Intracellular folate metabolism pathway.** Intracellular folate metabolism is shown. Folate provides methyl groups for nucleotide synthesis via 5,10-methylene-THF, or to methylation reactions via 5-methyl-THF, and S-adenosylmethionine (SAM). DHF, dihydrofolate; DHFR, dihydrofolate reductase (E.C.1.5.1.3); THF, tetrahydrofolate; SHMT, serine hydroxylmethyltransferase (E.C.2.1.2.1); MS, methionine synthase (E.C.2.1.1.13); dUMP, deoxyuridylate; dTMP, deoxythymidylate; TS, thymidylate synthase (E.C.2.1.1.45).

Another important role of folic acid is being involved in the metabolism of amino acids: homocysteine, methionine, serine, glycine, and histidine. Homocysteine and methionine are metabolized in the methylation cycle (Figure 13.4).[45] In this cycle, methionine synthase (MS) (E.C. 2.1.1.13) transfers methyl group from 5-methyl-THF on homocysteine, forming methionine and THF.[37,46] Some of this regenerated methionine is converted to S-adenosylmethionine (SAM) by the enzyme methionine adenosyltransferase (E.C. 2.5.1.6) that is an active form of methionine, and acts as a methyl donor to methyltransferases. The methylation of DNA is a common process, and influences on the epigenetic regulation of gene expression. Subsequently, SAM is converted to S-adenosyl-homocysteine (SAH) that is further hydrolyzed to homocysteine by the enzyme adenosylhomocysteinase (E.C. 3.3.1.1). 5,10-Methylene-THF is reduced to 5-methyl-THF in a NAD^+-dependent reaction by the methylenetetrahydrofolate reductase (MTHFR) (E.C. 1.5.1.20) and, in this way, it is channeled to the methylation cycle. In the liver, the methylation cycle also serves to degrade methionine. The excess of methionine is degraded to homocysteine that can be removed in three ways: by remethylation to methionine (folate cycle dependent), by reconversion to methionine via betaine (folate cycle independent), and by metabolism to cystathionine (transsulfuration pathway) that in the presence of vitamin B_6, forms cysteine, or further to the production of pyruvate (Figure 13.4).[34]

The important role in the amino acids processes is played by the 5,10-methylene-THF. Its main source within cells is a transformation of serine to glycine (Figure 13.2).[15] During this conversion, a hydroxymethyl group of serine is added to THF, creating 5,10-methylene-THF and glycine. This reaction is catalyzed by the enzyme vitamin-B_6-dependent glycine hydroxymethyltransferase (E.C. 2.1.2.1).[15,47] 5,10-Methylene-THF can be reduced to 5-methyl-THF, the most important form stored in the body.

Folate also plays an important role in the conversion of amino acid histidine to glutamic acid (Figure 13.5). During the breakdown of histidine, at first, formiminoglutamate (FIGLU) is formed that donates the formimino-group to THF, generating 5-formimino-THF and glutamate, through the enzyme glutamate formimidoyltransferase (E.C. 2.1.2.5) (Figures 13.2 and 13.5). The 5-formimino-THF may undergo desamination to 5-formyl-THF, or transformation to 5,10-methylene-THF. This reaction is limited by the lack or deficiency of folate within the cells, and then the FIGLU excess is excreted in the urine. The FIGLU test can be used to detect folate deficiency.[15,48] The glutamate is oxidized to α-ketoglutarate using glutamate dehydrogenase (E.C. 1.4.1.2).

Cellular folate metabolism can be affected by several known factors, such as polymorphism in the enzyme methylenetetrahydrofolate reductase that can reduce its activity to below 20%; a polymorphism in RFC that is a risk for fetal neural tube defects (NTD); the lack of PCFT;

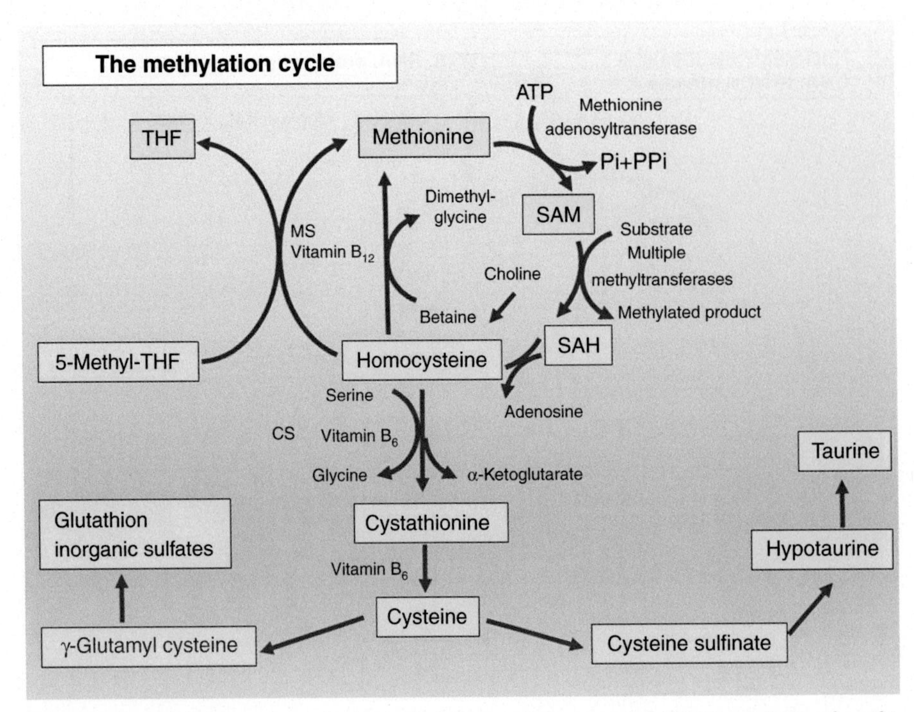

FIGURE 13.4 **The methylation cycle.** Folate cofactor in the methylation cycle is presented. Homocysteine and methionine are metabolized in this cycle. The excess of methionine is degraded to homocysteine that can be removed in three ways: by remethylation to methionine (folate cycle dependent), by reconversion to methionine via betaine (folate cycle independent), and by metabolism to cystathionine (transsulfuration pathway). Folate (5-methyl-THF) serves as substrate in the remethylation cycle, converting homocysteine to methionine and S-adenosylmethionine (SAM) that provides methyl groups for many reactions (e.g., methylation of DNA). THF, tetrahydrofolate; SAM, SAH, S-adenosylhomocysteine; MS, methionine synthase (E.C.2.1.1.13); CS, cystathionine beta-synthase (E.C.4.2.1.22).

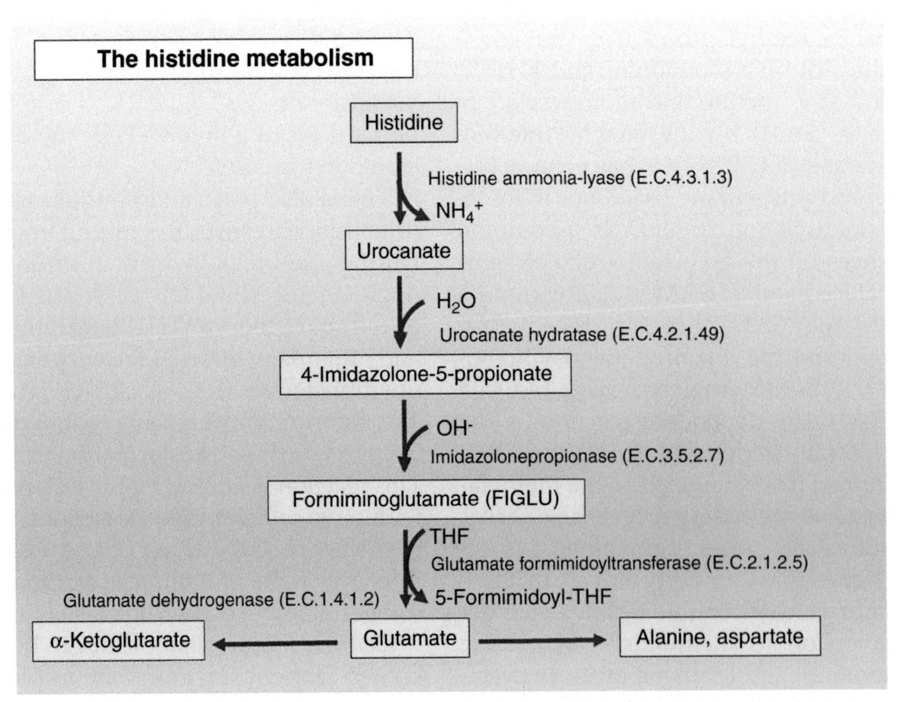

FIGURE 13.5 **The histidine metabolism.** Folate cofactor in the histidine metabolism is shown. Folate is involved in the conversion of formiminoglutamate (FIGLU) to glutamate, during the degradation of amino acid histidine. FIGLU donates the formimino-group to tetrahydrofolate (THF), forming 5-formimino-THF and glutamate.

and deficiency of vitamin B_{12} that reduces the activity of methionine synthase and leads to homocysteine accumulation in cells.[11] A number of conditions and factors can interfere with normal folate homeostasis and metabolism, including ethanol.

Folate and Alcoholism

Historic Background

The effect of long-term ethanol on folate metabolism has been studied since the 1960s. In 1964, Louis Sullivan and Victor Herbert wrote: "Anemia is a frequent finding in alcoholic patients for many reasons. One of the major causes is folate deficiency with associated macrocytosis and megaloblastic erythropoiesis."[49] The first observations indicated that the alcohol-induced fall in serum folate level is not a result of depletion of folate stores.[50] The proof of this should be an immediate decrease in serum folate level after alcohol ingestion; a rise when alcohol was stopped, and reduction when alcohol was resumed. In 1966, Chanarin and coworkers[51] found that hepatic folate in humans is 5 methyltetrahydrofolate, and the correlation between the hepatic folate and the urinary excretion of formiminoglutamic acid, as well as the correlation between the levels of hepatic folate and serum folate. The existence of a protein that binds folic acid in humans was suggested in 1970,[52] and the characteristics of folic acid-binding protein (FABP) were first described by Waxman and Schreiber.[53] According to this study, FABP elutes as a beta globulin and is recovered in the transferrin band region in polyacrylamide gel electrophoresis. FABP binds oxidized folyl-, mono-, and polyglutamates in preference to reduced folates. The attention to some factors that may interfere with the assay of FBP was approached Eichner et al.[54] They confirmed the existence of elevated serum levels of unsaturated FBP in some normal subjects, some women taking oral contraceptives, and most patients with uremia that produces false low results in folate radioassay. In 1975, McGuffin and coworkers, using an animal model, reported that the release of folate from liver stores was unimpaired by alcohol ingestion, and the liver folate store depletion rates were identical for alcoholic and folate starved animals.[55] According to these authors, the suppression of serum folate levels by alcohol ingestion must be sought in the internal metabolic sequences of folate, other than the delivery of folate stores to plasma. The further studies revealed that alcohol ingestion does not adversely affect tissue uptake of folates, but may retard the release of 5-methyltetrahydrofolate from tissue stores to plasma.[56] In the 1980s, Halsted summarized that the factors involved in the pathogenesis of folate deficiency in chronic alcoholism: dietary deficiency and decreased body stores of folate, intestinal malabsorption, serum binding, tissue affinity, renal excretion, and hepatic metabolism of folates.[57] Hepatic reduction, methylation, and formylation of reduced folate, and the synthesis of polyglutamyl folates are not affected by long-term ethanol (2 years) ingestion, and decreased levels of folate in the liver are due to a decreased ability to retain folates in the liver.[58] Another study suggested that folate deficiency in chronic alcoholism is caused by increased urinary and fecal excretion, and the result of decreased hepatic incorporation of exogenous folate in the liver.[59] An uptake of pteroylmonoglutamate by intestinal BBM is required for folate absorption, and this process is preceded by hydrolysis of pteroylpolyglutamate with hydrolase.[60] It has been reported that chronic ethanol exposure (11 months) inhibits jejunal folate hydrolase, and diminishes an uptake of pteroylmonoglutamate in the liver of animals. In a summary of the history of interaction of alcohol and folate, it should be stated that chronic alcoholism impairs folate absorption by inhibiting the expression of the reduced folate carrier, and decreasing the hepatic uptake, and renal conservation of circulating folate.[61]

Effect of Ethanol on Intestinal Folate Absorption

The initial step in the folate absorption in the proximal small intestine is the transport across the BBM of enterocytes, and then the exit of folate across the basolateral membrane into portal circulation. The folate uptake into cells and tissues occurs by transporters involving reduced folate carrier, proton-coupled folate transporter, and folate binding protein.[26]

It is well established that the primary effect of ethanol on folate metabolism is reflected in intestinal malabsorption. The mechanisms of regulation of folate malabsorption during chronic alcoholism involve impaired deconjugation, and transport of monoglutamic folate across the BBM. There is evidence that chronic alcohol ingestion alters the folate-binding kinetics in the intestinal BBM, that is, reduced maximal binding (B_{max}) of folic acid to the intestinal BBM.[62] The changes in B_{max} are attributed to site-specific binding to the extra vesicular sites. The best known folate transporter is the reduced folate carrier (RFC) that mediates the cellular uptake of reduced folates and antifolates.[63] The decreased intestinal BBM folate transport in alcoholism can be attributed to the increased K_m value, lowered V_{max} value, and decreased reduced folate carrier mRNA levels in jejunal tissue.[62] The changes in RFC protein levels in BBM were observed in all cell fractions representing villus tip, mid-villus, and crypt base. The altered folate-binding kinetics suggested that the affinity of the transporter and the number of transporter sites on BBM are reduced after chronic ethanol feeding. The expression of mRNA coding for RFC was reduced by more than threefold, whereas V_{max} by less than twofold. Other factors contributing to reduced intestinal transport of folate in chronic alcoholism are the disturbance of the SH group status at the binding site, and divalent and monovalent cation dependency. These observations may suggest the different effects of chronic ethanol feeding on the transcriptional and post-transcriptional regulation of RFC, or on another route of folate transport. A proton-coupled folate transporter has been found to play an important role in folate absorption in the intestine.[27] Both folate transporters are associated with specific plasma membrane domains called lipid rafts (LR), in the intestine and kidney.[64] Because chronic alcoholism is known to alter the lipid composition of membranes, it might disturb the localization of folate transporters in the membrane. The significant decreased association of folate transporters with lipid rafts, after chronic ethanol ingestion, was found to be responsible for folate malabsorption across the intestine. A downregulation in the RFC and PCFT protein levels in lipid rafts of BBM and BLM, in ethanol fed rats, corresponds to the reduced protein and mRNA expression. In summary, a deregulation in the folate uptake system in the intestine might provide reduced levels of folate in serum and blood red cells, during chronic alcoholism. The decreased folate uptake associated with downregulation of proton-coupled folate transporter and RFC expression at mRNA, and protein levels, was also observed across colon BLM.[65] The deregulation in the folate transport at brush border surface, during alcohol ingestion, was associated with cAMP-, PKA-, and PKC-mediated signaling pathways.[66] The decreased activity of cAMP-dependent protein kinase A upon alcohol ingestion, and then modulatory effect of PKC, altered the intestinal folate transport during alcoholism.

Effect of Ethanol on Renal Folate Reabsorption and Excretion

The kidneys play a significant role in folate homeostasis that is mediated by glomerular filtration, reabsorption, and secretion in the proximal tubules.[67] Earlier studies showed that an acute and chronic ethanol administration increased the urinary excretion of endogenous folate.[68,69] These data indicated that 5-methyltetrahydrofolic (5-MTHF) acid is rapidly taken up by the kidney, and metabolized to other folate and nonfolate forms. It was surprising that kidney incorporation of 5-MTHF was 10-fold greater than it was in the liver. The increase in the urinary excretion of 5-MTHF appeared to be due to altered reabsorption of this folate form.[68] The decreased tubular uptake of folic acid, during chronic alcoholism, is attributed to alter the binding and kinetic characteristics of folate transporters at the BBM and BLM of the kidney.[70,71] The increased K_m and lowered V_{max} values across the renal absorptive surface, in chronic ethanol ingestion, yields lower affinity of folate for transporters, and a lower number of folate transporters on the membrane. The final result of these changes is a decreased renal tubular reabsorption of folate, and an increased secretion in the proximal tubules. It is important to know that chronic ethanol ingestion increased urinary folate excretion only in ethanol-fed rats consuming folate-containing diet.[69] Increased urinary folate excretion has also been demonstrated in chronic alcoholic patients.[72,73]

Evidence from a recent study showed that the reduced activity of folate transport in renal tissue is due to the decreased expression of RFC and PCFT, at both mRNA and protein levels.[42,74] The downregulation in the expression of folate transporters in the kidney was seen in both BBM, as well as BLM. The decreased synthesis of folate transporters, after chronic alcoholism, is associated with the changes in the distribution of folate transporters in membrane lipid rafts. PCFT and RFC were found in the apical and basolateral side of renal membranes. Chronic ethanol ingestion disturbs the association between folate transporters and lipid rafts, in the BBM and BLM of kidney. Another folate transporter, a folate binding protein, has also been involved in folate reabsorption in renal proximal tubules.[75] It plays a major role in the regulation of physiological folate concentration because urinary folate excretion is regulated by more than 95% tubular reabsorption at the kidney BBM.[76] FBP is involved in transporting folate across the kidney BBM, independently or synergistically with RFC.[67] The decreased transport activity after chronic exposure to ethanol might be due to a reduced number of FBP molecules on the renal plasma membrane.[71]

Effect of Ethanol on Folate Metabolism in the Liver

Hepatic folate metabolism includes the transport of circulating 5-methyltetrahydrofolate across the hepatic membrane, binding to intracellular proteins, synthesis of polyglutamyl folates, and exit to the biliary ductules, or the hepatic vein. Any disturbances in the enterohepatic cycle of folate might lead to folate deficiency. The hepatocytes have separate systems for the uptake of naturally occurring reduced folates. There is strong evidence of an impaired enterohepatic recycling of folic acid in ethanol-fed rats.[77] Because hepatic uptake of folic acid in alcohol-fed monkeys was similar to that observed in the control animals, and the fecal excretion of labeled folic acid was increased in ethanol-fed animals, it has been supposed there exists a diminished enterohepatic recirculation of absorbed folate.[78] Long-term ethanol feeding resulted in significant decrease in hepatic folate levels, due to a decreased ability to retain folate in the liver, although hepatic folate metabolism as reduction, methylation and formylation of reduced folate, and synthesis of polyglutamyl folates, is not affected by long-term ethanol feeding.[79] Although the hepatic metabolism of folate seems to be unimpaired by chronic ethanol administration, an increased rate of pteroylpolyglutamate synthesis in the liver was observed.[80] Hepatic pteroylpolyglutamate is metabolized to polyglutamate forms, predominantly to pentaglutamates. The rate of this reaction in ethanol-fed animals is slightly faster than in the controls. Furthermore, the proportion of folate in the 5-methyltetrahydrofolate, a predominant derivative of folate in the liver, was slightly increased after ethanol ingestion. There is evidence that tissue folate levels in chronic ethanol-fed animals are decreased in its poly- and monoglutamated forms.[81] Because polyglutamate forms cannot cross biological membranes, these must be enzymatically hydrolyzed by γ-glutamyl hydrolase (GGH). In the enterocytes, but also in the pancreas monoglutamate form, it gets polyglutamate by the enzyme folylpolyglutamate synthase (FPGS). The activity and expression of GGH and FPGS play a significant role in folate homeostasis, regulating the intracellular folate levels.[81] In turn, these enzymes can influence the methylation of genes, especially genes of folate transporters, and enzymes involved in the regulation of folate levels in chronic alcoholism. The study showed that there is no change in the activity of folate hydrolyzing enzyme GGH, and significant downregulation in the expression of FPGS, in the intestine, colon, kidney, pancreas, and liver.[81] Thus, ethanol feeding leads to a decrease in the total and polyglutamated forms of folate in all tissues, while the initial deconjugation of polyglutamated folate by GGH is not impaired.

The folate uptake across the liver basolateral membrane, like in the small intestine and kidney, was carrier mediated. In addition to PCTF and RFC, liver express folate binding protein (FBP), another high affinity folate transporter.[82] The activity of these transporters plays an important role in regulating folate homeostasis. The diminished folate binding to the liver basolateral membrane, occurring via FBP and PCFT, can decrease the liver folate uptake in chronic alcoholism, and it resulted in the reduction of V_{max} and the unchanged K_m values in the ethanol-fed animals.[83] Thus, during chronic alcoholism, the number of carrier molecules on the liver BLM is reduced, but the affinity of transporters to folate remains unchanged. There is a minimal contribution of the RFC in the folate transport across the liver BLM. The decrease in the folate transport to the liver is due to decreased proteins and mRNA levels of folate transporters in the liver. The alteration of the lipid composition of the basolateral membrane in chronic alcoholism resulted in the disruption of lipid rafts of liver BLM, and decreased the association of folate transporters with LR microdomain of liver BLM. In conclusion, the folate uptake across the liver basolateral membrane is affected by chronic alcohol ingestion, and there exist many mechanisms that explain the folate deficiency in the liver in this condition.

The study demonstrated that chronic ethanol feeding alters the methylation pattern of folate transporters.[81] There was observed a significant hypomethylation in promoter regions of PCFT and RFC genes in the intestine, and hypermethylation of PCFT gene in the kidney and colon, hypermethylation of RFC in the liver, and hypermethylation of FPGS in the intestine and liver. Hypermethylation of the FPGS gene can be one of the mechanisms of the decreased mRNA levels of the enzyme in these tissues. The downregulation in the expression of FPGS might lead to decreased intracellular folate in its polyglutamylated form, thereby increasing the efflux of folate, and contributing to folate deficiency.

There is epidemiological evidence that decreases in serum folate concentration are related to metabolic pathways of ethanol in the liver. The reduction in serum folate levels is more marked in men with genetically inactive aldehyde dehydrogenase 2 (ALDH-2) that encoded mutant allele ALDH2*2, than those with active ALDH2 encoded by ALDH2*1/ALDH2*1 genotype.[84] These data indicate that acetaldehyde concentration plays a significant role in the development of decreased serum folate levels in alcohol drinking men.

Effect of Ethanol on the Folate Transport to Colon, Pancreas, and Fetus

The reduced folate levels during chronic alcoholism were also observed in the colon, pancreas, and human fetus. In the colon, folate undergoes the transport across the BBM into colonocytes, and then across the basolateral membrane to portal circulation. The study demonstrated that, in the exit of folate out of colonocytes, reduced folate carrier is mainly involved.[85] The decreased colon folate transport in chronic alcoholism should be attributed to decreased affinity, and the number of transport molecules at the colon BLM.[86] The decrease in folate uptake across colon BLM is associated with downregulation of folate transporters at protein and mRNA levels, and with decreases in the association of these transporters with lipid rafts.

Folate deficiency in the pancreas may be an additional effect imparted to the direct effect of excessive ethanol intake on pancreatic cells. Chronic ethanol ingestion in rats led to a significant decrease in pancreatic folate levels.[87] The kinetic characteristics of folate uptake across the pancreatic plasma membrane (PPM) is similar to folate uptake across intestinal, renal, and colon apical membranes. The folate absorption across PPM is attributed to folate transporters, PCFT and RFC. Similar to other organs, these folate transporters are associated with lipid rafts microdomain of the PPM. The decreased folic acid uptake by pancreatic acinar cells is due to reduced levels of folate transporter molecules in lipid rafts at the PPM, and to the decreased association of these proteins with lipid rafts.[87]

Folic acid is an essential nutrient that is required for the transfer of one-carbon unit during nucleic acid synthesis. For this reason, its cellular deficiency leads to disturbances in the normal physiology of the cell. Folate plays an essential role in the development of the central nervous system (CNS). The impaired transport of folate into the CNS is associated with the neurological defect in the developing embryo.[88] The results demonstrate that chronic and heavy alcohol use in pregnancies impairs folate transport to the fetus. Transport of folates across the placenta is mediated by placental folate receptors-α (FR-α), RFC, and PCFT at the microvillous membrane of the syncytiotrophoblast.[89] It is important that alcohol consumption in pregnancies creates an oxidative stress in the placenta and fetus that can be mitigated by folate. During pregnancies, folic acid acts as an antioxidant. A recent study documented that ethanol-induced upregulation of 10-formyltetrahydrofolate dehydrogenase (FDH) helps relieve ethanol-induced oxidative stress in embryos.[90] FDH catalyzes the reduction of 10-formyltetrahydrofolate to tetrahydrofolate. The upregulation of FDH was observed at mRNA and protein levels, and involved the transcription factor CEBPα. Folic acid is also critical for the detoxification of formic acid, a toxic metabolite of methanol that has been detected in umbilical cord blood from pregnancies with heavy alcohol consumption.[91] The neurotoxicity of formic acid can be mitigated by folic acid administration.

Ethanol, Folate, and Oxidative Stress

Alcoholic liver disease (ALD) is commonly associated with folate deficiency. In turn, the combination of ethanol feeding with folate deficiency activates hepatic endoplasmic reticulum stress.[92] The oxidative damage of the liver is activated through the activation of the cytochrome P-450 2E1(CYP2E1), and the induction of signal pathway for apoptosis and steatosis. The combination of ethanol feeding and a folate-deficient diet activates liver CYP2E1, and signals for apoptosis (glucose-regulated protein 78 (GRP78), caspase 12 and sterol regulatory element binding protein-1c (SREBP-1c)), and steatosis (the transcripts of fatty acid synthase, acetyl-CoA carboxylase and stearoyl-CoA desaturase). Folate, and its form 5-MTHF, is a central vitamin to methionine metabolism, therefore folate deficiency perturbs methionine metabolism. 5-MTHF is the substrate for methionine synthase reaction that generates homocysteine. Ethanol feeding, or a folate deficient diet, singly or in combination, promote oxidative liver injury, while increasing levels of homocysteine, SAH, and DNA strand breaks, and reducing levels of liver folate, reduced glutathione (GSH), SAM; and the ratio SAM to SAH.[93] Because SAH is an effective inhibitor of DNA methylation, the increase in the hypomethylation of DNA in relation to SAH and homocysteine was observed. In turn, SAM reduces the GSH by upregulation of cystathionine β synthase, and the transsulfuration of homocysteine to cystathionine. Additionally, SAM provides negative feedback to the conversion of 5,10-MTHF to 5-MTHF; therefore, SAM deficiency enhances the rate of this reaction. 5,10-MTHF is also a substrate for thymidine synthase, an enzyme that converts uracil for thymidine. Thus, SAM deficiency enhances the conversion of 5,10-MTHF for 5-MTHF, and decreases the availability of uracil for the thymidine synthase reaction that decreases the synthesis of thymidine. On the other hand, SAM supplementation may attenuate ALD by decreasing oxidative stress via upregulation of glutathione synthesis, reducing inflammation, and inhibiting the apoptosis of normal hepatocytes.[94]

It has been documented that methylenetetrahydrofolate reductase (MTHFR) gene polymorphism plays a role in the regulation of homocysteine status.[95] The polymorphism of MTHFR gene in the form of point mutation, 677C → T transition (MTHFR TT genotype), results in valine substitution for an alanine. Alcohol dependent subjects showed a significant decrease in MTHFR TT genotype prevalence than the controls.[96] Because the drinkers with this genotype exposed lower values of alcohol abusing markers, better liver function tests, a lower frequency of relapse to alcohol and no marked withdrawal symptoms, the MTHFR genotype could play a protective role against alcohol dependence.

CONCLUSIONS

These results strongly suggest the implication of ethanol ingestion in the folate status. In particular, folate deficiency resulted from intestinal malabsorption, increased renal excretion, impaired folate transport to colon, pancreas, and fetus, and altered hepatobiliary metabolism. Recent advances revealed that the responsible mechanism that involves the reduction in folate transport across biological membranes is attributed to downregulation of folate transporters at the protein and mRNA levels, and to disruption of lipid rafts microdomains on brush border and basolateral membranes of cells. The final effects of these disturbances are the low serum folate levels, and low folate stores in the organs.

Key Facts of Folate and Alcohol

- Folic acid (folate) belongs to the B vitamin group.
- The term folic acid refers to synthetically produced compounds that are used in fortified foods and supplements.
- Folic acid food fortification is mandatory in nearly 80 of countries worldwide.
- The term folate refers to the naturally occurring form of folic acid.
- Folate is not synthesized in the human body *de novo*, and must be supplied through the diet.

- An insufficient or lack of dietary folate can lead to folate deficiency.
- Chronic alcohol abuse leads to disturbances of folic acid metabolism.
- The diminished level of serum folic acid is present in 80% of alcoholics.
- Folate deficiency may lead to many serious health problems (i.e., cancer, cardiovascular disease, megaloblastic anemia).

Summary Points

- This chapter presents the physiology of folates in health and alcohol abuse conditions.
- Folates perform a very important function in the human body, as cofactors of a single carbon in the metabolic pathways named one-carbon metabolism.
- Folates are essential for the synthesis of purine and thymidine nucleotides needed for the support of DNA biosynthesis and repair, the synthesis of several amino acids, especially of methionine from homocysteine (methylation cycle).
- Alcohol abuse is a strong factor that implicates folate status, leading to deficiency of this vitamin, low serum concentrations, and low stores in the organs.
- Folate deficiency in alcoholism results from intestinal malabsorption, increased renal excretion, impaired folate transport to colon, pancreas, and fetus, and altered hepatobiliary metabolism.
- Folate deficiency leads to impairment of its metabolism causing hypomethylation, hyperhomocysteinemia, DNA damage, impaired cell proliferation, malignancies, and megaloblastic anemia.
- Folate status may be done by serum folate concentration, RBC levels and dietary folate intake.

Abbreviations

ABC ATP-binding cassette
ALD Alcoholic liver disease
ALDH Aldehyde dehydrogenase
ApABG *p*-Acetamidobenzoylglutamate
BBM Brush-border membrane
BLM Basolateral membrane
CEBPα CCAAT/enhancer-binding protein α
CYP2E1 Cytochrome P-450 2E1
dATP Deoxyadenoside triphosphate
dGTP Deoxyguanosine triphosphate
DHF Dihydrofolate
DHFR Dihydrofolate reductase
dTMP Deoxythymidine monophosphate
dUMP Deoxyuridylate monophosphate
FABP Folic acid-binding protein
FBP Folate-binding protein
FDH Formyltetrahydrofolate dehydrogenase
FIGLU Formiminoglutamic acid
FR Folate receptors
GCPII Glutamate carboxypeptidase II
GGH γ-Glutamyl hydrolase
GSH Reduced glutathione
MRP Multidrug resistance protein
MS Methionine synthase
5-MTHF 5-Methyltetrahydrofolate
MTHFR Methylenetetrahydrofolate reductase
OAT Organic anion transporters
***p*ABG** *p*-Aminobenzoylglutamate
PCFT Protein-coupled folate carrier
PKA Protein kinase A
PKC Protein kinase C
PPM Pancreatic plasma membrane
PteGlu Pteroylglutamate
PteGlu$_n$ Pteroylpolyglutamate
RFC Reduced-folate carrier
SAH S-adenosylhomocysteine
SAM S-adenosylmethionine
THF Tetrahydrofolate

References

1. Cho S, Johnson G, Song WO. Folate content of foods: comparison between databases compiled before and after new FDA fortification requirements. *J Food Comp Anal* 2002;**15**:293–307.
2. Stanger O. Physiology of folic acid in health and disease. *Curr Drug Metab* 2002;**3**:211–23.
3. Dietrich M, Brown CJ, Block G. The effect of folate fortification of cereal-grain products on blood folate status, dietary folate intake, and dietary folate sources among adult non-supplement users in the United States. *J Am Coll Nutr* 2005;**24**:266–74.
4. Krawinkel MB, Strohm D, Weissenborn A, et al. Revised D-A-CH intake recommendations for folate: how much is needed? *Eur J Clin Nutr* 2014;**68**:719–23.
5. Iron and folate supplementation. Integrated Management of Pregnancy and Childbirth (IMPAC). In: *Standards for maternal and neonatal care*, 1.8. Geneva: World Health Organization; 2006.
6. Stites TE, Bailey LB, Scott KC, Toth JP, Fisher WP, Gregory III JF. Kinetic modeling of folate metabolism through use of chronic administration of deuterium-labeled folic acid in men. *Am J Clin Nutr* 1997;**65**:53–60.
7. Gregory JF, Williamson J, Liao JF, Bailey LB, Toth JP. Urinary excretion of [2H4]folate by nonpregnant women following a single oral dose of [2H4]folic acid is a functional index of folate nutritional status. *J Nutr* 1998;**128**:1896–906.
8. Herbert V, Zalusky S. Interrelations of vitamin B12 and folic acid metabolism: folic acid clearance studies. *J Clin Invest* 1962;**41**:1263–76.
9. Subcommittee on the 10th Edition of the RDAs, Food and Nutrition Board, Commission on Life Sciences, National Research Council, editors. Water-soluble vitamins. In: *Recommended dietary allowances*. 10th ed. Washington, DC: National Academy Press; 1989. p. 115–173.
10. Clinical and Laboratory Standard Institute (CLSI). *Defining, establishing, and verifying reference intervals in the clinical laboratory: approved guideline.* 3rd ed. CLSI document C28-A3. Wayne, PA: CLSI; 2008.
11. Stover PJ. Physiology of folate and vitamin B12 in health and disease. *Nutr Rev* 2004;**62**:S3–S12.
12. Shane B. Folate chemistry and metabolism. In: Bailey LB, editor. *Folate in health and disease*. New York: Marcel Dekker Inc; 1995. p. 1–22.
13. Selhub J, Dhar GJ, Rosenberg IH. Gastrointestinal absorption of folates and antifolates. *Pharm Ther* 1983;**20**:397–418.

14. Selhub J, Powell GM, Rosenberg IH. Intestinal transport of 5-methyltetrahydrofolate. *Am J Physiol* 1984;**246**:G515–20.
15. Bailey LB, Gregory JF. Folate metabolism and requirements. *J Nutr* 1999;**129**:779–82.
16. Camilo E, Zimmerman J, Mason JB, et al. Folate synthesized by bacteria in the human upper small intestine is assimilated by the host. *Gastroenterology* 1996;**110**:991–8.
17. Herbert V. Folic acid. In: Shils ME, Olson JA, Shike M, Ross A, editors. *Modern nutrition in health and disease*. 9th ed. Philadelphia: Lippincott, Williams and Wilkins; 1999. p. 433–46.
18. Aufreiter S, Gregory III JF, Pfeiffer CM, et al. Folate is absorbed across the colon of adults: evidence from cecal infusion of (13) C-labeled [6S]-5-formyltetrahydrofolic acid. *Am J Clin Nutr* 2009;**90**:116–23.
19. Devlin AM, Ling E, Peerson JM, et al. Glutamate carboxypeptidase II: a polymorphism associated with lower levels of serum folate and hyperhomocysteinemia. *Hum Mol Gen* 2000;**9**:2837–44.
20. Galivan J, Ryan TJ, Chave K, Rhee M, Yao R, Yin D. Glutamyl hydrolase, pharmacological role and enzymatic characterization. *Pharmacol Ther* 2000;**85**:207–15.
21. Wang TT, Chandler CJ, Halsted CH. Intracellular pteroylpolyglutamate from human jejunal mucosa. Isolation and characterization. *J Biol Chem* 1986;**261**:13551–5.
22. Gregory JF. The bioavailability of folate. *Eur J Clin Nutr* 1997;**51**:554–9.
23. Chandler CJ, Wang TTY, Halsted CH. Pteroylpolyglutamate hydrolase from human jejunal brush borders. *J Biol Chem* 1986;**261**:928–33.
24. Balamurugan K, Said HM. Role of reduced folate carrier in intestinal folate uptake. *Am J Physiol Cell Physiol* 2006;**291**:C189–93.
25. Zhao R, Goldman ID. The proton-coupled folate transporter: physiological and pharmacological roles. *Curr Opin Pharmacol* 2013;**13**:875–80.
26. Zhao R, Matherly LH, Goldman ID. Membrane transporters and folate homeostasis: intestinal absorption and transport into systemic compartments and tissues. *Expert Rev Mol Med* 2009;**11**:e4.
27. Qiu A, Jansen M, Sakaris A, et al. Identification of an intestinal folate transporter and the molecular basis for hereditary folate malabsorption. *Cell* 2006;**127**:917–28.
28. Salazar MD, Ratnam M. The folate receptor: what does it promise in tissue-targeted therapeutics? *Cancer Metastasis Rev* 2007;**26**:141–52.
29. Kamen BA, Smith AK. A review of folate receptor alpha cycling and 5-methyltetrahydrofolate accumulation with an emphasis on cell models *in vitro*. *Adv Drug Deliv Rev* 2004;**56**:1085–97.
30. Scrimgeour KG. Biosynthesis of polyglutamates of folates. *Biochem Cell Biol* 1986;**64**:667–74.
31. Chanarin I, Perry J. Evidence of reduction and methylation of folate in the intestine during normal absorption. *Lancet* 1969;**2**:776–8.
32. Assaraf YG. The role of multidrug resistance efflux transporters in antifolate resistance and folate homeostasis. *Drug Resist Updat* 2006;**9**:227–46.
33. Kane MA, Elwood PC, Portillo RM, Antony AC, Kolhouse JF. The interrelationship of the soluble and membrane-associated folate binding proteins in human KB cells. *J Biol Chem* 1986;**261**:15625–31.
34. Krebs HA, Hems R, Tyler B. The regulation of folate and methionine metabolism. *Biochem J* 1976;**158**:341–53.
35. Gregory III JF. The bioavailability of folate. In: Bailey LB, editor. *Folate in health and disease*. New York: Marcel Dekker Inc; 1995. p. 135–95.
36. Gregory JF, Williamson J, Liao JF, Bailey LB, Toth JP. Kinetic model of folate metabolism in nonpregnant women consuming [H-2(2)] folic acid: isotopic labelling of urinary folate and the catabolite *para*-acetamidobenzoyloglutamate indices slow, intake-dependent, turnover of folate pools. *J Nutr* 1998;**128**:1896–906.
37. Wagner C. Biochemical role of folate in cellular metabolism. In: Bailey LB, editor. *Folate in health and disease*. New York: Marcel Dekker Inc; 1995. p. 23–32.
38. Steinberg SE, Campbell CL, Hillman RS. Kinetics of the normal folate enterohepatic cycle. *J Clin Invest* 1979;**64**:83–8.
39. Gregory III JF, Caudill MA, Opalko FJ, Bailey LB. Kinetics of folate turnover in pregnant women (second trimester) and nonpregnant controls during folic acid supplementation: stable-isotopic labelling of plasma folate, urinary folate and folate catabolites shows subtle effects of pregnancy on turnover of folate pools. *J Nutr* 2001;**131**:1928–37.
40. McPartlin J, Courtney G, McNulty H, Weir D, Scott J. The quantitative analysis of endogenous folate catabolites in human urine. *Anal Biochem* 1992;**206**:256–61.
41. Morshed KM, Ross DM, McMartin KE. Folate transport proteins mediate the bidirectional transport of 5-methyltetrahydrofolate in cultured human proximal tubule cells. *J Nutr* 1997;**127**:1137–47.
42. Hamid A, Kaur J. Decreased expression of transporters reduces folate uptake across renal absorptive surfaces in experimental alcoholism. *J Membr* 2007;**220**:69–77.
43. Kumar BA, Mohammed ZM, Vaziri ND, Said HM. Effect of folate over supplementation on folate uptake by human intestinal and renal epithelial cells. *Am J Clin Nutr* 2007;**86**:159–66.
44. Lin Y, Dueker SR, Follett JR, et al. Quantitation of *in vivo* human folate metabolism. *Am J Clin Nutr* 2004;**80**:680–91.
45. Witthoft CM, Arkbage K, Johansson M, et al. Folate absorption from folate-fortified and processed foods using a human ileostomy model. *Br J Nutr* 2006;**95**:181–7.
46. Young IS, Woodside JV. Folate and homocysteine. *Curr Opin Clin Nutr Metab Care* 2000;**3**:427–32.
47. Berg MJ. The importance of folic acid. *J Gend Specif Med* 1999;**2**:24–8.
48. Perry TL, Applegarth DA, Evans ME, Hansen S, Jellum E. Metabolic studies of a family with massive formiminoglutamic aciduria. *Pediatr Res* 1975;**9**:117–22.
49. Sullivan LW, Herbert V. Suppression hematopoiesis by ethanol. *J Clin Invest* 1964;**43**:2048–62.
50. Herbert V, Zalusky R, Davidson CS. Correlation of folate deficiency with alcoholism and associated macrocytosis, anemia, and liver disease. *Ann Intern Med* 1963;**58**:977–88.
51. Chanarin I, Hutchinson M, McLean A, Moule M. Hepatic folate in man. *Br Med J* 1966;**1**:396–9.
52. Markkanen T, Peltola O. Binding of folic acid activity by body fluids. *Acta Haematol* 1970;**43**:272–9.
53. Waxman S, Schreiber C. Characteristics of folic acid-binding protein in folate-deficient serum. *Blood* 1973;**42**:291–301.
54. Eichner ER, Hillman RS. Effect of alcohol on serum folate level. *J Clin Invest* 1973;**52**:584–91.
55. McGuffin R, Goff P, Hillman RS. The effect of diet and alcohol on the development of folate deficiency in the rat. *Br J Haematol* 1975;**31**:185–92.
56. Lane F, Goff P, McGuffin R, Eichner ER, Hillman RS. Folic acid metabolism in normal, folate deficient and alcoholic man. *Br J Haematol* 1976;**34**:489–500.
57. Halsted C. Folate deficiency in alcoholism. *Am J Clin Nutr* 1980;**33**:2736–40.
58. Tamura T, Romero JJ, Watson JE, Gong EJ, Halsted CH. Hepatic folate metabolism in the chronic alcoholic monkey. *J Lab Clin Med* 1981;**97**:654–61.
59. Tamura T, Halsted CH. Folate turnover in chronically alcoholic monkeys. *J Lab Clin Med* 1983;**101**:623–8.
60. Naughton CA, Chandler CJ, Duplantier RB, Halsted CH. Folate absorption in alcoholic pigs: *in vitro* hydrolysis and transport at the intestinal brush border membrane. *Am J Clin Nutr* 1989;**50**:1436–41.
61. Halsted CH, Villanueva JA, Devlin AM, Chandler CJ. Metabolic interactions of alcohol and folate. *J Nutr* 2002;**132**:2367S–72S.

62. Hamid A, Kaur J. Long-term alcohol ingestion alters the folate-binding kinetics in intestinal brush border membrane in experimental alcoholism. *Alcohol* 2007;**41**:441–6.
63. Sirotnak FM, Tolner B. Carrier-mediated membrane transport of folates in mammalian cells. *Annu Rev Nutr* 1999;**19**:91–122.
64. Wani NA, Thakur S, Najar RA, Nada R, Khanduja KL, Kaur J. Mechanistic insights of intestinal absorption and renal conservation of folate in chronic alcoholism. *Alcohol* 2013;**47**:121–30.
65. Wani NA, Hamid A, Khanduja KL, Kaur J. Folate malabsorption is associated with down-regulation of folate transporter expression and function at colon basolateral membrane in rats. *Br J Nutr* 2012;**107**:800–8.
66. Hamid A, Kaur J. Role of signaling pathways in the regulation of folate transport in ethanol-fed rats. *J Nutr Biochem* 2009;**20**:291–7.
67. Villanueva J, Ling EH, Chandler CJ, Halsted CH. Membrane and tissue distribution of folate binding protein in pig. *Am J Physiol* 1998;**275**:R1503–10.
68. Eisenga BH, Collins TD, McMartin KE. Effects of acute ethanol on urinary excretion of 5-methyltetrahydrofolic acid and folate derivatives in the rat. *J Nutr* 1989;**119**:1498–505.
69. McMartin KE, Collins TD, Eisenga BH, Fortney T, Bates WR, Bairnsfather L. Effects of chronic ethanol and diet treatment on urinary folate excretion and development of folate deficiency in the rat. *J Nutr* 1989;**119**:1490–7.
70. Hamid A, Kaur J. Kinetic characteristics of folate binding to rat renal brush border membrane in chronic alcoholism. *Mol Cell Biochem* 2005;**280**:219–25.
71. Hamid A, Kaur J. Chronic alcoholism alters the transport characteristics of folate in rat renal brush border membrane. *Alcohol* 2006;**38**:59–66.
72. Russell RM, Rosenberg IH, Wilson PD, et al. Increased urinary excretion and prolonged turnover time of folic acid during ethanol ingestion. *Am J Clin Nutr* 1983;**38**:64–70.
73. McMartin KE, Collins TD, Shiao CQ, Vidrine L, Redetzki HM. Study of dose-dependence and urinary folate excretion produced by ethanol in humans and rats. *Alcohol Clin Exp Res* 1986;**10**:419–24.
74. Wani NA, Thakur S, Najar RA, Nada R, Khanduja KL, Kaur J. Mechanistic insights of intestinal absorption and renal conservation of folate in chronic alcoholism. *Alcohol* 2013;**47**:121–30.
75. Birn H, Selhub J, Christensen EI. Internalization and intracellular transport of folate-binding protein in rat kidney proximal tubule. *Am J Physiol* 1993;**264**:C302–10.
76. Williams WM, Huang KC. Renal tubular transport of folic acid and methotrexate in the monkey. *Am J Physiol* 1982;**242**:F484–90.
77. Fernández O, Carreras O, Murillo ML. Intestinal absorption and enterohepatic circulation of folic acid: effect of ethanol. *Digestion* 1998;**59**:130–3.
78. Blocker DE, Thenen SW. Intestinal absorption, liver uptake, and excretion of 3H-folic acid in folic acid-deficient, alcohol-consuming nonhuman primates. *Am J Clin Nutr* 1987;**46**:503–10.
79. Tamura T, Romero JJ, Watson JE, Gong EJ, Halsted CH. Hepatic folate metabolism in the chronic alcoholic monkey. *J Lab Clin Med* 1981;**97**:654–61.
80. Wilkinson JA, Shane B. Folate metabolism in the ethanol-fed rat. *J Nutr* 1982;**112**:604–9.
81. Wani NA, Hamid A, Kaur J. Alcohol-associated folate disturbances result in altered methylation of folate-regulating genes. *Mol Cell Biochem* 2012;**363**:157–66.
82. Wani NA, Hamid A, Kaur J. Folate status in various pathophysiological conditions. *IUBMB Life* 2008;**60**:834–42.
83. Wani NA, Nada R, Khanduja KL, Kaur J. Decreased activity of folate transporters in lipid rafts resulted in reduced hepatic folate uptake in chronic alcoholism in rats. *Genes Nutr* 2013;**8**:209–19.
84. Yokoyama T, Saito K, Lwin H, et al. Epidemiological evidence that acetaldehyde plays a significant role in the development of decreased serum folate concentration and elevated mean corpuscular volume in alcohol drinkers. *Alcohol Clin Exp Res* 2005;**29**:622–30.
85. Dudeja PK, Kode A, Alnounou M, et al. Mechanism of folate transport across the human colonic basolateral membrane. *Am J Physiol Gastrointest Liver Physiol* 2001;**281**:G54–60.
86. Wani NA, Hamid A, Khanduja KL, Kaur J. Folate malabsorption is associated with down-regulation of folate transporter expression and function at colon basolateral membrane in rats. *Br J Nutr* 2012;**107**:800–8.
87. Wani NA, Nada R, Kaur J. Biochemical and molecular mechanisms of folate transport in rat pancreas; interference with ethanol ingestion. *PLoS One* 2011;**12**:e28599.
88. Hutson JR, Stade B, Lehotay DC, Collier CP, Kapur BM. Folic acid transport to the human fetus is decreased in pregnancies with chronic alcohol exposure. *PLoS One* 2012;**7**:e38057.
89. Solanky N, Requena Jimenez A, D'Souza SW, Sibley CP, Glazier JD. Expression of folate transporters in human placenta and implications for homocysteine metabolism. *Placenta* 2010;**31**:134–43.
90. Hsiao TH, Lin CJ, Chung YS, et al. Ethanol-induced upregulation of 10-formyltetrahydrofolate dehydrogenase helps relieve ethanol-induced oxidative stress. *Mol Cell Biol* 2014;**34**:498–509.
91. Kapur BM, Vandenbroucke AC, Adamchik Y, Lehotay DC, Carlen PL. Formic acid, a novel metabolite of chronic ethanol abuse, causes neurotoxicity, which is prevented by folic acid. *Alcohol Clin Exp Res* 2007;**31**:2114–20.
92. Esfandiari F, Villanueva JA, Wong DH, French SW, Halsted CH. Chronic ethanol feeding and folate deficiency activate hepatic endoplasmic reticulum stress pathway in micropigs. *Am J Physiol Gastrointest Liver Physiol* 2005;**289**:G54–63.
93. Halsted CH, Villanueva JA, Devlin AM, et al. Folate deficiency disturbs hepatic methionine metabolism and promotes liver injury in the ethanol-fed micropig. *Proc Natl Acad Sci USA* 2002;**99**:10072–7.
94. Purohit V, Abdelmalek MF, Barve S. Role of S-adenosylmethionine, folate, and betaine in the treatment of alcoholic liver disease: summary of a symposium. *Am J Clin Nutr* 2007;**86**:14–24.
95. Chiuve SE, Giovannucci EL, Hankinson SE. Alcohol intake and methylenetetrahydrofolate reductase polymorphism modify the relation of folate intake to plasma homocysteine. *Am J Clin Nutr* 2005;**82**:155–62.
96. Saffroy R, Benyamina A, Pham P. Protective effect against alcohol dependence of the thermolabile variant of MTHFR. *Drug Alcohol Depend* 2008;**96**:30–6.

CHAPTER

14

Effects of Acetaldehyde on Intestinal Barrier Function

Elhaseen E. Elamin, MD, MSc, PhD, Ad A. Masclee, MD, PhD, Daisy M. Jonkers, PhD

Division of Gastroenterology-Hepatology, School for Nutrition, Toxicology and Metabolism, Maastricht University Medical Center, Maastricht, the Netherlands

INTRODUCTION

Ethanol (i.e., ethyl alcohol) is widely consumed,[1] and has long been recognized to be associated with alcoholic liver disease (ALD).[2–4] In addition, excessive consumption can also affect all parts of the gastrointestinal tract (GIT), resulting in increased risk of oropharyngeal, esophageal, gastric, and colorectal cancer.[5,6] Injurious effects of ethanol are also caused by its metabolites, including altered motility, cytotoxic, and mutagenic effects.[6–8] Another major mechanistic consequence of ethanol consumption is gastrointestinal tract (GIT) epithelial barrier dysfunction.[9] Previous studies in humans,[10,11] as well as in animals,[12–14] have shown that both short- and long-term ethanol intake can increase intestinal permeability, thus enhancing permeation of bacteria and their byproducts, such as endotoxins, into the portal circulation.[15,16] In the liver, endotoxins stimulate Kupffer's cells and thereby releasing cytokines, chemokines, and reactive oxygen species (ROS) that can result in hepatocellular injury and, consequently, ALD.[17,18] However, increased intestinal permeability and subsequent translocation of substances can also induce intestinal inflammation, contributing to a vicious circle. For example, cytokines such as TNF-α and IL-1β have been shown to increase intestinal permeability *in vitro*.[19,20] Evidence from studies in patients and animals indicate that intestinal epithelial barrier disruption is a key event in the initiation and progression of a number of intestinal disorders, including inflammatory bowel diseases (IBD)[21] and irritable bowel syndrome (IBS).[22] Furthermore, epithelial barrier dysfunction is associated with increased susceptibility toward carcinogens, and can thus contribute to the increased risk of ethanol-related GIT cancers.[23–26]

Previous studies on modulating effects of ethanol on intestinal barrier function have focused on the small intestine. However, based on dosage, absorption, and metabolism, ethanol and its metabolites can also reach the large intestine.[27] Following oral intake, ethanol can be metabolized oxidatively and nonoxidatively, generating acetaldehyde, fatty acid ethyl esters and phosphatidylethanol, respectively.[28] Among these, acetaldehyde is a key metabolite that can be found at high concentrations in the oral cavity and colon lumen.[9,29] It is a highly reactive and carcinogenic compound, and its accumulation in the upper and lower GI tract is considered a major factor for disruption of epithelial barrier integrity and carcinogenesis.[30]

This chapter aims to summarize the principal observations, and the current state of understanding on effects of acetaldehyde on intestinal barrier function. First, intestinal barrier function will be outlined briefly. Next, the presence of acetaldehyde in the small and large intestine, involving generation by both the human host, and by the intestinal microbiota will be discussed. Later on, its effects on intestinal barrier function and involved mechanisms will be recited. Finally, promising nutrients that can prevent acetaldehyde-induced intestinal epithelial barrier function will be described.

THE INTESTINAL EPITHELIAL BARRIER AND APICAL JUNCTIONAL COMPLEX

The intestinal epithelium constitutes a physical and functional barrier between the host and the external environment, and represents the first defensive barrier against, for example, enteric pathogens and toxins.

Molecular Aspects of Alcohol and Nutrition. http://dx.doi.org/10.1016/B978-0-12-800773-0.00014-8

It is composed of a continuous monolayer of intestinal epithelial cells (IECs) that facilitate nutrient absorption, water, and ion fluxes, while restricting access of pathogenic substances and microorganisms by the means of transcellular and paracellular pathways.[31] The transcellular pathway is constituted by lipophobic pores located in the enterocytes' brush border membrane.[32] The paracellular pathway is regulated in large part by an apical intercellular junction, referred to as the tight junctions (TJs).[33] Besides, the adherens junctions (AJs) that are comprised mainly of E-cadherin and β-catenin, also contribute to the epithelial integrity and, together with the TJs, gap junctions, and desmosomes, constitute the apical junctional complex (AJC).[32] The TJs are composed of the transmembrane proteins, such as claudins, the integral membrane proteins occludin, and junctional adhesion molecules, as well as cytoplasmic zona occludens proteins (i.e., ZO-1, ZO-2, and ZO-3) that connect the TJ complex intracellularly with the actin cytoskeleton and the circumferential perijunctional F-actin ring.[34] The association between the TJs and the perijunctional F-actin ring is vital for maintaining its structure and function, and is regulated by various intra- and extracellular signaling molecules. Intracellular pathways involved in regulation of assembly and disassembly of the TJs include, for example, myosin light chain kinase (MLCK).[35] Rho GTPases,[36] protein kinase C,[37] mitogen activated protein kinases (MAPKs),[38] protein tyrosine kinase (PTK),[39] epithelial to mesenchymal transition (EMT),[40] zonulin,[41,42] and calcium.[43] Extracellularly, the TJs can be modulated via a large number of substances, including cytokines (e.g., IFNγ, TNF-α, and IL-1β), nutrients, and xenobiotics, such as nonsteroidal anti-inflammatory drugs (NSAIDs).[44–48] Both intracellular and extracellular stimuli can influence the apical F-actin organization and the AJC structure, thus modulating the highly dynamic paracellular permeability.

Apart from the epithelial integrity, the intestinal microbiota, by means of colonization resistance, the intestinal mucus layer, and immune cells contribute to an intact barrier function. In this chapter, we will focus on the epithelial integrity.

The integrity of the intestinal epithelial barrier can be examined by studying the structure and function of the TJs. Morphologically, the structure of the TJs can be examined in intestinal biopsies *ex vivo*, and cell lines *in vitro*, using freeze fracture electron microscopy. In this technique, the TJs are identified as chains of protein particles, or a network of anastomosing linear fibrils called TJ strands, sealing adjacent cell membranes,[49] and correlating positively with epithelial integrity.[50] In addition, the TJs can be evaluated by transmission electron microscopy in which fusion of membranes and electron density between cells can be observed.[51] More advanced techniques, such as fluorescence recovery after photobleaching (FRAP), have demonstrated that the TJs proteins occludin and ZO-1 are highly dynamic.[52,53] Furthermore, different techniques can be applied to investigate localization and expression of TJs, its associated proteins and coding genes, including immunohistochemistry and immunocytochemistry, Western blotting, and real-time polymerase chain reaction (PCR).[54,55]

TJs functionality is often determined using transepithelial electrical resistance (TEER) that measures flux of ions, or permeation of markers (e.g., fluorescein isothiocyanate-dextran, horseradish peroxidase, or sucralose) using intestinal mucosa mounted in Ussing chambers, or intestinal cell line monolayers (e.g., Caco-2).[56–59] In addition, recently, three dimensional culture of Caco-2 and T84 cells in synthetic extracellular matrix proteins have also demonstrated to be a reliable model for the assessment of intestinal epithelial barrier integrity *in vitro*.[60–62]

In vivo, the barrier function can be determined noninvasively by analyses of the urinary or plasma recovery of ingested test probes, such as sugars (monosaccharides and disaccharides, i.e., rhamnose and lactulose, respectively), polyethylene glycols (PEGs: 400, 1000, 4000), and radiolabeled chelators, such as ^{51}Cr-ethylenediaminetetraacetic acid (^{51}Cr-EDTA), and ^{99m}Tc-diethylenetriaminepentaacetic acid (^{99m}Tc-DTPA), and analysis of the sequelae of barrier dysfunction, such as serum/plasma and D-lactate and endotoxin levels.[63–66] To correct for differences in, for example, dilution by gastric secretions and emptying, intestinal transit, and renal function, combined test markers that differ in size and transport (i.e., paracellular or transcellular) can be used.[66]

GENERATION AND METABOLISM OF ACETALDEHYDE IN THE GASTROINTESTINAL TRACT

Ethanol Absorption, Distribution, and Elimination

Following oral intake, ethanol is absorbed from the GIT by simple diffusion, due to its small molecular size, moderate lipid, and excellent water solubility.[67] Its absorption begins in the mouth and esophagus, and about 20–70% of absorption takes place in the stomach and the proximal small intestine, respectively, indicating that the majority of ingested ethanol is absorbed before it reaches the colon.[68,69] Even so, ethanol can reach the terminal ileum and colon from the circulation.[27,70,71] There are large variations in absorption, distribution, and elimination of ethanol between individuals,[72,73] depending on, for example, ethanol dosage and concentration, gastric emptying, gender, genetic makeup, body mass index, and dietary intake.[74,75] Once absorbed, ethanol rapidly distributes throughout the body fluids, reaching various tissues and organs, based on body water content.[74] Since body water content is lower in females than in males, higher blood ethanol concentrations can be achieved after ingestion of a similar dose per kilogram body weight.[76,77]

Generation of Acetaldehyde and Metabolism

Acetaldehyde is generated from ethanol via oxidative pathways, including cytoplasmic alcohol dehydrogenase (ADH) in the liver, the microsomal ethanol oxidizing system cytochrome P450 2E1 (CYP2E1) in the microsomes, and by catalase in the peroxisomes (Figure 14.1).[28,69,78–81] Acetaldehyde is the first and most potent metabolite of ethanol, being highly toxic, mutagenic, and carcinogenic.[82] Although it is mainly metabolized in the liver, there is evidence that significant acetaldehyde production occurs in saliva, gastric juice, and in the colon lumen.[30] ADH is the main enzyme involved in acetaldehyde production, comprising 10 isoenzymes that have been grouped in five classes (based on tissue distributions, substrates specificities, and kinetic properties) (Figure 14.2).[28,69,83–85] Class I ADH enzymes have a low K_m for ethanol (1–2 mM), resulting in rapid conversion of ethanol into acetaldehyde, and are highly expressed in the liver as well as in the stomach.[79,86,87] Class IV ADH enzymes expressed in the oral cavity, esophagus, and stomach, are characterized by a high K_m for ethanol.[88–90] The esophagus has an ADH activity similar to that of the liver, and approximately four times higher than that of the stomach.[87] The acetaldehyde generated in the upper GI tract and the liver can also reach the small and large intestine via the vascular space. Furthermore, class I ADH isoenzymes are expressed in the small[91] and large intestine, predominately in the absorptive enterocytes of villous epithelium and, to a (much) lesser extent, in the cytoplasm of goblet cells and the *muscularis mucosae*.[82,91] They possess a low K_m (0.6–4 mM) and a high maximal velocity, thereby oxidizing ethanol at a constant rate and at low concentrations.[28,92] Interestingly, the overall activity of ADH in rectal mucosa was found to be comparable to gastric ADH, suggesting efficient ethanol metabolism in the lower GI tract.[80]

Besides ADH, the cytochrome P450 2E1 (CYP2E1)-dependent microsomal ethanol oxidizing system (MEOS) contributes to less than 10% of ethanol conversion into acetaldehyde.[93] However, its activity increases during long-term ethanol consumption,[94] resulting in generation of high acetaldehyde concentrations.[93,95]

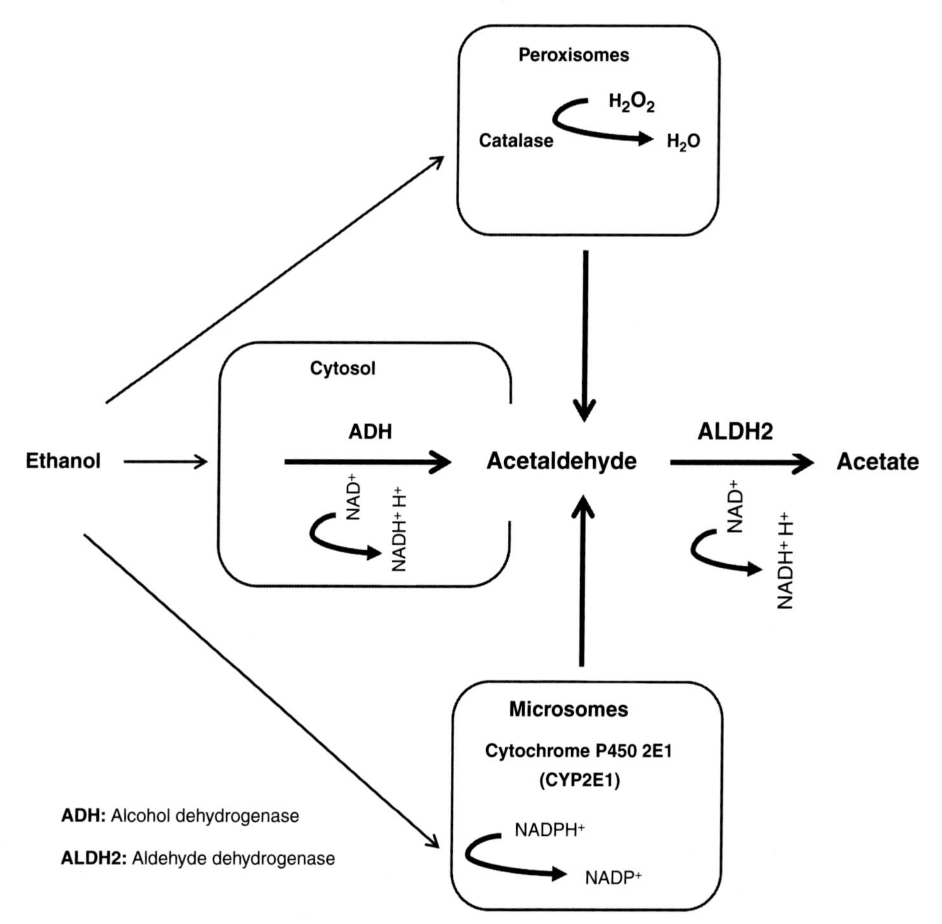

FIGURE 14.1 **Oxidative ethanol metabolism in the liver cells (hepatocytes).**

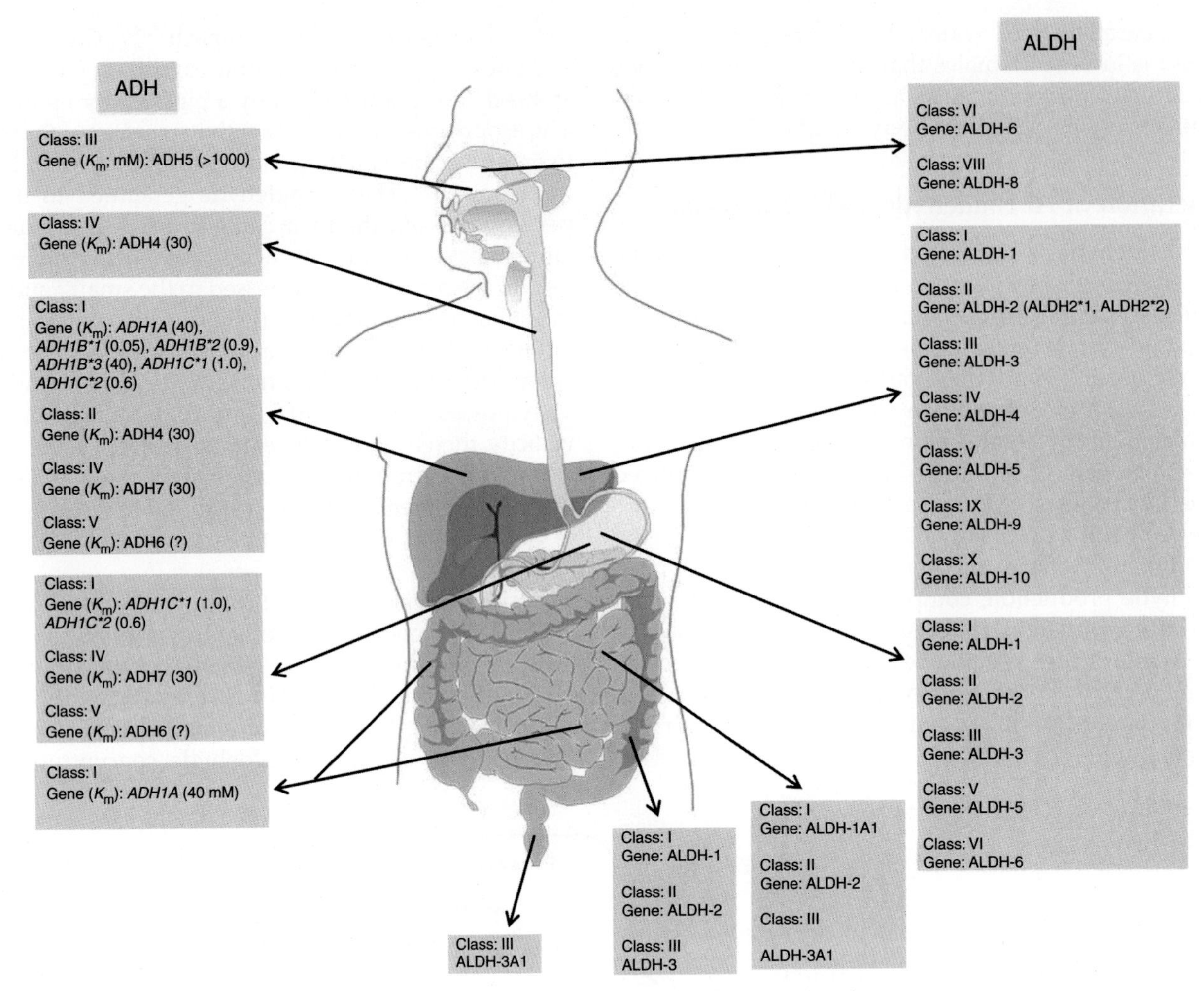

FIGURE 14.2 Distribution of ADH and ALDH isoenzymes among different body tissues and organs.

In rat liver, catalase has also been shown to oxidize ethanol into acetaldehyde in a H_2O_2-dependent reaction.[28] Its activity has been observed in human gastric and intestinal mucosae, but human data on its role in ethanol metabolism are lacking.[96] In addition to the oxidative metabolism, ethanol can be converted nonoxidatively into fatty acid ethyl esters (FAEEs),[97,98] and phosphatidylethanol (PEth).[28,98] Similar to acetaldehyde, intracellular accumulation of these metabolites can be injurious to different tissues, including intestinal epithelium.[99] Since effects of these metabolites are beyond the scope of this chapter, their damaging effects on the small and large intestine will not be further discussed.

Acetaldehyde Metabolism

Acetaldehyde is further metabolized mainly in the liver and, to a lesser extent, in the oral cavity, esophagus, and stomach, pancreas, and small and large intestine, through oxidation by aldehyde dehydrogenase (ALDH) into acetate (Figure 14.1),[82,88,89,91] and thereafter into CO_2 and H_2O_2.[100] To date, 10 ALDH isoenzymes have been identified (Figure 14.2).[101,102] Compared to ADH, ALDH is less active in the small and large intestinal mucosa, contributing to accumulation of acetaldehyde.[81,103]

Genetic Variations in Ethanol and Acetaldehyde Metabolism

There is a high individual variability in ethanol metabolism, mainly due to genetic variations and polymorphisms in the main ethanol and acetaldehyde metabolizing enzymes. For example, in Asians, the variant allele of ADH *ADH1C*1* is more frequent than in Caucasians and Africans. This enzyme oxidizes ethanol faster into acetaldehyde.[104,105] In Asians, also the variant allele of ALDH *ALDH2*2* is predominant (28–45%).[92] This enzyme encodes the inactive subunit of the enzyme ALDH2, and homozygous carriers of this allele experience readily the unpleasant symptoms associated with acetaldehyde accumulation.[106] These adverse effects are less severe in the heterozygous carriers, as they have 10–50% ALDH2 activity. However, these carriers are still at increased risk of developing ethanol-related GI cancers.[106–108] These genetic polymorphisms ascertain the level of acetaldehyde accumulation after ethanol consumption, thus having an impact on subject susceptibilities to both ethanol and acetaldehyde toxic and injurious effects.[109]

ROLE OF INTESTINAL MICROBIOTA IN PRODUCTION AND METABOLISM OF ACETALDEHYDE

In addition to ethanol ingestion and metabolism, the intestinal microbiota plays an important role in acetaldehyde generation and metabolism. There is evidence that small amounts of ethanol (i.e., 0.1–0.2 mg/L) can be produced in the GI tract by bacterial and yeast fermentation of carbohydrates,[110–112] later on absorbed and transferred via the portal vein to the liver, where it is metabolized into acetaldehyde.[113] The endogenously-produced ethanol and, indirectly, acetaldehyde, can reach high concentrations in patients with gut fermentation syndrome (i.e., auto-brewery syndrome),[114–116] and in conditions associated with bacterial overgrowth, such as jejunoileal bypass surgery and tropical sprue.[117,118] Examples of bacteria that can ferment carbohydrates to ethanol in the GIT[111] include, but are not limited to, *Helicobacter pylori,*[119] coliform bacteria such as *Klebsiella pneumoniae, Enterobacter cloacae,* and *Escherichia coli,*[117] in the small and/or large intestine, and *Clostridium* spp. in the colon.[120]

Both endogenous and exogenous ethanol can also be converted into acetaldehyde via the action of microbial ADH.[121,122] A number of commensal bacteria including, for example, *Streptococcus viridans,*[123] and *H. pylori*[124] and *E. coli*, but also *Candida* strains,[125] have been found capable of ethanol oxidation via an ADH-dependent reaction.[126–128] In addition, several bacteria do possess catalase, including members of Enterobacteriaceae family and Staphylococci.[129] It has been demonstrated that human colon contents can generate acetaldehyde via catalase-dependent activity.[130]

In addition to the role of microbiota in ethanol metabolism, there is evidence that ethanol intake can be associated with quantitative and qualitative changes in the intestinal microbiota. Long-term ethanol consumption has been shown to induce small intestinal bacterial overgrowth.[131,132] Studies using length heterogeneity PCR fingerprinting and multitag pyrosequencing showed that long-term ethanol intake also altered the composition of mucosa-associated microbiota in sigmoid biopsies in alcoholics, compared to healthy controls, resulting in lower numbers of Bacteroidetes and higher Proteobacteria counts.[133] In line with these data, a reduction in Bacteroidetes and increase in Proteobacteria have also been demonstrated in patients with hepatitis B and ethanol-related liver cirrhosis, compared to healthy controls using 454 pyrosequencing. Changes have also been observed on the family level, including, for example, increased numbers of Enterobacteriaceae and Streptococcaceae, and a reduction of Lachnospiraceae.[134]

In summary, acetaldehyde can reach high concentrations in the upper and lower GIT tract, resulting either from ethanol intake, or locally mediated by activity of mucosal and bacterial ethanol-metabolizing enzymes. However, its highly volatile character makes its measurement in tissues rather difficult. Systemic acetaldehyde levels, even after a high dose of ethanol, are very low (1–5 μM), due to rapid liver metabolism.[135] However, local acetaldehyde levels in the oral cavity (i.e., saliva) can differ between subjects, reaching concentrations varying between 143 μM and 400 μM, after a moderate dose (i.e., 20 g, equaling two standard drinks) of ethanol administration.[136] Theoretically, it has been estimated that acetaldehyde in saliva could be up to 450 μM at blood ethanol levels of 44 mM.[136] In the lower GI tract, acetaldehyde levels have been assessed by Jokelainen et al., showing that incubation of human colon contents with ethanol concentrations measured *in vivo,* after moderate ethanol consumption (i.e., 22 and 44 mM), can result in generation of 58 and 238 μM, respectively.[137] Since the capacity of the microbiota[108,109] and intestinal mucosa[138] to further metabolize acetaldehyde into acetate by ALDH is low, acetaldehyde levels can accumulate in the colon lumen.[103,137] Resulting high concentrations predispose the colorectal mucosa to acetaldehyde-induced barrier disruption, and its pathological consequences such as alcoholic liver disease, small and large intestinal inflammation, and carcinogenesis.[139]

EFFECTS OF ACETALDEHYDE ON INTESTINAL BARRIER EPITHELIAL INTEGRITY

The majority of data on effects of acetaldehyde on intestinal permeability have emerged from a limited number of *in vitro* studies using Caco-2 cell monolayers. An overview of key studies is provided in Table 14.1. Rao et al. were the first to demonstrate that luminal exposure of acetaldehyde, at concentrations ranging from 25 μM to 760 μM (0.11–3.3 mg/dL), can decrease transepithelial electrical resistance (TEER) and increase paracellular permeability of the permeation marker inulin.[56,140] Subsequent studies in Caco-2 cell monolayers have shown comparable data (Table 14.1).[38,39,129,141–144] In a three-dimensional Caco-2 cell culture model, basolateral exposure to acetaldehyde, at concentrations varying between 25 μM and 200 μM, has also been shown to increase the paracellular permeability to fluorescein isothiocyanate-labeled dextran 4 KD (FITC-D4) in a dose-dependent manner.[61] The study also showed that acetaldehyde can act additively with ethanol to increase permeability.[61]

TABLE 14.1 *In Vitro* and *Ex Vivo* Studies Exploring the Effects of Acetaldehyde on Intestinal Epithelial Barrier Integrity

Model	Acetaldehyde concentration (μM)	Time of exposure (h)	Significant findings	Possible mechanism(s) involved	References
In Vitro ***CELL CULTURE***					
Caco-2	100–760	4	Reduction of TEER Increase in mannitol permeability	–	[56,140]
Caco-2	650	6	Decrease in TEER Increase in inulin permeability Tyrosine phosphorylation of ZO-1, E-cadherin, and β-catenin	Inhibition of protein tyrosine phosphatase	[39]
Caco-2	100–600	4	Decrease in TEER Increase in inulin and endotoxin permeability Dissociation of ZO-1, occludin, E-cadherin, and β-catenin	–	[143]
Caco-2	100–600	4	Decrease in TEER Increase in inulin and endotoxin permeability Reorganization of occludin, ZO-1, E-cadherin, and β-catenin Reorganization of actin cytoskeleton	–	[141]
Caco-2	100–600	3–6	Redistribution, tyrosine phosphorylation, and reduction in ZO-1, occludin, E-cadherin, and β-catenin protein levels	Inhibition of protein tyrosine phosphatase	[145]
Caco-2	400	0.5	Redistribution and tyrosine phosphorylation of E-cadherin and β-catenin Abolishment of interaction of β-catenin with E-cadherin	Tyrosine kinase activation	[146]
Caco-2	1000	5	Decrease in TEER Increase in mannitol and sucrose permeability	–	[142]
Caco-2	100–760	5	Decrease in TEER Increase in inulin permeability Redistribution of ZO-1, occludin, E-cadherin, and β-catenin Reorganization of actin cytoskeleton Tyrosine phosphorylation of occludin, ZO-1, claudin-3, and E-cadherin	Tyrosine kinase activation	[38]
Caco-2	300	5	Decrease in TEER and increase in inulin permeability Redistribution of occludin, ZO-1, E-cadherin, and β-catenin	Tyrosine-phosphorylation	[38]
Caco-2	200–600	1	Increase in inulin permeability Redistribution of occludin and ZO-1	Protein phosphatase 2A activation	[144]
Caco-2	25–100	3	Increase in FITC-D4 permeation Disruption of ZO-1 and occludin integrity Hyperacetylation of microtubules	Hyperacetylation of microtubules	[61]
Caco-2	25	3	Increase in FITC-D4 permeation Disruption of ZO-1 and occludin, E-cadherin and β-catenin integrity	Activation of EMT transcription factor Snail	[129]
Ex Vivo ***ANALYSES***					
Left human colon mucosal biopsies	100–600	3–6	Reduction in occludin, ZO-1, E-cadherin, and β-catenin protein levels Increase in tyrosine phosphorylation of occludin, E-cadherin, and β-catenin	Inhibition of protein tyrosine phosphatase	[145]
Rat colon strips	50–160	1	Increased FITC-D4 permeation	–	[57]

TEER, Transepithelial electrical resistance; FITC-D4, fluorescein isothiocyanate-labeled dextran 4 kDa; ZO-1, zona occludens-1; EMT, epithelial to mesenchymal transition.

The concentrations of acetaldehyde used in the previous studies are in the range to be expected in the intestine after moderate and high ethanol consumption. Furthermore, *ex vivo* incubation of rat colon strips, mounted in Ussing chambers with acetaldehyde (i.e., 50–160 μM for 1 h), dose-dependently increased FITC-D4 permeability, whereas ethanol up to 18 mM failed to induce barrier dysfunction, indicating that the oxidation of ethanol into acetaldehyde in colon lumen is important for barrier disruption.[57] In the same study, ethanol administration in rats resulted in endotoxemia that was dependent on ethanol metabolism into acetaldehyde, by the intestinal resident microbiota.[57] This confirms earlier observations that acetaldehyde is potentially more effective than ethanol in inducing intestinal barrier dysfunction.[56] Data on direct effects of acetaldehyde on intestinal barrier function in humans and animals are lacking, most likely due to its carcinogenic potentials.

In addition to generation of acetaldehyde, an altered intestinal microbiota can contribute to disruption of the TJs, either directly through induction of changes in TJs protein expression and distribution,[147] or by generating high levels of LPS.[148] Long- and short-term ethanol consumption has been shown to induce changes in the composition of intestinal microbiota,[133] and to enhance LPS blood levels.[131,149,150] On intestinal epithelium, LPS exerts several injurious effects, including disruption of the TJs integrity.[151] Therefore, presence of acetaldehyde and LPS in combination may represent a two-hit insult on intestinal epithelium that can result in a vicious cycle of barrier dysfunction.

MECHANISMS OF ACETALDEHYDE-INDUCED INTESTINAL BARRIER DYSFUNCTION

So far, a number of mechanisms underlying acetaldehyde-induced barrier dysfunction have been identified, including loss of TJs and/or AJs integrity, and direct epithelial cell damage, each of which will be discussed in succeeding sections.

Effects on TJs and/or AJs Integrity

The main mechanisms by which acetaldehyde disrupts barrier function are through direct and/or indirect effects on TJs integrity (Table 14.1). Interactions between AJ proteins (E-cadherin and β-catenin), TJ proteins (ZO-1 and occludin), and cytoskeletal proteins are crucial for the organization of the TJ complex, and for subsequent maintenance of intestinal epithelial barrier.[39,140,141,143,145,146] Acetaldehyde has been demonstrated to disrupt ZO-1, occludin, E-cadherin, and β-catenin in Caco-2 cell monolayers,[39,140,141,143,146] and to dissociate these proteins from the actin cytoskeleton, resulting in loss of TJs and AJs integrity, and increased paracellular permeability.[145,152]

Cell signaling pathways, such as protein kinases,[153] and protein phosphatases 2A and 1 (PP2A and PP1), regulate the integrity of TJs in different epithelia.[154] To explore the role of the signaling elements in acetaldehyde-induced TJ disruption, a number of signaling pathways involved in regulating epithelial barrier function have been examined. Using Caco-2 cell monolayers, protein tyrosine kinase (PTK),[153–157] and protein tyrosine phosphatase (PTP)[154] are demonstrated to be key cell signaling players in acetaldehyde-induced loss of TJs integrity. Activation of PTK and inhibition of PTP by acetaldehyde have been shown to induce protein tyrosine phosphorylation of E-cadherin and β-catenin, resulting in loss of (or decreased) interaction between these proteins, and to disrupt the interactions between the AJ proteins and PTP1B,[158] respectively, and consequently barrier dysfunction.[141] Furthermore, Dunagan et al. have demonstrated that PP2A activation is required for acetaldehyde-induced TJs disruption, mediated by PP2A translocation to the TJs, and dephosphorylation of occludin on threonine residues.[144] Their study also reported a cross talk between PTK and PP2A pathways, demonstrating that PTK-dependent association of PP2A with occludin and threonine dephosphorylation.

One study was published validating the disruption of the TJs and AJs by *ex vivo* exposure of acetaldehyde on human colon biopsies, and confirmed that acetaldehyde can induce protein tyrosine phosphatase activation and, subsequently, tyrosine phosphorylation of occludin, E-cadherin, and β-catenin resulting in redistribution of TJ and AJ proteins from the intercellular junctions, and the dissociation of these proteins from the actin cytoskeleton and, consequently, barrier dysfunction.[145]

Additional to phosphorylation, protein hyperacetylation can interfere with TJs integrity.[159] Using a three dimensional Caco-2 cell culture model, it has been demonstrated that exposure to acetaldehyde (25 μM ~ 110 μg/dL) for 3 h can increase paracellular permeability, through mechanisms involving hyperacetylation of the microtubular protein α-tubulin.[160]

Although ethanol has been shown to stimulate activation of the transcription factor Snail, and biomarkers of epithelial to mesenchymal transition (TEM) resulting in disruption of the TJs proteins, studies investigating effects of acetaldehyde on TEM are scarce.[161] However, in one study, exposure to acetaldehyde at 25 μM concentration (0.25 mM ~ 0.11 mg/dL) for 3 h has recently been reported to activate the epithelial-to-mesenchymal transition (EMT) factor Snail in Caco-2 monolayers, resulting in redistribution and reduction of TJs and AJs protein levels and thus, intestinal epithelial barrier disruption.[129] The role of hyperacetylation and EMT in acetaldehyde-induced barrier disruption in humans has not yet been studied. Therefore, confirmation of these findings in human small and large intestinal mucosa merits further investigations. In addition, the role of pathways and cell signaling molecules known to mediate ethanol-induced intestinal barrier disruption, such as Rho kinase,[162] myosin light chain kinase,[163] Ca^{2+},[162] and hepatocyte nuclear factor-4α (HNF-4α),[164] as well as nuclear factor κB (NF-κB)[165] on acetaldehyde-induced TJs disruption has to be examined.

Oxidative stress has been a major research focus in understanding the mechanism of intestinal epithelial barrier dysfunction. A number of studies have shown that oxidative stress-induced proto-oncogene tyrosine-protein kinase Src (c-Src), PP2A, and PTK activation leading to tyrosine phosphorylation of the TJ proteins ZO-1 and occludin, and the AJs proteins E-cadherin and β-catenin, is an important mechanism of loss of intestinal barrier integrity.[153,155,166] There is a compelling evidence to indicate the involvement of cellular oxidative stress in mediating ethanol and acetaldehyde-induced intestinal barrier dysfunction.[13,167] Ethanol-induced CYP2E1 activation, and subsequent increase in oxidative stress, has been shown to upregulate the circadian clock genes CLOCK and PER2 in rats and Caco-2 cell monolayers, resulting in intestinal barrier dysfunction.[168,169] Since acetaldehyde is generated from ethanol, and has strong oxidative potential, its involvement in circadian clock-mediated loss of barrier integrity cannot be excluded, but data are not yet available.

So far, only one study has investigated the direct role of oxidative stress in acetaldehyde-induced intestinal barrier dysfunction, showing that oxidative stress mediates acetaldehyde-induced activation of the EMT transcription factor Snail, resulting in disruption of the TJs and AJs proteins and, consequently, loss of intestinal epithelial integrity.[129] Since acetaldehyde is a highly potent oxidant and increases ROS production,[170] further studies are needed to explore the effects of acetaldehyde on circadian clock, and to delineate the mechanism involved in EMT-activation and intestinal barrier disruption.

Direct Damage to Epithelial Cells

In addition to the TJs, intact IECs, including enterocytes and the secretory goblet cells, are also important for maintaining the barrier function of the epithelium.[31] Histological studies on rectal mucosa of heavy ethanol consumers, where high levels of acetaldehyde are to be expected, have shown a number of ultrastructural changes, including distorted mitochondria, dilated endoplasmic reticulum and decreased number of mucin-secreting goblet cells.[171] Apart from these histological human data, most observations on IECs damage, as well as further mechanistic insight, stems from *in vitro* experiments. In Caco-2 cell monolayers, short- and long-term luminal (i.e., apical) exposure to acetaldehyde (500–1000 μM ~ 2.2–4.4 mg/dL) has been shown to inhibit and stimulate cell proliferation, respectively, to decrease sucrase activity, disrupt differentiation, and reduce cell adhesion to type I and IV collagens.[33] This indicates that colon intraluminal acetaldehyde can be associated with disturbed barrier integrity.

Acetaldehyde up to 100 μM has not been shown to induce cytotoxicy in Caco-2 cells.[61] Effects of higher concentrations on enterocytes have not been studied. However, in a recent study, effects of short-term acetaldehyde (0, 25, 50, 75, 100 μM ~ 0, 0.11, 0.22, 0.33, 0.44 mg/dL) on intestinal goblet-like cells (LS174T) have been investigated.[112] Data have demonstrated that acetaldehyde largely increased reactive oxygen species generation, decreased mitochondrial function, and decreased ATP levels in dose-dependent fashion. In addition, acetaldehyde induced intramitochondrial Ca^{2+} accumulation and cell apoptosis.[112] Since goblet cells contribute to the intestinal barrier function through mucin production, the cytotoxic effects induced by these clinically relevant concentrations may contribute to acetaldehyde-induced intestinal barrier dysfunction. Further studies investigating the effects of acetaldehyde on goblet cells, mucin production and glycosylation, and mucin barrier function, during long-term of exposure, are warranted.

MODULATION OF ACETALDEHYDE-INDUCED INTESTINAL EPITHELIAL BARRIER FUNCTION

Several nutritional and pharmaceutical strategies have been applied in an attempt to counteract the detrimental effects of ethanol and acetaldehyde on intestinal barrier function. Most of the studies have been carried out *in vitro* and in animals, focusing on ethanol. However, these studies are also relevant to acetaldehyde as being the first and very potent ethanol metabolite.

In a mouse model of ALD, zinc supplementation has been shown to attenuate blood levels of endotoxin, serum alanine aminotransferase (ALT) activity, and hepatic TNF-α level, and thereby preventing ethanol- and, indirectly, acetaldehyde-induced liver injury. Attenuation of ethanol-induced transfer of endotoxin of the intestine to the circulation has been ascribed to preservation of intestinal morphology and permeability,[16,172] and zinc antioxidant effects.[173] These promoting effects on barrier function have been supported by evidence showing that zinc deficiency can induce ileal oxidative stress in mice, sensitizing epithelial cells to ethanol injury, and inhibiting HNF-4α, resulting in loss of TJs integrity.[174,175] Since zinc deficiency is common in long-term ethanol consumers,[176] its supplementation may prevent mucosal oxidative stress induced by ethanol and acetaldehyde and, thus, preserving barrier integrity.

In addition to zinc, Tang et al. and Keshavarzian et al. have demonstrated that oat supplementation, which can result in increased short chain fatty acids (SCFAs) levels in the intestinal lumen, can prevent increased intestinal hyperpermeability and steatohepatitis in long-term ethanol-fed rats, via antioxidant-dependent mechanisms.[14,177,178] In line with these data, pretreatment of Caco-2 monolayers with clinically relevant concentration of butyrate, propionate, and acetate (2, 4, 8 mmol/L) was found to ameliorate ethanol-induced barrier dysfunction, by reducing oxidative stress, improving mitochondrial function, and increasing intracellular ATP levels, indicating the antioxidant potential of these compounds.[179] The study has further demonstrated that the promotive effects of the SCFA on barrier function are adenosine monophosphate-activated protein kinase (AMPK)-dependent.[179] Previous studies in Caco-2 cells have also shown that butyrate improves epithelial barrier integrity by enhancing TJs assembly, by activating AMPK.[180]

The Caco-2 cells used in this study do not express ADH,[181] and *in vitro* data on effects of SCFA after exposure to acetaldehyde are lacking. However, since acetaldehyde has oxidative potency and can accumulate in the colon, where high levels of SCFAs are available, future studies should focus on beneficial effects of SCFAs on acetaldehyde-induced barrier disruption.

Since oxidative stress mediates intestinal barrier dysfunction,[153] protective effects of the antioxidant L-cysteine on acetaldehyde-induced loss of intestinal TJs integrity have been examined. Treatment of Caco-2 monolayers with the antioxidants L-cysteine has been shown to attenuate acetaldehyde-induced oxidative stress, and subsequent EMT activation, and barrier disruption.[182] Human studies on effects of antioxidant on acetaldehyde-induced intestinal barrier dysfunction are lacking. However, in one study, slow-releasing buccal L-cysteine has been shown to eliminate up to two-thirds of acetaldehyde from saliva, after ethanol intake.[183]

Beneficial effects of different cytoprotective and cytotrophic factors that are known to promote proliferation and differentiation of gastrointestinal epithelium, such as epidermal growth factor (EGF) and glutamine, have also been investigated.[184] In Caco-2 monolayers, EGF improved the barrier function, and decreased paracellular permeability to inulin and LPS, by preventing acetaldehyde-induced reorganization of ZO-1, occluding, E-cadherin, and β-catenin, and enhancing their interactions with the actin cytoskeleton.[141] Comparable *ex vivo* data have also demonstrated that EGF can protect intestinal epithelium against acetaldehyde-induced TJs and AJs disruption, in human colon mucosal biopsies.[145] Further investigations by Suzuki et al. in Caco-2 cell monolayers have shown that these protective effects are mediated by mechanisms involving a number of cell signaling pathways, including phospholipase Cγ1 (PLCγ1), calcium, phosphokinase CβI (PKCβI), and protein kinase Cε (PKCε).[152] Moreover, Samak et al. have demonstrated in an elegant study that the EGF-mediated protection against acetaldehyde-induced loss of TJs integrity requires activity of the isoform of mitogen-activated protein kinase (MAPK) extracellular signal-regulated kinase 1/2 (ERK1/2), and that EGF-mediated protection of AJs is independent of MAPK activities.[38]

L-Glutamine has also been shown to protect intestinal epithelium from acetaldehyde-induced barrier dysfunction.[143] It increases TEER, and decreases inulin and lipopolysaccharide permeation, by inhibiting acetaldehyde-induced redistribution of ZO-1, occludin, E-cadherin, and β-catenin, and their dissociation from actin cytoskeleton, using Caco-2 cell monolayers. In a structural study using human colon mucosa, L-glutamine has also been demonstrated to prevent acetaldehyde-induced TJs and AJs disruption.[145] Evidence has also shown that L-glutamine exerts its barrier, promoting effects by activating EGF receptor, indicating a cross talk between L-glutamine and EGF in preventing acetaldehyde-induced barrier disruption.

Another potential biotherapeutic agents that may promote intestinal barrier function are probiotics. Probiotics have the potential to ameliorate intestinal barrier dysfunction, to reduce inflammation, and to change the composition of the gut microbiota.

For instance, probiotic bacteria including *Lactobacillus plantarum* WCFS1,[185] *E. coli* Nissle 1917,[186,187] *Bifidobacterium infantis,* [188] and *L. plantarum* MB452,[189] have been shown to promote intestinal barrier integrity *in vitro,* by increasing ZO-1, ZO-2, occludin, and/or decreasing claudin-2 protein expression, and were found to protect against phorbol ester-induced dislocation of ZO-1 and occludin. Furthermore, inhibiting growth of Gram-negative bacteria in the intestine may reduce the amount of endotoxin that, in turn, may attenuate endotoxemia-associated intestinal and liver injury.[190]

In animal models, effects of several probiotic strains on ethanol-induced intestinal barrier dysfunction have been examined. For example, Nanji et al. have shown that feeding of rats with the probiotic *Lactobacillus* GG can reduce plasma levels of endotoxin, and severity of liver injury.[18] Using the same strain, it has been shown, in mice[191,192] and in rats, that *Lactobacillus* GG administration significantly blunts ethanol-induced oxidative stress, and improves ethanol-induced gut leakiness.[193] In humans, Karczewski et al. have shown that intraduodenal administration of *L. plantarum* WCFS1, in healthy human volunteers, can increase ZO-1 and occludin expression in duodenal biopsies.[185] So far, the potential beneficial effect of probiotics in ethanol-induced liver diseases has been investigated in only one human pilot study, showing that short-term oral supplementation with *Bifidobacterium bifidum* and *L. plantarum* 8PA3 can restore the intestinal microbiota and attenuate ethanol-induced liver injury.[194] Effect on barrier function or expression of TJ proteins was not examined.

Probiotics can modulate ethanol and acetaldehyde generation and metabolism. *In vitro*, Nosova et al. have demonstrated that *Bifidobacterium* spp. And, to a greater extent, *Lactobacillus* GG are weak acetaldehyde generators, but have a relatively high acetaldehyde-metabolizing capacity.[195] By reducing the acetaldehyde-generating microbiota, and by enhancing acetaldehyde metabolism, probiotics may reduce acetaldehyde accumulation in colon. However, further studies in humans are required to confirm this possibility, and to investigate whether probiotic treatment can attenuate acetaldehyde-induced intestinal barrier disruption, and ethanol-related liver diseases.

CONCLUSIONS

In summary, there is strong evidence that acetaldehyde can reach the entire GI tract, including the small, as well as the large, intestine. Acetaldehyde, in addition to ethanol itself, can increase intestinal permeability resulting in endotoxemia, that is one of the key factors involved in the pathogenesis of ALD, but can also contribute to intestinal inflammation and carcinogenesis. In addition to being a source of endotoxin, intestinal microbiota is involved in generation and accumulation of acetaldehyde in the colon lumen, thereby influencing epithelial barrier function.

Several mechanisms can be involved in acetaldehyde-induced barrier dysfunction, including disruption of AJs and TJs integrity, mediated via protein phosphorylation and acetylation. Key mechanisms involved are summarized in Figure 14.3. The effects of acetaldehyde on barrier function are mediated by several cell

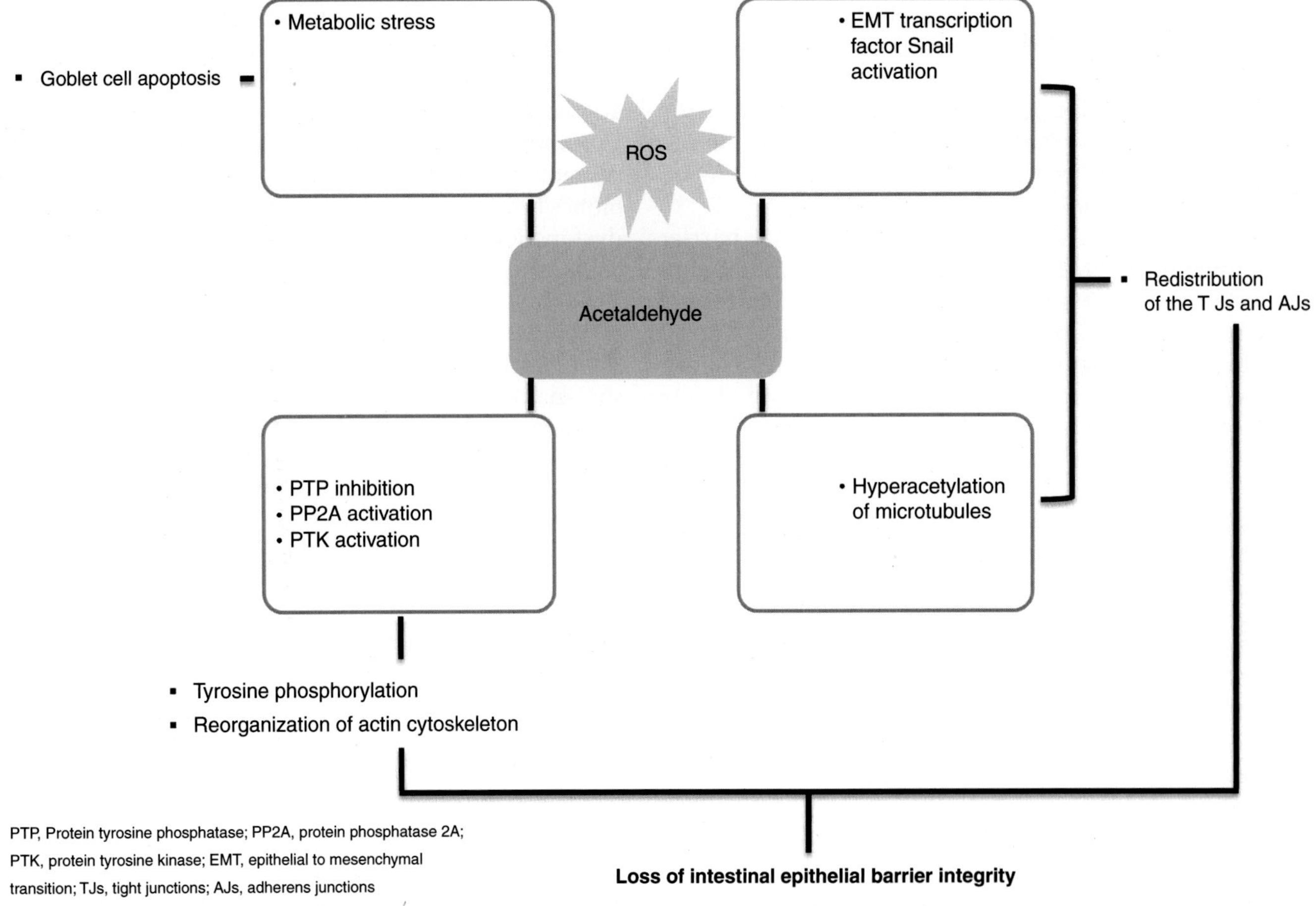

FIGURE 14.3 Summary of potential mechanisms and pathways involved in acetaldehyde-induced disruption of intestinal epithelial integrity, based on literature findings. Acetaldehyde induces inhibition of PTP and activation of PP2A and PTK, resulting in increased tyrosine phosphorylation of the TJs and AJs, and reorganizes cell cytoskeleton that in turn disrupts the TJs. Acetaldehyde also induces oxidative stress and increases ROS generation, resulting in apoptosis of goblet cells, activation of EMT and hyperacetylation of microtubules and, consequently, disruption of the TJs and AJs. These modulations can result in loss of intestinal epithelial integrity and thereby increasing permeability.

signaling pathways, including activation of PTK, PP2A, and the EMT transcription factor Snail, and inhibition of PTP. Together with the strong oxidative potency resulting in oxidative stress, these pathways can act in an additive or a synergistic way to disrupt the barrier function. Thus, future studies to understand precisely the role of interaction between ROS and these pathways in acetaldehyde-induced intestinal barrier disruption are warranted. Acetaldehyde also exerts deleterious effects on intestinal barrier components, including inhibition and stimulation of proliferation and differentiation, respectively in enterocytes, and induction of apoptosis in mucin-secreting goblet cells.

The injurious effects of acetaldehyde on intestinal barrier could potentially be prevented by a number of cytoprotective factors, including SCFAs, zinc, and probiotics, as well as EGF and glutamine. Further understanding and delineation of the mechanisms involved in the actions of these protective factors may provide leads for the development of therapeutic strategies that can prevent acetaldehyde-induced barrier disruption and its consequences.

Summary Points

- This chapter focuses on acetaldehyde, a potent toxic and carcinogenic ethanol metabolite that is injurious to the liver and intestinal epithelium. It contributes to alcoholic liver disease (ALD), as well as the development of cancers in the gastrointestinal tract.
- Acetaldehyde is produced by oxidation of ethanol, involving alcohol dehydrogenase (ADH) in hepatocytes cytosol, the microsomal ethanol oxidizing system (MEOS) cytochrome P450 2E1 (CYP2E1) in microsomes, and catalase in the peroxisomes.
- Intestinal barrier disruption with subsequent increased intestinal permeability to endotoxins is considered a major contributor to ALD and ethanol-related intestinal damage.
- Evidence indicates that acetaldehyde at 25–600 μM can disrupt intestinal epithelial tight and adherens junctions, and thereby increasing paracellular permeability.
- Acetaldehyde-induced activation of a number of cell signaling pathways including tyrosine kinases, posttranslational modification of cytoskeletal proteins, and induction of epithelial-to-mesenchymal transition have been suggested as possible mechanisms involved in acetaldehyde-induced intestinal barrier dysfunction.
- More research focusing on injurious effects of ethanol and acetaldehyde on intestinal epithelium, including potential mechanisms involved and modulators thereof is indispensable to identify leads for potentially preventive and/or therapeutic targets.

References

1. WHO Global Status Report on Alcohol 2004. Geneva: Department of Mental Health and Substance Abuse: World Health Organisation; 2004.
2. Mann RE, Smart RG, Govoni R. The epidemiology of alcoholic liver disease. *Alcohol Res Health* 2003;**27**(3):209–19.
3. Saunders JB, Latt N. Epidemiology of alcoholic liver disease. *Baillieres Clin Gastroenterol* 1993;**7**(3):555–79.
4. Deltenre P, Mathurin P. Epidemiology of alcoholic liver disease and new challenges. *Gastroenterol Clin Biol* 2009;**33**(12):1147–50.
5. Chiang CP, Wu CW, Lee SP, et al. Expression pattern, ethanol-metabolizing activities, and cellular localization of alcohol and aldehyde dehydrogenases in human pancreas: implications for pathogenesis of alcohol-induced pancreatic injury. *Alcohol Clin Exp Res* 2009;**33**(6):1059–68.
6. Salaspuro MP. Alcohol consumption and cancer of the gastrointestinal tract. *Best Pract Res Clin Gastroenterol* 2003;**17**(4):679–94.
7. Robles EA, Mezey E, Halsted CH, Schuster MM. Effect of ethanol on motility of the small intestine. *Johns Hopkins Med J* 1974;**135**(1):17–24.
8. Asai K, Buurman WA, Reutelingsperger CP, Schutte B, Kaminishi M. Low concentrations of ethanol induce apoptosis in human intestinal cells. *Scand J Gastroenterol* 2003;**38**(11):1154–61.
9. Rao R. Endotoxemia and gut barrier dysfunction in alcoholic liver disease. *Hepatology* 2009;**50**(2):638–44.
10. Bjarnason I, Peters TJ, Wise RJ. The leaky gut of alcoholism: possible route of entry for toxic compounds. *Lancet* 1984;**1**(8370):179–82.
11. Robinson GM, Orrego H, Israel Y, Devenyi P, Kapur BM. Low-molecular-weight polyethylene glycol as a probe of gastrointestinal permeability after alcohol ingestion. *Dig Dis Sci* 1981;**26**(11):971–7.
12. Mathurin P, Deng QG, Keshavarzian A, Choudhary S, Holmes EW, Tsukamoto H. Exacerbation of alcoholic liver injury by enteral endotoxin in rats. *Hepatology* 2000;**32**(5):1008–17.
13. Keshavarzian A, Farhadi A, Forsyth CB, et al. Evidence that chronic alcohol exposure promotes intestinal oxidative stress, intestinal hyperpermeability and endotoxemia prior to development of alcoholic steatohepatitis in rats. *J Hepatol* 2009;**50**(3):538–47.
14. Keshavarzian A, Choudhary S, Holmes EW, et al. Preventing gut leakiness by oats supplementation ameliorates alcohol-induced liver damage in rats. *J Pharmacol Exp Ther* 2001;**299**(2):442–8.
15. Parlesa k A, Schafer C, Schutz T, Bode JC, Bode C. Increased intestinal permeability to macromolecules and endotoxemia in patients with chronic alcohol abuse in different stages of alcohol-induced liver disease. *J Hepatol* 2000;**32**(5):742–7.
16. Lambert JC, Zhou Z, Wang L, Song Z, McClain CJ, Kang YJ. Prevention of alterations in intestinal permeability is involved in zinc inhibition of acute ethanol-induced liver damage in mice. *J Pharmacol Exp Ther* 2003;**305**(3):880–6.
17. Adachi Y, Bradford BU, Gao W, Bojes HK, Thurman RG. Inactivation of Kupffer cells prevents early alcohol-induced liver injury. *Hepatology* 1994;**20**(2):453–60.
18. Nanji AA, Khettry U, Sadrzadeh SM. *Lactobacillus* feeding reduces endotoxemia and severity of experimental alcoholic liver (disease). *Proc Soc Exp Biol Med* 1994;**205**(3):243–7.
19. Ma TY, Boivin MA, Ye D, Pedram A, Said HM. Mechanism of TNF-{alpha} modulation of Caco-2 intestinal epithelial tight junction barrier: role of myosin light-chain kinase protein expression. *Am J Physiol Gastrointest Liver Physiol* 2005;**288**(3):G422–30.
20. Al-Sadi R, Ye D, Dokladny K, Ma TY. Mechanism of IL-1beta-induced increase in intestinal epithelial tight junction permeability. *J Immunol* 2008;**180**(8):5653–61.
21. Djamali A, Reese S, Yracheta J, Oberley T, Hullett D, Becker B. Epithelial-to-mesenchymal transition and oxidative stress in chronic allograft nephropathy. *Am J Transplant* 2005;**5**(3):500–9.

22. Welcker K, Martin A, Kolle P, Siebeck M, Gross M. Increased intestinal permeability in patients with inflammatory bowel disease. *Eur J Med Res* 2004;**9**(10):456–60.
23. Squier CA, Cox P, Hall BK. Enhanced penetration of nitrosonornicotine across oral mucosa in the presence of ethanol. *J Oral Pathol* 1986;**15**(5):276–9.
24. Howie NM, Trigkas TK, Cruchley AT, Wertz PW, Squier CA, Williams DM. Short-term exposure to alcohol increases the permeability of human oral mucosa. *Oral Dis* 2001;**7**(6):349–54.
25. Squier CA, Kremer MJ, Wertz PW. Effect of ethanol on lipid metabolism and epithelial permeability barrier of skin and oral mucosa in the rat. *J Oral Pathol Med* 2003;**32**(10):595–9.
26. Wight AJ, Ogden GR. Possible mechanisms by which alcohol may influence the development of oral cancer – a review. *Oral Oncol* 1998;**34**(6):441–7.
27. Halsted CH, Robles EA, Mezey E. Distribution of ethanol in the human gastrointestinal tract. *Am J Clin Nutr* 1973;**26**(8):831–4.
28. Zakhari S. Overview: how is alcohol metabolized by the body? *Alcohol Res Health* 2006;**29**(4):245–54.
29. Purohit V, Bode JC, Bode C, et al. Alcohol, intestinal bacterial growth, intestinal permeability to endotoxin, and medical consequences: summary of a symposium. *Alcohol* 2008;**42**(5):349–61.
30. Salaspuro MP. Acetaldehyde, microbes, and cancer of the digestive tract. *Crit Rev Clin Lab Sci* 2003;**40**(2):183–208.
31. Peterson LW, Artis D. Intestinal epithelial cells: regulators of barrier function and immune homeostasis. *Nat Rev Immunol* 2014;**14**(3):141–53.
32. Turner JR. Intestinal mucosal barrier function in health and disease. *Nat Rev Immunol* 2009;**9**(11):799–809.
33. Koivisto T, Salaspuro M. Acetaldehyde alters proliferation, differentiation and adhesion properties of human colon adenocarcinoma cell line Caco-2. *Carcinogenesis* 1998;**19**(11):2031–6.
34. Van Itallie CM, Anderson JM. The molecular physiology of tight junction pores. *Physiology* 2004;**19**:331–8.
35. Blair SA, Kane SV, Clayburgh DR, Turner JR. Epithelial myosin light chain kinase expression and activity are upregulated in inflammatory bowel disease. *Lab Invest* 2006;**86**(2):191–201.
36. Gopalakrishnan S, Raman N, Atkinson SJ, Marrs JA. Rho GTPase signaling regulates tight junction assembly and protects tight junctions during ATP depletion. *Am J Physiol* 1998;**275**(3 Pt. 1): C798–809.
37. Jain S, Suzuki T, Seth A, Samak G, Rao R. Protein kinase Cζ phosphorylates occludin and promotes assembly of epithelial tight junctions. *Biochem J* 2011;**437**(2):289–99.
38. Samak G, Aggarwal S, Rao RK. ERK is involved in EGF-mediated protection of tight junctions, but not adherens junctions, in acetaldehyde-treated Caco-2 cell monolayers. *Am J Physiol Gastrointest Liver Physiol* 2011;**301**(1):G50–9.
39. Atkinson KJ, Rao RK. Role of protein tyrosine phosphorylation in acetaldehyde-induced disruption of epithelial tight junctions. *Am J Physiol Gastrointest Liver Physiol* 2001;**280**(6):G1280–8.
40. Elamin E, Masclee A, Troost F, Dekker J, Jonkers D. Activation of the epithelial-to-mesenchymal transition factor snail mediates acetaldehyde-induced intestinal epithelial barrier disruption. *Alcohol Clin Exp Res* 2014;**38**:344–53.
41. Fasano A. Zonulin and its regulation of intestinal barrier function: the biological door to inflammation, autoimmunity, and cancer. *Physiol Rev* 2011;**91**(1):151–75.
42. Gonzalez-Mariscal L, Tapia R, Chamorro D. Crosstalk of tight junction components with signaling pathways. *Biochim Biophys Acta* 2008;**1778**(3):729–56.
43. Samak G, Narayanan D, Jaggar JH, Rao R. CaV1.3 channels and intracellular calcium mediate osmotic stress-induced N-terminal c-Jun kinase activation and disruption of tight junctions in Caco-2 CELL MONOLAYERS. *J Biol Chem* 2011; **286**(34):30232–43.
44. Lichtenberger LM, Zhou Y, Dial EJ, Raphael RM. NSAID injury to the gastrointestinal tract: evidence that NSAIDs interact with phospholipids to weaken the hydrophobic surface barrier and induce the formation of unstable pores in membranes. *J Pharm Pharmacol* 2006;**58**(11):1421–8.
45. Watson CJ, Hoare CJ, Garrod DR, Carlson GL, Warhurst G. Interferon-gamma selectively increases epithelial permeability to large molecules by activating different populations of paracellular pores. *J Cell Sci* 2005;**118**(Pt. 22):5221–30.
46. Wang F, Graham WV, Wang Y, Witkowski ED, Schwarz BT, Turner JR. Interferon-gamma and tumor necrosis factor-alpha synergize to induce intestinal epithelial barrier dysfunction by up-regulating myosin light chain kinase expression. *Am J Pathol* 2005;**166**(2):409–19.
47. Al-Sadi R, Ye D, Said HM, Ma TY. Cellular and molecular mechanism of interleukin-1beta modulation of Caco-2 intestinal epithelial tight junction barrier. *J Cell Mol Med* 2011;**15**(4):970–82.
48. Nusrat A, Turner JR, Madara JL. Molecular physiology and pathophysiology of tight junctions. IV. Regulation of tight junctions by extracellular stimuli: nutrients, cytokines, and immune cells. *Am J Physiol Gastrointest Liver Physiol* 2000;**279**(5):G851–7.
49. Staehelin LA, Mukherjee TM, Williams AW. Freeze-etch appearance of the tight junctions in the epithelium of small and large intestine of mice. *Protoplasma* 1969;**67**(2):165–84.
50. Claude P, Goodenough DA. Fracture faces of zonulae occludentes from "tight" and "leaky" epithelia. *J Cell Biol* 1973;**58**(2):390–400.
51. Goodenough DA, Revel JP. A fine structural analysis of intercellular junctions in the mouse liver. *J Cell Biol* 1970;**45**(2):272–90.
52. Shen L, Weber CR, Turner JR. The tight junction protein complex undergoes rapid and continuous molecular remodeling at steady state. *J Cell Biol* 2008;**181**(4):683–95.
53. Riesen FK, Rothen-Rutishauser B, Wunderli-Allenspach H. A ZO1-GFP fusion protein to study the dynamics of tight junctions in living cells. *Histochem Cell Biol* 2002;**117**(4):307–15.
54. Tang VW. Proteomic and bioinformatic analysis of epithelial tight junction reveals an unexpected cluster of synaptic molecules. *Biol Direct* 2006;**1**:37.
55. Noth R, Lange-Grumfeld J, Stuber E, et al. Increased intestinal permeability and tight junction disruption by altered expression and localization of occludin in a murine graft versus host disease model. *BMC Gastroenterol* 2011;**11**:109.
56. Rao RK. Acetaldehyde-induced increase in paracellular permeability in Caco-2 cell monolayer. *Alcohol Clin Exp Res* 1998;**22**(8):1724–30.
57. Ferrier L, Berard F, Debrauwer L, et al. Impairment of the intestinal barrier by ethanol involves enteric microflora and mast cell activation in rodents. *Am J Pathol* 2006;**168**(4):1148–54.
58. Shen L, Weber CR, Raleigh DR, Yu D, Turner JR. Tight junction pore and leak pathways: a dynamic duo. *Annu Rev Physiol* 2011;**73**:283–309.
59. Hidalgo IJ. Cultured intestinal epithelial cell models. *Pharm Biotechnol* 1996;**8**:35–50.
60. Juuti-Uusitalo K, Klunder LJ, Sjollema KA, et al. Differential effects of TNF (TNFSF2) and IFN-gamma on intestinal epithelial cell morphogenesis and barrier function in three-dimensional culture. *PLoS One* 2011;**6**(8):e22967.
61. Elamin E, Jonkers D, Juuti-Uusitalo K, et al. Effects of ethanol and acetaldehyde on tight junction integrity: *in vitro* study in a three dimensional intestinal epithelial cell culture model. *PLoS One* 2012;**7**(4):e35008.
62. Elamin E, Masclee A, Juuti-Uusitalo K, et al. Fatty acid ethyl esters induce intestinal epithelial barrier dysfunction via a reactive oxygen species-dependent mechanism in a three-dimensional cell culture model. *PLoS One* 2013;**8**(3):e58561.
63. Menzies IS. Intestinal permeability in coeliac disease. *Gut* 1972;**13**(10):847.

64. Menzies IS, Laker MF, Pounder R, et al. Abnormal intestinal permeability to sugars in villous atrophy. *Lancet* 1979;**2**(8152): 1107–9.
65. Grootjans J, Thuijls G, Verdam F, Derikx JP, Lenaerts K, Buurman WA. Non-invasive assessment of barrier integrity and function of the human gut. *World J Gastrointest Surg* 2010;**2**(3):61–9.
66. Bjarnason I, MacPherson A, Hollander D. Intestinal permeability: an overview. *Gastroenterology* 1995;**108**(5):1566–81.
67. Crabb DW, Bosron WF, Li TK. Ethanol metabolism. *Pharmacol Ther* 1987;**34**(1):59–73.
68. Levitt MD, Li R, DeMaster EG, Elson M, Furne J, Levitt DG. Use of measurements of ethanol absorption from stomach and intestine to assess human ethanol metabolism. *Am J Physiol* 1997;**273**(4 Pt. 1):G951–7.
69. Norberg A, Jones AW, Hahn RG, Gabrielsson JL. Role of variability in explaining ethanol pharmacokinetics: research and forensic applications. *Clin Pharmacokinet* 2003;**42**(1):1–31.
70. Levitt MD, Doizaki W, Levine AS. Hypothesis: metabolic activity of the colonic bacteria influences organ injury from ethanol. *Hepatology* 1982;**2**(5):598–600.
71. Jones AW. Distribution of ethanol between saliva and blood in man. *Clin Exp Pharmacol Physiol* 1979;**6**(1):53–9.
72. Hachet-Haas M, Converset N, Marchal O, et al. FRET and colocalization analyzer – a method to validate measurements of sensitized emission FRET acquired by confocal microscopy and available as an ImageJ Plug-in. *Microsc Res Tech* 2006;**69**(12):941–56.
73. Chen Y, Lu Q, Schneeberger EE, Goodenough DA. Restoration of tight junction structure and barrier function by down-regulation of the mitogen-activated protein kinase pathway in ras-transformed Madin–Darby canine kidney cells. *Mol Biol Cell* 2000;**11**(3):849–62.
74. Eckardt MJ, File SE, Gessa GL, et al. Effects of moderate alcohol consumption on the central nervous system. *Alcohol Clin Exp Res* 1998;**22**(5):998–1040.
75. Oneta CM, Simanowski UA, Martinez M, et al. First pass metabolism of ethanol is strikingly influenced by the speed of gastric emptying. *Gut* 1998;**43**(5):612–9.
76. Goist Jr KC, Sutker PB. Acute alcohol intoxication and body composition in women and men. *Pharmacol Biochem Behav* 1985;**22**(5):811–4.
77. Lammers SM, Mainzer DE, Breteler MH. Do alcohol pharmacokinetics in women vary due to the menstrual cycle? *Addiction* 1995;**90**(1):23–30.
78. Agarwal DP, Goedde HW. Pharmacogenetics of alcohol dehydrogenase (ADH). *Pharmacol Ther* 1990;**45**(1):69–83.
79. Seitz HK, Oneta CM. Gastrointestinal alcohol dehydrogenase. *Nutr Rev* 1998;**56**(2 Pt. 1):52–60.
80. Seitz HK, Egerer G, Oneta C, et al. Alcohol dehydrogenase in the human colon and rectum. *Digestion* 1996;**57**(2):105–8.
81. Vaglenova J, Martinez SE, Porte S, Duester G, Farres J, Pares X. Expression, localization and potential physiological significance of alcohol dehydrogenase in the gastrointestinal tract. *Eur J Biochem* 2003;**270**(12):2652–62.
82. Yin SJ, Liao CS, Lee YC, Wu CW, Jao SW. Genetic polymorphism and activities of human colon alcohol and aldehyde dehydrogenases: no gender and age differences. *Alcohol Clin Exp Res* 1994;**18**(5):1256–60.
83. Jörnvall H, Danielsson O, Hjelmqvist L, Persson B, Shafqat J. The alcohol dehydrogenase system. *Adv Exp Med Biol* 1995;**372**:281–94.
84. Jörnvall H, Hoog JO. Nomenclature of alcohol dehydrogenases. *Alcohol Alcohol* 1995;**30**(2):153–61.
85. Jelski W, Szmitkowski M. Alcohol dehydrogenase (ADH) and aldehyde dehydrogenase (ALDH) in the cancer diseases. *Clin Chim Acta* 2008;**395**(1–2):1–5.
86. Fisher OZ, Peppas NA. Quantifying tight junction disruption caused by biomimetic pH-sensitive hydrogel drug carriers. *J Drug Deliv Sci Technol* 2008;**18**(1):47–50.
87. Woods A, Cheung PC, Smith FC, et al. Characterization of AMP-activated protein kinase beta and gamma subunits. Assembly of the heterotrimeric complex *in vitro*. *J Biol Chem* 1996;**271**(17):10282–90.
88. Dong YJ, Peng TK, Yin SJ. Expression and activities of class IV alcohol dehydrogenase and class III aldehyde dehydrogenase in human mouth. *Alcohol* 1996;**13**(3):257–62.
89. Yin SJ, Chou FJ, Chao SF, et al. Alcohol and aldehyde dehydrogenases in human esophagus: comparison with the stomach enzyme activities. *Alcohol Clin Exp Res* 1993;**17**(2):376–81.
90. Pares X, Cederlund E, Moreno A, Saubi N, Hoog JO, Jornvall H. Class IV alcohol dehydrogenase (the gastric enzyme). Structural analysis of human sigma sigma-ADH reveals class IV to be variable and confirms the presence of a fifth mammalian alcohol dehydrogenase class. *FEBS Lett* 1992;**303**(1):69–72.
91. Chiang CP, Wu CW, Lee SP, et al. Expression pattern, ethanol-metabolizing activities, and cellular localization of alcohol and aldehyde dehydrogenases in human small intestine. *Alcohol Clin Exp Res* 2012;**36**(12):2047–58.
92. Li TK, Bosron WF. Genetic variability of enzymes of alcohol metabolism in human beings. *Ann Emerg Med* 1986;**15**(9):997–1004.
93. Lieber CS. Cytochrome P-4502E1: its physiological and pathological role. *Physiol Rev* 1997;**77**(2):517–44.
94. Vanhoutvin SA, Troost FJ, Hamer HM, et al. Butyrate-induced transcriptional changes in human colonic mucosa. *PLoS One* 2009;**4**(8):e6759.
95. Ingelman-Sundberg M, Ronis MJ, Lindros KO, Eliasson E, Zhukov A. Ethanol-inducible cytochrome P4502E1: regulation, enzymology and molecular biology. *Alcohol Alcohol Suppl* 1994;**2**:131–9.
96. Beno I, Volkovová K, Staruchová M, Koszeghyová L. The activity of Cu/Zn-superoxide dismutase and catalase of gastric mucosa in chronic gastritis, and the effect of alpha-tocopherol. *Bratisl Lek Listy* 1994;**95**(1):9–14.
97. Laukoetter MG, Nava P, Nusrat A. Role of the intestinal barrier in inflammatory bowel disease. *World J Gastroenterol* 2008;**14**(3):401–7.
98. Best CA, Laposata M. Fatty acid ethyl esters: toxic non-oxidative metabolites of ethanol and markers of ethanol intake. *Front Biosci* 2003;**8**:e202–17.
99. Pannequin J, Delaunay N, Darido C, et al. Phosphatidylethanol accumulation promotes intestinal hyperplasia by inducing ZONAB-mediated cell density increase in response to chronic ethanol exposure. *Mol Cancer Res* 2007;**5**(11):1147–57.
100. Lieber CS. Metabolism of alcohol. *Clin Liver Dis* 2005;**9**(1):1–35.
101. Agarwal DP. Molecular genetic aspects of alcohol metabolism and alcoholism. *Pharmacopsychiatry* 1997;**30**(3):79–84.
102. Bosron WF, Li TK. Genetic polymorphism of human liver alcohol and aldehyde dehydrogenases, and their relationship to alcohol metabolism and alcoholism. *Hepatology* 1986;**6**(3):502–10.
103. Salaspuro M. Bacteriocolonic pathway for ethanol oxidation: characteristics and implications. *Ann Med* 1996;**28**(3):195–200.
104. Li TK, Yin SJ, Crabb DW, O'Connor S, Ramchandani VA. Genetic and environmental influences on alcohol metabolism in humans. *Alcohol Clin Exp Res* 2001;**25**(1):136–44.
105. Freudenheim JL, Ambrosone CB, Moysich KB, et al. Alcohol dehydrogenase 3 genotype modification of the association of alcohol consumption with breast cancer risk. *Cancer Causes Control* 1999;**10**(5):369–77.
106. Crabb DW, Edenberg HJ, Bosron WF, Li TK. Genotypes for aldehyde dehydrogenase deficiency and alcohol sensitivity. The inactive ALDH2(2) allele is dominant. *J Clin Invest* 1989;**83**(1):314–6.
107. Baan R, Straif K, Grosse Y, et al. Carcinogenicity of alcoholic beverages. *Lancet Oncol* 2007;**8**(4):292–3.
108. Seitz HK, Stickel F. Molecular mechanisms of alcohol-mediated carcinogenesis. *Nat Rev Cancer* 2007;**7**(8):599–612.

109. Elamin EE, Masclee AA, Dekker J, Jonkers DM. Ethanol metabolism and its effects on the intestinal epithelial barrier. *Nutr Rev* 2013;**71**(7):483–99.
110. Blomstrand R. Observations of the formation of ethanol in the intestinal tract in man. *Life Sci II* 1971;**10**(10):575–82.
111. Remedio RN, Castellar A, Barbosa RA, Gomes RJ, Caetano FH. Morphological analysis of colon goblet cells and submucosa in type I diabetic rats submitted to physical training. *Microsc Res Tech* 2012;**75**(6):821–8.
112. Geertinger P, Bodenhoff J, Helweg-Larsen K, Lund A. Endogenous alcohol production by intestinal fermentation in sudden infant death. *Z Rechtsmed* 1982;**89**(3):167–72.
113. Krebs HA, Perkins JR. The physiological role of liver alcohol dehydrogenase. *Biochem J* 1970;**118**(4):635–44.
114. Kaji H, Asanuma Y, Yahara O, et al. Intragastrointestinal alcohol fermentation syndrome: report of two cases and review of the literature. *J Forensic Sci Soc* 1984;**24**(5):461–71.
115. Logan BK, Jones AW. Endogenous ethanol 'auto-brewery syndrome' as a drunk-driving defence challenge. *Med Sci Law* 2000;**40**(3):206–15.
116. Logan BK, Jones AW. Endogenous ethanol production in a child with short gut syndrome. *J Pediatr Gastroenterol Nutr* 2003;**36**(3):419–20 author's reply 420–421.
117. Klipstein FA, Engert RF. Enterotoxigenic intestinal bacteria in tropical sprue. III. Preliminary characterization of *Klebsiella pneumoniae* enterotoxin. *J Infect Dis* 1975;**132**(2):200–3.
118. Mezey E, Imbembo AL, Potter JJ, Rent KC, Lombardo R, Holt PR. Endogenous ethanol production and hepatic disease following jejunoileal bypass for morbid obesity. *Am J Clin Nutr* 1975;**28**(11):1277–83.
119. Roine RP, Salmela KS, Salaspuro M. Alcohol metabolism in *Helicobacter pylori*-infected stomach. *Ann Med* 1995;**27**(5):583–8.
120. Lin Y, Tanaka S. Ethanol fermentation from biomass resources: current state and prospects. *Appl Microbiol Biotechnol* 2006;**69**(6):627–42.
121. Still JL. Alcohol enzyme of Bact. coli. *Biochem J* 1940; **34**(8–9):1177–82.
122. Baraona E, Julkunen R, Tannenbaum L, Lieber CS. Role of intestinal bacterial overgrowth in ethanol production and metabolism in rats. *Gastroenterology* 1986;**90**(1):103–10.
123. Kurkivuori J, Salaspuro V, Kaihovaara P, et al. Acetaldehyde production from ethanol by oral streptococci. *Oral Oncol* 2007;**43**(2):181–6.
124. Roine RP, Salmela KS, Hook-Nikanne J, Kosunen TU, Salaspuro M. Alcohol dehydrogenase mediated acetaldehyde production by *Helicobacter pylori* – a possible mechanism behind gastric injury. *Life Sci* 1992;**51**(17):1333–7.
125. Tillonen J, Homann N, Rautio M, Jousimies-Somer H, Salaspuro M. Role of yeasts in the salivary acetaldehyde production from ethanol among risk groups for ethanol-associated oral cavity cancer. *Alcohol Clin Exp Res* 1999;**23**(8):1409–15.
126. Jokelainen K, Siitonen A, Jousimies-Somer H, Nosova T, Heine R, Salaspuro M. *In vitro* alcohol dehydrogenase-mediated acetaldehyde production by aerobic bacteria representing the normal colonic flora in man. *Alcohol Clin Exp Res* 1996;**20**(6):967–72.
127. Salaspuro V, Nyfors S, Heine R, Siitonen A, Salaspuro M, Jousimies-Somer H. Ethanol oxidation and acetaldehyde production *in vitro* by human intestinal strains of *Escherichia coli* under aerobic, microaerobic, and anaerobic conditions. *Scand J Gastroenterol* 1999;**34**(10):967–73.
128. Kishore R, Hill JR, McMullen MR, Frenkel J, Nagy LE. ERK1/2 and Egr-1 contribute to increased TNF-alpha production in rat Kupffer cells after chronic ethanol feeding. *Am J Physiol Gastrointest Liver Physiol* 2002;**282**(1):G6–G15.
129. Foster T. Staphylococcus. In: Baron S, editor. *Medical microbiology*. 4th ed. University of Texas Medical Branch, Department of Microbiology (January 28, 1997) Galveston, TX; 1996.
130. Tillonen J, Kaihovaara P, Jousimies-Somer H, Heine R, Salaspuro M. Role of catalase in *in vitro* acetaldehyde formation by human colonic contents. *Alcohol Clin Exp Res* 1998;**22**(5): 1113–9.
131. Bode JC, Bode C, Heidelbach R, Durr HK, Martini GA. Jejunal microflora in patients with chronic alcohol abuse. *Hepatogastroenterology* 1984;**31**(1):30–4.
132. Casafont Morencos F, de las Heras Castano G, Martin Ramos L, Lopez Arias MJ, Ledesma F, Pons Romero F. Small bowel bacterial overgrowth in patients with alcoholic cirrhosis. *Dig Dis Sci* 1996;**41**(3):552–6.
133. Mutlu EA, Gillevet PM, Rangwala H, et al. Colonic microbiome is altered in alcoholism. *Am J Physiol Gastrointest Liver Physiol* 2012;**302**(9):G966–78.
134. Rossi MA, Zucoloto S. Effect of chronic ethanol ingestion on the small intestinal ultrastructure in rats. *Beitr Pathol* 1977;**161**(1):50–61.
135. Stowell AR. An improved method for the determination of acetaldehyde in human blood with minimal ethanol interference. *Clin Chim Acta* 1979;**98**(3):201–5.
136. Homann N, Jousimies-Somer H, Jokelainen K, Heine R, Salaspuro M. High acetaldehyde levels in saliva after ethanol consumption: methodological aspects and pathogenetic implications. *Carcinogenesis* 1997;**18**(9):1739–43.
137. Jokelainen K, Roine RP, Vaananen H, Farkkila M, Salaspuro M. *In vitro* acetaldehyde formation by human colonic bacteria. *Gut* 1994;**35**(9):1271–4.
138. Koivisto T, Salaspuro M. Aldehyde dehydrogenases of the rat colon: comparison with other tissues of the alimentary tract and the liver. *Alcohol Clin Exp Res* 1996;**20**(3):551–5.
139. Salaspuro M. Interrelationship between alcohol, smoking, acetaldehyde and cancer. *Novartis Found Symp* 2007;**285**:80–9 discussion 89–96, 198–199.
140. Rao RK. Acetaldehyde-induced barrier disruption and paracellular permeability in Caco-2 cell monolayer. *Methods Mol Biol* 2008;**447**:171–83.
141. Sheth P, Seth A, Thangavel M, Basuroy S, Rao RK. Epidermal growth factor prevents acetaldehyde-induced paracellular permeability in Caco-2 cell monolayer. *Alcohol Clin Exp Res* 2004;**28**(5):797–804.
142. Fisher SJ, Swaan PW, Eddington ND. The ethanol metabolite acetaldehyde increases paracellular drug permeability in vitro and oral bioavailability in vivo. *J Pharmacol Exp Ther* 2010;**332**(1):326–33.
143. Seth A, Basuroy S, Sheth P, Rao RK. L-Glutamine ameliorates acetaldehyde-induced increase in paracellular permeability in Caco-2 cell monolayer. *Am J Physiol Gastrointest Liver Physiol* 2004;**287**(3):G510–7.
144. Dunagan M, Chaudhry K, Samak G, Rao RK. Acetaldehyde disrupts tight junctions in Caco-2 cell monolayers by a protein phosphatase 2A-dependent mechanism. *Am J Physiol Gastrointest Liver Physiol* 2012;**303**:G1356–64.
145. Basuroy S, Sheth P, Mansbach CM, Rao RK. Acetaldehyde disrupts tight junctions and adherens junctions in human colonic mucosa: protection by EGF and L-glutamine. *Am J Physiol Gastrointest Liver Physiol* 2005;**289**(2):G367–75.
146. Sheth P, Seth A, Atkinson KJ, et al. Acetaldehyde dissociates the PTP1B-E-cadherin-beta-catenin complex in Caco-2 cell monolayers by a phosphorylation-dependent mechanism. *Biochem J* 2007;**402**(2):291–300.
147. Shifflett DE, Clayburgh DR, Koutsouris A, Turner JR, Hecht GA. Enteropathogenic *E. coli* disrupts tight junction barrier function and structure *in vivo*. *Lab Invest* 2005;**85**(10):1308–24.
148. Hietbrink F, Besselink MG, Renooij W, et al. Systemic inflammation increases intestinal permeability during experimental human endotoxemia. *Shock* 2009;**32**(4):374–8.

149. Tsukita S, Furuse M. Occludin and claudins in tight-junction strands: leading or supporting players? *Trends Cell Biol* 1999;**9**(7):268–73.
150. Fukui H, Brauner B, Bode JC, Bode C. Plasma endotoxin concentrations in patients with alcoholic and non-alcoholic liver disease: reevaluation with an improved chromogenic assay. *J Hepatol* 1991;**12**(2):162–9.
151. Forsythe RM, Xu DZ, Lu Q, Deitch EA. Lipopolysaccharide-induced enterocyte-derived nitric oxide induces intestinal monolayer permeability in an autocrine fashion. *Shock* 2002;**17**(3):180–4.
152. Suzuki T, Seth A, Rao R. Role of phospholipase Cgamma-induced activation of protein kinase Cepsilon (PKCepsilon) and PKCbetaI in epidermal growth factor-mediated protection of tight junctions from acetaldehyde in Caco-2 cell monolayers. *J Biol Chem* 2008;**283**(6):3574–83.
153. Basuroy S, Sheth P, Kuppuswamy D, Balasubramanian S, Ray RM, Rao RK. Expression of kinase-inactive c-Src delays oxidative stress-induced disassembly and accelerates calcium-mediated reassembly of tight junctions in the Caco-2 cell monolayer. *J Biol Chem* 2003;**278**(14):11916–24.
154. Seth A, Sheth P, Elias BC, Rao R. Protein phosphatases 2A and 1 interact with occludin and negatively regulate the assembly of tight junctions in the CACO-2 cell monolayer. *J Biol Chem* 2007;**282**(15):11487–98.
155. Rao RK, Basuroy S, Rao VU, Karnaky Jr KJ, Gupta A. Tyrosine phosphorylation and dissociation of occludin-ZO-1 and E-cadherin-beta-catenin complexes from the cytoskeleton by oxidative stress. *Biochem J* 2002;**368**(Pt. 2):471–81.
156. Rao RK, Li L, Baker RD, Baker SS, Gupta A. Glutathione oxidation and PTPase inhibition by hydrogen peroxide in Caco-2 cell monolayer. *Am J Physiol Gastrointest Liver Physiol* 2000;**279**(2): G332–40.
157. Sheth P, Basuroy S, Li C, Naren AP, Rao RK. Role of phosphatidylinositol 3-kinase in oxidative stress-induced disruption of tight junctions. *J Biol Chem* 2003;**278**(49):49239–45.
158. Atkinson KJ, Rao RK. Role of protein tyrosine phosphorylation in acetaldehyde-induced disruption of epithelial tight junctions. *Am J Physiol Gastrointest Liver Physiol* 2001;**280**(6):G1280–8.
159. Ivanov AI, McCall IC, Babbin B, Samarin SN, Nusrat A, Parkos CA. Microtubules regulate disassembly of epithelial apical junctions. *BMC Cell Biol* 2006;**7**:12.
160. Elhaseen E, Daisy J, Freddy T, et al. S1792 effects of ethanol and acetaldehyde on epithelial integrity in a three dimensional (3D) epithelial cell culture model. *Gastroenterology* 2009;**138**(Suppl. 1):S-275.
161. Forsyth CB, Tang Y, Shaikh M, Zhang L, Keshavarzian A. Role of snail activation in alcohol-induced iNOS-mediated disruption of intestinal epithelial cell permeability. *Alcohol Clin Exp Res* 2011;**35**:1635–43.
162. Elamin E, Masclee A, Dekker J, Jonkers D. Ethanol disrupts intestinal epithelial tight junction integrity through intracellular calcium-mediated Rho/ROCK activation. *Am J Physiol Gastrointest Liver Physiol* 2014;**306**(8):G677–85.
163. Zahs A, Bird MD, Ramirez L, Turner JR, Choudhry MA, Kovacs EJ. Inhibition of long myosin light-chain kinase activation alleviates intestinal damage after binge ethanol exposure and burn injury. *Am J Physiol Gastrointest Liver Physiol* 2012;**303**(6):G705–12.
164. Zhong W, Zhao Y, McClain CJ, Kang YJ, Zhou Z. Inactivation of hepatocyte nuclear factor-4{alpha} mediates alcohol-induced downregulation of intestinal tight junction proteins. *Am J Physiol Gastrointest Liver Physiol* 2010;**299**(3):G643–51.
165. Banan A, Keshavarzian A, Zhang L, et al. NF-kappaB activation as a key mechanism in ethanol-induced disruption of the F-actin cytoskeleton and monolayer barrier integrity in intestinal epithelium. *Alcohol* 2007;**41**(6):447–60.
166. Sheth P, Samak G, Shull JA, Seth A, Rao R. Protein phosphatase 2A plays a role in hydrogen peroxide-induced disruption of tight junctions in Caco-2 cell monolayers. *Biochem J* 2009;**421**(1):59–70.
167. Varella Morandi Junqueira-Franco M, Ernesto Troncon L, Garcia Chiarello P, do Rosario Del Lama Unamuno M, Afonso Jordao A, Vannucchi H. Intestinal permeability and oxidative stress in patients with alcoholic pellagra. *Clin Nutr* 2006;**25**(6):977–83.
168. Forsyth CB, Voigt RM, Shaikh M, et al. Role for intestinal CYP2E1 in alcohol-induced circadian gene-mediated intestinal hyperpermeability. *Am J Physiol Gastrointest Liver Physiol* 2013;**305**:G185–95.
169. Swanson G, Forsyth CB, Tang Y, et al. Role of intestinal circadian genes in alcohol-induced gut leakiness. *Alcohol Clin Exp Res* 2011;**35**(7):1305–14.
170. Caro AA, Cederbaum AI. Oxidative stress, toxicology, and pharmacology of CYP2E1. *Annu Rev Pharmacol Toxicol* 2004;**44**:27–42.
171. Brozinsky S, Fani K, Grosberg SJ, Wapnick S. Alcohol ingestion-induced changes in the human rectal mucosa: light and electron microscopic studies. *Dis Colon Rectum* 1978;**21**(5):329–35.
172. Lambert JC, Zhou Z, Wang L, Song Z, McClain CJ, Kang YJ. Preservation of intestinal structural integrity by zinc is independent of metallothionein in alcohol-intoxicated mice. *Am J Pathol* 2004;**164**(6):1959–66.
173. Li J, Hu W, Baldassare JJ, et al. The ethanol metabolite, linolenic acid ethyl ester, stimulates mitogen-activated protein kinase and cyclin signaling in hepatic stellate cells. *Life Sci* 2003;**73**(9):1083–96.
174. Zhong W, McClain CJ, Cave M, Kang YJ, Zhou Z. The role of zinc deficiency in alcohol-induced intestinal barrier dysfunction. *Am J Physiol Gastrointest Liver Physiol* 2010;**298**(5):G625–33.
175. Zhong W, Zhao Y, McClain CJ, Kang YJ, Zhou Z. Inactivation of hepatocyte nuclear factor-4{alpha} mediates alcohol-induced downregulation of intestinal tight junction proteins. *Am J Physiol Gastrointest Liver Physiol* 2010;**299**(3):G643–51.
176. McClain CJ, Su LC. Zinc deficiency in the alcoholic: a review. *Alcohol Clin Exp Res* 1983;**7**(1):5–10.
177. Tang Y, Forsyth CB, Banan A, Fields JZ, Keshavarzian A. Oats supplementation prevents alcohol-induced gut leakiness in rats by preventing alcohol-induced oxidative tissue damage. *J Pharmacol Exp Ther* 2009;**329**:952–8.
178. Tang Y, Forsyth CB, Banan A, Fields JZ, Keshavarzian A. Oats supplementation prevents alcohol-induced gut leakiness in rats by preventing alcohol-induced oxidative tissue damage. *J Pharmacol Exp Ther* 2009;**329**(3):952–8.
179. Hamer HM, Jonkers DM, Bast A, et al. Butyrate modulates oxidative stress in the colonic mucosa of healthy humans. *Clin Nutr* 2009;**28**(1):88–93.
180. Peng L, Li ZR, Green RS, Holzman IR, Lin J. Butyrate enhances the intestinal barrier by facilitating tight junction assembly via activation of AMP-activated protein kinase in Caco-2 cell monolayers. *J Nutr* 2009;**139**(9):1619–25.
181. Koivisto T, Salaspuro M. Effects of acetaldehyde on brush border enzyme activities in human colon adenocarcinoma cell line Caco-2. *Alcohol Clin Exp Res* 1997;**21**(9):1599–605.
182. Elamin E, Masclee A, Troost F, Dekker J, Jonkers D. Activation of the epithelial-to-mesenchymal transition factor snail mediates acetaldehyde-induced intestinal epithelial barrier disruption. *Alcohol Clin Exp Res* 2013;**38**:344–53.
183. Salaspuro VJ, Hietala JM, Marvola ML, Salaspuro MP. Eliminating carcinogenic acetaldehyde by cysteine from saliva during smoking. *Cancer Epidemiol Biomarkers Prev* 2006;**15**(1):146–9.
184. Rao RK. Biologically active peptides in the gastrointestinal lumen. *Life Sci* 1991;**48**(18):1685–704.
185. Karczewski J, Troost FJ, Konings I, et al. Regulation of human epithelial tight junction proteins by *Lactobacillus plantarum in vivo*

and protective effects on the epithelial barrier. *Am J Physiol Gastrointest Liver Physiol* 2010;**298**(6):G851–9.

186. Zyrek AA, Cichon C, Helms S, Enders C, Sonnenborn U, Schmidt MA. Molecular mechanisms underlying the probiotic effects of *Escherichia coli* Nissle 1917 involve ZO-2 and PKCzeta redistribution resulting in tight junction and epithelial barrier repair. *Cell Microbiol* 2007;**9**(3):804–16.
187. Ukena SN, Singh A, Dringenberg U, et al. Probiotic *Escherichia coli* Nissle 1917 inhibits leaky gut by enhancing mucosal integrity. *PLoS One* 2007;**2**(12):e1308.
188. Ewaschuk JB, Diaz H, Meddings L, et al. Secreted bioactive factors from *Bifidobacterium infantis* enhance epithelial cell barrier function. *Am J Physiol Gastrointest Liver Physiol* 2008;**295**(5):G1025–34.
189. Anderson RC, Cookson AL, McNabb WC, et al. *Lactobacillus plantarum* MB452 enhances the function of the intestinal barrier by increasing the expression levels of genes involved in tight junction formation. *BMC Microbiol* 2010;**10**:316.
190. Nardone G, Rocco A. Probiotics: a potential target for the prevention and treatment of steatohepatitis. *J Clin Gastroenterol* 2004;**38**(6 Suppl.):S121–2.
191. Wang Y, Kirpich I, Liu Y, et al. *Lactobacillus rhamnosus* GG treatment potentiates intestinal hypoxia-inducible factor, promotes intestinal integrity and ameliorates alcohol-induced liver injury. *Am J Pathol* 2011;**179**(6):2866–75.
192. Wang Y, Liu Y, Sidhu A, Ma Z, McClain C, Feng W. *Lactobacillus rhamnosus* GG culture supernatant ameliorates acute alcohol-induced intestinal permeability and liver injury. *Am J Physiol Gastrointest Liver Physiol* 2012;**303**(1):G32–41.
193. Forsyth CB, Farhadi A, Jakate SM, Tang Y, Shaikh M, Keshavarzian A. *Lactobacillus* GG treatment ameliorates alcohol-induced intestinal oxidative stress, gut leakiness, and liver injury in a rat model of alcoholic steatohepatitis. *Alcohol* 2009;**43**(2):163–72.
194. Kirpich IA, Solovieva NV, Leikhter SN, et al. Probiotics restore bowel flora and improve liver enzymes in human alcohol-induced liver injury: a pilot study. *Alcohol* 2008;**42**(8):675–82.
195. Nosova T, Jousimies-Somer H, Jokelainen K, Heine R, Salaspuro M. Acetaldehyde production and metabolism by human indigenous and probiotic *Lactobacillus* and *Bifidobacterium* strains. *Alcohol Alcohol* 2000;**35**(6):561–8.

CHAPTER

15

Cholesterol Regulation by Leptin in Alcoholic Liver Disease

Balasubramaniyan Vairappan, PhD

Department of Biochemistry, Jawaharlal Institute of Postgraduate Medical Education and Research (JIPMER), Dhanvantari Nagar, Pondicherry, India

INTRODUCTION

Alcohol abuse is an acute health problem throughout the world. The word alcohol is derived from Arabic *al-kohl* meaning "subtle" (finely divided spirit). Alcohol is an organic compound possessing hydroxyl group (polar) attached to an alkyl group (nonpolar). Chronic heavy alcohol consumption can lead to alcoholic liver disease (ALD) that is a major cause of morbidity and mortality worldwide, and the prevalence is currently alarming high in India.[1] The pathophysiology of this illness is poorly understood, and two-third of patients with advanced ALD die within 2 years of diagnosis.[2] Hence, it is important to understand the pathogenetic mechanisms, and to identify urgently new targets for therapy of this debilitating disease. Clinical manifestations of ALD include fatty liver (also called hepatic steatosis), steatohepatitis (steatosis coupled with inflammation), fibrosis (deposition of collagen and extracellular matrix), and cirrhosis (fibrosis with nodular regeneration) (Figure 15.1). Furthermore, chronic heavy alcohol consumption is an established risk factor for the development of hepatocellular carcinoma (HCC) in cirrhosis.[3] In the developed nations, ALD is a main consequence of end stage ailments that require liver transplantation. The majority of chronic heavy drinkers develop fatty liver; the condition is associated with the accumulation of lipid droplets (mainly, triglyceride (TG)) primarily in the hepatocytes,[4] and is reversible if a person reduces alcohol consumption.[5] It is estimated that daily ethanol consumption exceeding 40–80 g/day for males and 20–40 g/day for females, for a decade, may cause ALD.[6,7] Furthermore, chronic heavy alcohol consumption is also associated with other disorders unrelated to the liver, such as infections, malignancies, metabolic syndrome, cardiovascular events, neurological disorders, acute pancreatitis, and kidney injury.[8] In contrast, moderate alcohol consumption has a protective effect on coronary heart disease.[9]

ALCOHOL METABOLISM

Alcohol metabolism changes the redox state of the liver that leads to alterations in hepatic lipid, carbohydrate, protein, lactate, and uric acid metabolism.[10] Indeed, the molecular mechanisms that account for these alterations are not completely well understood.[6] Ingested alcohol is eliminated principally through its metabolic degradation, via multiple enzymatic pathways, such as alcohol dehydrogenase (ADH), cytochrome P4502E1 (CYP2E1), and catalase. Although many organs show ethanol-metabolizing properties, more than 90% of ethanol is metabolized into acetaldehyde in the liver, primarily, in the area near the central vein.[11] Acetaldehyde, a reactive aldehyde, can lead to oxidation of lipids and nucleic acids, as well as the formation of protein adducts, and is subsequently oxidized into acetate by aldehyde dehydrogenase (ALDH) in the liver.[12] Indeed, the most part of acetate is converted to CO_2 and H_2O, before leaving the hepatocytes. Furthermore, CYP2E1 mediated alcohol metabolism leads to the generation of reactive oxygen species (ROS) that elicit multiple deleterious effects on hepatocytes, including dysregulation of fatty acid (FA) synthesis and oxidation.[11]

EFFECTS OF ALCOHOL ON LIPID METABOLISM

Alcohol, particularly in large doses, is a direct hepatotoxin, but it can also act as a "permissive" agent that causes liver injury in association with a wide variety of

Molecular Aspects of Alcohol and Nutrition. http://dx.doi.org/10.1016/B978-0-12-800773-0.00015-X

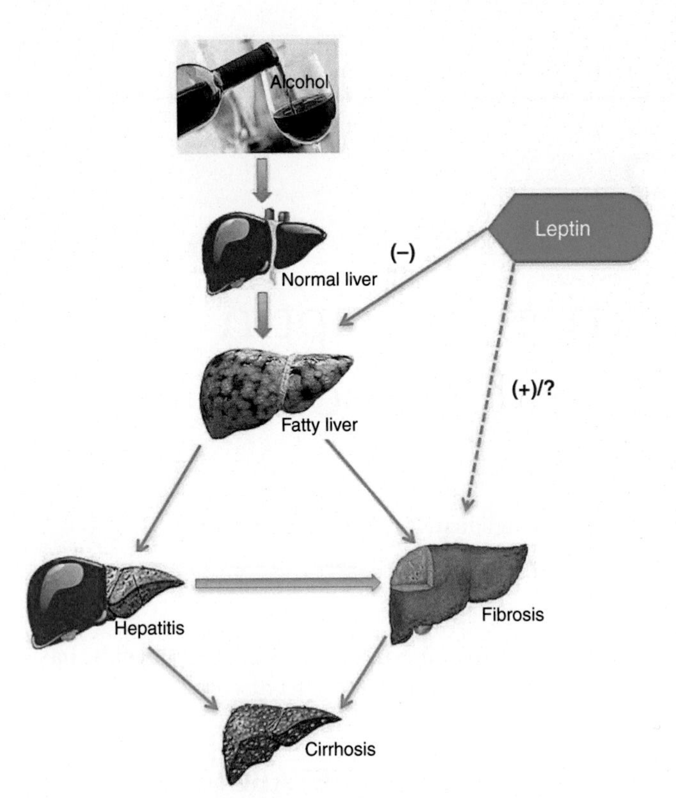

FIGURE 15.1 Schematic representation of the effect of leptin on inhibiting alcohol-induced fatty liver disease. Note: (+) denotes stimulation; (−) denotes inhibition.

other factors, such as various forms of cancer, brain damage, and fetal injuries during pregnancy. The accumulation of fat in the liver, in response to chronic ethanol consumption, can lead to more severe forms of liver injury.[13] This has resulted in an increase in the incidence of alcohol-related disorders, and is also a cause of increased morbidity and mortality in alcoholics.

Alcohol intake results in a wide spectrum of hepatic dysfunction in humans and experimental animals. Although the association between heavy alcohol intake and ALD has been well established, the precise molecular mechanisms that underlie the progression of liver injury are still not completely understood. However, enhanced hepatic lipogenesis has been proposed as an important biochemical mechanism. Lipids are a group of naturally occurring small hydrophobic molecules that play a multitude of crucial roles. Furthermore, lipids act as structural elements in biological membranes they store energy, and they function as signaling molecules in cellular pathways.[14] Lipid homeostasis is maintained by balanced lipid synthesis, catabolism (mainly, through β-oxidation), and secretion.[6] Hepatic lipid uptake is a function of substrate delivery and transport into the hepatocyte, and several genetic models exemplify this aspect of hepatic lipid metabolism.

The accumulation of fat within the hepatocytes, following heavy alcohol intake, leads to the most common disruptions of lipid metabolism. Alcohol-induced fatty liver disease is often associated with hyperlipidemia. The accumulation of lipids, both in the liver and blood, are multifactorial phenomena, and develop gradually during the first month of ethanol administration, and persisted thereafter for at least 1 year for a rat,[15] and 3 years for a baboon.[16] Our earlier studies have also shown evidence that chronic alcohol supplementation to mice, over a period of 7–8 consecutive weeks, developed fatty liver (it increased both blood and tissue lipid levels), with mild inflammation, when compared to isocaloric glucose fed control mice.[17–19] In this context, nutritional competence of the given diets to control mice was exhibited by their ability to maintain normal growth and liver function. Furthermore, alcohol-fed animals have also shown to gain weight, though the growth-promoting ability of ethanol calories was smaller than that of carbohydrate.[20] In fact, isocaloric replacement of carbohydrate by fat or administration of carbohydrate-deficient diets did not reproduce the effect of ethanol.[21] Thus, the development of fatty liver is due to ethanol exposure, and not to the manipulation of the other sources of calories.[21]

Ethanol is a powerful inducer of hyperlipidemia, including hypertriglyceridemia and hypercholesterolemia, in both animals and humans.[17–19] Ethanol supplementation to mice also causes changes in lipoprotein metabolism.[17–19] The interaction of ethanol with lipid metabolism is complex. When ethanol is present, it becomes the preferred fuel for the liver, and displaces fat as a source of energy, blocking fat oxidation and favoring fat accumulation.[22] In this context, our earlier report shows that plasma and tissue lipid levels increased, following alcohol supplementation to mice,[17] consistent with other reports, where the administration of ethanol to rats changes the metabolism of serum and tissue lipids.[23] Our observations also showed the significant increase of plasma, liver, brain, and kidney lipid profile in alcohol-treated mice.[17]

TRIGLYCERIDES

TGs are the predominant molecules of energy storage in eukaryotes. Elevated blood TG has been identified as an independent risk factor for many cardiovascular diseases, and in patients with ALD. Substantial TG accumulation in the hepatocytes is strongly associated with fatty liver, and the subsequent progression of fibrosis/cirrhosis and HCC. TG synthesis is primarily catalyzed by the acyl-CoA-diacylglycerol acyltransferase (DGAT EC; 2.3.120).[24,25] Two isoforms of DGAT (1 and 2) have been identified from a distinct gene family, and are widely expressed in human and rodent liver.[24,25] Hence, it is of great importance to have a better understanding of the molecular mechanism of DGAT in health and disease. The evidence suggested that the two enzymes play different roles in TG metabolism, with DGAT2 participating in steatosis,[26] and DGAT1 in VLDL synthesis.[27] Overexpression of liver-specific DGAT2 in mice results in hepatic steatosis.[26] Furthermore, upregulation of hepatic DGAT2 by chronic alcohol exposure contributes to the development of ALD.[26] Conversely, inhibition of DGAT2 with antisense oligonucleotides reverses hepatic steatosis in ob/ob mice, and in mice challenged with high-fat diet,[28,29] suggesting that this enzyme plays a critical role in the development of alcohol induced fatty liver disease. Support for this view comes from the observation that our unpublished previously analyzed global gene expression profile (DNA microarray analysis) of control and fibrotic mice, showing altered *Dgat1* and *Dgat2* genes in fibrotic mice (0.6 and 2.61 log fold changes, compared to control mice, respectively). Neutral fat (mainly TG) deposition is the initial stage of ALD; it plays a critical role in disease progression.[26] We found significantly increased plasma and tissue concentrations of TG, following chronic heavy alcohol supplementation to mice, compared to control mice.[17] Surplus TG storage is the hallmark of obesity and other cardiovascular metabolic diseases. A tight regulation between TG synthesis, hydrolysis, secretion, and FA oxidation is mandatory to prevent lipid accumulation, as well as lipid depletion from hepatocytes.[26] Therefore, understanding the TG metabolic pathway is considered vital for the development of new therapies to attenuate pathophysiological conditions associated with excessive hepatic TG accumulation, in ALD.

FATTY ACIDS

FAs are the key components of most lipids of biological importance. Among the FAs, saturated FAs are an essential source of energy, while polyunsaturated fatty acids (PUFA) that are abundant in phospholipids play an important role in determining the functional and structural integrity of the cell membrane. Any change in the composition of these PUFA has a direct relationship in determining the structural integrity of the cell membrane. Alcohol toxicity facilitates esterification of the accumulated FA to TG and cholesterol esters (CE), all of which accumulate in the liver, and are disposed of, in part, as serum lipoprotein, resulting in moderate hyperlipemia.[21] Indeed, the precise molecular mechanisms remain obscure, and it is usually accepted that elevated liver *de novo* FA synthesis has a prime role in the progression of fatty liver. Obesity, in parallel with insulin resistance, is thought to be crucial for its development. Moreover, an increased supply of free fatty acids (FFA) due to obesity and enhanced *de novo* lipogenesis, from extra-hepatic insulin resistance, led to hepatic fat accumulation.[30] Hepatic steatosis results from an imbalance between intrahepatic TG production and removal. Both uptake of FA to the liver and *de novo* synthesis contribute to hepatic TG production, whereas FA β-oxidation, and formation of very low density lipoprotein (VLDL) particles, contribute to hepatic TG removal. Mavrelis et al. observed higher levels of FFA in the liver of both alcoholics and obese subjects.[30] They also found that unsaturated FA predominated in the TG fractions, while saturated FA predominated in the FFA fractions, and postulated that these FFA could be hepatotoxic, and contribute to the liver damage of obese individuals and alcoholics.[30] This agrees with our previous studies showing evidence of significantly increased FFA accumulation in the plasma and liver of alcohol fed mice, compared to controls.[17]

Alcohol abuse is also known to induce alterations in the FA composition in the plasma and tissues.[31] It has been shown that short-term ethanol administration decreased linoleic acid (18:2) and arachidonic acid (20:4), but increased oleic acid (18:1) in the liver.[32] In addition, chronic alcohol consumption increases erythrocyte stearic acid (18:0) and palmitic acid (16:0).[33] Our study showed significant increase in the percentage of palmitic acid (16:0), stearic acid (18:0), oleic acid (18:1), and docosapentaenoic acid (22:5) levels, observed in ethanol supplemented mice, whereas the percentage of palmitoleic acid (16:1) and arachidonic acid (20:4) were significantly decreased in the liver of ethanol-treated mice (unpublished data). Ethanol modifies the membrane bilayer by interacting with phospholipids, thus increasing the membrane fluidity. Acetaldehyde, the major metabolite of ethanol, is more liposoluble than ethanol, and thus capable of producing membrane perturbations. This alteration in membrane structure is due to a reduction of polyunsaturated fatty acid content.[34] Moreover, studies have shown that ethanol ingestion alters the composition of liver FA, depending on the FA composition of the dietary fat.[35] In this context, Nanji et al. have observed that the changes in the FA composition in ethanol fed rats have been influenced by a high fat diet containing linoleic acid.[36] These changes in FA correlate with the mitochondrial damage in ALD.

HEPATIC LIPID METABOLISM IS CONTROLLED BY SEVERAL MASTER TRANSCRIPTION FACTORS

PPAR-α

PPAR-α is a nuclear hormone receptor involved in regulating lipid metabolism and transport, FA oxidation, and glucose homeostasis.[37] PPAR-α is predominantly expressed in cells or tissues capable of oxidizing FA, including hepatocytes, but not in Kupffer cells and hepatic stellate cells (HSC).[38,39] When activated, it binds as a heterodimer with retinoid X receptor (RXR) to peroxisome proliferator response elements, in genes involved in the FA oxidation pathways.[40] Moreover, activation of PPAR-α target genes, such as liver type fatty acid binding protein (L-FABP) and acyl-CoA dehydrogenase (ACAD), results in increased uptake and oxidation of FFA, increased TG hydrolysis, and upregulation of ApoA I&II.[37] The net effect is increased FA oxidation, decreased serum TG, a rise in high-density lipoprotein (HDL), and an increase in cholesterol efflux.[6] Nakajima et al. found that PPAR-α knockout mice fed ethanol developed marked hepatomegaly, steatohepatitis, liver cell death and proliferation, and portal fibrosis.[41] Moreover, liver lesions induced by alcohol in these mice also reflect a reduced ability to catabolize very-long-chain FAs and their metabolites, due to an inability to upregulate the FA oxidation systems in the liver, in the absence of PPAR-α.[42] Ethanol infusion also causes downregulation of PPAR-α, and ethanol fed PPAR-α null mice developed hepatocyte damage that is not found in wild type mice.[43] In this context, PPAR-α agonists, such as fibrate drugs, lower blood TG and glucose level, and are used as therapeutic agents in the treatment of metabolic syndromes by stimulating its receptor activity.[44] Furthermore, treatment of ethanol fed animals with PPAR-α agonists reduces the toxic effects of ethanol, and reverses hepatic fat accumulation.[43] In fact, fish oil contains *n*-3 FA, a known PPAR-α activator shown to prevent acute ethanol induced fatty liver.[45]

AMPK

AMP-activated protein kinase (AMPK) is an energy sensor that senses cellular stresses (oxidative stress and reduced energy charge), increases the activity of the major energy-generating pathways (glycolysis and FA oxidation), and downregulates energy-demanding processes (FA and cholesterol synthesis, protein synthesis).[40] Activation of AMPK increases FA oxidation and inhibits synthesis, whereas inhibition of AMPK blocks FA oxidation and promotes FA synthesis.[40,46] Its activation also suppresses the expression of key lipogenic enzymes, such as acetyl-CoA carboxylase (ACC), and fatty acid synthase (FASN).[47] ACC is a key regulatory enzyme also involved in FA synthesis.[48] Phosphorylation of hepatic ACC *in vivo* is mostly achieved by AMPK, rendering the enzyme inactive.[48,49] Alteration of ACC activity is also essential for the control of carnitine palmitoyltransferase I (CPT-I), a key enzyme of hepatic ketogenesis. CPT I inhibition is made by malonyl-CoA, the product of the reaction of ACC, so that coordinate control of synthesis and oxidation of FA is achieved.[50] Malonyl-CoA is degraded by malonyl-CoA decarboxylase that is activated by AMPK. In this context, Guzman et al. reported that ethanol feeding to rats reversibly decreases hepatic CPT-1 activity, and increases enzyme sensitivity to malonyl-CoA.[51] Furthermore, it regulates lipid synthesis both by direct effects on sterol regulatory element-binding protein (SREBP)-1c, and through phosphorylation and inhibition of ACC and 3-hydroxy-3-methylglutaryl (HMG)-CoA reductase.[52] Thus, compounds that specifically inhibit the SREBP pathway, while activating AMPK, can decrease the biosynthesis of both cholesterol and FA, and can be useful in the treatment of alcoholic fatty liver. Further, it has been reported that chronic ethanol ingestion impaired the regulation of FA metabolism by decreased activity of AMPK, and reduced enzyme sensitivity to variation of the AMP/ATP ratio in the liver.[53]

Sterol Regulatory Element-Binding Proteins

SREBPs belong to the basic-helix–loop–helix leucine zipper class of transcription factors. Low levels of sterols induce their cleavage, forming a water-soluble N-terminal domain containing a bHLH-Zip motif that is translocated to the nucleus.[54] SREBPs regulate the transcription of more than 30 genes that encode participants in the biosynthesis of cholesterol, FAs, and TGs.[54] Two isoforms of SREBPs (1 and 2) were identified, with distinct properties. Although SREBP-1a regulates genes related to lipid and cholesterol synthesis, its activity is regulated by sterol levels. SREBP-1c regulates genes required for glucose metabolism, and FA and lipid production.[54] Furthermore, hepatic and adipose over-expression of SREBP-2 led to a significant increase in the accumulation of cholesterol.[55] Earlier studies noted that ethanol induces transcription of SREBP-regulated promoters, via increased concentrations of mature SREBP-1 protein.[56,57] Karasawa et al. observed that SREBP-1 deficiency in the model of accelerated atherosclerosis in low-density lipoprotein (LDL) receptor-deficient ($ldlr^{-/-}$) mice prevented Western diet-induced hyperlipidemia, and attenuated atherosclerosis without altering liver TG and cholesterol levels.[58] SREBP-2 synthesized as a 125 kDa protein that regulates genes involved in cholesterol metabolism (e.g., those encoding HMG-CoA synthase, HMG-CoA reductase, and squalene synthase, and the LDL receptor).[59] Chronic alcohol feeding resulted in enhanced SREBP-2 activation, manifested by an increased mature form of SREBP-2 proteins in the liver of alcohol-fed rats, and this was associated with increased HMG-CoA gene expression.[56] Our data also showed evidence that increased SREBP-2 protein expression that was found in the liver of ethanol treated mice, correlated with increased hepatic HMG-CoA reductase activity,[60] and this indicated that alcohol enhanced the intracellular cholesterol biosynthesis pathway.

CHOLESTEROL

Cholesterol originates from circulating plasma lipoproteins that contain both unesterified (free) cholesterol (FC), and cholesteryl ester (CE). Cholesterol is an essential component of cell membranes that regulates their structure and function, and plays an important role in signal transduction. Mitochondria are cholesterol-poor organelles with estimates ranging from 0.5% to 3% of the content found in other cellular membranes.[61] The limited amount of cholesterol in mitochondrial membranes, however, plays an important physiological role in specific tissues where cholesterol is used for steroidogenesis and synthesis of steroid hormones.[62] The liver plays a central role in the regulation of cholesterol homeostasis. HMG-CoA reductase, a rate-limiting enzyme, is primarily involved in cholesterol biosynthesis. In nonalcoholic steatohepatitis (NASH) patients, both HMG-CoA reductase activity and /or expression were shown to be significantly increased, and were related to hepatic FC accumulation and the severity of liver histology.[63] The increased HMG-CoA reductase expression in NASH also derived from enhanced gene transcription by SREBP-2, the principal transcriptional activator of HMG-CoA reductase. In the hepatocyte, cholesterol exists as FC, and as CE. The conversion of FC to CE, and hepatic esterification of FC, plays a significant role in protecting the hepatocytes from FC accumulation.[64] Previous studies have shown evidence that ethanol feeding increased hepatic esterified cholesterol levels in rats, and this was associated with both enhanced cholesterol biosynthesis, and decreased BA excretion.[65] Liver cholesterol esterification is primarily catalyzed by the enzyme Acyl-CoA cholesterol O-acyl transferase (ACAT). The majority of this enzyme activity was found to be localized to RNA rich microsomes.[66] Thus, the newly formed CE is stored in smooth microsomes and the cytoplasm, thought to represent an inert storage pool of cellular cholesterol.[67] ACAT 2, a key tissue cholesterol esterifying enzyme, is primarily found within lipoprotein producing cells, including hepatocytes.[68] ACAT2 fosters cholesterol absorption by the intestine, and the secretion of CE-enriched VLDL by the liver.[69] Previous studies have shown that ACAT2 plays a critical role in the production of atherogenic apoB-containing lipoproteins,[70] and its specific inhibitors are extremely effective in preventing atherosclerosis. However, ACAT2-deficient mice show unexpected hypertriglyceridemia, thus the alteration of increased production or decreased clearance of VLDL TG occurred in these mice.[71] CE hydrolysis also represents a potential regulatory process that would affect cellular FC concentrations. The intracellular free and ester cholesterol concentrations are regulated by two enzymes, cholesterol ester synthase (CES) and cholesterol ester hydrolase (CEH). CES is responsible for the intracellular esterification of cholesterol and, thus, its storage. CEH hydrolyzes the cholesterol esters, liberating free cholesterol that may be *in situ* incorporated into the lipoproteins (VLDL, HDL), or may serve as a substrate for bile acid synthesis. We previously found elevated activity of CES, and reduced activity of CEH in ethanol fed mice, resulting in increased esterification and decreased intracellular hydrolysis of EC,[60] and this may be responsible for the reduction in plasma FC concentrations in these mice. Hence, cellular cholesterol under the influence of ACAT and CEH can potentially undergo a constant cycling between the free and ester forms.

Effect of Ethanol on Cholesterol Metabolism

ALD is a major cause of end-stage liver disease that requires liver transplantation in developed countries, as well as developing nations. Over the past few decades, significant advancement has been made in our understanding of the molecular mechanisms underlying the pathogenesis of ALD. Importantly, induction of hepatic cholesterol accumulation promoted steatohepatitis and fibrosis,[72] while reduction of hepatic cholesterol overload attenuated liver disease severity.[63] Thus, proper coordination of cholesterol biosynthesis and trafficking is essential to human health. Hepatic lipid abnormalities seen after chronic alcohol intoxication induce increased levels of FC, CE, and FA esters particularly – the fatty acyl composition of membrane phospholipids.[73] Moreover, increase in plasma ACAT activity after ethanol feeding correlated well with increased levels of esterified cholesterol.[74] Our earlier study has shown evidence that chronic ethanol feeding to mice resulted in significant elevation of hepatic cholesterol and plasma total, and of free and ester cholesterol concentrations.[60] Chronic heavy ethanol consumption also affects multiple lipid metabolic pathways in the liver, such as stimulating *de novo* lipogenesis, enhancing FA uptake, and suppressing FA oxidation.[17,75,76] ACAT and HMG-CoA reductase were significantly increased in the ethanol fed rat liver.[77] Thus, chronic ethanol administration caused a marked accumulation of hepatic CE, and was associated with a significant increase in the activities of enzymes ACAT and HMG-CoA in the liver.

In NAFLD patients, the development of NASH and fibrosis paralleled hepatic FC accumulation.[63,78] Moreover, inhibition of the catabolism of cholesterol to bile salt may contribute to hepatic accumulation and hypercholesterolemia.[6] In rabbits, dietary cholesterol has been suggested to cause stellate cell activation, leading to perisinusoidal fibrosis.[79] A previous human study has shown evidence that dietary cholesterol is a critical factor in the development of steatohepatitis.[78] In this context, the National Health and Nutrition Examination Survey demonstrated that higher dietary cholesterol consumption independently predicted a higher risk of cirrhosis.[80] Increased hepatic cholesterol accumulation, following alcohol toxicity, plays a key role in the development and progression of ALD. Furthermore, elevated cholesterol synthesis and secretion in human hepatomal cells was observed following ethanol administration.[81] Wang et al. reported that chronic alcohol consumption disrupted cholesterol homeostasis in rats.[82] It has been previously reported that ethanol fed rats show increased hepatic esterified cholesterol levels that is associated with both enhanced cholesterol biosynthesis, and decreased bile acid excretion.[65]

Bile Acids

Bile is the main route for cholesterol elimination from the body, and reverse cholesterol transport (RCT) is a key metabolic pathway for the movement of extra cholesterol from peripheral tissues to the liver, for biliary secretion. Bile acids (BA) are vital metabolic signaling molecules, derived during hepatic cholesterol metabolism. They can modulate lipid, glucose, and energy metabolism.[83] The key rate limiting enzyme cholesterol 7α hydroxylase (CYP7A1) is involved in the conversion of hepatic cholesterol into BA, and thus leads to cholesterol elimination from the human body. The CYP7A1 gene is highly regulated via numerous signaling pathways.[84] Previously, it has been documented that upregulation of hepatic CYP7A1 expression, following short-term cholesterol feeding to rodents, resulted in an increased BA formation, and normal hepatic cholesterol homeostasis.[85] Increased hepatic CYP7A1, in response to dietary cholesterol, is mediated by the oxysterol sensor, liver X receptor α (LXR α). Further, LXR α-deficient mice fail to upregulate hepatic CYP7A1 expression, in response to dietary cholesterol leading, thus, to massive hepatic cholesterol accumulation.[86] Interestingly, rodent, but not human, CYP7A1 gene has the promoter region where the LXR response element can bind and activate, so humans may not upregulate hepatic CYP7A1 in response to dietary cholesterol.[87] In this context, Xie et al. reported that ethanol consumption altered expression of genes related to BA metabolism, and BA transport in the liver and the ileum.[88] Furthermore, ethanol-induced fatty liver is accompanied by alterations in the BA profile and the composition of the gut microbiome.[89] These changes in the enterohepatic circulation of BA are important not only for feedback inhibition of BA synthesis, but also for whole-body lipid homeostasis.[90]

Humans synthesize two primary bile acids, such as cholic acid (CA), and chenodeoxycholic acid (CDCA). These BA are synthesized by the neutral and the acidic pathways.[91,92] The initial step of the neutral pathway is a hydroxylation at the 7a-position of cholesterol, a reaction catalyzed by the rate-limiting enzyme CYP7A1. The acidic pathway begins with the oxidation of the cholesterol side chain, in which the first step is catalyzed by sterol 27-hydroxylase (CYP27A1).[93] Both *in vivo* and *in vitro* studies have shown that the "acidic" pathway may be responsible for up to 50% of bile acid synthesis.[94,95] Previous studies have documented that expression of CYP7A1 for the "neutral" pathway of BA synthesis in the liver was significantly up regulated in ethanol-treated rats.[88] Furthermore, ethanol ingestion in a normal healthy person was shown to increase plasma CYP7A1 and 7a-hydroxy-4-cholesten-3-one levels.[96] Ethanol supplementation to rats for 4–7 weeks also showed elevated HMG-CoA reductase activity, while hepatic CYP7A1 activity was decreased.[97] Xie et al. also stated that ethanol caused increased expression of genes involved in BA biosynthesis, efflux transport, and reduced expression of genes regulating bile acid influx transport in the liver.[88]

OBESITY, LEPTIN, AND ALD

Obesity

Human obesity is associated with increased rates of hepatic cholesterol and BA synthesis,[98] both of which are decreased in obese subjects during weight loss.[99] Obesity is known to predispose individuals to liver disease by increasing hepatic sensitivity to endotoxin.[100] An earlier study pointed out that obesity is strongly associated with steatohepatitis that often progresses to cirrhosis, in alcoholic individuals, giving a clue that the progression of hepatic ailments related to obesity might be similar to that of ALD.[100] Although obese people show obvious hyperleptinemia,[101] it is still unclear whether an increase in systemic leptin concentrations is involved in the progression of liver diseases. Clinical studies have disclosed obesity as a risk factor in the pathogenesis of chronic hepatic diseases, including ALD.

Leptin

Leptin (from the Greek, *leptos* meaning thin) was identified by positional cloning in 1994[102] as a key hormone in the regulation of body weight and energy balance. Leptin is a 167 amino acid secreted protein, encoded by the ob gene. It is predominantly expressed by adipocytes, and its plasma levels correlate well with the body fat mass.[103] Wang et al. observed that leptin is expressed in several areas, including liver.[104] Leptin was also reported to have intense effects on lipid and carbohydrate metabolism. Compelling evidence also indicates that leptin suppresses glucose production and *de novo* lipogenesis, and also induces FA oxidation in the liver.[105–107] In addition, it also induces glucose uptake and FA oxidation in the skeletal muscle, thereby preventing lipid accumulation in nonadipose tissues, and increasing insulin sensitivity.[106,108] Recombinant leptin has been used in clinical trials for common types of obesity.[109,110] Furthermore, leptin therapy is increasingly being used in human disorders of congenital and acquired leptin deficiency, with, in some cases, dramatic therapeutic impact on dyslipidemia.[111,112] Leptin treatment also promotes biliary cholesterol elimination, during weight loss in ob/ob mice,[113] and impaired biliary lipid secretion in obese Zucker rats.[114] In this context, our earlier study has shown evidence that leptin administration significantly attenuated alcohol induced hyperlipidemia.[17–19,60] Further evidence also pointed out that hepatic steatosis and plasma dyslipidemia induced by sucrose feeding is rapidly downregulated, following leptin administration.[115] Our results were also supporting the previous notion that leptin regulates alcohol induced fatty liver by virtue of its lipid lowering effect.[17–19,60] Although leptin has a profound effect on regulating alcohol induced or other fatty liver diseases, it was also shown to have profibrogenic responses in rodents exposed with hepatotoxin; however, the mechanisms are not entirely understood (Figure 15.1).

Prime Involvement of Leptin on Regulation of Lipid Metabolism

Lipid metabolism is regulated by several hormones, and leptin is considered one of them. It is a balance between lipid synthesis and degradation that determines fat mass. Over 90% of total energy reserves are stored in adipocytes, such as TG, that can be hydrolyzed (lipolysis) following hormonal stimulation to release FA. FA has two possible fates: β-oxidation to produce ATP, or re-esterification back into TG. Many studies have shown that leptin has a direct autocrine or paracrine mode of action on lipid metabolism.[104,107,116] Leptin appears to mediate FA metabolism by changing enzyme mRNA levels and concentration. For example, the presence of leptin inhibits the expression of acetyl-CoA carboxylase (ACC) in adipocytes,[117] a rate-limiting enzyme for long chain FA synthesis, and is essential for the conversion of carbohydrates to FA, and caloric storage as TG. This stimulation of FA oxidation is probably the key event for the tissue lipid lowering and insulin-sensitizing effects of leptin. This was demonstrated recently to occur through direct or indirect (via either the central nervous system, or a putative inhibition of stearoyl-CoA desaturase-1 activity) stimulation of AMP that inactivates ACC, and decreases malonyl-CoA concentration, thus stimulating CPT 1 mediated FA oxidation in the mitochondria.[116,118] Thus, the lipolytic effect of leptin was well documented and has been regarded as an antisteatotic hormone (Tables 15.1 and 15.2).

TABLE 15.1 Effects of Leptin on Different Key Enzymes of Lipogenesis, Lipolysis, and FA Oxidation and Cholesterol Metabolism

Effects	Cell type or tissue	References
DECREASED LIPOGENESIS		
Inhibition of acetyl-CoA carboxylase	Muscle	[106]
Acetyl-CoA carboxylase mRNA↓	Preadipocytes, aortic endothelial cells	[119]
FA acid synthase mRNA↓	Adipocytes, liver	[104,120,121]
Glycerol-3-phosphate acyltransferase (GPAT) mRNA↓	Pancreatic islets	[122,123]
INCREASED LIPOLYSIS		
Hormone-sensitive lipase mRNA↑	Adipocytes	[121]
INCREASED FA OXIDATION		
Acyl-CoA oxidase mRNA↑	Adipocytes, pancreatic islets	[104,122,123]
Carnitine palmitoyl transferase-1 mRNA↑	Adipocytes, pancreatic islets, aortic endothelial cells	[104,106,119–121]
Indirect disinhibition of carnitine palmitoyltransferase (its inhibitor malonyl-CoA decreases as a consequence of inhibition of acetyl-CoA carboxylase)	Muscle	[104,106,119–121]
LIPOPROTEIN, CHOLESTEROL, AND BILE ACID SYNTHESIS		
Lecithin cholesterol acyl transferase↑	Plasma	[17–19,60]
Lipoprotein lipase↑	Heart, liver	
HMG-CoA reductase↓	Heart, liver	
Cholesterol ester synthase↓	Heart, liver	
Cholesterol ester hydrolase		
Cholesterol 7a hydroxylase		

TABLE 15.2 Effect of Leptin and Ethanol on Hepatic Gene or Enzymes Involved in Cholesterol and Bile Acids Synthesis

Gene/enzymes involved in cholesterol and bile acid metabolism	Abbreviations	Effect of ethanol	Effect of leptin	References
3-Hydroxy-3-methylglutaryl-coenzyme A reductase	HMG-CoA reductase	Increased	Decreased	[60]
Cholesterol ester synthase	CES	Increased	Decreased	[60]
Cholesterol ester hydrolase	CEH	Decreased	Increased	[60]
Acyl-CoA cholesterol O-acyl transferase	ACAT	Upregulated	Unchanged	[74,113]
Sterol regulatory element-binding protein	SREBP-2	Increased	Decreased	[60]
Cholesterol 7α hydroxylase	CYP7A1	Increased/decreased	Upregulated	[96,97,124]
Sterol 27-hydroxylase	CYP27A1	Unchanged	Upregulated	[88,93,114,124]

Effect of Exogenous Leptin on Ethanol-Induced Hypercholesterolemia

Leptin treatment administered to ethanol intoxicated mice resulted in significant lowering of hepatic and plasma cholesterol formation. Emerging data now indicates that leptin may be an important regulator of RCT.[125] Considering that adipose tissue is a major cholesterol storage depot in the body,[126] depletion of TG by leptin necessitates transport of excess cholesterol to the liver, for elimination via bile.[113] Furthermore, leptin has shown to increase the activity of CYP7A1, a rate-limiting enzyme involved in the conversion of cholesterol into bile.[125] In our previous study, we found leptin administration results in the significant decrease of CES activity, whereas CEH activity was found significantly upregulated in ethanol fed mice, compared to ethanol alone fed mice.[60] Furthermore, we also observed that increased hepatic HMG-CoA reductase activity, and elevated hepatic SREBP-2 protein expression in alcohol fed mice was significantly decreased, following treatment with leptin, implementing the regulatory role of leptin in cholesterol metabolism.[60] Considering that leptin promotes hepatic uptake of plasma cholesterol,[127] partly due to upregulation of scavenger receptor-BI (SR-BI),[125] an increased flux of cholesterol into the liver would provide a mechanistic explanation for the observed decrease in HMG-CoA reductase activity in the liver of alcohol fed mice, after leptin administration. It is also important to note that hepatocytes express leptin receptors[118,128] that mediate insulin like signaling.[118] It has also been shown, in primary hepatocyte cultures, that leptin enhances the inhibitory effects of insulin on glycogenolysis and hepatic glucose production.[128] Moreover, in hepatocytes, insulin treatment has shown to downregulate CYP7A1, while upregulating HMG-CoA reductase.[108,115,125] Thus, exogenous leptin administration may indirectly downregulate cholesterol synthesis through insulin-mediated action. Van Patten et al. have shown evidence that leptin administration to mice resulted in a marked decrease of both BA pool size and hydrophobicity, accompanied by a pronounced decrease in cholesterol absorption.[114] It is postulated that leptin induces the catabolism of cholesterol in the liver, thus reducing the plasma cholesterol concentration, which further confirmed, with other findings, that leptin enhances the activity of CYP7A1 and stimulates BA production.[124] Thus, administration of exogenous leptin markedly reduced alcohol-induced hepatic hypercholesterolemia by virtue of decreased synthesis and increased catabolism of cholesterol, and could be a new potential therapeutic target for the management of alcohol induced hepatic hypercholesterolemia (Figure 15.2).

Possible Other Nutritional Therapies for Alcohol-Induced Hypercholesterolemia

Folic acid supplementation was shown to increase bile flow, BA synthesis from cholesterol, and BA excretion via feces, thus provoking a decrease in serum and hepatic cholesterol. Thus, folic acid supplementation is beneficial to alcoholic patients.[89] Moreover, Ojeda et al. have shown, in ethanol induced hypercholesterolaemic rats, that folic acid supplementation reduced HMG-CoA reductase activity and associated cholesterol synthesis, and increased catabolism.[129] Moreover, the other naturally occurring nutritious approach using persimmon, a fruit rich in vitamin C and tannins, in combination with vinegar, prevents metabolic disorders induced by chronic administration of alcohol.[130] In this context, it was shown previously in NAFLD, that oral administration of atorvastatin + vitamin C + vitamin E reduces HMG-CoA reductase activity.[131] (Table 15.3). Germinated brown rice extracts, which contain a high level of γ-aminobutyric acid (GABA) were shown to protect from ethanol-induced chronic liver diseases by increasing serum and liver high-density lipoprotein cholesterol (HDL-C) concentrations, and decreasing ethanol-induced elevated hepatic TG and TC levels.[132] In addition, Marmillot et al. studied the effect of dietary

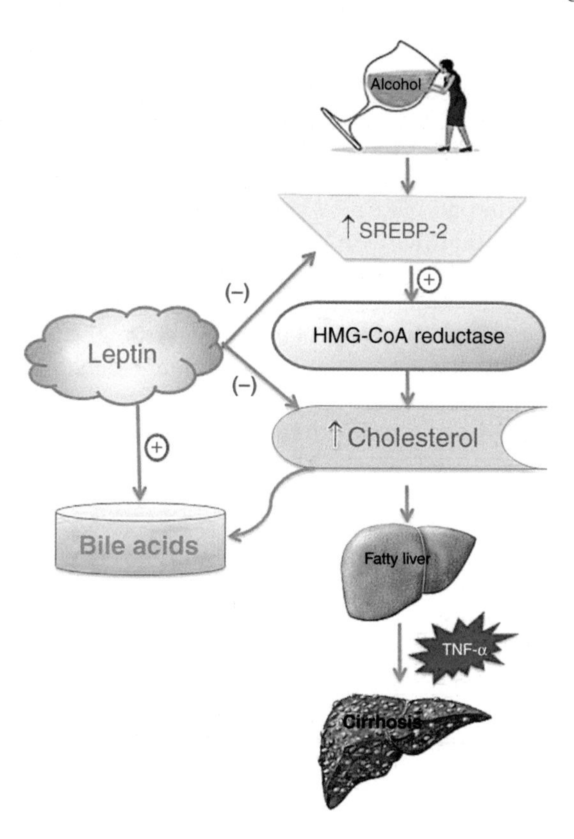

FIGURE 15.2 **Proposed role of leptin in regulating cholesterol synthesis in alcohol-induced fatty liver disease.**[60] Chronic alcohol consumption results in elevated hepatic master transcription factor, such as SREBP-2 that increases hepatic cholesterol concentration by stimulating HMG-CoA reductase activity, and thus causes fatty liver. On the background of inflammation (TNF-α), fatty liver develops steatohepatitis that eventually progresses to cirrhosis, and subsequent HCC. Leptin administration to ethanol-fed rodents resulted in downregulation of SREBP-2 protein expression, and subsequently decreased HMG-CoA reductase activity, and thus reduction of cholesterol levels and improved hepatic bile acids concentration; this, in turn, decreases alcohol induced fatty liver disease. Note: (+) indicates stimulation; (−) indicates inhibition; ↑ indicates increase; HCC, hepatocellular carcinoma; TNF, tumor necrosis factor.

TABLE 15.3 Nutritious Agents Used in the Treatment of Alcohol-Induced Hypercholesterolemia

Nutritional therapy	Administration route	Action/effect of the drug	Overall effect	References
Folic acid	Oral	Cholesterol levels↓ HMG-CoA reductase↓	Decreased cholesterol synthesis and increased catabolism in ALD	[89,129]
Persimmon-vinegar	Oral	Plasma and liver total cholesterol↓ Hepatic ACC mRNA↓ Hepatic CPT-I mRNA↑	Alcohol induced metabolic syndrome	[130]
Germinated brown rice extracts	Oral	Liver HDL-C↑ Liver total cholesterol↓	Alcohol related disorders	[132]
Omega-3 FAs	Oral	HDL-C↑	Reverse cholesterol transport	[133]
Taurine		Hepatic lipid content↓	Alcohol induced hepatic steatosis	[134]
Resveratrol	Oral	AMPK↑	Alcoholic fatty liver disease	[135]
Lipid-lowering statins: atorvastatin + vitamin C + vitamin E	Oral	Inhibit HMG-CoA reductase	NASH	[131]
Luteolin	Oral	SREBP↓ AMPK↑ SREBP-1c↑	Alcoholic steatosis	[46]
Traditional herbal medicine: *Schisandra chinensis*	Oral	Phospho AMPK↑ PPAR α↑	Alcoholic fatty liver	[139]

Note: ↓ indicates decrease; ↑ indicates increase.

omega-3 FAs and chronic ethanol consumption on RCT in rats.[133] Furthermore, taurine, one of the naturally occurring sulfonic acids and a well known antioxidant, was shown to reverse ethanol-induced hepatic steatosis and lipid peroxidation in rats,[134] and may have a role in reducing alcohol induced hyperlipidemia. Furthermore, resveratrol, a dietary polyphenol found in grape skins, has been identified as a potent antioxidant and activator for both sirtuin 1 and AMPK, and decreased alcoholic fatty liver disease in rodents.[135] The other antioxidants, L-carnitine and β-carotene, were shown to attenuate ethanol induced oxidative stress, and may have an impact in reducing ethanol-induced hypercholesterolemia.[136,137] In fact, green tea polyphenol treatment administered to ethanol fed rats, results in attenuating erythrocyte membrane oxidative stress.[138] Last but not least, luteolin, a flavonoid, was shown to reduce SREBP and improve AMPK and SREBP-1c gene expression in ALD.[46] Collectively, these nutritional approaches are encouraging, and may become available in the future for the reduction of hypercholesterolemia in patients with ALD.

CONCLUSIONS

In conclusion, this chapter revealed a plausible mechanism by which exogenous leptin treatment administered to alcohol fed mice ameliorated the levels of hepatic lipid profile, including cholesterol and the enzymes responsible for its synthesis and degradation. Furthermore, leptin treatment was shown to elevate the enzymes involved in BA synthesis. Thus, administration of exogenous leptin markedly reduced alcohol-induced hepatic hypercholesterolemia, by virtue of decreased synthesis, and increased catabolism of cholesterol, and could be a new potential therapeutic target for the management of early alcohol induced fatty liver disease. Indeed, further molecular studies are needed to classify the potential beneficial effect of leptin on ALD.

Key Facts

- Alcohol administration to both humans and animals resulted in elevated lipid profile, such as TGs, free fatty acids (FFA), phospholipids, and cholesterol in the liver and blood.
- Dyslipidemia is an abnormal level of lipids (e.g., cholesterol, TGs, and lipoproteins) in the blood.
- Lipids are a large and distinct group of naturally occurring organic compounds, combined by an ester linkage, primarily from alcohol and FAs. This group of molecules includes fats, waxes, oils, fat-soluble vitamins, hormones phospholipids, steroids (like cholesterol), TGs, and other related compounds.
- Cholesterol is a key molecule for all animal life; its increased concentration following heavy alcohol consumption leads to hepatic hypercholesterolemia.
- Chronic alcohol supplementation increased the intracellular cholesterol biosynthetic pathway, by enhancing hepatic SREBP-2 protein expression; it also increased HMG-CoA reductase activity.
- Cholesterol 7α hydroxylase (CYP7A1) is involved in the conversion of hepatic cholesterol into bile acids (BA); thus, cholesterol elimination from the body was shown to be decreased following alcohol consumption.
- Leptin administration to ethanol intoxicated mice markedly reduced hepatic hypercholesterolemia by virtue of decreased synthesis, and increased catabolism of cholesterol.
- Several agents that regulate hypercholesterolemia are in the pipeline, but their clinical efficacy and safety are yet to be established.
- Antioxidants are chemical compounds or natural substances that neutralize free radicals (unstable molecules) produced by oxidation in the human body; they protect the cell from the damage.

Summary Points

- Alcohol induced fatty liver disease was shown to be associated with the progression of cirrhosis and hepatocellular carcinoma (HCC).
- Leptin, the ob gene product, is a key hormone in regulating body weight and energy balance.
- This chapter deals with the involvement of leptin on alcohol induced lipid metabolism.
- Previous studies have shown evidence that exogenous leptin administration is associated with decreased lipid synthesis.
- Leptin administration was also shown to regulate alcohol induced hepatic cholesterol synthesis.
- Several other antioxidants and lipid lowering agents are discussed in this chapter, all of which have been shown to ameliorate alcohol induced hyperlipidemia.

Acknowledgment

Supported by the Department of Biotechnology (DBT) Ramalingaswami Re-entry Fellowship 5-year grant from the Government of India.

Abbreviations

ACAT Acyl-CoA cholesterol acyltransferase
ACC Acetyl-CoA carboxylases
ALD Alcoholic liver disease
AMPK AMP-activated protein kinases
BA Bile acids
CE Cholesterol esters

CPT	Carnitine palmitoyltransferase
CVD	Cardiovascular diseases
CYP2E1	Cytochrome P4502E1
DAG	Diacylglycerol
DGAT	Diacylglycerol acyltransferase
ER	Endoplasmic reticulum
FA	Fatty acid
FABP	Fatty acid binding protein
FC	Free cholesterol
FFA	Free fatty acid
HCC	Hepatocellular carcinoma
LXR	Liver X receptor
MAPK	Mitogen activated protein kinase
MUFA	Mono unsaturated fatty acid
NAFLD	Nonalcoholic fatty liver disease
NEFA	Nonesterified fatty acid
PPAR	Peroxisome proliferator activated receptor
PUFA	Polyunsaturated fatty acid
ROS	Reactive oxygen species
SCD	Steroyl-CoA desaturases
SREBP-1c	Sterol regulatory element binding protein 1c
STAT	Signal transducer and activator of transcription
TC	Total cholesterol
TG	Triglyceride
UDCA	Ursodeoxycholic acid
UFA	Unsaturated fatty acids
VLDL	Very low-density lipoprotein

References

1. Dutta AK. Genetic factors affecting susceptibility to alcoholic liver disease in an Indian population. *Ann Hepatol* 2013;**12**(6):901–7.
2. Leake I. Alcoholic liver disease: ASMase implicated in alcoholic liver disease. *Nat Rev Gastroenterol Hepatol* 2013;**10**(7):384.
3. Morgan TR, Mandayam S, Jamal MM. Alcohol and hepatocellular carcinoma. *Gastroenterology* 2004;**127**(5 Suppl. 1):S87–96.
4. Borowsky SA, Perlow W, Baraona E, Lieber CS. Relationship of alcoholic hypertriglyceridemia to stage of liver disease and dietary lipid. *Dig Dis Sci* 1980;**25**(1):22–7.
5. Altamirano J, Bataller R. Alcoholic liver disease: pathogenesis and new targets for therapy. *Nat Rev Gastroenterol Hepatol* 2011;**8**(9): 491–501.
6. Gyamfi MA, Wan YJ. Pathogenesis of alcoholic liver disease: the role of nuclear receptors. *Exp Biol Med (Maywood)* 2010;**235**(5): 547–60.
7. Teli MR, Day CP, Burt AD, Bennett MK, James OF. Determinants of progression to cirrhosis or fibrosis in pure alcoholic fatty liver. *Lancet* 1995;**346**(8981):987–90.
8. Gao B, Seki E, Brenner DA, Friedman S, Cohen JI, Nagy L, et al. Innate immunity in alcoholic liver disease. *Am J Physiol Gastrointest Liver Physiol* 2011;**300**(4):G516–25.
9. O'Keefe JH, Bhatti SK, Bajwa A, DiNicolantonio JJ, Lavie CJ. Alcohol and cardiovascular health: the dose makes the poison... or the remedy. *Mayo Clin Proc* 2014;**89**(3):382–93.
10. Zakhari S, Li TK. Determinants of alcohol use and abuse: impact of quantity and frequency patterns on liver disease. *Hepatology* 2007;**46**(6):2032–9.
11. Barnes MA, Roychowdhury S, Nagy LE. Innate immunity and cell death in alcoholic liver disease: role of cytochrome P4502E1. *Redox Biol* 2014;**2**:929–35.
12. Ramchandani VA, Bosron WF, Li TK. Research advances in ethanol metabolism. *Pathol Biol (Paris)* 2001;**49**(9):676–82.
13. Seth D, Haber PS, Syn WK, Diehl AM, Day CP. Pathogenesis of alcohol-induced liver disease: classical concepts and recent advances. *J Gastroenterol Hepatol* 2011;**26**(7):1089–105.
14. Finkelstein J, Heemels MT, Shadan S, Weiss U. Lipids in health and disease. *Nature* 2014;**510**(7503):47.
15. Lieber CS. Interactions of ethanol, drug and lipid metabolism; adaptive changes after ethanol consumption. *Clin Sci* 1970;**39**(3): 8P–9P.
16. Feinman L, Lieber CS. Hepatic collagen metabolism: effect of alcohol consumption in rats and baboons. *Science* 1972;**176**(4036):795.
17. Balasubramaniyan V, Manju V, Nalini N. Effect of leptin administration on plasma and tissue lipids in alcohol induced liver injury. *Hum Exp Toxicol* 2003;**22**(3):149–54.
18. Balasubramaniyan V, Nalini N. The potential beneficial effect of leptin on an experimental model of hyperlipidemia, induced by chronic ethanol treatment. *Clin Chim Acta* 2003;**337**(1–2):85–91.
19. Balasubramaniyan V, Nalini N. Intraperitoneal leptin regulates lipid metabolism in ethanol supplemented *Mus musculas* heart. *Life Sci* 2006;**78**(8):831–7.
20. Pirola RC, Lieber CS. The energy cost of the metabolism of drugs, including ethanol. *Pharmacology* 1972;**7**(3):185–96.
21. Baraona E, Lieber CS. Effects of ethanol on lipid metabolism. *J Lipid Res* 1979;**20**(3):289–315.
22. Lieber CS, Savolainen M. Ethanol and lipids. *Alcohol Clin Exp Res* 1984;**8**(4):409–23.
23. Remla A, Menon PV, Kurup PA, Kumari S, Saravarghese S. Effect of ethanol administration on metabolism of lipids in heart and aorta in isoproterenol induced myocardial infarction in rats. *Indian J Exp Biol* 1991;**29**(3):244–8.
24. Liu Q, Siloto RM, Lehner R, Stone SJ, Weselake RJ. Acyl-CoA:diacylglycerol acyltransferase: molecular biology, biochemistry and biotechnology. *Prog Lipid Res* 2012;**51**(4):350–77.
25. Cases S, Smith SJ, Zheng YW, Myers HM, Lear SR, Sande E, et al. Identification of a gene encoding an acyl CoA:diacylglycerol acyltransferase, a key enzyme in triacylglycerol synthesis. *Proc Natl Acad Sci USA* 1998;**95**(22):13018–23.
26. Wang Z, Yao T, Song Z. Involvement and mechanism of DGAT2 upregulation in the pathogenesis of alcoholic fatty liver disease. *J Lipid Res* 2010;**51**(11):3158–65.
27. Yamazaki T, Sasaki E, Kakinuma C, Yano T, Miura S, Ezaki O. Increased very low density lipoprotein secretion and gonadal fat mass in mice overexpressing liver DGAT1. *J Biol Chem* 2005;**280**(22): 21506–14.
28. Choi CS, Savage DB, Kulkarni A, Yu XX, Liu ZX, Morino K, et al. Suppression of diacylglycerol acyltransferase-2 (DGAT2), but not DGAT1, with antisense oligonucleotides reverses diet-induced hepatic steatosis and insulin resistance. *J Biol Chem* 2007;**282**(31): 22678–88.
29. Yu XX, Murray SF, Pandey SK, Booten SL, Bao D, Song XZ, et al. Antisense oligonucleotide reduction of DGAT2 expression improves hepatic steatosis and hyperlipidemia in obese mice. *Hepatology* 2005;**42**(2):362–71.
30. Mavrelis PG, Ammon HV, Gleysteen JJ, Komorowski RA, Charaf UK. Hepatic free fatty acids in alcoholic liver disease and morbid obesity. *Hepatology* 1983;**3**(2):226–31.
31. Gomez-Tubio A, Pita ML, Tavares E, Murillo ML, Delgado MJ, Carreras O. Changes in the fatty acid profile of plasma and adipose tissue in rats after long-term ethanol feeding. *Alcohol Clin Exp Res* 1995;**19**(3):747–52.
32. Cunnane SC, Manku MS, Horrobin DF. Effect of ethanol on liver triglycerides and fatty acid composition in the golden Syrian hamster. *Ann Nutr Metab* 1985;**29**(4):246–52.
33. Shiraishi K, Matsuzaki S, Itakura M, Ishida H. Abnormality in membrane fatty acid compositions of cells measured on erythrocyte in alcoholic liver disease. *Alcohol Clin Exp Res* 1996;**20**(1 Suppl.): 56A–9A.
34. Littleton JM, John G. Synaptosomal membrane lipids of mice during continuous exposure to ethanol. *J Pharm Pharmacol* 1977;**29**(9): 579–80.

35. Reitz RC. Dietary fatty acids and alcohol: effects on cellular membranes. *Alcohol Alcohol* 1993;**28**(1):59–71.
36. Nanji AA, Sadrzadeh SM, Dannenberg AJ. Liver microsomal fatty acid composition in ethanol-fed rats: effect of different dietary fats and relationship to liver injury. *Alcohol Clin Exp Res* 1994;**18**(4):1024–8.
37. Yu S, Rao S, Reddy JK. Peroxisome proliferator-activated receptors, fatty acid oxidation, steatohepatitis and hepatocarcinogenesis. *Curr Mol Med* 2003;**3**(6):561–72.
38. Peters JM, Rusyn I, Rose ML, Gonzalez FJ, Thurman RG. Peroxisome proliferator-activated receptor alpha is restricted to hepatic parenchymal cells, not Kupffer cells: implications for the mechanism of action of peroxisome proliferators in hepatocarcinogenesis. *Carcinogenesis* 2000;**21**(4):823–6.
39. Hellemans K, Michalik L, Dittie A, Knorr A, Rombouts K, De Jong J, et al. Peroxisome proliferator-activated receptor-beta signaling contributes to enhanced proliferation of hepatic stellate cells. *Gastroenterology* 2003;**124**(1):184–201.
40. Sozio M, Crabb DW. Alcohol and lipid metabolism. *Am J Physiol Endocrinol Metab* 2008;**295**(1):E10–6.
41. Nakajima T, Kamijo Y, Tanaka N, Sugiyama E, Tanaka E, Kiyosawa K, et al. Peroxisome proliferator-activated receptor alpha protects against alcohol-induced liver damage. *Hepatology* 2004;**40**(4):972–80.
42. Rao MS, Reddy JK. PPARalpha in the pathogenesis of fatty liver disease. *Hepatology* 2004;**40**(4):783–6.
43. Nanji AA, Dannenberg AJ, Jokelainen K, Bass NM. Alcoholic liver injury in the rat is associated with reduced expression of peroxisome proliferator-alpha (PPARalpha)-regulated genes and is ameliorated by PPARalpha activation. *J Pharmacol Exp Ther* 2004;**310**(1):417–24.
44. Larter CZ, Yeh MM, Van Rooyen DM, Brooling J, Ghatora K, Farrell GC. Peroxisome proliferator-activated receptor-alpha agonist, Wy 14,643, improves metabolic indices, steatosis and ballooning in diabetic mice with non-alcoholic steatohepatitis. *J Gastroenterol Hepatol* 2012;**27**(2):341–50.
45. Wada S, Yamazaki T, Kawano Y, Miura S, Ezaki O. Fish oil fed prior to ethanol administration prevents acute ethanol-induced fatty liver in mice. *J Hepatol* 2008;**49**(3):441–50.
46. Liu G, Zhang Y, Liu C, Xu D, Zhang R, Cheng Y, et al. Luteolin alleviates alcoholic liver disease induced by chronic and binge ethanol feeding in mice. *J Nutr* 2014;**144**(7):1009–15.
47. Zhou G, Myers R, Li Y, Chen Y, Shen X, Fenyk-Melody J, et al. Role of AMP-activated protein kinase in mechanism of metformin action. *J Clin Invest* 2001;**108**(8):1167–74.
48. Kim KH. Regulation of mammalian acetyl-coenzyme A carboxylase. *Annu Rev Nutr* 1997;**17**:77–99.
49. Hardie DG. Regulation of fatty acid and cholesterol metabolism by the AMP-activated protein kinase. *Biochim Biophys Acta* 1992;**1123**(3):231–8.
50. McGarry JD, Brown NF. The mitochondrial carnitine palmitoyltransferase system. From concept to molecular analysis. *Eur J Biochem* 1997;**244**(1):1–14.
51. Guzman M, Castro J. Alterations in the regulatory properties of hepatic fatty acid oxidation and carnitine palmitoyltransferase I activity after ethanol feeding and withdrawal. *Alcohol Clin Exp Res* 1990;**14**(3):472–7.
52. Tomita K, Tamiya G, Ando S, Kitamura N, Koizumi H, Kato S, et al. AICAR, an AMPK activator, has protective effects on alcohol-induced fatty liver in rats. *Alcohol Clin Exp Res* 2005;**29**(12 Suppl.):240S–5S.
53. García-Villafranca J, Guillén A, Castro J. Ethanol consumption impairs regulation of fatty acid metabolism by decreasing the activity of AMP-activated protein kinase in rat liver. *Biochimie* 2008;**90**(3):460–6.
54. Liu Y, Major AS, Zienkiewicz J, Gabriel CL, Veach RA, Moore DJ, et al. Nuclear transport modulation reduces hypercholesterolemia, atherosclerosis, and fatty liver. *J Am Heart Assoc* 2013;**2**(2):e000093.
55. Horton JD, Goldstein JL, Brown MS. SREBPs: activators of the complete program of cholesterol and fatty acid synthesis in the liver. *J Clin Invest* 2002;**109**(9):1125–31.
56. Ji C, Chan C, Kaplowitz N. Predominant role of sterol response element binding proteins (SREBP) lipogenic pathways in hepatic steatosis in the murine intragastric ethanol feeding model. *J Hepatol* 2006;**45**(5):717–24.
57. Liu J. Ethanol and liver: recent insights into the mechanisms of ethanol-induced fatty liver. *World J Gastroenterol* 2014;**20**(40):14672–85.
58. Karasawa T, Takahashi A, Saito R, Sekiya M, Igarashi M, Iwasaki H, et al. Sterol regulatory element-binding protein-1 determines plasma remnant lipoproteins and accelerates atherosclerosis in low-density lipoprotein receptor-deficient mice. *Arterioscler Thromb Vasc Biol* 2011;**31**(8):1788–95.
59. Shimano H, Horton JD, Hammer RE, Shimomura I, Brown MS, Goldstein JL. Overproduction of cholesterol and fatty acids causes massive liver enlargement in transgenic mice expressing truncated SREBP-1a. *J Clin Invest* 1996;**98**(7):1575–84.
60. Balasubramaniyan VM, Nalini G. Leptin administration regulates hepatic cholesterol synthesis in a mouse model of alcoholic fatty liver disease. *J Hepatol* 2013;**58**(Suppl. 1):S226.
61. Jefcoate C. High-flux mitochondrial cholesterol trafficking, a specialized function of the adrenal cortex. *J Clin Invest* 2002;**110**(7):881–90.
62. Fernández A, Colell A, Garcia-Ruiz C, Fernandez-Checa JC. Cholesterol and sphingolipids in alcohol-induced liver injury. *J Gastroenterol Hepatol* 2008;**23**(Suppl. 1):S9–S15.
63. Min HK, Kapoor A, Fuchs M, Mirshahi F, Zhou H, Maher J, et al. Increased hepatic synthesis and dysregulation of cholesterol metabolism is associated with the severity of nonalcoholic fatty liver disease. *Cell Metab* 2012;**15**(5):665–74.
64. Spector AA, Mathur SN, Kaduce TL. Role of acylcoenzyme A: cholesterol o-acyltransferase in cholesterol metabolism. *Prog Lipid Res* 1979;**18**(1):31–53.
65. Lefevre AF, DeCarli LM, Lieber CS. Effect of ethanol on cholesterol and bile acid metabolism. *J Lipid Res* 1972;**13**(1):48–55.
66. Erickson SK, Shrewsbury MA, Brooks C, Meyer DJ. Rat liver acyl-coenzyme A:cholesterol acyltransferase: its regulation *in vivo* and some of its properties *in vitro*. *J Lipid Res* 1980;**21**(7):930–41.
67. Hashimoto S, Fogelman AM. Smooth microsomes. A trap for cholesteryl ester formed in hepatic microsomes. *J Biol Chem* 1980;**255**(18):8678–84.
68. Parini P, Davis M, Lada AT, Erickson SK, Wright TL, Gustafsson U, et al. ACAT2 is localized to hepatocytes and is the major cholesterol-esterifying enzyme in human liver. *Circulation* 2004;**110**(14):2017–23.
69. Alger HM, Brown JM, Sawyer JK, Kelley KL, Shah R, Wilson MD, et al. Inhibition of acyl-coenzyme A:cholesterol acyltransferase 2 (ACAT2) prevents dietary cholesterol-associated steatosis by enhancing hepatic triglyceride mobilization. *J Biol Chem* 2010;**285**(19):14267–74.
70. Lee RG, Kelley KL, Sawyer JK, Farese Jr RV, Parks JS, Rudel LL. Plasma cholesteryl esters provided by lecithin:cholesterol acyltransferase and acyl-coenzyme a:cholesterol acyltransferase 2 have opposite atherosclerotic potential. *Circ Res* 2004;**95**(10):998–1004.
71. Willner EL, Tow B, Buhman KK, Wilson M, Sanan DA, Rudel LL, et al. Deficiency of acyl CoA:cholesterol acyltransferase 2 prevents atherosclerosis in apolipoprotein E-deficient mice. *Proc Natl Acad Sci USA* 2003;**100**(3):1262–7.
72. Savard C, Tartaglione EV, Kuver R, Haigh WG, Farrell GC, Subramanian S, et al. Synergistic interaction of dietary cholesterol and dietary fat in inducing experimental steatohepatitis. *Hepatology* 2013;**57**(1):81–92.
73. Hungund BL, Zheng Z, Lin L, Barkai AI. Ganglioside GM1 reduces ethanol induced phospholipase A2 activity in synaptosomal preparations from mice. *Neurochem Int* 1994;**25**(4):321–5.

74. Field FJ, Boydstun JS, LaBrecque DR. Effect of chronic ethanol ingestion on hepatic and intestinal acyl coenzyme a:cholesterol acyltransferase and 3-hydroxy-3-methylglutaryl coenzyme a reductase in the rat. *Hepatology* 1985;**5**(1):133–8.
75. You M, Crabb DW. Molecular mechanisms of alcoholic fatty liver: role of sterol regulatory element-binding proteins. *Alcohol* 2004;**34**(1):39–43.
76. Crabb DW, Galli A, Fischer M, You M. Molecular mechanisms of alcoholic fatty liver: role of peroxisome proliferator-activated receptor alpha. *Alcohol* 2004;**34**(1):35–8.
77. Sanchez-Amate MC, Zurera JM, Carrasco MP, Segovia JL, Marco C. Ethanol and lipid metabolism. Differential effects on liver and brain microsomes. *FEBS Lett* 1991;**293**(1–2):215–8.
78. Caballero F, Fernández A, De Lacy AM, Fernandez-Checa JC, Caballería J, García-Ruiz C. Enhanced free cholesterol, SREBP-2 and StAR expression in human NASH. *J Hepatol* 2009;**50**(4):789–96.
79. Wanless IR, Belgiorno J, Huet PM. Hepatic sinusoidal fibrosis induced by cholesterol and stilbestrol in the rabbit: 1. Morphology and inhibition of fibrogenesis by dipyridamole. *Hepatology* 1996;**24**(4):855–64.
80. Ioannou GN, Morrow OB, Connole ML, Lee SP. Association between dietary nutrient composition and the incidence of cirrhosis or liver cancer in the United States population. *Hepatology* 2009;**50**(1):175–84.
81. Visioli F, Monti S, Colombo C, Galli C. Ethanol enhances cholesterol synthesis and secretion in human hepatomal cells. *Alcohol* 1998;**15**(4):299–303.
82. Wang Z, Yao T, Song Z. Chronic alcohol consumption disrupted cholesterol homeostasis in rats: down-regulation of low-density lipoprotein receptor and enhancement of cholesterol biosynthesis pathway in the liver. *Alcohol Clin Exp Res* 2010;**34**(3):471–8.
83. Watanabe M, Houten SM, Mataki C, Christoffolete MA, Kim BW, Sato H, et al. Bile acids induce energy expenditure by promoting intracellular thyroid hormone activation. *Nature* 2006;**439**(7075): 484–9.
84. Chiang JY. Bile acids: regulation of synthesis. *J Lipid Res* 2009; **50**(10):1955–66.
85. Janowski BA, Willy PJ, Devi TR, Falck JR, Mangelsdorf DJ. An oxysterol signalling pathway mediated by the nuclear receptor LXR alpha. *Nature* 1996;**383**(6602):728–31.
86. Peet DJ, Turley SD, Ma W, Janowski BA, Lobaccaro JM, Hammer RE, et al. Cholesterol and bile acid metabolism are impaired in mice lacking the nuclear oxysterol receptor LXR alpha. *Cell* 1998; **93**(5):693–704.
87. Tiemann M, Han Z, Soccio R, Bollineni J, Shefer S, Sehayek E, et al. Cholesterol feeding of mice expressing cholesterol 7alpha-hydroxylase increases bile acid pool size despite decreased enzyme activity. *Proc Natl Acad Sci USA* 2004;**101**(7):1846–51.
88. Xie G, Zhong W, Li H, Li Q, Qiu Y, Zheng X, et al. Alteration of bile acid metabolism in the rat induced by chronic ethanol consumption. *FASEB J* 2013;**27**(9):3583–93.
89. Delgado-Villa MJ, Ojeda ML, Rubio JM, Murillo ML, Sánchez OC. Beneficial role of dietary folic acid on cholesterol and bile acid metabolism in ethanol-fed rats. *J Stud Alcohol Drugs* 2009;**70**(4): 615–22.
90. Dumas ME, Barton RH, Toye A, Cloarec O, Blancher C, Rothwell A, et al. Metabolic profiling reveals a contribution of gut microbiota to fatty liver phenotype in insulin-resistant mice. *Proc Natl Acad Sci USA* 2006;**103**(33):12511–6.
91. Chiang JY. Regulation of bile acid synthesis: pathways, nuclear receptors, and mechanisms. *J Hepatol* 2004;**40**(3):539–51.
92. Russell DW. The enzymes, regulation, and genetics of bile acid synthesis. *Annu Rev Biochem* 2003;**72**:137–74.
93. Nilsson LM, Sjovall J, Strom S, Bodin K, Nowak G, Einarsson C, et al. Ethanol stimulates bile acid formation in primary human hepatocytes. *Biochem Biophys Res Commun* 2007;**364**(4):743–7.
94. Stravitz RT, Vlahcevic ZR, Russell TL, Heizer ML, Avadhani NG, Hylemon PB. Regulation of sterol 27-hydroxylase and an alternative pathway of bile acid biosynthesis in primary cultures of rat hepatocytes. *J Steroid Biochem Mol Biol* 1996;**57**(5–6):337–47.
95. Hall EA, Ren S, Hylemon PB, Rodriguez-Agudo D, Redford K, Marques D, et al. Detection of the steroidogenic acute regulatory protein, StAR, in human liver cells. *Biochim Biophys Acta* 2005;**1733**(2–3):111–9.
96. Crouse JR, Grundy SM. Effects of alcohol on plasma lipoproteins and cholesterol and triglyceride metabolism in man. *J Lipid Res* 1984;**25**(5):486–96.
97. Maruyama S, Murawaki Y, Hirayama C. Effects of chronic ethanol administration on hepatic cholesterol and bile acid synthesis in relation to serum high density lipoprotein cholesterol in rats. *Res Commun Chem Pathol Pharmacol* 1986;**53**(1):3–21.
98. Stahlberg D, Rudling M, Angelin B, Bjorkhem I, Forsell P, Nilsell K, et al. Hepatic cholesterol metabolism in human obesity. *Hepatology* 1997;**25**(6):1447–50.
99. Nishina PM, Lowe S, Wang J, Paigen B. Characterization of plasma lipids in genetically obese mice: the mutants obese, diabetes, fat, tubby, and lethal yellow. *Metabolism* 1994;**43**(5):549–53.
100. Yang SQ, Lin HZ, Lane MD, Clemens M, Diehl AM. Obesity increases sensitivity to endotoxin liver injury: implications for the pathogenesis of steatohepatitis. *Proc Natl Acad Sci USA* 1997;**94**(6): 2557–62.
101. Considine RV, Sinha MK, Heiman ML, Kriauciunas A, Stephens TW, Nyce MR, et al. Serum immunoreactive-leptin concentrations in normal-weight and obese humans. *N Engl J Med* 1996;**334**(5): 292–5.
102. Zhang Y, Proenca R, Maffei M, Barone M, Leopold L, Friedman JM. Positional cloning of the mouse obese gene and its human homologue. *Nature* 1994;**372**(6505):425–32.
103. Maffei M, Halaas J, Ravussin E, Pratley RE, Lee GH, Zhang Y, et al. Leptin levels in human and rodent: measurement of plasma leptin and ob RNA in obese and weight-reduced subjects. *Nat Med* 1995;**1**(11):1155–61.
104. Wang MY, Lee Y, Unger RH. Novel form of lipolysis induced by leptin. *J Biol Chem* 1999;**274**(25):17541–4.
105. Huang W, Dedousis N, Bandi A, Lopaschuk GD, O'Doherty RM. Liver triglyceride secretion and lipid oxidative metabolism are rapidly altered by leptin *in vivo*. *Endocrinology* 2006;**147**(3): 1480–7.
106. Minokoshi Y, Kim YB, Peroni OD, Fryer LG, Muller C, Carling D, et al. Leptin stimulates fatty-acid oxidation by activating AMP-activated protein kinase. *Nature* 2002;**415**(6869):339–43.
107. Reidy SP, Weber J. Leptin: an essential regulator of lipid metabolism. *Comp Biochem Physiol A Mol Integr Physiol* 2000;**125**(3): 285–98.
108. Gallardo N, Bonzon-Kulichenko E, Fernandez-Agullo T, Molto E, Gomez-Alonso S, Blanco P, et al. Tissue-specific effects of central leptin on the expression of genes involved in lipid metabolism in liver and white adipose tissue. *Endocrinology* 2007;**148**(12): 5604–10.
109. Fogteloo AJ, Pijl H, Frolich M, McCamish M, Meinders AE. Effects of recombinant human leptin treatment as an adjunct of moderate energy restriction on body weight, resting energy expenditure and energy intake in obese humans. *Diabetes Nutr Metab* 2003;**16**(2):109–14.
110. Westerterp-Plantenga MS, Saris WH, Hukshorn CJ, Campfield LA. Effects of weekly administration of pegylated recombinant human OB protein on appetite profile and energy metabolism in obese men. *Am J Clin Nutr* 2001;**74**(4):426–34.
111. Farooqi IS, Matarese G, Lord GM, Keogh JM, Lawrence E, Agwu C, et al. Beneficial effects of leptin on obesity, T cell hyporesponsiveness, and neuroendocrine/metabolic dysfunction of human congenital leptin deficiency. *J Clin Invest* 2002;**110**(8):1093–103.

112. Prieur X, Tung YC, Griffin JL, Farooqi IS, O'Rahilly S, Coll AP. Leptin regulates peripheral lipid metabolism primarily through central effects on food intake. *Endocrinology* 2008;**149**(11):5432–9.
113. Hyogo H, Roy S, Paigen B, Cohen DE. Leptin promotes biliary cholesterol elimination during weight loss in ob/ob mice by regulating the enterohepatic circulation of bile salts. *J Biol Chem* 2002;**277**(37):34117–24.
114. Vanpatten S, Karkanias GB, Rossetti L, Cohen DE. Intracerebroventricular leptin regulates hepatic cholesterol metabolism. *Biochem J* 2004;**379**(Pt. 2):229–33.
115. Huang W, Dedousis N, O'Doherty RM. Hepatic steatosis and plasma dyslipidemia induced by a high-sucrose diet are corrected by an acute leptin infusion. *J Appl Physiol (1985)* 2007;**102**(6): 2260–5.
116. Unger RH, Zhou YT, Orci L. Regulation of fatty acid homeostasis in cells: novel role of leptin. *Proc Natl Acad Sci USA* 1999;**96**(5):2327–32.
117. Bai Y, Zhang S, Kim KS, Lee JK, Kim KH. Obese gene expression alters the ability of 30A5 preadipocytes to respond to lipogenic hormones. *J Biol Chem* 1996;**271**(24):13939–42.
118. Cohen B, Novick D, Rubinstein M. Modulation of insulin activities by leptin. *Science* 1996;**274**(5290):1185–8.
119. Yamagishi SI, Edelstein D, Du XL, Kaneda Y, Guzman M, Brownlee M. Leptin induces mitochondrial superoxide production and monocyte chemoattractant protein-1 expression in aortic endothelial cells by increasing fatty acid oxidation via protein kinase A. *J Biol Chem* 2001;**276**(27):25096–100.
120. Fukuda H, Iritani N, Sugimoto T, Ikeda H. Transcriptional regulation of fatty acid synthase gene by insulin/glucose, polyunsaturated fatty acid and leptin in hepatocytes and adipocytes in normal and genetically obese rats. *Eur J Biochem* 1999;**260**(2):505–11.
121. Sarmiento U, Benson B, Kaufman S, Ross L, Qi M, Scully S, et al. Morphologic and molecular changes induced by recombinant human leptin in the white and brown adipose tissues of C57BL/6 mice. *Lab Invest* 1997;**77**(3):243–56.
122. Bryson JM, Phuyal JL, Swan V, Caterson ID. Leptin has acute effects on glucose and lipid metabolism in both lean and gold thioglucose-obese mice. *Am J Physiol* 1999;**277**(3 Pt. 1):E417–22.
123. Zhou YT, Shimabukuro M, Koyama K, Lee Y, Wang MY, Trieu F, et al. Induction by leptin of uncoupling protein-2 and enzymes of fatty acid oxidation. *Proc Natl Acad Sci USA* 1997;**94**(12):6386–90.
124. VanPatten S, Ranginani N, Shefer S, Nguyen LB, Rossetti L, Cohen DE. Impaired biliary lipid secretion in obese Zucker rats: leptin promotes hepatic cholesterol clearance. *Am J Physiol Gastrointest Liver Physiol* 2001;**281**(2):G393–404.
125. Lundasen T, Liao W, Angelin B, Rudling M. Leptin induces the hepatic high density lipoprotein receptor scavenger receptor B type I (SR-BI) but not cholesterol 7alpha-hydroxylase (Cyp7a1) in leptin-deficient (ob/ob) mice. *J Biol Chem* 2003;**278**(44):43224–8.
126. Angel A, Farkas J. Regulation of cholesterol storage in adipose tissue. *J Lipid Res* 1974;**15**(5):491–9.
127. Silver DL, Jiang XC, Tall AR. Increased high density lipoprotein (HDL), defective hepatic catabolism of ApoA-I and ApoA-II, and decreased ApoA-I mRNA in ob/ob mice. Possible role of leptin in stimulation of HDL turnover. *J Biol Chem* 1999;**274**(7):4140–6.
128. Zhao AZ, Shinohara MM, Huang D, Shimizu M, Eldar-Finkelman H, Krebs EG, et al. Leptin induces insulin-like signaling that antagonizes cAMP elevation by glucagon in hepatocytes. *J Biol Chem* 2000;**275**(15):11348–54.
129. Ojeda ML, Delgado-Villa MJ, Llopis R, Murillo ML, Carreras O. Lipid metabolism in ethanol-treated rat pups and adults: effects of folic acid. *Alcohol Alcohol* 2008;**43**(5):544–50.
130. Moon YJ, Cha YS. Effects of persimmon-vinegar on lipid metabolism and alcohol clearance in chronic alcohol-fed rats. *J Med Food* 2008;**11**(1):38–45.
131. Arendt BM, Allard JP. Effect of atorvastatin, vitamin E and C on nonalcoholic fatty liver disease: is the combination required? *Am J Gastroenterol* 2011;**106**(1):78–80.
132. Oh SH, Soh JR, Cha YS. Germinated brown rice extract shows a nutraceutical effect in the recovery of chronic alcohol-related symptoms. *J Med Food* 2003;**6**(2):115–21.
133. Marmillot P, Rao MN, Liu QH, Chirtel SJ, Lakshman MR. Effect of dietary omega-3 fatty acids and chronic ethanol consumption on reverse cholesterol transport in rats. *Metabolism* 2000;**49**(4): 508–12.
134. Kerai MD, Waterfield CJ, Kenyon SH, Asker DS, Timbrell JA. Reversal of ethanol-induced hepatic steatosis and lipid peroxidation by taurine: a study in rats. *Alcohol Alcohol* 1999;**34**(4):529–41.
135. Ajmo JM, Liang X, Rogers CQ, Pennock B, You M. Resveratrol alleviates alcoholic fatty liver in mice. *Am J Physiol Gastrointest Liver Physiol* 2008;**295**(4):G833–42.
136. Augustyniak A, Skrzydlewska E. L-Carnitine in the lipid and protein protection against ethanol-induced oxidative stress. *Alcohol* 2009;**43**(3):217–23.
137. Lin WT, Huang CC, Lin TJ, Chen JR, Shieh MJ, Peng HC, et al. Effects of beta-carotene on antioxidant status in rats with chronic alcohol consumption. *Cell Biochem Funct* 2009;**27**(6):344–50.
138. Dobrzynska I, Szachowicz-Petelska B, Ostrowska J, Skrzydlewska E, Figaszewski Z. Protective effect of green tea on erythrocyte membrane of different age rats intoxicated with ethanol. *Chem Biol Interact* 2005;**156**(1):41–53.
139. Park HJ, Lee SJ, Song Y, Jang SH, Ko YG, Kang SN, et al. *Schisandra chinensis* prevents alcohol-induced fatty liver disease in rats. *J Med Food* 2014;**17**(1):103–10.

CHAPTER

16

The Corticotropin Releasing Factor System and Alcohol Consumption

*Andrey E. Ryabinin, PhD**, *William J. Giardino, PhD***

*Department of Behavioral Neuroscience, Oregon Health and Science University, Portland, OR, USA

**Department of Psychiatry & Behavioral Sciences, Stanford University, Stanford, CA, USA

INTRODUCTION: ALCOHOL AND THE BRAIN

Alcohol use disorder is a devastating disease characterized by compulsive use of alcohol (ethanol; EtOH). It can arise from adaptations within neural circuits, leading to a persistent dysregulation of drug-seeking.[1,2] Efforts to characterize the maladaptive changes underlying this phenomenon have identified numerous brain regions, neurotransmitter systems, and genes that work in concert to drive voluntary, repeated intake of intoxicating doses of EtOH.[3–5]

EtOH is hypothesized to interact directly with receptors for gamma-amino butyric acid (GABA) and glutamate.[6,7] Consistent with this idea, voluntary EtOH consumption causes plastic changes in GABA, and glutamate transmission within the mesolimbic dopamine (DA) reward pathway, including the ventral tegmental area (VTA), and nucleus accumbens (NAc).[8,9] Furthermore, EtOH dependence alters GABA and glutamate physiology within stress-related circuits of the extended amygdala, including the central nucleus of the amygdala (CeA), and bed nucleus of stria terminalis (BNST).[10–12] Extrasynaptic GABA-A receptors mediate behavioral and physiological effects of EtOH,[13] and these receptors are also modulated by neuroactive steroids. Several lines of evidence support a sex-specific role for endogenous neurosteroids in EtOH sensitivity, drinking, and withdrawal.[14,15]

The conditioned rewarding effects of EtOH rely on DA, GABA, glutamate, and endogenous opioid signaling within mesolimbic, amygdalar, and cortical brain areas.[16–21] Perhaps related to their effects on reward, these neurotransmitter systems also drive "excessive" EtOH consumption, in which rodent subjects exceed the National Institutes of Health (NIH) criterion for binge drinking (blood ethanol concentrations >80 mg/dL).[22–24]

While neurotransmitter systems directly targeted by ETOH, or involved in conditioned rewarding effects of ETOH, are clearly important for regulation of alcohol consumption, another extremely important system contributing to the addictive properties of this drug is the corticotropin releasing factor (CRF) system. This system is crucial for regulating an organism's response to various stressors, as well as ETOH's interactions with these stressors, and ETOH self-administration itself. This chapter focuses on the role of this system in moderate and excessive alcohol consumption.

CRF, UROCORTINS, AND THE HPA-AXIS

The CRF system is also known as the corticotropin-releasing hormone system. The mammalian CRF system consists of three additional ligands (urocortin 1, urocortin 2, and urocortin 3), two main forms of receptors (CRF1 and CRF2), and the CRF-binding protein (CRFBP).[25] Components of the system have distinct abbreviations, when referring to the gene versus the protein form, although these terms (and several variations) are used interchangeably in the CRF system literature. The nomenclature used in this review is presented at the end of this chapter (Table 16.1).

CRF has high affinity for CRF1 receptor, but not CRF2 receptor. Urocortin 1 (Ucn1) has high affinity for both CRF1 and CRF2 receptors. Ucn2 and Ucn3 have high affinity for CRF2 receptor, but not CRF1.[26] CRF1 is expressed widely throughout the brain, with major sites in the extended amygdala, septal nuclei, hypothalamus, cortex, and mesolimbic pathway, while CRF2 expression

Molecular Aspects of Alcohol and Nutrition. http://dx.doi.org/10.1016/B978-0-12-800773-0.00016-1

TABLE 16.1 CRF System Nomenclature

Name	Gene	Protein
Corticotropin-releasing factor	*Crh*	CRF
Urocortin-1	*Ucn*	Ucn1
Urocortin-2	*Ucn2*	Ucn2
Urocortin-3	*Ucn3*	Ucn3
Urocortin peptides (any two, or all three)	*Ucns*	Ucns
CRF binding protein	*Crhbp*	CRFBP
CRF type-1 receptor	*Crhr1*	CRF1
CRF type-2 receptor	*Crhr2*	CRF2

When discussing mutant mouse lines, we referred to the deleted gene in its abbreviated form, but to the mouse line itself with the protein abbreviation, as to acknowledge that both the gene and protein are dysfunctional in the mutant mice.

is expressed primarily in the lateral septum (LS), raphe nuclei, extended amygdala, ventromedial hypothalamus (VMH), and brainstem.[27–29] Alternatively spliced forms of CRF receptors, as well as CRF2/dopamine D1 heteromer receptors, have also been described, but their properties and distributions are not well studied.[30,31]

The best-characterized site of CRF expression is paraventricular nucleus of hypothalamus (PVN), but CRF-expressing neurons are also present in the cortex, areas of the extended amygdala, medial septum, thalamus, cerebellum, autonomic midbrain and hindbrain.[29,32] CRF released from the PVN stimulates CRF1 in the anterior pituitary gland, and causes secretion of adrenocorticotropin hormone (ACTH), resulting in release of corticosterone or cortisol (CORT) from the adrenal glands.[33,34] Together, this system is referred to as the hypothalamic-pituitary-adrenal (HPA) axis, the primary neuroendocrine response to stress. CORT exerts negative and positive feedback on components of the central CRF system in a site-specific manner, via signaling through mineralocorticoid and glucocorticoid receptors.[35] Outside of the HPA-axis, the CeA and BNST are two key sites of CRF expression that influence stress- and addiction-related behavior.[36–38] Extrahypothalamic CRF neurons coordinate the neurobehavioral response to stress via projections to the locus coeruleus (LC), raphe nuclei, and extended amygdala.[29,39]

The main site of Ucn1 expression is the centrally projecting Edinger–Westphal nucleus (EWcp). Smaller populations of Ucn1-positive neurons are also present in the lateral superior olive (LSO), and the supraoptic nucleus (SON).[40–44] EWcp-Ucn1 neurons project to the lateral septum (LS) and dorsal raphe nucleus (DRN), among other areas of the brain and spinal cord.[45,46,42,47] Ucn2 is expressed in the LC, PVN, SON, and arcuate nucleus.[48,49] Ucn2 projections have not yet been characterized, but several lines of evidence suggest that Ucn2 regulates the hypothalamic-pituitary-gonadal axis, likely via release from the PVN and/or SON.[28,50–52] Ucn3-containing cell bodies are localized mainly to the medial amygdala, BNST, and hypothalamus.[53–55] Ucn3 pathways are somewhat well characterized, with major projections to the BNST, LS, and VMH.[55–57]

Ucns are influenced by HPA-axis tone. Levels of Ucn2 and Ucn3 in the hypothalamus and amygdala are directly regulated by stress and glucocorticoids.[49,58] In addition, levels of EWcp-*Ucn* messenger ribonucleic acid (mRNA) fluctuate in a circadian rhythm, opposite to plasma CORT levels, suggesting HPA regulation.[59]

The largely distinct (yet partially redundant) patterns of CRF and Ucns expression, across key limbic brain areas, likely underlie the complex contributions of these systems to mobilization and recovery of the stress response. These contributions were difficult to assess using standard pharmacological techniques, because administration of each particular peptide into a specific region could mimic effects of a different family member endogenously expressed in the same neuroanatomical site. Therefore, these roles were better revealed by genetic studies focusing on each component of the system.

FUNCTION OF CRF SYSTEM COMPONENTS REVEALED BY GENETIC KNOCKOUTS

The hypothesized roles of the CRF system in regulation of HPA-axis activity and stress-related behavior were largely confirmed upon examination of mice containing targeted mutations in individual genes of the CRF system. For example, deletion of *Crhr1* produced an anxiolytic behavioral phenotype associated with low basal and stress-induced levels of CORT and ACTH.[60,61] The results from three independently generated CRF2 KO mouse lines were varied, with deletions of *Crhr2* causing increased, decreased, or no change in anxiety-like behavior and CORT/ACTH levels.[62–64] Mice lacking CRFBP displayed increased anxiety-like behavior, despite no change in basal or stress-induced CORT and ACTH levels.[65]

Deletion of *Crh* reduced levels of CORT and ACTH as expected,[66] although anxiety-like behavior in CRF KO mice was comparable to wild-type (WT) littermate control mice.[67] The results from three independently-generated Ucn1 KO mouse lines were varied, with *Ucn* deletion causing either increased, decreased, or no change in anxiety-like behavior and CORT/ACTH levels.[43,68,69] Interestingly, Vetter et al.[43] described auditory deficits in Ucn1 KO mice, perhaps reflecting a consequence of the loss of Ucn1 in the LSO (an auditory brainstem nucleus).

Deletion of *Ucn2* produced a female-specific antidepressant-like phenotype, despite enhancements in plasma CORT and ACTH.[70] More recently, an independently generated Ucn2 KO mouse line displayed normal HPA-axis activity, and reduced aggressive behavior.[71] The first neural and behavioral characterization of Ucn3 KO and WT mice focused on the enhanced social recognition abilities, observed following deletion of *Ucn3*.[55]

Double and triple deletions of *Ucns* (*Ucn/Ucn2* and *Ucn/Ucn2/Ucn3*) produced decreases and increases in anxiety-like behavior, respectively. These effects were accompanied by complex effects on the HPA-axis and stress reactivity.[72,73] Double and triple Ucns KO mice also showed substantial alterations within the 5-hydroxytryptamine (serotonin; 5-HT) system.[74] Complementing the link between Ucns and 5-HT, a recent report observed decreased EWcp-*Ucn* mRNA in 5-HT transporter KO versus WT mice.[75] These data suggest that Ucn1 signaling in the raphe nuclei is a potential mechanism by which stress regulates 5-HT tone and accompanying mood. Indeed, an extensive literature documents the interactions between CRF/Ucns signaling and raphe nuclei 5-HT transmission, within the context of stress and addiction.[76–79]

Multiple site-specific compensations in expression of CRF system components have been identified in CRF system KO mice.[25] These alterations, as well as KO effects on anxiety-like behavior and HPA-axis activity, are summarized in Table 16.2. Taken together, these studies indicate that different components of the CRF systems differentially regulate anxiety-, mood-, and stress-related responses in a stressor- and context-specific way. This is important for the current review, as alcohol and addiction-related behaviors also influence, and are influenced, by stress.

EFFECTS OF EtOH ON THE HPA-AXIS

Like all drugs of abuse, EtOH can activate the HPA-axis.[80] Dr Catherine Rivier and her coworkers at the Salk Institute performed seminal studies in rats, showing that intraperitoneal (i.p.) administration of EtOH (1–3 grams per kilogram body weight (g/kg)) significantly increased CORT and ACTH levels. The effects were attenuated by either i.c.v. immunoneutralization of CRF,[81] or electrolytic lesion of the PVN,[82] suggesting the importance of PVN-CRF and the HPA-axis. A further study revealed that a nonselective CRF receptor antagonist with minimal actions in the pituitary also reduced the EtOH-induced ACTH response.[83] This result indicated that *extrahypothalamic* CRF components were also important for driving the effects of EtOH on the HPA-axis.[84]

Despite the HPA-activating effects of EtOH in naïve rats, rats pretreated with i.p. EtOH showed a blunted ACTH response to acute stress, or i.c.v. CRF administration.[85] Consistent with this finding, rats with prior EtOH vapor exposure also showed an attenuated ACTH response to intravenous CRF, and footshock stress.[86] Others have replicated the dampened neuroendocrine state observed following EtOH vapor dependence.[87] However, the HPA-blunting effects of EtOH appeared to differ across development, as 21-day-old rat pups born to EtOH vapor-exposed dams displayed an *accentuated* ACTH response to stress, relative to control rats.[88] Indeed, several prenatal exposure studies observed a hyperresponsive HPA phenotype in rodents born to EtOH-treated dams.[89]

In several cases, *Crh* mRNA was elevated in the PVN of EtOH-exposed versus control rats.[86,88,90] However, other studies reported no influence of EtOH on PVN-CRF expression,[91–94] suggesting that effects differed, depending on the method of exposure and the timecourse of analysis.

EFFECTS OF EtOH ON THE EXTRAHYPOTHALAMIC CRF SYSTEM

Elevations in *Crh* mRNA or CRF-immunoreactivity (IR) were observed in the CeA, following long-term EtOH vapor or diet exposure,[95–97] and following short-term binge drinking.[98] However, other studies reported decreases in amygdalar *Crh* mRNA and CRF-IR, following voluntary drinking,[99–101] or withdrawal from liquid EtOH diet.[91] In yet a few other cases, authors reported no significant effects of EtOH on CeA-CRF expression.[90,102]

Following chronic EtOH vapor, *Crhr1* levels in the amygdala were increased, while amygdala *Crhr2* levels were decreased.[96] In contrast, voluntary EtOH drinking reduced *Crhr1* mRNA in the CeA and the NAcc of selectively-bred high-preferring rats.[103] Although repeated i.p. EtOH, or voluntary EtOH intake, had no effect on EWcp-Ucn1 protein expression, EtOH drinking significantly reduced the number of Ucn1 fibers present in the LS.[104,105] Furthermore, repeated i.p. EtOH increased CRF2 binding in the LS and DRN, fact that could reflect either increased or decreased release of Ucn1 from the EWcp.[106]

Given the dynamics of gene transcription, and the differing rates of peptide synthesis, release, and binding, it can be difficult to interpret mRNA findings without corresponding data at the protein level, and vice versa. Nevertheless, the overall picture indicates that acute EtOH exposure reduces CRF system activity, while abstinence from chronic EtOH increases it. Supporting the interpretation of enhanced CRF release during EtOH abstinence, *in vivo* microdialysis studies found increased extracellular CRF content in the amygdala and BNST, following 6–12 h of forced abstinence from liquid EtOH diet.[107,108]

TABLE 16.2 CRF System KO Effects

	Crhr1	*Crhr2*	*Crh*	*Crhbp*	Ucn	Ucn2	Ucn3	Ucn + Ucn2	Ucn + Ucn2 + Ucn3
Anxiety	↓	↑?	–	↑	↑?	–	–	↓	↑x
CORT	↓	↑x?	↓	–	↑	↑x?	–	↑x	↓x
ACTH	↓	↑x?	↓	–	↑	↑x?	–	n/a	n/a
PVN-*Crh*	↑	↑?	n/a	n/a	–	–	n/a	↑	↑
CeA-*Crh*	↑	↑	n/a	n/a	n/a	↑	n/a	↑	n/a
EWcp-*Ucn*	n/a	↑?	↑	n/a	n/a	–	n/a	n/a	n/a
LS-*Crhr2*	–	n/a	n/a	n/a	↓	↑	n/a	↓x	↑
DRN-*Crhr2*	–	n/a	n/a	n/a	n/a	↑	n/a	↑	↑x

The top row refers to the deleted gene (or genes). The left-hand column lists the behavior, HPA-axis marker, or component of the CRF system assessed. Up arrow, increased in KO versus WT; down arrow, decreased in KO versus WT; horizontal line, assessed, but no significant effects; n/a, not assessed or not applicable; ?, conflicting effects reported; x, direction of effect depends on the sex, circadian timepoint, or stress level; x?, direction of effect may depend on sex, circadian timepoint, or stress levels, but conflicting effects reported.

While the studies above used candidate gene and candidate mechanisms approaches, studies using nonbiased brain activity mapping approaches also suggested extreme sensitivity of some of the components of the CRF system to ETOH. One of such approaches mapped the induction of inducible transcription factors (ITFs) throughout the brain.[109] In the initial neural mapping studies, experimenters forcibly administered EtOH to rodents, and noted increased protein expression of the prototypical ITF c-Fos in several brain areas, including mesolimbic, amygdalar systems, and the EWcp.[110–113] Further experiments improved the face validity of the neural mapping approach by implementing immunohistochemistry (IHC) for ITFs, following *voluntary* consumption of EtOH. Across several different drinking paradigms, and among numerous rodent strains and species, the Ucn1-containing EWcp was the only brain area that consistently displayed significantly elevated c-Fos expression, following oral EtOH consumption.[104,114–120] The preferential sensitivity of EWcp over other regions to self-administered ETOH suggested involvement of the Ucn1 neurocircuit to the regulation of alcohol drinking.

EFFECTS OF CRF RECEPTORS ON EtOH DRINKING

Over the years, researchers have used various EtOH drinking and exposure paradigms to investigate the involvement of the CRF system in EtOH consumption. Disruption of CRF1 by pharmacological and genetic methods attenuated the enhancements in EtOH drinking, and operant self-administration observed in EtOH-dependent rodents.[97,121–124] Additional studies that performed intra-CeA microinfusions of a nonselective CRF receptor antagonist found similar effects.[125,126] After publication of these findings, some argued that CRF1 antagonists might selectively inhibit EtOH intake in dependent subjects.[127] However, short-term studies provided evidence that CRF1 signaling drives binge drinking, even in a nondependent state,[128–131] possibly via the CeA as well,[98] or the VTA.[132] Disruption of CRF1 signaling by pharmacological blockade or genetic deletion also decreased EtOH drinking in the long-term intermittent procedure,[133–135] and the 2-bottle choice continuous procedure.[136,137] Others have argued that CRF1 manipulations are effective at decreasing drinking only when subjects reach binge levels of EtOH intake,[138,139] but data supporting this hypothesis are lacking.

While most results suggest that CRF1 signaling facilitates EtOH intake, central administration of CRF and Ucn1 unexpectedly *decreased* EtOH drinking.[140–142] These findings are perhaps related to CRF and Ucn1's abilities to decrease food and H_2O intake when administered i.c.v.[143] Moreover, the contribution of CRF1 to alcohol drinking, in the short term limited access procedure, is not specific to ETOH, as consumption of water and food is also decreased by KO of CRF1, or administration of CRF1 antagonists.[128] It can be argued that the CRF1 system contributes to basal consummatory behaviors, whereas ETOH dependence sensitizes this system to manipulation of the CRF1 mechanisms.

The CRF system can influence EtOH-related behavior via inherent genetic differences in the function of its components, and their interactions with stress-related environmental factors. The first identification of a *Crhr1* variant associated with binge drinking in humans was described in 2006,[144] and further studies revealed *Crhr1* genotypes that interacted with prior stress history to influence excessive EtOH intake.[145–148]

In a nonhuman primate model, Dr Christina Barr and coworkers reported that early life stress (maternal separation) increased EtOH consumption during adulthood, but only in monkeys with a functional variant in the promoter of the *Crh* gene that conferred increased sensitivity to glucocorticoids, and stress-induced HPA-axis activation.[149] In a similar vein, the laboratory of Dr Markus Heilig identified an allelic variant in the *Crhr1* promoter that differed in frequency between rats genetically selected for high EtOH preference, and their control line.[150] The variant was associated with *Crhr1* expression in the NAcc and amygdala (upregulated in high-preferring rats versus controls), and conferred increased sensitivity to the effects of CRF1 blockade on stress- and EtOH-related behaviors.[151] Accordingly, Dr Heilig advocated a pharmacogenetic approach, in which genetic sequence analysis may aid in selection of appropriate pharmacotherapy for clinical management of alcoholic patients.[152]

With regard to the CRF2 receptor, deletion of *Crhr2* increased EtOH intake in one limited-access procedure, but had no effect in another.[129,153] Moreover, a quantitative trait analysis of high alcohol preferring (P) and nonpreferring (NP) rat lines identified a chromosomal locus, containing the *Crhr2* gene associated with differences in alcohol consumption and body weight.[154,155] Further genetic analysis revealed an insertion into the *Crhr2* promoter associated with lower CRF2 binding in the amygdala of inbred P rats versus inbred NP rats.[156] Studies focusing on CRF2 ligand Ucn3 found that i.c.v. administration decreased EtOH drinking in nondependent mice,[130,157] but intra-CeA administration had bidirectional effects on EtOH intake, in dependent versus nondependent rats.[158,159] Overall, the data indicate that CRF1 facilitates EtOH consumption, while CRF2 inhibits it. However, there are several exceptions to this framework that require further study.

Ucn1 AND EtOH DRINKING

Since Ucn1 acts on both CRF1 and CRF2 receptors, its contribution to EtOH drinking needs to be reviewed separately. This need is further highlighted by the preferential sensitivity of neural activity in EWcp, the primary source of Ucn1 in the brain, to self-administration of EtOH (Section "Effects of EtOH on the Extrahypothalamic CRF System"). In agreement with a unique role of Ucn1 in regulation of EtOH consumption, a series of studies found higher Ucn1 immunoreactivity in EWcp-Ucn1, in several lines of mice and rats selectively bred for high EtOH consumption.[104,160,161] Further analysis, comparing mRNA in EWcp alcohol-preferring C57BL/6J, and alcohol-avoiding DBA/2J mice, identified a number of transcripts, including Ucn1 mRNA, expressed higher in the preferring lines.[162] These studies suggested that EWcp Ucn1 neurons regulate predisposition to alcohol drinking. To test a functional role for the EWcp in voluntary 2-bottle choice EtOH drinking, the Ryabinin laboratory measured EtOH consumption in high-drinking C57BL/6J mice that received either electrolytic lesion of the EWcp, or sham control surgery. Surgical ablation of the EWcp attenuated intake of, and preference for, 3, 6, and 10% EtOH. In contrast, it did not attenuate preference for sucrose, saccharin, or saline over water,[47,163] supporting the hypothesis that EWcp neurons play an important and selective role in voluntary EtOH consumption.[164] Among lesioned mice, Ucn1-IR fibers were decreased in the LS and DRN, suggesting that these brain areas might mediate Ucn1's effects on preference drinking. However, a functional link between Ucn1 expression and EtOH drinking remained to be determined, as Ucn1 is only one of several neuropeptide system components that are highly enriched within the EWcp.[162] Furthermore, Ucn1 neurons are intermingled with DAergic neurons of the adjacent rostral linear nucleus of the raphe (RLi), suggesting that the EWcp could be regulated by local DA release.[165,166] Indeed, inhibition of VTA neurons, via site-specific activation of the GABA-A receptor, or the autoinhibitory dopamine type-2 receptor (Drd2) increased c-Fos expression in the EWcp.[167] Thus, additional techniques were required in order to determine whether EWcp lesions affected EtOH drinking specifically, via disruption of Ucn1 function.

One approach to address the specificity of Ucn1's contribution to EtOH drinking was to combine EWcp lesions with the genetic deletion of Ucn1. A study of 2-bottle choice drinking, performed in Ucn1 KO and WT littermate mice, showed that both deletion of Ucn1 and lesions of EWcp attenuated EtOH preference over water. Moreover, the lesions did not further decrease EtOH preference in the Ucn1 KO mice, indicating that this phenotype is mediated by Ucn1 from EWcp. The effect on ethanol intake was more complex. Specifically, lesions of EWcp attenuated intake of 10% EtOH, in both Ucn1 KO and WT littermates,[168] an effect suggesting that other neurotransmitter systems, besides Ucn1 in EWcp, also contribute to regulation of EtOH intake. Further experiments showed that genetic deletion of *Ucn* attenuated ethanol-induced conditioned place preference, but not ethanol-induced conditioned place aversion, indicating that Ucn1 contributes to regulation of EtOH drinking, by promoting rewarding properties of EtOH, and not through nonspecific effects on EtOH sensitivity, or amnestic effects. A similar attenuation of ethanol-induced conditioned place preference was also observed in CRF2 KO mice, suggesting that Ucn1 promotes rewarding properties of EtOH, by acting on CRF2 receptors.[168]

INTERACTIONS OF CRF SYSTEM, STRESS, EtOH DRINKING AND DEPENDENCE

Effects of stress on EtOH drinking are bidirectional, depending on the stressor length and the developmental timepoint, as reviewed elsewhere in detail.[169] With respect to the CRF system, either genetic KO or pharmacological blockade of CRF1 blunted the increased EtOH intake, observed following forced swim stress, or social defeat stress.[137,170,171] Furthermore, mice lacking both CRF1 and CRF2 receptors showed a blunted response to the delayed effects of repeated swim stress on increased EtOH drinking in the 2-BC CA procedure.[137] One study found that mice lacking CRF1 displayed an *enhanced* stress-induced increase in EtOH drinking,[172] although this effect was later explained by loss of CRF1 in the pituitary, rather than the brain.[170] Overall, these data suggest that stress can be a motivating factor for excessive EtOH drinking, and that the underlying mechanisms likely rely on complex contributions from the CRF system.

In 2002, the laboratory of Dr George Koob reported that, relative to air-exposed control rats, EtOH vapor-dependent rats displayed increased anxiety-like behavior at both acute (2 h), and protracted (5 weeks) timepoints of withdrawal.[173] This state of heightened anxiety was also observed 6 weeks following cessation of a liquid EtOH diet, and was reversed by i.c.v. administration of either a nonselective CRF receptor antagonist, or the CRF2 agonist Ucn3.[159,174] Observing similarities between the effects of CRF system manipulations on enhanced EtOH drinking, and enhanced anxiety-like behavior in the postdependent state, Dr Koob hypothesized that EtOH dependence produces adaptations in the CRF system that allow a negative affective state to predominate during EtOH withdrawal. In this model, EtOH dependence recruits CRF and other anxiogenic neuropeptide systems, thus producing a stress-like state that permits negative reinforcement processes to drive compulsive EtOH-seeking.[1,2,38]

The transition to EtOH dependence may be characterized by the appearance of physiological adaptations in neural circuit function (i.e., neuroplastic changes).[175,176] Dr Marisa Roberto and her coworkers characterized the effects of chronic EtOH exposure on neuropeptide-mediated plasticity in the CeA.[177] CRF increased GABA release from CeA interneurons, and this effect was potentiated in EtOH vapor-dependent rats via a CRF1-dependent mechanism.[97] Furthermore, baseline GABA release was enhanced in the CeA of high EtOH-preferring versus control rats, and retrograde tracing revealed that EtOH specifically activated CRF1-containing, BNST-projecting neurons in the CeA.[178,179] In the BNST, several studies observed effects of EtOH on electrophysiological interactions between CRF, DA, and glutamate.[10,180,181]

Alcohol dependence contributes to relapse in patients trying to maintain sobriety. Attempts to study EtOH relapse-like behavior in rodents focused on the reinstatement model, in which EtOH-seeking was assessed following acquisition and extinction of operant EtOH self-administration. Studies in rats from the laboratory of Dr A.D. Le revealed that behavioral (electric footshock) or pharmacological stressors (i.c.v. CRF or i.p. yohimbine (alpha-2 adrenoreceptor antagonist)) increased operant behavioral responding on an active lever, previously associated with EtOH availability[182]

Stress-induced reinstatement of operant EtOH-seeking was significantly reduced by blockade of CRF1, but not by removal of the adrenal glands,[183–185] suggesting limited involvement of the HPA-axis. Focused on extrahypothalamic CRF systems, the Le group observed significant decreases in stress-induced EtOH reinstatement, following i.c.v. and intramedian raphe nucleus (MRN) administration of a nonselective CRF receptor antagonist.[184–186]

Footshock, CRF, and yohimbine all increased CRF mRNA in the BNST,[187] and the BNST projects to the MRN.[188] Thus, one could speculate that the BNST and MRN are critical nodes in the circuit underlying stress-induced reinstatement of EtOH-seeking. However, neurons in the MRN express both CRF1 and CRF2,[28] and intra-MRN CRF2 antagonists have not yet been tested in the stress-induced reinstatement model. Thus, it remains possible that stress-induced EtOH relapse is mediated via Ucns/CRF2 signaling in the MRN. Even if MRN-mediated effects on reinstatement were mediated entirely by CRF1, the underlying ligand could be either CRF or Ucn1, as both bind CRF1, and both directly innervate the MRN (from the BNST and EWcp, respectively).[44,47,189]

CONCLUSIONS

Taken together, a plethora of studies implicate the CRF system in the regulation of alcohol consumption. However, the contribution of this system is probably not as straightforward as previously suggested. In particular, stress-induced CRF acting on CRF1 receptors, and activating the HPA axis cannot be the only contributor leading to increased alcohol consumption. An intricate play between CRF, Ucns, CRF receptors, and CRFBP, regulates alcohol consumption in a brain region- and context-specific way. Current pharmacological approaches attempting to manipulate the CRF system, in order to combat alcohol addiction, may be too crude for this delicately balanced neurocircuitry regulating many aspects of mental and physiological functions. As novel molecular tools allow specific components of this system to be manipulated in precisely defined populations of neurons, these neurocircuits may soon be clarified, and more sophisticated approaches to treat alcohol addiction may be developed.

Key Facts of the Corticotropin Releasing Factor System

- The mammalian corticoptropin releasing factor (CRF) system includes four peptides: CRF, urocortin (Ucn)1, Ucn2, and Ucn3.
- CRF and Ucns target CRF1 receptor, CRF2 receptor, and CRF-binding protein (CRFBP).
- CRF has high affinity for CRF1 and CRFBP, but not CRF2.
- Ucn1 has high affinity for CRF1, CRF2, and CRFBP.
- Ucn2 and Ucn3 have high affinity for CRF2.

Summary Points

- This chapter focuses on the contribution of the corticotropin releasing factor (CRF) peptide system to the regulation of alcohol consumption.
- The CRF system is more complex than previously thought, and includes four ligands of CRF receptors: CRF, Ucn1, Ucn2, and Ucn3.
- These four ligands are differentially expressed in brain, and involved in regulation of anxiety, mood, and stress-related responses.
- CRF and Ucn1-containing neurons are sensitive to alcohol administration, self-administration, and/or withdrawal from alcohol.
- Manipulations of the CRF and CRF1 affect alcohol consumption in dependent animals, but also can nonspecifically modulate general consummatory behaviors.
- Manipulations of Ucn1 can regulate alcohol consumption in nondependent animals, and this regulation appears to be ethanol-specific.

References

1. Koob GF, Le Moal M. Review. Neurobiological mechanisms for opponent motivational processes in addiction. *Philos Trans R Soc Lond B Biol Sci* 2008;**363**(1507):3113–23.
2. Koob GF. A role for brain stress systems in addiction. *Neuron* 2008;**59**(1):11–34.
3. Koob GF, Roberts AJ, Schulteis G, Parsons LH, Heyser CJ, Hyytia P, et al. Neurocircuitry targets in ethanol reward and dependence. *Alcohol Clin Exp Res* 1998;**22**(1):3–9.
4. Crabbe JC, Phillips TJ, Harris RA, Arends MA, Koob GF. Alcohol-related genes: contributions from studies with genetically engineered mice. *Addict Biol* 2006;**11**(3–4):195–269.
5. Crabbe JC, Harris RA, Koob GF. Preclinical studies of alcohol binge drinking. *Ann N Y Acad Sci* 2011;**1216**:24–40.
6. Harris RA, Trudell JR, Mihic SJ. Ethanol's molecular targets. *Sci Signal* 2008;**1**(28):re7.
7. Howard RJ, Slesinger PA, Davies DL, Das J, Trudell JR, Harris RA. Alcohol-binding sites in distinct brain proteins: the quest for atomic level resolution. *Alcohol Clin Exp Res* 2011;**35**(9):1561–73.
8. Stuber GD, Hopf FW, Hahn J, Cho SL, Guillory A, Bonci A. Voluntary ethanol intake enhances excitatory synaptic strength in the ventral tegmental area. *Alcohol Clin Exp Res* 2008;**32**(10):1714–20.
9. Seif T, Chang SJ, Simms JA, Gibb SL, Dadgar J, Chen BT, et al. Cortical activation of accumbens hyperpolarization-active NMDARs mediates aversion-resistant alcohol intake. *Nat Neurosci* 2013;**16**(8):1094–100.
10. Silberman Y, Winder DG. Emerging role for corticotropin releasing factor signaling in the bed nucleus of the stria terminalis at the intersection of stress and reward. *Front Psychiatry* 2013;**4**:42.
11. Wills TA, Klug JR, Silberman Y, Baucum AJ, Weitlauf C, Colbran RJ, et al. GluN2B subunit deletion reveals key role in acute and chronic ethanol sensitivity of glutamate synapses in bed nucleus of the stria terminalis. *Proc Natl Acad Sci USA* 2012;**109**(5):E278–87.
12. Roberto M, Gilpin NW, Siggins GR. The central amygdala and alcohol: role of gamma-aminobutyric acid, glutamate, and neuropeptides. *Cold Spring Harb Perspect Med* 2012;**2**(12):a012195.
13. Lobo IA, Harris RA. GABA(A) receptors and alcohol. *Pharmacol Biochem Behav* 2008;**90**(1):90–4.
14. Helms CM, Rossi DJ, Grant KA. Neurosteroid influences on sensitivity to ethanol. *Front Endocrinol (Lausanne)* 2012;**3**:10.
15. Finn DA, Beckley EH, Kaufman KR, Ford MM. Manipulation of GABAergic steroids: Sex differences in the effects on alcohol drinking- and withdrawal-related behaviors. *Horm Behav* 2010;**57**(1):12–22.
16. Cunningham CL, Howard MA, Gill SJ, Rubinstein M, Low MJ, Grandy DK. Ethanol-conditioned place preference is reduced in dopamine D2 receptor-deficient mice. *Pharmacol Biochem Behav* 2000;**67**(4):693–9.
17. Bechtholt AJ, Cunningham CL. Ethanol-induced conditioned place preference is expressed through a ventral tegmental area dependent mechanism. *Behav Neurosci* 2005;**119**(1):213–23.
18. Gremel CM, Cunningham CL. Roles of the nucleus accumbens and amygdala in the acquisition and expression of ethanol-conditioned behavior in mice. *J Neurosci* 2008;**28**(5):1076–84.
19. Gremel CM, Cunningham CL. Involvement of amygdala dopamine and nucleus accumbens NMDA receptors in ethanol-seeking behavior in mice. *Neuropsychopharmacology* 2009;**34**(6):1443–53.
20. Gremel CM, Young EA, Cunningham CL. Blockade of opioid receptors in anterior cingulate cortex disrupts ethanol-seeking behavior in mice. *Behav Brain Res* 2011;**219**(2):358–62.
21. Young EA, Dreumont SE, Cunningham CL. Role of nucleus accumbens dopamine receptor subtypes in the learning and expression of alcohol-seeking behavior. *Neurobiol Learn Mem* 2014;**108**:28–37.
22. Tanchuck MA, Yoneyama N, Ford MM, Fretwell AM, Finn DA. Assessment of GABA-B, metabotropic glutamate, and opioid receptor involvement in an animal model of binge drinking. *Alcohol* 2011;**45**(1):33–44.
23. Rice OV, Patrick J, Schonhar CD, Ning H, Ashby Jr CR. The effects of the preferential dopamine D(3) receptor antagonist S33138 on ethanol binge drinking in C57BL/6J mice. *Synapse* 2012;**66**(11):975–8.
24. Sabino V, Kwak J, Rice KC, Cottone P. Pharmacological characterization of the 20% alcohol intermittent access model in sardinian alcohol-preferring rats: a model of binge-like drinking. *Alcohol Clin Exp Res* 2013;**37**(4):635–43.
25. Bale TL, Vale WW. CRF and CRF receptors: role in stress responsivity and other behaviors. *Annu Rev Pharmacol Toxicol* 2004;**44**:525–57.
26. Giardino WJ, Ryabinin AE. Corticotropin-releasing factor: innocent until proven guilty. *Nat Rev Neurosci* 2012;**13**(1):70.
27. Chalmers DT, Lovenberg TW, De Souza EB. Localization of novel corticotropin-releasing factor receptor (CRF2) mRNA expression to specific subcortical nuclei in rat brain: comparison with CRF1 receptor mRNA expression. *J Neurosci* 1995;**15**(10):6340–50.
28. Van Pett K, Viau V, Bittencourt JC, Chan RK, Li HY, Arias C, et al. Distribution of mRNAs encoding CRF receptors in brain and pituitary of rat and mouse. *J Comp Neurol* 2000;**428**(2):191–212.

29. Reul JM, Holsboer F. Corticotropin-releasing factor receptors 1 and 2 in anxiety and depression. *Curr Opin Pharmacol* 2002;**2**(1):23–33.
30. Chen A, Perrin M, Brar B, Li C, Jamieson P, Digruccio M, et al. Mouse corticotropin-releasing factor receptor type 2alpha gene: isolation, distribution, pharmacological characterization and regulation by stress and glucocorticoids. *Mol Endocrinol* 2005;**19**(2):441–58.
31. Fuenzalida J, Galaz P, Araya KA, Slater PG, Blanco EH, Campusano JM, et al. Dopamine D1 and corticotrophin releasing hormone type-2alpha receptors assemble into functionally interacting complexes in living cells. *Br J Pharmacol* 2014;**171**(24):5650–64.
32. Swanson LW, Sawchenko PE, Rivier J, Vale WW. Organization of ovine corticotropin-releasing factor immunoreactive cells and fibers in the rat brain: an immunohistochemical study. *Neuroendocrinology* 1983;**36**(3):165–86.
33. Elliott E, Ezra-Nevo G, Regev L, Neufeld-Cohen A, Chen A. Resilience to social stress coincides with functional DNA methylation of the Crf gene in adult mice. *Nat Neurosci* 2010;**13**(11):1351–3.
34. Bonfiglio JJ, Inda C, Refojo D, Holsboer F, Arzt E, Silberstein S. The corticotropin-releasing hormone network and the hypothalamic-pituitary-adrenal axis: molecular and cellular mechanisms involved. *Neuroendocrinology* 2010;**94**(1):12–20.
35. Makino S, Hashimoto K, Gold PW. Multiple feedback mechanisms activating corticotropin-releasing hormone system in the brain during stress. *Pharmacol Biochem Behav* 2002;**73**(1):147–58.
36. Regev L, Tsoory M, Gil S, Chen A. Site-specific genetic manipulation of amygdala corticotropin-releasing factor reveals its imperative role in mediating behavioral response to challenge. *Biol Psychiatry* 2012;**71**(4):317–26.
37. Regev L, Neufeld-Cohen A, Tsoory M, Kuperman Y, Getselter D, Gil S, et al. Prolonged and site-specific over-expression of corticotropin-releasing factor reveals differential roles for extended amygdala nuclei in emotional regulation. *Mol Psychiatry* 2011;**16**(7):714–28.
38. Koob GF. The role of CRF and CRF-related peptides in the dark side of addiction. *Brain Res* 2010;**1314**:3–14.
39. Koob GF. Corticotropin-releasing factor, norepinephrine, and stress. *Biol Psychiatry* 1999;**46**(9):1167–80.
40. Vaughan J, Donaldson C, Bittencourt J, Perrin MH, Lewis K, Sutton S, et al. Urocortin, a mammalian neuropeptide related to fish urotensin I and to corticotropin-releasing factor. *Nature* 1995;**378**(6554):287–92.
41. Kozicz T, Yanaihara H, Arimura A. Distribution of urocortin-like immunoreactivity in the central nervous system of the rat. *J Comp Neurol* 1998;**391**(1):1–10.
42. Bittencourt JC, Vaughan J, Arias C, Rissman RA, Vale WW, Sawchenko PE. Urocortin expression in rat brain: evidence against a pervasive relationship of urocortin-containing projections with targets bearing type 2 CRF receptors. *J Comp Neurol* 1999;**415**(3):285–312.
43. Vetter DE, Li C, Zhao L, Contarino A, Liberman MC, Smith GW, et al. Urocortin-deficient mice show hearing impairment and increased anxiety-like behavior. *Nat Genet* 2002;**31**(4):363–9.
44. Weitemier AZ, Tsivkovskaia NO, Ryabinin AE. Urocortin 1 distribution in mouse brain is strain-dependent. *Neuroscience* 2005;**132**(3):729–40.
45. Loewy AD, Saper CB, Yamodis ND. Re-evaluation of the efferent projections of the Edinger-Westphal nucleus in the cat. *Brain Res* 1978;**141**(1):153–9.
46. Loewy AD, Saper CB. Edinger-Westphal nucleus: projections to the brain stem and spinal cord in the cat. *Brain Res* 1978;**150**(1):1–27.
47. Bachtell RK, Weitemier AZ, Ryabinin AE. Lesions of the Edinger-Westphal nucleus in C57BL/6J mice disrupt ethanol-induced hypothermia and ethanol consumption. *Eur J Neurosci* 2004;**20**(6):1613–23.
48. Reyes TM, Lewis K, Perrin MH, Kunitake KS, Vaughan J, Arias CA, et al. Urocortin II: a member of the corticotropin-releasing factor (CRF) neuropeptide family that is selectively bound by type 2 CRF receptors. *Proc Natl Acad Sci USA* 2001;**98**(5):2843–8.
49. Tanaka Y, Makino S, Noguchi T, Tamura K, Kaneda T, Hashimoto K. Effect of stress and adrenalectomy on urocortin II mRNA expression in the hypothalamic paraventricular nucleus of the rat. *Neuroendocrinology* 2003;**78**(1):1–11.
50. Kageyama K, Li C, Vale WW. Corticotropin-releasing factor receptor type 2 messenger ribonucleic acid in rat pituitary: localization and regulation by immune challenge, restraint stress, and glucocorticoids. *Endocrinology* 2003;**144**(4):1524–32.
51. Nemoto T, Iwasaki-Sekino A, Yamauchi N, Shibasaki T. Regulation of the expression and secretion of urocortin 2 in rat pituitary. *J Endocrinol* 2007;**192**(2):443–52.
52. Nemoto T, Iwasaki-Sekino A, Yamauchi N, Shibasaki T. Role of urocortin 2 secreted by the pituitary in the stress-induced suppression of luteinizing hormone secretion in rats. *Am J Physiol Endocrinol Metab* 2010;**299**(4):E567–75.
53. Lewis K, Li C, Perrin MH, Blount A, Kunitake K, Donaldson C, et al. Identification of urocortin III, an additional member of the corticotropin-releasing factor (CRF) family with high affinity for the CRF2 receptor. *Proc Natl Acad Sci USA* 2001;**98**(13):7570–5.
54. Li C, Vaughan J, Sawchenko PE, Vale WW. Urocortin III-immunoreactive projections in rat brain: partial overlap with sites of type 2 corticotrophin-releasing factor receptor expression. *J Neurosci* 2002;**22**(3):991–1001.
55. Deussing JM, Breu J, Kuhne C, Kallnik M, Bunck M, Glasl L, et al. Urocortin 3 modulates social discrimination abilities via corticotropin-releasing hormone receptor type 2. *J Neurosci* 2010;**30**(27):9103–16.
56. Wittmann G, Fuzesi T, Liposits Z, Lechan RM, Fekete C. Distribution and axonal projections of neurons coexpressing thyrotropin-releasing hormone and urocortin 3 in the rat brain. *J Comp Neurol* 2009;**517**(6):825–40.
57. Cavalcante JC, Sita LV, Mascaro MB, Bittencourt JC, Elias CF. Distribution of urocortin 3 neurons innervating the ventral premammillary nucleus in the rat brain. *Brain Res* 2006;**1089**(1):116–25.
58. Jamieson PM, Li C, Kukura C, Vaughan J, Vale W. Urocortin 3 modulates the neuroendocrine stress response and is regulated in rat amygdala and hypothalamus by stress and glucocorticoids. *Endocrinology* 2006;**147**(10):4578–88.
59. Gaszner B, Van Wijk DC, Korosi A, Jozsa R, Roubos EW, Kozicz T. Diurnal expression of period 2 and urocortin 1 in neurones of the non-preganglionic Edinger-Westphal nucleus in the rat. *Stress* 2009;**12**(2):115–24.
60. Timpl P, Spanagel R, Sillaber I, Kresse A, Reul JM, Stalla GK, et al. Impaired stress response and reduced anxiety in mice lacking a functional corticotropin-releasing hormone receptor 1. *Nat Genet* 1998;**19**(2):162–6.
61. Smith GW, Aubry JM, Dellu F, Contarino A, Bilezikjian LM, Gold LH, et al. Corticotropin releasing factor receptor 1-deficient mice display decreased anxiety, impaired stress response, and aberrant neuroendocrine development. *Neuron* 1998;**20**(6):1093–102.
62. Coste SC, Kesterson RA, Heldwein KA, Stevens SL, Heard AD, Hollis JH, et al. Abnormal adaptations to stress and impaired cardiovascular function in mice lacking corticotropin-releasing hormone receptor-2. *Nat Genet* 2000;**24**(4):403–9.
63. Kishimoto T, Radulovic J, Radulovic M, Lin CR, Schrick C, Hooshmand F, et al. Deletion of crhr2 reveals an anxiolytic role for corticotropin-releasing hormone receptor-2. *Nat Genet* 2000;**24**(4):415–9.
64. Bale TL, Contarino A, Smith GW, Chan R, Gold LH, Sawchenko PE, et al. Mice deficient for corticotropin-releasing hormone receptor-2 display anxiety-like behaviour and are hypersensitive to stress. *Nat Genet* 2000;**24**(4):410–4.

65. Karolyi IJ, Burrows HL, Ramesh TM, Nakajima M, Lesh JS, Seong E, et al. Altered anxiety and weight gain in corticotropin-releasing hormone-binding protein-deficient mice. *Proc Natl Acad Sci USA* 1999;**96**(20):11595–600.
66. Muglia L, Jacobson L, Dikkes P, Majzoub JA. Corticotropin-releasing hormone deficiency reveals major fetal but not adult glucocorticoid need. *Nature* 1995;**373**(6513):427–32.
67. Weninger SC, Dunn AJ, Muglia LJ, Dikkes P, Miczek KA, Swiergiel AH, et al. Stress-induced behaviors require the corticotropin-releasing hormone (CRH) receptor, but not CRH. *Proc Natl Acad Sci USA* 1999;**96**(14):8283–8.
68. Wang X, Su H, Copenhagen LD, Vaishnav S, Pieri F, Shope CD, et al. Urocortin-deficient mice display normal stress-induced anxiety behavior and autonomic control but an impaired acoustic startle response. *Mol Cell Biol* 2002;**22**(18):6605–10.
69. Zalutskaya AA, Arai M, Bounoutas GS, Abou-Samra AB. Impaired adaptation to repeated restraint and decreased response to cold in urocortin 1 knockout mice. *Am J Physiol Endocrinol Metab* 2007;**293**(1):E259–63.
70. Chen A, Zorrilla E, Smith S, Rousso D, Levy C, Vaughan J, et al. Urocortin 2-deficient mice exhibit gender-specific alterations in circadian hypothalamus-pituitary-adrenal axis and depressive-like behavior. *J Neurosci* 2006;**26**(20):5500–10.
71. Breu J, Touma C, Holter SM, Knapman A, Wurst W, Deussing JM. Urocortin 2 modulates aspects of social behaviour in mice. Behav. *Brain Res* 2012;**233**(2):331–6.
72. Neufeld-Cohen A, Evans AK, Getselter D, Spyroglou A, Hill A, Gil S, et al. Urocortin-1 and -2 double-deficient mice show robust anxiolytic phenotype and modified serotonergic activity in anxiety circuits. *Mol Psychiatry* 2010;**15**(4):426–41 339.
73. Neufeld-Cohen A, Tsoory MM, Evans AK, Getselter D, Gil S, Lowry CA, et al. A triple urocortin knockout mouse model reveals an essential role for urocortins in stress recovery. *Proc Natl Acad Sci USA* 2010;**107**(44):19020–5.
74. Kozicz T. The missing link; the significance of urocortin 1/urocortin 2 in the modulation of the dorsal raphe serotoninergic system. *Mol Psychiatry* 2010;**15**(4):340–1.
75. Fabre V, Massart R, Rachalski A, Jennings K, Brass A, Sharp T, et al. Differential gene expression in mutant mice overexpressing or deficient in the serotonin transporter: a focus on urocortin 1. *Eur Neuropsychopharmacol* 2011;**21**(1):33–44.
76. Bethea CL, Lima FB, Centeno ML, Weissheimer KV, Senashova O, Reddy AP, et al. Effects of citalopram on serotonin and CRF systems in the midbrain of primates with differences in stress sensitivity. *J Chem Neuroanat* 2011;**41**(4):200–18.
77. Valentino RJ, Lucki I, Van Bockstaele E. Corticotropin-releasing factor in the dorsal raphe nucleus: Linking stress coping and addiction. *Brain Res* 2010;**1314**:29–37.
78. Vuong SM, Oliver HA, Scholl JL, Oliver KM, Forster GL. Increased anxiety-like behavior of rats during amphetamine withdrawal is reversed by CRF2 receptor antagonism. *Behav Brain Res* 2010;**208**(1):278–81.
79. Lukkes JL, Forster GL, Renner KJ, Summers CH. Corticotropin-releasing factor 1 and 2 receptors in the dorsal raphe differentially affect serotonin release in the nucleus accumbens. *Eur J Pharmacol* 2008;**578**(2–3):185–93.
80. Armario A. Activation of the hypothalamic-pituitary-adrenal axis by addictive drugs: different pathways, common outcome. *Trends Pharmacol Sci* 2010;**31**(7):318–25.
81. Rivier C, Bruhn T, Vale W. Effect of ethanol on the hypothalamic-pituitary-adrenal axis in the rat: role of corticotropin-releasing factor (CRF). *J Pharmacol Exp Ther* 1984;**229**(1):127–31.
82. Rivest S, Rivier C. Lesions of hypothalamic PVN partially attenuate stimulatory action of alcohol on ACTH secretion in rats. *Am J Physiol* 1994;**266**(2 Pt. 2):R553–8.
83. Rivier C, Rivier J, Lee S. Importance of pituitary and brain receptors for corticotrophin-releasing factor in modulating alcohol-induced ACTH secretion in the rat. *Brain Res* 1996;**721**(1–2):83–90.
84. Rivier C. Alcohol stimulates ACTH secretion in the rat: mechanisms of action and interactions with other stimuli. *Alcohol Clin Exp Res* 1996;**20**(2):240–54.
85. Rivier C, Vale W. Interaction between ethanol and stress on ACTH and beta-endorphin secretion. *Alcohol Clin Exp Res* 1988;**12**(2):206–10.
86. Rivier C, Imaki T, Vale W. Prolonged exposure to alcohol: effect on CRF mRNA levels, and CRF- and stress-induced ACTH secretion in the rat. *Brain Res* 1990;**520**(1–2):1–5.
87. Richardson HN, Lee SY, O'Dell LE, Koob GF, Rivier CL. Alcohol self-administration acutely stimulates the hypothalamic-pituitary-adrenal axis, but alcohol dependence leads to a dampened neuroendocrine state. *Eur J Neurosci* 2008;**28**(8):1641–53.
88. Lee S, Imaki T, Vale W, Rivier C. Effect of prenatal exposure to ethanol on the activity of the hypothalamic-pituitary-adrenal axis of the offspring: Importance of the time of exposure to ethanol and possible modulating mechanisms. *Mol Cell Neurosci* 1990;**1**(2):168–77.
89. Hellemans KG, Sliwowska JH, Verma P, Weinberg J. Prenatal alcohol exposure: fetal programming and later life vulnerability to stress, depression and anxiety disorders. *Neurosci Biobehav Rev* 2010;**34**(6):791–807.
90. Ogilvie KM, Lee S, Rivier C. Role of arginine vasopressin and corticotropin-releasing factor in mediating alcohol-induced adrenocorticotropin and vasopressin secretion in male rats bearing lesions of the paraventricular nuclei. *Brain Res* 1997;**744**(1):83–95.
91. Wills TA, Knapp DJ, Overstreet DH, Breese GR. Interactions of stress and CRF in ethanol-withdrawal induced anxiety in adolescent and adult rats. *Alcohol Clin Exp Res* 2010;**34**(9):1603–12.
92. Zhou Y, Franck J, Spangler R, Maggos CE, Ho A, Kreek MJ. Reduced hypothalamic POMC and anterior pituitary CRF1 receptor mRNA levels after acute, but not chronic, daily "binge" intragastric alcohol administration. *Alcohol Clin Exp Res* 2000;**24**(10):1575–82.
93. Lee S, Rivier C. Alcohol increases the expression of type 1, but not type 2 alpha corticotropin-releasing factor (CRF) receptor messenger ribonucleic acid in the rat hypothalamus. *Brain Res Mol Brain Res* 1997;**52**(1):78–89.
94. Rivier C, Lee S. Acute alcohol administration stimulates the activity of hypothalamic neurons that express corticotropin-releasing factor and vasopressin. *Brain Res* 1996;**726**(1–2):1–10.
95. Zorrilla EP, Valdez GR, Weiss F. Changes in levels of regional CRF-like-immunoreactivity and plasma corticosterone during protracted drug withdrawal in dependent rats. *Psychopharmacology (Berl)* 2001;**158**(4):374–81.
96. Sommer WH, Rimondini R, Hansson AC, Hipskind PA, Gehlert DR, Barr CS, et al. Upregulation of voluntary alcohol intake, behavioral sensitivity to stress, and amygdala crhr1 expression following a history of dependence. *Biol Psychiatry* 2008;**63**(2):139–45.
97. Roberto M, Cruz MT, Gilpin NW, Sabino V, Schweitzer P, Bajo M, et al. Corticotropin releasing factor-induced amygdala gamma-aminobutyric Acid release plays a key role in alcohol dependence. *Biol Psychiatry* 2010;**67**(9):831–9.
98. Lowery-Gionta EG, Navarro M, Li C, Pleil KE, Rinker JA, Cox BR, et al. Corticotropin releasing factor signaling in the central amygdala is recruited during binge-like ethanol consumption in C57BL/6J mice. *J Neurosci* 2012;**32**(10):3405–13.
99. Karanikas CA, Lu YL, Richardson HN. Adolescent drinking targets corticotropin-releasing factor peptide-labeled cells in the central amygdala of male and female rats. *Neuroscience* 2013;**249**:98–105.

100. Gilpin NW, Karanikas CA, Richardson HN. Adolescent binge drinking leads to changes in alcohol drinking, anxiety, and amygdalar corticotropin releasing factor cells in adulthood in male rats. *PLoS One* 2012;**7**(2):e31466.
101. Falco AM, Bergstrom HC, Bachus SE, Smith RF. Persisting changes in basolateral amygdala mRNAs after chronic ethanol consumption. *Physiol Behav* 2009;**96**(1):169–73.
102. Walker BM, Drimmer DA, Walker JL, Liu T, Mathe AA, Ehlers CL. Effects of prolonged ethanol vapor exposure on forced swim behavior, and neuropeptide Y and corticotropin-releasing factor levels in rat brains. *Alcohol* 2010;**44**(6):487–93.
103. Hansson AC, Cippitelli A, Sommer WH, Ciccocioppo R, Heilig M. Region-specific down-regulation of Crhr1 gene expression in alcohol-preferring msP rats following ad lib access to alcohol. *Addict Biol* 2007;**12**(1):30–4.
104. Bachtell RK, Weitemier AZ, Galvan-Rosas A, Tsivkovskaia NO, Risinger FO, Phillips TJ, et al. The Edinger-Westphal-lateral septum urocortin pathway and its relationship to alcohol consumption. *J Neurosci* 2003;**23**(6):2477–87.
105. Bachtell RK, Tsivkovskaia NO, Ryabinin AE. Strain differences in urocortin expression in the Edinger-Westphal nucleus and its relation to alcohol-induced hypothermia. *Neuroscience* 2002;**113**(2):421–34.
106. Weitemier AZ, Ryabinin AE. Brain region-specific regulation of urocortin 1 innervation and corticotropin-releasing factor receptor type 2 binding by ethanol exposure. *Alcohol Clin Exp Res* 2005;**29**(9):1610–20.
107. Merlo Pich E, Lorang M, Yeganeh M, Rodriguez de Fonseca F, Raber J, Koob GF, et al. Increase of extracellular corticotropin-releasing factor-like immunoreactivity levels in the amygdala of awake rats during restraint stress and ethanol withdrawal as measured by microdialysis. *J Neurosci* 1995;**15**(8):5439–47.
108. Olive MF, Koenig HN, Nannini MA, Hodge CW. Elevated extracellular CRF levels in the bed nucleus of the stria terminalis during ethanol withdrawal and reduction by subsequent ethanol intake. *Pharmacol Biochem Behav* 2002;**72**(1–2):213–20.
109. Morgan JI, Cohen DR, Hempstead JL, Curran T. Mapping patterns of c-fos expression in the central nervous system after seizure. *Science* 1987;**237**(4811):192–7.
110. Chang SL, Patel NA, Romero AA. Activation and desensitization of Fos immunoreactivity in the rat brain following ethanol administration. *Brain Res* 1995;**679**(1):89–98.
111. Ryabinin AE, Criado JR, Henriksen SJ, Bloom FE, Wilson MC. Differential sensitivity of c-Fos expression in hippocampus and other brain regions to moderate and low doses of alcohol. *Mol Psychiatry* 1997;**2**(1):32–43.
112. Knapp DJ, Braun CJ, Duncan GE, Qian Y, Fernandes A, Crews FT, et al. Regional specificity of ethanol and NMDA action in brain revealed with FOS-like immunohistochemistry and differential routes of drug administration. *Alcohol Clin Exp Res* 2001;**25**(11):1662–72.
113. Murphy NP, Sakoori K, Okabe C. Lack of evidence of a role for the neurosteroid allopregnanolone in ethanol-induced reward and c-fos expression in DBA/2 mice. *Brain Res* 2006;**1094**(1):107–18.
114. Anacker AM, Loftis JM, Kaur S, Ryabinin AE. Prairie voles as a novel model of socially facilitated excessive drinking. *Addict Biol* 2011;**16**(1):92–107.
115. Bachtell RK, Wang YM, Freeman P, Risinger FO, Ryabinin AE. Alcohol drinking produces brain region-selective changes in expression of inducible transcription factors. *Brain Res* 1999;**847**(2):157–65.
116. Ryabinin AE, Bachtell RK, Freeman P, Risinger FO. ITF expression in mouse brain during acquisition of alcohol self-administration. *Brain Res* 2001;**890**(1):192–5.
117. Ryabinin AE, Galvan-Rosas A, Bachtell RK, Risinger FO. High alcohol/sucrose consumption during dark circadian phase in C57BL/6J mice: involvement of hippocampus, lateral septum and urocortin-positive cells of the Edinger-Westphal nucleus. *Psychopharmacology (Berl)* 2003;**165**(3):296–305.
118. Sharpe AL, Tsivkovskaia NO, Ryabinin AE. Ataxia and c-Fos expression in mice drinking ethanol in a limited access session. *Alcohol Clin Exp Res* 2005;**29**(8):1419–26.
119. Topple AN, Hunt GE, McGregor IS. Possible neural substrates of beer-craving in rats. *Neurosci Lett* 1998;**252**(2):99–102.
120. Weitemier AZ, Woerner A, Backstrom P, Hyytia P, Ryabinin AE. Expression of c-Fos in Alko alcohol rats responding for ethanol in an operant paradigm. *Alcohol Clin Exp Res* 2001;**25**(5):704–10.
121. Richardson HN, Zhao Y, Fekete EM, Funk CK, Wirsching P, Janda KD, et al. MPZP: a novel small molecule corticotropin-releasing factor type 1 receptor (CRF1) antagonist. *Pharmacol Biochem Behav* 2008;**88**(4):497–510.
122. Chu K, Koob GF, Cole M, Zorrilla EP, Roberts AJ. Dependence-induced increases in ethanol self-administration in mice are blocked by the CRF1 receptor antagonist antalarmin and by CRF1 receptor knockout. *Pharmacol Biochem Behav* 2007;**86**(4):813–21.
123. Gehlert DR, Cippitelli A, Thorsell A, Le AD, Hipskind PA, Hamdouchi C, et al. 3-(4-Chloro-2-morpholin-4-yl-thiazol-5-yl)-8-(1-ethylpropyl)-2,6-dimethyl-imidazo [1,2-b]pyridazine: a novel brain-penetrant, orally available corticotropin-releasing factor receptor 1 antagonist with efficacy in animal models of alcoholism. *J Neurosci* 2007;**27**(10):2718–26.
124. Funk CK, Zorrilla EP, Lee MJ, Rice KC, Koob GF. Corticotropin-releasing factor 1 antagonists selectively reduce ethanol self-administration in ethanol-dependent rats. *Biol Psychiatry* 2007;**61**(1):78–86.
125. Finn DA, Snelling C, Fretwell AM, Tanchuck MA, Underwood L, Cole M, et al. Increased drinking during withdrawal from intermittent ethanol exposure is blocked by the CRF receptor antagonist D-Phe-CRF(12-41). *Alcohol Clin Exp Res* 2007;**31**(6):939–49.
126. Funk CK, O'Dell LE, Crawford EF, Koob GF. Corticotropin-releasing factor within the central nucleus of the amygdala mediates enhanced ethanol self-administration in withdrawn, ethanol-dependent rats. *J Neurosci* 2006;**26**(44):11324–32.
127. Heilig M, Koob GF. A key role for corticotropin-releasing factor in alcohol dependence. *Trends Neurosci* 2007;**30**(8):399–406.
128. Giardino WJ, Ryabinin AE. CRF1 Receptor Signaling Regulates Food and Fluid Intake in the Drinking-in-the-Dark Model of Binge Alcohol Consumption. *Alcohol Clin Exp Res* 2013;**37**(7):1161–70.
129. Kaur S, Li J, Stenzel-Poore MP, Ryabinin AE. Corticotropin-releasing factor acting on corticotropin-releasing factor receptor type 1 is critical for binge alcohol drinking in mice. *Alcohol Clin Exp Res* 2012;**36**(2):369–76.
130. Lowery EG, Spanos M, Navarro M, Lyons AM, Hodge CW, Thiele TE. CRF-1 antagonist and CRF-2 agonist decrease binge-like ethanol drinking in C57BL/6J mice independent of the HPA axis. *Neuropsychopharmacology* 2010;**35**(6):1241–52.
131. Sparta DR, Sparrow AM, Lowery EG, Fee JR, Knapp DJ, Thiele TE. Blockade of the corticotropin releasing factor type 1 receptor attenuates elevated ethanol drinking associated with drinking in the dark procedures. *Alcohol Clin Exp Res* 2008;**32**(2):259–65.
132. Sparta DR, Hopf FW, Gibb SL, Cho SL, Stuber GD, Messing RO, et al. Binge ethanol-drinking potentiates corticotropin releasing factor R1 receptor activity in the ventral tegmental area. *Alcohol Clin Exp Res* 2013;**37**(10):1680–7.
133. Hwa LS, Debold JF, Miczek KA. Alcohol in excess: CRF(1) receptors in the rat and mouse VTA and DRN. *Psychopharmacology (Berl)* 2013;**225**(2):313–27.
134. Simms JA, Nielsen CK, Li R, Bartlett SE. Intermittent access ethanol consumption dysregulates CRF function in the hypothalamus and is attenuated by the CRF-R1 antagonist, CP-376395. *Addict Biol* 2014;**19**(4):606–11.

135. Cippitelli A, Damadzic R, Singley E, Thorsell A, Ciccocioppo R, Eskay RL, et al. Pharmacological blockade of corticotropin-releasing hormone receptor 1 (CRH1R) reduces voluntary consumption of high alcohol concentrations in non-dependent Wistar rats. *Pharmacol Biochem Behav* 2011;**100**(3):522–9.
136. Lodge DJ, Lawrence AJ. The CRF1 receptor antagonist antalarmin reduces volitional ethanol consumption in isolation-reared fawn-hooded rats. *Neuroscience* 2003;**117**(2):243–7.
137. Pastor R, Reed C, Burkhart-Kasch S, Li N, Sharpe AL, Coste SC, et al. Ethanol concentration-dependent effects and the role of stress on ethanol drinking in corticotropin-releasing factor type 1 and double type 1 and 2 receptor knockout mice. *Psychopharmacology (Berl)* 2011;**218**(1):169–77.
138. Thiele TE. Commentary: studies on binge-like ethanol drinking may help to identify the neurobiological mechanisms underlying the transition to dependence. *Alcohol Clin Exp Res* 2012;**36**(2):193–6.
139. Lowery EG, Thiele TE. Pre-clinical evidence that corticotropin-releasing factor (CRF) receptor antagonists are promising targets for pharmacological treatment of alcoholism. *CNS Neurol Disord Drug Targets* 2010;**9**(1):77–86.
140. Bell SM, Reynolds JG, Thiele TE, Gan J, Figlewicz DP, Woods SC. Effects of third intracerebroventricular injections of corticotropin-releasing factor (CRF) on ethanol drinking and food intake. *Psychopharmacology (Berl)* 1998;**139**(1–2):128–35.
141. Thorsell A, Slawecki CJ, Ehlers CL. Effects of neuropeptide Y and corticotropin-releasing factor on ethanol intake in Wistar rats: interaction with chronic ethanol exposure. *Behav Brain Res* 2005;**161**(1):133–40.
142. Ryabinin AE, Yoneyama N, Tanchuck MA, Mark GP, Finn DA. Urocortin 1 microinjection into the mouse lateral septum regulates the acquisition and expression of alcohol consumption. *Neuroscience* 2008;**151**(3):780–90.
143. Spina M, Merlo-Pich E, Chan RK, Basso AM, Rivier J, Vale W, et al. Appetite-suppressing effects of urocortin, a CRF-related neuropeptide. *Science* 1996;**273**(5281):1561–4.
144. Treutlein J, Kissling C, Frank J, Wiemann S, Dong L, Depner M, et al. Genetic association of the human corticotropin releasing hormone receptor 1 (CRHR1) with binge drinking and alcohol intake patterns in two independent samples. *Mol Psychiatry* 2006;**11**(6):594–602.
145. Ray LA, Sehl M, Bujarski S, Hutchison K, Blaine S, Enoch MA. The CRHR1 gene, trauma exposure, and alcoholism risk: a test of G x E effects. *Genes Brain Behav* 2013;**12**(4):361–9.
146. Nelson EC, Agrawal A, Pergadia ML, Wang JC, Whitfield JB, Saccone FS, et al. H2 haplotype at chromosome 17q21.31 protects against childhood sexual abuse-associated risk for alcohol consumption and dependence. *Addict Biol* 2010;**15**(1):1–11.
147. Schmid B, Blomeyer D, Treutlein J, Zimmermann US, Buchmann AF, Schmidt MH, et al. Interacting effects of CRHR1 gene and stressful life events on drinking initiation and progression among 19-year-olds. *Int J Neuropsychopharmacol* 2010;**13**(6):703–14.
148. Blomeyer D, Treutlein J, Esser G, Schmidt MH, Schumann G, Laucht M. Interaction between CRHR1 gene and stressful life events predicts adolescent heavy alcohol use. *Biol Psychiatry* 2008;**63**(2):146–51.
149. Barr CS, Dvoskin RL, Gupte M, Sommer W, Sun H, Schwandt ML, et al. Functional CRH variation increases stress-induced alcohol consumption in primates. *Proc Natl Acad Sci USA* 2009;**106**(34):14593–8.
150. Hansson AC, Cippitelli A, Sommer WH, Fedeli A, Bjork K, Soverchia L, et al. Variation at the rat Crhr1 locus and sensitivity to relapse into alcohol seeking induced by environmental stress. *Proc Natl Acad Sci USA* 2006;**103**(41):15236–41.
151. Ayanwuyi LO, Carvajal F, Lerma-Cabrera JM, Domi E, Bjork K, Ubaldi M, et al. Role of a genetic polymorphism in the corticotropin-releasing factor receptor 1 gene in alcohol drinking and seeking behaviors of marchigian sardinian alcohol-preferring rats. *Front Psychiatry* 2013;**4**:23.
152. Heilig M, Goldman D, Berrettini W, O'Brien CP. Pharmacogenetic approaches to the treatment of alcohol addiction. *Nat Rev Neurosci* 2011;**12**(11):670–84.
153. Sharpe AL, Coste SC, Burkhart-Kasch S, Li N, Stenzel-Poore MP, Phillips TJ. Mice deficient in corticotropin-releasing factor receptor type 2 exhibit normal ethanol-associated behaviors. *Alcohol Clin Exp Res* 2005;**29**(9):1601–9.
154. Carr LG, Foroud T, Bice P, Gobbett T, Ivashina J, Edenberg H, et al. A quantitative trait locus for alcohol consumption in selectively bred rat lines. *Alcohol Clin Exp Res* 1998;**22**(4):884–7.
155. Spence JP, Lai D, Shekhar A, Carr LG, Foroud T, Liang T. Quantitative trait locus for body weight identified on rat chromosome 4 in inbred alcohol-preferring and -nonpreferring rats: potential implications for neuropeptide Y and corticotrophin releasing hormone 2. *Alcohol* 2013;**47**(1):63–7.
156. Yong W, Spence JP, Eskay R, Fitz SD, Damadzic R, Lai D, et al. Alcohol-preferring rats show decreased corticotropin-releasing hormone-2 receptor expression and differences in HPA activation compared to alcohol-nonpreferring rats. *Alcohol Clin Exp Res* 2014;**38**:1275–83.
157. Sharpe AL, Phillips TJ. Central urocortin 3 administration decreases limited-access ethanol intake in nondependent mice. *Behav Pharmacol* 2009;**20**(4):346–51.
158. Funk CK, Koob GF. A CRF(2) agonist administered into the central nucleus of the amygdala decreases ethanol self-administration in ethanol-dependent rats. *Brain Res* 2007;**1155**:172–8.
159. Valdez GR, Sabino V, Koob GF. Increased anxiety-like behavior and ethanol self-administration in dependent rats: reversal via corticotropin-releasing factor-2 receptor activation. *Alcohol Clin Exp Res* 2004;**28**(6):865–72.
160. Turek VF, Tsivkovskaia NO, Hyytia P, Harding S, Le AD, Ryabinin AE. Urocortin 1 expression in five pairs of rat lines selectively bred for differences in alcohol drinking. *Psychopharmacology (Berl)* 2005;**181**(3):511–7.
161. Fonareva I, Spangler E, Cannella N, Sabino V, Cottone P, Ciccocioppo R, et al. Increased perioculomotor urocortin 1 immunoreactivity in genetically selected alcohol preferring rats. *Alcohol Clin Exp Res* 2009;**33**(11):1956–65.
162. Giardino WJ, Cote DM, Li J, Ryabinin AE. Characterization of genetic differences within the centrally projecting Edinger-Westphal nucleus of C57BL/6J and DBA/2J mice by expression profiling. *Front Neuroanat* 2012;**6**:5.
163. Weitemier AZ, Ryabinin AE. Lesions of the Edinger-Westphal nucleus alter food and water consumption. *Behav Neurosci* 2005;**119**(5):1235–43.
164. Ryabinin AE, Weitemier AZ. The urocortin 1 neurocircuit: ethanol-sensitivity and potential involvement in alcohol consumption. *Brain Res Rev* 2006;**52**(2):368–80.
165. Kozicz T. Axon terminals containing tyrosine hydroxylase- and dopamine-beta-hydroxylase immunoreactivity form synapses with galanin immunoreactive neurons in the lateral division of the bed nucleus of the stria terminalis in the rat. *Brain Res* 2001;**914**(1–2):23–33.
166. Bachtell RK, Tsivkovskaia NO, Ryabinin AE. Alcohol-induced c-Fos expression in the Edinger-Westphal nucleus: pharmacological and signal transduction mechanisms. *J Pharmacol Exp Ther* 2002;**302**(2):516–24.
167. Ryabinin AE, Cocking DL, Kaur S. Inhbition of VTA neurons activates the centrally projecting Edinger-Westphal nucleus: Evidence of a stress-reward link? *J Chem Neuroanat* 2013;**54**:57–61.
168. Giardino WJ, Cocking DL, Kaur S, Cunningham CL, Ryabinin AE. Urocortin-1 within the Centrally-Projecting Edinger-Westphal Nucleus is Crtical for Ethanol Preference. *PLoS One* 2011;**6**(10):e26997.

169. Becker HC, Lopez MF, Doremus-Fitzwater TL. Effects of stress on alcohol drinking: a review of animal studies. *Psychopharmacology (Berl)* 2011;**218**(1):131–56.
170. Molander A, Vengeliene V, Heilig M, Wurst W, Deussing JM, Spanagel R. Brain-specific inactivation of the Crhr1 gene inhibits post-dependent and stress-induced alcohol intake, but does not affect relapse-like drinking. *Neuropsychopharmacology* 2012;**37**(4):1047–56.
171. Lowery EG, Sparrow AM, Breese GR, Knapp DJ, Thiele TE. The CRF-1 receptor antagonist, CP-154,526, attenuates stress-induced increases in ethanol consumption by BALB/cJ mice. *Alcohol Clin Exp Res* 2008;**32**(2):240–8.
172. Sillaber I, Rammes G, Zimmermann S, Mahal B, Zieglgansberger W, Wurst W, et al. Enhanced and delayed stress-induced alcohol drinking in mice lacking functional CRH1 receptors. *Science* 2002;**296**(5569):931–3.
173. Valdez GR, Roberts AJ, Chan K, Davis H, Brennan M, Zorrilla EP, et al. Increased ethanol self-administration and anxiety-like behavior during acute ethanol withdrawal and protracted abstinence: regulation by corticotropin-releasing factor. *Alcohol Clin Exp Res* 2002;**26**(10):1494–501.
174. Valdez GR, Zorrilla EP, Roberts AJ, Koob GF. Antagonism of corticotropin-releasing factor attenuates the enhanced responsiveness to stress observed during protracted ethanol abstinence. *Alcohol* 2003;**29**(2):55–60.
175. Koob GF, Le Moal M. Plasticity of reward neurocircuitry and the 'dark side' of drug addiction. *Nat Neurosci* 2005;**8**(11):1442–4.
176. McCool BA. Ethanol modulation of synaptic plasticity. *Neuropharmacology* 2011;**61**(7):1097–108.
177. Gilpin NW, Roberto M. Neuropeptide modulation of central amygdala neuroplasticity is a key mediator of alcohol dependence. *Neurosci Biobehav Rev* 2012;**36**(2):873–88.
178. Herman MA, Contet C, Justice NJ, Vale W, Roberto M. Novel subunit-specific tonic GABA currents and differential effects of ethanol in the central amygdala of CRF receptor-1 reporter mice. *J Neurosci* 2013;**33**(8):3284–98.
179. Herman MA, Kallupi M, Luu G, Oleata CS, Heilig M, Koob GF, et al. Enhanced GABAergic transmission in the central nucleus of the amygdala of genetically selected Marchigian Sardinian rats: Alcohol and CRF effects. *Neuropharmacology* 2013;**67**:337–48.
180. Kash TL, Nobis WP, Matthews RT, Winder DG. Dopamine enhances fast excitatory synaptic transmission in the extended amygdala by a CRF-R1-dependent process. *J Neurosci* 2008;**28**(51):13856–65.
181. Silberman Y, Matthews RT, Winder DG. A corticotropin releasing factor pathway for ethanol regulation of the ventral tegmental area in the bed nucleus of the stria terminalis. *J Neurosci* 2013;**33**(3):950–60.
182. Le A, Shaham Y. Neurobiology of relapse to alcohol in rats. *Pharmacol Ther* 2002;**94**(1–2):137–56.
183. Marinelli PW, Funk D, Juzytsch W, Harding S, Rice KC, Shaham Y, et al. The CRF1 receptor antagonist antalarmin attenuates yohimbine-induced increases in operant alcohol self-administration and reinstatement of alcohol seeking in rats. *Psychopharmacology (Berl)* 2007;**195**(3):345–55.
184. Le AD, Harding S, Juzytsch W, Fletcher PJ, Shaham Y. The role of corticotropin-releasing factor in the median raphe nucleus in relapse to alcohol. *J Neurosci* 2002;**22**(18):7844–9.
185. Le AD, Harding S, Juzytsch W, Watchus J, Shalev U, Shaham Y. The role of corticotrophin-releasing factor in stress-induced relapse to alcohol-seeking behavior in rats. *Psychopharmacology (Berl)* 2000;**150**(3):317–24.
186. Le AD, Funk D, Coen K, Li Z, Shaham Y. Role of corticotropin-releasing factor in the median raphe nucleus in yohimbine-induced reinstatement of alcohol seeking in rats. *Addict Biol* 2011;**18**(3):448–51.
187. Funk D, Li Z, Le AD. Effects of environmental and pharmacological stressors on c-fos and corticotropin-releasing factor mRNA in rat brain: Relationship to the reinstatement of alcohol seeking. *Neuroscience* 2006;**138**(1):235–43.
188. Behzadi G, Kalen P, Parvopassu F, Wiklund L. Afferents to the median raphe nucleus of the rat: retrograde cholera toxin and wheat germ conjugated horseradish peroxidase tracing, and selective D-[3H]aspartate labelling of possible excitatory amino acid inputs. *Neuroscience* 1990;**37**(1):77–100.
189. Dong HW, Swanson LW. Projections from bed nuclei of the stria terminalis, dorsomedial nucleus: implications for cerebral hemisphere integration of neuroendocrine, autonomic, and drinking responses. *J Comp Neurol* 2006;**494**(1):75–107.

CHAPTER

17

Metabolic Profiling Approaches for Biomarkers of Ethanol Intake

*Helen G. Gika, PhD**, *Ian D. Wilson, DSc***

*Department of Chemical Engineering, Aristotle University of Thessaloniki, Thessaloniki, Greece

**Department of Surgery & Cancer, Imperial College, London, UK

INTRODUCTION

The pleasures and problems associated with alcohol consumption by humans are well known, with the most pressing problems involving either excessive acute exposures (so called "binge" drinking) or the chronic, long term, excess alcohol consumption that leads to alcoholism.[1]

In attempting to understand ethanol-related toxicity and disease, there is arguably a requirement for the discovery of new biomarkers that can be used to monitor the effects of long-term exposure, to assess the effectiveness of therapy, and guide strategies for rehabilitation. Some of the most useful and easily implement biomarkers of alcohol exposure are the longer-lived ethanol metabolites, such as ethyl glucuronide (and related conjugates), and various modified lipids (e.g., ethyl oleate, arachidonate, phosphatidylethanol, etc.) that can be found in the circulation and urine,[2–8] and even hair.[9] It is now well established that alcohol consumption affects the function of many organs, biochemical processes, and signaling pathways and, therefore, the use of "omics" based approaches, such as metabonomics/metabolomics, to examine the effects of this toxin might reasonably be expected to provide novel insights into mechanisms of toxicity. Indeed, moving toward a more holistic "systems biology" approach could well have advantages,[10] by providing a more global overview of ethanol-induced metabolic perturbations. Here we describe the current state of global metabolic profiling research, as it relates to the effects of alcohol in animals and humans.

METABOLIC PROFILING

Metabolic profiling (metabolomics/metabonomics) is the measurement of the complement of low-molecular-weight metabolites and their intermediates in biological systems that reflect the response to any stimuli (e.g., drug treatment, disease, genetic modification, etc.).[11] The measurement and interpretation of the endogenous metabolite profile of the biological matrix analyzed (typically urine, serum, or tissue extract) provides an insight into the changes induced by external stimuli, and can enhance our knowledge of inherent biological variation within subpopulations. Such global metabolic profiling studies have already been shown to offer promise, by highlighting potential mechanistic biomarkers, in toxicology investigations in preclinical drug development.[12,13]

ANALYTICAL TECHNOLOGIES FOR GLOBAL METABOLIC PROFILING

The principal technologies that have been developed to obtain comprehensive endogenous metabolic profiles are based on nuclear magnetic resonance (NMR) spectroscopy and mass spectrometry (MS).[14] Both of these techniques are capable of providing data on a wide range of analytes present in biological samples, thus generating a metabolic phenotype that can be mined for further information on metabolite composition. ^{1}H-NMR spectroscopy, although of limited sensitivity, has been demonstrated to be a robust and reliable technique that

Molecular Aspects of Alcohol and Nutrition. http://dx.doi.org/10.1016/B978-0-12-800773-0.00017-3

is well suited to the analysis of large numbers of samples, with good coverage of different classes of metabolite, and ready quantification. MS provides a complementary technique to NMR spectroscopy for metabolite profiling, offering often greater, but analyte-dependent, sensitivity, and allowing only qualitative analysis in the absence of isotopically labeled internal standards. MS-based profiling can be performed using direct infusion of the sample into the ion source of the spectrometer, but is usually applied by linking the mass spectrometer to a separation technique (gas chromatography (GC), liquid chromatography (LC), or capillary electrophoresis (CE)). Both GC-MS and LC-MS are important global metabolic profiling tools, and recent advances in LC, such as ultra (high) performance liquid chromatography (UPLC/UHPLC-MS) have ensured that LC-MS MS technology has become the dominant metabolic profiling approach.[15] Compared to conventional HPLC analysis, U(H)PLC offers increased speed and sensitivity and, as a result, increases the number of metabolites detected, thereby enhancing metabolome coverage.[16] In Figure 17.1, an example is given of the metabolite profiles (3D-chromatograms) obtained by HPLC-MS and UPLC-MS, for the same urine sample, illustrating both the complexity of the acquired data, and the dependence of the profile on the resolution of the chromatographic method. It is accepted that, currently, there is no single analytical method that can detect and monitor all of the thousands of small molecules present in biological samples. The selection of the most suitable analytical tool is, in general, a compromise between sensitivity and selectivity. In Figure 17.2, the relative sensitivity and selectivity of different analytical techniques applied for metabolite profiling is shown.

Furthermore, the various analytical methodologies have different requirements for samples, in terms of quantities, and the degree of sample preparation

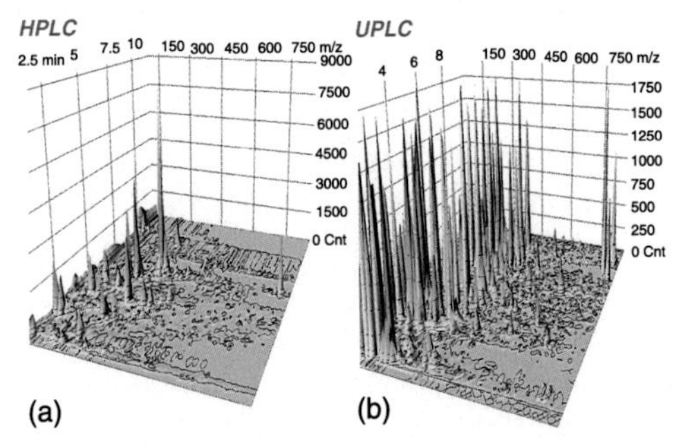

FIGURE 17.1 3D chromatograms of (a) HPLC-MS and (b) UPLC-MS mouse urine metabolic profiles, showing retention time (*X* axis), *m/z* (*Z* axis), and intensity (*Y* axis). From this, it is obvious that the metabolic fingerprint of a sample is dependent on the profiling method applied. The higher number of metabolites the method can detect, with higher sensitivity, the better. *Reproduced from Ref. [16] with permission from ACS Publications.*

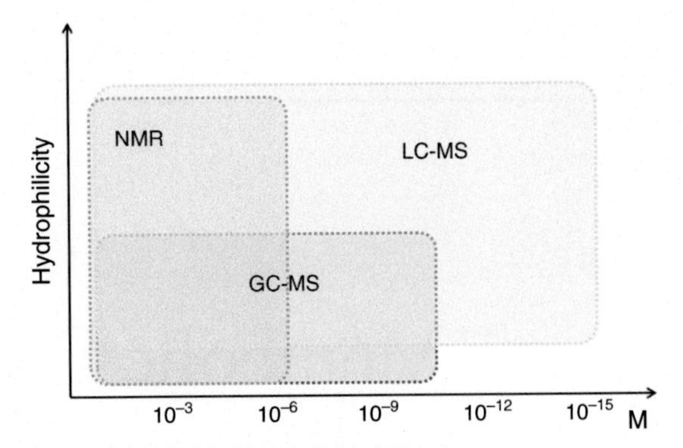

FIGURE 17.2 Relative sensitivity and selectivity of the three analytical techniques mostly applied for metabolite profiling.

required prior to analysis. Self-evidently, sample preparation should be kept to the absolute minimum required to avoid "editing" the subsequent profile, via the unintended selective concentration/loss of metabolites. In this context, where liquid samples, such as biological fluids, are to be analyzed by ^{1}H-NMR spectroscopy, the removal of particulates and dilution in a suitable D_2O-based buffer is all that is required.[17] In the case of LC-MS-based analysis methods, apart for urine where, like NMR, all that is needed is to dilute the sample, centrifuge, and inject,[18] the situation can be somewhat more complex. So, with the LC-MS metabolic profiling of blood-derived samples, such as plasma or serum, the removal of proteins, generally by precipitation with an excess of an organic solvent, such as methanol or acetonitrile, is absolutely essential to prevent the rapid degradation of the analytical system.[19] The application of GC-MS, unless it is limited to the analysis of volatiles (such as those found in breath, etc.), of course requires derivatization, via, for example, trimethylsilylation and methoximation,[20] such that involatile metabolites (that represent the majority) are made amenable to analysis. Clearly, when it comes to the analysis of tissues, much more is needed in the way of sample preparation. At its simplest, tissue analysis involves homogenization, extraction into a suitable solvent, and then often the use of selective extraction to separate lipophilic and polar hydrophilic metabolites. Chloroform or dichloromethane are suitable for the extraction of hydrophobic metabolites, with the hydrophilic metabolites extracted using water/methanol mixtures.[21] Many aspects of the preparation and analysis of samples for metabolic phenotyping are now reasonably well defined (e.g., the following protocols providing guidance for this type of work[17–21]).

Currently, it seems that the most popular metabolic profiling technique used for studies on the effects of ethanol administration (chronic or acute) is LC-MS,[22–31] with a significant contribution from ^{1}H-NMR spectroscopy.[32–38] A GC-MS-based application has also been reported.[39]

ETHANOL EXPOSURE STUDIES ON ANIMAL MODELS

Most of the investigations of the effects of alcohol on metabolic phenotypes in animal models have been undertaken in rodents, where dosing has generally been oral, either via gavage, in the diet, via the drinking water, or by providing the animals with the "Lieber–DeCarli" liquid diet.[40] A more invasive model, based on intragastric feeding, with the advantage of offering a more controlled means of alcohol delivery, has also been used. There follows a description of the studies conducted either by NMR or by MS-based profiling methods in rodents treated by these methods of administration.

^{1}H-NMR spectroscopy has been used to study the metabolic perturbations associated with alcohol abuse[32] in adult male rats, dosed orally with ethanol at 5 g/kg. The animals were sacrificed 3 h after dosing, and samples of liver, serum, and brain were taken. In a further study, another group of rats was dosed for 4 days using a protocol that produces a similar physical dependence and neuropathology to that seen in humans, with half of the animals also given the antioxidant butylated hydroxytoluene (BHT). Liver samples were collected within 1 h of the final dose. The control groups for both studies comprised two groups of animals, receiving either water, or a sucrose solution that provided the same amount of calories as the ethanol dose. This study demonstrated that ethanol administration in both single or multiple dose treatments resulted in liver profiles that differed from the control because of changes in choline-containing compounds, betaine, lactate, alanine, glucose, glycogen, and ethyl glucuronide (EtG), with larger changes following multiple dose treatments. From the ^{1}H-NMR profiles there was no evidence of an increase in hepatic glucose or glycogen levels following a single dose, while a decrease in both was detected following multiple doses of ethanol. In addition, there was evidence of reduced formation of EtG in the livers of rats receiving BHT. For serum obtained after the single dose decreased, concentrations of lactate and alanine, coupled to increased acetate and acetoacetate, were noted. There was no detectable effect on the metabolic profiles seen for brain extracts.

In a study in male C57Bl/6J mice, the metabolic phenotypes were produced when a high-fat corn oil version of the Lieber-DeCarli diet was used, in an attempt to more accurately mimic the liver injury seen in humans, resulting from alcohol abuse.[33] Urine and liver extracts were analyzed using ^{1}H-NMR spectroscopy, and Fourier transform ion cyclotron resonance mass spectrometry (FTICR-MS). The mice were fed *ad libitum* for up to 36 days, with interim urine collections for days 22, 30, and 36 (analyzed using ^{1}H-NMR spectroscopy). As well as this group, liver samples were harvested from a parallel group of mice, after 28 days of feeding. Aqueous liver extracts prepared from these samples that were observed to show steatosis, were then profiled using ^{1}H-NMR spectroscopy. Organic extract metabolic profiles were also acquired, using direct infusion (DI) into the FTICR-MS. The ^{1}H-NMR-generated profiles for urine, and the aqueous liver extracts of the dosed mice, were significantly different from those of the controls, with 11 metabolites altered in urine, and 9 in liver extracts, as a result of exposure to alcohol. Ethanol administration caused the amounts of *N*-acetylglutamine and *N*-acetyglycine to be increased in urine, while taurine concentrations were lower. This last change was thought to result from the activation of the glutathione pathway, in response to alcohol toxicity. The organic extracts showed differences in the liver metabolic profiles of alcohol treated animals, compared to the controls with 14 compounds, seen to be significantly changed, including free fatty acids such as arachidonic, palmitic, linoleic, stearic, eicosanoic, and 11,14-eicosadienoic acids. Interestingly, these compounds had previously been seen to be altered in serum samples from male human drinkers.[41] The authors also noted changes in the profiles that probably reflected liver injury, rather than being ethanol-specific. They concluded that the model resulted in a characteristic metabolic phenotype that could be differentiated from other, acute models.

LC-MS-based studies of the effects of ethanol exposure on the metabolic phenotypes of rats, undertaken in order to study the protective effects of the traditional Chinese medicine Yin Chen Hao Tang (YCHT),[22] employed a complex study design, with samples obtained from five groups of animals receiving various treatments. So, one treatment consisted of rats being given 4 days of ethanol administration (5 g/kg, orally), while another group, also exposed to ethanol for 4 days, received YCHT. The YHCT animals were subdivided into three treatment regimens, receiving either 0.6, 2.8, or 4.8 g/kg/day (PO), beginning 3 days before dosing with ethanol. Controls were dosed with water. Hepatotoxicity was shown for the ethanol-only treated group using histopathology, with protective effects shown for YCHT treated animals. PCA of the UPLC-MS urinary metabolite profiling data revealed that the responses of the control, ethanol treated, and ethanol + YCHT-treated rats were different. One of the metabolites seen to have changed was ceramide that the authors suggested indicated an involvement of the sphingomyelin pathway in the hepatic effects of ethanol consumption. From the PCA results, the authors considered that treatment with YCHT had a positive effect on the livers of the ethanol-treated mice, by improving metabolic processes affected by exposure to ethanol.

Another study, in female rats, examined the effects of chronic Sake consumption on the brain and liver,[34] as part of an investigation that included both proteomic and transcriptomic profiling. The animals had free access to Sake (15% alcohol) for 1 year. The resulting brain and liver samples were ground in liquid nitrogen, extracted with methanol-water-chloroform, and the aqueous phases were profiled by ^{1}H-NMR spectroscopy. Analysis of the resulting data showed alterations in the concentrations of the metabolites contained in these tissues, compared to the controls, with, for example, brain extracts showing significant reductions in the amounts of gamma aminobutyric acid, isoleucine, *N*-acetyl aspartate, taurine, and glutamate, coupled with increased quantities of valine, arginine, ornithine, alanine, glutamine, and choline. Liver from Sake-treated animals showed reductions in bile acids and glucose, and raised glycine and other metabolites, albeit with low statistical significance ($p < 0.1$, or $p < 0.2$ by Student's *t*-test). Given the potential of long-term ethanol exposure to affect lipid profiles, it would perhaps also have been valuable to have performed a lipidomic analysis on the hydrophobic fractions, as well.

In a study to find potential early prognostic biomarkers of alcoholic liver disease (ALD), an investigation was undertaken in wild-type and P*para*-null mice,[23] as the peroxisome proliferator-activated receptor alpha (PPAR-α) has been shown to protect against alcohol-induced liver damage.[42] The animals were maintained for 6 months on either a 4% alcohol liquid (Lieber–DeCarli) diet or, in the case of the controls, a maltose dextran diet that provided an equivalent calorific intake, with 24 h urine samples collected each month. The resulting metabolic signatures were found to depend on the strain of mouse used with several ions, seen to be up- or downregulated in the urine of wild type animals after 2 months exposure to ethanol, but later in *Ppara*-null-derived mice (after 3 or 4 months), and to a smaller extent. Changes were noted for 4-hydroxyphenylacetic acid, and its sulfate conjugate and indol-3-lactic acid, both of which were upregulated after exposure to ethanol, while 2-hydroxyphenylacetic, adipic, and pimelic acids were all reduced in quantity. Many other ions were also seen to change, depending on alcohol administration, and in only one or other of the two strains of mouse. The *Ppara*-null mice also excreted lower amounts of ethyl glucuronide and ethyl sulfate than the wild type animals during the early stages of alcohol exposure, and the downregulation seen for pimelic and adipic acids was also greater in these animals, indicating differences in fatty acid oxidation, compared to the WT. Tryptophan metabolism in *Ppara*-null mice appeared to result in the conversion of tryptophan into NAD^+, and the build-up of NADH. This was thought to be because the oxidation of ethanol helps to reduce indole-3-pyruvic acid to indole-3-lactic acid, to restore the redox balance. Interestingly, indole-3-lactic acid was found to be increased in this study, particularly in *Ppara*-null mice, and might have potential as an early stage biomarker of ALD. A further investigation using the *Ppara*-null mouse model, expressed in C57BL/6 and 129/SvJ mice, was used to examine if there was an effect of the genetic background on the changes in urinary metabolic phenotypes with ALD.[24] This study confirmed that increased amounts of indole-3-lactic acid and phenyllactic acid appeared to represent a specific metabolic response for alcohol-treated *Ppara*-null mice, irrespective of their genetic background, with the liver pathologies seen in these animals similar to the early stages of human ALD. As such, indole-3-lactic acid and phenyllactic acid may represent potential biomarkers of the early stages of ALD in humans.

The lipid profiling of livers from rats and mice dosed with ethanol intragastrically, for 1 month (with equivalent caloric intake for both control and treated animals) has been investigated using LC-MS.[25] In this study, the ethanol intake was gradually raised from 24.3% to 34.4% of the total coloric intake. Extracts of the liver samples, taken from control and the alcohol-fed rodents, showed quite different profiles for ethanol-treated versus control animals (Figure 17.3a). In Figure 17.3b, the S-plot generated from the PCA model data for these mouse liver samples is illustrated; it assisted in finding the significantly altered ions between the control and alcohol treated groups. Ions were then identified using molecular formula prediction tools, MS/MS spectra, and search in databases. Metabolites seen to be present in larger amounts following ethanol exposure included octadecatrienoic and eicosapentaenoic acids, the ethyl esters of arachidonic, docosahexaenoic, linoleic, and oleic acids, and phosphatidylethanol homologues. While lipid profiles were broadly similar for both species, there were metabolites that gave different responses in rats, compared to mice, and vice-versa. Thus, glycerophospholipids showed both increases and decreases in alcohol-administered rats and mice, while fatty acyl metabolites (e.g., docosapentaenoic acid, arachidonic acid, and adrenic acid) were elevated in rat, but decreased in mouse liver extracts, respectively. Compared to rat, the amounts of retinol and cholesterol were markedly decreased in the mouse liver extracts. Such differences should be kept in mind, when seeking to extrapolate results obtained in rodent models to man. The UPLC-MS analysis of urine and plasma from the same study has also been reported in a further study,[26] with samples collected after 2 and 4 weeks of ethanol administration. In Figure 17.4, the PCA scores plot, derived from the urinary LC-MS data, is given. As with the liver metabolite profiles,[25] differences were seen between rats and mice, over the time course of ethanol administration. For example, changes in the amounts of some tryptophan metabolites, as well as arginine, hydroxyproline, taurine, 4-hydroxy-phenylacetic acid,

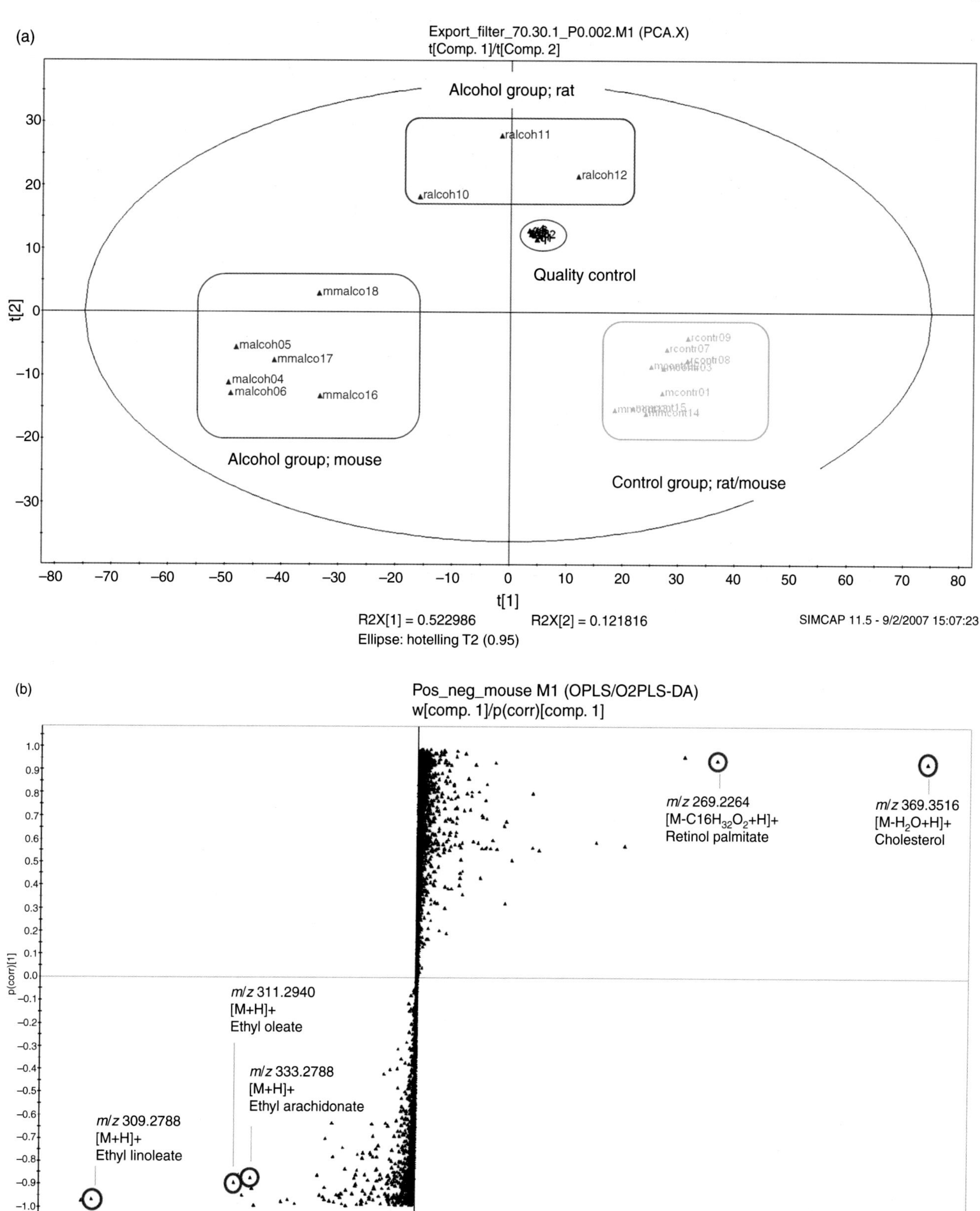

FIGURE 17.3 (a) Principal components analysis plot, based on the detected ions by the applied metabolic profiling technique, is able to cluster the liver extracts samples of ethanol treated animals separately from controls, and according to species. (b) S-plot generated on the PCA model data for mouse liver samples is illustrated; it assisted in finding the significantly altered ions between the control group and alcohol treated group. *Reproduced from Ref. [25] with permission from ACS Publications.*

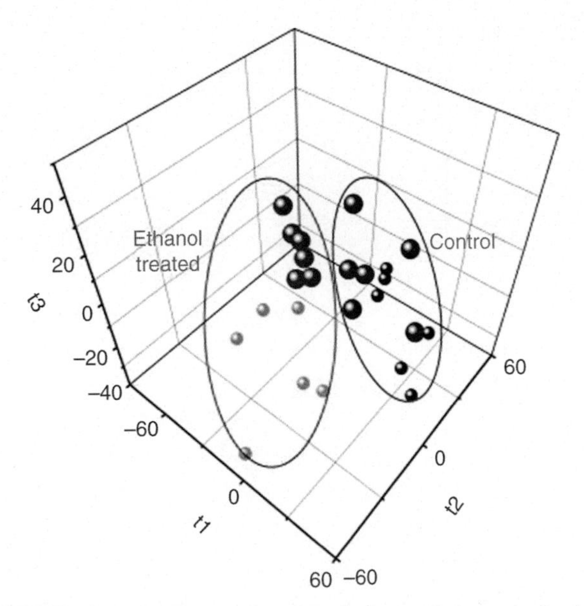

FIGURE 17.4 3D PCA scores plot of mouse plasma samples (+ve ESI UPLC TOF MS profiles) after 2 and 4 weeks of ethanol treatment. Bigger symbols correspond to 4-week samples, and smaller to 2-week samples. t1, t2, and t3 represent the three major principal components. *Reproduced from Ref. [26] with permission from Elsevier.*

together with its sulfate and acetylglycine conjugates, were seen between ethanol dosed and control mice. These changes were not all linearly correlated with time as, in some cases, metabolites were seen to rise initially, only to decrease in subsequent time samples (e.g., eicosatrienoic acid and urso- or cheno-deoxycholic acid sulfate in mouse urine). For plasma, the relative amounts of glycerophosphocholines and glycerol phosphoserines decreased in the ethanol treated, compared to the control, mice.

The use of ^{1}H- and ^{31}P-NMR spectroscopy has been described for the lipid profiling of liver extracts from male Fischer 344 rats, in a study where the determination of alcohol-driven metabolic changes in liver and plasma samples was investigated.[35] In the case of plasma, ethanol caused the concentrations of PC/lysophosphatidyl choline and the aliphatic methyl esters to decrease, while it also altered the concentrations of cholesterol, triglycerides, fatty acids, and phospholipids in the liver. When ^{31}P- and ^{1}H-NMR spectroscopy was undertaken on the liver extracts, the spectra showed changes in hepatic phospholipid profiles. The fact that similar changes in the plasma and liver lipids profiles were seen after exposure to ethanol suggests that plasma has the potential to reveal early-stage ALD. A repeat of the study covering 2 and 3 months of dosing[36] gave a similar outcome, with decreases in the amounts of phosphatidylcholine observed for both liver and plasma, together with increased fatty acid, and triglyceride concentrations in liver versus decreased amounts of fatty acids detected in plasma. An increase in the level of unsaturation of fatty acyl chains in liver lipids was also seen, in contrast to plasma. Perhaps unsurprisingly, these changes became more pronounced as ethanol was dosed for longer.

Serum LC-MS metabolic profiles of mice with hepatocellular carcinomas (HCC), control nude mice, and those dosed daily by gavage with ethanol (with the dose increased over 4 weeks from 5% to 40% ethanol) were compared in an attempt to find correlations in the hepatic liver injury induced by ethanol and hepatic carcinomas.[27] It was found that the alcohol-treated mice shared some similarities in metabolite profiles with the xenograft-bearing animals that were not seen in the normal animals. In particular, five metabolic pathways were affected in the tumor-bearing and ethanol dosed mice (phenylalanine and tyrosine metabolism, leucine degradation, tryptophan metabolism, and sphingolipid and glycerophospholipid metabolism), in comparison to the controls. There were, however, disease-specific changes in the lysophosphatidylcholines (LPCs), with those containing saturated or monounsaturated fatty acids lowered for both models, but LPCs containing polyunsaturated fatty acids increased in samples obtained from the mice exposed to ethanol.

A study designed to explore the effect of ethanol administration and genetic differences on the urinary metabolite profiles of Sprague–Dawley (SD) and Wistar rats has also been undertaken.[24,39] After intragastric administration of ethanol, for 4 days, SD rats appeared to show a greater tolerance, and ability to recover from, dosing with ethanol, compared to Wistar rats. Capillary GC-MS of the samples showed changes in a range of metabolites, including short-chain dicarboxylic acids, TCA cycle intermediates, and amino acids, with increased amounts of malonic and ethylmalonic acid in Wistar, compared to SD rats, considered to be related to a greater degree of hepatic injury. Strain related differences were also noted for a number of gut-microbial cometabolites, such as *p*-cresol, 4-hydroxy-phenylacetic acid, etc., with the downregulation of hippuric acid combined with an upregulation of the amounts of glycine and benzoic acid, seen for Wistar rats. Ethanol also caused a decrease in tryptophan and indole-3-acetic acid concentrations, suggesting changes in tryptophan metabolism.

The effects of the presence or absence of CYP2E1, an enzyme that can oxidize ethanol, generating reactive oxygen species during the production of acetaldehyde, have been investigated in control and Cyp2e1-null mice, maintained on a Lieber-DeCarli diet for 21 days, with urine samples collected on days 7, 14, and 21. UPLC-MS analysis of these samples showed that urinary concentrations of *N*-acetyltaurine (NAT) had increased dramatically, in response to alcohol exposure.[31] The increased concentrations of NAT were correlated with the expression of CYP2E1. It was also observed that acetate concentrations were increased, and those of taurine decreased, with ethanol intake as an effect of NAT biosynthesis.

Studies on septic peritonitis were undertaken in a mouse model, with ^{1}NMR spectroscopy used for the metabolic profiling of pancreatic tissue. Mice were administered either alcohol (*ad libitum*), or water, for 12 weeks, followed by cecal ligation and puncture.[38] The analysis of the aqueous extracts of the pancreas showed that, in alcohol-fed mice, acetate, adenosine, xanthine, acetoacetate, 3-hydroxybutyrate, and betaine were higher, and cytidine, uracil, fumarate, creatine phosphate, creatine, and choline were lower, than in the equivalent control mice. These changes were seen in the absence of histological differences.

HPLC-MS-based metabolic profiling has been used to study the effects of ethanol on the metabolic profiles of various regions of the gastrointestinal (GI) tract of rats, administered ethanol for 8 weeks.[29] Profiles were obtained from the GI contents of the stomach, duodenum, jejunum, ileum, cecum, colon, and rectum. The samples were first extracted with cold water, and then with cold methanol, prior to LC-MS analysis (with a targeted GC-MS analysis also used for short chain acids, and the amino acids leucine, isoleucine, and valine). Significant changes in metabolite profiles, compared to control rats, were seen for all the regions of the GI tract, sampled for alcohol-fed animals, with ethanol intake resulting in decreased amounts of all of the amino acids, in all regions examined. Concentrations of steroids were lower in the stomach, duodenum, and jejunum. The fatty acids were found to be elevated, while branched chain amino acids, as well as carnitines, and other metabolites involved in lipid metabolism, were present in lower amounts, compared to controls. The study also found that concentrations of acetic acid were elevated in stomach, ileum, and cecum, but propanoic, 2-methylpropanoic, butyric, 2-methylbutyric, 3-methylbutyric acid, propionic, isobutyric, and heptanoic acids were all lowered in the ileum, colon, cecum, and rectum. Such changes may reflect alterations in the intestinal microbiome that is known to be affected in alcoholism.[43]

ETHANOL EXPOSURE STUDIES IN MAN

Given the potential of metabolic profiling, it is perhaps surprising that few studies have been undertaken in humans, but that is the case. Reversed-phase UPLC-MS has been applied to the analysis of serum obtained from normal controls, and subjects suffering from alcoholic cirrhosis or hepatitis B (HBV-induced cirrhosis).[30] The study was performed with the aim of finding biomarkers that could distinguish between the alcohol and HBV-induced hepatic cirrhosis, thereby enabling clinical diagnosis of two conditions. The results suggested that some markers of hepatic cirrhosis (e.g., the lysophosphatidylcholines (LPCs)) were common to both types of liver disease. Thus, the LPCs were lower, while certain bile acids, hypoxanthine, and stearamide appeared to be increased in all patients. In the case of alcoholic cirrhosis, both oleamide and myristamide concentrations were raised, in comparison to the controls (and decreased in serum sample from HBV-induced cirrhosis patients), suggesting a possible means of differential diagnosis, if these observations can be confirmed in further studies.

Another study aimed to investigate the beneficial effect of drinking wine, compared to other alcoholic beverages, with respect to reduced alcohol toxicity.[37] Analysis by ^{1}H-NMR spectroscopy was performed on the urine of 61 subjects, in 3 groups, administered either red wine, nonalcoholic red wine, or gin, for 28 days. Concluding that the study demonstrated the ability of metabolomics to be applied to nutritional intervention studies, it was noted that gut microfloral cometabolites, such as hippurate and 4-hydroxyphenylacetate, showed an effect, albeit with a larger increase in the amounts present in urine from subjects who received the nonalcoholic wine.

The investigation of changes in metabolite concentrations related to alcohol intake, in just over 3000 subjects taking part in epidemiological studies (from the KORA F4, KORA F3, and the TwinsUK adult bioresource studies), was undertaken using targeted metabolomic analysis (Direct Infusion MS with ESI and tandem MS), using the AbsoluteIDQ p150 assay.[31] When corrected for factors such as age, BMI, smoking, HDL, and triglycerides, etc., analysis of the profiling data highlighted a range of metabolites that differed between moderate-to-heavy drinkers, and light drinkers, in the KORA F4 study, with 40 altered in males and 18 in females. A subset of 10 metabolites in males, and 5 in females, could be used to distinguish between light drinkers and moderate to heavy ones, with good sensitivity and specificity. A number of these proposed alcohol-related metabolic changes were also seen for samples obtained in the KORA F3 and TwinsUK studies. In Figure 17.5, the correlation of metabolite concentrations with ethanol consumption is shown by clusterograms, where the samples are classified into two groups, light and moderate to heavy drinkers. On the basis of these data, it was suggested that profiles using diacylphosphatidylcholines, lysophosphatidylcholines, ether lipids, and sphingolipids, provided a new class of biomarkers for determining excess alcohol intake, with potential for use in both clinical and epidemiological investigations, while also acknowledging the limitations of the targeted analysis used, in terms of limited metabolome coverage.

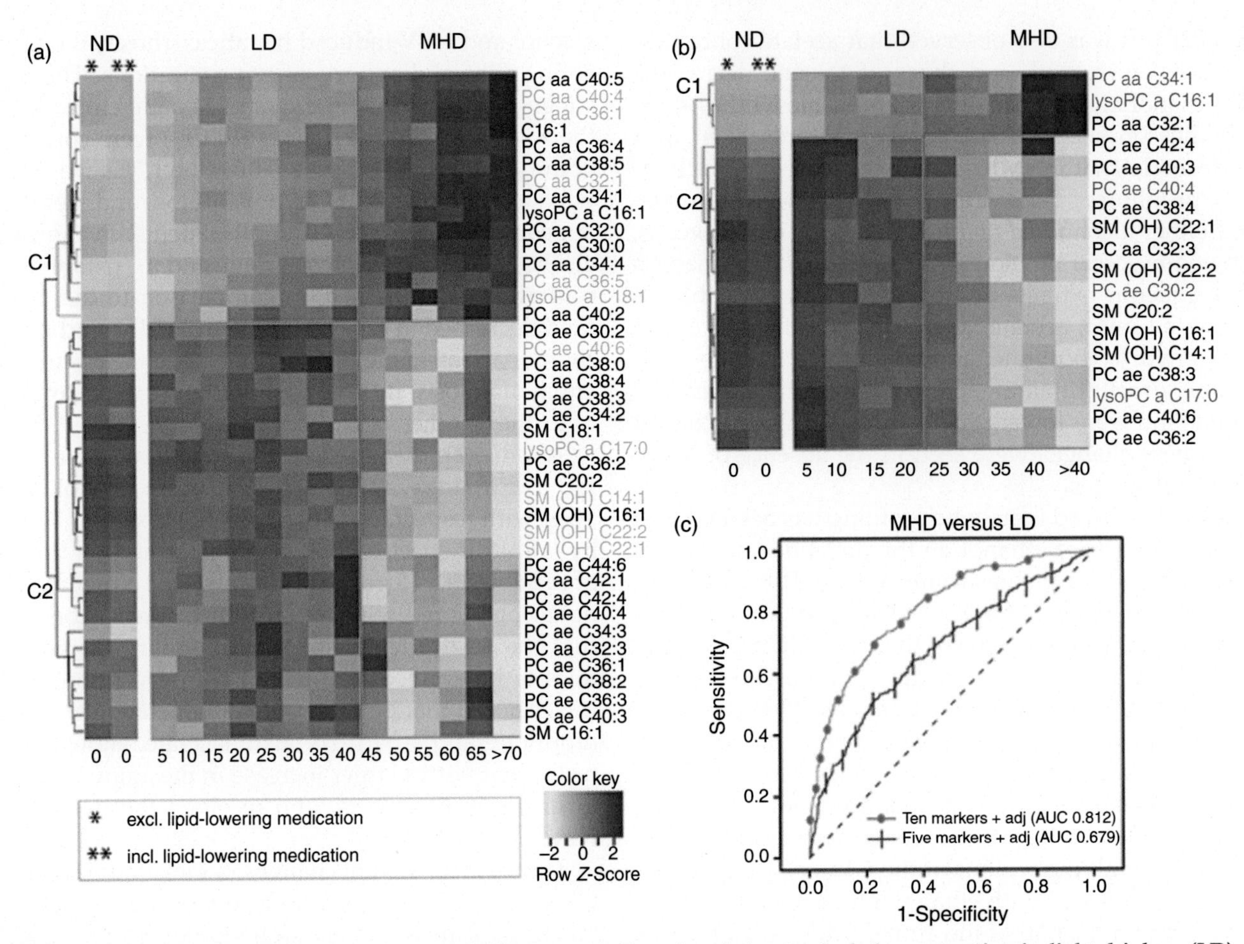

FIGURE 17.5 Heatmaps showing 40 and 18 metabolite concentrations in relation to alcohol consumption in light drinkers (LD), and moderate-to-heavy drinkers (MHD), in (a) males, and (b) females, respectively. The additional two-columns show the effect of lipid-lowering medication (i.e., statins, fibrates, herbal-based lipid-lowering agents) on metabolite concentrations in nondrinkers (ND). Relative concentration of metabolites are represented by *x*-fold s.d. from overall mean concentrations, for groups of alcohol consumption of 5 g/day. Horizontal axis displays the alcohol concentration in gram per day, while vertical axis represent hierarchical clustering. C1 consists of metabolites that increase in concentration with increasing alcohol consumption (high in MHD and low in LD). In contrast, C2 consist of metabolites that decrease in concentration with increasing alcohol consumption (low in MHD and high in LD). The 10/5 most significant metabolites separating MHD from LD in males/females are highlighted in blue and pink (c). *Reproduced from Ref. [31] with permission from Nature Publishing Group Limited.*

CONCLUSIONS

Studies in rodents have demonstrated the potential of metabonomics/metabolomics to reveal the consequences of ethanol administration on endogenous metabolism (via the effects on tryptophan metabolism, or lipid profiles), resulting in the detection of novel biomarkers of its toxicity, such as indole-3-lactic acid.

While currently limited in scope, the results of these studies are encouraging, and suggest that further investigations, using these metabolic phenotyping approaches, in carefully conducted studies in humans, are warranted, and would provide valuable new insights into the short and long term effects of alcohol exposure. It is to be hoped that, in this way, new and mechanistic biomarkers would be discovered, with potential for use in monitoring the effects of ethanol consumption and therapy, for ALD and alcoholism, in man.

Key Facts

Alcohol Consumption

- Alcohol, according to WHO, is a sedative, hypotonic with effects similar to barbiturates. Apart from social effects of use, alcohol intoxication may result in poisoning, or even death. Long term heavy use may result in dependence, or in a wide variety of physical and organic mental disorders.
- Alcoholism is defined by WHO as the continual drinking or periodic consumption of alcohol, characterized by impaired control over drinking, frequent episodes of intoxication, and preoccupation with alcohol and the use of alcohol, despite adverse consequences.
- Moderate consumption is up to 1 drink/day for women, and up to 2 drinks/day for men.

- Binge drinking is a pattern of drinking that brings blood alcohol concentration (BAC) levels to 0.08 g/dL. This typically occurs after four drinks for women, and five drinks for men, in about 2 h.
- Heavy drinking is considered to take place when drinking five or more drinks on the same occasion, on each of 5 or more days, in the past 30 days.

Ethanol Toxicity

- Ethanol diffuses across membranes, and distributes through cells and tissues, affecting cell function by interacting with certain proteins and cell membranes.
- Ethanol metabolism results in the generation of acetaldehyde, a highly reactive and toxic byproduct that damages tissue, and forms reactive oxygen species.
- Acetate is produced from the oxidation of acetaldehyde, and escapes the liver to the blood, where it is oxidized to carbon dioxide in heart, skeletal muscle, and brain cells. It increases blood flow into the liver, and depresses the central nervous system, as well as affecting various metabolic processes.

Global Metabolic Profiling

- Metabolic profiling or metabolomics is a system biology approach that incorporates a holistic approach for the characterization of metabolites produced in metabolic pathways.
- The analytical technologies that give the potential to obtain a holistic view of the biological sample are NMR spectroscopy, and LC-MS/GC-MS.
- Most of the published work on metabolomics, for the study of alcohol related disorders, were applied to urine or liver samples of rodents. There are also metabolomics studies on brain, serum or plasma and, recently, on pancreas tissue and gastrointestinal track content.

Summary Points

- Alcohol abuse is a growing social problem that is related to many major health problems, such as liver cirrhosis, cancer, and physiological dependence.
- Alcohol induced disorders involve many interrelated mechanisms; thus, a global approach, such as systems biology, would be an ideal way to study alcohol effects at different levels, from metabolites to genes.
- Metabolomics can provide a global view on ethanol induced metabolic perturbations, and shed light on mechanisms of toxicity and related disorders.
- Metabolic profiling analysis performed on urine, blood, etc., gives information on the changes induced in the metabolic content of biological systems, after an intervention with the aim of discovering mechanisms and metabolic pathways affected.
- NMR, LC-MS, and GC-MS metabolic profiling methods have been applied to study the metabolic perturbations associated with alcohol abuse and consumption, in animal models and humans.
- Studies have demonstrated the alteration of metabolic fingerprint with ethanol consumption, and their ability to reveal effects on various metabolic pathways, for example, on tryptophan and lipid metabolism, but also novel biomarkers of its toxicity, such as indole-3-lactic acid.

References

1. Zakhari S, Li T-K. Determinants of alcohol use and abuse: impact of quantity and frequency patterns on liver disease. *Hepatology* 2007;**46**(6):2032–9.
2. Laposata EA, Lange LG. Presence of nonoxidative ethanol metabolism in human organs commonly damaged by ethanol abuse. *Science* 1986;**231**(4737):497–9.
3. Jaakonmaki PI, Knox KL, Horning EC, Horning MG. The characterization by gas–liquid chromatography of ethyl β-D-glucosiduronic acid as a metabolite of ethanol in rat and man. *Eur J Pharmacol* 1967;**1**:63–70.
4. Schmitt G, Aderjan R, Keller T, Wu M. Ethyl glucuronide: an unusual ethanol metabolite in humans. Synthesis, analytical data, and determination in serum and urine. *J Anal Toxicol* 1995;**19**(2):91–4.
5. Bicker W, Lämmerhofer M, Keller T, Schuhmacher R, Krska R, Lindner W. Validated method for the determination of the ethanol consumption markers ethyl glucuronide, ethyl phosphate, and ethyl sulfate in human urine by reversed-phase/weak anion exchange liquid chromatography-tandem mass spectrometry. *Anal Chem* 2006;**78**(16):5884–92.
6. Dresen S, Weinmann W, Wurst FM. Forensic confirmatory analysis of ethyl sulfate – a new marker for alcohol consumption – by liquid-chromatography/electrospray ionization/tandem mass spectrometry. *J Am Soc Mass Spectrom* 2004;**15**(11):1644–8.
7. Laposata M. Fatty acid ethyl esters: short-term and long-term serum markers of ethanol intake. *Clin Chem* 1997;**43**(8 Pt. 2):1527–34.
8. Varga A, Hansson P, Lundqvist C, Alling C. Phosphatidylethanol in blood as a marker of ethanol consumption in healthy volunteers: comparison with other markers. *Alcohol Clin Exp Res* 1998;**22**(8):1832–7.
9. Politi L, Morini L, Leone F, Polettini A. Ethyl glucuronide in hair: is it a reliable marker of chronic high levels of alcohol consumption? *Addiction* 2006;**101**(10):1408–12.
10. Guo QM, Zakhari S. Commentary: systems biology and its relevance to alcohol research. *Alcohol Res Health* 2008;**31**(1):5–11.
11. Nicholson JK, Lindon JC, Holmes E. Metabonomics: understanding the metabolic responses of living systems to pathophysiological stimuli via multivariate statistical analysis of biological NMR spectroscopic data. *Xenobiotica* 1999;**29**(11):1181–9.
12. Robertson DG, Watkins PB, Reily MD. Metabolomics in toxicology: preclinical and clinical applications. *Toxicol Sci* 2011;**120**(S1):S146–70.
13. Clarke J, Haselden NJ C. Metabolic profiling as a tool for understanding mechanisms of toxicity. *Toxicol Pathol* 2008;**36**(1):140–7.
14. Lenz EM, Wilson ID. Analytical strategies in metabonomics. *J Proteome Res* 2007;**6**(2):443–58.
15. Theodoridis G, Gika HG, Wilson ID. Mass spectrometry-based holistic analytical approaches for metabolite profiling in systems biology studies. *Mass Spectrom Rev* 2011;**30**(5):884–906.

16. Wilson ID, Nicholson JK, Castro-Perez J, et al. High resolution "ultra performance" liquid chromatography coupled to oa-TOF mass spectrometry as a tool for differential metabolic pathway profiling in functional genomic studies. *J Proteome Res* 2005;**4**(2):591–8.
17. Beckonert O, Keun HC, Ebbels TMD, et al. Metabolic profiling, metabolomic and metabonomic procedures for NMR spectroscopy of urine, plasma, serum and tissue extracts. *Nat Protocols* 2007;**2**(11):2692–703.
18. Want EJ, Wilson ID, Gika H, et al. Global metabolic profiling procedures for urine using UPLC–MS. *Nat Protocols* 2010;**5**(6):1005–18.
19. Gika H, Theodoridis G. Sample preparation prior to the LC–MS-based metabolomics/metabonomics of blood-derived samples. *Bioanalysis* 2011;**3**(14):1647–61.
20. Dunn WB, Broadhurst D, Begley P, et al. Procedures for large-scale metabolic profiling of serum and plasma using gas chromatography and liquid chromatography coupled to mass spectrometry. *Nat Protocols* 2011;**6**(7):1060–83.
21. Lin CY, Wu H, Tjeerdema RS, Viant MR. Evaluation of metabolite extraction strategies from tissue samples using NMR. *Metabolomics* 2007;**3**(1):55–67.
22. Wang X, Lv H, Sun H, et al. Metabolic urinary profiling of alcohol hepatotoxicity and intervention effects of Yin Chen Hao Tang in rats using ultra-performance liquid chromatography/electrospray ionization quadruple time-of-flight mass spectrometry. *J Pharm Biomed Anal* 2008;**48**(4):1161–8.
23. Manna SK, Patterson AD, Yang Q, et al. Identification of noninvasive biomarkers for alcohol-induced liver disease using urinary metabolomics and the Ppara-null mouse. *J Proteome Res* 2010;**9**(8):4176–88.
24. Manna SK, Patterson AD, Yang Q, et al. UPLC–MS-based urine metabolomics reveals indole-3-lactic acid and phenyllactic acid as conserved biomarkers for alcohol-induced liver disease in the Ppara-null mouse model. *J Proteome Res* 2011;**10**(9):4120–33.
25. Loftus N, Barnes A, Ashton S, et al. Metabonomic investigation of liver profiles of nonpolar metabolites obtained from alcohol-dosed rats and mice using high mass accuracy MSn analysis. *J Proteome Res* 2011;**10**(2):705–13.
26. Gika HG, Ji C, Theodoridis GA, Michopoulos F, Kaplowitz N, Wilson ID. Investigation of chronic alcohol consumption in rodents via ultra-high-performance liquid chromatography-mass spectrometry based metabolite profiling. *J Chromatogr A* 2012;**1259**:128–37.
27. Li S, Liu H, Jin Y, Lin S, Cai Z, Jiang Y. Metabolomics study of alcohol-induced liver injury and hepatocellular carcinoma xenografts in mice. *J Chromatogr B Analyt Technol Biomed Life Sci* 2011;**879**(24):2369–75.
28. Shi X, Yao D, Chen C. Identification of *N*-acetyltaurine as a novel metabolite of ethanol through metabolomics-guided biochemical analysis. *J Biol Chem* 2012;**287**(9):6336–49.
29. Xie G, Zhong W, Zheng X, et al. Chronic ethanol consumption alters mammalian gastrointestinal content metabolites. *J Proteome Res* 2013;**12**(7):3297–306.
30. Lian J, Liu W, Hao S, et al. A serum metabonomic study on the difference between alcohol- and HBV-induced liver cirrhosis by ultraperformance liquid chromatography coupled to mass spectrometry plus quadrupole time-of-flight mass spectrometry. *Chin Med J* 2011;**124**(9):1367–73.
31. Jaremek M, Yu Z, Mangino M, et al. Alcohol-induced metabolomic differences in humans. *Transl Psychiatry* 2013;**3**(7):e276.
32. Nicholas PC, Kim D, Crews FT, Macdonald JM. 1H NMR-based metabolomic analysis of liver, serum, and brain following ethanol administration in rats. *Chem Res Toxicol* 2008;**21**(2):408–20.
33. Bradford BU, O'Connell TM, Han J, et al. Metabolomic profiling of a modified alcohol liquid diet model for liver injury in the mouse uncovers new markers of disease. *Toxicol Appl Pharmacol* 2008;**232**(2):236–43.
34. Masuo Y, Imai T, Shibato J, et al. Omic analyses unravels global molecular changes in the brain and liver of a rat model for chronic Sake (Japanese alcoholic beverage) intake. *Electrophoresis* 2009;**30**(8):1259–75.
35. Fernando H, Kondraganti S, Bhopale KK, et al. 1H and 31P NMR lipidome of ethanol-induced fatty liver. *Alcohol Clin Exp Res* 2010;**34**(11):1937–47.
36. Fernando H, Bhopale KK, Kondraganti S, et al. Lipidomic changes in rat liver after long-term exposure to ethanol. *Toxicol Appl Pharmacol* 2011;**255**(2):127–37.
37. Vázquez-Fresno R, Llorach R, Alcaro F, et al. (1)H-NMR-based metabolomic analysis of the effect of moderate wine consumption on subjects with cardiovascular risk factors. *Electrophoresis* 2012;**33**(15):2345–54.
38. Yoseph BP, Breed E, Overgaard CE, et al. Chronic alcohol ingestion increases mortality and organ injury in a murine model of septic peritonitis. *PLoS One* 2013;**8**(5):e62792.
39. Gao X, Zhao A, Zhou M, et al. GC/MS-based urinary metabolomics reveals systematic differences in metabolism and ethanol response between Sprague–Dawley and Wistar rats. *Metabolomics* 2011;**7**(3):363–74.
40. Lieber CS, DeCarli LM. Ethanol oxidation by hepatic microsomes: adaptive increase after ethanol feeding. *Science* 1968;**162**(3856):917–8.
41. Simon JA, Fong J, Bemert JT, Browner WS. Relation of smoking and alcohol consumption to serum fatty acids. *Am J Epidemiol* 1996;**144**(4):325–34.
42. Nakajima T, Kamijo Y, Tanaka N, et al. Peroxisome proliferator-activated receptor alpha protects against alcohol-induced liver damage. *Hepatology* 2004;**40**(4):972–80.
43. Mutlu EA, Gillevet PM, Rangwala H, et al. Colonic microbiome is altered in alcoholism. *Am J Physiol Gastrointest Liver Physiol* 2012;**302**(9):G966–78.

SECTION III

GENETIC MACHINERY AND ITS FUNCTION

18 *Gene Expression in Alcoholism: An Overview* 225

19 Cytochrome P4502E1 *Gene Polymorphisms and the Risks of Ethanol-Induced Health Problems in Alcoholics* 231

20 *Genes Associated with Alcohol Withdrawal* 247

21 *Alcohol and Epigenetic Modulations* 261

22 *The miRNA and Extracellular Vesicles in Alcoholic Liver Disease* 275

23 *Alcohol Metabolism and Epigenetic Methylation and Acetylation* 287

24 *Molecular Mechanisms of Alcohol-Associated Carcinogenesis* 305

25 *Molecular Link Between Alcohol and Breast Cancer: the Role of Salsolinol* 315

26 *Ethanol Impairs Phospholipase D Signaling in Astrocytes* 325

27 *Metabolic Changes in Alcohol Gonadotoxicity* 337

28 *Molecular Effects of Alcohol on Iron Metabolism* 355

CHAPTER

18

Gene Expression in Alcoholism: An Overview

Reem Ghazali, MSc,**, Vinood B. Patel, PhD***

*Department of Clinical Biochemistry, King Abdul Aziz University, Jeddah, Saudi Arabia

**Department of Biomedical Sciences, Faculty of Science & Technology, University of Westminster, London, UK

INTRODUCTION

Disorders relating to heavy alcohol use typically manifest into two different syndromes: alcohol dependence and alcohol abuse. The latter is considered milder than alcohol dependence. However, in reality, both syndromes lead down the same path, and over time, alcohol abuse results in the same health and behavioral problems as alcohol dependence.[1] Chronic alcohol consumption is associated with serious negative social behavior and community problems. Moreover, alcohol consumption correlates strongly with the incidence of various diseases, such as liver disease, diabetes, cancer, cardiovascular disease, overweight, and obesity.[2] Alcoholism is a heterogeneous and complex disorder, and is influenced to some extent by environmental and genetic/familial factors.[1,3] However, it is generally believed that genetic factors are responsible for about 50–60% of alcoholism cases.[4] The aim of this chapter is to illustrate the wide range of gene expression involved in alcohol dependence.

BRAIN SEROTONIN AND ALCOHOLISM

Wang et al.[4] subdivided alcohol dependence into three different categories: alcoholism based on depression and anxiety; pure and antisocial alcoholism, the latter characterized by an antisocial personality character; and high familial risk giving rise to the early onset of alcoholism. These authors demonstrated that antisocial alcoholism is highly correlated with various genetic polymorphisms such as the serotonin (5-hydroxytryptamine (5-HT)) receptor 1B A-161T.[4] Serotonin (5-HT) is a neurotransmitter, and is believed to be linked to many behavior disorders, such as alcoholism, suicidality, and aggression.[5] 5-HT1B is situated in the postsynaptic and presynaptic terminals, and mediates serotonin release by functioning as a heteroreceptor or autoreceptor.[4] The interaction between the serotonin transporter (5-HTTLPR) and the serotonin 1B receptor (5-HT1B) is associated with vulnerability in individuals to antisocial alcoholism that may occur when brain serotonin function is reduced.[4] Furthermore, the 5-HTT-linked promoter region (5-HTTLPR) polymorphism can affect either the function or the expression of serotonin transport, and that the short allele (s-allele) of this gene displays reduced serotonin uptake and serotonin transport expression, as well as decreased transcriptional activity *in vitro*.[4] Although many studies have found that there is an association between the polymorphism of the 5-HTTLPR and the s-allele of this gene, in terms of the risk of antisocial alcoholism, as well as an association between the polymorphism of 5-HT1B G861C and the occurrence of the same disorder,[6,7] the polymorphism of neither gene alone is responsible for antisocial alcoholism, rather as mentioned before, the interaction between the transporter and the receptor is the risk factor for this alcoholic disorder.[4]

The 5-HT3 receptor (5-HT3R) has also been found to contribute to the risk of antisocial alcoholism. It consists of two subunits, 5-HT3A and 5-HT3B, and there is a similarity of approximately 44% between the amino acid sequences of both subunits. These two subunits are encoded by the HTR3A and HTR3B genes that are colocalized on chromosome 11q23.1.[8] Disrupted 5-HT3R activity may be a characteristic linked to the early onset of alcoholism.[9,10] Ducci et al.[8] found that HTR3B genetic variation correlates with antisocial alcoholism, and that HTR3B SNP (rs3782025) can predict the risk of this disorder.[8]

Molecular Aspects of Alcohol and Nutrition. http://dx.doi.org/10.1016/B978-0-12-800773-0.00018-5

NMDA RECEPTOR AND ALCOHOLISM

The *N*-methyl-D-aspartate (NMDA) subtype of glutamate receptor is thought to play an important role in alcoholism.[11] The NMDA receptor (NMDAR) is composed of various heterodimer complexes, containing NR1 together with subunits from NR2, and subunits from NR3 with perhaps a person's susceptibility to alcoholism dependent on NR1 and NR2 submit composition.[12]. Earlier studies by Lovinger et al.[13] showed that acute alcohol is generally inhibitory on NMDAR activity, with increasing alcohol concentrations from 5 mM to 100 mM, whereas in animal studies chronic alcohol exposure leads to upregulation of NMDA receptors.[14] However, decreased NMDAR expression and synaptic function has also been reported, following chronic alcohol exposure,[15] whereas alcohol withdrawal leads to an increase in glutamate release.[16] These studies indicate the complexity of issues surrounding the effects of alcohol dosage on NMDA receptor activation or inhibition, and consequently glutamate metabolism. While therapeutic drugs have targeted the NMDAR, it is possible selective antagonism of particular NMDA subunits may have more beneficial effects.[17]

FETAL ALCOHOL SYNDROME

Fetal alcohol syndrome (FAS) is considered one of the primary causes of mental retardation in children whose mothers consumed excessive alcohol during pregnancy. Manifestations of this disease include facial feature deformation, brain injury, and fetal growth defects that can affect the development of the fetal brain, at any stage of gestation, and cause defects in the proliferation of neuron migration and position.[18] Fetal alcohol syndrome is considered to be the most severe of all fetal alcohol spectrum disorders. There are various factors that affect the severity of these disorders, including the overall amount of alcohol consumed by pregnant women and the duration of consumption, malnutrition, and genetic factors. All of these factors have become the focus of today's researchers.[19]

GSC is an organizer-specific and homeobox gene that is expressed and involved in the formation of the endoderm, as well as in the morphogenesis of the axial mesendoderm. It is believed that the GSC gene is overexpressed after alcohol exposure, and deregulated in the early stages of embryo differentiation by nodal signaling that may contribute to fetal alcohol syndrome.[20] Nodal is considered to be a member of a cytokine family known as TGF-β, and plays a vital role in the development of many processes, including early development of the fetus. It has also been found that the expression of this gene increases after ethanol consumption, and that nodal is a target point in the pathway of nodal signaling that consequently contributes to the regulation of GSC expression in the early stages of embryogenesis.[20]

There are many other genes that play a central role in the early phase of fetal development, whose expression can be also altered by excessive alcohol consumption. One of those genes is aldolase C (ALDOC or zebrin II). This gene is a member of the brain aldolase family, and is considered to be a type of brain isozyme that is involved in glycolysis. It is heterogeneously expressed in the cerebral cortex and plays a vital role in the development of the central nervous system.[21,22] Halder et al.[20] demonstrated that the expression of ALDOC can be affected and decreased after high ethanol consumption, and this is believed to affect and inhibit the development of the fetus.[20]

SWEET PREFERENCE AND ALCOHOL DEPENDENCE

Many studies have linked the association between sweet preference and the genetic risk of alcohol dependence.[23] Regardless of alcoholic status, sweet preference is more associated with individuals who have a family history of alcohol dependence, specifically, paternal history, than those who have nonalcoholism paternal history.[23,24] It has also been found that alcoholic individuals prefer sweet substances by 65% compared to nonalcoholic subjects, who account only for 16%.[24]

It seems that bulimic as well as obese individuals have a sweet taste preference, and medicate themselves (self-reward) with sugar and sweets that, in turn, activates both reward pathways (beta-endorphin and dopamine in the nucleus accumbens) similar to alcohol dependence, and consequently, stimulates their release. Dopamine is a neurotransmitter that plays a key role in controlling the brain's pleasure and reward centers. The dopamine receptor D2 gene (DRD2) is considered to be a pleasure determinant and is located in the 11q22.3-q23 chromosome; therefore, any deficiency in the numbers of these receptors will indicate the risk of alcoholism due to the association between this gene and alcohol disorders. One of the alleles of the dopamine receptor D2, known as the Al allele, has contributed to many psychological and addictive disorders, such as alcohol dependence, bulimia, and obesity. The presence of this allele is strongly associated with the diminished density of the DRD2 gene that explains the decreased number of dopamine receptors in obese individuals, compared to nonobese people.[3,25–28] Therefore, it seems that the correlation between preference for high-sucrose products and alcohol intake may be due to the similarity of the impact of both substances on neurotransmitter systems, such as the serotonergic and dopaminergic systems.[28]

Ankyrin repeat and kinase domain containing 1 (ANKK1) is another gene believed to play a central role, via its signal transduction in the dopaminergic reward system. The ANKK1gene has a Taq 1A polymorphism situated within the exon of this gene,[3,28] and that the A1 allele of Taq 1A is considered to be the determinant of sweet preference, whereas the A2 allele suppresses this feature.[28] Therefore, the A1A1 and A1A2 genotypes are significantly expressed in people with a sweet preference, while the A2A2 genotype is expressed highly in those with no sweet preference. This suggests that the ANKK1 gene and the polymorphism of Taq 1A play a central etiological role in sweet taste preference that can pave the way for the risk of alcohol dependence.

STEM CELL THERAPY AND ALCOHOLISM

As mentioned previously in this chapter, chronic alcohol consumption is a central causative component that can lead to a number of associated diseases. Alcoholic liver disease (ALD) is one such disease. It results from prolonged high consumption of ethanol, and consequently leads to a spectrum of ALD stages, originating with steatosis, followed by steatohepatitis that is characterized by fibrosis, leading to cirrhosis and, in some cases, hepatocellular cancer.[29] It is known that chronic liver disease ends with liver cirrhosis that can be cured currently by means of liver transplantation, one of the most effective therapies.[30] However, it has now been found that this stage may also be treated with stem cells derived from bone marrow, such as CD34+ stem cells that have the ability to differentiate to hepatocytes. CD34+ is also believed to be a promising therapy based on its role in the proliferation of the liver, following any serious loss of hepatic mass, and in the repair of either chronic or acute liver injury.[29–31] Moreover, it has been reported that autologous CD133+ stem cell portal administration has a potential role in accelerating the regeneration of the liver, and ameliorating hepatic injury, as well as improving the function of the liver.[31] Terai et al.[30] showed that bone marrow stem cell infusion in patients with liver cirrhosis induced the proliferation of hepatocytes, as well as improved hepatic function.[30]

GENETIC RISKS OF ALCOHOLISM

Various research studies have been conducted to study genetic influences with respect to the risk of alcohol dependence. Alcohol can act on many neurotransmitter genes that subsequently end up affecting various cellular processes. However, few genes with functional loci have shown efficacy in terms of treating alcoholism and reversing its impact. Alcohol dependence genes have been mapped by the analysis of whole genomes in a region on chromosome 4 that contains gene clusters for the alcohol dehydrogenase enzyme, as well as for α-synuclein (SNCA).[32] SNCA is a polymorphic gene that contributes to brain plasticity modulation, neurotransmission, and neurogenesis. The pathological effect of this gene can be instigated from any abnormal level; for instance, a reduced level of SNCA results in lower oxidative stress protection, while an increased level may play a role in degenerative neuron diseases such as Parkinson's. However, from an alcoholism perspective, the SNCA gene plays a role in stimulating alcohol cravings. Therefore, low levels of the SNCA gene can predispose the individual to alcohol cravings that may be a result of increased neurotransmitter activity, leading to excessive alcohol consumption. This, in turn, increases and upregulates the gene's expression in alcoholic dependents.[32,33] GFAP (glial fibrillary acidic protein) is a gene involved in processes such as cell communication, cellular adhesion, and neuron interaction. GFAP expression is decreased in the brains of alcoholics, yet elevated in individuals predisposed to alcoholism.[33] It is believed that excessive and chronic alcohol consumption leads to various types of organ damage in the body, including brain damage, by attenuating the central nervous system, and causing brain atrophy. This reduces the weight of the brain, which explains the reduced level of the gene. The GFAP gene has a vital role to play in maintaining the white matter located in the frontal lobe of the brain. Therefore, any acute or high alcohol consumption may lead to a reduction in white matter, and affect the expression of the GFAP gene.[33,34]

MicroRNA AND ALCOHOLISM

MicroRNAs (miRNA) are small endogenous nonprotein-coding RNA molecules of approximately 21–23 bases that can potentially regulate hundreds or thousands of protein coding mRNA species, at the translational level or via increased degradation, and are now known to be involved in psychiatric disorders,[35] and forms of addiction, such as cocaine.[36] In terms of alcoholism, while miRNAs are involved in liver disease,[37] their role in addiction may have equal importance. The study by Lewohl et al.[38] showed significant upregulation of 35 miRNAs found in the prefrontal cortex of alcoholics, with the authors indicating this may contribute to the deterioration and concomitant adaptation of neuronal functioning in alcoholism, whereas, in a model of alcohol dependence, 41 miRNAs were altered in the rat prefrontal cortex,[39] with genes involved in neurotransmission, neuroadaptation, and synaptic plasticity affected.[39] One particular miRNA of interest is miR-9. This was upregulated, following acute alcohol treatment in rat striatal neurons, causing a decrease in splice variants mRNA encoding the large-conductance calcium- and voltage-activated potassium channel (BK) sensitive to alcohol potentiation, thus supporting a role for alcohol tolerance.[40] Although, 5-day treatment at higher alcohol concentrations (70 mM) showed downregulation of miR-9,[41] a fact that again highlights the complexity of varied responses in different models and doses of alcohol. Nevertheless, these studies indicate the potential biological role of miRNAs in regulating genes involved in alcoholism.

CONCLUSIONS

This chapter has reviewed a select number of the studies conducted to elucidate, and reveal new approaches relating to new genes that may contribute to addressing the risk of alcoholism. However, none of the data collected regarding those genes appears to be promising in terms of their potential use, either as therapy, or as analytical markers. For that reason, further studies and experimentation are still needed to uncover novel gene(s) for treating alcoholism, and to assess the impact of alcohol on the expression of the many genes implicated in the risk of alcoholism.

Key Facts

- According to the World Health Organization, the estimated number of alcoholics worldwide is 140 million people.
- In 2013, 139,000 deaths resulted from alcohol misuse.
- The prevalence of alcoholism is higher among males, than females.
- The risk of alcohol dependence is correlated to the genetic variation among different racial groups.

Summary Points

- Alcoholism is a worldwide issue with a deleterious effect on health and behavior.
- Genes are either up/downregulated by chronic alcohol consumption.
- Many genes contribute in predisposing the person to alcohol craving.
- Neurotransmitter genes, serotonin (5-HT) polymorphism is highly associated with alcoholism, and other behavioral disorders.
- Aldolase C and GSC genes can negatively affect fetus development, causing fetal alcohol syndrome.
- The association between sweet preference and alcoholism is highly correlated to the deficiency of dopamine receptor D2 gene (DRD2), and the polymorphism of ANKK1.
- Stem cells infusion can be a promising therapy for alcoholic liver cirrhosis.

References

1. Kendler KS, Myers J. Clinical indices of familial alcohol use disorder. *Alcohol Clin Exp Res* 2012;**36**(12):2126–31.
2. Pettigrew S, Jongenelis M, Chikritzhs T, Slevin T, Pratt I, Glance D, Liang W. Developing cancer warning statements for alcoholic beverages. *BMC Public Health* 2014;**14**(1):786.
3. Meyers JL. The association between DRD2/ANKK1 and genetically informed measures of alcohol use and problems. *Addict Biol* 2013;**18**(3):523.
4. Wang T, Lee S, Chen S, Chang Y, Chen S, Chu C, Huang S, Tzeng N, Wang C, Lee I, Yeh T, Yang Y, Lu R. Interaction between serotonin transporter and serotonin receptor 1 B genes polymorphisms may be associated with antisocial alcoholism. *Behav Brain Funct* 2012;**8**:18.
5. Best J, Nijhout F, Reed M. Serotonin synthesis, release and reuptake in terminals: a mathematical model. *Theor Biol Med Model* 2010;**7**(1):34.
6. Lappalainen J, Long J, Eggert M, Ozaki N, Robin R, Brown G, Naukkarinen H, Virkkunen M, Linnoila M, Goldman D. Linkage of antisocial alcoholism to the serotonin 5-HT1B receptor gene in 2 populations. *Arch Gen Psychiatry* 1998;**55**(11):989–94.
7. Parsian A, Cloninger CR. Serotonergic pathway genes and subtypes of alcoholism: association studies. *Psychiatr Genet* 2001;**11**(2):89–94.
8. Ducci F, Enoch M, Yuan Q, Shen P, White KV, Hodgkinson C, Albaugh B, Virkkunen M, Goldman D. HTR3B is associated with alcoholism with antisocial behavior and alpha EEG power – an intermediate phenotype for alcoholism and co-morbid behaviors. *Alcohol* 2009;**43**(1):73–84.
9. Campbell AD, McBride WJ. Serotonin- 3 receptor and ethanol-stimulated dopamine release in the nucleus accumbens. *Pharmacol Biochem Behav* 1995;**51**(4):835–42.
10. Johnson B, Roache J, Ait-Daoud N, Zanca N, Velazquez M. Ondansetron reduces the craving of biologically predisposed alcoholics. *Psychopharmacology* 2002;**160**(4):408–13.
11. Ron D, Wang J. The NMDA receptor and alcohol addiction. In: Van Dongen AM, editor. *Biology of the NMDA receptor*. Boca Raton, FL: CRC Press; 2009.
12. Kaniakova M, Krausova B, Vyklicky V, Korinek M, Lichnerova K, Vyklicky L, Horak M. Key amino acid residues within the third membrane domains of NR1 and NR2 subunits contribute to the regulation of the surface delivery of NMDA receptors. *J Biol Chem* 2012;**287**:26423–34.
13. Lovinger DM, White G, Weight FF. Ethanol inhibits NMDA-activated ion current in hippocampal neurons. *Science* 1989;**243**:1721–4.
14. Lovinger DM, Roberto M. Synaptic effects induced by alcohol. *Curr Top Behav Neurosci* 2013;**13**:31–86.
15. Holmes A, Fitzgerald PJ, MacPherson KP, DeBrouse L, Colacicco G, Flynn SM, Masneuf S, Pleil KE, Li C, Marcinkiewcz CA, Kash TL, Gunduz-Cinar O, Camp M. Chronic alcohol remodels prefrontal neurons and disrupts NMDAR-mediated fear extinction encoding. *Nat Neurosci* 2012;**10**:1359–61.
16. Kumari M, Anji A. An old story with new twist: do NMDAR1 mRNA binding protein regulate expression of the NMDAR1 receptor in the presence of alcohol? *Ann NY Acad Sci* 2005;**1053**:311–8.
17. Holmes A, Spanagel R, Krystal JH. Glutamatergic targets for new alcohol medications. *Psychopharmacology* 2013;**229**:539–54.
18. El Fatimy R, Miozzo F, Mouël A, Abane R, Schwendimann L, Sabéran-Djoneidi D, Thonel A, Massaoudi I, Paslaru L, Hashimoto-Torii K, Christians E, Rakic P, Gressens P, Mezger V. Heat shock factor 2 is a stress- responsive mediator of neuronal migration defects in models of fetal alcohol syndrome. *EMBO Mol Med* 2014;**6**(8):1043–61.
19. McCarthy N, Eberhart JK. Gene-ethanol interactions underlying fetal alcohol spectrum disorders. *Cell Mol Life Sci* 2014;**71**(14):2699.

20. Halder D, Park JH, Choi MR, Chai JC, Lee YS, Mandal C, Jung K, Chai YG. Chronic ethanol exposure increases goosecoid (GSC) expression in human embryonic carcinoma cell differentiation. *J Appl Toxicol* 2014;**34**(1):66–75.
21. Skala-Rubinson H, Vinh J, Labas V, Kahn A, Tuy F. Novel target sequences for Pax-6 in the brain-specific activating regions of the rat aldolase C gene. *J Biol Chem* 2002;**277**(49):47190–6.
22. Fujita H, Aoki H, Ajioka I, Yamazaki M, Abe M, Oh-Nishi A, Sakimura K, Sugihara I. Detailed expression pattern of aldolase C (Aldoc) in the cerebellum, retina and other areas of the CNS studied in Aldoc-Venus knock-in mice. *PLoS One* 2014;**9**(1):e86679.
23. Fortuna J. Sweet preference, sugar addiction and the familial history of alcohol dependence: shared neural pathways and genes. *J Psychoactive Drugs* 2010;**42**(2):147–51.
24. Kampov-Polevoy AB, Ziedonis D, Steinberg ML, Pinsky I, Krejci J, Eick C, Boland G, Khalitov E, Crews FT. Association between sweet preference and paternal history of alcoholism in psychiatric and substance abuse patients. *Alcohol Clin Exp Res* 2003;**27**(12):1929–36.
25. Stice E, Spoor S, Bohon C, Small DM. Relation between obesity and blunted striatal response to food is moderated by TaqIA A1 allele. *Science* 2008;**322**(5900):449–52.
26. Nisoli E, Brunani A, Borgomainerio E, Tonello C, Dioni L, Briscini L, Redaelli G, Molinari E, Cavagnini F, Carruba M. D2 dopamine receptor (DRD2) gene Taq1A polymorphism and the eating-related psychological traits in eating disorders (anorexia nervosa and bulimia) and obesity. *Eat Weight Disord* 2007;**12**(2):91–6.
27. Davis C, Levitan RD, Kaplan AS, Carter J, Reid C, Curtis C, Patte K, Hwang R, Kennedy JL. Reward sensitivity and the D2 dopamine receptor gene: a case-control study of binge eating disorder. *Prog Neuropsychopharmacol Biol Psychiatry* 2008;**32**(3):620–8.
28. Jabłoński M, Jasiewicz A, Kucharska-Mazur J, Samochowiec J, Bienkowski P, Mierzejewski P, Samochowiec A. The effect of selected polymorphisms of the dopamine receptor gene DRD2 and the ANKK-1 on the preference of concentrations of sucrose solutions in men with alcohol dependence. *Psychiatr Danub* 2013;**25**(4):371–8.
29. Gao B, Bataller R. Alcoholic liver disease: pathogenesis and new therapeutic targets. *Gastroenterology* 2011;**141**(5):1572–85.
30. Terai S, Ishikawa T, Omori K, Aoyama K, Marumoto Y, Urata Y, Yokoyama Y, Uchida K, Yamasaki T, Fujii Y, Okita K, Sakaida I. Improved liver function in liver cirrhosis patients after autologous bone marrow cell infusion (ABMI) therapy. *Hepatology* 2006;**44**(4):402A.
31. Am Esch J, Knoefel W, Klein M, Ghodsizad A, Fuerst G, Poll L, Piechaczek C, Burchardt E, Feifel N, Stoldt V, Stockschläder M, Stoecklein N, Tustas R, Eisenberger C, Peiper M, Häussinger D, Hosch S. Portal application of autologous CD133+ bone marrow cells to the liver: a novel concept to support hepatic regeneration. *Stem Cells* 2005;**23**(4):463–70.
32. Janeczek P, Mackay RK, Lea RA, Dodd P, Lewohl JM. Reduced expression of α-synuclein in alcoholic brain: influence of SNCA-Rep1 genotype. *Addict Biol* 2014;**19**(3):509–15.
33. Levey D, Le-Niculescu H, Frank J, Ayalew M, Jain N, Kirlin B, Learman R, Winiger E, Rodd Z, Shekhar A, Schork N, Kiefe F, Wodarz N, Müller-Myhsok B, Dahmen N, Nöthen M, Sherva R, Farrer L, Smith A, Kranzler H, Rietschel M, Gelernter J, Niculescu A. Genetic risk prediction and neurobiological understanding of alcoholism. *Transl Psychiatr* 2014;**4**(5):e391.
34. Lewohl JM, Wixey J, Harper C, Dodd PR. Expression of MBP, PLP, MAG, CNP, and GFAP in the human alcoholic brain. *Alcohol Clin Exp Res* 2005;**29**(9):1698–705.
35. Kolshus E, Dalton VS, Ryan KM, McLoughlin DM. When less is more – microRNAs and psychiatric disorders. *Acta Psychiatr Scand* 2014;**129**:241–56.
36. Jonkman S, Kenny PJ. Molecular, cellular, and structural mechanisms of cocaine addiction: a key role for microRNAs. *Neuropsychopharmacology* 2013;**38**:198–211.
37. Dolganiuc A, Petrasek J, Kodys K, Catalano D, Mandrekar P, Velayudham A, Szabo G. MicroRNA expression profile in Lieber-DeCarli diet-induced alcoholic and methionine choline deficient diet-induced nonalcoholic steatohepatitis models in mice. *Alcohol Clin Exp Res* 2009;**33**:1704–10.
38. Lewohl JM, Nunez YO, Dodd PR, Tiwari GR, Harris RA, Mayfield RD. Up-regulation of microRNAs in brain of human alcoholics. *Alcohol Clin Exp Res* 2011;**35**:1928–37.
39. Tapocik JD, Solomon M, Flanigan M, Meinhardt M, Barbier E, Schank JR, Schwandt M, Sommer WH, Heilig M. Coordinated dysregulation of mRNAs and microRNAs in the rat medial prefrontal cortex following a history of alcohol dependence. *Pharmacogenomics J* 2013;**13**:286–96.
40. Pietrzykowski AZ, Friesen RM, Martin GE, Puig SI, Nowak CL, Wynne PM, Siegelmann HT, Treistman SN. Posttranscriptional regulation of BK channel splice variant stability by miR-9 underlies neuroadaptation to alcohol. *Neuron* 2008;**59**:274–87.
41. Sathyan P, Golden HB, Miranda RC. Competing interactions between micro-RNAs determine neural progenitor survival and proliferation after ethanol exposure: evidence from an *ex vivo* model of the fetal cerebral cortical neuroepithelium. *J Neurosci* 2007;**27**:8546–57.

CHAPTER

19

Cytochrome P4502E1 Gene Polymorphisms and the Risks of Ethanol-Induced Health Problems in Alcoholics

Tao Zeng, PhD, Ke-Qin Xie, PhD

Institute of Toxicology, School of Public Health, Shandong University, Jinan City, China

The relationship between heavy drinking and risks of health problems, such as alcohol dependence (AD) and alcoholic organ damage, is well known. However, not everyone who drinks chronically and heavily develops these diseases; this indicates that genetic variations may be responsible for interindividual differences in the susceptibility to AD and alcohol-related organ damage. Accumulating evidences have suggested that these differences may be associated with the polymorphisms of many genes, such as those encoding enzymes involved in ethanol metabolism.

Cytochrome P450 2E1 (CYP2E1) is a major component of the microsomal ethanol-oxidizing system (MEOS), and is responsible for about 10% of total ethanol metabolism.[1–3] CYP2E1 can catalyze the conversion of ethanol to acetaldehyde that will then be metabolized to acetate, by acetaldehyde dehydrogenase. CYP2E1-mediated ethanol oxidation is a major pathway for ethanol-induced oxidative stress that has been demonstrated to play critical roles in the development of ethanol-induced liver and pancreas damage.[4,5] In this context, the interindividual difference in the susceptibility to various ethanol-induced diseases may be related to the difference in the catalytic activity of CYP2E1. In fact, *in vitro* studies with human microsomes demonstrated remarkable interindividual variation in levels of both protein, and catalytic activity of CYP2E1.[6,7] *In vivo* studies using chlorzoxazone as probe drug revealed about four- to five-fold variations in CYP2E1 activity in healthy subjects, or in ethanol-induced alcoholics.[8,9] These interindividual variations might be related with the functional polymorphisms of *CYP2E1* gene. Therefore, the functional polymorphisms of *CYP2E1* gene might be an important factor in determining the relative risks of ethanol-induced diseases.

In this chapter, we describe the current knowledge about the reported genetic polymorphisms of *CYP2E1* gene, and summarize the available human epidemiological studies about the relationship between *CYP2E1* gene polymorphisms, and risks of ethanol-induced health problems: AD, alcoholic liver diseases (ALD), and alcoholic pancreatitis (AP).

THE POLYMORPHISMS OF *CYP2E1* GENE AND THE NOMENCLATURES OF *CYP2E1* ALLELES

The human *CYP2E1* gene is located on the tenth chromosome and consists of nine exons and eight introns. *CYP2E1* gene contains a typical TATA-box, and occupies 11413 bp of genomic DNA.[10] A number of *CYP2E1* gene polymorphic loci have been reported so far that can be found on the human *CYP450* allele nomenclature website (http://www.imm.ki.se/CYPalleles/cyp2e1.htm). These *CYP2E1* allelic variants are located in the 5′-flanking region, in introns, or in the transcribed regions (exons), and were initially detected by restriction fragment length polymorphism (RFLP) analysis, single-strand conformational polymorphism (SSCP) analysis, or DNA sequencing. The nomenclature for *CYP2E1* alleles, the site of mutation, nucleotide, and amino acid changes, the RFLP sits, and the effects on enzyme activity, are summarized in Table 19.1.

Molecular Aspects of Alcohol and Nutrition. http://dx.doi.org/10.1016/B978-0-12-800773-0.00019-7

TABLE 19.1 Nomenclature of *CYP2E1* Gene Polymorphisms According to the Human *Cytochrome P450* (*CYP*) Allele Nomenclature Committee

Allele	Protein	Site of mutation	Nucleotide change	Amino acid change	RFLP	Enzyme activity *in vitro*	Enzyme activity *in vivo*
*CYP2E1*1A*	CYP2E1*1		None	None		Wild type (wt)	Wild type (wt)
*CYP2E1*1B*	CYP2E1*1	Intron 7	9896C > G	None	*TaqI*–	Similar to wt[11]	Unknown
*CYP2E1*1C*	CYP2E1*1	5′-Flanking region	insertion 6 tandem repeats	None		Unknown	Unknown
*CYP2E1*1D*	CYP2E1*1	5′-Flanking region	insertion 8 tandem repeats	None	*Dra*I; *Xba*I	Increase the activity in Hela cells, not in B16A2 cells[12,13]	Increase activity after alcohol exposure and in obese subjects[14]
*CYP2E1*2*	CYP2E1*2	Exon 2	−1132G > A	R76H		Reduced protein level by 37% and activity by 36%[15]	Unknown
*CYP2E1*3*	CYP2E1*3	Exon 8	−10023G > A	V389I		Similar to wild type[15]	Unknown
*CYP2E1*4*	CYP2E1*4	Exon 4	−4768G > A	V179I		Similar to wild type[16]	Unknown
*CYP2E1*5A*	CYP2E1*1	5′-Flanking region	−1293G > C; −1053C > T; 7632T > A	None	*Pst*I + ; *Rsa*I–; *Dra*I–	Similar to wt using human liver microsomes[11]	Less inducible[9]
*CYP2E1*5B*	CYP2E1*1	5′-Flanking region	−1293G > C; −1053C > T;	None	*Pst*I + ; *Rsa*I–	Increased the transcription and activity[17–19]; similar to wt[11,20]	Decreased activity[9,21]
*CYP2E1*6*	CYP2E1*1	Intron 6	7632 T > A	None	*Dra*I–	Similar to wt[11]	Decreased activity[8]
*CYP2E1*7A*	CYP2E1*1	5′-Flanking region	−333T > A	None		Unknown	Unknown
*CYP2E1*7B*	CYP2E1*1	5′-Flanking region	−71G > T; −333T > A	None		Unknown	Unknown
*CYP2E1*7C*	CYP2E1*1	5′-Flanking region	−333T > A; −352A > G	None		Unknown	Unknown

RFLP, restriction fragment length polymorphism; R, H, V, I, represents Arginine, Histidine, valine, and Isoleucide, respectively.

The Influence of *CYP2E1* Gene Polymorphisms on the Catalytic Activity of the Enzyme

In comparison to other *CYP450* genes involved in drug metabolism, the human *CYP2E1* gene is relatively well conserved.[15] Most of the *CYP2E1* polymorphic sites are located in introns or in the 5′-flanking regulatory region that make the protein usually intact. Hitherto, only three polymorphic loci including *CYP2E1*2*, **3*, **4* in exon 2, 8, and 4, respectively, have been reported to result in amino acid modification of the CYP2E1 protein.[15,16] However, only the *CYP2E1*2* allele might influence the catalytic activity of the enzyme.[15] Although all the other polymorphic loci are located in the noncoding regions, some polymorphisms, including the *CYP2E1*1C*, **1D*, **5A*, **5B* in the 5′-flanking region, and the *CYP2E1*6* in intron 6, may also modulate the expression of *CYP2E1*.

The extensively studied *CYP2E1 Rsa I* polymorphism (in *CYP2E1*5A* and **5B*) is in complete linkage disequilibrium with the *Pst I* polymorphism.[17,18] Allele with *Rsa I* site (*Rsa I +* , *Pst I* –) is defined as the wild type (c1 allele), while the variant allele (*Rsa I*–, *Pst I* +) was designated as c2 allele. *In vitro* study showed that homozygous c2c2 genotype was associated with 10 time increase in *CYP2E1* gene transcription.[17] Further studies with microsomes of Japanese liver samples showed that the mRNA levels of subjects with c1c2 genotype were higher than those with c1c1 genotype, thus also demonstrating that transcriptional activity of the c2 allele was stronger than that of the c1 allele.[18,19] However, other *in vitro* studies did not find significant increase of CYP2E1 activity in c2 carriers, compared with c1 carriers.[11,20] Similarly, *in vivo* studies using chlorzoxazone as the probe drug also failed to confirm the initially observed *in vitro* results. For example, Kim et al. found there was no association between c2 allele and higher activity of CYP2E1 in Caucasians,[8] while the study by Marchand et al. showed that Japanese with c2c2 genotype had significantly reduced chlorzoxazone clearance, compared with those with the c1c1 genotype.[21] In another study, decreased activity of CYP2E1 in alcoholic patients with c1c2 genotype was observed, compared with those with c1c1 genotype in Caucasians; this suggested that the c2 allele might be associated with lower inducibility in CYP2E1 activity.[9] These contradictory results might be related to the status of the enrolled subjects, as it has been well documented that CYP2E1 activity could be influenced by many factors, including the diets (high fat, low-carbohydrates, garlic, etc.) and the physiological condition (obesity, diabetes, etc).[21,22] The interethnic difference might be another contributor to the contradictory results. It has been demonstrated that the total CYP content of Japanese subjects was consistently lower than those of European origin (0.26 nmol/mg protein versus 0.43 nmol/mg protein).[6] The CYP2E1 catalytic activity was also significantly lower in Japanese than in Caucasians.[23] Furthermore, the frequency of c2 allele in Japanese is higher than that of Caucasians.[11] These data suggest pronounced interethnic differences might exist in CYP expression, and also in the genetic polymorphisms of *CYP2E1* gene.

The polymorphism detectable with *Dra I* in intron 6 (*CYP2E1*6*) produces three genotypes: the homozygous CC and DD, and the heterozygous CD. It has been demonstrated that the subjects with CD genotype had statistically significant lower catalytic activity, than those with the DD genotype, although no statistical difference was noted between the homozygous genotypes (CC versus DD).[8] Inoue et al. investigated the associations between the *CYP2E1*1A*, **5A*, **5B*, **6* and the protein expression and enzyme catalytic activities in 39 Japanese and 45 Caucasians, but did not find any significant association.[11]

The insertion mutation in the 5'-flanking region between -2270 and -1672 bp on the human *CYP2E1* region was initially identified by using DNA restriction endonuclease *Xba I*.[14] The presence of the insertion appeared to be associated with elevated CYP2E1 activity, but only in obese patients and chronic alcoholic subjects.[14] Hu et al. characterized the polymorphic repeat sequence of this region, and identified two alleles: *CYP2E1*1C* (six sequence repeats) and *CYP2E1*1D* (eight sequence repeats).[24] *In vitro* studies showed that constitutive reporter gene activity produced by the *CYP2E1*1D* allele was significantly greater than that produced by the *CYP2E1*1C* allele in Hela cells, but not in human hepatoblastoma cell (B16A2 cells).[12,13]

The other polymorphisms in non-coding mutations (*CYP2E1*1B*, **7A*, **7B*, and **7C*) are relatively rarely studied. The *CYP2E1*1B* polymorphism, located on intron 7, could be detected by restriction enzyme *Taq I*. The *CYP2E1*7A*, **7B*, and **7C* alleles are all associated with a -333T > A nucleotide change. The **7B* and **7C* alleles have an additional –71G > T nucleotide change, and –352A > G nucleotide change, respectively. Higher levels of acrylonitrile-hemoglobin adducts in persons with *CYP2E1*7C* allele was observed in a cohort of 59 persons involved in the industrial handling of low levels of acrylonitrile. Higher adduct levels would be compatible with a slower CYP2E1-mediated metabolism of acrylonitrile, although the results were not statistically significant.[25]

Population Distribution of *CYP2E1* Gene Polymorphisms

Striking interethnic difference in the c2 allele frequency of *CYP2E1 Rsa I/Pst I* polymorphism was reported in different populations. The c2 allele frequency was about 2–5% in Caucasians,[9,26–28] 5% in African-Americans,[28] and 20–28% in Asians.[18,28] The C allele frequency for *CYP2E1*6* polymorphism was about 8–11% in Caucasians and African Americans,[9,26,27] and 24–29% in Asians. The *CYP2E1*1D* frequency was about 10–15% for African Americans, 15–23% in Asians, and 1–6.9% in Caucasians.[14,24,29]

THE POLYMORPHISMS OF *CYP2E1* GENE AND INDIVIDUAL SUSCEPTIBILITY TO ALCOHOL DEPENDENCE (AD)

The Terminology of Alcohol Dependence, Abuse, Misuse, and Alcoholism

AD is a substance related disorder in which an individual is addicted to alcohol either physically or mentally, and continues to use alcohol despite significant areas of dysfunction, evidence of physical dependence, and/or related hardship. Alcohol misuse is generally considered an umbrella term for both "alcohol dependence" and "alcohol abuse." AD is differentiated from alcohol abuse by two main classification systems: the Diagnostic and Statistical Manual of Mental Disorders (DSM-IV), and the World Health Organization (WHO)'s International Classification of Diseases (ICD-10).[30] The diagnosis of AD demands that at least three of seven of the following criteria during a 12 month period: tolerance, withdrawal, persistent consumption despite harmful consequences, and a loss of interest in alternative activities or former interests.[31] The term "alcoholism" is a commonly used, but poorly defined, and less specific term. The inexactness of the term led a 1979 WHO Expert Committee to disfavor it, preferring the narrower formulation of alcohol dependence syndrome as one among a wide range of alcohol related problems. However, alcoholism is still widely used as a diagnostic and descriptive term, despite its ambiguous meaning. In 1992, the National Council on Alcoholism and Drug Dependence (NCADD) and the American Society of Addiction Medicine defined alcoholism as a primary, chronic disease with genetic, psychosocial, and environmental factors influencing its development and manifestations.[32] In professional and research contexts, the term "alcoholism" often encompasses both alcohol abuse and AD, and sometimes is considered equivalent to AD.

Potential Links Between *CYP2E1* Gene Polymorphisms and Individual Susceptibility to AD

Acetaldehyde has been demonstrated to play important roles in the pathogenesis of AD.[33] Acetaldehyde possesses reinforcing properties, a fact that suggests that some of the behavioral pharmacological effects induced by ethanol may be results of the formation of acetaldehyde, and supports the involvement of acetaldehyde in AD.[34,35] Ethanol can be metabolized to acetaldehyde through three pathways, that is, the alcohol dehydrogenase (ADH), catalase, and MEOS (mainly CYP2E1). The accumulation of acetaldehyde is influenced by the catalytic activities of these enzymes. As ADH is not physiologically active in the brain, the CYP2E1 should play important roles in the conversion of ethanol to acetaldehyde in the rodent brains.[35] Therefore, the *CYP2E1* gene polymorphisms may influence individual susceptibility to AD by altering the metabolism of ethanol to acetaldehyde in the brain.

Human Epidemiological Studies About the *CYP2E1* Gene Polymorphisms and Individual Susceptibility to AD

A number of epidemiological studies have been performed, trying to elucidate the roles of polymorphisms of *CYP2E1* gene, and the individual susceptibility to AD (Table 19.2). The results of these studies are conflicting, and there is no consensus on the roles of the *CYP2E1* gene polymorphisms and risks of AD, although some evidence suggests that some polymorphic loci might be risk factors for AD.

The most studied is the *CYP2E1 Rsa I/Pst I* polymorphism, as the mutant c2 allele has been suggested to increase enzymatic activity *in vitro*.[17] It would be plausible that the c2 allele frequency in AD patient is higher than that of control subjects. Indeed, a significantly higher frequency of c2 allele was observed in AD patients, compared with nonalcoholics, in male Koreans.[48] Similarly, the study by Konishi et al. demonstrated that the percentage of subjects who carried the c2 alleles was significantly higher in alcoholics than the controls in Mexican Americans.[40] However, there were also studies in which the c2 allele frequencies in AD patients were similar, or even lower, than those in controls. For example, no significant difference in c2 allele frequency was found between AD patients and nonalcoholic controls in Japanese.[42,44,45] A study performed in Chinese also revealed null association between c2 allele and AD risks.[43] Furthermore, studies in Caucasians (including Polish, Spanish, Italian, and British) demonstrated there were no significant differences in c2 allele frequencies or genotype distributions between AD patients and nonalcoholics.[36,38,39,41,46] These contradictory results may be related with the gene–gene interactions or the gene–environment interactions. For example, in the study by Konishi et al., the authors found that the c2 allele frequency in alcoholics carrying *ADH3*1/*1* allele was almost three times higher than that in nonalcoholics carrying *ADH3*1/*1* allele, which was only about 1.2 fold when *ADH3* polymorphism was not considered.[40] Furthermore, some studies suggested that *CYP2E1* c2 allele might be associated with age of AD onset. The persons with c1c2 genotypes became alcohol dependent at a considerably younger age than the subjects with c1c1 genotypes.[36,40]

In addition to the *CYP2E1 Rsa I/Pst I* polymorphism, the associations between the *CYP2E1 Dra I* polymorphism, the *CYP2E1 Taq I* polymorphism, and the susceptibility to AD have also been investigated, but the results were also inconsistent.[37–41,47] For the *CYP2E1 Dra I* polymorphism, one study showed that C allele was associated with decreased risks of AD,[37] finding that appears to be consistent with the concept that the mutant C allele produces a less inducible form of CYP2E1.[9] In another study, the frequency of C allele in AD patients was significantly higher than that in controls, suggesting that the C allele might be related to the risk of developing AD.[47] However, studies performed in Italians and Spanish showed null association between *CYP2E1 Dra I* polymorphism and risks of AD.[38,39,41] In regard with *CYP2E1 Taq I* polymorphism, one study showed significant association between *CYP2E1 Taq I* polymorphism and AD risk,[37] while another study showed null association.[40]

Meta-Analysis of the *CYP2E1* Gene Polymorphisms and Susceptibility to AD

Meta-analysis is a quantitative method that focuses on contrasting and combining results from different studies. Meta-analysis can increase the statistical power by increasing the sample size. At present, no meta-analysis

TABLE 19.2 Summary of the Studies About the Association Between *CYP2E1* Gene Polymorphisms and Susceptibility to Alcohol Dependence

Polymorphisms	Study	Country	No. of AD	No. of controls	Allele frequency in AD		Allele frequency in controls		Association (yes or no)
					Mutant	Wild	Mutant	Wild	
CYP2E1 Rsa I/Pst I polymorphism									
	Cichoz-Lach et al.[36]	Poland	204	172	2.21	97.79	1.16	98.84	No
	Montano Loza et al.[37]	Mexico	59	59	23.73	76.27	29.66	70.34	No
	Lorenzo et al.[38]	Spain	85	42	4.71	95.29	2.38	97.62	No
	Vidal et al.[39]	Spain	64	264	5.47	94.53	3.79	96.21	No
	Kim et al.[48]	Korea	100	128	35.00	65.00	19.92	80.08	Yes
	Konishi et al.[40]	Mexico	101	104	17.82	82.18	11.54	88.46	Yes
	Pastorelli et al.[41]	Italy	60	64	6.67	93.33	10.16	89.84	No
	Nakamura et al.[42]	Japan	53	97	22.64	77.36	18.04	81.96	No
	Carr et al.[43]	China	80	135	26.88	73.13	27.94	72.06	No
	Maezawa et al.[44]	Japan	96	60	19.27	80.73	18.33	81.67	No
	Iwahashi et al.[45]	Japan	53	75	18.87	81.13	20.00	80.00	No
	Ball et al.[46]	UK	38	54	1.28	98.72	2.78	97.22	No
CYP2E1 Dra I polymorphism									
	Montano et al.[37]	Mexico	59	59	24.57	75.42	39.83	60.17	Yes
	Lorenzo et al.[38]	Spain	85	42	18.24	81.76	9.52	90.48	No
	Vidal et al.[39]	Spain	64	264	11.36	88.64	15.63	84.38	No
	Konishi et al.[40]	Mexico	101	104	18.81	81.19	15.78	84.13	No
	Pastorelli et al.[41]	Italy	60	64	2.50	97.5	4.69	95.31	No
	Iwahashi et al.[47]	Japan	35	130	48.57	51.43	27.69	72.31	Yes
CYP2E1 Taq I polymorphism									
	Montano et al.[37]	Mexico	59	59	40.68	59.32	21.19	78.81	Yes
	Konishi et al.[40]	Mexico	101	104	15.35	84.65	15.38	84.62	No

on the association between *CYP2E1* gene polymorphisms and AD susceptibility was found in the literature by searching the PubMed database (http://www.ncbi.nlm.nih.gov/pubmed/guide/). Herein, we performed a meta-analysis trying to make a more definitive conclusion between *CYP2E1 Rsa I/Pst I* polymorphism and AD susceptibility. The combined results showed that c1c2/c2c2 genotype was not significantly associated with the individual susceptibility to AD (c1c2/c2c2 vs. c1c1, OR = 1.18, 95% CI 0.95–1.47). Subgroup analyses revealed no significant association in Caucasians (c1c2/c2c2 vs. c1c1, OR = 1.12, 95% CI 0.66–1.90) or Asians (c1c2/c2c2 vs. c1c1, OR = 1.22, 95% CI 0.92–1.62). Although a significant difference in the c2 allele frequency between AD patients and the controls was found, when all studies were combined (c2 versus c1, OR = 1.23, 95%CI 1.02–1.47), subgroup analyses showed no significant association in Caucasians (c2 vs. c1, OR = 1.15, 95% CI 0.70–1.90) or in Asians (c2 vs. c1, OR = 1.21, 95% CI 0.85–1.71) (Figure 19.1). The results of the meta-analyses of studies about the *CYP2E1 Dra I* polymorphism and AD susceptibility also revealed that this polymorphism may be not significantly associated with the susceptibility to AD (CC/CD vs. DD, OR = 1.03, 95% CI 0.46–2.32; C vs. D, OR = 1.08, 95% CI 0.84–1.38).

CYP2E1 GENE POLYMORPHISMS AND SUSCEPTIBILITY TO ALCOHOLIC LIVER DISEASE (ALD)

ALD encompasses a series of progressive liver damage including hepatic steatosis, hepatitis, fibrosis, and finally cirrhosis.[49] Alcoholic steatosis (fatty liver) is characterized by the accumulation of fat in the liver cells,

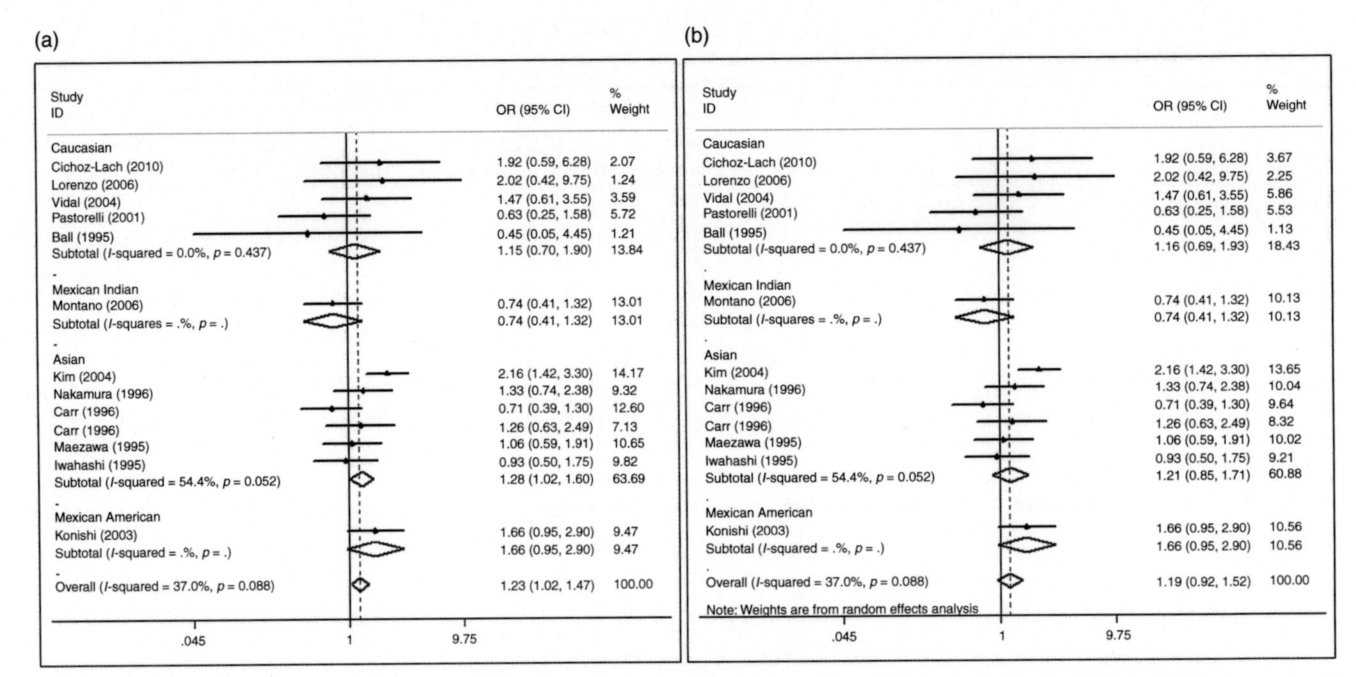

FIGURE 19.1 Meta-analysis for *CYP2E1 Rsa I/Pst I* polymorphism and the susceptibility to alcoholic dependence (allele c2 versus c1). (a) Fixed effect model; (b) random effect model. Each study was shown by a point estimate of the effect size (OR), and its 95% confidence interval (95% CI) (horizontal lines). The white diamond denotes the pooled OR.

while alcoholic hepatitis is shown by the inflammation of hepatocytes. Alcoholic cirrhosis is a late stage of serious liver disease marked by inflammation, fibrosis, and damaged membranes preventing detoxification of chemicals in the body, ending in scarring and necrosis.

Potential Links Between CYP2E1 Gene Polymorphisms and Susceptibility to ALD

It has been well documented that oxidative stress plays critical roles in the initiation and progression of ALD.[5,50–53] Many pathways have been demonstrated to be associated with ethanol-induced oxidative stress. Among them, the activation of CYP2E1 appears to be a major contributor. CYP2E1 can catalyze the metabolism of ethanol to acetaldehyde, causing free radicals and reactive oxygen species (ROS) overproduction.[54–58] CYP2E1 possesses a unique ability to reduce molecular oxygen to highly reactive compounds: superoxide anion radical ($O_2^{\bullet}$), single oxygen (O_2^1), hydrogen peroxide (H_2O_2), and hydroxyl radical ($OH\cdot$), even in the absence of substrate.[59] It has been demonstrated that ethanol-induced liver injury and lipid peroxidation was correlated well with the CYP2E1 levels.[60–62] Furthermore, CYP2E1 inhibitors significantly blocked the lipid peroxidation and ameliorated the pathologic changes in ethanol-treated animals,[63–66] while CYP2E1 overexpressing mice displayed higher transaminase activities and histological features of liver injury, when compared with the control mice.[67] All these studies support the critical roles of CYP2E1 in the etiology of ALD. Therefore, the functional polymorphisms of *CYP2E1* gene might be associated with the individual susceptibility to ALD.

Human Epidemiological Studies About the *CYP2E1* Gene Polymorphisms and Individual Susceptibility to ALD

Up to now, a large number of epidemiological studies have been conducted trying to reveal the associations between individual susceptibility to ALD and *CYP2E1* gene polymorphisms (Table 19.3). These studies were performed in different countries, including China, Japan, Korea, Span, Italy, United Kingdom, France, America, Brazil, India, Mexico, etc., and mainly focused on *CYP2E1 Rsa I/Pst I* polymorphism[38,39,43,46,68–90] and *CYP2E1 Dra I* polymorphism.[38,39,75,79,81,85,90–92] In these studies, the genotype distributions and allele frequencies were compared between ALD patients and alcoholics without ALD, and/or between ALD patients and nonalcoholics (healthy subjects). All the ALD patients have unambiguous liver diseases that were diagnosed by clinical, laboratory, ultrasound examination, and/or liver biopsy, with exclusion of other causes, such as hepatitis virus infection, inherited and autoimmune liver diseases, and liver cancer. Alcoholics without ALD were those heavy drinkers without evidences of liver diseases, while nonalcoholics were healthy subjects who drank less than 10 g/day ethanol.

TABLE 19.3 Summary of Studies for the Association Between *CYP2E1 Rsa I/Pst I* Polymorphism and Susceptibility to Alcoholic Liver Diseases (ALD)

Study	Country	Histological types of ALD	ALD patients (*n*)			Alcoholics without ALD (*n*)			Nonalcoholics (*n*)			ALD patients (%)		Alcoholics without ALD (%)		Nonalcoholics (%)	
			c1c1	c1c2	c2c2	c1c1	c1c2	c2c2	c1c1	c1c2	c2c2	c2	c1	c2	c1	c2	c1
Liu et al.[68]	China	AC/AH/AFL/HCC	203	106	44	271	19	10	325	24	11	27.5	72.5	6.5	93.5	6.4	93.6
Garcia-Banuelos et al.[69]	Mexico	AC	25	16	0				73	15	2	19.5	80.5			10.6	89.4
Khan et al.[70]	India	AC	161	14	0	137	3	0	250	5	0	4.0	96.0	1.1	98.9	1.0	99.0
Lorenzo et al.[38]	Spain	AC/AH/AFL	53	2	3	27	0	0	40	2	0	6.9	93.1	0.0	100	2.4	97.6
Cichoz-Lach et al.[71]	Poland	AC	53	4	0	43	0	0	54	0	0	3.5	96.5	0.0	100	0.0	100
Vidal et al.[39]	Spain	AC/AH/AFL	94	5	0	42	4	1	57	7	0	2.5	97.5	6.4	93.6	5.5	94.5
Kim et al.[72]	Korea	AC	17	4	0				51	34	15	13.6	86.4			32.0	68.0
Burim et al.[73]	Brazil	AC	59	6	0	37	4	0	197	23	1	4.6	95.4	4.9	95.1	5.7	94.3
Kee et al.[74]	Korea	AC	17	10	3	4	7	1	23	15	0	26.7	73.3	37.5	62.5	19.7	80.3
Frenzer et al.[75]	Australia	AC	56	1	0	54	3	0	188	12	0	0.9	99.1	2.6	97.4	3.0	97.0
Monzoni et al.[76]	Italy	AC/AH/AFL	66	14	1	85	7	0				9.9	90.1	3.8	96.5		
Lee et al.[77]	Korea	AC	34	21	1	32	19	1	41	22	1	20.5	79.5	20.2	79.8	18.8	81.3
Zhang et al.[78]	China	AC/AH/HCC	2	50	3				17	9	0	50.9	49.1			17.3	82.7
Wong et al.[79]	UK	AH/AF/AC	59	2	0				350	25	0	1.6	98.4			3.3	96.7
Rodrigo et al.[80]	Spain	AC	112	8	0	28	2	0	183	17	0	3.3	96.7	3.3	96.7	4.3	95.8
Parsian et al.[81]	USA	AC	43	0	0				na	na	na	0.0	100			1.2	98.8
Grove et al.[82]	UK	AC/AF/AH	226	14	0				117	4	0	2.9	97.1			1.7	98.3
Tanaka et al.[83]	Japan	AC/AF/AH	13	9	4	30	11	1				32.7	67.3	15.5	84.5		
Chao et al.[84]	China	AC	42	29	4	12	5	2	56	38	6	24.7	75.3	23.7	76.3	25.0	75.0
Lucas et al.[85]	France	AC	101	9	0	188	12	2	248	11	1	4.1	95.9	4.0	96.0	2.5	97.5
Carr et al.[43]	China	AC	18	10	2	28	18	0	52	45	3	23.3	76.7	19.6	80.4	25.5	74.5

(*Continued*)

TABLE 19.3 Summary of Studies for the Association Between *CYP2E1 Rsa I/Pst I* Polymorphism and Susceptibility to Alcoholic Liver Diseases (ALD) (*cont.*)

Study	Country	Histological types of ALD	ALD patients (*n*)			Alcoholics without ALD (*n*)			Nonalcoholics (*n*)			ALD patients (%)		Alcoholics without ALD (%)		Nonalcoholics (%)	
			c1c1	c1c2	c2c2	c1c1	c1c2	c2c2	c1c1	c1c2	c2c2	c2	c1	c2	c1	c2	c1
Agundez et al.[86]	Spain	AC	56	2	0				130	7	0	1.7	98.3			2.6	97.4
Yamauchi et al.[87]	Japan	AC	47	26	0				40	18	2	17.8	82.2			18.3	81.7
Pirmohamed et al.[88]	UK	AC/AH	77	17	1	55	2	1	97	3	0	10	90	3.4	96.6	1.5	98.5
Carr et al.[89]	USA	AC/AH	49	3	1	35	4	0	31	1	0	4.7	95.3	5.1	94.9	1.6	98.4
Ball et al.[46]	UK	AC	34	3	0				102	6	0	4.1	95.9			2.8	97.2
Ingelman-Sundberg et al.[90]	Italy	AC	53	3	0				104	10	0	2.7	97.3			4.4	95.6

ALD, alcoholic liver disease; AC, alcoholic cirrhosis; AH, alcoholic hepatitis; AFL, alcoholic fatty liver; AF, alcoholic fibrosis; HCC, hepatic carcinoma; na, data not available.

Unfortunately, there is still no consensus as these epidemiological studies provided highly inconsistent results. For example, the study by Liu et al. found that the frequency of the mutant c2 allele in ALD patients was significantly higher than that in alcoholics without ALD, and in nonalcoholics in the Han, Mongol, and Chaoxian nationalities in the northeast of China.[68] However, another study performed in Chinese showed that there was no significant difference in the c2 allele frequency among ALD patients, alcoholics without ALD, and nonalcoholics,[84] while the study in Koreans showed that the c2 allele frequency in ALD patients was significantly lower than that in nonalcoholics.[72] Conflicting results also existed in studies performed in Caucasians. Monzoni et al. found that the c2 allele frequency in ALD patients was significantly higher than that in alcoholic patients without ALD, in Italians.[76] However, no significant association between *CYP2E1 Rsa I/Pst I* polymorphism and susceptibility to ALD was found in French and in Spanish patients.[39,85]

Meta-Analysis of the *CYP2E1* Gene Polymorphisms and Susceptibility to ALD

Zintzaras et al. performed a meta-analysis to evaluate the association between the polymorphisms of ethanol-metabolizing enzymes, including *CYP2E1 Rsa I/Pst I* polymorphism, and ALD risks.[93] They combined eight studies, and the results of the meta-analysis showed that c2 allele was not significantly associated with the risk of alcoholic cirrhosis (worldwide population: OR = 1.13, 95% CI 0.76–1.68; Caucasians: OR = 1.58, 95% CI 0.76–3.28; Asians: OR = 0.97, 95% CI 0.58–1.62).[93] In an updated meta-analysis, a total of 27 studies and nine studies were included for the analyses of the association between *CYP2E1 Rsa I/Pst I* polymorphism and ALD risks, and between *CYP2E1 Dra I* polymorphism and ALD risks, respectively.[94] The combined results showed a significant association between c2c2 genotype and ALD risks, when ALD patients were compared with nonalcoholics (c2c2 vs. c1c1: OR = 3.12, 95% CI 1.91–5.11). Subgroup analyses showed that the c2c2 genotype was associated with increased risk of ALD in Asians (c2c2 vs. c1c1: OR = 4.11, 95% CI 2.32–7.29) and in Caucasians (c1c2 vs. c1c1: OR = 1.63, 95% CI 1.05–2.53; c2c2/c1c2 vs. c1c1: OR = 1.58, 95% CI 1.04–2.42), when ALD patients were compared with alcoholics without ALD. However, no significant association was observed in Asians or Caucasians, when ALD patients were compared with nonalcoholics. The subgroup meta-analyses also revealed that *CYP2E1 Pst I/Rsa I* polymorphism was not significantly associated with the risks of alcoholic cirrhosis, neither in Asians, nor in Caucasians. However, the pooled results of the studies in which cases were composed by mixed types of ALD patients (steatosis, hepatitis, fibrosis, and cirrhosis) showed significant association in Asians (c2 vs. c1, OR = 4.95, 95% CI 3.55–6.89; c2c2/c2c1 vs. c1c1, OR = 4.63, 95% CI 1.75–12.26) and in Caucasians (c2 vs. c1, OR = 2.58, 95% CI 1.42–4.67; c2c2/c2c1 vs. c1c1, OR = 2.58, 95% CI 1.37–4.87), suggesting that *CYP2E1 Pst I/Rsa I* polymorphism might be significantly associated with the earlier liver damages, such as alcoholic steatosis and hepatitis, but not with the advanced ALD (alcoholic cirrhosis).[94]

The pooled results of the studies for *CYP2E1 Dra I* polymorphism and the risks of ALD showed no significant association (ALD patients vs. alcoholics without ALD: C vs. D, OR = 1.09, 95% CI 0.80–1.48; CC/CD vs. DD, OR = 1.00, 95% CI 0.75–1.33; ALD patients vs. nonalcoholics: C vs. D, OR = 1.13, 95% CI 0.77–1.66; CC/CD vs. DD: OR = 1.13, 95% CI 0.76–1.70). Subgroup analyses showed there was no significant association between *CYP2E1 Dra I* polymorphism and risks of alcoholic cirrhosis.

CYP2E1 GENE POLYMORPHISM AND RISKS OF ALCOHOLIC PANCREATITIS (AP)

Potential Links Between *CYP2E1* Gene Polymorphisms and Susceptibility to AP

Alcohol abuse is an important risk factor for AP.[95] There is now sufficient evidence that the pancreas has the capacity to metabolize ethanol via both the oxidative and nonoxidative pathways. The resulting byproducts (oxygen radicals) exert toxicity to both the acinar cells and pancreatic stellate cells, that could lead to acute and chronic pancreatitis.[96] CYP2E1 is one of the possible mechanisms for ethanol-induced oxidative stress in the pancreas, and the associated oxidative stress might play important roles in the pathogenesis of AP.[4] Thus, *CYP2E1* functional polymorphisms might be also associated with the risks of AP.

Human Epidemiological Studies About the *CYP2E1* Gene Polymorphisms and Individual Susceptibility to AP

A number of studies have investigated the relationship between *CYP2E1 Rsa I/Pst I* polymorphism, *CYP2E1 Dra I* polymorphism, and the individual susceptibility to AP (Table 19.4). In a study performed in Polish patients, it was found that the c2 allele frequency was about 2.3% in AP patients, while c2 allele was not detected in alcoholics without AP or in nonalcoholics.[71] The study by Burim et al. showed an increased frequency of the c1c2 genotype in AP patients, compared with the alcoholics without AP or nonalcoholics (21.4% vs. 9.8% or 10.8%) in Brazilians. Although the statistical analyses showed no significant difference

TABLE 19.4 Summary of Studies About the Association Between *CYP2E1 Rsa I /Pst I* Polymorphism and Susceptibility to Alcoholic Pancreatitis (AP)

		AP patients (*n*)			Alcoholics without AP (*n*)			Nonalcoholics (*n*)			AP patients (%)		Alcoholics without AP (%)		Nonalcoholics (%)	
Study	**Country**	**c1c1**	**c1c2**	**c2c2**	**c1c1**	**c1c2**	**c2c2**	**c1c1**	**c1c2**	**c2c2**	**c1**	**c2**	**c1**	**c2**	**c1**	**c2**
Cichoż-Lach et al.[71]	Poland	42	2	0	43	0	0	54	0	0	97.7	2.3	100	0	100	0
Verlaan et al.[97]	Netherlands	75	7	0	88	5	0	122	6	0	95.7	4.3	97.3	2.7	97.7	2.3
Kim et al.[72]	Korea	17	11	1				51	34	15	77.6	22.4			68.0	32.0
Burim et al.[73]	Brazil	11	3	0	37	4	0	197	23	1	89.3	10.7	95.1	4.9	94.3	5.7
Frenzer et al.[75]	Australia	65	6	0	54	3	0	188	12	0	95.8	4.2	97.4	2.6	97.0	3.0
Yang et al.[98]	UK	55	2	0	42	4	0	150	5	0	98.2	1.8	95.7	4.3	98.4	1.6
Maruyama et al.[99]	Japan	30	21	2	30	15	1				76.4	23.6	81.5	18.5		
Chao et al.[84]	China	30	15	3	12	5	2	56	38	6	78.1	21.9	76.3	23.7	75.0	25.0
Matsumoto et al.[100]	Japan	9	2	0	39	21	2				90.9	9.1	79.8	20.2		

among the genotypes distribution or allele frequencies among different groups, the promotive effects of the c2 allele to AP could not be completely excluded, as a significant difference might be detected if increasing the sample size.[73] Results of other studies did not support hypothesis of correlation between c2 allele and chronic AP. The study by Verlaan et al. showed that the frequencies of c2 alleles did not differ among AP patients, alcoholics without AP, and nonalcoholics, while a weak positive association between the homozygous DD genotype (*CYP2E1 Dra I* polymorphism) and the development of AP in Dutch.[97] The relationships between *CYP2E1 Rsa I* / Pst I polymorphism, *CYP2E1 Dra I* polymorphism and susceptibility to AP were also investigated in Australians by Frenzer et al., who found that the genotypes distributions and allele frequencies of both polymorphisms were similar in AP patients, alcoholics without AP, and healthy blood donors.[75] Yang et al. investigated the relationship between four *CYP2E1* polymorphisms (*−35G > T, Rsa I/Pst I, Dra I,* and **3* polymorphisms) and the susceptibility to AP, in British patients. The results showed that none of the above *CYP2E1* polymorphisms was likely to be involved in the susceptibility and pathogenesis of AP.[98]

The roles of *CYP2E1* gene polymorphisms in the susceptibility to AP were also investigated in Asians. Although a study performed in Koreans showed that the c2 allele frequency was significantly lower than that in nonalcoholics (22.4% vs. 32.0%), they did not compare the frequency of allele between AP patients and alcoholics without AP.[72] In a study performed in Chinese, there was no significant difference in genotype distribution or allele frequencies of *CYP2E1 Rsa I/Pst I* polymorphism among AP patients, alcoholics without AP, and nonalcoholics.[84] Two studies investigated the associations between *CYP2E1 Rsa I/Pst I* polymorphism and AP risks in Japanese. The study by Maruyama et al. showed a slightly higher c2 allele frequency in AP patients than in alcoholics without AP (23.6% vs. 18.5%), while the study by Matsumoto et al. showed a slightly lower frequency of c2 allele in AP patients than that in alcoholics without AP (9.8% vs. 20.2%).[99,100]

Meta-Analysis of the *CYP2E1* Gene Polymorphisms and Susceptibility to AP

Null association between *CYP2E1 Rsa I/Pst I* polymorphism and AP risks was found when AP patients compared with alcoholics without AP (c1c2/c2c2 vs. c1c1, OR = 1.21, 95% CI 0.78–1.89), and when AP patients were compared with nonalcoholics (c1c2/c2c2 vs. c1c1, OR = 1.09, 95% CI 0.73–1.62). Similarly, no association between c2 allele frequency and AP risks was detected (AP vs. alcoholics without AP: OR = 1.14, 95% CI 0.77–1.70; AP vs. nonalcoholics: OR = 0.98, 95% CI 0.69–1.39). The subgroup analyses by ethnicity also revealed no significant association (Figure 19.2). Collectively, the results of the meta-analyses showed that *CYP2E1 Rsa I/Pst I* polymorphism was not significantly associated with individual susceptibility to AP.

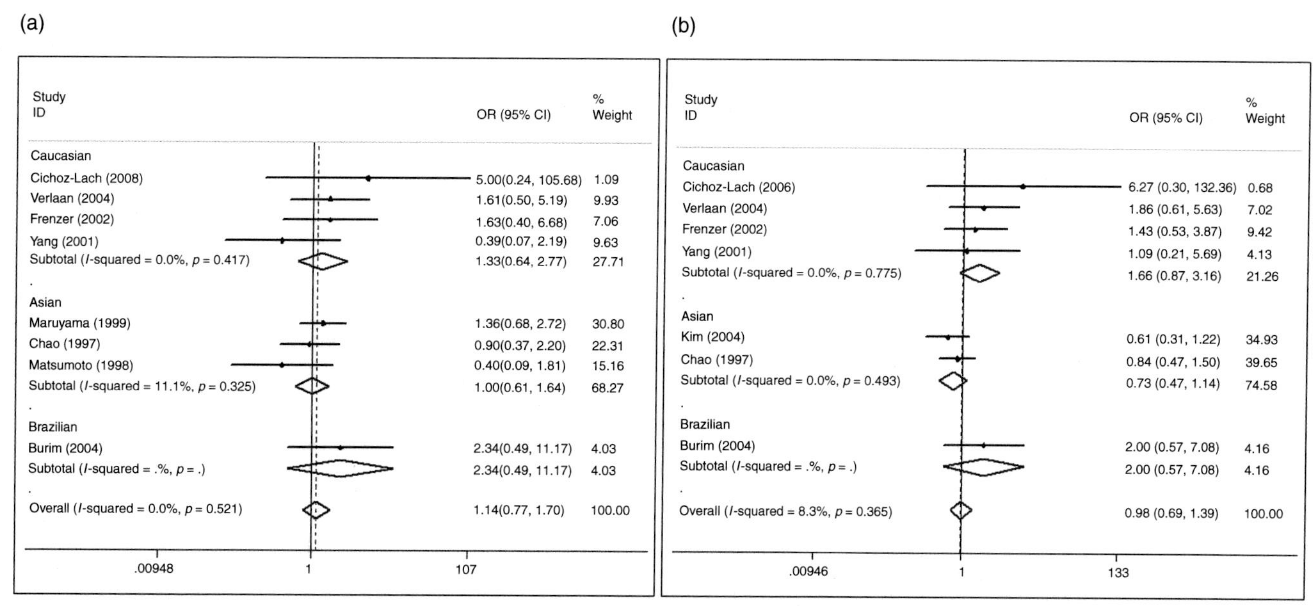

FIGURE 19.2 **Meta-analysis for *CYP2E1 Rsa I/Pst I* polymorphism and the susceptibility to alcoholic pancreatitis (AP) (allele c2 versus c1).** (a) AP patients versus alcoholics without AP; (b) AP patients versus nonalcoholics. Each study was shown by a point estimate of the effect size (OR), and its 95% confidence interval (95% CI) (horizontal lines). The white diamond denotes the pooled OR.

CONCLUSIONS

Alcohol abuse can lead to damage to different organs, especially the liver, the brain, and the pancreas. Since not all the heavy drinkers develop such diseases, there is an urgent task to identify the potential risk factors that may confer individual susceptibility to these diseases. CYP2E1 is a major component of the MEOS, and a key contributor to ethanol-induced oxidative stress. Thus, the knowledge of functional polymorphisms of *CYP2E1* gene might help to identify individuals who are vulnerable, to a smaller or larger extent, to these diseases. However, the genetic predisposition of *CYP2E1* polymorphisms to AD, ALD, and AP, is a still an area of debate, and there is no consensus on the roles of the enzymes, although a number of studies have been performed trying to elucidate the roles of *CYP2E1* polymorphisms in the susceptibility and pathogenesis of these diseases.

The conflicting results of different studies might be related to many potential factors. First, most of the population studies suffer from poor statistical power, due to the low frequencies of mutant alleles (such as the c2 allele in Caucasians), and the small sample size. Second, high interethnic difference might be another contributor to the contradictory results. Third, CYP2E1 activity could be influenced by many factors, including the diets, the physiological condition, the amount of ethanol intake, the type of ethanol, and the pattern of alcohol consumption. However, no studies considered the data on dietary and physiological factors in their studies. The roles of *CYP2E1* gene polymorphisms in ethanol-induced organ damage might be confounded by these factors. Last, AD, ALD, and AP are all multifactor diseases that include many genetic factors and environmental factors. However, the effects of gene–gene and gene–environment interactions were not addressed in most of these studies.

In conclusion, a number of polymorphic loci have been found in the *CYP2E1* gene. Some functional polymorphisms are theoretically to be associated with ethanol-induced diseases, such as AD, ALD, and AP. However, the results of the available studies could not provide solid connections between the *CYP2E1* gene polymorphisms and individual susceptibility to these diseases. In order to clearly delineate the association between *CYP2E1* gene polymorphisms and the susceptibility of alcoholics to various ethanol-induced diseases, large, well-designed epidemiological studies are warranted.

Key Facts

Heavy Drinking

- Heavy drinking, moderate drinking, and binge drinking are three concepts describing the drinking pattern.
- Heavy drinking is a rough concept without clear definition. Adult men with ethanol consumption > 60 g/day and adult women with ethanol consumption > 40 g/day are usually considered heavy drinkers.
- Moderate drinking means no more than three-four standard drinks (one standard drink is about 14 g alcohol) per drinking, no more than nine drinks per week for women and 14 for men.
- Binge drinking means drinking so much in a short time that blood alcohol concentration levels reach 0.08 g/dL.
- Heavy drinking and binge drinking may lead to health problems, while moderate drinking is usually considered beneficial to health.

Oxidative Stress

- Oxidative stress refers to an imbalance status of the production of reactive oxygen species (ROS), and the ability to detoxify them by the antioxidant system.
- The antioxidant system contains glutathione and antioxidant enzymes, such as superoxide dismutase, catalase, glutathione peroxidase, heme oxygenase-1, etc.
- CYP2E1 and NADPH oxidase are two major sources of ROS production in alcoholic liver disease.
- Oxidative stress plays important roles in the pathogenesis of alcoholic liver disease (ALD)

Meta-Analysis

- Meta-analysis is a statistical method to combine results of different studies, especially those with small sample size or with conflicting results.
- Meta-analysis is often an important component of systematic reviews.
- Literature search is the first step, and is very important for meta-analysis, as incomplete literature search may bring incorrect results.
- Meta-analysis is done by identifying a common statistical measure that is shared among studies, and calculating a weighted average of that common measure.
- The statistics of meta-analysis could be conducted with software such as Stata or Review manager (RevMan).

Pancreatitis

- The pancreas is a glandular organ in the digestive system and endocrine system of vertebrates.
- Pancreas can produce several important hormones, including insulin, glucagon, somatostatin, and pancreatic polypeptide.
- Pancreatitis is characterized by irreversible morphological and functional alterations of the pancreas.

- There are two main types of pancreatitis: acute pancreatitis, and chronic pancreatitis.
- The symptoms of pancreatitis include upper abdominal pain, nausea, vomiting, as well as, exocrine and endocrine insufficiencies.
- Alcohol abuse is one of the most common risk factors for pancreatitis.

Summary Points

- This chapter describes the current knowledge about the *CYP2E1* gene polymorphisms, and its relationship with the individual susceptibility to alcohol-induced health problems.
- CYP2E1 is a major component of the microsomal ethanol-oxidizing system (MEOS) that could be activated after chronic alcohol consumption.
- CYP2E1 is believed to be a major contributor to alcohol-induced oxidative stress.
- A number of *CYP2E1* gene polymorphic loci in the 5′-flanking region, introns and exons, have been reported.
- The functional polymorphisms of *CYP2E1* gene may influence the interindividual variation of the CYP2E1 catalytic activity.
- A series of epidemiological studies have been conducted to explore the potential association between *CYP2E1* polymorphism, and susceptibility to alcohol dependence and alcohol-induced organ damage, but obtained conflicting results.
- The conflicting results may be related to some factors, such as ethnic difference, diets, and physiological conditions of study population, the amount, type and pattern of alcohol, etc.
- Meta-analysis of available studies shows no association between *CYP2E1* polymorphism and alcohol dependence, or alcohol-induced organ damage.

References

1. Lieber CS. Microsomal ethanol-oxidizing system (MEOS): the first 30 years (1968-1998)--a review. *Alcohol Clin Exp Res* 1999;**23**(6):991–1007.
2. Lieber CS. The discovery of the microsomal ethanol oxidizing system and its physiologic and pathologic role. *Drug Metab Rev* 2004;**36**(3–4):511–29.
3. Takahashi T, Lasker JM, Rosman AS, Lieber CS. Induction of cytochrome P-4502E1 in the human liver by ethanol is caused by a corresponding increase in encoding messenger RNA. *Hepatol* 1993;**17**(2):236–45.
4. Apte MV, Pirola RC, Wilson JS. Molecular mechanisms of alcoholic pancreatitis. *Dig Dis* 2005;**23**(3–4):232–40.
5. Cederbaum AI, Lu Y, Wu D. Role of oxidative stress in alcohol-induced liver injury. *Arch Toxicol* 2009;**83**(6):519–48.
6. Shimada T, Yamazaki H, Mimura M, Inui Y, Guengerich FP. Interindividual variations in human liver cytochrome P-450 enzymes involved in the oxidation of drugs, carcinogens and toxic chemicals: studies with liver microsomes of 30 Japanese and 30 Caucasians. *J Pharmacol Exp Ther* 1994;**270**(1):414–23.
7. Forrester LM, Henderson CJ, Glancey MJ, et al. Relative expression of cytochrome P450 isoenzymes in human liver and association with the metabolism of drugs and xenobiotics. *Biochem J* 1992;**281**(Pt. 2):359–68.
8. Kim RB, O'Shea D. Interindividual variability of chlorzoxazone 6-hydroxylation in men and women and its relationship to CYP2E1 genetic polymorphisms. *Clin Pharmacol Ther* 1995;**57**(6):645–55.
9. Lucas D, Menez C, Girre C, et al. Cytochrome P450 2E1 genotype and chlorzoxazone metabolism in healthy and alcoholic Caucasian subjects. *Pharmacogenetics* 1995;**5**(5):298–304.
10. Umeno M, McBride OW, Yang CS, Gelboin HV, Gonzalez FJ. Human ethanol-inducible P450IIE1: complete gene sequence, promoter characterization, chromosome mapping, and cDNA-directed expression. *Biochemistry* 1988;**27**(25):9006–13.
11. Inoue K, Yamazaki H, Shimada T. Characterization of liver microsomal 7-ethoxycoumarin O-deethylation and chlorzoxazone 6-hydroxylation activities in Japanese and Caucasian subjects genotyped for CYP2E1 gene. *Arch Toxicol* 2000;**74**(7):372–8.
12. Nomura F, Itoga S, Uchimoto T, et al. Transcriptional activity of the tandem repeat polymorphism in the 5′-flanking region of the human CYP2E1 gene. *Alcohol Clin Exp Res* 2003;**27**(8 Suppl):42S–6S.
13. Uchimoto T, Itoga S, Nezu M, Sunaga M, Tomonaga T, Nomura F. Role of the genetic polymorphisms in the 5′-flanking region for transcriptional regulation of the human CYP2E1 gene. *Alcohol Clin Exp Res* 2007;**31**(1 Suppl):S36–42.
14. McCarver DG, Byun R, Hines RN, Hichme M, Wegenek W. A genetic polymorphism in the regulatory sequences of human CYP2E1: association with increased chlorzoxazone hydroxylation in the presence of obesity and ethanol intake. *Toxicol Appl Pharmacol* 1998;**152**(1):276–81.
15. Hu Y, Oscarson M, Johansson I, et al. Genetic polymorphism of human CYP2E1: characterization of two variant alleles. *Mol Pharmacol* 1997;**51**(3):370–6.
16. Fairbrother KS, Grove J, de Waziers I, et al. Detection and characterization of novel polymorphisms in the CYP2E1 gene. *Pharmacogenetics* 1998;**8**(6):543–52.
17. Hayashi S, Watanabe J, Kawajiri K. Genetic polymorphisms in the 5′-flanking region change transcriptional regulation of the human cytochrome P450IIE1 gene. *J Biochem* 1991;**110**(4):559–65.
18. Watanabe J, Hayashi S, Kawajiri K. Different regulation and expression of the human CYP2E1 gene due to the RsaI polymorphism in the 5′-flanking region. *J Biochem* 1994;**116**(2):321–6.
19. Tsutsumi M, Wang JS, Takase S, Takada A. Hepatic messenger RNA contents of cytochrome P4502E1 in patients with different P4502E1 genotypes. *Alcohol Alcohol Suppl* 1994;**29**(1):29–32.
20. Powell H, Kitteringham NR, Pirmohamed M, Smith DA, Park BK. Expression of cytochrome P4502E1 in human liver: assessment by mRNA, genotype and phenotype. *Pharmacogenetics* 1998;**8**(5):411–21.
21. Marchand LL, Wilkinson GR, Wilkens LR. Genetic and dietary predictors of CYP2E1 activity: a phenotyping study in Hawaii Japanese using chlorzoxazone. *Cancer Epidemiol Biomarkers Prev* 1999;**8**(6):495–500.
22. Khemawoot P, Yokogawa K, Shimada T, Miyamoto K. Obesity-induced increase of CYP2E1 activity and its effect on disposition kinetics of chlorzoxazone in Zucker rats. *Biochem Pharmacol* 2007;**73**(1):155–62.
23. Kim RB, Yamazaki H, Chiba K, et al. *In vivo* and *in vitro* characterization of CYP2E1 activity in Japanese and Caucasians. *J Pharmacol Exp Ther* 1996;**279**(1):4–11.
24. Hu Y, Hakkola J, Oscarson M, Ingelman-Sundberg M. Structural and functional characterization of the 5′-flanking region of the rat

and human cytochrome P450 2E1 genes: identification of a polymorphic repeat in the human gene. *Biochem Biophys Res Commun* 1999;**263**(2):286–93.
25. Bolt HM, Roos PH, Thier R. The cytochrome P-450 isoenzyme CYP2E1 in the biological processing of industrial chemicals: consequences for occupational and environmental medicine. *Int Arch Occup Environ Health* 2003;**76**(3):174–85.
26. Persson I, Johansson I, Bergling H, et al. Genetic polymorphism of cytochrome P4502E1 in a Swedish population. Relationship to incidence of lung cancer. *FEBS Lett* 1993;**319**(3):207–11.
27. Tsutsumi M, Takada A, Wang JS. Genetic polymorphisms of cytochrome P4502E1 related to the development of alcoholic liver disease. *Gastroenterology* 1994;**107**(5):1430–5.
28. Kato S, Shields PG, Caporaso NE, et al. Cytochrome P450IIE1 genetic polymorphisms, racial variation, and lung cancer risk. *Cancer Res* 1992;**52**(23):6712–5.
29. Fritsche E, Pittman GS, Bell DA. Localization, sequence analysis, and ethnic distribution of a 96-bp insertion in the promoter of the human CYP2E1 gene. *Mutat Res* 2000;**432**(1–2):1–5.
30. Saunders JB. Substance dependence and non-dependence in the Diagnostic and Statistical Manual of Mental Disorders (DSM) and the International Classification of Diseases (ICD): can an identical conceptualization be achieved? *Addiction* 2006;**101**(Suppl. 1):48–58.
31. Laramee P, Kusel J, Leonard S, Aubin HJ, Francois C, Daeppen JB. The economic burden of alcohol dependence in Europe. *Alcohol Alcohol* 2013;**48**(3):259–69.
32. Morse RM, Flavin DK. The definition of alcoholism. The Joint Committee of the National Council on Alcoholism and Drug Dependence and the American Society of Addiction Medicine to Study the Definition and Criteria for the Diagnosis of Alcoholism. *JAMA* 1992;**268**(8):1012–4.
33. Deng XS, Deitrich RA. Putative role of brain acetaldehyde in ethanol addiction. *Curr Drug Abuse Rev* 2008;**1**(1):3–8.
34. Quertemont E, Tambour S, Tirelli E. The role of acetaldehyde in the neurobehavioral effects of ethanol: a comprehensive review of animal studies. *Prog Neurobiol* 2005;**75**(4):247–74.
35. Zimatkin SM, Pronko SP, Vasiliou V, Gonzalez FJ, Deitrich RA. Enzymatic mechanisms of ethanol oxidation in the brain. *Alcohol Clin Exp Res* 2006;**30**(9):1500–5.
36. Cichoz-Lach H, Celinski K, Wojcierowski J, Slomka M, Lis E. Genetic polymorphism of alcohol-metabolizing enzyme and alcohol dependence in Polish men. *Braz J Med Biol Res* 2010;**43**(3):257–61.
37. Montano Loza AJ, Ramirez Iglesias MT, Perez Diaz I, et al. Association of alcohol-metabolizing genes with alcoholism in a Mexican Indian (Otomi) population. *Alcohol* 2006;**39**(2):73–9.
38. Lorenzo A, Auguet T, Vidal F, et al. Polymorphisms of alcohol-metabolizing enzymes and the risk for alcoholism and alcoholic liver disease in Caucasian Spanish women. *Drug Alcohol Depend* 2006;**84**(2):195–200.
39. Vidal F, Lorenzo A, Auguet T, et al. Genetic polymorphisms of ADH2, ADH3, CYP4502E1 Dra-I and Pst-I, and ALDH2 in Spanish men: lack of association with alcoholism and alcoholic liver disease. *J Hepatol* 2004;**41**(5):744–50.
40. Konishi T, Calvillo M, Leng AS, et al. The ADH3*2 and CYP2E1 c2 alleles increase the risk of alcoholism in Mexican American men. *Exp Mol Pathol* 2003;**74**(2):183–9.
41. Pastorelli R, Bardazzi G, Saieva C, et al. Genetic determinants of alcohol addiction and metabolism: a survey in Italy. *Alcohol Clin Exp Res* 2001;**25**(2):221–7.
42. Nakamura K, Iwahashi K, Matsuo Y, Miyatake R, Ichikawa Y, Suwaki H. Characteristics of Japanese alcoholics with the atypical aldehyde dehydrogenase 2*2. I. A comparison of the genotypes of ALDH2, ADH2, ADH3, and cytochrome P-4502E1 between alcoholics and nonalcoholics. *Alcohol Clin Exp Res* 1996;**20**(1):52–5.
43. Carr LG, Yi IS, Li TK, Yin SJ. Cytochrome P4502E1 genotypes, alcoholism, and alcoholic cirrhosis in Han Chinese and Atayal Natives of Taiwan. *Alcohol Clin Exp Res* 1996;**20**(1):43–6.
44. Maezawa Y, Yamauchi M, Toda G, Suzuki H, Sakurai S. Alcohol-metabolizing enzyme polymorphisms and alcoholism in Japan. *Alcohol Clin Exp Res* 1995;**19**(4):951–4.
45. Iwahashi K, Matsuo Y, Suwaki H, Nakamura K, Ichikawa Y. CYP2E1 and ALDH2 genotypes and alcohol dependence in Japanese. *Alcohol Clin Exp Res* 1995;**19**(3):564–6.
46. Ball DM, Sherman D, Gibb R, et al. No association between the c2 allele at the cytochrome P450IIE1 gene and alcohol induced liver disease, alcohol Korsakoff's syndrome or alcohol dependence syndrome. *Drug Alcohol Depend* 1995;**39**(3):181–4.
47. Iwahashi K, Ameno S, Ameno K, et al. Relationship between alcoholism and CYP2E1 C/D polymorphism. *Neuropsychobiol* 1998;**38**(4):218–21.
48. Kim SA, Kim JW, Song JY, Park S, Lee HJ, Chung JH. Association of polymorphisms in nicotinic acetylcholine receptor alpha 4 subunit gene (CHRNA4), mu-opioid receptor gene (OPRM1), and ethanol-metabolizing enzyme genes with alcoholism in Korean patients. *Alcohol* 2004;**34**(2–3):115–20.
49. O'Shea RS, Dasarathy S, McCullough AJ. Alcoholic liver disease. *Hepatol* 2010;**51**(1):307–28.
50. Albano E. Oxidative mechanisms in the pathogenesis of alcoholic liver disease. *Mol Aspects Med* 2008;**29**(1–2):9–16.
51. Dey A, Cederbaum AI. Alcohol and oxidative liver injury. *Hepatol* 2006;**43**(2 Suppl. 1):S63–74.
52. Koch OR, Pani G, Borrello S, et al. Oxidative stress and antioxidant defenses in ethanol-induced cell injury. *Mol Aspects Med* 2004;**25**(1–2):191–8.
53. Zeng T, Zhang CL, Song FY, Zhao XL, Xie KQ. CMZ Reversed Chronic Ethanol-Induced Disturbance of PPAR-alpha Possibly by Suppressing Oxidative Stress and PGC-1alpha Acetylation, and Activating the MAPK and GSK3beta Pathway. *PLoS One* 2014;**9**(6):e98658.
54. Liu LG, Yan H, Yao P, et al. CYP2E1-dependent hepatotoxicity and oxidative damage after ethanol administration in human primary hepatocytes. *World J Gastroenterol* 2005;**11**(29):4530–5.
55. Bansal S, Liu CP, Sepuri NB, et al. Mitochondria-targeted cytochrome P450 2E1 induces oxidative damage and augments alcohol-mediated oxidative stress. *J Biol Chem* 2010;**285**(32):24609–19.
56. Cederbaum AI. Role of CYP2E1 in ethanol-induced oxidant stress, fatty liver and hepatotoxicity. *Dig Dis* 2010;**28**(6):802–11.
57. Lu Y, Cederbaum AI. CYP2E1 and oxidative liver injury by alcohol. *Free Radic Biol Med* 2008;**44**(5):723–38.
58. Kessova I, Cederbaum AI. CYP2E1: biochemistry, toxicology, regulation and function in ethanol-induced liver injury. *Curr Mol Med* 2003;**3**(6):509–18.
59. Danko IM, Chaschin NA. Association of CYP2E1 gene polymorphism with predisposition to cancer development. *Exp Oncol* 2005;**27**(4):248–56.
60. Castillo T, Koop DR, Kamimura S, Triadafilopoulos G, Tsukamoto H. Role of cytochrome P-450 2E1 in ethanol-, carbon tetrachloride- and iron-dependent microsomal lipid peroxidation. *Hepatol* 1992;**16**(4):992–6.
61. Ronis MJ, Huang J, Crouch J, et al. Cytochrome P450 CYP 2E1 induction during chronic alcohol exposure occurs by a two-step mechanism associated with blood alcohol concentrations in rats. *J Pharmacol Exp Ther* 1993;**264**(2):944–50.
62. French SW, Morimoto M, Reitz RC, et al. Lipid peroxidation, CYP2E1 and arachidonic acid metabolism in alcoholic liver disease in rats. *J Nutr* 1997;**127**(5 Suppl):907S–11S.
63. Morimoto M, Reitz RC, Morin RJ, Nguyen K, Ingelman-Sundberg M, French SW. CYP-2E1 inhibitors partially ameliorate the changes in hepatic fatty acid composition induced in rats by chronic administration of ethanol and a high fat diet. *J Nutr* 1995;**125**(12):2953–64.

64. Albano E, Clot P, Morimoto M, Tomasi A, Ingelman-Sundberg M, French SW. Role of cytochrome P4502E1-dependent formation of hydroxyethyl free radical in the development of liver damage in rats intragastrically fed with ethanol. *Hepatol* 1996;**23**(1):155–63.
65. Gouillon Z, Lucas D, Li J, et al. Inhibition of ethanol-induced liver disease in the intragastric feeding rat model by chlormethiazole. *Proc Soc Exp Biol Med* 2000;**224**(4):302–8.
66. Zeng T, Zhang CL, Song FY, Zhao XL, Xie KQ. Garlic oil alleviated ethanol-induced fat accumulation via modulation of SREBP-1, PPAR-alpha, and CYP2E1. *Food Chem Toxicol* 2012;**50**(3–4):485–91.
67. Morgan K, French SW, Morgan TR. Production of a cytochrome P450 2E1 transgenic mouse and initial evaluation of alcoholic liver damage. *Hepatol* 2002;**36**(1):122–34.
68. Liu Y, Zhou LY, Meng XW. Genetic polymorphism of two enzymes with alcoholic liver disease in Northeast China. *Hepatogastroenterol* 2012;**59**(113):204–7.
69. Garcia-Banuelos J, Panduro A, Gordillo-Bastidas D, et al. Genetic polymorphisms of genes coding to alcohol-metabolizing enzymes in western Mexicans: association of CYP2E1*c2/CYP2E1*5B allele with cirrhosis and liver function. *Alcohol Clin Exp Res* 2012;**36**(3):425–31.
70. Khan AJ, Husain Q, Choudhuri G, Parmar D. Association of polymorphism in alcohol dehydrogenase and interaction with other genetic risk factors with alcoholic liver cirrhosis. *Drug Alcohol Depend* 2010;**109**(1–3):190–7.
71. Cichoż-Lach H, Partycka J, Nesina I, Wojcierowski J, Słomka M, Celiński K. Genetic polymorphism of CYP2E1 and digestive tract alcohol damage among Polish individuals. *Alcohol Clin Exp Res* 2006;**30**(5):878–82.
72. Kim MS, Lee DH, Kang HS, et al. Genetic polymorphisms of alcohol-metabolizing enzymes and cytokines in patients with alcohol induced pancreatitis and alcoholic liver cirrhosis. *Korean J Gastroenterol* 2004;**43**(6):355–63.
73. Burim RV, Canalle R, Martinelli Ade L, Takahashi CS. Polymorphisms in glutathione S-transferases GSTM1, GSTT1 and GSTP1 and cytochromes P450 CYP2E1 and CYP1A1 and susceptibility to cirrhosis or pancreatitis in alcoholics. *Mutagenesis* 2004;**19**(4):291–8.
74. Kee JY, Kim MO, You IY, et al. Effects of genetic polymorphisms of ethanol-metabolizing enzymes on alcohol drinking behaviors. *Korean J Hepatol* 2003;**9**(2):89–97.
75. Frenzer A, Butler WJ, Norton ID, et al. Polymorphism in alcohol-metabolizing enzymes, glutathione S-transferases and apolipoprotein E and susceptibility to alcohol-induced cirrhosis and chronic pancreatitis. *J Gastroenterol Hepatol* 2002;**17**(2):177–82.
76. Monzoni A, Masutti F, Saccoccio G, Bellentani S, Tiribelli C, Giacca M. Genetic determinants of ethanol-induced liver damage. *Mol Med* 2001;**7**(4):255–62.
77. Lee HC, Lee HS, Jung SH, et al. Association between polymorphisms of ethanol-metabolizing enzymes and susceptibility to alcoholic cirrhosis in a Korean male population. *J Korean Med Sci* 2001;**16**(6):745–50.
78. Zhang S, Liu S, Liu H, Zhu W. Genotyping cytochrome P450IIE1 in alcoholic liver diseases and its significance. *Zhonghua Gan Zang Bing Za Zhi* 2000;**8**(6):338.
79. Wong NA, Rae F, Simpson KJ, Murray GD, Harrison DJ. Genetic polymorphisms of cytochrome p4502E1 and susceptibility to alcoholic liver disease and hepatocellular carcinoma in a white population: a study and literature review, including meta-analysis. *Mol Pathol* 2000;**53**(2):88–93.
80. Rodrigo L, Alvarez V, Rodriguez M, Perez R, Alvarez R, Coto E. N-acetyltransferase-2, glutathione S-transferase M1, alcohol dehydrogenase, and cytochrome P450IIE1 genotypes in alcoholic liver cirrhosis: a case-control study. *Scand J Gastroenterol* 1999;**34**(3):303–7.
81. Parsian A, Cloninger CR, Zhang ZH. Association studies of polymorphisms of CYP2E1 gene in alcoholics with cirrhosis, antisocial personality, and normal controls. *Alcohol Clin Exp Res* 1998;**22**(4):888–91.
82. Grove J, Brown AS, Daly AK, Bassendine MF, James OF, Day CP. The RsaI polymorphism of CYP2E1 and susceptibility to alcoholic liver disease in Caucasians: effect on age of presentation and dependence on alcohol dehydrogenase genotype. *Pharmacogenetics* 1998;**8**(4):335–42.
83. Tanaka F, Shiratori Y, Yokosuka O, Imazeki F, Tsukada Y, Omata M. Polymorphism of alcohol-metabolizing genes affects drinking behavior and alcoholic liver disease in Japanese men. *Alcohol Clin Exp Res* 1997;**21**(4):596–601.
84. Chao YC, Young TH, Tang HS, Hsu CT. Alcoholism and alcoholic organ damage and genetic polymorphisms of alcohol metabolizing enzymes in Chinese patients. *Hepatol* 1997;**25**(1):112–7.
85. Lucas D, Menez C, Floch F, et al. Cytochromes P4502E1 and P4501A1 genotypes and susceptibility to cirrhosis or upper aerodigestive tract cancer in alcoholic Caucasians. *Alcohol Clin Exp Res* 1996;**20**(6):1033–7.
86. Agúndez J, Ladero J, Díaz-Rubio M, Benítez J. Rsa I polymorphism at the cytochrome P4502E1 locus is not related to the risk of alcohol-related severe liver disease. *Liver* 1996;**16**(6):380–3.
87. Yamauchi M, Maezawa Y, Mizuhara Y, et al. Polymorphisms in alcohol metabolizing enzyme genes and alcoholic cirrhosis in Japanese patients: a multivariate analysis. *Hepatol* 1995;**22**(4 Pt 1): 1136–42.
88. Pirmohamed M, Kitteringham NR, Quest LJ, et al. Genetic polymorphism of cytochrome P4502E1 and risk of alcoholic liver disease in Caucasians. *Pharmacogenetics* 1995;**5**(6):351–7.
89. Carr LG, Hartleroad JY, Liang Y, Mendenhall C, Moritz T, Thomasson H. Polymorphism at the P450IIE1 locus is not associated with alcoholic liver disease in Caucasian men. *Alcohol Clin Exp Res* 1995;**19**(1):182–4.
90. Ingelman-Sundberg M, Johansson I, Yin H, et al. Ethanol-inducible cytochrome P4502E1: genetic polymorphism, regulation, and possible role in the etiology of alcohol-induced liver disease. *Alcohol* 1993;**10**(6):447–52.
91. Khan AJ, Ruwali M, Choudhuri G, Mathur N, Husain Q, Parmar D. Polymorphism in cytochrome P450 2E1 and interaction with other genetic risk factors and susceptibility to alcoholic liver cirrhosis. *Mutat Res* 2009;**664**(1–2):55–63.
92. Savolainen VT, Pajarinen J, Perola M, Penttila A, Karhunen PJ. Polymorphism in the cytochrome P450 2E1 gene and the risk of alcoholic liver disease. *J Hepatol* 1997;**26**(1):55–61.
93. Zintzaras E, Stefanidis I, Santos M, Vidal F. Do alcohol-metabolizing enzyme gene polymorphisms increase the risk of alcoholism and alcoholic liver disease? *Hepatol* 2006;**43**(2):352–61.
94. Zeng T, Guo FF, Zhang CL, Song FY, Zhao XL, Xie KQ. Roles of cytochrome P4502E1 gene polymorphisms and the risks of alcoholic liver disease: a meta-analysis. *PLoS One* 2013;**8**(1):e54188.
95. Yadav D, Lowenfels AB. The epidemiology of pancreatitis and pancreatic cancer. *Gastroenterol* 2013;**144**(6):1252–61.
96. Vonlaufen A, Wilson JS, Pirola RC, Apte MV. Role of alcohol metabolism in chronic pancreatitis. *Alcohol Res Health* 2007;**30**(1):48–54.
97. Verlaan M, Te Morsche RH, Roelofs HM, et al. Genetic polymorphisms in alcohol-metabolizing enzymes and chronic pancreatitis. *Alcohol Alcohol* 2004;**39**(1):20–4.
98. Yang B, O'Reilly DA, Demaine AG, Kingsnorth AN. Study of polymorphisms in the CYP2E1 gene in patients with alcoholic pancreatitis. *Alcohol* 2001;**23**(2):91–7.
99. Maruyama K, Takahashi H, Matsushita S, et al. Genotypes of alcohol-metabolizing enzymes in relation to alcoholic chronic pancreatitis in Japan. *Alcohol Clin Exp Res* 1999;**23**(4 Suppl): 85S–91S.
100. Matsumoto M, Takahashi H, Maruyama K, et al. Genotypes of alcohol-metabolizing enzymes and the risk for alcoholic chronic pancreatitis in Japanese alcoholics. *Alcohol Clin Exp Res* 1996;**20**(9 Suppl):289A–92A.

CHAPTER

20

Genes Associated with Alcohol Withdrawal

Kesheng Wang, PhD, Liang Wang, MD, PhD, MPH

Department of Biostatistics and Epidemiology, College of Public Health, East Tennessee State University, Johnson City, USA

INTRODUCTION

Worldwide, alcohol is the third leading risk factor for disease burden, while its harmful use leads to 2.5 million deaths every year.[1] In the United States in 2012, 71% of people aged 18 or older reported that they drank in the past year; while 24.6% of people aged 18 or older reported that they engaged in binge drinking in the past month, and 7.1% reported that they engaged in heavy drinking in the past month.[2] Alcohol dependence (AD) is a complex disease, with devastating effects on individuals, families, and society. The fourth edition of the Diagnostic and Statistical Manual (DSM–IV), published by the American Psychiatric Association, described two distinct disorders—alcohol abuse and AD—with specific criteria for each. The fifth edition, DSM–V, integrates the two DSM–IV disorders, alcohol abuse and AD, into a single disorder called alcohol use disorder (AUD), with mild, moderate, and severe subclassifications.[3] It has been reported that approximately 17 million adults aged 18 and older (7.2% of this age group) in the US had an AUD in 2012, including 11.2 million men (9.9% of men) and 5.7 million women (4.6% of women).[2] Furthermore, AD frequently coexists with other addictions, such as illicit substance abuse (cocaine dependence and marijuana dependence) and nicotine dependence.[4,5] Besides other addictions, AD also coexists frequently with other psychiatric diseases, including both internalizing disorders (e.g., depression and anxiety) and externalizing disorders (e.g., antisocial personality disorder, conduct disorder, and attention deficit hyperactivity disorder).[6,7] Alcohol consumption is increasing in many countries, and is an important cause of cancer worldwide.[8–10] Besides acting as a major risk factor for cardiac disease[11,12] and liver disease,[13,14] AD could damage the brain through many mechanisms, such as alcohol-induced brain dysfunctions and cognitive impairment.[15,16]

Alcohol withdrawal or alcohol withdrawal symptom (AWS) refers to a cluster of symptoms that may occur when a heavy drinker suddenly stops or significantly reduces their alcohol intake. Common symptoms include anxiety or nervousness, depression, fatigue, irritability, shakiness, mood swings, nightmares, headache, insomnia, loss of appetite, nausea and vomiting, rapid heart rate, sweating and tremor of the hands or other body parts. Some symptoms typical of withdrawal include seizures and delirium tremens.[17] The severity of alcohol withdrawal depends on various factors, such as age, genetics, degree of alcohol intake, length of time of alcohol cessation, and number of previous detoxifications.[18,19] The severity of withdrawal can vary from mild symptoms, such as sleep disturbances and anxiety, to severe symptoms, such as (1) alcohol hallucinosis: patients have problems in visual, auditory, or hallucinations, but are otherwise coherent,[20–31] (2) withdrawal seizures: seizures occur within 48 h of alcohol cessations,[21–24] (3) delirium tremens: patients have problems such as hyperadrenergic state, disorientation, tremors, diaphoresis, impaired attention/consciousness, and visual and auditory hallucinations.[20–23] These symptoms make alcohol abstinence difficult, and increase the risk of relapse in recovering alcoholics.[25] Withdrawal symptoms, the presence of which can indicate AD with a "physiological component," as defined by DSM-IV, appear to have particular clinical relevance, and may identify an important subpopulation of alcohol dependent individuals with a more severe clinical course.[26,27]

Family, twin, and adoption studies have indicated that genetic and environmental factors and their interactions contribute to the development of AD, with

Molecular Aspects of Alcohol and Nutrition. http://dx.doi.org/10.1016/B978-0-12-800773-0.00020-3

a heritability of more than 0.5.[28–33] Understanding the genetic basis of AD and AWS is a crucial step for developing efficient prevention strategies and personalized treatments. Animal models, whole-genome linkage, and candidate gene association studies have successfully identified several chromosome regions and genes that are related to AWS. For example, several candidate genes might be susceptible to AWS, such as CCK,[34] ADH1B and ADH4,[35] NPY gene,[36,37] CNR1,[38] 5-HT1B,[39] NQO2,[40] MTHFR,[41,42] 5-HT1A,[43] and SLC29A1.[44] The genome-wide association study (GWAS) has been used successfully in identifying single nucleotide polymorphisms (SNPs) associated with several common diseases.[45] Recently, a GWAS identified that two SNPs (rs7590720, $p = 9.72 \times 10^{-9}$; rs1344694, $p = 1.69 \times 10^{-8}$) had genome-wide significant associations with AD in German samples.[46] Bierut et al. analyzed 1M Illumina Human SNPs from The Study of Addiction: Genetics and Environment (SAGE), and identified 15 SNPs associated with AD ($p < 10^{-5}$);[33] meanwhile Edenberg et al. identified 199 SNPs ($p \leq 2.1 \times 10^{-4}$) using a GWAS of the Collaborative Study on the Genetics of Alcoholism (COGA) sample, and a cluster of genes on chromosome 11, and confirmed a number of genes associated with AD, including CPE, DNASE2B, SLC10A2, ARL6IP5, ID4, GATA4, SYNE1, and ADCY3.[47] Lind et al. conducted a pooling-based GWAS and meta-analysis, using a case-control design.[48] Wang et al. performed a meta-analysis of AD using the COGA and SAGE samples, and identified three new loci (KIAA0040, THSD7B and NRD1) and confirmed the previous association of PKNOX2 with AD.[49] More recently, GWASs have been conducted with alcohol consumption,[50,51] event-related brain oscillations,[52] and quantitative consumption, and diagnostic and quantitative dependence.[53] The GWAS is a powerful tool for unlocking the genetic basis of complex diseases such as AD. Recently, Wang et al. conducted the first GWAS to examine novel genetic factors of AWS using the COGA data.[54] This chapter reviews the recent findings in genetic studies of AWS.

CANDIDATE GENE STUDIES

Previous animal models, linkage study, candidate gene association study, gene expression analysis, and epigenetic studies have provided several candidate genes for AWS. Part of the candidate genes are presented in Table 20.1.

ADH1B and ADH4

McPherson et al. used a combination of somatic cell hybrid DNA analysis and *in situ* hybridization to localize the alcohol dehydrogenase 4 (ADH4) gene (also known as ADH-2; HEL-S-4) to human chromosome 4q22-4q23, in the cluster of alcohol dehydrogenase (ADH) genes.[69] Using a genome-wide linkage study, Reich et al. supported the genetic linkage between alcoholism and the region of chromosome 4 that includes the ADH genes;[70] while Long et al. supported the genetic linkage between AD and a nearby region on chromosome 4.[71] The ADH4 gene encodes alcohol dehydrogenase 4 that belongs to class II of the ADH genes. Alcohol dehydrogenase 1B, beta polypeptide (ADH1B) gene (also known as ADH2; HEL-S-117) is located at 4q23.[72] According to Smith et al., ADH1B is expressed in the lung in early fetal life and remains active in this tissue throughout life. It is also active in liver, after about the first trimester, and gradually increases in activity. The enzyme is also active in the adult kidney.[73] One recent study using a Native American community sample showed that one SNP within ADH1B (rs2066702; ADH1B*3) and another SNP at the 5′ end of ADH4 (rs3762894) were significantly associated with AWS ($p = 0.0018$ and 0.0012, respectively).[35] The minor alleles of both SNPs (C allele of rs3762894; T allele of rs2066702) were found to have a protective effect on the development of AWS, suggesting that variants in the ADH1B and ADH4 genes may be protective against the development of some symptoms associated with AD. Specifically, both rs2066702[74–77] and rs3762894 have been associated with AD.[75,78]

CCK

The cholecystokinin (CCK) gene is located at 3p22.1 and is a brain/gut peptide. In the gut, it induces the release of pancreatic enzymes and the contraction of the gallbladder.[79] The CCK is an important neurotransmitter that gives the influences on firings, anxiety, notiception, and dopamine-related behavior; however, the CCK coexists in the dopaminergic neurons, interacting with dopamine serotonin, gamma-aminobutyric acid, substance P, and enkephalins. It has been suggested that allelic mutation in the promoter region of the CCK gene might be susceptible to delirium tremens caused by alcohol abuse[34] and hallucination[56] in Japanese samples.

CNR1

The cannabinoid receptor 1 (brain) (CNR1) gene (also known as CB1; CNR; CB-R; CB1A; CB1R; CANN6; CB1K5) is located at 6q14-q15.[80] Animal model suggests that CB1 is involved in the motivational properties of opiates and in the development of physical dependence, and extended the concept of an interconnected role of CB1 and opiate receptors in the brain areas mediating addictive behavior.[81] Another study suggests that the endogenous cannabinoid system provides an on-demand protection against acute excitotoxicity in central nervous

TABLE 20.1 Candidate Genes for Alcohol Withdrawal Symptoms

CHR*	Gene location**	Gene	Gene full name	References
1	1p13.1	NGF	Nerve growth factor	Heberlein et al.[55]
1	1p36.3	MTHFR	Methylenetetrahydrofolate reductase	Lutz et al.[41]
3	3p22.1	CCK	Cholecystokinin	Okubo et al.[56]
4	4q22	ADH4	Alcohol dehydrogenase 4	Gizer et al.[35]
4	4q23	ADH1B	Alcohol dehydrogenase 1B (class I), beta polypeptide	Gizer et al.[35]
5	5q12.3	5-HT1A	5-Hydroxytryptamine (serotonin) receptor 1A	Lee et al.[43]
5	5p15	DAT	Sopamine transporter	Gorwood et al.[57]
6	6p21.1	RNT1	Solute carrier family 29 (equilibrative nucleoside transporter), member 1	Kim et al.[44]
6	6p21.3	GABABR1	Gamma-aminobutyric acid (GABA) B receptor, 1	Köhnke et al.[58]
6	6p21.31	FKBP5	FK506-binding protein 5	Huang et al.[59]
6	6p25	NQO2	NRH-quinone oxidoreductase 2	Okubo et al.[40]
6	6q13	HTR1B	5-Hydroxytryptamine (serotonin) receptor 1B	Sun et al.[39]
6	6q14-q15	CNR1	Cannabinoid receptor	Schmidt et al.[38]
7	7p15.1	NPY	Neuropeptide Y	Koehnke et al.[60]; Wetherill et al.[37]
7	7q31	LEP	Leptin	Hillemacher et al.[61]
8	8q11.2	OPRK1	Opioid receptor, kappa 1	Wang et al.[62]
9	9q34.3	GRIN1	Glutamate receptor, ionotropic, *N*-methyl D-aspartate 1	Rujescu et al.[63]
10	10q26.2	NPS	Neuropeptide S	Ruggeri et al.[64]
11	11p23	DRD2	Dopamine receptor D2	Grzywacz et al.[65]
11	11p15.5	TH	Tyrosine hydroxylase	Sander et al.[66]; Hack et al.[67]
12	12p13.1	NR2B	Glutamate receptor, ionotropic, *N*-methyl D-aspartate 2B	Biermann et al.[68]
17	17q11	SLC6A4	Solute carrier family 6 (neurotransmitter transporter), member 4	Bleich et al.[42]

** Chromosome;*
*** Cytogenetic locations.*

system neurons.[82] Schmidt et al. suggests that the homozygous genotype CNR1 1359A/A confers vulnerability to alcohol withdrawal delirium, in a Caucasian sample.[38]

DAT

The dopamine transporter (DAT) gene (also known as DAT1; PKDYS) is located at 5p15.3, and it plays a key role in homeostatic regulation of dopaminergic neurotransmission; it could also be involved in the variability of two severe AWS, alcohol-withdrawal seizure and delirium tremens.[83,84] The A9 allele of the DAT1 gene was associated with more severe effects of alcohol withdrawal. The possible mechanism may be due to modifications of the brain's capacity to compensate for long-term effects of ethanol on cerebral function.[85] Gorwood et al. provided convergent data in favor of a significant role of the DAT gene in the risk for some severe withdrawal symptoms.[57] Another study suggested that the A9 allele of the DAT gene was involved in vulnerability to alcohol withdrawal complications in women, but these complications may differ from those associated with this polymorphism in alcohol-dependent men.[86]

DRD2

The dopamine receptor D2 (DRD2) gene (also known as D2R; D2DR) is located at 11q23.[87] Previous study showed the influence of DRD2 on tiapride efficacy in alcohol withdrawal.[88] However, another study had negative results.[89] The D2 dopamine receptor is a G protein-coupled receptor located on postsynaptic dopaminergic neurons that is centrally involved in reward-mediating mesocorticolimbic pathways.[90] A recent study from Poland showed statistically significant associations between SNP in exon 8 A/G (rs6276) in the DRD2 gene and AWS with seizures, suggesting that DRD2 polymorphism was associated with AWS.[65]

ENT1

The type 1 equilibrative nucleoside transporter (ENT1) gene is also known as solute carrier family 29 (equilibrative nucleoside transporter), member 1 (SLC29A1) and is located at 6p21.1.[91] The ENT1 that regulates adenosine levels is known to regulate ethanol sensitivity and preference. Analyses of the combined data set in subjects of European ancestry, recruited from two independent sites, showed an association of the 647C variant and AD with withdrawal seizures at the nominally significant level. These results suggest a potential contribution of a genetic variant of ENT1 to the development of alcoholism, with increased risk of alcohol withdrawal-induced seizures in humans.[44]

FKBP5

FK506-binding protein 5 (FKBP5) (also known as P54; AIG6; FKBP51; FKBP54; PPIase; Ptg-10) is located at 6p21.31. This gene may play a role in immunoregulation and basic cellular processes involving protein folding and trafficking. Baughman et al. showed that the murine Fkbp51 gene was expressed in all tissues, most abundantly in T lymphocytes and in the thymus.[92] Alcohol withdrawal was associated with hypothalamic-pituitary-adrenal (HPA) axis dysfunction. The FKBP5 gene codes for a cochaperone, FK506-binding protein 5, that exerts negative feedback on HPA axis function. Six FKBP5 SNPs (rs3800373, rs9296158, rs3777747, rs9380524, rs1360780, and rs9470080) were found to be associated with AWS in a human sample. Such findings suggest that FKBP5 variants may trigger different adaptive changes in HPA axis regulation, during alcohol withdrawal, with concomitant effects on withdrawal severity.[59]

GABABR1

The gamma-aminobutyric acid (GABA) B receptor, 1 (GABABR1) gene (also known as GB1; GPRC3A; GABABR1; GABBR1-3; dJ271M21.1.1; dJ271M21.1.2) is involved in the GABAergic neurotransmission of the mammalian central nervous system. The metabotropic GABA-B receptors are coupled to G proteins, and modulate synaptic transmission through intracellular effector systems. GABA-B receptors function by inhibiting presynaptic transmitter release, or by increasing the potassium conductance responsible for long-lasting inhibitory postsynaptic potentials.[93,94] By genomic sequence analysis, Goei et al. (1998) mapped the human GABABR1 gene to chromosome 6p21.3.[94] One animal model showed that the GABAB receptor may play a role in mediating convulsions during alcohol withdrawal.[95] Another animal model study mapped 3 loci on mouse chromosomes 1, 4, and 11, which contain genes that influence alcohol withdrawal severity. Candidate genes, in proximity to the chromosome 11 locus, include genes encoding the alpha1, alpha6, and gamma2 subunits of type-A receptors for the inhibitory neurotransmitter, GABA.[96] One recent study genotyped one SNP within GABABR1 (rs29230). This study compared the allele and genotype frequencies between a group of alcoholics with a history of AWS ($n = 69$), and those with only mild withdrawal symptoms ($n = 97$), and did not find an association between the SNP and AWS.[58] Due to the fact that the sample size in this study is small, further studies are needed to evaluate the role of GABABR1 gene on AWS.

GRIN1

The glutamate receptor, ionotropic, *N*-methyl D-aspartate 1 gene (also known as NR1; MRD8; GluN1; NMDA1; NMDAR1) is located at 9q34.3.[97] It has been reported that the expression and alternative splicing of the obligatory NR1 subunit is altered by alcohol exposure, emphasizing the involvement of the NR1 subunit, that is coded by the GRIN1 gene, in alcohol-mediated effects. One human study using a German sample showed that the GRIN1 locus may modify the susceptibility to seizures during alcohol withdrawal.[63]

5-HT1A

The 5-hydroxytryptamine (serotonin) receptor 1A (5-HT1A) gene (also known as G-21; 5HT1a; PFMCD; 5-HT1A; 5-HT-1A; ADRBRL1; ADRB2RL1) is located at 5q12.3.[98] Recently, one Korean study investigated the role of serotonergic genes in the development of AD and the manifestation of AWS, using 97 Korean male inpatients with AD, and 76 Korean healthy male subjects. This study assessed the patients' AWS using the Clinical Institute Withdrawal Assessment for Alcohol (CIWA-Ar) scale, and found that the CIWA-Ar subscale scores of nausea, anxiety, and headache, and total CIWA-Ar scale scores were significantly higher in G+ genotypes (CG+ GG), than in G- genotype. The results suggest that the genetic polymorphism of the 5-HT1A receptor may play a role in AD, and that polymorphisms of serotonergic genes may be important in withdrawal symptoms of patients with AD.[43]

5-HTR1B

The 5-hydroxytryptamine (serotonin) receptor 1B (5-HTR1B) gene (also known as S12; 5-HT1B; HTR1D2; HTR1DB; 5-HT1DB) is located at 6q13.[99] The neurotransmitter serotonin (5-hydroxytryptamine; 5-HT) exerts a wide variety of physiologic functions through a multiplicity of receptors, and may be involved in human neuropsychiatric disorders such as anxiety, depression, or migraine. These receptors consist of several main groups, subdivided into several distinct subtypes, on the basis of their pharmacologic characteristics, coupling to intracellular second messengers, and distribution within the nervous system.[100] One study using a Taiwanese Han population screened for genetic variation in the coding, promoter, and partial 3′ untranslated regions of the HTR1B locus of 158 AD cases with withdrawal symptoms, and 149 control subjects.[39] The results show positive associations between variant A-161T and AWS, at both the allelic and genotypic levels, and suggest that the HTR1B A-161T polymorphism may be valuable both as a functional, and as an anonymous genetic marker for HTR1B.

LEP

Leptin (LEP) (also known as OB; OBS; LEPD) is located at 7q31.[101] Leptin is a 16-kD protein that plays a critical role in the regulation of body weight, by inhibiting food intake and stimulating energy expenditure; while defects in leptin production cause severe hereditary obesity in rodents and humans.[102–105] There is evidence that higher serum levels of leptin are positively linked to the extent of craving, during intoxication treatment.[106–108] Leptin has been shown to be involved in the effects of the androgen receptor and steroid 5-alpha reductase II on craving, during alcohol withdrawal.[109] A recent study showed that low methylation status was associated with increasing serum leptin levels, and elevation of craving for alcohol in the referring patients group. These findings point toward a pathophysiological relevance of changes in DNA methylation of the LEP gene promoter region in AD.[61]

MTHFR

The methylenetetrahydrofolate reductase (NAD(P)H) (MTHFR) gene is located at 1p36.3.[110] Genetic variation in this gene influences susceptibility to occlusive vascular disease, neural tube defects, colon cancer, and acute leukemia. Mutations in this gene are associated with methylenetetrahydrofolate reductase deficiency. It has been suggested that the elevated homocysteine plasma levels are considered as a risk factor for the occurrence of seizures, during alcohol withdrawal; while tomocysteine plasma concentrations seemed to be influenced by the methylenetetrahydrofolate reductase (MTHFR) C677T-polymorphism. Lutz et al. investigated whether the T-allele of the MTHFR C677T-polymorphism is associated with AD and alcohol withdrawal seizure, using a group of 102 healthy controls and 221 alcoholic patients, including 97 patients with a history of mild withdrawal symptoms, and 70 patients with a history of alcohol withdrawal seizure.[41] They found that the T-allele was significantly associated with withdrawal seizure, by comparing alcoholic patients with a history of withdrawal seizure ($p = 0.03$). These results suggest an influence of the MTHFR C677T-polymorphism on the etiology of AWS and AD in men, in a western European population.

NGF

Nerve growth factor (NGF) gene (also known as NGFB; HSAN5; Beta-NGF) is located at 1p13.1. NGF is a polypeptide involved in the regulation of growth and differentiation of sympathetic and certain sensory neurons.[111] One study suggests that the NGF levels may be a trait marker for the development of AD.[112] One recent study investigated the Cytosin-phosphatidyl-Guanin (CpG) island promoter methylation of the NGF gene in the blood of alcohol-dependent patients (57 male patients), during withdrawal, and found that the NGF serum levels were significantly associated with the mean methylation of the investigated CpG-sites. These results imply an epigenetic regulation of the NGF gene during alcohol withdrawal.[55]

NPY

The neuropeptide Y (NPY) gene (also known as PYY4) is located at 7p15.1,[113] and exists in both the central and peripheral nervous system. It is thought to modulate many functions, such as feeding behavior, anxiety-associated behavior, circadian rhythm, seizure modulation, and hormone secretion. Recent studies have revealed that NPY influences alcohol consumption in mice, and that alcohol-preferring rats had lower concentrations of NPY-like immunoreactivity, compared with alcohol-nonpreferring rats, in several brain regions. Kauhanen et al. analyzed 889 middle-aged men from eastern Finland for the leu7-to-pro polymorphism of NPY, and suggested that alcohol preference in humans may be regulated by the NPY system.[114]

One study analyzed the whole coding region and 5′-untranslating region of the NPY gene for 163 Japanese male alcoholics, with different withdrawal symptoms (93 with delirium tremens, 71 with seizures, and 49 with hallucinations), and 98 Japanese male controls. It has been found that frequency of the T allele and frequency of the genotype that possessed T alleles (CT, TT) at the 5671 locus were significantly higher in patients with seizure, than in those without seizure ($p < 0.05$), that suggested that a C to T substitution at the 5671 locus of the NPY gene may be associated with seizure during alcohol withdrawal.[36] Another study tested the hypothesis that the T1128C polymorphism is associated with the diagnosis of alcoholism, or with severe forms of alcohol withdrawal, using two groups of alcoholics with severe withdrawal symptoms (delirium tremens, $n = 83$; withdrawal seizures, $n = 65$), and one group of alcoholics with mild withdrawal symptoms ($n = 97$). The frequency of the C-allele in the individuals with severe withdrawal symptoms was higher; however, not reaching statistical significance.[60] Lappalainen et al. suggested that the NPY pro7 allele is a risk factor for AD.[115] A recent study using part of the Collaborative Studies on the Genetics of Alcoholism (COGA) study, showed that sequence variations in NPY receptor genes are associated with AD, particularly a severe subtype of AD characterized by withdrawal symptoms, comorbid alcohol and cocaine dependence, and cocaine dependence.[37]

NQO2

The NRH-quinone oxidoreductase 2 (NQO2) gene (also known as QR2; DHQV; DIA6; NMOR2) is involved in phase II detoxification reactions. It is thought to be important for detoxification of catechol-o-quinones in the central nervous system. By fluorescence *in situ* hybridization, Jaiswal et al. narrowed the mapping of NQO2 to chromosome 6p25.[116] One study investigated the association between one polymorphism of NQO2 gene and AWS (such as delirium tremens, hallucination, and seizure) using 247 Japanese male alcoholic patients with AWS or without the symptoms, and 134 age-matched Japanese male controls.[40] They found significant differences between alcoholic patients and controls in genotype frequency at an insertion/deletion site in the promoter region of the NQO2 gene ($p = 0.0014$). They found that the frequency of the homozygous genotype for the D allele, at this locus, was significantly higher in delirium tremens-positive patients ($p = 0.0004$) and in hallucination-positive patients ($p = 0.0001$), as well as in patients displaying both delirium tremens and hallucination ($p = 0.0002$), than in controls. These results suggested that a polymorphism in the promoter region of the NQO2 gene may play an important role in the pathogenesis of alcoholism and AWS.[40]

NPS

Neuropeptide S (NPS) gene is located at 10q26.2. It has been reported that the NPSR mRNA is widely distributed in the brain, including the amygdala and the midline thalamic nuclei. Central administration of NPS increases locomotor activity in mice, and decreases paradoxical rapid eye movement sleep and slow wave sleep in rats. Xu et al. concluded that NPS may mediate arousal and anxiety.[117] The NPS may play a role in the modulation of ethanol drinking, and possibly anxiety-like behavior in rats selectively bred for high alcohol drinking.[118] In a rat model, NPS receptor mRNA expression is increased in different brain areas of postdependent rats; as shown in the DB test, this expression change is functionally relevant, suggesting that NPS receptor (NPSR) gene expression is altered during withdrawal.[64]

NR2B

The *N*-methyl-D-aspartate (NMDA) receptor is a glutamate-activated ion channel permeable to Na+, K+, and Ca(2+). It is found at excitatory synapses throughout the brain. Glutamate receptor, ionotropic, *N*-methyl D-aspartate 2B (GRIN2B) gene (also known as MRD6; NR2B; hNR3; EIEE27; GluN2B; NMDAR2B) is a subunit of NMDA. Mandich et al. localized the human NMDAR2B gene to 12p12;[119] while Endele et al. noted that the GRIN2B gene maps to chromosome 12p13.1.[120] The NR2 subunit acts as the agonist binding site for glutamate. This receptor is the predominant excitatory neurotransmitter receptor in the mammalian brain. *N*-Methyl-D-aspartate (NMDA) receptor-mediated glutamatergic neurotransmission may play a central role in the development of alcohol dependence, and this alteration is supposed to be due to a differential upregulation of the NR2B type of subunits. NR2B subunit selective NMDA antagonists inhibit the neurotoxic effect of alcohol-withdrawal in primary cultures of rat cortical neurons.[121] However, neither the analyzed SNPs (rs1806201 and rs1806191), nor any of their haplotypes likely modify susceptibility to AD or withdrawal-related phenotypes.[122] A methylation analysis was conducted using bisulfite sequencing of a fragment of the NR2B promoter region, and results showed that the expression of the NR2B receptor increased significantly during the first 24 hours of withdrawal treatment (day 1; $t = 4.1$, $p = 0.001$), and also on and day 3 ($t = 2.4$; $p = 0.029$). These findings might explain the observation of an impact of alcohol consumption patterns on the gravity of withdrawal symptoms.[68]

OPRK1

The opioid receptor, κ1 (OPRK1) gene (also known as KOR; OPRK; KOR-1; K-OR-1) is located at 8q11.2.[123,124] A recent study tested the hypotheses that the genetic polymorphisms in "The opioid receptor, k1 (OPRK1)" in the OPRK1 gene are associated with methadone treatment responses in a Taiwan methadone maintenance treatment cohort. They found that the haplotype rs10958350-rs7016778-rs12675595 was associated with gooseflesh skin (global $p < 0.0001$), yawning (global $p = 0.0001$), and restlessness (global $p < 0.0001$) withdrawal symptoms. These findings suggested that genetic polymorphisms in the genetic polymorphisms in OPRK1 gene region were associated with body weight, alcohol use, and opioid withdrawal symptoms in methadone maintenance treatment cohort patients.[62]

SLC6A4

Solute carrier family 6 (neurotransmitter transporter), member 4 (SLC6A4) gene (also known as HTT; 5HTT; OCD1; SERT; 5-HTT; SERT1; HSERT; 5-HTTLPR) is a neurotransmitter in the central and peripheral nervous systems. By somatic cell hybrid and *in situ* hybridization studies, Ramamoorthy et al. mapped a single gene encoding the human 5-HT transporter to chromosome 17q11.1-q12.[125] Feinn et al. conducted a meta-analysis of the association of the functional serotonin transporter promoter polymorphism with AD.[126] One study investigated the association of one 5-HTTLPR polymorphism with obsessive-compulsive alcohol craving in 124 male Caucasian patients admitted for alcohol detoxification treatment. The results suggests that the 5-HTTLPR polymorphism was associated with higher compulsive alcohol craving at the beginning of alcohol withdrawal.[42]

TH

The tyrosine hydroxylase (TH) (also known as TYH; DYT14; DYT5b) is located at 11p15.5.[127,128] The protein encoded by this gene is involved in the conversion of tyrosine to dopamine. It is the rate-limiting enzyme in the synthesis of catecholamines, hence it plays a key role in the physiology of adrenergic neurons. The rare allele containing 10-repeats (A10) of a polymorphic tetranucleotide motif in the first intron of the tyrosine hydroxylase (TH) gene was significantly increased in the alcoholics with withdrawal delirium (3.2%), compared with that in the controls using a German sample, fact suggesting that the TH gene may mediate vulnerability to alcohol-withdrawal delirium in a small proportion of alcohol-dependent subjects.[66] Using a case-control study, a recent study identified association of one SNP in TH (rs11564717) with the withdrawal factor score ($p = 0.0094$).[67]

TABLE 20.2 Top 25 SNPs in the Genome-Wide Association STUDY of Alcohol Withdrawal Symptoms in the COGA Sample

CHR	SNP	Position*	Gene	Allele**	P_{COGA}†
5	rs770182	107146039	Near EFNA5	C	3.65E-06
8	rs4129295	117544778	Near TRPS1	T	3.92E-06
20	rs8126263	23714277	Near CST2	T	5.66E-06
9	rs10975990	7016716	KDM4C	A	7.15E-06
3	rs3903132	115183416	KIAA1407	A	1.12E-05
5	rs2416371	113803769	KCNN2	T	1.32E-05
5	rs12523436	43312226	NIM1	T	1.70E-05
14	rs10141191	97957263	Near BCL11B	A	1.78E-05
17	rs17674853	64263617	Near ABCA8	G	2.46E-05
5	rs4705665	113771552	KCNN2	G	2.55E-05
3	rs7649435	117836222	BZW1L1	G	2.58E-05
9	rs10758821	7020268	KDM4C	T	2.79E-05
14	rs7161221	97962636	Near BCL11B	C	2.86E-05
3	rs324556	115213977	KIAA1407	T	3.16E-05
1	rs11804222	11859101	Near PLOD1	A	3.23E-05
14	rs17412116	32036099	AKAP6	A	3.25E-05
13	rs1543660	28753650	MTUS2	C	3.62E-05
10	rs2854992	73256591	PSAP	T	3.81E-05
2	rs16827526	186749074	Near FSIP2	G	4.30E-05
3	rs1286665	25550679	RARB	T	4.34E-05
11	rs10834476	24731082	LUZP2	C	4.77E-05
2	rs1251207	129698342	Near POTEF	C	4.79E-05
9	rs1407862	7022776	KDM4C	T	4.93E-05
4	rs2466990	171230346	AADAT	G	4.94E-05
10	rs11250672	1655466	ADARB2	A	4.99E-05

* *Physical position (bp).*
** *Minor allele.*
† *P-value based on logistic regression.*
Adapted from Ref. [54].

GENOME-WIDE ASSOCIATION STUDY

Recently, Wang et al. reported the first GWAS of AWS using the COGA sample that contains 1M Illumina SNPs (1,069,796 SNPs) based on 1025 cases with AD, and 569 controls.[47,54] In this study, 879 Caucasian individuals (466 AD cases with AWS and 413 controls) were included for genome-wide association analysis of AWS.[54] Phenotypes include AD and AWS as a binary trait, according to DSM-IV diagnosis.[37] In this replication study of AWS,[54] family-based association analysis was performed for the Australian twin-family study of alcohol use disorder (OZALC study) sample with 825 individuals (318 AD cases with AWS and 507 controls) from 273 families.[48,129,130] Parts of the results of GWAS results of AWS[54] are presented in Table 20.2. The first top SNP rs770182 is between EFNA5 and FBXL17 at 5q21, closer to EFNA5. Ephrin-A5 is a protein encoded by the EFNA5 gene (also known as AF1, EFL5, RAGS, EPLG7, GLC1M, LERK7) in humans. The EPH and EPH-related receptors have been implicated in mediating developmental events, particularly in the nervous system.[131–133] The next interesting locus is KDM4C (also known as GASC1, JHDM3C, JMJD2C, FLJ25949, KIAA0780, bA146B14.1) at 9p24.1. Nagase et al. cloned JMJD2C that they designated KIAA0780, and detected JMJD2C expression in all tissues examined. Highest expression was in brain, and lowest was in skeletal muscle and spleen.[134] The third replicated locus was at 2q32.1 near FSIP2 (the top SNP in the COGA sample is rs16827526, $p = 4.30 \times 10^{-5}$). Several flanking SNPs of the top hits in the COGA sample (Table 20.2) demonstrated borderline associations with AWS in the OZALC sample (rs3386600, rs963660, rs10834486, and rs1874995 for KCNN2, AADAT, LUZP2 and PSAP, respectively). In addition, several SNPs in KIAA1407, 8q24.12 (near TRPS1) and 20p11.21 (near CST2) were associated with AWS ($p < 1.5\ 10^{-5}$) in the COGA sample. For GABRG1, there were 14 SNPs associated with AWS (the top SNP is rs10032631 with $p = 0.0044$) while for GABRG3, 19 SNPs were associated AWS (the top SNP is rs7484955 with $p = 0.0061$). Furthermore, rs11576001 within GABRA1 gene revealed association with AWS ($p = 0.0091$).

LIMITATIONS

This chapter focuses on candidate genes that might be susceptible to AWS. However, part of genes may just show suggestive evidence for AWS. Furthermore, for most genes, the results are limited to one sample or population. Therefore, the results need to be confirmed in other samples or populations. The GWAS of AWS provided some genes or loci associated with AWS,[54] but further functional studies and replication studies are warranted.

CONCLUSIONS

This chapter provides candidate genes with common variants that might be susceptible to AWS. It has been suggested that both the common and rare variants are involved in complex diseases.[135,136] It is also possible that a large number (e.g., >100) of susceptibility polymorphisms of small effects working together may provide useful individual-level risk prediction.[137,138] Furthermore, complex diseases, such as AD and AWS, result from the interplay of many genetic and environmental factors. If some of the unexplained heritability in GWAS is due to interactions, then one goal might be to use interactions to discover novel genes/regions.[139,140] In order to create a comprehensive catalogue of common and rare variants in individuals with AD and AWS, it is useful to combine the results of GWAS, gene–gene and gene–environment interactions, gene expression, and epigenetics with the recent rapid technological advancements in the next-generation sequencing (NGS), including whole exome sequencing, transcriptome sequencing, and whole genome sequencing.

Key Facts

Complex Disease

- Any disease that cannot be defined by a Mendelian pattern that is controlled by a single locus.
- Several genes in combination with environmental factors cause the disease.
- No single gene or environmental factor is sufficient to cause the disease.
- Complex disease is present with a higher population frequency and, therefore, leading to a greater public health concern.

Alcohol Withdrawal

- Worldwide, alcohol is the third leading risk factor for disease burden, while its harmful use leads to 2.5 million deaths every year.
- Alcohol use disorder (AUD), including alcohol abuse and alcohol dependence (AD), are medical conditions that doctors diagnose when a patient's drinking causes distress or harm, based on the Diagnostic and Statistical Manual of Mental Disorders (DSM)-V.
- Alcohol withdrawal or alcohol withdrawal symptom (AWS) refers to a cluster of symptoms that may occur when a heavy drinker suddenly stops or significantly reduces their alcohol intake.
- Withdrawal symptoms, the presence of which can indicate AD with a "physiological component" as defined by DSM-IV, appear to have particular clinical relevance and may identify an important subpopulation of AD individuals with a more severe clinical course.
- Numerous family, twin, and adoption studies have suggested that AD represents a heritable condition, with approximately 50–60% of the variance in the development of AD being explained by genetic influences.

Gene Discovery Methods

- Linkage study is to find genetic regions that may include candidate genes for complex diseases.
- Candidate gene association study is to test pre-specified genes of interest that may affect phenotypes or disease states.
- The genome-wide association studies (GWAS) is to search for genetic variants in the whole genome.
- Epigenetics involves genetic control by factors other than an individual's DNA sequence.
- Next-generation sequencing (NGS) is a promising tool to identify copy number variation, novel genetic variants, and genes related to complex diseases.
- Animal models have been used for linkage and gene functional studies.

ADH1B Gene

- Alcohol dehydrogenase 1B, beta polypeptide (ADH1B) gene (also known as ADH2) encodes the beta subunit of class I alcohol dehydrogenase (ADH), an enzyme that catalyzes the rate-limiting step for ethanol metabolism. The ADH1B gene is located at 4q22-4q23.
- ADH1B is expressed in the lung in early fetal life, and remains active in this tissue throughout life.
- ADH1B is also active in liver in some period of the life, and in the adult kidney.
- Polymorphisms of ADH1B have been associated with AD, AWS, and some cancers.

Summary Points

- Family, twin, and adoption studies have indicated that genetic and environmental factors and their interactions contribute to the development of alcohol dependence (AD).
- Alcohol withdrawal or alcohol withdrawal symptom (AWS) refers to a cluster of symptoms that may occur when a heavy drinker suddenly stops or significantly reduces their alcohol intake.
- AWS can start as early as two hours after the last drink, persist for weeks, and range from mild anxiety and shakiness to severe complications, such as seizures and delirium tremens.
- Understanding the genetic basis of AD and AWS is essential to develop efficient prevention strategies and personalized treatments of AD and AWS.
- This chapter reviews recently discovered AWS candidate genes by using animal models, linkage analysis, candidate gene study, genome-wide association study (GWAS), gene expression study, and methylation analyses.
- To better understand the genetic etiology of AWS, it is useful to combine the results of GWAS, gene-gene and gene-environment interactions, gene expression, and epigenetics, with the recent rapid technological advancements in the next-generation sequencing (NGS).

Acknowledgments

Funding support for the CIDR-COGA Study was provided through the Center for Inherited Disease Research (CIDR) and the Collaborative Study on the Genetics of Alcoholism (COGA). The CIDR-COGA Study is a genome-wide association studies funded as part of the Collaborative Study on the Genetics of Alcoholism (COGA). Assistance with phenotype harmonization and genotype cleaning, as well as with general study coordination, was provided by the COGA. Assistance with data cleaning was provided by the National Center for Biotechnology Information. Support for collection of datasets and samples was provided by the Collaborative Study on the Genetics of Alcoholism (COGA; U10 AA008401). Funding support for genotyping, that was performed at the Johns Hopkins University Center for Inherited Disease Research, was provided by the NIH GEI (U01HG004438), the National Institute on Alcohol Abuse and Alcoholism, and the NIH contract "High throughput genotyping for studying the genetic contributions to human disease." (HHSN268200782096C). The datasets used for the analyses described in this manuscript were obtained from dbGaP at http://www.ncbi.nlm.nih.gov/sites/entrez?Db=gap through dbGaP accession number: phs000125.v1.p1. The dataset for replication study was obtained from the CIDA database found at http://www.ncbi.nlm.nih.gov/projects/gap/ through the dbGAP accession number Study Accession: phs000181.v1.p1. Funding support for the (CIDR-OZALC GWAS) was provided through the Center for Inherited Disease Research (CIDR) and the National Institute on Alcohol Abuse and Alcoholism (NIAAA) (Study Accession: phs000181.v1.p1). CIDR-OZALC GWAS is a genome-wide association studies funded as part of the NIAAA grant 5 R01 AA013320-04. Assistance with phenotype harmonization and genotype cleaning, as well as with general study coordination, was provided by the CIDR-OZALC GWAS. Assistance with data cleaning was provided by the National Center for Biotechnology Information. Support for collection of datasets and samples was provided by the MARC: Risk Mechanisms in Alcoholism and Comorbidity (MARC; P60 AA011998-11). Funding support for genotyping, that was performed at the Johns Hopkins University Center for Inherited Disease Research, was provided by the NIH GEI (U01HG004438), the National Institute on Alcohol Abuse and Alcoholism, and the NIH contract "High throughput genotyping for studying the genetic contributions to human disease." (HHSN268200782096C).

Abbreviations

AD Alcohol dependence
AUD Alcohol use disorder
AWS Alcohol withdrawal symptoms
DNA Deoxyribonucleic acid
GWAS Genome-wide association study
NGS Next-generation sequencing
SNP Single-nucleotide polymorphism

References

1. WHO. *Draft global strategy to reduce the harmful use of alcohol*. Geneva, Switzerland: World Health Organization; 2010.
2. Substance Abuse and Mental Health Services Administration (SAMHSA). National survey on drug use and health (NSDUH). Rockville, MD. Available from: http://www.samhsa.gov/data/NSDUH/2012SummNatFindDetTables/DetTabs/NSDUH-DetTabsSect5peTabs1to56-2012.htm#Tab5.8A; 2012
3. http://pubs.niaaa.nih.gov/publications/dsmfactsheet/dsmfact.pdf.
4. Swan GE, Carmelli D, Cardon LR. Heavy consumption of cigarettes, alcohol and coffee in male twins. *J Stud Alcohol* 1997;**58**:182–90.
5. Hopfer CJ, Stallings MC, Hewitt JK. Common genetic and environmental vulnerability for alcohol and tobacco use in a volunteer sample of older female twins. *J Stud Alcohol* 2001;**62**:717–23.
6. Grant BF, Stinson FS, Dawson DA, Chou SP, Dufour MC, Compton W, Pickering RP, Kaplan K. Prevalence and co-occurrence of substance use disorders and independent mood and anxiety disorders: results from the National Epidemiologic Survey on Alcohol and Related Conditions. *Arch Gen Psychiatry* 2004;**61**:807–16.
7. Grant BF, Stinson FS, Dawson DA, Chou SP, Ruan WJ, Pickering RP. Co-occurrence of 12-month alcohol and drug use disorders and personality disorders in the United States: results from the National Epidemiologic Survey on Alcohol and Related Conditions. *Arch Gen Psychiatry* 2004;**61**:361–8.
8. Longnecker MP. Alcohol consumption and risk of cancer in humans: an overview. *Alcohol* 1995;**12**:87–96.

9. Boffetta P, Hashibe M. Alcohol and cancer. *Lancet Oncol* 2006;**7**: 149–56.
10. Druesne-Pecollo N, Tehard B, Mallet Y, Gerber M, Norat T, Hercberg S, Latino-Martel P. Alcohol and genetic polymorphisms: effect on risk of alcohol-related cancer. *Lancet Oncol* 2009;**10**:173–80.
11. Ettinger PO, Wu CF, De La Cruz C, Weisse AB, Ahmed SS, Regan TJ. Arrhythmias and the "Holiday Heart": alcohol-associated cardiac rhythm disorders. *Am Heart J* 1978;**95**:555–62.
12. Dolara A, Marascio D. Heart-alcohol relations: effective dosage. *G Ital Cardiol* 1982;**12**:740–2.
13. Kuller LH, May SJ, Perper JA. The relationship between alcohol, liver disease, and testicular pathology. *Am J Epidemiol* 1978;**108**: 192–9.
14. Menon KV, Gores GJ, Shah VH. Pathogenesis, diagnosis, and treatment of alcoholic liver disease. *Mayo Clinic Proc* 2001;**76**:1021–9.
15. Lovinger DM. Serotonin's role in alcohol's effects on the brain. *Alcohol Health Res World* 1997;**21**:114–20.
16. Savage LM, Candon PM, Hohmann HL. Alcohol-induced brain pathology and behavioral dysfunction: using an animal model to examine sex differences. *Alcohol Clin Exp Res* 2000;**24**:465–75.
17. Hughes JR. Alcohol withdrawal seizures. *Epilepsy Behav* 2009;**15**: 92–7.
18. Liskow BI, Rinck C, Campbell J, DeSouza C. Alcohol withdrawal in the elderly. *J Stud Alcohol* 1989;**50**:414–21.
19. Becker HC. Kindling in Alcohol Withdrawal. *Alcohol Health Res World* 1998;**22**(1) http://pubs.niaaa.nih.gov/publications/arh22-1/25-34.pdf (accessed 14.11.2010).
20. Bayard M, McIntyre J, Hill KR, Woodside J. Alcohol withdrawal syndrome. *Am Fam Physician* 2004;**69**:1443–50.
21. Jennifer F, Wilson, et al. In the clinic. Alcohol use. *Ann Intern Med* 2009;**150** ITC3-1-ITC3-15.
22. Schuckit MA. Alcohol-use disorders. *Lancet* 2009;**373**:492–501.
23. O'Connor PG. Alcohol abuse and dependence. In: Goldman L, Schafer AI, editors. *Cecil medicine*. 24th ed. Philadelphia, Pa: Saunders Elsevier; 2011 [chapter 32].
24. Manasco A, Chang S, Larriviere J, Hamm LL, Glass M. Alcohol withdrawal. *South Med J* 2012;**105**:607–12.
25. Economidou D, Cippitelli A, Stopponi S, Braconi S, Clementi S, Ubaldi M, Martin-Fardon R, Weiss F, Massi M, Ciccocioppo R. Activation of brain NOP receptors attenuates acute and protracted alcohol withdrawal syndrome in the rat. *Alcohol Clin Exp Res* 2011;**35**:747–55.
26. Schuckit MA, Smith TL, Daeppen JB, Eng M, Li TK, Hesselbrock VM, Nurnberger Jr JI, Bucholz KK. Clinical relevance of the distinction between alcohol dependence with and without a physiological component. *Am J Psychiatry* 1998;**155**:733–40.
27. Langenbucher J, Martin CS, Labouvie E, Sanjuan PM, Bavly L, Pollock NK. Toward the DSM-V: the Withdrawal-Gate Model versus the DSM-IV in the diagnosis of alcohol abuse and dependence. *J Consult Clin Psychol* 2000;**68**:799–809.
28. Heath AC, Bucholz KK, Madden PA, Dinwiddie SH, Slutske WS, Bierut LJ, Statham DJ, Dunne MP, Whitfield JB, Martin NG. Genetic and environmental contributions to alcohol dependence risk in a national twin sample: consistency of findings in women and men. *Psychol Med* 1997;**27**:1381–96.
29. McGue. M. A behavioral-genetic perspective on children of alcoholics. *Alcohol Health Res World* 1997;**21**:210–7.
30. Schuckit MA. Genetics of the risk for alcoholism. *Am J Addict* 2000;**9**:103–12.
31. Goldman D, Oroszi G, Ducci F. The genetics of addictions: uncovering the genes. *Nat Rev Genet* 2005;**6**:521–32.
32. Kalsi G, Prescott CA, Kendler KS, Riley BP. Unraveling the molecular mechanisms of alcohol dependence. *Trends Genet* 2009;**25**:49–55.
33. Bierut LJ, Agrawal A, Bucholz KK, Doheny KF, Laurie C, Pugh E, Fisher S, Fox L, Howells W, Bertelsen S, Hinrichs AL, Almasy L, Breslau N, Culverhouse RC, Dick DM, Edenberg HJ, Foroud T, Grucza RA, Hatsukami D, Hesselbrock V, Johnson EO, Kramer J, Krueger RF, Kuperman S, Lynskey M, Mann K, Neuman RJ, Nothen MM, Nurnberger Jr JI, Porjesz B, Ridinger M, Saccone NL, Saccone SF, Schuckit MA, Tischfield JA, Wang JC, Rietschel M, Goate AM, Rice JP. A genome-wide association study of alcohol dependence. *Proc Natl Acad Sci USA* 2010;**107**:5082–7.
34. Okubo T, Harada S, Higuchi S, Matsushita S. Genetic association between alcohol withdrawal syndromes and polymorphism of CCK gene promoter. *Alcohol Clin Exp Res* 1999;**23**(4 Suppl.):11S–2S.
35. Gizer IR, Edenberg HJ, Gilder DA, Wilhelmsen KC, Ehlers CL. Association of alcohol dehydrogenase genes with alcohol-related phenotypes in a Native American community sample. *Alcohol Clin Exp Res* 2008;**35**:2008–18.
36. Okubo T, Harada S. Polymorphism of the neuropeptide Y gene: an association study with alcohol withdrawal. *Alcohol Clin Exp Res* 2001;**25**(6 Suppl.):59S–62S.
37. Wetherill L, Schuckit MA, Hesselbrock V, Xuei X, Liang T, Dick DM, Kramer J, Nurnberger Jr JI, Tischfield JA, Porjesz B, Edenberg HJ, Foroud T. Neuropeptide Y receptor genes are associated with alcohol dependence, alcohol withdrawal phenotypes, and cocaine dependence. *Alcohol Clin Exp Res* 2008;**32**:2031–40.
38. Schmidt LG, Samochowiec J, Finckh U, Fiszer-Piosik E, Horodnicki J, Wendel B, Rommelspacher H, Hoehe MR. Association of a CB1 cannabinoid receptor gene (CNR1) polymorphism with severe alcohol dependence. *Drug Alcohol Depend* 2002;**65**:221–4.
39. Sun HF, Chang YT, Fann CS, Chang CJ, Chen YH, Hsu YP, Yu WY, Cheng AT. Association study of novel human serotonin 5-HT(1B) polymorphisms with alcohol dependence in Taiwanese Han. *Biol Psychiatry* 2002;**51**:896–901.
40. Okubo T, Harada S, Higuchi S, Matsushita S. Association analyses between polymorphisms of the phase II detoxification enzymes (GSTM1, NQO1, NQO2) and alcohol withdrawal syndromes. *Alcohol Clin Exp Res* 2003;**27**(8 Suppl.):68S–71S.
41. Lutz UC, Batra A, Kolb W, Machicao F, Maurer S, Köhnke MD. Methylenetetrahydrofolate reductase C677T-polymorphism and its association with alcohol withdrawal seizure. *Alcohol Clin Exp Res* 2006;**30**:1966–71.
42. Bleich S, Bönsch D, Rauh J, Bayerlein K, Fiszer R, Frieling H, Hillemacher T. Association of the long allele of the 5-HTTLPR polymorphism with compulsive craving in alcohol dependence. *Alcohol Alcohol* 2007;**42**:509–12.
43. Lee YS, Choi SW, Han DH, Kim DJ, Joe KH. Clinical manifestation of alcohol withdrawal syndromes related to genetic polymorphisms of two serotonin receptors and serotonin transporter. *Eur Addict Res* 2009;**15**:9–46.
44. Kim JH, Karpyak VM, Biernacka JM, Nam HW, Lee MR, Preuss UW, Zill P, Yoon G, Colby C, Mrazek DA, Choi DS. Functional role of the polymorphic 647 T/C variant of ENT1 (SLC29A1) and its association with alcohol withdrawal seizures. *PLoS One* 2011;**6**:e16331.
45. Wellcome Trust Case Control Consortium. Genome-wide association study of 14,000 cases of seven common diseases and 3,000 shared controls. *Nature* 2007;**447**:661–78.
46. Treutlein J, Cichon S, Ridinger M, Wodarz N, Soyka M, Zill P, Maier W, Moessner R, Gaebel W, Dahmen N, Fehr C, Scherbaum N, Steffens M, Ludwig KU, Frank J, Wichmann HE, Schreiber S, Dragano N, Sommer WH, Leonardi-Essmann F, Lourdusamy A, Gebicke-Haerter P, Wienker TF, Sullivan PF, Nothen MM, Kiefer F, Spanagel R, Man K, Rietschel M. Genome-wide association study of alcohol dependence. *Arch Gen Psychiatry* 2009;**66**:773–84.
47. Edenberg HJ, Koller DL, Xuei X, Wetherill L, McClintick JN, Almasy L, Bierut LJ, Bucholz KK, Goate A, Aliev F, Dick D, Hesselbrock V, Hinrichs A, Kramer J, Kuperman S, Nurnberger Jr JI, Rice JP, Schuckit MA, Taylor R, Todd Webb B, Tischfield JA, Porjesz B, Foroud T. Genome-wide association study of alcohol dependence implicates a region on chromosome 11. *Alcohol Clin Exp Res* 2010;**34**:840–52.

48. Lind PA, Macgregor S, Vink JM, Pergadia ML, Hansell NK, de Moor MH, Smit AB, Hottenga JJ, Richter MM, Heath AC, Martin NG, Willemsen G, de Geus EJ, Vogelzangs N, Penninx BW, Whitfield JB, Montgomery GW, Boomsma DI, Madden PA. A genomewide association study of nicotine and alcohol dependence in Australian and Dutch populations. *Twin Res Hum Genet* 2010;**13**:10–29.
49. Wang KS, Liu XF, Zhang QY, Pan Y, Aragam N, Zeng M. Meta-analysis of two genome-wide association studies identifies 3 new loci for alcohol dependence. *J Psychiatr Res* 2011;**45**:1419–25.
50. Baik I, Cho NH, Kim SH, Han BG, Shin C. Genome-wide association studies identify genetic loci related to alcohol consumption in Korean men. *Am J Clin Nutr* 2011;**93**:809–16.
51. Schumann G, Coin LJ, Lourdusamy A, Charoen P, Berger KH, Stacey D, Desrivières S, Aliev FA, Khan AA, Amin N, Aulchenko YS, Bakalkin G, Bakker SJ, Balkau B, Beulens JW, Bilbao A, de Boer RA, Beury D, Bots ML, Breetvelt EJ, Cauchi S, Cavalcanti-Proença C, Chambers JC, Clarke TK, Dahmen N, de Geus EJ, Dick D, Ducci F, Easton A, Edenberg HJ, Esko T, Fernández-Medarde A, Foroud T, Freimer NB, Girault JA, Grobbee DE, Guarrera S, Gudbjartsson DF, Hartikainen AL, Heath AC, Hesselbrock V, Hofman A, Hottenga JJ, Isohanni MK, Kaprio J, Khaw KT, Kuehnel B, Laitinen J, Lobbens S, Luan J, Mangino M, Maroteaux M, Matullo G, McCarthy MI, Mueller C, Navis G, Numans ME, Núñez A, Nyholt DR, Onland-Moret CN, Oostra BA, O'Reilly PF, Palkovits M, Penninx BW, Polidoro S, Pouta A, Prokopenko I, Ricceri F, Santos E, Smit JH, Soranzo N, Song K, Sovio U, Stumvoll M, Surakk I, Thorgeirsson TE, Thorsteinsdottir U, Troakes C, Tyrfingsson T, Tönjes A, Uiterwaal CS, Uitterlinden AG, van der Harst P, van der Schouw YT, Staehlin O, Vogelzangs N, Vollenweider P, Waeber G, Wareham NJ, Waterworth DM, Whitfield JB, Wichmann EH, Willemsen G, Witteman JC, Yuan X, Zhai G, Zhao JH, Zhang W, Martin NG, Metspalu A, Doering A, Scott J, Spector TD, Loos RJ, Boomsma DI, Mooser V, Peltonen L, Stefansson K, van Duijn CM, Vineis P, Sommer WH, Kooner JS, Spanagel R, Heberlein UA, Jarvelin MR, Elliott P. Genome-wide association and genetic functional studies identify autism susceptibility candidate 2 gene (AUTS2) in the regulation of alcohol consumption. *Proc Natl Acad Sci USA* 2011;**108**:7119–24.
52. Zlojutro M, Manz N, Rangaswamy M, Xuei X, Flury-Wetherill L, Koller D, Bierut LJ, Goate A, Hesselbrock V, Kuperman S, Nurnberger Jr J, Rice JP, Schuckit MA, Foroud T, Edenberg HJ, Porjesz B, Almasy L. Genome-wide association study of theta band event-related oscillations identifies serotonin receptor gene HTR7 influencing risk of alcohol dependence. *Am J Med Genet B Neuropsychiatr Genet* 2011;**156B**(1):44–58.
53. Heath AC, Whitfield JB, Martin NG, Pergadia ML, Goate AM, Lind PA, McEvoy BP, Schrage AJ, Grant JD, Chou YL, Zhu R, Henders AK, Medland SE, Gordon SD, Nelson EC, Agrawal A, Nyholt DR, Bucholz KK, Madden PA, Montgomery GW. A quantitative-trait genome-wide association study of alcoholism risk in the community: findings and implications. *Biol Psychiatry* 2011;**70**:513–8.
54. Wang KS, Liu X, Zhang Q, Wu LY, Zeng M. Genome-wide association study identifies 5q21 and 9p24.1 (KDM4C) loci associated with alcohol withdrawal syndromes. *J Neural Transm* 2012;**119**:425–33.
55. Heberlein A, Muschler M, Frieling H, Behr M, Eberlein C, Wilhelm J, Gröschl M, Kornhuber J, Bleich S, Hillemacher T. Epigenetic down regulation of nerve growth factor during alcohol withdrawal. *Addict Biol* 2013;**18**:508–10.
56. Okubo T, Harada S, Higuchi S, Matsushita S. Genetic polymorphism of the CCK gene in patients with alcohol withdrawal syndromes. *Alcohol Clin Exp Res* 2000;**24**(4 Suppl.):2S–4S.
57. Gorwood P, Limosin F, Batel P, Hamon M, Adès J, Boni C. The A9 allele of the dopamine transporter gene is associated with delirium tremens and alcohol-withdrawal seizure. *Biol Psychiatry* 2003;**53**:85–92.
58. Köhnke M, Schick S, Lutz U, Köhnke A, Vonthein R, Kolb W, Batra A. The polymorphism GABABR1 T1974C[rs29230] of the GABAB receptor gene is not associated with the diagnosis of alcoholism or alcohol withdrawal seizures. *Addict Biol* 2006;**11**:152–6.
59. Huang MC, Schwandt ML, Chester JA, Kirchhoff AM, Kao CF, Liang T, Tapocik JD, Ramchandani VA, George DT, Hodgkinson CA, Goldman D, Heilig M. FKBP5 moderates alcohol withdrawal severity: human genetic association and functional validation in knockout mice. *Neuropsychopharmacol* 2014;**39**:2029–38.
60. Koehnke MD, Schick S, Lutz U, Willecke M, Koehnke AM, Kolb W, Gaertner I. Severity of alcohol withdrawal syndromes and the T1128C polymorphism of the neuropeptide Y gene. *J Neural Transm* 2002;**109**:1423–9.
61. Hillemacher T, Weinland C, Lenz B, Kraus T, Heberlein A, Glahn A, Muschler MA, Bleich S, Kornhuber J, Frieling H. DNA methylation of the LEP gene is associated with craving during alcohol withdrawal. *Psychoneuroendocrinol* 2015;**51**:371–7.
62. Wang SC, Tsou HH, Chung RH, Chang YS, Fang CP, Chen CH, Ho IK, Kuo HW, Liu SC, Shih YH, Wu HY, Huang BH, Lin KM, Chen AC, Hsiao CF, Liu YL. The association of genetic polymorphisms in the κ-opioid receptor 1 gene with body weight, alcohol use, and withdrawal symptoms in patients with methadone maintenance. *J Clin Psychopharmacol* 2014;**34**:205–11.
63. Rujescu D, Soyka M, Dahmen N, Preuss U, Hartmann AM, Giegling I, Koller G, Bondy B, Möller HJ, Szegedi A. GRIN1 locus may modify the susceptibility to seizures during alcohol withdrawal. *Am J Med Genet B Neuropsychiatr Genet* 2005;**133B**(1):85–7.
64. Ruggeri B, Braconi S, Cannella N, Kallupi M, Soverchia L, Ciccocioppo R, Ubaldi M. Neuropeptide S receptor gene expression in alcohol withdrawal and protracted abstinence in postdependent rats. *Alcohol Clin Exp Res* 2010;**34**:90–7.
65. Grzywacz A, Jasiewicz A, Małecka I, Suchanecka A, Grochans E, Karakiewicz B, Samochowiec A, Bieńkowski P, Samochowiec J. Influence of DRD2 and ANKK1 polymorphisms on the manifestation of withdrawal syndrome symptoms in alcohol addiction. *Pharmacol Rep* 2012;**64**:1126–34.
66. Sander T, Harms H, Rommelspacher H, Hoehe M, Schmidt LG. Possible allelic association of a tyrosine hydroxylase polymorphism with vulnerability to alcohol-withdrawal delirium. *Psychiatr Genet* 1998;**8**:13–7.
67. Hack LM, Kalsi G, Aliev F, Kuo PH, Prescott CA, Patterson DG, Walsh D, Dick DM, Riley BP, Kendler KS. Limited associations of dopamine system genes with alcohol dependence and related traits in the Irish Affected Sib Pair Study of Alcohol Dependence (IASPSAD). *Alcohol Clin Exp Res* 2011;**35**:376–85.
68. Biermann T, Reulbach U, Lenz B, Frieling H, Muschler M, Hillemacher T, Kornhuber J, Bleich S. N-methyl-D-aspartate 2b receptor subtype (NR2B) promoter methylation in patients during alcohol withdrawal. *J Neural Transm* 2009;**116**:615–22.
69. McPherson JD, Smith M, Wagner C, Wasmuth JJ, Hoog J-O. Mapping of the class II alcohol dehydrogenase gene locus to 4q22 (Abstract). *Cytogenet Cell Genet* 1989;**51**:1043.
70. Reich T, Edenberg HJ, Goate A, Williams JT, Rice JP, Van Eerdewegh P, Foroud T, Hesselbrock V, Schuckit MA, Bucholz K, Porjesz B, Li TK. 11 others. Genome-wide search for genes affecting the risk for alcohol dependence. *Am J Med Genet* 1998;**81**:207–15.
71. Long JC, Knowler WC, Hanson RL, Robin RW, Urbanek M, Moore E, Bennett PH, Goldman D. Evidence for genetic linkage to alcohol dependence on chromosomes 4 and 11 from an autosome-wide scan in an American Indian population. *Am J Med Genet* 1998;**81**:216–21.
72. Osier MV, Pakstis AJ, Soodyall H, Comas D, Goldman D, Odunsi A, Okonofua F, Parnas J, Schulz LO, Bertranpetit J, Bonne-Tamir B, Lu R-B, Kidd JR, Kidd KK. A global perspective on genetic variation at the ADH genes reveals unusual patterns of linkage disequilibrium and diversity. *Am J Hum Genet* 2002;**71**:84–99.

73. Smith M, Hopkinson DA, Harris H. Studies on the subunit structure and molecular size of the human dehydrogenase isozymes determined by the different loci, ADH(1), ADH(2), and ADH(3). *Ann Hum Genet* 1973;**36**:401–14.
74. Wall TL, Carr LG, Ehlers CL. Protective association of genetic variation in alcohol dehydrogenase with alcohol dependence in Native American Mission Indians. *Am J Psychiatry* 2003;**160**:41–6.
75. Edenberg HJ, Xuei X, Chen HJ, Tian H, Wetherill LF, Dick DM, Almasy L, Bierut L, Bucholz KK, Goate A, Hesselbrock V, Kuperman S, Nurnberger J, Porjesz B, Rice J, Schuckit M, Tischfield J, Begleiter H, Foroud T. Association of alcohol dehydrogenase genes with alcohol dependence: a comprehensive analysis. *Hum Mol Genet* 2006;**15**:1539–49.
76. Luo X, Kranzler HR, Zuo L, Wang S, Schork NJ, Gelernter J. Diplotype trend regression analysis of the ADH gene cluster and the ALDH2 gene: multiple significant associations with alcohol dependence. *Am J Hum Genet* 2006;**78**:973–87.
77. Ehlers CL, Montane-Jaime K, Moore S, Shafe S, Joseph R, Carr LG. Association of the ADHIB*3 allele with alcohol-related phenotypes in Trinidad. *Alcohol Clin Exp Res* 2007;**31**:216–20.
78. Macgregor S, Lind PA, Bucholz KK, Hansell NK, Madden PA, Richter MM, Montgomery GW, Martin NG, Heath AC, Whitfield JB. Associations of ADH and ALDH2 gene variation with self report alcohol reactions, consumption and dependence: an integrated analysis. *Hum Mol Genet* 2009;**18**:580–93.
79. Takahashi Y, Fukushige S, Murotsu T, Matsubara K. Structure of human cholecystokinin gene and its chromosomal location. *Gene* 1986;**50**:353–60.
80. Modi WS, Bonner TI. Localization of the cannabanoid (sic) receptor locus using non-isotopic in situ hybridization. (Abstract). *Cytogenet Cell Genet* 1991;**58**:1915.
81. Ledent C, Valverde O, Cossu G, Petitet F, Aubert J-F, Beslot F, Bohme GA, Imperato A, Pedrazzini T, Roques BP, Vassart G, Fratta W, Parmentier M. Unresponsiveness to cannabinoids and reduced addictive effects of opiates in CB(1) receptor knockout mice. *Science* 1999;**283**:401–4.
82. Marsicano G, Goodenough S, Monory K, Hermann H, Eder M, Cannich A, Azad SC, Cascio MG, Gutierrez SO, van der Stelt M, Lopez-Rodriguez ML, Casanova E, Schutz G, Zieglgansberger W, Di Marzo V, Behl C, Lutz B. CB1 cannabinoid receptors and on-demand defense against excitotoxicity. *Science* 2003;**302**:84–8.
83. Giros B, El Mestikawy S, Godinot N, Zheng K, Han H, Yang-Feng T, Caron MG. Cloning, pharmacological characterization, and chromosome assignment of the human dopamine transporter. *Molec Pharm* 1992;**42**:383–90.
84. Vandenbergh DJ, Persico AM, Uhl GR. A human dopamine transporter cDNA predicts reduced glycosylation, displays a novel repetitive element and provides racially-dimorphic TaqI RFLPs. *Molec Brain Res* 1992;**15**:161–6.
85. Schmidt LG, Harms H, Kuhn S, Rommelspacher H, Sander T. Modification of alcohol withdrawal by the A9 allele of the dopamine transporter gene. *Am J Psychiatry* 1998;**155**:474–8.
86. Limosin F, Loze JY, Boni C, Fedeli LP, Hamon M, Rouillon F, Adès J, Gorwood P. The A9 allele of the dopamine transporter gene increases the risk of visual hallucinations during alcohol withdrawal in alcohol-dependent women. *Neurosci Lett* 2004;**362**:91–4.
87. Eubanks JH, Djabali M, Selleri L, Grandy DK, Civelli O, McElligott DL, Evans GA. Structure and linkage of the D2 dopamine receptor and neural cell adhesion molecule genes on human chromosome 11q23. *Genomics* 1992;**14**:1010–8.
88. Lucht MJ, Kuehn KU, Schroeder W, Armbruster J, Abraham G, Schattenberg A, Gaensicke M, Barnow S, Tretzel H, Herrmann FH, Freyberger HJ. Influence of the dopamine D2 receptor (DRD2) exon 8 genotype on efficacy of tiapride and clinical outcome of alcohol withdrawal. *Pharmacogenetics* 2001;**11**:647–53.
89. Schmidt LG, Sander T. Genetics of alcohol withdrawal. *Eur Psychiatry* 2000;**15**:135–9.
90. Neville MJ, Johnstone EC, Walton RT. Identification and characterization of ANKK1: a novel kinase gene closely linked to DRD2 on chromosome band 11q23.1. *Hum Mutat* 2004;**23**:540–5.
91. Coe IR, Griffiths M, Young JD, Baldwin SA, Cass CE. Assignment of the human equilibrative nucleoside transporter (hENT1) to 6p21.1-p21. 2. *Genomics* 2007;**45**:459–60.
92. Baughman G, Wiederrecht GJ, Campbell NF, Martin MM, Bourgeois S. FKBP51, a novel T-cell-specific immunophilin capable of calcineurin inhibition. *Mol Cell Biol* 1995;**15**:4395–402.
93. Bittiger H, Froestl W, Mickel SJ, Olpe H-R. GABA-B receptor antagonists: from synthesis to therapeutic applications. *Trends Pharm Sci* 1993;**14**:391–4.
94. Goei VL, Choi J, Ahn J, Bowlus CL, Raha-Chowdhury R, Gruen JR. Human gamma-aminobutyric acid B receptor gene: complementary DNA cloning, expression, chromosomal location, and genomic organization. *Biol Psychiat* 1998;**44**:659–66.
95. Humeniuk RE, White JM, Ong J. The effects of GABAB ligands on alcohol withdrawal in mice. *Pharmacol Biochem Behav* 1994;**49**:561–6.
96. Buck KJ, Metten P, Belknap JK, Crabbe JC. Quantitative trait loci involved in genetic predisposition to acute alcohol withdrawal in mice. *J Neurosci* 1997;**17**:3946–55.
97. Karp SJ, Masu M, Eki T, Ozawa K, Nakanishi S. Molecular cloning and chromosomal localization of the key subunit of the human N-methyl-D-aspartate receptor. *J Biol Chem* 1993;**268**:3728–33.
98. Kobilka BK, Frielle T, Collins S, Yang-Feng T, Kobilka TS, Francke U, Lefkowitz RJ, Caron MG. An intronless gene encoding a potential member of the family of receptors coupled to guanine nucleotide regulatory proteins. *Nature* 1987;**329**:75–9.
99. Jin H, Oksenberg D, Ashkenazi A, Peroutka SJ, Duncan AMV, Rozmahel R, Yang Y, Mengod G, Palacios JM, O'Dowd BF. Characterization of the human 5-hydroxytryptamine(1B) receptor. *J Biol Chem* 1992;**267**:5735–8.
100. Zifa E, Fillion G. 5-Hydroxytryptamine receptors. *Pharm Rev* 1992;**44**:401–58.
101. Friedman JM, Leibel RL, Siegel DS, Walsh J, Bahary N. Molecular mapping of the mouse ob mutation. *Genomics* 1991;**11**:1054–62.
102. Miller SG, De Vos P, Guerre-Millo M, Wong K, Hermann T, Staels B, Briggs MR, Auwerx J. The adipocyte specific transcription factor C/EBPalpha modulates human ob gene expression. *Proc Natl Acad Sci USA* 1996;**93**:5507–11.
103. Paracchini V, Pedotti P, Taioli E. Genetics of leptin and obesity: a HuGE review. *Am J Epidemiol* 2005;**162**:101–14.
104. Kalra SP. Central leptin insufficiency syndrome: an interactive etiology for obesity, metabolic and neural diseases and for designing new therapeutic interventions. *Peptides* 2008;**29**(1):127–38.
105. Kim JH, Choi JH. Pathophysiology and clinical characteristics of hypothalamic obesity in children and adolescents. *Ann Pediatr Endocrinol Metab* 2013;**18**:161–7.
106. Kiefer F, Jahn H, Kellner M, Naber D, Wiedemann K. Leptin as a possible modulator of craving for alcohol. *Arch Gen Psychiatry* 2001;**58**:509–10.
107. Hillemacher T, Bleich S, Frieling H, Schanze A, Wilhelm J, Sperling W, Kornhuber J, Kraus T. Evidence of an association of leptin serum levels and craving in alcohol dependence. *Psychoneuroendocrinology* 2007;**32**:87–90.
108. Kiefer F, Jahn H, Otte C, Demiralay C, Wolf K, Wiedemann K. et al. Increasing leptin precedes craving and relapse during pharmacological abstinence maintenance treatment of alcoholism. *J Psychiatr Res* 2005;**39**:545–51.
109. Lenz B, Schopp E, Muller CP, Bleich S, Hillemacher T, Kornhuber J. Association of V89L SRD5A2 polymorphism with craving and serum leptin levels in male alcohol addicts. *Psychopharmacology (Berl.)* 2012;**224**:421–9.

110. Goyette P, Sumner JS, Milos R, Duncan AMV, Rosenblatt DS, Matthews RG, Rozen R. Human methylenetetrahydrofolate reductase: isolation of cDNA, mapping and mutation identification. *Nature Genet* 1994;**7**:195–200.
111. Levi-Montalcini R. The nerve growth factor thirty-five years later. *Science* 1987;**237**:1154–62.
112. Yoon SJ, Roh S, Lee H, Lee JY, Lee BH, Kim YK, Kim DJ. Possible role of nerve growth factor in the pathogenesis of alcohol dependence. *Alcohol Clin Exp Res* 2006;**30**:1060–5.
113. Baker E, Hort YJ, Ball H, Sutherland GR, Shine J, Herzog H. Assignment of the human neuropeptide Y gene to chromosome 7p15.1 by nonisotopic in situ hybridization. *Genomics* 1995;**26**: 163–4.
114. Kauhanen J, Karvonen MK, Pesonen U, Koulu M, Tuomainen T-P, Uusitupa MIJ, Salonen JT. Neuropeptide Y polymorphism and alcohol consumption in middle-aged men. *Am J Med Genet* 2000;**93**:117–21.
115. Lappalainen J, Kranzler HR, Malison R, Price LH, Van Dyck C, Rosenheck RA, Cramer J, Southwick S, Charney D, Krystal J, Gelernter J. A functional neuropeptide Y leu7pro polymorphism associated with alcohol dependence in a large population sample from the United States. *Arch Gen Psychiat* 2002;**59**:825–31.
116. Jaiswal AK, Bell DW, Radjendirane V, Testa JR. Localization of human NQO1 gene to chromosome 16q22 and NQO2-6p25 and associated polymorphisms. *Pharmacogenetics* 1999;**9**:413–8.
117. Xu Y-L, Reinscheid RK, Huitron-Resendiz S, Clark SD, Wang Z, Lin SH, Brucher FA, Zeng J, Ly NK, Henriksen SJ, de Lecea L, Civelli O. Neuropeptide S: a neuropeptide promoting arousal and anxiolytic-like effects. *Neuron* 2004;**43**:487–97.
118. Badia-Elder NE, Henderson AN, Bertholomey ML, Dodge NC, Stewart RB. The effects of neuropeptide S on ethanol drinking and other related behaviors in alcohol-preferring and -nonpreferring rats. *Alcohol Clin Exp Res* 2008;**32**:1380–7.
119. Mandich P, Schito AM, Bellone E, Antonacci R, Finelli P, Rocchi M, Ajmar F. Mapping of the human NMDAR2B receptor subunit gene (GRIN2B) to chromosome 12p12. *Genomics* 1994;**22**:216–8.
120. Endele S, Rosenberger G, Geider K, Popp B, Tamer C, Stefanova I, Milh M, Kortum F, Fritsch A, Pientka FK, Hellenbroich Y, Kalscheuer VM. 16 others. Mutations in GRIN2A and GRIN2B encoding regulatory subunits of NMDA receptors cause variable neurodevelopmental phenotypes. *Nature Genet* 2010;**42**: 1021–6.
121. Nagy J, Horváth C, Farkas S, Kolok S, Szombathelyi Z. NR2B subunit selective NMDA antagonists inhibit neurotoxic effect of alcohol-withdrawal in primary cultures of rat cortical neurones. *Neurochem Int* 2004;**44**:17–23.
122. Tadic A, Dahmen N, Szegedi A, Rujescu D, Giegling I, Koller G, Anghelescu I, Fehr C, Klawe C, Preuss UW, Sander T, Toliat MR, Singer P, Bondy B, Soyka M. Polymorphisms in the NMDA subunit 2B are not associated with alcohol dependence and alcohol withdrawal-induced seizures and delirium tremens. *Eur Arch Psychiatry Clin Neurosci* 2005;**255**:129–35.
123. Yasuda K, Espinosa III R, Takeda J, Le Beau MM, Bell GI. Localization of the kappa opioid receptor gene to human chromosome band 8q11.2. *Genomics* 1994;**19**:596–7.
124. Simonin F, Gaveriaux-Ruff C, Befort K, Matthes H, Lannes B, Micheletti G, Mattei M-G, Charron G, Bloch B, Kieffer B. Kappa-opioid receptor in humans: cDNA and genomic cloning, chromosomal assignment, functional expression, pharmacology, and expression pattern in the central nervous system. *Proc Natl Acad Sci USA* 1995;**92**:7006–10.
125. Ramamoorthy S, Bauman AL, Moore KR, Han H, Yang-Feng T, Chang AS, Ganapathy V, Blakely RD. Antidepressant- and cocaine-sensitive human serotonin transporter: molecular cloning, expression, and chromosomal localization. *Proc Natl Acad Sci USA* 1993;**90**: 2542–6.
126. Feinn R, Nellissery M, Kranzler HR. Meta-analysis of the association of a functional serotonin transporter promoter polymorphism with alcohol dependence. *Am J Med Genet (Neuropsychiat Genet)* 2005;**133B**:79–84.
127. Craig SP, Buckle VJ, Lamouroux A, Mallet J, Craig I. Localization of the human tyrosine hydroxylase gene to 11p15: gene duplication and evolution of metabolic pathways. *Cytogenet Cell Genet* 1986;**42**:29–32.
128. Xue F, Kidd JR, Pakstis AJ, Castiglione CM, Mallet J, Kidd KK. Tyrosine hydroxylase maps to the short arm of chromosome 11 proximal to the insulin and HRAS1 loci. *Genomics* 1988;**2**:288–93.
129. Nelson EC, Heath AC, Bucholz KK, Madden PA, Fu Q, Knopik V, Lynskey MT, Lynskey MT, Whitfield JB, Statham DJ, Martin NG. Genetic epidemiology of alcohol-induced blackouts. *Arch Gen Psychiatry* 2004;**61**:257–63.
130. Grant JD, Agrawal A, Bucholz KK, Madden PA, Pergadia ML, Nelson EC, Lynskey MT, Todd RD, Todorov AA, Hansell NK, Whitfield JB, Martin NG, Heath AC. Alcohol consumption indices of genetic risk for alcohol dependence. *Biol Psychiatry* 2009;**66**: 795–800.
131. Cerretti DP, Copeland NG, Gilbert DJ, Jenkins NA, Kuefer MU, Valentine V, Shapiro DN, Cui X, Morris SW. The gene encoding LERK-7 (EPLG7, Epl7), a ligand for the Eph-related receptor tyrosine kinases, maps to human chromosome 5 at band q21 and to mouse chromosome 17. *Genomics* 1996;**35**:376–9.
132. Kozlosky CJ, VandenBos T, Park L, Cerretti DP, Carpenter MK. LERK-7: a ligand of the Eph-related kinases is developmentally regulated in the brain. *Cytokine* 1997;**9**:540–9.
133. Wilkinson DG. Multiple roles of EPH receptors and ephrins in neural development. *Nat Rev Neurosci* 2001;**2**:155–64.
134. Nagase T, Ishikawa K, Suyama M, Kikuno R, Miyajima N, Tanaka A, Kotani H, Nomura N, Ohara O. Prediction of the coding sequences of unidentified human genes. XI. The complete sequences of 100 new cDNA clones from brain which code for large proteins in vitro. *DNA Res* 1998;**5**:277–86.
135. Iyengar SK, Elston RC. The genetic basis of complex traits: rare variants or "common gene, common disease"? *Methods Mol Biol* 2007;**376**:71–84.
136. Edenberg HJ. Common and rare variants in alcohol dependence. *Biol Psychiatry* 2011;**70**:498–9.
137. Janssens AC, Moonesinghe R, Yang Q, Steyerberg EW, van Duijn CM, Khoury MJ. The impact of genotype frequencies on the clinical validity of genomic profiling for predicting common chronic diseases. *Genet Med* 2007;**9**:528–53.
138. Craddock N, Sklar P. Genetics of bipolar disorder: successful start to a long journey. *Trends Genet* 2009;**25**:99–105.
139. Kraft P, Yen YC, Stram DO, Morrison J, Gauderman WJ. Exploiting gene-environment interaction to detect genetic associations. *Hum Hered* 2007;**63**:111–9.
140. Thomas D. Methods for investigating gene-environment interactions in candidate pathway and genome-wide association studies. *Annu Rev Public Health* 2010;**31**:21–36.

CHAPTER

21

Alcohol and Epigenetic Modulations

Claudio D'Addario, PhD,**, Mauro Maccarrone, MS, PhD†,‡*

*Faculty of Bioscience and Technology for Food, Agriculture and Environment, University of Teramo, Teramo, Italy

**Department of Clinical Neuroscience, Center for Molecular Medicine, Karolinska Institutet, Stockholm, Sweden

†School of Medicine and Center of Integrated Research, Campus Bio-Medico University of Rome, Rome, Italy

‡European Center for Brain Research (CERC)/Santa Lucia Foundation, Rome, Italy

INTRODUCTION

Alcoholism is an etiologically and clinically heterogeneous disorder, in which compulsive alcohol seeking and use represent core symptoms. Exposure to alcohol is, of course, a necessary precondition; however, environment and heritability factors can also play a dramatic role in controlling the individual vulnerability to develop alcohol abuse. Recently, molecular research has paved the basis for understanding the interaction between alcohol exposure, environmental stress, and heritable factors in the development and progression of alcoholism. Indeed, if on one hand, genetic determinants play a pivotal role in innate vulnerability to develop alcohol abuse problems, on the other hand, it has been recognized that environment (i.e., stress, conditioning factors) and exposure to alcohol may activate epigenetic mechanisms, which are further able to influence disease progression.[1,2] It is well known that these mechanisms, including histone modifications (e.g., acetylation, phosphorylation, methylation, ubiquitination, and ADP-ribosylation) and DNA methylation can evoke transient changes in gene expression, involving chemical changes of chromatin and DNA, without affecting the actual DNA sequence of the organism. Thus, the gene expression is somewhat temporary (i.e., the DNA is not permanently altered). Although how long it actually lasts depends on the specific process. The groundbreaking discovery of epigenetic regulation mechanisms is opening new perspectives in the understanding of psychiatric diseases, including addiction, and the impact that environment has on their progression.[3] Understanding the nature of genetic and epigenetic (gene × environment) interaction in regulating the individual risk of becoming an alcohol abuser, is therefore of relevance for the development of preventive strategies, or new pharmacotherapeutic remedies. In line with this, recent findings have already pointed to enzymes involved in epigenetic mechanisms, such as histone deacetylases, histone acetyltransferases, and DNA methyltransferases, as possible novel therapeutic targets for the treatment of alcoholism.

Several genes have been found by genetic studies to be of relevance in alcohol abuse disorders, in both humans and animal models.[4–6] Additionally, alterations in gene expression evoked by exposure to ethanol (acutely and chronically) have been observed in specific neuronal circuits associated with the development of tolerance and dependence.[7] Epigenetic mechanisms are known to be able to regulate these changes in gene expression.[8] Thus, the positive- and negative-affective states in alcoholism are characterized by the development of different behaviors, such as withdrawal and relapse symptoms, to which chromatin remodeling in the neuronal circuits might take part.[9]

Specific conformational state of chromatin can be changed through enzyme-mediated covalent modifications of DNA and structural chromatin proteins (i.e., histones).[10] Epigenetic marks include not only DNA methylation and histone modifications, as described in the aforementioned section, but also noncoding RNAs (miRNAs), which are now considered to be epigenetic means for regulating gene expression.

Molecular Aspects of Alcohol and Nutrition. http://dx.doi.org/10.1016/B978-0-12-800773-0.00021-5

EPIGENETICS

The concept of epigenetics was introduced in 1942 by Waddington, as "the branch of biology which studies the causal interactions between genes and their products, which bring the phenotype into being."[11]

Identical DNA sequences might result in different gene expression profiles through epigenetic mechanisms, filling the gap between genotype and phenotype.

Reversible heritable epigenetic mechanisms, such as DNA methylation and posttranslational modifications of histone tails (acetylation, methylation, phosphorylation, etc.), remodel chromatin structure and open up the DNA template, making it accessible to various transcription factors, coactivators and/or corepressors, thus regulating gene expression.[12]

The interaction among these events, together with the regulation by miRNAs, contribute to the epigenetic status of the cells.

Methylation of DNA

DNA methylation, the most widely studied epigenetic mark, is a covalent modification that occurs in mammals through the addition of a $-CH_3$ group at cytosines, primarily in CpG dinucleotides (CpG sites), leading to the formation of 5-methyl cytosine (5mC). CpG sites are usually rare (~1%) in mammalian genomes, and partly clustered into CpG islands.[13] The latter are regions of the genome with at least 200 bp, with a GC percentage that is higher than 50%, and with a CpG ratio between observed/expected residues, which are at least equal to 0.6.

Sixty percent of human genes have CpG islands in the promoter region or first exon, and DNA methylation in promoter regions is often associated with transcriptional silencing[14] achieved by repressing the binding of transcription factors, or by recruiting methyl-DNA binding proteins, like MeCP2. The latter, in turn, recruits histone-modifying enzymes that induce the formation of compact heterochromatin.[15] Enzymes known as DNA methyltransferases (DNMTs) catalyze the methylation reaction,[14] a process reversed by demethylases (DDM). DNMT1 is the "maintenance" DNMT able to regenerate the methylcytosine marks on the newly synthesized complementary DNA strand arising from DNA replication.[16] DNMT3a and DNMT3b add methyl groups *de novo*.[17]

It should be also noted that DNA methylation does not occur exclusively at CpG islands, and that there are CpG sites in regulatory regions outside the promoters. Tissue-specific DNA methylation has been found 1–2 kb downstream or upstream, at CpG island "shores," and are strongly related to gene expression inactivation.[18]

Finally, hydroxylation of methylcytosine leading to 5-hydroxymethylcytosine (5hmC) has been recognized as a marker for gene activity, that counteracts the role of transcriptional repressors targeting 5mC.[19] A role for 5hmC has also been proposed as an intermediate in DNA demethylation.[20] 5hmC is present in mammalian DNA at physiologically relevant levels, and in a tissue-specific manner.[21] Ten–eleven translocation 1 (TET) mammalian enzymes (TET1, TET2, and TET3) have been identified as 5mC dioxygenases responsible for catalyzing the conversion of 5mC–5hmC[22] (Figure 21.1 for a summary).

Histone Modifications

Genomic DNA is packaged into a highly compact structure to form chromatin, which is made up of nucleosomes consisting of short stretches of DNA (147 bp), wrapped around histone octamers (two H3, H4, H2A, and H2B).[23] The latter are joined together by linker DNA and histone H1, interacting with the nucleosome core and the linker DNA (Figure 21.2).

Furthermore, there are histone variants of different types (e.g., H3.1, H3.2, and H3.3; H2A1-6 and H2A.7), some of which are associated with the persistence of distinct states of the active gene.[24]

Regulation of chromatin structure and transcription is driven by posttranslational modifications, primarily in the N-terminal tails of histone proteins,[25] including acetylation at lysine, methylation at lysine and arginine, phosphorylation at serine and threonine, ubiquitination, ADP addition, and ribosylation at lysine.[26]

Acetylation, associated with transcriptional activation,[27] occurs mainly at different positions of lysine (K) residues on histone H3 (K4, K9, K14, and K28) and histone H4 (K5, K8, K12, and K16).[25] Acetylation and deacetylation depend on the balance between histone acetyltransferases (HATs),[28] comprising of five families (GNATs, MYST, p300/CBP Transcription factor HATs, and nuclear hormone-related HATs), and histone deacetylases (HDACs)[29] that are divided in to four classes (class I–IV). Each family or class has additional members. Instead, histone methylation, depending on the sites of the modification, can either be activating or inhibiting of gene expression.[30] Methylations of H3K9, H3K27, and H4K20 have been associated with gene silencing, whereas H3K4, K3–K36, and H3-K79 methylation leads to gene induction.[31] Histone K can be mono-, di-, or trimethylated. The trimethylation of K has been considered to be involved in longterm epigenetic memory. Histone methylations are mediated by histone methyltransferases, and the methyl group is donated by S-adenosyl-menthionine (SAM). For a long time histone methylation has been considered as a permanent and irreversible epigenetic mark, responsible, in concert with DNA methylation for chromatin remodelling.[32] However, it is now clear that there are enzymes, which can demethylate the methylated histone K residues as well as methylated arginines, via amine oxidation, hydroxylation, or deamination.[33]

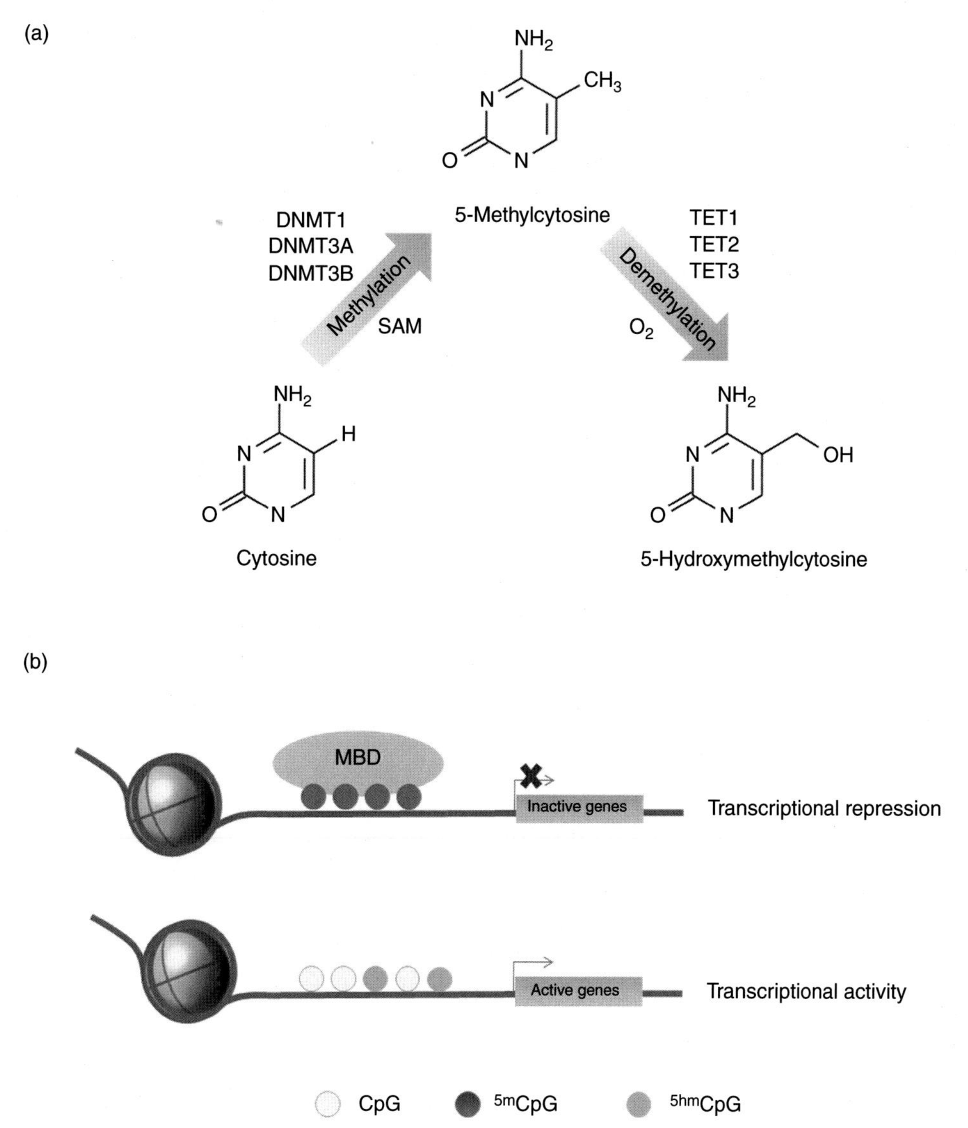

FIGURE 21.1 Schematic representation of genomic DNA methylation. (a) Conversion of cytosine in 5-methylcytosine by DNMTs, and demethylation by TET to produce 5-hydroxymethylcytosine. (b) Role of DNA methylation at CpG sites in transcriptional repression and activation. Methyl-CpG-binding domain (MBD) binds to DNA-containing methylated CpGs.

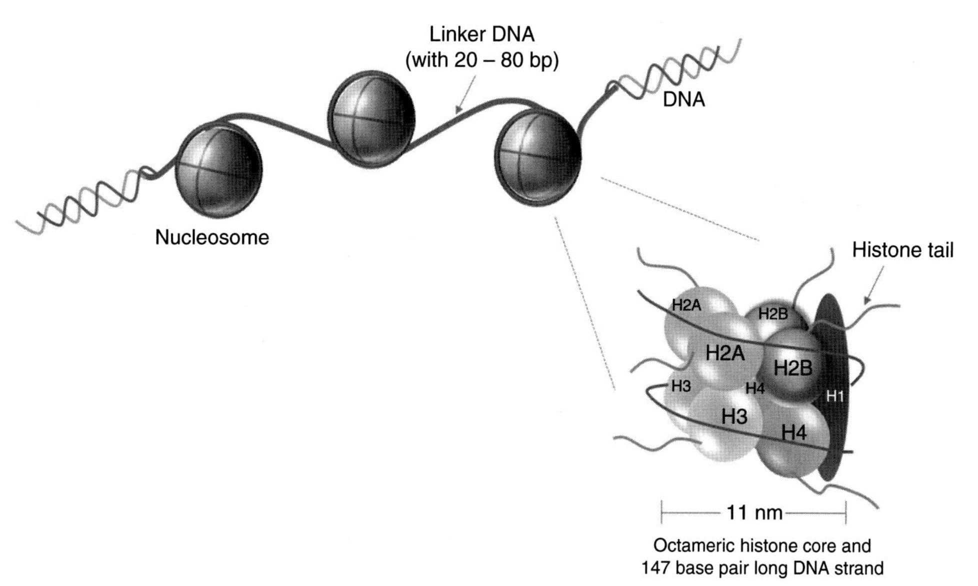

FIGURE 21.2 Schematic diagram of nucleosomes. Each comprising 147 bp of DNA wrapped around an octamer of histones (the core nucleosome), made of two molecules each of H2A, H2B, H3, and H4. DNA linker connects nucleosomes.

The different combinations of histone modification patterns support a "histone code" hypothesis,[34] referring to an epigenetic system that mediates distinct downstream events in eukaryotic genomes (Figure 21.3).

The scenario can be even more complex when considering that the same K residue (e.g., K4 or K9) might be both acetylated and methylated, or that a subtype of histone variants (e.g., H3.1, H3.2, and H3.3) might be modified differently.

Figure 21.4 summarizes the epigenetic markers of open and condensed chromatin.

microRNAs

Epigenetic regulators have been recently extended to microRNAs (miRNAs), which are able to alter the transcriptional potential of a gene without changing the DNA sequence. miRNAs are 21–23 nucleotides long, single-stranded RNA molecules, encoded by genes and transcribed, but not translated, into proteins (noncoding RNA). The discovery of miRNAs and their profound effect in controlling gene expression is revolutionizing our understanding of gene regulation.[35] Their binding to miRNA-recognition elements in target genes, generally results in either suppression of translation or degradation of mRNAs. These miRNAs are highly abundant in the brain and they play important roles in multiple biological processes, such as, neuronal differentiation,[36] brain development,[37] synapse formation and plasticity,[38] and neurodegeneration.[39] Besides, miRNAs also appear to mediate the cellular adaptations induced by exposure to a number of drugs of abuse, including alcohol.[40–43] See Figure 21.5 for details on miRNA biogenesis and alcohol influence.

ALCOHOL EFFECTS ON EPIGENETIC MECHANISMS

Ethanol and/or its metabolites (acetaldehyde, acetate, acetyl-CoA, and reactive-oxygen species)[44] can induce epigenetic changes through several mechanisms, including alterations in the activity of epigenetic enzymes, availability of substrates for histone acetylation, or DNA and histone methylation, as well as by influencing miRNA production. *For alcohol effects on DNA methylation please refer "Chapter 23. Alcohol Metabolism and Epigenetic Methylation" by Professor F.C. Zhou.*

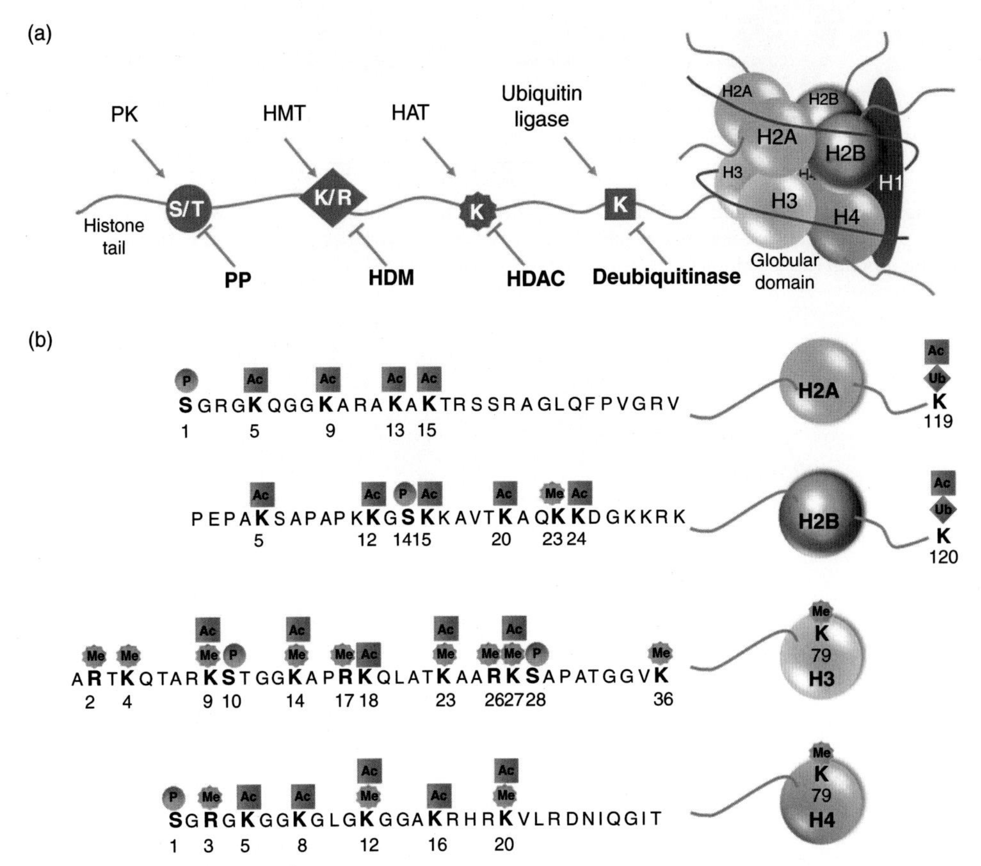

FIGURE 21.3 **Schematics.** (a) The enzymes responsible for adding and removing marks on histone tails. (b) The amino-terminal tails of core histones, with indication of the amino acid position and the posttranslational modification (ac, acetylated; HAT, histone acetyltransferase; HDAC, histone deacetylase; HDM, histone demethylase; HMT, histone methyltransferase; me, methylated; ph, phosphorylated; ub, ubiquitinilation).

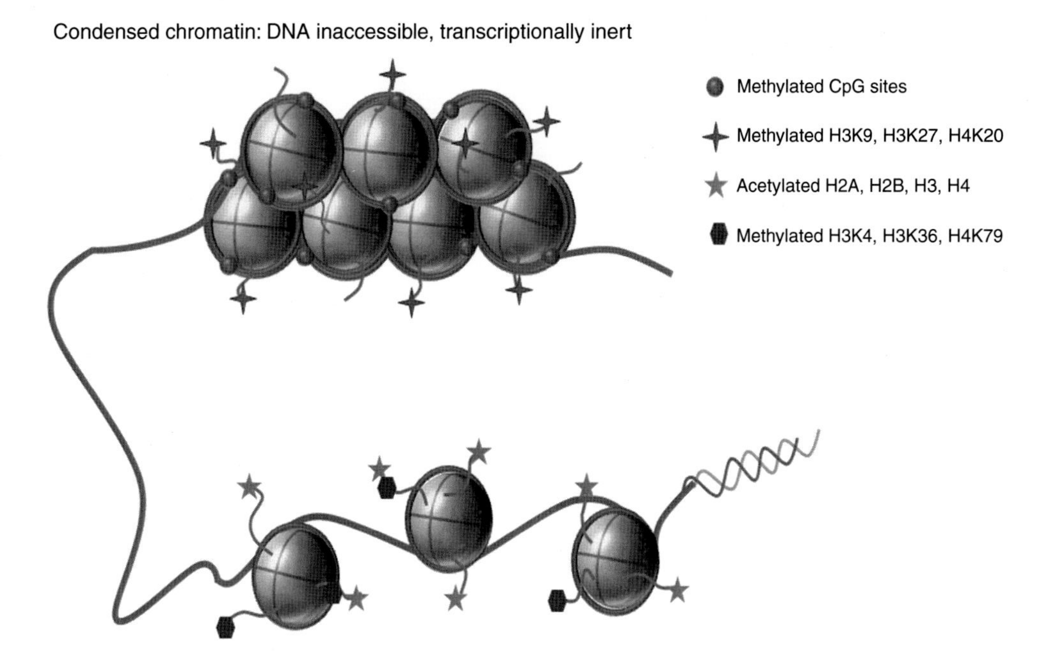

FIGURE 21.4 **Epigenetic marks of open and condensed chromatin.**

Alcohol Effects on Histone Modifications

Different studies indicate that histone modifications are involved in alcohol-related events,[9,45] and many research efforts are trying to clarify the role of individual modifications and/or combinatorial effects of those modifications. Of great interest in this frame, is the study of how these changes at histone tails are established, recognized, and inherited.

Histone Acetylation

Most evidence on histone modifications induced by alcohol has focused on histone acetylation, often by analyzing the activities of the enzymes responsible for adding (HAT) or removing (HDAC) acetyl groups.

HDAC inhibitors are effective on different alcohol-related behaviors, including withdrawal-related anxiety,[9] locomotor sensitization,[46] alcohol consumption,[47] conditioned-place aversion,[48] and rapid tolerance.[49] For alcohol dependence, Pandey et al.[9] proposed a chromatin remodeling induced by HDAC in the amygdala of rats. Acute ethanol increased H3K9 and H4K8 acetylation, whereas during withdrawal, after chronic alcohol, a decrease in this acetylation was observed, associated with anxiety-like behaviors. Moreover, they also showed that these anxiety-like behaviors could be reversed by treatment with the HDAC inhibitor trichostatin A (TSA).[9] TSA treatment was also shown to reverse rapid tolerance to anxiolytic effects, induced by alcohol.[49] Moreover, TSA increased alcohol consumption in mice having continuous access to both water and alcohol.[47] Sodium butyrate, another HDAC inhibitor, was able to selectively alter some alcohol-related behaviors (e.g., enhanced ethanol-induced locomotor sensitization), without any effect on others (e.g., ethanol tolerance or withdrawal).[46] Another study found, in the central and medial nucleus of amygdale, that alcohol-preferring rats innately display higher nuclear HDAC activity and, among the different HDAC isoforms, higher protein levels of HDAC2, as well as lower acetylation of H3K9, but not of H3K14, when compared with nonpreferring rats.[50] Other studies focused on HAT activity. For instance, rats fed with alcohol through intragastric tube, showed increased acetylation of H3K18[51] and H3K9, thus enhancing the activity of a HAT, called p300, in the liver.[52] Incidentally, ethanol metabolism induces an increase in acetyl-CoA, which is used in histone acetylation by HATs.[53]

It is known that acetylation of different H3 and H4 residues influences gene expression at different time points,[54] and, when it is related to alcohol, impacts on many genes, such as cyclic-AMP responsive element binding protein (CBP), neuropeptide Y (*NPY)*,[9] *FosB*[48] in selected animal brain regions, as well as *NR2B* in primary cortical neurons,[55] and ADH1 in the liver.[56] Thus, histone acetylation is tissue-, brain region-, and cell type-specific. Indeed, following a single dose of ethanol into the stomach, it increases in the liver, lungs, and testes, but not in other tissues, like whole brains of rats.[57] In the brain, alteration in H3 or H4, acetylation was observed in central and medial, but not basolateral, nuclei of the amygdala,[9,49] and was specific for neurons.[49]

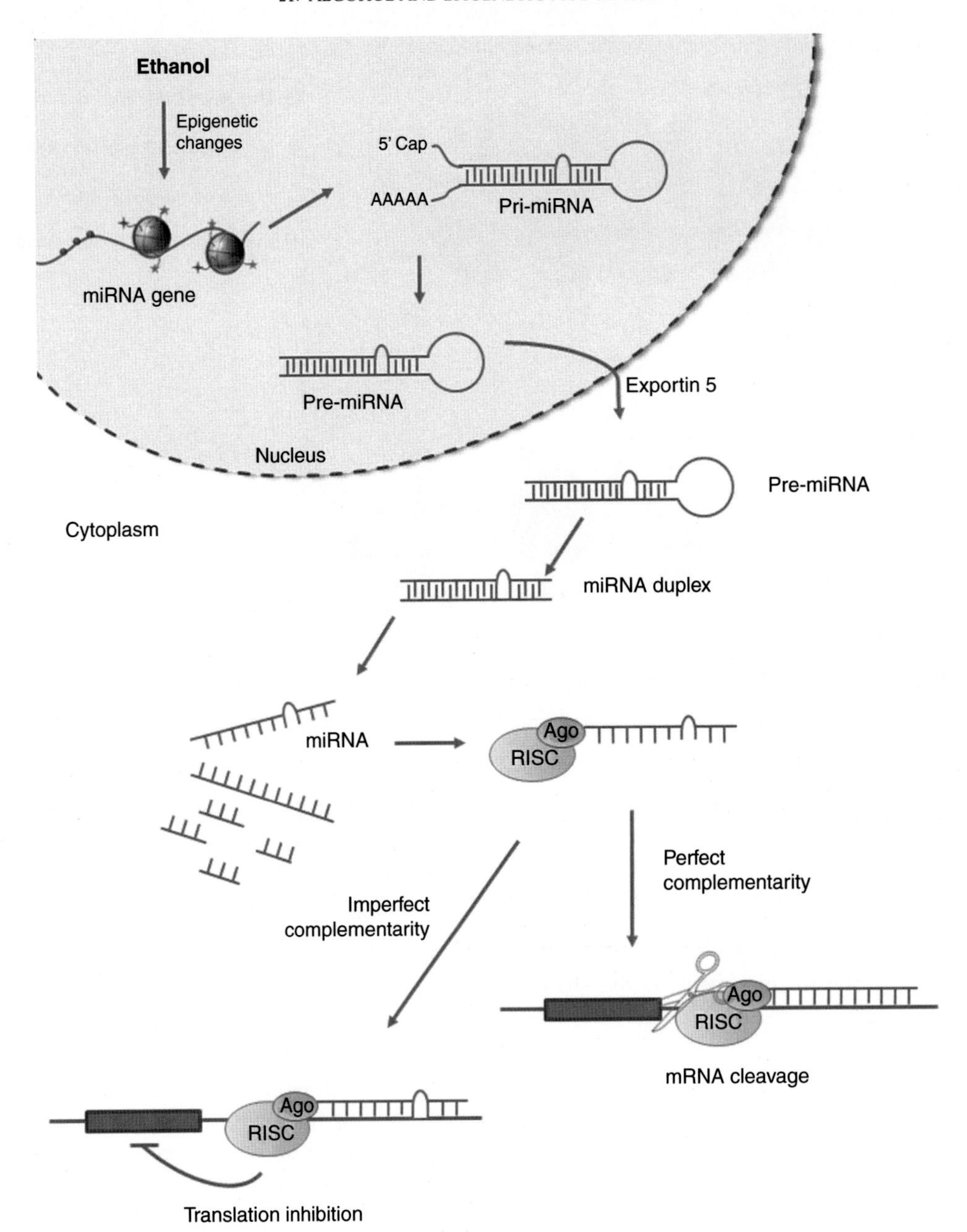

FIGURE 21.5 **Possible alcohol effects on miRNAs pathways by evoking epigenetic changes at miRNA genes.** Within the nucleus, primary miRNA (pri-miRNA) becomes precursor miRNA (pre-miRNA), which is then exported to the cytoplasm, where the Dicer protein cleaves it into mature miRNA. miRNAs are thus loaded on a silencing complex, called RNA-induced silencing complex (RISC), and including also an Argonaut protein (Ago). Then, they target mRNAs by selective base-pairing, primarily in the 3′-UTR, and either inhibit their expression, or speed up their degradation.

Moreover, alcohol effects on histone acetylation might be different also depending on species, genotype, age, dose, route of administration, and duration of exposure. Chronic ethanol vapor increases global and gene-specific histone acetylation in the ventral midbrain, during withdrawal, that peaked around 10 h postethanol.[58] Adolescent rats show more changes in histone acetylation, following intermittent alcohol exposure, when compared to adult rats.[48,59] In addition, time-course is also crucial, since histone acetylation measured 24 h after the last of repeated alcohol injections was increased in some brain areas (e.g., frontal cortex and nucleus accumbens), decreased in others (e.g., striatum), and unchanged in yet others (e.g., hippocampus).[59] Rat hepatocytes exposed to ethanol exhibited a maximal and selective increase in H3K9ac levels after 24 h treatment with the highest ethanol concentration tested.[60] Acetylation of other H3 lysines (i.e., K14, K18, and K23) was not affected by ethanol.[56,57,60]

A link has been suggested between H3 acetylation and phosphorylation,[61] since some HATs have preferences for phosphorylated histone H3.[62] Alcohol evokes H3 phosphorylation at Ser10 and Ser28 mediated by p38 MAPK in primary culture of rat hepatocytes.[63]

One mechanism by which histone acetylation might promote transcriptional activation is via activation of CREB, that recruits CBP with intrinsic HAT.[2] In alcoholics, the downregulation of CREB and CBP genes was observed,[64] as well as the reduction in pCREB and CBP levels in the amygdala of rats undergoing withdrawal after chronic ethanol exposure,[9] and again in pCREB in the rat cerebellum, following chronic ethanol.[65] Moreover, alcohol-induced neurodegeneration was associated with decreased CREB transcription.[66]

Histone Methylation

As yet, histone methylation changes by alcohol have been less explored, and some studies have evaluated exposure to ethanol of different cell types. Alcohol exposure induced reductions in H3K27me3 and H3K4me3 at promoters of genes involved in neural precursor cell identity and differentiation.[67] It was also observed in rat hepatocytes that ethanol caused decreased and increased methylation of H3K9 and H3K4, respectively.[68] In the latter study, histone changes were associated with genes expression regulation, supporting an epigenetic regulation by alcohol.

However, this is not always the case. Indeed, others observed alcohol-induced increases in H3K4me3, a mark of actively transcribed genes, in alcoholics at the global and gene-specific levels in the brain cortex,[64] and either increased or decreased levels in specific genes promoters in the hippocampus.[69] These alterations did not correlate with differences in gene expression. These findings are in line with another report on the effects of ethanol and its metabolite acetaldehyde on various chromatin marks, and the transcription of prodynorphin (PDYN) gene in a human neuroblastoma cell line. This study showed increased H3K4me3 after 72 h of ethanol exposure that did not result in initiation of PDYN transcription, but kept the gene in a poised state for later reactivation.[70] Later on, the same authors also reported a decrease of the repressive mark H3K27me3, and an upregulation of PDYN mRNA in rats treated for 1 day with alcohol.[71]

Another study focused on ethanol effects on HMT, showing that alcohol exposure during postnatal day 7, comparable to the third trimester of human pregnancy, evoked neuronal cell loss in mice, partially through enhanced activity of lysine dimethyltransferase G9a, and increased levels of H3K9me2 and H3K27me2.[72]

In Figure 21.6, remodeling of chromatin that depends on the degree of alcohol consumption or on the development of withdrawal is reported.

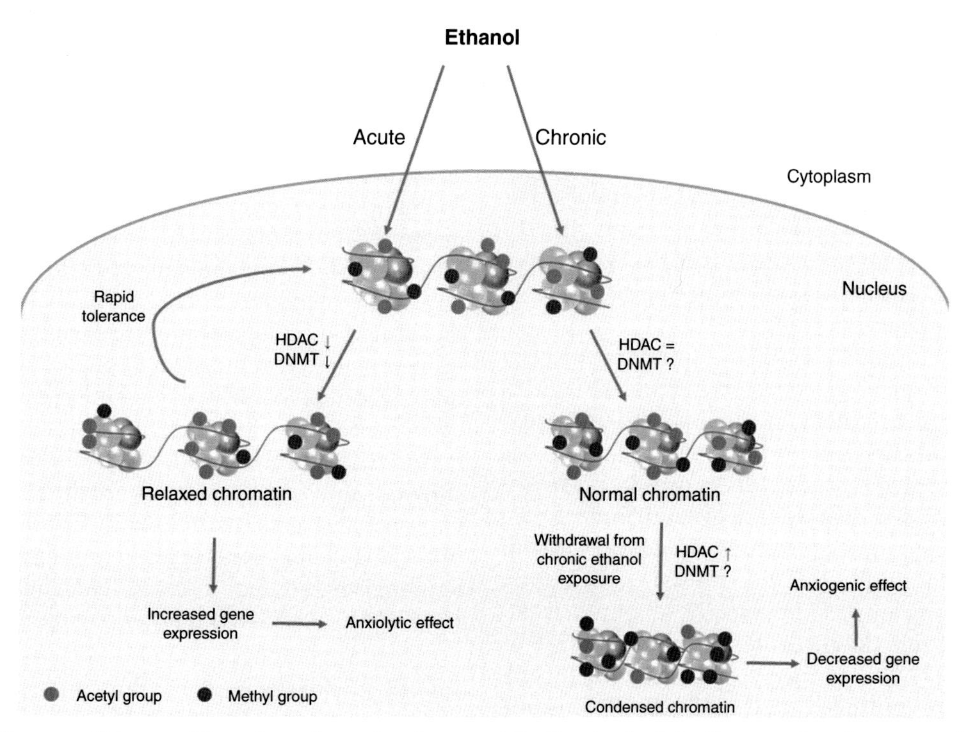

FIGURE 21.6 Diagram of the possible epigenetic mechanisms contributing to alcohol effects, as already suggested by Pandey et al.,[9] and proposed by Starkman et al.[7] Acute ethanol exposure can inhibit both histone deacetylase (HDAC) and DNA methyltransferase (DNMT) activities, resulting in relaxed chromatin structure and increased gene expression. Chronic ethanol exposure evokes neuroadaptations that do not induce relevant changes in the histone code. However, withdrawal after chronic ethanol exposure increases HDAC activity, and chromatin might become more condensed and, thus, less accessible.

Alcohol Effects on miRNA Expression

Alcohol induces alterations in miRNA expression in different cells and tissues: neuronal cells from rodents,[40] human cells,[73,74] human,[43,75] rat,[41,76,77] and mouse[78,79] brain samples, and developing zebrafish (where they were correlated with neurobehavioral and skeletal abnormalities).[80,81] Downregulation of several miRNAs (miR-9, miR-21, miR-153, and miR-335) was evoked by alcohol in fetal cultures of mouse cerebral cortical neuroepithelium.[40] Moreover, alcohol upregulates miR-9 expression in rat's brain, resulting in downregulation of BK channel variants, with high sensitivity to alcohol.[41] Using microarray, it has been reported that 35 human miRNAs are upregulated in alcoholic brain samples.[43] Among these, miR-203 is the one with most mRNA targets.[43] The expression of six of these 35 miRNAs (miR-7, miR-153, miR-152, miR-15B, miR-203, and miR-144) has been also studied upon exposure to ethanol in three human cell lines, HEK293T, SHSY5Y, and 1321 N1. Chronic alcohol exposure followed by 5 days of withdrawal induced the upregulation of several miRNAs in each of these cell lines, similarly to expression changes identified in postmortem human brain.[73] Others showed selective changes evoked by alcohol in miRNAs expression in SHSY5Y cells, where 28 miRNAs were altered by a long-term exposure to a low dose of ethanol (0.5% (v/v) for 72 h), and 13 miRNAs by a shortterm exposure of a high dose (2.5% (v/v) for 4 h).[74] Among these miRNA, the authors showed that miR-497 regulates expression of the antiapoptotic factor BCL2, and of cyclin D2, having a crucial role in neurogenesis maintenance, whereas miR-302b regulates expression of cyclin D2. Different miRNAs were also upregulated in fetal mouse brains after prenatal ethanol exposure, where an association between miR-10 and ethanol-induced teratogenesis, and also a negative correlation of miR-10a expression with that of its target gene Hoxa1 was observed.[78]

Moreover, approximately 41 miRNAs and 165 mRNAs in the rat medial prefrontal cortex (mPFC) were significantly altered after a history of alcohol dependence.[76] Among these, miR-206, upregulated upon alcohol dependence, was deeply studied because it represses BDNF expression, and was found to contribute to escalated alcohol consumption in the case of a history of dependence.[77] Among the multiple miRNAs aberrantly expressed in the nucleus accumbens (NAc) of rats treated with alcohol, the downregulation of miR-382 was found to target D1 dopamine receptor *in vivo* (NAc), as well as *in vitro* (cultured neuronal cells).[82] Chronic alcohol induces miR-155 in mouse cerebellum, thus regulating proinflammatory cytokines such as tumor necrosis factor-α (TNF-α) and monocyte chemotactic protein-1 (MCP1).[79]

ALCOHOL METABOLISM INFLUENCES NUTRITIONAL STATUS VIA EPIGENETIC MECHANISMS

Alcohol is known to influence nutritional status with different effects, also at the metabolic level in relation to the drinking patterns. Moderate alcohol intake seems to have no effect on body composition or metabolism,[83] and ethanol can be used as energy source.[84] However, in heavy drinkers, alcohol represents a risk factor that leads to many nutritional disturbances and deficiencies. Nutrients can dramatically affect gene expression, and alcohol-induced nutrient imbalance may be a major contributor to alterations in gene expression. Again, these changes could be driven by epigenetic mechanisms evoked especially during alcohol metabolism in several manners, such as via SAM deficiency, increased NADH/NAD$^+$ ratio, and acetate formation. Here, we focus on how nutrient disturbances during alcohol metabolism can impact histone modifications. *For the effects on DNA methylation refer "Chapter 23. Alcohol Metabolism and Epigenetic Methylation" by Professor F.C. Zhou.*

SAM Deficiencies Induced by Alcohol Affected Histone Methylation

Mammals cannot synthesize folate, choline, or methionine, whose dietary intake is essential for normal metabolic homeostasis. Methionine synthase depends on vitamin B12 and uses methyl-5,6,7,8-tetrahydrofolate for transmethylation. Methionine synthase is inhibited by acetaldehyde, the first product of alcohol metabolism in liver and brain. It was observed in rats that diets deficient in B vitamins (e.g., folic acid and choline) increased consumption of ethanol, whereas vitamin-enriched diets decreased it.[85] Folates and other B vitamins are critical for one-carbon metabolism and the synthesis of SAM, which is essential in numerous cellular processes, and is the principal biological methyl donor with a central role in the epigenetic regulation of genes.[86] Alcohol effects on one-carbon metabolism can result in the lack of availability of methionine, the endogenous precursor of SAM, and hence in hepatic SAM deficiency.[87,88] A few studies already showed how SAM treatment might affect histone methylation[51] and gene expression.[89] Chronic administration of SAM, together with ethanol, intragastrically to rats attenuated the ethanol-induced liver injury, and increased H3K27Me3, which is a marker of gene repression.[51]

Alcohol Metabolism Evokes an Increase in the Hepatic NADH/NAD$^+$ Ratio, which Inhibits HDAC Activity

Alcohol metabolism utilizes NAD$^+$ when alcohol dehydrogenase converts alcohol to acetaldehyde, and when acetaldehyde dehydrogenase further converts it to acetate. In both reactions, NAD^+ is reduced to NADH. The NADH/NAD^+ ratio influences gene expressions through several pathways. One of these involves sirtuins (SIRTs), NAD^+-dependent enzymes that have HDAC activity,[90] recognizes histones, and the transcription factor p53 as substrates.[91] SIRTs, or Class 3 HDACs, are activated only in the presence of NAD^+, which is hydrolyzed into nicotinamide, a potent inhibitor of HDAC activity of SIRTs. It has been hypothesized that ethanol-induced inhibition of HDACs is due to the depletion of NAD^+ caused during its metabolism.[92]

Acetate Produced by Ethanol Metabolism Induces Hat Activity

The end-product of ethanol metabolism in the liver is free acetate,[93] which is incorporated into acetyl-CoA,[94] the substrate for histone acetylation. Thus, it has been suggested that acetate might induce HAT activity by increasing substrate availability for the reaction. Moreover, acetate is also the product of the deacetylation reaction, thus free acetate might cause a feedback inhibition of HDACs.[95]

FUTURE DIRECTIONS

Different approaches have been used to identify genes related to alcoholism risk and, among candidate genes, several encode neurotransmitters and their receptors (*National Institute on Alcohol Abuse and Alcoholism (NIAAA). The genetics of alcoholism. Alcohol Alert No. 60. Rockville, MD: NIAAA, 2003*). Genes shape how an individual experiences (and is susceptible to) alcohol, and the alterations in many of these genes that influence the risk of developing alcoholism have been identified.[96] Remarkably, knowing how exposure to alcohol can change gene expression should provide insights into the brain mechanisms evoked by alcohol itself, and hence it could identify potential targets for therapeutic intervention.

Alcohol activates brain circuits and induces neuronal plasticity that converts alcohol-induced signals into long-term alterations; at the molecular level, the latter are coordinated by complex gene expression mechanisms. Considerable evidence suggests that alcohol activates pathways of the dopaminergic transmission in the mesocorticolimbic-brain circuits.[97] However, it is now apparent that other neuronal systems, such as opioid,[98] endocannabinoid,[99] GABAergic, and glutamatergic,[100] as well as serotonergic[101] systems, are involved. As yet, the local control of mRNA translation of multiple neurotransmitters remains partly unclear. However, recent evidence suggests that epigenetic regulation of the endocannabinoid and the opioid systems, for which functional interactions have been demonstrated in drug addiction, might as well play a role in alcoholism.[102] As for the endogenous opioid system, alterations in epigenetic mechanisms evoked by alcohol have been reported *in vivo* and *in vitro* for precursors of the major genes: PDYN,[70,71] proenkephalin (PENK), proopiomelanocortin[103] peptide precursors, as well as mu (MOP),[104] and nociceptin (NOP) receptor.[105]

To date, the epigenetic regulation of endocannabinoid system genes in response to alcohol remains completely unexplored, apart from one study showing that, in a mouse model of fetal alcohol spectrum disorders, the reduction in brain cannabinoid receptor 1 (*Cnr1*) expression is coupled with an increased complementary micro-RNA (miR-26b).[106] Consistently, an earlier study reported that *Cnr1* is involved in the neuropharmacological effects of alcohol,[107] and that gene variations or expression alterations are associated with mood disorders.[108] Thus, it appears of relevance to further investigate the regulation of endocannabinoid system elements by epigenetic mechanisms.[109]

The endocannabinoid system controls lipid signaling pathways by acting both in the central nervous system, and in peripheral tissues,[110] and a deeper understanding of its regulation might lead to the development of new clinical strategies for different human pathologies, including alcoholism.[111]

Ethanol is also a dietary constituent, and the relevance of the endocannabinoid system in nutrition should be stressed, because of its role in controlling food intake and energy balance through multiple central and peripheral mechanisms.[112] Based on these observations, the possibility of an "epigenetic therapy," applicable to endocannabinoid signaling in alcohol research, seems to hold promise. As reported, the epigenetic modifications evoked by alcohol might differ, depending on the degree of exposure (acute vs. chronic). In the case of chronic alcohol exposure, a strong correlation has been made with nutrient deficiencies, and understanding the role of nutrients in the regulation of epigenetic modifications of relevant genes will provide new insights into potential dietary supplementation in alcoholics.

In conclusion, different epigenetic processes control alcohol-induced changes in brain gene expression; however, these global changes are not completely consistent across the whole genome, because many genes show opposite epigenetic changes in their promoters. Thus, the study of new possible targets (i.e., endocannabinoid system genes), together with the possibility to reverse the epigenetic signature, would offer a new approach for more effective treatments of alcohol dependence.

CONCLUSIONS

Key Facts

Epigenetic Modulation

- Epigenetic mechanisms can regulate gene expression, involving chemical modifications of DNA, without affecting the actual DNA sequence of the organism.
- Alterations in gene expressions are transient, although for how long they actually last, is unclear.
- Epigenetic mechanisms include histone modifications, DNA methylation, and microRNA expression.
- Epigenetic marks are reversible, leading research to focus on epigenetic therapy for different diseases.
- Environmental factors modulate the establishment and maintenance of epigenetic modifications, thus influencing phenotype.
- Molecular research has paved way for the understanding of how environment and heritability factors might interact to facilitate the progression of different diseases, including alcoholism, pointing at the role of epigenetic mechanisms.

Summary Points

- This chapter focuses on the epigenetic modulation evoked by alcohol.
- Epigenetic mechanisms are posttranslational modifications regulating gene expression, without causing variation in the DNA sequence.
- Recently, molecular research explored how exposure to alcohol may activate epigenetic mechanisms without further influencing disease progression.
- Alcohol exposure is correlated with nutrient deficiencies, and the influences of nutritional status via epigenetics, is of relevance.
- New possible alcohol targets based on the modulation of the epigenetic signature have been suggested to offer new approaches for alcohol-dependence treatment.

Acknowledgments

This work was supported by the Italian Ministry of Education, University and Research, under the grants FIRB-RBFR12DELS to CDA and PRIN 2010-11 to MM. We would like to thank Dr Mariangela Pucci, for her kind assistance with the artwork.

References

1. Zimmermann US, Blomeyer D, Laucht M, Mann KF. How gene-stress-behavior interactions can promote adolescent alcohol use: the roles of predrinking allostatic load and childhood behavior disorders. *Pharmacol Biochem Behav* 2007;**86**:246–62.
2. Moonat S, Starkman BG, Sakharkar A, Pandey SC. Neuroscience of alcoholism: molecular and cellular mechanisms. *Cell Mol Life Sci* 2010;**67**:73–88.
3. Ptak C, Petronis A. Epigenetic approaches to psychiatric disorders. *Dialogues Clin Neurosci* 2010;**12**:25–35.
4. Crabbe JC, Phillips TJ, Harris RA, Arends MA, Koob GF. Alcohol-related genes: contributions from studies with genetically engineered mice. *Addict Biol* 2006;**11**:195–269.
5. Pignataro L, Varodayan FP, Tannenholz LE, Harrison NL. The regulation of neuronal gene expression by alcohol. *Pharmacol Ther* 2009;**124**:324–35.
6. Spanagel R, Bartsch D, Brors B, Dahmen N, Deussing J, Eils R, Ende G, Gallinat J, Gebicke-Haerter P, Heinz A, Kiefer F, Jäger W, Mann K, Matthäus F, Nöthen M, Rietschel M, Sartorius A, Schütz G, Sommer WH. An integrated genome research network for studying the genetics of alcohol addiction. *Addict Biol* 2010;**15**: 369–79.
7. Starkman BG, Sakharkar AJ, Pandey SC. Epigenetics-beyond the genome in alcoholism. *Alcohol Res* 2012;**34**:293–305.
8. Urdinguio RG, Sanchez-Mut JV, Esteller M. Epigenetic mechanisms in neurological diseases: genes, syndromes, and therapies. *Lancet Neurol* 2009;**8**:1056–72.
9. Pandey SC, Ugale R, Zhang H, Tang L, Prakash A. Brain chromatin remodeling: a novel mechanism of alcoholism. *J Neurosci* 2008;**28**:3729–37.
10. Borrelli E, Nestler EJ, Allis CD, Sassone-Corsi P. Decoding the epigenetic language of neuronal plasticity. *Neuron* 2008;**60**:961–74.
11. Waddington C. The epigenotype. *Endeavour* 1942;**1**:18–20.
12. Hsieh J, Gage FH. Chromatin remodeling in neural development and plasticity. *Curr Opin Cell Biol* 2005;**17**:664–71.
13. Esteller M. Epigenetics in evolution and disease. *Lancet* 2008;**372**: 90–6.
14. Miranda TB, Jones PA. A methylation: the nuts and bolts of repression. *J Cell Physiol* 2007;**213**:384–90.
15. Dhasarathy A, Wade PA. The MBD protein family-reading an epigenetic mark? *Mutat Res* 2008;**647**:39–43.
16. Leonhardt H, Page AW, Weier HU, Bestor TH. A targeting sequence directs DNA methyltransferase to sites of DNA replication in mammalian nuclei. *Cell* 1992;**71**:865–73.
17. Okano M, Bell DW, Haber DA, Li E. DNA methyltransferases Dnmt3a and Dnmt3b are essential for *de novo* methylation and mammalian development. *Cell* 1999;**99**:247–57.
18. Doi A, Park IH, Wen B, Murakami P, Aryee MJ, Irizarry R, Herb B, Ladd-Acosta C, Rho J, Loewer S, Miller J, Schlaeger T, Daley GQ, Feinberg AP. Differential methylation of tissue- and cancer-specific CpG island shores distinguishes human induced pluripotent stem cells, embryonic stem cells and fibroblasts. *Nat Genet* 2009;**41**:1350–3.
19. Kriaucionis S, Heintz N. The nuclear DNA base 5-hydroxymethylcytosine is present in Purkinje neurons and the brain. *Science* 2009;**324**:929–30.
20. Wu SC, Zhang Y. Active DNA demethylation: many roads lead to Rome. *Nat Rev Mol Cell Biol* 2010;**11**:607–20.
21. Jin SG, Wu X, Li AX, Pfeifer GP. Genomic mapping of 5-hydroxymethylcytosine in the human brain. *Nucleic Acids Res* 2011;**39**:5015–24.
22. Ito S, D'Alessio AC, Taranova OV, Hong K, Sowers LC, Zhang Y. Role of Tet proteins in 5mC to 5hmC conversion, ES-cell self-renewal and inner cell mass specification. *Nature* 2010;**466**: 1129–33.
23. Luger K, Rechsteiner TJ, Richmond TJ. Preparation of nucleosome core particle from recombinant histones. *Methods Enzymol* 1999; **304**:3–19.
24. Ng RK, Gurdon JB. Epigenetic inheritance of cell differentiation status. *Cell Cycle* 2008;**7**:1173–7.
25. Turner BM. Cellular memory and the histone code. *Cell* 2002; **111**:285–91.
26. Spencer VA, Davie JR. Role of covalent modifications of histones in regulating gene expression. *Gene* 1999;**240**:1–12.

27. Kouzarides T. SnapShot: histone-modifying enzymes. *Cell* 2007; **131**:822.
28. Shahbazian MD, Grunstein M. Functions of site-specific histone acetylation and deacetylation. *Annu Rev Biochem* 2007;**76**: 75–100.
29. Yang XJ, Seto E. HATs and HDACs: from structure, function and regulation to novel strategies fortherapy and prevention. *Oncogene* 2007;**26**:5310–8.
30. Martin C, Zhang Y. The diverse functions of histone lysine methylation. *Nat Rev Mol Cell Biol* 2005;**6**:838–49.
31. Jenuwein T. The epigenetic magic of histone lysine methylation. *FEBS J* 2006;**273**:3121–35.
32. Henckel A, Nakabayashi K, Sanz LA, Feil R, Hata K, Arnaud P. Histone methylation is mechanistically linked to DNA methylation at imprinting control regions in mammals. *Hum Mol Genet* 2009;**18**:3375–83.
33. Cloos PA, Christensen J, Agger K, Helin K. Erasing the methyl mark: histone demethylases at the center of cellular differentiation and disease. *Genes Dev* 2008;**22**:1115–40.
34. Strahl BD, Allis CD. The language of covalent histone modifications. *Nature* 2000;**403**:41–5.
35. Ambros V. microRNAs: tiny regulators with great potential. *Cell* 2001;**107**:823–6.
36. Cheng LC, Pastrana E, Tavazoie M, Doetsch F. miR-124 regulates adult neurogenesis in the subventricular zone stem cell niche. *Nat Neurosci* 2009;**12**:399–408.
37. Fiore R, Siegel G, Schratt G. MicroRNA function in neuronal development, plasticity and disease. *Genet* 2008;**9**:102–14.
38. Schratt GM, Tuebing F, Nigh EA, Kane CG, Sabatini ME, Kiebler M, Greenberg ME. A brain-specific microRNA regulates dendritic spine development. *Nature* 2006;**439**:283–9.
39. Schaefer A, O'Carroll D, Tan CL, Hillman D, Sugimori M, Llinas R, Greengard P. Cerebellar neurodegeneration in the absence of microRNAs. *J Exp Med* 2007;**204**:1553–8.
40. Sathyan P, Golden HB, Miranda RC. Competing interactions between micro-RNAs determine neural progenitor survival and proliferation after ethanol exposure: evidence from an ex vivo model of the fetal cerebral cortical neuroepithelium. *J Neurosci* 2007;**27**: 8546–57.
41. Pietrzykowski AZ, Friesen RM, Martin GE, Puig SI, Nowak CL, Wynne PM, Siegelmann HT, Treistman SN. Posttranscriptional regulation of BK channel splice variant stability by miR-9 underlies neuroadaptation to alcohol. *Neuron* 2008;**59**:274–87.
42. Miranda RC, Pietrzykowski AZ, Tang Y, Sathyan P, Mayfield D, Keshavarzian A, Sampson W, Hereld D. MicroRNAs: master regulators of ethanol abuse and toxicity? *Alcohol Clin Exp Res* 2010;**34**: 575–87.
43. Lewohl JM, Nunez YO, Dodd PR, Tiwari GR, Harris RA, Mayfield RD. Up-regulation of microRNAs in brain of human alcoholics. *Alcohol Clin Exp Res* 2011;**35**:1928–37.
44. Choudhury M, Shukla SD. Surrogate alcohols and their metabolites modify histone H3 acetylation: involvement of histone acetyl transferase and histone deacetylase. *Alcohol Clin Exp Res* 2008;**32**: 829–39.
45. Weaver IC, Meaney MJ, Szyf M. Maternal care effects on the hippocampal transcriptome and anxiety-mediated behaviors in the offspring that are reversible in adulthood. *Proc Natl Acad Sci USA* 2006;**103**:3480–5.
46. Sanchis-Segura C, Lopez-Atalaya JP, Barco A. Selective boosting of transcriptional and behavioral responses to drugs of abuse by histone deacetylase inhibition. *Neuropsychopharmacology* 2009;**34**: 2642–54.
47. Wolstenholme JT, Warner JA, Capparuccini MI, Archer KJ, Shelton KL, Miles MF. Genomic analysis of individual differences in ethanol drinking: evidence for non-genetic factors in C57BL/6 mice. *PLoS One* 2011;**6**:e21100.
48. Pascual M, Do Couto BR, Alfonso-Loeches S, Aguilar MA, Rodriguez-Arias M, Guerri C. Changes in histone acetylation in the prefrontal cortex of ethanol-exposed adolescent rats are associated with ethanol-induced place conditioning. *Neuropharmacology* 2012;**62**:2309–19.
49. Sakharkar AJ, Zhang H, Tang L, Shi G, Pandey SC. Histone deacetylases (HDAC)-induced histone modifications in the amygdala: a role in rapid tolerance to the anxiolytic effects of ethanol. *Alcohol Clin Exp Res* 2012;**36**:61–71.
50. Moonat S, Sakharkar AJ, Zhang H, Tang L, Pandey SC. Aberrant histone deacetylase 2-mediated histone modifications and synaptic plasticity in the amygdala predisposes to anxiety and alcoholism. *Biol Psychiatry* 2013;**73**:763–73.
51. Bardag-Gorce F, Li J, Oliva J, Lu SC, French BA, French SW. The cyclic pattern of blood alcohol levels during continuous ethanol feeding in rats: the effect of feeding *S*-adenosylmethionine. *Exp Mol Pathol* 2010;**88**:380–7.
52. Bardag-Gorce F, French BA, Joyce M, Baires M, Montgomery RO, Li J, French S. Histone acetyltransferase p300 modulates gene expression in an epigenetic manner at high blood alcohol levels. *Exp Mol Pathol* 2007;**82**:197–202.
53. Yamashita H, Kaneyuki T, Tagawa K. Production of acetate in the liver and its utilization in peripheral tissues. *Biochim Biophys Acta* 2001;**1532**:79–87.
54. Renthal W, Nestler EJ. Histone acetylation in drug addiction. *Semin Cell Dev Biol* 2009;**20**:387–94.
55. Qiang M, Denny A, Lieu M, Carreon S, Li J. Histone H3K9 modifications are a local chromatin event involved in ethanol-induced neuroadaptation of the NR2B gene. *Epigenetics* 2011;**6**: 1095–104.
56. Park PH, Lim RW, Shukla SD. Involvement of histone acetyltransferase (HAT) in ethanol-induced acetylation of histone H3 in hepatocytes: potential mechanism for gene expression. *Am J Physiol Gastrointest Liver Physiol* 2005;**289**:G1124–36.
57. Kim JS, Shukla SD. Acute *in vivo* effect of ethanol (binge drinking) on histone H3 modifications in rat tissues. *Alcohol Alcohol* 2006; **41**:126–32.
58. Shibasaki M, Mizuno K, Kurokawa K, Ohkuma S. Enhancement of histone acetylation in midbrain of mice with ethanol physical dependence and its withdrawal. *Synapse* 2011;**65**: 1244–50.
59. Pascual M, Boix J, Felipo V, Guerri C. Repeated alcohol administration during adolescence causes changes in the mesolimbic dopaminergic and glutamatergic systems and promotes alcohol intake in the adult rat. *J Neurochem* 2009;**108**:920–31.
60. Park PH, Miller R, Shukla SD. Acetylation of histone H3 at lysine 9 by ethanol in rat hepatocytes. *Biochem Biophys Res Commun* 2003; **306**:501–4.
61. Grant PA. A tale of histone modifications. *Genome Biol* 2001;**2** (4): reviews 0003.1–0003.6.
62. Cheung P, Tanner KG, Cheung WL, Sassone-Corsi P, Denu JM, Allis CD. Synergistic coupling of histone H3 phosphorylation and acetylation in response to epidermal growth factor stimulation. *Mol Cell* 2000;**5**:905–15.
63. Lee YJ, Shukla SD. Histone H3 phosphorylation at serine 10 and serine 28 is mediated by p38 MAPK in rat hepatocytes exposed to ethanol and acetaldehyde. *Eur J Pharmacol* 2007;**573**: 29–38.
64. Ponomarev I, Wang S, Zhang L, Harris RA, Mayfield RD. Gene coexpression networks in human brain identify epigenetic modifications in alcohol dependence. *J Neurosci* 2012;**32**:1884–97.
65. Yang X, Horn K, Wand GS. Chronic ethanol exposure impairs phosphorylation of CREB and CRE-binding activity in rat striatum. *Alcohol Clin Exp Res* 1998;**22**:382–90.
66. Crews FT, Nixon K. Mechanisms of neurodegeneration and regeneration in alcoholism. *Alcohol Alcohol* 2009;**44**:115–27.

67. Veazey KJ, Carnahan MN, Muller D, Miranda RC, Golding MC. Alcohol-induced epigenetic alterations to developmentally crucial genes regulating neural stemness and differentiation. *Alcohol Clin Exp Res* 2013;**37**:1111–22.
68. Pal-Bhadra M, Bhadra U, Jackson DE, Mamatha L, Park PH, Shukla SD. Distinct methylation patterns in histone H3 at Lys-4 and Lys-9 correlate with up- and down-regulation of genes by ethanol in hepatocytes. *Life Sci* 2007;**81**:979–87.
69. Zhou Z, Yuan Q, Mash DC, Goldman D. Substance-specific and shared transcription and epigenetic changes in the human hippocampus chronically exposed to cocaine and alcohol. *Proc Natl Acad Sci USA* 2011;**108**:6626–31.
70. D'Addario C, Johansson S, Candeletti S, Romualdi P, Ögren SO, Terenius L, Ekström TJ. Ethanol and acetaldehyde exposure induces specific epigenetic modifications in the prodynorphin gene promoter in a human neuroblastoma cell line. *FASEB J* 2011;**25**:1069–75.
71. D'Addario C, Caputi FF, Ekström TJ, Di Benedetto M, Maccarrone M, Romualdi P, Candeletti S. Ethanol induces epigenetic modulation of prodynorphin and pronociceptin gene expression in the rat amygdala complex. *J Mol Neurosci* 2013;**49**:312–9.
72. Subbanna S, Shivakumar M, Umapathy NS, Saito M, Mohan PS, Kumar A, Nixon RA, Verin AD, Psychoyos D, Basavarajappa BS. G9a-mediated histone methylation regulates ethanol-induced neurodegeneration in the neonatal mouse brain. *Neurobiol Dis* 2013;**54**:475–85.
73. Van Steenwyk G, Janeczek P, Lewohl JM. Differential effects of chronic and chronic-Intermittent ethanol treatment and its withdrawal on the expression of miRNAs. *Brain Sci* 2013;**3**:744–56.
74. Yadav S, Pandey A, Shukla A, Talwelkar SS, Kumar A, Pant AB, Parmar D. miR-497 and miR-302b regulate ethanol-induced neuronal cell death through BCL2 protein and cyclin D2. *J Biol Chem* 2011;**286**:37347–57.
75. Manzardo AM, Gunewardena S, Butler MG. Over-expression of the miRNA cluster at chromosome 14q32 in the alcoholic brain correlates with suppression of predicted target mRNA required for oligodendrocyte proliferation. *Gene* 2013;**526**:356–63.
76. Tapocik JD, Solomon M, Flanigan M, Meinhardt M, Barbier E, Schank JR, Schwandt M, Sommer WH, Heilig M. Coordinated dysregulation of mRNAs and microRNAs in the rat medial prefrontal cortex following a history of alcohol dependence. *Pharmacogenomics J* 2013;**13**:286–96.
77. Tapocik JD, Barbier E, Flanigan M, Solomon M, Pincus A, Pilling A, Sun H, Schank JR, King C, Heilig M. microRNA-206 in rat medial prefrontal cortex regulates BDNF expression and alcohol drinking. *J Neurosci* 2014;**34**:4581–8.
78. Wang LL, Zhang Z, Li Q, Yang R, Pei X, Xu Y, Wang J, Zhou SF, Li Y. Ethanol exposure induces differential microRNA and target gene expression and teratogenic effects which can be suppressed by folic acid supplementation. *Hum Reprod* 2009;**24**:562–79.
79. Lippai D, Bala S, Csak T, Kurt-Jones EA, Szabo G. Chronic alcohol-induced microRNA-155 contributes to neuroinflammation in a TLR4-dependent manner in mice. *PLoS One* 2013;**8**:e70945.
80. Soares AR, Pereira PM, Ferreira V, Reverendo M, Simões J, Bezerra AR, Moura GR, Santos MA. Ethanol exposure induces upregulation of specific microRNAs in zebrafish embryos. *Toxicol Sci* 2012;**127**:18–28.
81. Tal TL, Franzosa JA, Tilton SC, Philbrick KA, Iwaniec UT, Turner RT, Waters KM, Tanguay RL. MicroRNAs control neurobehavioral development and function in zebrafish. *FASEB J* 2012;**26**:1452–61.
82. Li J, Li J, Liu X, Qin S, Guan Y, Liu Y, Cheng Y, Chen X, Li W, Wang S, Xiong M, Kuzhikandathil EV, Ye JH, Zhang C. MicroRNA expression profile and functional analysis reveal that miR-382 is a critical novel gene of alcohol addiction. *EMBO Mol Med* 2013;**5**:1402–14.
83. Cordain L, Bryan ED, Melby CL, Smith MJ. Influence of moderate daily wine consumption on body weight regulation and metabolism in healthy free-living males. *J Am Coll Nutr* 1997;**16**:134–9.
84. Lieber CS. Hepatic, metabolic and toxic effects of ethanol: 1991 update. *Alcohol Clin Exp Res* 1991;**15**:573–92.
85. Williams RJ, Berry LJ, Beerstecher E. Individual metabolic patterns, alcoholism, genetotrophic diseases. *Proc Natl Acad Sci USA* 1949;**35**:265–71.
86. Lu SC, Mato JM. S-Adenosylmethionine in cell growth, apoptosis and liver cancer. *J Gastroenterol Hepatol* 2008;**23**:S73–7.
87. Blasco C, Caballería J, Deulofeu R, Lligoña A, Parés A, Lluis JM, Gual A, Rodés J. Prevalence and mechanisms of hyperhomocysteinemia in chronic alcoholics. *Alcohol Clin Exp Res* 2005;**29**:1044–8.
88. Hamid A, Wani NA, Kaur J. New perspectives on folate transport in relation to alcoholism-induced folatemalabsorption--association with epigenome stability and cancer development. *FEBS J* 2009;**276**:2175–91.
89. Li J, Bardag-Gorce F, Oliva J, Dedes J, French BA, French SW. Gene expression modifications in the liver caused by binge drinking and S-adenosylmethionine feeding. The role of epigenetic changes. *Genes Nutr* 2009;**5**:169–79.
90. Imai S, Armstrong CM, Kaeberlein M, Guarente L. Transcriptional silencing and longevity protein Sir2 is an NAD-dependent histone deacetylase. *Nature* 2000;**403**:795–800.
91. Vaziri H, Dessain SK, Ng Eaton E, Imai SI, Frye RA, Pandita TK, Guarente L, Weinberg RA. hSIR2(SIRT1) functions as an NAD-dependent p53 deacetylase. *Cell* 2001;**107**:149–59.
92. Osna NA. Histone modifications and alcohol-induced liver disease: are altered nutrients the missing link? *World J Gastroenterol* 2011;**17**:2465–72.
93. Lieber CS. Metabolism of alcohol. *Clin Liver Dis* 2005;**9**:1–35.
94. Fujino TIY, Osborne TF, Takahashi S, Yamamoto TT, Sakai J. Sources of acetyl-coA: acetyl-coA synthetase 1 and 2. *Curr Med Chem Immunol Endocr Metab Agents* 2003;**3**:207–10.
95. Kendrick SF, O'Boyle G, Mann J, Zeybel M, Palmer J, Jones DE, Day CP. Acetate, the key modulator of inflammatory responses in acute alcoholic hepatitis. *Hepatology* 2010;**51**:1988–97.
96. Worst TJ, Vrana KE. Alcohol and gene expression in the central nervous system. *Alcohol Alcohol* 2005;**40**:63–75.
97. Söderpalm B, Löf E, Ericson M. Mechanistic studies of ethanol's interaction with the mesolimbic dopamine reward system. *Pharmacopsychiatry* 2009;**42**(Suppl. 1):S87–94.
98. Méndez M, Morales-Mulia M. Role of mu and delta opioid receptors in alcohol drinking behaviour. *Curr Drug Abuse Rev* 2008;**1**:239–52.
99. Pava MJ, Woodward JJ. A review of the interactions between alcohol and the endocannabinoid system: implications for alcohol dependence and future directions for research. *Alcohol* 2012;**46**:185–204.
100. Colombo G, Addolorato G, Agabio R, Carai MA, Pibiri F, Serra S, Vacca G, Gessa GL. Role of GABA(B) receptor in alcohol dependence: reducing effect of baclofen on alcohol intake and alcohol motivational properties in rats and amelioration of alcohol withdrawal syndrome and alcohol craving in human alcoholics. *Neurotox Res* 2004;**6**:403–14.
101. Johnson BA. Role of the serotonergic system in the neurobiology of alcoholism: implications for treatment. *CNS Drugs* 2004;**18**:1105–18.
102. López-Moreno JA, López-Jiménez A, Gorriti MA, de Fonseca FR. Functional interactions between endogenous cannabinoid and opioid systems: focus on alcohol, genetics and drug-addicted behaviors. *Curr Drug Targets* 2010;**11**:406–28.
103. Muschler MA, Hillemacher T, Kraus C, Kornhuber J, Bleich S, Frieling H. DNA methylation of the POMC gene promoter is associated with craving in alcohol dependence. *J Neural Transm* 2010;**117**:513–9.

104. Zhang H, Herman AI, Kranzler HR, Anton RF, Simen AA, Gelernter J. Hypermethylation of OPRM1 promoter region in European Americans with alcohol dependence. *J Hum Genet* 2012; **57**:670–5.
105. Zhang H, Wang F, Kranzler HR, Zhao H, Gelernter J. Profiling of childhood adversity-associated DNA methylation changes in alcoholic patients and healthy controls. *PLoS One* 2013;**8**:e65648.
106. Stringer RL, Laufer BI, Kleiber ML, Singh SM. Reduced expression of brain cannabinoid receptor 1 (Cnr1) is coupled with an increased complementary micro-RNA (miR-26b) in a mouse model of fetal alcohol spectrum disorders. *Clin Epigenetics* 2013;**5**:14.
107. Adermark L, Jonsson S, Ericson M, Soderpalm B. Intermittent ethanol consumption depresses endocannabinoid-signaling in the dorsolateral striatum of rat. *Neuropharmacology* 2011;**5**: 1160–5.
108. Dubreucq S, Kambire S, Conforzi M, Metna-Laurent M, Cannich A, Soria-Gomez E, Richard E, Marsicano G, Chaouloff F. Cannabinoid type 1 receptors located on single-minded 1-expressing neurons control emotional behaviors. *Neurosci* 2012;**5**:230–44.
109. D'Addario C, Di Francesco A, Pucci M, Finazzi-Agrò A, Maccarrone M. Epigenetic mechanisms and endocannabinoid signalling. *FEBS J* 2013;**280**:1905–17.
110. Galve-Roperh I, Chiurchiù V, Díaz-Alonso J, Bari M, Guzmán M, Maccarrone M. Cannabinoid receptor signaling in progenitor/stem cell proliferation and differentiation. *Prog Lipid Res* 2013;**52**:633–50.
111. Pacher P, Kunos G. Cannabinoids and endocannabinoids in human health/disease. *FEBS J* 2013;**280**:1918–43.
112. Maccarrone M, Gasperi V, Catani MV, Diep TA, Dainese E, Hansen HS, Avigliano L. The endocannabinoid system and its relevance for nutrition. *Annu Rev Nutr* 2010;**30**:423–40.

CHAPTER

22

The miRNA and Extracellular Vesicles in Alcoholic Liver Disease

Fatemeh Momen-Heravi, DDS, MPH, Shashi Bala, PhD

Department of Medicine, University of Massachusetts Medical School, Worcester, MA, USA

INTRODUCTION

Alcoholic Liver Disease

Alcoholic liver disease (ALD) is a major global health concern, and its spectrum ranges from fatty liver to alcoholic hepatitis and cirrhosis, while, in some cases, it leads to hepatocellular carcinoma (HCC).[1,2] Increased oxidative stress, circulating endotoxin, induction of TNF-α signaling, and subsequent epigenetic events are the main contributing factors of ALD.[3–5] Alcohol metabolism increases the production of reactive oxygen species (ROS) via multiple mechanisms, including the mitochondrial electron transport chain and cytochrome P450 2E1 (Cyp2e1)-mediated alcohol metabolism.[6,7] The pathologic mechanisms of ALD involve complex interactions between the direct effects of alcohol, and its toxic metabolites on various cell types in the liver and gut.[8]

MicroRNAs

Recent advances in analytical methods to explore transcriptome have resulted in numerous improvements in identifying and understanding noncoding RNAs. Among noncoding RNAs, the regulatory roles of microRNAs (miRNAs) in cellular processes are now being revealed in detail. This chapter discusses the insights into the biology of miRNA in ALD.

miRNAs are short noncoding RNA molecules (21–23 nucleotides) that do not encode for any protein. The first miRNA, lin-4 was discovered in 1993 in *Caenorhabditis elegans*.[9] Since 1993, microRNA field has seen a tremendous growth (Figure 22.1). Biogenesis of miRNA is initiated in the nucleus followed by processing into the mature miRNA (18–21 nucleotides) in the cytoplasm.[10] To execute their function, mature miRNAs are loaded onto an RNA inducing silencing complex and, depending on their 3′ untranslated region complementarity to target mRNA, binding of miRNA leads to either target mRNA degradation, or translational repression (Figure 22.2).[10] One miRNA can target more than 100 genes, leading to complex changes in genetic networks. Till date, the number of annotated miRNA genes in humans is approximately 2000 (as per Sanger Institute).[11] Various studies have revealed new roles for miRNA, such as in the regulation of DNA methylation, and target regulation at the 5′ UTR.[12]

microRNA Expression in Alcoholic Liver Disease

Alcoholic hepatitis, alcoholic fatty liver, and cirrhosis are the major segments of ALD worldwide. Although the toxic effects of alcohol likely result from complex interactions between genes and the environment, the molecular mechanisms of alcohol-induced liver damage remains undefined.

Various miRNAs have been affected in ALD, however, till date, the precise role of few miRNAs are shown in ALD. First microarray study in a mouse model of ALD (Lieber-DeCarli alcohol diet) demonstrated that alcohol upregulates 1% and downregulates 1% of known miRNAs, in the livers of alcohol-fed mice.[13] miRNA, such as miR-705 and miR-1224, were increased, whereas another set of miRNAs, including miR-182, miR-183, and miR-199a-3p, were decreased in mice fed with alcohol diet.[13] Gut plays a crucial role in the pathogenesis of ALD, and ethanol increases miR-212 levels that then regulates ZO1 expression in intestinal epithelial cells, and in alcoholic patients.[14] EtOH-induced miR-199 downregulation in rat liver sinusoidal endothelial cells and human

Molecular Aspects of Alcohol and Nutrition. http://dx.doi.org/10.1016/B978-0-12-800773-0.00022-7

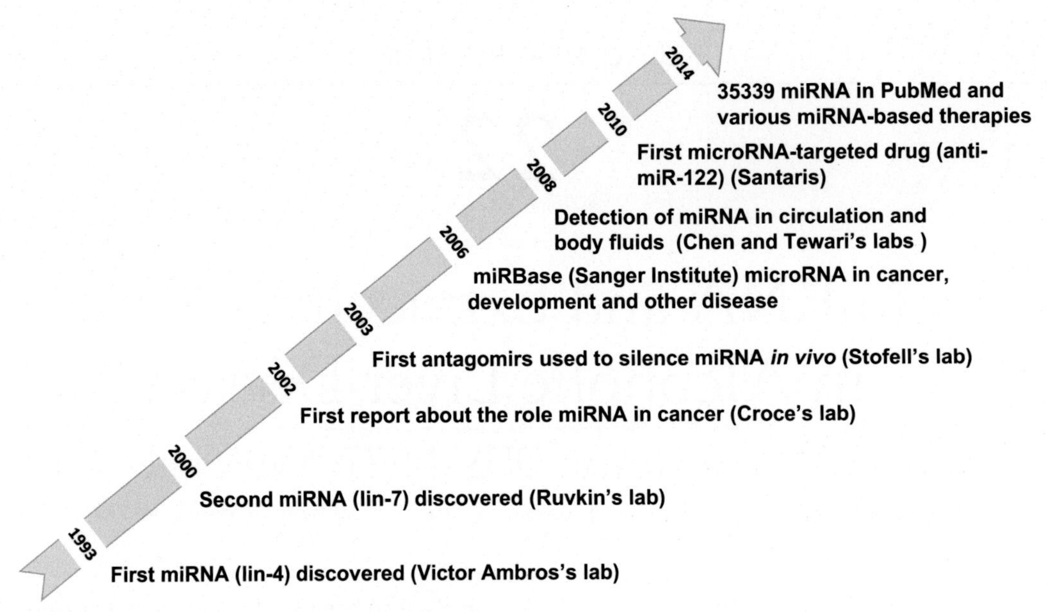

FIGURE 22.1 **Major events in miRNA history; 21 years of journey.**

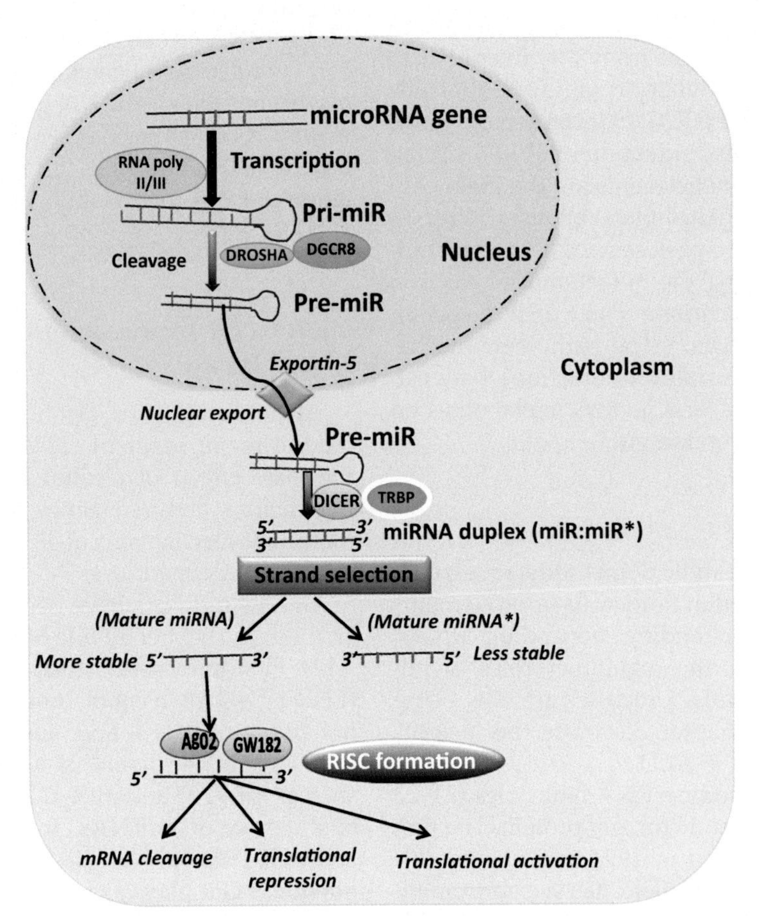

FIGURE 22.2 **Biogenesis of miRNA.** miRNAs are transcribed from miRNA gene via RNA polymerase II or III as primary (pri) miRNA transcript in the nucleus. Pri-miRNA undergoes cleavage by Drosha-DGCR8 complex as precursor (pre) miRNA. Pre-miRNA is exported to cytoplasm via exportin-5 complex where further processing takes place. Dicer, along with TRBP, cleaves pre-miRNA to mature form (miRNA duplex) and strand selection takes place. Depending upon stability, functional strand is loaded together with Ago2 and GW182 into the RNA-inducing silencing complex (RISC). The less stable strand of miRNA gets degraded. Depending on matching of seed region of mature miRNA to the 3′ UTR of target mRNA gene, target mRNA either undergoes for mRNA cleavage, translational repression, or activation. *, star strand

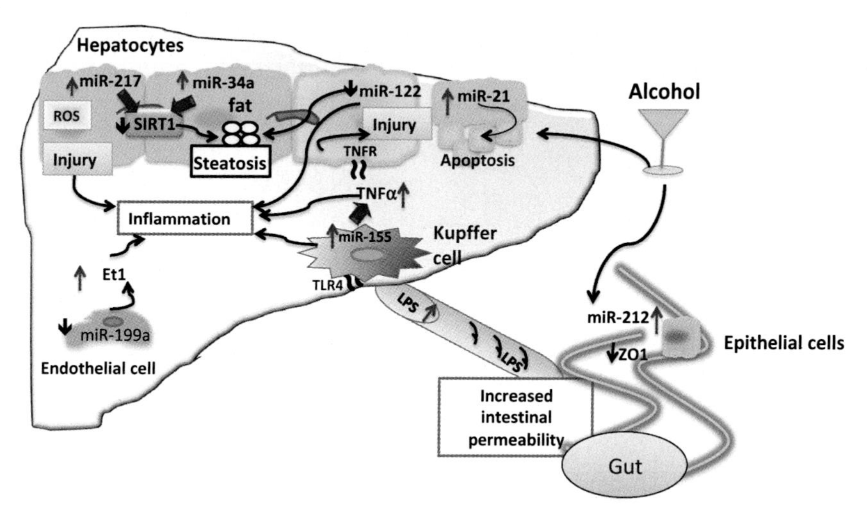

FIGURE 22.3 **Role of miRNAs in alcoholic liver disease.** Alcohol and its metabolites increase miR-212 levels in the gut epithelial cells. miR-212 targets tight junction protein, ZO1. Decreased ZO1 disrupts gut integrity and results in increased intestinal permeability. The flow of microbial translocation into intestinal lumen increases, and causes an increase in lipopolysaccharide (LPS) in the portal vein. LPS in the liver affects immune (Kupffer cells), parenchymal (hepatocytes), and nonimmune cells. This results in the activation of Kupffer cells and induction of miR-155 and TNF-α. Increased TNF-α causes injury to hepatocytes, and damaged hepatocytes release danger molecules that are recognized by various immune cells resulting in inflammation. Alcohol induces miR-21 and increases apoptosis of hepatocytes. Alcohol metabolism increases oxidative stress that results in the induction of miR-34a and miR-217, and decrease in miR-122 in hepatocytes. Dysregulation of these miRNAs result in steatosis via targeting genes involved in fatty acid metabolism, including SIRT1 and LIPIN1. Oxidative stress decreases miR-199a in the endothelial cells that results in increase of Et1, and contributes to liver inflammation.

endothelial cells contributed to augmented HIF-1α and ET-1 expression.[15] Figure 22.3 shows a schematic representation of miRNA-regulated pathways in ALD.

miRNA-122

miR-122 is highly expressed in hepatocytes, representing about 70% of total miRNAs pool in the liver.[4,16] Approximately 66,000 copies per cell of miR-122 are present in hepatocytes, making it one of the highly expressed miRNAs in any tissue.[4,16] miR-122 expression is developmentally regulated, and its levels increase in the liver over the course of embryonic development.[16] miR-122 is a highly studied miRNA due to its role in cholesterol metabolism,[17] and HCC,[17,18] as well as its vital role in promoting hepatitis C virus (HCV) replication.[19,20] miR-122 is derived from a single genomic locus on chromosome 18, in humans. Inhibition of miR-122 by anti-miRNA strategies caused a reduction in plasma cholesterol levels in nonhuman primates and in chimpanzee, suggesting miR-122 plays an imperative role in maintaining liver homeostasis.[21,22]

Alcohol administration in mice for 5 weeks resulted in decrease in miR-122 levels in the liver.[23] miR-122 inhibition with AAV8 TUD vector expressing anti-miR-122 resulted in an increase in MCP1, TNF-α, and IL-1β in a mouse model of ALD (Lieber-DeCarli alcohol diet).[24] This study suggests a role of miR-122 in regulation of inflammation and fat accumulation.[24] However, further studies are needed to reveal if this phenotype is due to either direct role of miR-122 in hepatocytes, or via hepatocytes-mediated immune cell activation. The underlying mechanisms of decreased miR-122 in ALD, and its definite role in various liver cell types, have to be defined. Other studies have also revealed the role of miR-122 in liver inflammation, as miR-122 deficient mice had increased Ccl2 and proinflammatory cytokines, such as IL-6 and TNF-α.[17]

Prolonged chronic alcohol use in humans leads to liver fibrosis, and miR-122 expression is decreased in the fibrotic livers.[25,26] Hepatic levels of miR-122 are also reduced in diet-induced obesity (NASH or NAFLD),[27] suggesting a broad-spectrum role of miRNA-122 in different liver diseases. Moreover, genetic deletion of miR-122 in mice results in the progression of steatohepatitis, fibrosis, and HCC.[17] The phenotype was somewhat expected, as several miR-122 targets are associated with tumorigenesis, including ADAM10, cyclin G1, and ADAM17.

miRNA-155

miR-155 is a major regulator of inflammation,[28] and alcohol induced TNF-α and other cytokines, and contributes to inflammation. miR-155 is a highly studied immune miRNA, and most of the studies are focused on the role of miR-155 in immune cells, such as monocytes, macrophages, dendritic cells (DCs), and T cells. Increased levels of miR-155 have been found in various inflammatory diseases, including RA, neuro- or autoimmune-inflammation, and various cancers.[28–32]

Alcohol induced miR-155 expression in Kupffer cells (KCs) isolated from mice fed with alcohol for 5 weeks.[33] miR-155 expression correlated with TNF-α in KCs and RAW 264.7 macrophages. Functionally, inhibition of miR-155 was associated with decreases in TNF-α levels, suggesting a role of miR-155 in TNF-α regulation. One miRNA can regulate multiple genes of a pathway, therefore it is possible that miR-155 might regulate other genes that are directly or indirectly involved in TNF-α regulation. A recent miRNA array analysis showed increase of miR-155 in the livers of alcoholic hepatitis patients, and was correlated positively with the severity of the disease.[34]

miRNA-21

A miRNA array study performed on the livers from mice fed with intergastric feeding for 4 weeks revealed the significant increase of miR-34a, miR-21, miR-882, and decrease of miR-122, miR-192, miR-181b, miR-181a, and let7a. In total, 46 miRNAs were significantly altered in the livers of alcohol treated mice, compared to normal.[35,36]

miR-21 was a highly induced miRNA in the liver after alcohol feeding, and was also increased in ethanol-treated N-Heps, HepG2, and HSCs *in vitro*, and this increase was associated with higher apoptosis.[36] This study revealed that miR-21 is increased frequently in human alcoholic liver injury. miR-21 is a well known survival factor in liver injury and HCC development. The miR-21 gene is located on chromosome 17, close to the location of p53 and p53 regulates miR-21.[36] These findings suggest a key role of miR-21 in the regulation of survival and transformation of human hepatic cells during alcoholic liver injury. Activation of IL-6/Stat3 signaling resulted in increase in miR-21 *in vivo,* and miR-21 modulated processes, such as cell proliferation, apoptosis, and survival, and some of these effects were mediated through DR5 and FASLG, the well-characterized regulator genes of apoptosis.[36] These findings support a functional role of miR-21 in promoting liver tissue repair and liver fibrosis, in the development of ALD.

miRNA-34a

Alcohol induces miR-34a levels, and regulates Sirt1 and caspase-2.[35] In general, p53 is involved in the activation of miR-34a. miR-34a regulates cell survival, migration, and remodeling properties via repression of target genes, and modulation of downstream signaling pathways.[35] miR-34a levels were also increased in ethanol-treated N-Heps and HiBECs, and was associated with higher rate of apoptosis.[35] CASP2 and SIRT1 were found to be targets of miR-34a in these cells.[35] Further, transfection of human hepatocytes with miR-34a precursor increased MMP-2 and MMP-9 activity. This study suggests miR-34a as a putative mediator of tissue remodeling, and it might contribute to liver reconstruction and fibrosis in ALD.[35]

miRNA-217

miR-217 induction was found in the livers of chronically ethanol-fed mice and in AML-12 hepatocytes.[37] Overexpression of miR-217 in AML-12 cells resulted in the decrease of SIRT1 that caused fat accumulation in hepatocytes. Further, miR-217 impaired the function of lipin-1 in AML-12 hepatocytes. Lipin-1 regulates lipid metabolism, and acts as a phosphatidic acid phosphohydrolase type enzyme in the cytosol, and a transcriptional coactivator in the nucleus.[37] The mechanism through which alcohol induces miR-217 is not known.

Cell-Specific Effects of Alcohol on miRNAs

The concept of cell-specific effects of miRNA is emerging, however, the cell-specific role of miRNAs in ALD is yet to be determined. Alcohol induced miR-155 and miR-132 levels in the Kupffer cells and hepatocytes isolated after 5 weeks of feeding mice.[38] miR-125b levels were decreased in hepatocytes after alcohol feeding.[38] Decreased miR-125b levels were found in HCC, and also associated with increased placenta growth factor.[38] The physiological relevance of these miRNA in hepatocytes of ALD awaits further investigation.

Mechanism of Abnormal miRNA Expression

Recent developments indicate that ethanol induces epigenetic alterations, particularly acetylation, methylation of histones, and hypo- and hypermethylation of DNA.[39] All these changes modulate miRNA expression. However, the precise mechanism through which epigenetic changes affect miRNA in ALD remains to be determined.

EXTRACELLULAR VESICLE-ASSOCIATED miRNAs AS POTENTIAL BIOMARKERS FOR ALD

Extracellular Vesicles: Types and Terminology

Extracellular vesicles (EVs), including exosomes, microvesicles (MVs), and apoptotic bodies, are membranous, cell-derived, mixed population of vesicles approximately 4000–5000 nm in diameter, which are released by a variety of cells into biofluids, and intercellular microenvironments.[40,41] The terminologies used for referring to EVs have changed tremendously over the past 10 years. First, isolated EVs were named based on the cellular origin and their size. These concepts led to the development of different nomenclatures, such as ectosomes (vesicles secreted by monocytes and neutrophils), oncosomes (vesicles derived from tumor cells), exosome-like vesicles, microparticles, prostasomes (vesicles derived from seminal fluid), microparticles (mostly referred to the vesicles shed from platelets in blood), apoptotic bodies, exosomes, nanoparticles, and MVs. Particularly, this variation in terminology, incomplete understanding of each sub population of EVs biogenesis, nonstandardized isolation methods, and downstream characterization steps led to using the words exosomes and MVs interchangeably.[40–43] Toward an effective resolution, a recent consensus in the scientific community was achieved by categorizing EVs based on their biogenesis that places them in three groups of exosomes, MVs, and apoptotic bodies. The two most famous groups of EVs are exosomes and MVs. They attained vast attention in scientific literature because of their roles in disease pathogenesis, and biomarker discovery. Both exosomes and MVs carry various biomolecules, including proteins, mRNAs, miRNAs, and lipid molecules. Exosomes originate from multivesicular bodies, and are enriched in tetraspanins protein family like CD63 and CD81. MVs pitch off from the plasma membrane, and are enriched in phosphatidylserine and annexin V.[40,42] In this chapter, we use the term EVs to refer to both exosomes and MVs.

Extracellular Vesicles as Natural Carriers of miRNAs: Physiological and Pathological Roles

EVs act as multifunctional signaling complexes for cell-to-cell communication, control pivotal cellular, and physiological functions. EVs apply their biological effects in various manners: activating cell surface receptors via their surface marker, or/and exposing their intravesicular content, delivering different kinds of RNAs, including mRNAs and noncoding regulatory RNAs (such as miRNAs), transferring cytoplasmic or membranous proteins, and transmitting infectious pathogens into the recipient cells.[42,44–46] Their fundamental roles in physiological cross-talk are reported in several biological processes, including B-cell activation, stem cell activation, injury repair, immune tolerance induction, immunosuppression, activation of monocytes and NK cells, cell phenotype modulation, and synaptic plasticity. The potential roles of EVs in the process of various liver diseases, such as alcoholic hepatitis, fibrosis, portal hyper tension, and HCC were presented previously.[47] A recent study showed that EVs act as conveyors of miRNA-214 to primary mouse hepatocytes and LX-2 cells. The miRNA-214 in LX-2 cells was shuttled by exosomes to recipient LX-2 cells or human HepG2 hepatocytes, and suppresses connective tissue growth factor (CCN2) drives fibrogenesis in hepatic stellate cells (HSC). However, in the presence of fibrosis-inducing stimuli, the expression of miRNA-214 was decreased. Overall, diverse EV contents, and their close relation to intercellular signaling, position them as great candidate biomarkers of specific liver diseases.

Extracellular Vesicles-Associated miRNA as Liquid Biopsies

Circulating EVs have been detected in various types of biofluids, including serum, plasma, saliva, urine, amniotic fluid, semen, and breast milk.[40,42,48] Elevated levels of various subpopulations of circulating EVs were reported in patients with various liver diseases, including ALD,[49] hepatitis C,[50] and acute liver failure.[51] It has been shown that alcohol consumption can trigger EV production in mice.[52] Circulating miRNA can be in the form of EV associated,[53] bind to HDL/LDL,[54] and associated with RNA-binding proteins, such as Argonaute 2 and Argonaute 1.[55] Although, traditionally, most of the research on circulating miRNAs in the context of biomarker discovery and developing companion diagnostic test was focused on total circulating miRNAs, instead of considering different types of circulating miRNAs, such as EV-associated miRNAs. Analyzing EV rich fraction of serum and plasma is getting more attention in biomarkers discovery, since it has a higher rate of reproducibility and more stability.[42,56] Interestingly, EV-enriched fraction of miRNAs was useful for the prediction of inflammation and fibrosis in nonalcoholic steatohepatitis (NASH) and HCV, as well as categorizing patients with 96.59% accuracy.[56] Consistently, EV-enriched fraction of miRNAs in alcohol-fed mice was more stable and informative in prediction of ethanol induced liver injury, compared to measuring alanine aminotransferase (ALT) alone.[23]

With the emergence of new concepts in the domain of biomarker discovery, such as "liquid biopsy," there is a huge paradigm shift from liver biopsy to identify the extent of liver injury. In the context of ALD, single biopsy of liver is an invasive procedure, and provides a spatially and transient limited snapshot of the liver damage. In contrast, liquid biopsy is a noninvasive approach that depicts a landscape of liver function (Figure 22.4).

EV-Associated miRNAs as Biomarkers in Alcoholic Hepatitis and Other Liver Diseases

Although identifying the origin of EVs in plasma and serum is a challenge, it has been shown that circulating EVs consist of a mixed population of different EVs, originating from various cell types.[57] Interestingly, it has been shown that circulating EVs contain liver-specific mRNAs and miRNAs, under both physiological and diseased conditions.[23,56–58] miRNAs are released from damaged liver tissue to the circulation after various

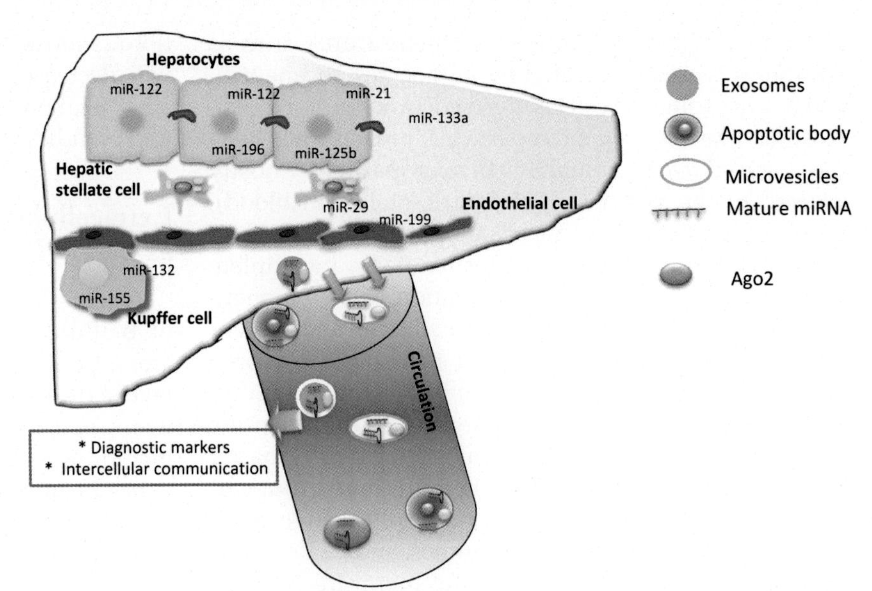

FIGURE 22.4 Extracellular miRNAs as new biomarkers of alcoholic liver disease. Liver injury caused by alcohol results in the release of miRNAs into circulation. The circulating miRNAs can be associated with either EV, such as exosomes, or with RNA inducing silencing complex proteins, such as Ago proteins.

liver injuries, including alcoholic hepatitis.[23,48] As we discussed earlier, miRNAs are present in two distinct forms, such as those of EV-associated miRNAs and free-floating miRNAs. However, the free-floating form of miRNAs is prone to degradation in the blood. This issue could be at least partially overcome by using miRNAs packed into the exosomes.[23,56–58]

Our group previously showed that, liver-specific miRNA-122 and inflammation-associated miRNA-155 gets significantly increased in EVs isolated from serum and plasma of alcohol-fed mice. Particularly, using EV associated miRNA-122 and miRNA-155 levels were more informative biomarkers of alcohol induced liver injury, than ALT alone.[23] Moreover, in mouse models of lipopolysaccharide-induced inflammatory liver injury (LILI), EV-associated miRNA-122 and miRNA-155 were remarkably increased in circulation.[23] It seems that a detectable miRNA-122 level in EVs could be an indicator of liver injury, since miRNA-122 was not detectable in EVs isolated from sera of human healthy controls.[59]

Intriguingly, the miRNA expression profile in human blood EVs was also found to be changed in patients with chronic hepatitis C, NASH, and nonalcoholic fatty liver disease (NAFLD); such changes were associated with the degree and severity of liver damage, and provided high diagnostic accuracy,[56,58] suggesting that miRNA profiling is a promising alternative to diagnosing liver disease. In a study assessing the diagnostic value of EV-associated miRNA in distinguishing various chronic liver diseases, NASH and chronic hepatitis C were compared.[56] This study demonstrated that, by using EV-associated miRNA profile, the stage of chronic liver disease (NASH and chronic hepatitis C in their model) can be determined, and it is correlated with the clinical status of the liver diseases. While these results suggest promising potentials for using EV-associated miRNAs for liver disease diagnosis, future studies, especially in ALD patients in a large cohort, are warranted to fully elucidate the diagnostic potential of EV-associated miRNAs profile. A list of identified deregulated EV-associated miRNAs in different liver diseases is presented in Table 22.1.

Employing Suitable EV Isolation and Characterization Methods to Increase Diagnostic Accuracy

Proper isolation and characterization of EVs and their cargos in body fluids are the prerequisite for potential downstream biomarker discovery. To find candidate biomarkers in EVs or subpopulation of the EVs, special attention must be paid to isolate EVs, or each subpopulation (exosomes and MVs) in a reproducible manner that increases the specificity of the assay.

The procedure starts from EV isolation. Although the gold standard and most frequently used procedure for EV isolation/purification is differential centrifugation, the efficiency was reported to be low, and it is not capable of differentiating between different subpopulation of EVs.[60] Ultracentrifugation involves several centrifugation and ultracentrifugation steps, while protocols vary across users, and this may lead to inconsistencies in recovery of EVs. In our experience, coupling the ultracentrifugation with microfiltration techniques, reduces both isolation time and increases purity of isolated EVs.[40] Moreover, considering the rotor k factor and normalizing viscosity is also among the additional steps that can increase EV recovery.[60,61]

Another limitation of using differential centrifugation for isolating EVs is coprecipitation of protein aggregates, apoptotic bodies, or nucleosomal fragments that could lead to less specificity of isolated EVs, and confounding

TABLE 22.1 Differential EV-Associated miRNAs in Liver Diseases

EV-associated miRNAs	Liver diseases	Species	Deregulation patterns	Detected in biological fluids	References
miRNA-122	ALD, LILI, NAFLD, and CHCV	Mouse, mouse, mouse, and human	Upregulated, upregulated, and upregulated	Serum/plasma, serum/plasma, serum, and serum	[23,56,58]
miRNA-192	NAFLD	Mouse	Upregulated	Serum	[58]
miRNA-155	ALD, LILI	Mouse	Upregulated	Serum	[23]
miRNA-320c	CHC	Human	Upregulated	Serum	[56]
miRNA-451	CHC	Human	Downregulated	Serum	[56]
miRNA-762	CHC	Human	Downregulated	Serum	[56]
miRNA1207-5p	CHC	Human	Upregulated	Serum	[56]
miRNA1225-5p	CHC	Human	Upregulated	Serum	[56]
miRNA-1275	CHC	Human	Upregulated	Serum	[56]
miRNA-486	NASH	Human	Downregulated	Serum	[56]
miRNA-762	NASH	Human	Downregulated	Serum	[56]

ALD, alcoholic liver disease; NAFLD, nonalcoholic fatty liver disease; LILI, lipopolysaccharide-induced inflammatory liver injury; CHC, chronic HCV; NASH, nonalcoholic steatohepatitis.

the interpretation of the assay. One way to address these issues is to use a sucrose gradient, which separates vesicles based on their different flotation densities, and subsequently increases downstream analysis precision. Another approach that can lead to more specificity and sensitivity of EV isolation is using immune magnetic beads coupled with surface antibodies for surface markers of different subpopulation of EVs. This strategy is not only a specific and sensitive method, but it can also be coupled with EV-condensation strategies to decrease cost and increase efficiency of the isolation.

Future Direction of EV-Associated miRNAs for ALD Diagnosis

EV-associated miRNAs have shown promising potential as biomarkers for the diagnosis and monitoring liver damage. However, human studies in the field of ALD are lacking. In the coming years, investigators have to validate the use of EV-associated miRNA as biomarker for ALD, in large cohorts of patients. Diagnostic value of EV-associated miRNAs should be compared and combined with other potential biomarkers of ALD. In fact, using expression levels of adequate numbers of miRNAs by using high throughput screening methods, and combining different biomarkers for identifying diseases, will increase diagnostic ability and accuracy, compared to using a single target miRNA. By recent strides in exosome isolation and characterization technologies, investigators and industries have to develop new reproducible assays that can routinely facilitate EV-associated miRNAs studies in different settings.

miRNA-TARGETED THERAPIES FOR ALD

The discussion above highlighted the fact that both tissue and circulating miRNAs play a role in the initiation and progression of ALD, suggesting their potential for serving as therapeutic targets in the future. In the following section, we will summarize recent processes and approaches relevant to miRNA-targeted therapy in general, and also in ALD (Figure 22.5).

miRNA-Level Therapeutics Using Conventional Gene Delivery Vehicles

The aforementioned evidence shows that miRNA plays pivotal roles in the progression of ALD. Thus, modifying the function of miRNA by using miRNA inhibitors, or augmenting miRNA biological functions using miRNA inhibitors, mimics, and precursors, seems an attractive approach to control pathological alterations in the ALD pathogenesis (Figure 22.5). Delivery of therapeutic genetic molecules in naked forms is very difficult, due to rapid clearance,[62] presence of RNases that reduce the half-life of small interfering RNAs,[63] and the lack of controllable cell specific distribution.[64] Thus, the cellular uptake of genetic biomacromolecules in general forms is very low, even after successful systemic delivery. For overcoming those limitations, gene delivery vehicles (GDVs) were introduced to fulfill the expectation of clinical nucleic acid-based therapy. Two most-studied types of GDVs are viral vectors and cationic liposomes.[65,66] Viral vectors are classified into adeno-associated viral vectors, retroviral/lentiviral vectors, and adenoviral

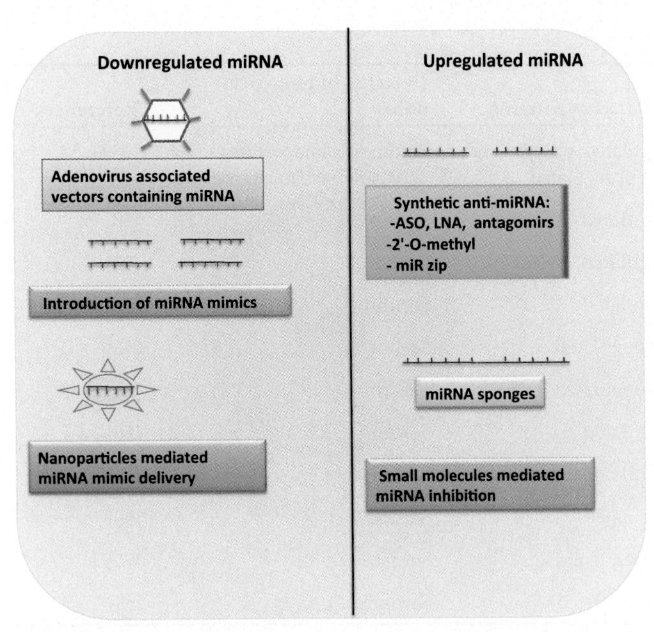

FIGURE 22.5 **miRNA-based therapeutic delivery to treat alcoholic liver disease.** Downregulated miRNA levels can be increased by delivering miRNA, using either adenoviral associated vectors, nanoparticles, or miRNA mimics. Upregulated miRNA function can be inhibited using synthetic antimiRs, such as antisense oligonucleotides (ASO), lock nucleic acid (LNA), antagomiRs, miR ZIp, 2′-O-methyl, or by delivering miRNA sponges or small molecules mediated miRNA inhibition.

vectors.[67] All three types of viral vectors can induce robust immune response against vehicles and transgenes. Adeno-associated viral vectors have drawbacks, including low probability of integration, reduced efficiency of repeated administration, and lack of ability to transfer RNA-based cargos. Limitations of retroviral/lentiviral vectors include high rate of insertional mutagenesis, and reduced efficiency over time. Cationic liposomes are usually cytotoxic, trigger immune reactions, exhibit low transduction efficiency, compared to viral vectors, and can be rapidly cleared from the circulation.[64,66] The immune-activation properties of both viral vectors and cationic liposomes necessitate use of concomitant immunosuppressive strategies to enhance uptake, and reduce the adverse immune responses. The considerable risk of current GDVs limited their use to life-threatening acute diseases in which benefits of therapy immediately outweigh the risks. However, for chronic diseases like ALD, where treatment requires repeated administration, a gene therapy vehicle (GDVs) with lower risk profile and ability to be administrated repetitively is needed. One of the new classes of biological GDVs is exosomes.

miRNA-Targeted Therapeutics Using Exosomes

Exosomes are a subset of EV in the 50–150 nm range. The fact that exosomes naturally transport mRNA, miRNA, and proteins between cells, has stimulated the interest for using exosomes as GVDs. Here, we highlight the opportunities and challenges associated with harnessing exosomes for miRNA therapeutics, as well as, practical workflow in alcoholic hepatitis (Figure 22.6).

Exosomes can be loaded with the miRNA-targeted therapeutic molecules, either prior to their release, by transacting the exosome producing cells, or after the release.[68] Although alternative methods, such as chemical transfection of exosomes, were used to load exosomes with biomolecules of interest,[69] electroporation is regarded as a more reliable method, and frequently reported as a method to load exosomes with miRNA or siRNA. Short interfering RNA (siRNA) was delivered via exosomes to the brain, in mouse model of Alzheimer's disease.[70] Specific targeting was achieved by generating exosomes, which expressed fused Lamp2b and neuron-specific RVG peptide. Purified exosomes were loaded with exogenous siRNA against β-secratase1 by electroporation. Intravenously, injected RVG-targeted exosomes delivered siRNA specifically to neurons, microglia, and oligodendrocytes in the brain, resulting in a specific gene knockdown with 62% protein knockdown of β-secratase1, a therapeutic target in Alzheimer's disease, in wild type mice. Another group showed that human plasma exosomes could be loaded with siRNA using electroporation.[69] Consequently, the siRNA-loaded exosomes can be internalized by human monocytes and lymphocytes, which lead to a knockdown effect of the siRNA targeted proteins in the recipient cells. In a recent study done by authors in the domain of alcoholic hepatitis,[71] miRNA-155 mimics and miRNA-155 inhibitors were successfully loaded into B cell-derived exosomes via electroporation, and delivered to the primary hepatocytes and RAW macrophages, respectively. Chemical-based transfection of exosomes using commercial transfection reagents has shown less efficiency of siRNA loading into the exosomes, when HiPerFect transfection reagent was used, compared to electroporation.[69,72] Together, those studies suggest that exosomes can be loaded by electroporation and deliver functional cargo to the recipient cells. Another approach to produce exosomes harboring therapeutic miRNA-targeted molecules is transfection of parental cells with therapeutic sequence, and subsequently they package and release the therapeutic molecule within exosomes. In a recent study, cells were transfected to express let-7a, and released exosomes from transfected cells were collected and shown to contain let-7a, which significantly suppressed the growth of xenograft breast cancer tumors in mice.[73]

Opportunities and Suitable Potentials of Exosomes for miRNA-Level Therapeutic

The recognized biological function of exosomes as natural conveyors of intercellular communications has

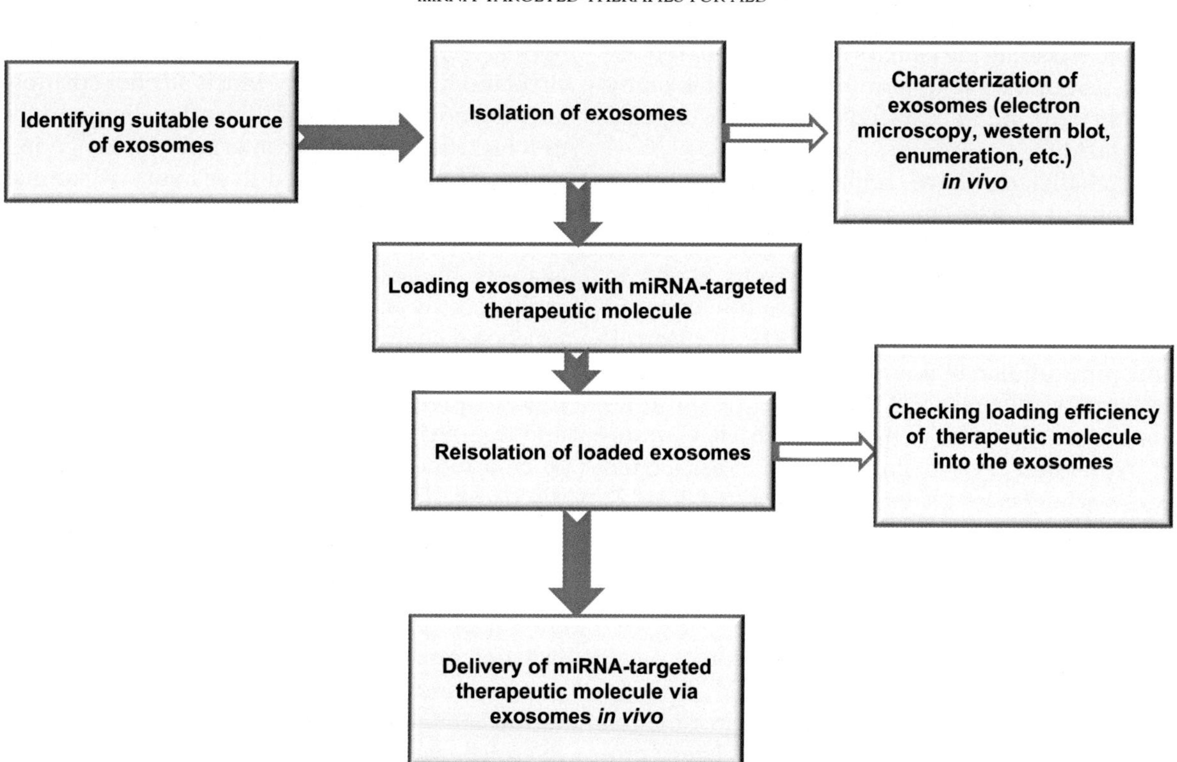

FIGURE 22.6 **Workflow of miRNA-based therapeutic delivery using exosomes to treat alcoholic liver disease.**

been further explored in the delivery of genetic molecules. Results of several studies showed that exosomes can be harnessed to establish a natural delivery mechanism, that is, safe and efficient.[71,73,74] One great benefit of exosomes for acting as biological GDVs, is their potential to mediate gene delivery without inducing adverse immune reactions.[71,74] Repeated intravenous (IV) administration of autologous exosomes derived from DCs did not induce any immune activation in mice.[69] Allogenic exosomes derived from BALB/c DC did not induce immune reaction in mice; splenic DCs subsequently isolated from the recipient mice did not show maturation markers indicative of immune activation.[75] Immune tolerance of exosomes appears to be present between species.[76] For instance, exosomes derived from human mesenchymal stem cells and human HEK293 cells were tolerated and functional, in the recipient mice.[77,78] However, those studies did not investigate the risk of acute inflammation, the potential of adverse effect in repeated administration, and long-term immunological effects. From the cytotoxicity perspective, exosomes from different cell types showed minimal baseline toxicity in the recipient cells, and were considered to be a safe treatment.[71]

Although exosomes are well suited to deliver a variety of biomacromolecules, including proteins, miRNAs, mRNAs, genomic DNAs, and noncoding RNAs, the efficiency of exosome-mediated delivery varies based on the cell type, cell status of parental cells, and recipient cell type. To obtain optimal treatment efficiency, each exosome-mediated delivery protocol has to be optimized, based on the recipient cell type, special cargo, and exosome source.[71] Thus, using exosomes as delivery vehicles for miRNA-targeted therapeutics needs identification and isolation of suitable exosomes, characterization of exosomes, and optimization of delivery methods. Moreover, the efficiency of miRNA-targeted therapy should be compared to other delivery vehicles.

Although developing exosome-mediated delivery for targeting miRNAs that play roles in alcoholic hepatitis treatment still remained a challenge, our recent study[71] showed successful delivery of functionally active miRNA-155 inhibitor, both *in vitro* and *in vivo*. Both exogenous miRNA-155 mimic and miRNA-155 inhibitor were successfully introduced into murine B cell-derived exosomes via electroporation. Loaded exosomes were capable of delivering miRNA-155 mimic and miRNA-155 inhibitor to primary mouse hepatocytes and RAW 264.7 macrophages, respectively. Exosomes showed significantly increased functional delivery of miRNA-155 inhibitor into the RAW 264.7 cells, causing inhibition of miRNA-155, and functionally led to statistically significant decrease of TNF-α protein level, compared to transfection reagents. In the *in vivo* setting, injection of miRNA-155 loaded exosomes were able to mediate successful delivery of miRNA-155 mimic to the liver and hepatocytes of miRNA-155 knockout mice, as early as 10 min after IV injection. However, further studies at various time points are needed to elucidate the kinetics of exosome-based delivery, and determined long-term efficiency. This study is the only *in vitro* and *in vivo* study of miRNA-targeted treatment for ALD.

Several challenges have remained during clinical development of exosome mediated miRNA-based therapeutics for alcoholic hepatitis. One of the major challenges is that miRNA target numerous genes, in diverse type of cells, and their long-term function and biological response in complex networks of intercellular and intracellular biological pathways remained unclear. For example, miRNA-122 plays different roles in various liver diseases. With respect to HCC, a decreased expression level of miRNA-122 is associated withHCC,[79] whereas an increased level of miRNA-122 plays an important positive role in regulation of hepatitis C virus replication.[46] The fact that each miRNA is involved in various pathways and biological processes, makes miRNA-targeted therapeutics complex, and requires the fine control of miRNA expression.

Another challenge in exosome-mediated delivery of miRNA-target therapies is the difficulty in isolation methods and quantification of the exosomes. Every protocol of using exosomes as delivery vehicles should be first meticulously optimized, based on the source of exosomes, cargo of the exosomes, recipient cells or organs, and therapeutic targets.

CONCLUSIONS

In this chapter, we summarized recent findings and perspectives on the role of miRNAs in the development and progression of ALD, clinical application of miRNAs as biomarkers, and potential therapeutic targets. The differential expression of miRNAs in the liver and circulation appears to reflect the complex and extensive signaling potential of these regulatory noncoding RNAs in ALD. Researchers are now beginning to unravel the complex roles of miRNAs in ALD. Although both *in vitro* and *in vivo* data clearly demonstrate the role of certain deregulated miRNAs in ALD, the driving force behind miRNA research was not only understanding the role of miRNA in the pathophysiology of ALD, but also the potential for using miRNAs a noninvasive biomarker, as well as targets of finely tuned therapeutics. In spite of this, there are still some challenges that need to be overcome before the provision of miRNAs as biomarkers and biological therapeutic targets. A critical challenge in miRNA research has been the poor reproducibility in different settings, and the lack of standard normalizer particular to each condition, and standard assay methods.

Perhaps the greatest challenge in introducing miRNAs as biomarkers is understanding the balance between normal and pathological deregulation of miRNA level in circulation, the degree to which alcohol corrupts or ablates the normal state of miRNA signaling, and the extent to which these signaling interactions change in various phases of alcohol-induced liver injury. What makes this especially challenging is the multifunctional and multidimensional nature of miRNA signaling. In the functional analysis, the ability of miRNAs to concurrently and simultaneously signal in various types of cells, makes functional analysis difficult.

As we mentioned earlier, there are two types of miRNA present in circulation: EV-associated miRNA, and free-floating miRNA. EV-associated miRNAs seem to be a more viable and reliable source for biomarker discovery, due to more stability, specificity, and reproducibility, compared to free floating miRNA. For the miRNAs that are packaged into the EVs, the EVs must be first isolated using a highly effective method and reproducible protocol. Although various methods of EV isolation exist, most of the methods are time-consuming, need various quality control processes, and expert technicians. Moreover, isolation techniques may serve to enrich exosomes subpopulations that have to be taken into account when comparing different studies, and may lead to different results. Nonetheless, due to significant challenges that are being undertaken by many groups showing strong scientific interest in unraveling the miRNAs, and validating studies and assays, the authors envision the emergence of miRNA diagnostic platforms for ALD that can improve individualized patient care and prognosis.

For developing clinical miRNA-targeted therapeutics, and to fully appreciate the signaling consequence of targeting one miRNA, further studies will need to investigate the result of enhancing miRNA activities or suppressing miRNA activities in not only pathological phenotypes, but also in other long-term compensatory feedback loops. Dissecting and exploring the contributions of deregulated miRNA component in ALD, and introducing GVDs and exosomes to elicit the required therapeutic response, will undoubtedly pave the translational way in the administration of miRNA-targeted therapies. It is for these reasons, and the many diverse reasons discussed in this article, that miRNAs might prove to be one of the most useful biomarkers and therapeutic targets so far identified in ALD.

Key Facts

- miRNAs are short noncoding RNA molecules that do not encode for any protein.
- Various miRNAs have been affected in alcoholic liver disease.
- Extracellular miRNAs are emerging as new biomarkers in ALD.

- EVs (exosomes, MVs, and apoptotic bodies) are released by various cells into the intercellular microenvironment and circulation.
- EV associated miRNA is more stable to RNase activity, and regarded as a more reproducible source of miRNA for biomarker discovery.

Summary Points

- In this chapter, we describe some of the recent advances on the role of miRNAs (miRNAs) in alcoholic liver disease (ALD).
- miRNAs play pivotal roles in the initiation, progression, and prognosis of ALD.
- miRNAs such as miRNA-122, miRNA-155, miRNA-21, miRNA-217, and miRNA-34a are deregulated in ALD.
- miRNAs has potential as new therapeutic targets to alleviate ALD.
- EVs play roles in the pathogenesis of different liver diseases, including ALD.
- EVs carry a variety of biomacromolecules and nucleic acids, including miRNA that can be used for biomarker discovery and disease monitoring.
- The smallest subpopulation of EVs, called exosomes, is the best studied subpopulation, and can be used for RNA inferences delivery and miRNA-based therapeutics.

Acknowledgment

Competing financial statement: authors declare no competing financial interests.

References

1. Adachi M, Brenner DA. Clinical syndromes of alcoholic liver disease. *Dig Dis* 2005;**23**:255–63.
2. Gao B, et al. Innate immunity in alcoholic liver disease. *Am J Physiol Gastrointest Liver Physiol* 2011;**300**:G516–25.
3. Bala S, Marcos M, Gattu A, Catalano D, Szabo G. Acute binge drinking increases serum endotoxin and bacterial DNA levels in healthy individuals. *PLoS One* 2014;**9**:e96864.
4. Szabo G, Bala S. MicroRNAs in liver disease. *Nat Rev Gastroenterol Hepatol* 2013;**10**:542–52.
5. Ambade A, Mandrekar P. Oxidative stress and inflammation: essential partners in alcoholic liver disease. *Int J Hepatol* 2012; **2012**:853175.
6. Cederbaum AI. CYP2E1 potentiates toxicity in obesity and after chronic ethanol treatment. *Drug Metabol Drug Interact* 2012;**27**:125–44.
7. Hartmann P, Chen WC, Schnabl B. The intestinal microbiome and the leaky gut as therapeutic targets in alcoholic liver disease. *Front Physiol* 2012;**3**:402.
8. Szabo G, Bala S. Alcoholic liver disease and the gut-liver axis. *World J Gastroenterol* 2010;**16**:1321–9.
9. Lee RC, Feinbaum RL, Ambros V. The *C. elegans* heterochronic gene lin-4 encodes small RNAs with antisense complementarity to lin-14. *Cell* 1993;**75**:843–54.
10. Bartel DP. MicroRNAs: genomics, biogenesis, mechanism, and function. *Cell* 2004;**116**:281–97.
11. Friedlander MR, et al. Evidence for the biogenesis of more than 1,000 novel human microRNAs. *Genome Biol* 2014;**15**:R57 2014-15-4-r57.
12. Bala S, Marcos M, Szabo G. Emerging role of microRNAs in liver diseases. *World J Gastroenterol* 2009;**15**:5633–40.
13. Dolganiuc A, et al. MicroRNA expression profile in Lieber-DeCarli diet-induced alcoholic and methionine choline deficient diet-induced nonalcoholic steatohepatitis models in mice. *Alcohol Clin Exp Res* 2009;**33**:10.
14. Tang Y, et al. Effect of alcohol on miR-212 expression in intestinal epithelial cells and its potential role in alcoholic liver disease. *Alcohol Clin Exp Res* 2008;**32**:355–64.
15. Yeligar S, Tsukamoto H, Kalra VK. Ethanol-induced expression of ET-1 and ET-BR in liver sinusoidal endothelial cells and human endothelial cells involves hypoxia-inducible factor-1alpha and microrNA-199. *J Immunol* 2009;**183**:5232–43.
16. Jopling C. Liver-specific microRNA-122: biogenesis and function. *RNA Biol* 2012;**9**:137–42.
17. Hsu SH, et al. Essential metabolic, anti-inflammatory, and anti-tumorigenic functions of miR-122 in liver. *J Clin Invest* 2012;**122**: 2871–83.
18. Tsai WC, et al. MicroRNA-122 plays a critical role in liver homeostasis and hepatocarcinogenesis. *J Clin Invest* 2012;**122**:2884–97.
19. Janssen HL, et al. Treatment of HCV infection by targeting microRNA. *N Engl J Med* 2013;**368**:1685–94.
20. Conrad KD, et al. microRNA-122 dependent binding of Ago2 protein to hepatitis C virus RNA is associated with enhanced RNA stability and translation stimulation. *PLoS One* 2013;**8**:e56272.
21. Lanford RE, et al. Therapeutic silencing of microRNA-122 in primates with chronic hepatitis C virus infection. *Science* 2010;**327**: 198–201.
22. Elmen J, et al. LNA-mediated microRNA silencing in non-human primates. *Nature* 2008;**452**:896–9.
23. Bala S, et al. Circulating microRNAs in exosomes indicate hepatocyte injury and inflammation in alcoholic, drug-induced and inflammatory liver diseases. *Hepatol* 2012;**56**:1946–57.
24. Satishchandran A, Bala S, Szabo G. The role of miR-122 in the pathogenesis of chronic alcoholic liver disease. *Alcohol* 2013; **47**:574.
25. Li J, et al. miR-122 regulates collagen production via targeting hepatic stellate cells and suppressing P4HA1 expression. *J Hepatol* 2013;**58**:522–8.
26. Trebicka J, et al. Hepatic and serum levels of miR-122 after chronic HCV-induced fibrosis. *J Hepatol* 2013;**58**:234–9.
27. Csak T, et al. MicroRNA-122 regulates hypoxia-inducible factor-1 and vimentin in hepatocytes and correlates with fibrosis in diet-induced steatohepatitis. *Liver Int* 2014;**35**:532–41.
28. O'Connell RM, et al. MicroRNA-155 promotes autoimmune inflammation by enhancing inflammatory T cell development. *Immunity* 2010;**33**:607–19.
29. Kong W, et al. MicroRNA-155 regulates cell survival, growth, and chemosensitivity by targeting FOXO3a in breast cancer. *J Biol Chem* 2010;**285**:17869–79.
30. Kurowska-Stolarska M, et al. MicroRNA-155 as a proinflammatory regulator in clinical and experimental arthritis. *Proc Natl Acad Sci USA* 2011;**108**:11193–8.
31. Tili E, et al. Mutator activity induced by microRNA-155 (miR-155) links inflammation and cancer. *Proc Natl Acad Sci USA* 2011;**108**: 4908–13.
32. Lippai D, Bala S, Csak T, Kurt-Jones EA, Szabo G. Chronic alcohol-induced microRNA-155 contributes to neuroinflammation in a TLR4-dependent manner in mice. *PLoS One* 2013;**8**:e70945.
33. Bala S, et al. Up-regulation of microRNA-155 in macrophages contributes to increased tumor necrosis factor {alpha} (TNF{alpha}) production via increased mRNA half-life in alcoholic liver disease. *J Biol Chem* 2011;**286**:1436–44.

34. Biaya D, Affò S, Rodrigo-Torres D, Morales-Ibanez O, Altamirano J, Coll M, Vila M, Lozano JJ, Arroyo V, Caballería J, Bataller R, Ginés P, Sancho-Bru P. Integrative microrna profiling identifies mir-21 and mir-155 as potential regulators in alcoholic hepatitis. *J Hepatol* 2014;**60**(1 Suppl.):S59.
35. Meng F, et al. Epigenetic regulation of miR-34a expression in alcoholic liver injury. *Am J Pathol* 2012;**181**:804–17.
36. Francis H, et al. Regulation of the extrinsic apoptotic pathway by microRNA-21 in alcoholic liver injury. *J Biol Chem* 2014;**289**: 27526–39.
37. Yin H, et al. MicroRNA-217 promotes ethanol-induced fat accumulation in hepatocytes by down-regulating SIRT1. *J Biol Chem* 2012;**287**:9817–26.
38. Bala S, Szabo G. MicroRNA signature in alcoholic liver disease. *Int J Hepatol* 2012;**2012**:498232.
39. Shukla SD, et al. Emerging role of epigenetics in the actions of alcohol. *Alcohol Clin Exp Res* 2008;**32**:1525–34.
40. Momen-Heravi F, et al. Current methods for the isolation of extracellular vesicles. *Biol Chem* 2013;**394**:1253–62.
41. Simpson RJ, Kalra H, Mathivanan S. ExoCarta as a resource for exosomal research. *J Extracell Vesicles* 2012;**1**.
42. Jia S, et al. Emerging technologies in extracellular vesicle-based molecular diagnostics. *Expert Rev Mol Diagn* 2014;**14**:307–21.
43. Kalra H, et al. Vesiclepedia: a compendium for extracellular vesicles with continuous community annotation. *PLoS Biol* 2012;**10**: e1001450.
44. Bashratyan R, Sheng H, Regn D, Rahman MJ, Dai YD. Insulinoma-released exosomes activate autoreactive marginal zone-like B cells that expand endogenously in prediabetic NOD mice. *Eur J Immunol* 2013;**43**:2588–97.
45. Ramachandran S, Palanisamy V. Horizontal transfer of RNAs: exosomes as mediators of intercellular communication. *Wiley Interdiscip Rev RNA* 2012;**3**:286–93.
46. Bukong TN, Momen-Heravi F, Kodys K, Bala S, Szabo G. Exosomes from hepatitis C infected patients transmit HCV infection and contain replication competent viral RNA in complex with Ago2-miR122-HSP90. *PLoS Pathog* 2014;**10**:e1004424.
47. Chen L, et al. Epigenetic regulation of connective tissue growth factor by microRNA-214 delivery in exosomes from mouse or human hepatic stellate cells. *Hepatology* 2014;**59**:1118–29.
48. Lemoinne S, et al. The emerging roles of microvesicles in liver diseases. *Nat Rev Gastroenterol Hepatol* 2014;**11**:350–61.
49. Ogasawara F, et al. Platelet activation in patients with alcoholic liver disease. *Tokai J Exp Clin Med* 2005;**30**:41–8.
50. Brodsky SV, et al. Dynamics of circulating microparticles in liver transplant patients. *J Gastrointestin Liver Dis* 2008;**17**:261–8.
51. Agarwal B, et al. Evaluation of coagulation abnormalities in acute liver failure. *J Hepatol* 2012;**57**:780–6.
52. Chen Y, Davis-Gorman G, Watson RR, McDonagh PF. Platelet CD62p expression and microparticle in murine acquired immune deficiency syndrome and chronic ethanol consumption. *Alcohol Alcohol* 2003;**38**:25–30.
53. Valadi H, et al. Exosome-mediated transfer of mRNAs and microRNAs is a novel mechanism of genetic exchange between cells. *Nat Cell Biol* 2007;**9**:654–9.
54. Vickers KC, Palmisano BT, Shoucri BM, Shamburek RD, Remaley AT. MicroRNAs are transported in plasma and delivered to recipient cells by high-density lipoproteins. *Nat Cell Biol* 2011;**13**:423–33.
55. Arroyo JD, et al. Argonaute2 complexes carry a population of circulating microRNAs independent of vesicles in human plasma. *Proc Natl Acad Sci USA* 2011;**108**:5003–8.
56. Murakami Y, et al. Comprehensive miRNA expression analysis in peripheral blood can diagnose liver disease. *PLoS One* 2012;**7**:e48366.
57. Yang X, Weng Z, Mendrick DL, Shi Q. Circulating extracellular vesicles as a potential source of new biomarkers of drug-induced liver injury. *Toxicol Lett* 2014;**225**:401–6.
58. Povero D, et al. Circulating extracellular vesicles with specific proteome and liver microRNAs are potential biomarkers for liver injury in experimental fatty liver disease. *PLoS One* 2014;**9**:e113651.
59. Hunter MP, et al. Detection of microRNA expression in human peripheral blood microvesicles. *PLoS One* 2008;**3**:e3694.
60. Momen-Heravi F, et al. Impact of biofluid viscosity on size and sedimentation efficiency of the isolated microvesicles. *Front Physiol* 2012;**3**:162.
61. Cvjetkovic A, Lotvall J, Lasser C. The influence of rotor type and centrifugation time on the yield and purity of extracellular vesicles. *J Extracell Vesicles* 2014;**3**:10.3402/jev.v3.23111.
62. Takakura Y, Nishikawa M, Yamashita F, Hashida M. Development of gene drug delivery systems based on pharmacokinetic studies. *Eur J Pharm Sci* 2001;**13**:71–6.
63. Soutschek J, et al. Therapeutic silencing of an endogenous gene by systemic administration of modified siRNAs. *Nature* 2004;**432**: 173–8.
64. Seow Y, Wood MJ. Biological gene delivery vehicles: beyond viral vectors. *Mol Ther* 2009;**17**:767–77.
65. Gehrig S, Sami H, Ogris M. Gene therapy and imaging in preclinical and clinical oncology: recent developments in therapy and theranostics. *Ther Deliv* 2014;**5**:1275–96.
66. Awada A, et al. A randomized controlled phase II trial of a novel composition of paclitaxel embedded into neutral and cationic lipids targeting tumor endothelial cells in advanced triple-negative breast cancer (TNBC). *Ann Oncol* 2014;**25**:824–31.
67. Serguera C, Bemelmans AP. Gene therapy of the central nervous system: general considerations on viral vectors for gene transfer into the brain. *Rev Neurol (Paris)* 2014;**170**:727–38.
68. Lasser C. Exosomes in diagnostic and therapeutic applications: biomarker, vaccine and RNA interference delivery vehicle. *Expert Opin Biol Ther* 2015;**15**:103–17.
69. Wahlgren J, et al. Plasma exosomes can deliver exogenous short interfering RNA to monocytes and lymphocytes. *Nucleic Acids Res* 2012;**40**:e130.
70. Alvarez-Erviti L, et al. Delivery of siRNA to the mouse brain by systemic injection of targeted exosomes. *Nat Biotechnol* 2011;**29**: 341–5.
71. Momen-Heravi F, Bala S, Bukong T, Szabo G. Exosome-mediated delivery of functionally active miRNA-155 inhibitor to macrophages. *Nanomed* 2014;**7**:1517–1527.
72. Shtam TA, et al. Exosomes are natural carriers of exogenous siRNA to human cells *in vitro*. *Cell Commun Signal* 2013;**11**:88.
73. Ohno S, et al. Systemically injected exosomes targeted to EGFR deliver antitumor microRNA to breast cancer cells. *Mol Ther* 2013;**21**: 185–91.
74. Banizs AB, et al. *In vitro* evaluation of endothelial exosomes as carriers for small interfering ribonucleic acid delivery. *Int J Nanomed* 2014;**9**:4223–30.
75. Morelli AE, et al. Endocytosis, intracellular sorting, and processing of exosomes by dendritic cells. *Blood* 2004;**104**:3257–66.
76. Marcus ME, Leonard JN. FedExosomes: engineering therapeutic biological nanoparticles that truly deliver. *Pharmaceuticals (Basel)* 2013;**6**:659–80.
77. Arslan F, et al. Mesenchymal stem cell-derived exosomes increase ATP levels, decrease oxidative stress and activate PI3K/Akt pathway to enhance myocardial viability and prevent adverse remodeling after myocardial ischemia/reperfusion injury. *Stem Cell Res* 2013;**10**:301–12.
78. Lee C, et al. Exosomes mediate the cytoprotective action of mesenchymal stromal cells on hypoxia-induced pulmonary hypertension. *Circulation* 2012;**126**:2601–11.
79. Wu Q, et al. Decreased expression of hepatocyte nuclear factor 4alpha (Hnf4alpha)/microRNA-122 (miR-122) axis in hepatitis B virus-associated hepatocellular carcinoma enhances potential oncogenic GALNT10 protein activity. *J Biol Chem* 2015;**290**:1170–85.

CHAPTER

23

Alcohol Metabolism and Epigenetic Methylation and Acetylation

Marisol Resendiz, BA, BS, Sherry Chiao-Ling Lo, PhD**,†, Jill K. Badin, BS*, Ya-Jen Chiu, BS**,‡, Feng C. Zhou, PhD*,**,†,§*

*Department of Cellular & Integrative Physiology, Stark Neuroscience Research Institute, Indiana University-Purdue University at Indianapolis, Indianapolis, IN, USA

**Department of Anatomy and Cell Biology, Indiana University-Purdue University at Indianapolis, Indianapolis, IN, USA

†Indiana Alcohol Research Center, Indiana University-Purdue University at Indianapolis, Indianapolis, IN, USA

‡Department of Life Science, National Taiwan Normal University, Taipei, Taiwan

§Indiana University School of Medicine, and Department of Psychology, Indiana University-Purdue University at Indianapolis, Indianapolis, IN, USA

INTRODUCTION

Beside multiple known physiological effects, which are not the focus of this chapter, alcohol has now been identified as an effectual epigenetic modifier through its unique chemical structure and metabolites, and its multifactorial impacts. One of the most critical is perhaps its ability to alter methylation and acetylation biogenesis, ultimately altering DNA and histone structure – the backbone of epigenetics. After ingestion, alcohol is oxidized in the liver by a series of enzymes, and converted to acetaldehyde, which in turn is converted to acetate and, further, to acetyl-CoA. Meanwhile, alcohol is a powerful inhibitor of the one-carbon metabolic cycle and, therefore, a means by which to influence the methyl donor pool required for many transmethylation reactions. Both of the above processes alter cellular methyl and acetyl availability, consequently impacting normal DNA methylation and histone tail codes (as well as microRNA, though not discussed here). As such, alcohol has a direct role as a major modifier of epigenetics (Sections "Alcohol Metabolism and Epigenetic Acetylation" and "Alcohol and Methyl Metabolism"). Alcohol has also demonstrated its capability to affect a wide range of epigenetic enzymes to further affect DNA methylation and demethylation, as well as histone methyl and acetyl transfer in a substrate and donor independent manner. The nature of alcohol to act on the enzymes required for epigenetic modification only serves to further implicate alcohol as a keen environmental effector of epigenetics (Section "Alcohol Affects the Epigenetics Converting Enzymes"). The range of epigenetic changes and the consequential transcriptional change brought about by alcohol is vast. Through this epigenetic pathway, alcohol has been reported to alter cellular functions, neurochemistry, receptor structure, neurodevelopment, and organ formation. Novel hypotheses on the pathology and mechanism of alcohol in fetal alcohol syndrome, cancer, liver dysfunction, and mental dysfunction are currently being investigated with promising supporting evidence (Section "Epigenetic Footprint of Alcohol Induced Dysfunction and Disease"). Alcohol affects DNA methylation on multiple levels. We advocate that alcohol can hyper- and hypomethylate DNA at different chromatic regions, and can alter both canonical 5-methylcytosine (5mC), and its derivatives (such as 5-hydroxylmethylcytosine (5hmC)) that are increasingly being found to be functionally diverse.[1] Furthermore, while acute alcohol altered DNA methylation has been reported to be reversible, chronic alcohol may have longer range, or even transgenerational consequences (Section "Epigenetic Footprint of Alcohol Induced Dysfunction and Disease"). Alcohol has

Molecular Aspects of Alcohol and Nutrition. http://dx.doi.org/10.1016/B978-0-12-800773-0.00023-9

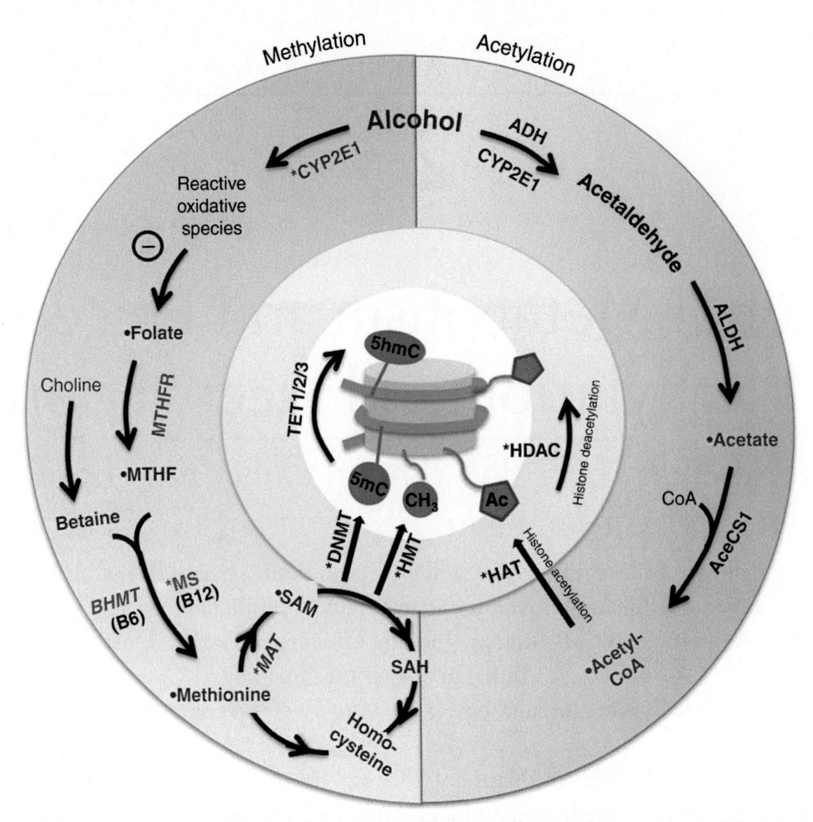

FIGURE 23.1 **Alcohol's acetyl and methyl metabolism on epigenetics.** On the acetylation side (red hemisphere), alcohol is metabolized by ADH and CYP2E1 into acetaldehyde. This process produces many reactive oxygen species that pour over to affect folate metabolism. Acetaldehyde is further metabolized to acetate by ALDH enzymes. Acetate is then converted to acetyl-CoA, the acetyl donor for histone acetylation enzymes acting on the amino acids of variable histone proteins (center). Once established, histone deacetylases can remove the acetyl group. On the methylation side (blue hemisphere), alcohol and its generated reactive oxidative species exert negative effect on folate and choline metabolism. dietary folate is metabolized by MTHFR to methyl tetrahydrofolate (MTHF). Meanwhile, dietary choline is converted to betaine. Betaine and MTHF both serve as the methyl donor for homocysteine conversion to methionine by the enzyme methionine synthase (MS). Methionine next becomes "activated" by methyl adenosyltransferase enzymes into the final methyl donor form, known as S-adenosylmethionine (SAM). SAM is utilized by both DNA methyltransferases and histone methyl transferases to the 5′ carbon of cytosine bases or histone tail residues, respectively (center). Solid dots, substrate; stars, enzymes, which are known to be affected by alcohol.

been associated with humans as a beverage of consumption for over 7 thousand years, though it may have even longer ties (millions of years before civilization).[2] Today, two thirds of adults consume alcohol on a regular basis, in the western world.[3] Thus, alcohol's epigenetic memory may be well recorded in human DNA. Since methylated DNA has higher frequency of mutation, alcohol may, through DNA methylation, affect the genomic make-up that would render alcohol an epigenomic–genomic interaction. A better understanding of epigenetic pathogenic mechanisms could enable development of novel treatments of alcohol-related diseases. Further, since alcohol-altered DNA methylation is recoded like a memory in some tissues, a diagnostic approach can be designed to exploit epigenetic markers for alcohol exposure or pathology (Section "Treatment and Biomarkers of Alcohol Diseases Through Epigenetic Pathway"). Though it is clear that alcohol plays a role in altering epigenetic schedules tied to normal transcriptional patterns, and that alcohol (at least partially) exerts its deleterious effects on mammalian systems via the exploitation of epigenetic mechanisms, many questions remain to be answered in the resolution of alcohol-epigenetic dynamics, and the transduction of environmentally regulated disease.

ALCOHOL METABOLISM AND EPIGENETIC ACETYLATION

Acetaldehyde

To understand how alcohol affects epigenetic histone modifications and DNA methylation, it is necessary to begin with an appreciation of alcohol metabolism (Figure 23.1). When alcohol enters the body, it is oxidized into acetaldehyde by one of three different classes of enzymes: alcohol dehydrogenase (ADH), cytochrome P450 isoform 2E1 (CYP2E1), and catalase. Acetaldehyde is a strong teratogenic agent that is suspected to cause several of the adverse effects associated with alcohol, through the production of reactive oxidation species (ROS). Acetaldehyde is present at much lower levels in the blood than alcohol, after alcohol consumption.[4] Chronic consumption of alcohol induces the activity of CYP2E1 that is found in the mitochondria of hepatocytes and, similar to ADH, oxidizes alcohol into acetaldehyde (Figure 23.1).

Histone acetylation is associated with gene activation while histone deacetylation is associated with gene silencing.[5] Modifying the enzymes responsible for histone acetylation and deacetylation (histone acetyltransferases (HATs) and histone deacetylases (HDACs), respectively) represents one way that alcohol and its metabolites can regulate gene expression. It has recently been hypothesized that histone acetylation may lead to chromatin instability, resulting in the transcription of cryptic promoters, and subsequent aberrant peptide synthesis.[6] The oxidative stress produced by ROS can mediate histone acetylation and deacetylation by increasing HAT activity, and decreasing HDAC activity.[7] These enzymes will be discussed in greater detail in Section "Alcohol Affects the Epigenetics Converting Enzymes." Choudhury et al.[8] demonstrated that ROS production plays a vital role in histone acetylation, particularly on histone H3 at lysine 9 (H3AcK9), in rat hepatocytes. Cells treated with alcohol and an ROS inhibitor exhibited reduced levels of ROS and H3AcK9, as compared to cells treated with just alcohol, whereas cells treated with alcohol and an ROS inducer exhibited higher levels of ROS and H3AcK9.[8] Therefore, acetaldehyde and subsequent ROS production from alcohol metabolism may influence chromatin remodeling. CYP2E1 has also been shown to directly reduce histone deacetylase 2 (HDAC2) activity in HepG2 cells, leading to increased histone acetylation.[9] The effects of acetaldehyde on DNA methylation will be discussed further in (Section "Alcohol and Methyl Metabolism").

Acetate

Acetate is the second major downstream metabolite of alcohol and can directly and indirectly effect epigenetic targets (i.e., histones) through acetylation (see Figure 23.1). Downstream of acetaldehyde, aldehyde dehydrogenase (ALDH) catalyzes the irreversible oxidation of acetaldehyde to acetate.[10] The irreversibility of the reaction helps to maintain a relatively low concentration of acetaldehyde in the body.[11] In healthy men, an average of 77% of ingested ethanol is converted into acetate, making it the primary metabolite of alcohol.[12] In rats chronically exposed to alcohol, about 90% of acetate produced in the liver is released into the bloodstream, and blood acetate levels peak around 1.5–2 mM.[11] Acetate may then systemically alter the cellular milieu of acetyl-CoA, affecting histone acetylation as acetyl-CoA is formed through the binding of acetate to coenzyme A, by acetyl-coenzyme A synthetase (AceCS1) (Figure 23.1). The acetyl-CoA produced then either enters the Krebs cycle, to supply energy to the cell, or, through histone acetyl transferase (HAT) activity, donates an acetyl group to a histone, preferentially acetylating lysine 9 residues on histone 3 proteins (H3K9Ac).[13,14] Acetate decreases class I HDAC activity in hepatic cells and neurons, but does not show any significant changes in HAT activity in those cell types.[15] Other enzymes involved in the acetyl-CoA pathway (as well as acetyl-CoA itself) play a role in histone acetylation. In a study by Takahashi et al.[16] it was shown that when AceCS1 is inactivated, there is a rapid deacetylation of histones. This suggests that either AceCS1 activity or the amount of acetyl-CoA in the system affect histone acetylation.[16] In another study, it was found that acetate treatment produced a higher level of acetyl-CoA in cultured rat hepatocytes.[17] This could contribute to the higher histone acetylation observed after alcohol consumption.

Alcohol can also alter the activities of several metabolic enzymes (Table 23.1). Alcohol exposure, for example, causes about a 10-fold increase in ADH and CYP2E1 activity in cultured liver Hep-G2 cells.[18] Alcohol also causes a significant decrease in ALDH enzyme activity in chronic alcoholics, causing a build-up of acetaldehyde.[19] A build-up of acetaldehyde increases the formation of harmful acetaldehyde-DNA adducts, and produces higher levels of ROS that contribute to histone acetylation alterations.[20] However, in a study by Chrostek et al.[21] investigating ALDH activity in rat liver, there was a decrease in ALDH activity in older rats (aged 24 months), signifying that activity levels are a complex process, likely controlled by several different variables.[21]

One of the variables that can affect the activity of alcohol-metabolizing enzymes is intrinsic expression. Translationally, there are many enzyme variants suspected, among different ethnic groups that may affect alcohol metabolism and pose potentially significant influences on epigenetic dynamics. This interesting finding begs the question of whether there can be meaningful differences in the magnitude of DNA methylation and histone modification among different ethnicities, established long before the onset of alcohol consumption occurs. Theoretically, since alcohol and its major metabolites are known to cause various epigenomic changes, the ethnic diversity of alcohol metabolism should yield differential epigenetic responses that may be lasting, and of clinical relevance (see Box 23.1). However, there has not been literature published to date that collectively explores this hypothesis in depth. This would be an interesting and novel investigation that could elucidate how alcohol affects different populations and diverse epigenetic profiles.

ALCOHOL AND METHYL METABOLISM

Alcohol and alcohol metabolism have direct ties to dietary processes regulating epigenetic mechanisms. Perhaps the most pertinent is alcohol's involvement in

TABLE 23.1 Alcohol Affects Three Levels of Metabolic Enzymes

	Alcohol effect on activity/expression	Animal/tissue	References
1. ALCOHOL METABOLISM ENZYMES			
ADH	Increased activity	Human liver HEP-G2 cell (*in vitro*)	[18]
ALDH	Decreased activity	Rat liver	[21]
CYP2E1	Increased activity	Human liver HEP-G2 cell (*in vitro*)	[18]
2. METHYL METABOLISM ENZYMES			
Methionine synthase	Activity inhibited	Rat liver	[22,23]
MTHFR	Activity reduced	Chick brain	[24]
BHMT	Acute ethanol increased activity	Rat liver	[23]
MAT	Decreased expression and activity	Rat liver	[25]
3. EPIGENETIC ENZYMES			
DNMT1	Decreased expression and activity	Mouse brain	[26]
DNMT3a/3b	Decreased expression	Alcoholic patient blood	[27]
HDAC	Decreased expression and activity	Mouse brain, liver	[28,29]
CBP/p300	Decreased expression	Rat FASD brain	[30]
CBP/p300	Increased expression	Rat brain, acute administration	[29]
Ga9 (HMT)	Increased activity	Mouse brain	[31]

(1) Alcohol metabolic, (2) methyl metabolism, and (3) epigenetic enzyme activity and/or expression, are affected by alcohol in multiple alcohol exposure models.

BOX 23.1

GENETIC VARIANTS AFFECTING ALCOHOL METABOLISM MAY CONTRIBUTE TO DIFFERENTIAL EPIGENETIC VULNERABILITY

There are several subtypes of ADH enzymes associated with ethanol metabolism. ADH1A, ADH1B, and ADH1C are responsible for about 70% of total ethanol oxidation in the liver, and ADH4 is responsible for the remaining 30%.[32] There are also several different subtypes of the ALDH enzyme that are involved in alcohol metabolism. ALDH1A2, primarily found in the cytosol, is responsible for converting retinol into retinoic acid-metabolites of vitamin A.[33] ALDH2 is the other major enzyme that converts acetaldehyde to acetate in the mitochondria.[32]

Different variants of these enzymes exist within the population and have different relative activities with regard to the rate of alcohol metabolism. ADH enzymes with higher activity will generate acetaldehyde quicker, and ALDH enzymes with lower activity will convert acetaldehyde to acetate more slowly.[34] The differing rates of acetaldehyde production and breakdown, as a result of these variants, will cause either an increase or decrease in the epigenetic modifications associated with alcohol. These variant enzymes are seen at disproportionate frequencies in certain races and ethnic groups. An example of these variants is ADH1B*2 and ADH1B*3 that have 40-and 30-fold higher activity, respectively, than the wild-type ALDH1B*1 enzyme, therefore producing higher levels of acetaldehyde.[35,36] These enzymes are present in 4–7% of Native Americans and 2–13% of Mexican Americans.[35] Almost all European Americans, however, express the wild type ADH1B*1 allele.[37] Furthermore, the defective ALDH2*2 allele is common in East Asian populations and causes several adverse effects after drinking, such as facial flushing.[34,36] This is due to a build-up of acetaldehyde in individuals homozygous for ALDH2*2, with blood acetaldehyde peaking at levels 3.3-fold higher than individuals heterozygous for ALDH2*1/*2 after moderate drinking.[36] This variant is extremely common among East Asians, with some studies reporting up to 50% ALDH2*2 prevalence in the East Asian population.[34,36] Although common in the East Asian population, this variant has not been detected in any other ethnic group.[34,36] As this chapter describes, alcohol metabolism and epigenetic regulation are closely tied. Existing genetic variation affecting alcohol metabolism is likely then to result in differential epigenetic outcomes that may serve to influence the transcriptome and the cellular and functional properties of the affected individual throughout life.

methionine metabolism and transmethylation enzymes. The metabolism of dietary methionine precursors is critical for the activation of the methyl donor, S-adenosylmethionine (SAM), required for DNA methyltransferase enzymes (DNMT) and, ultimately, for DNA and histone methylation (see Figure 23.1). The availability of SAM is of paramount importance to DNA and histone methylation, purported to affect lasting transcriptional and physiological paradigms (see Section "Epigenetic Footprint of Alcohol Induced Dysfunction and Disease"). Much work has been reported on the interference of alcohol in several facets of methionine metabolism. This work compliments the many reports of alcohol's distinct effect on DNA and histone methylation.

Alcohol and Folate

Folate is the primary dietary source of methionine, and alcohol has shown a multitude of negative effects on folate intake. Folate deficiency, in turn, has been shown to have a broad range of detrimental consequences, such as neuronal cell death and inhibited neurogenesis.[38] Very early studies of human alcoholism correlated alcohol ingestion with dramatic falls in serum folate.[39] A rat model of alcoholism similarly showed reductions in serum 5-methyltetrahydrofolate (5-MTHF), an intermediate metabolite of folate.[40] Alcohol-induced folate deficiency is corroborated by increased excretion of urinary folate in chronically-exposed rats.[41] Additionally, MTHF reductase (MTHFR), the enzyme responsible for the conversion of dietary folate to the donor 5-MTHF, demonstrated a significant reduction in the brain after ethanol-treatment.[24] Heavy alcohol drinkers in one study, who expressed the MTHFR polymorphism CT677 (renders the enzyme thermo-labile and <50% ineffective), were also more likely to suffer from hyper-homocystemia, a common indicator of inhibited methionine metabolism.[42] Alcohol may also induce oxidative stress on methionine metabolism, via the metabolism of alcohol to acetaldehyde (see Section "Alcohol Metabolism and Epigenetic Acetylation") that causes the release of super oxides shown to aggravate the cleavage of folate *in vivo*.[43]

MTHF, an intermediate of folate and methionine, is utilized as a methyl donor for homocysteine in the conversion to methionine by the enzyme methionine synthase (MS). MS has demonstrated decreased activity in an animal model of alcoholism.[22] Moreover, Barak et al. reported that, in a rat alcohol liquid diet model, MS is decreased by up to 50% by chronic exposure, though it has since been believed that this inhibition is not directly due to ethanol, but rather the formation of adducts by acetaldehyde.[44]

Alcohol exposure during fetal stages has also demonstrated detrimental potential in the progress of methionine metabolism and, as such, may play a role in altering fetal DNA and histone signatures. Folate levels are critical, for normal neural development and, unsurprisingly, fetal alcohol offspring mirror many phenotypes of folate deficiency.[45] Fetal alcohol is reported to decrease the transport of folic acid from mother to offspring,[46] though earlier reports have demonstrated maternal-fetal folate transport is actually increased in the presence of ethanol or folate deficiency.[47] The absorption of jejunal free folate appears to be increased in ethanol exposed pups, despite a measured decrease in milk folate levels, possibly hinting at an alternative mechanism of milk folate absorption adapted by ethanol-exposed pups.[48] While the investigation of maternal-fetal folate dynamics continues, folate deficiency *in utero* continues to prove detrimental not only for neural development, but for other congenital diseases, several of which have also been correlated with fetal alcohol disease.[49] In fact, a promising therapeutic approach to alcohol-related fetal disease has been the dietary supplementation of folate and other methionine-based factors (see Section "Treatment and Biomarkers of Alcohol Diseases Through Epigenetic Pathway").

Alcohol and Choline/Betaine

The other major methyl source for homocysteine-methionine conversion is choline. Upon oxidative conversion to betaine, choline acts as an alternate methyl donor for homocysteine–methionine conversion via the enzyme betaine-homocysteine methyltransferase (BHMT). Alcohol effects on betaine have been described as initially overactivating BHMT, consequently depleting hepatic betaine levels,[23] and over long periods of time, contributing to reduced liver SAM levels. This appears to describe an initial compensation of the betaine-methionine pathway in the presence of alcohol inhibition of folate metabolism, possibly accounting for why SAM is not depleted by acute alcohol exposures, but is decreased in chronic exposure paradigms.[23] While SAM is eventually depleted, betaine supplementation is capable of normalizing SAM:SAH (S-adenosylhomocysteine) ratios, preventing the upregulation of nitric oxide (NO) that has been shown to induce mitochondrial DNA damage and alcoholic liver disease.[50]

Like folate, normal choline metabolism is thought to play an important developmental role, particularly for cognitive ability,[51] implying that methyl metabolism and adjacent applications may be a crucial means of regulating developmental processes. Perinatal choline supplementation has been shown to mitigate various phenotypes associated with fetal alcohol syndrome.[52] Moreover, choline supplementation in fetal alcohol models has demonstrated the capacity to normalize molecular parameters of methionine metabolism, such as DNA and histone methylation.[53] While choline-alcohol effects remain to be expanded, it is clear that choline is a critical contributor

TABLE 23.2 Genes Epigenetically Affected by Alcohol with Hypomethylation or Active Histone Marks are Listed in Functional Clusters

Epigenetic marks related to gene activation						
Gene	**Description**	**Epigenetic effect**	**Model**	**Cell type**	**Function**	**References**
ADH1A	Alcohol dehydrogenase 1A	Hypomethylated	Human alcohol dependent	Lymphocyte	Metabolism	[60]
ADH7	Alcohol dehydrogenase 7	Hypomethylated	Human alcohol dependent	Lymphocyte	Metabolism	[60]
ADH1	Alcohol dehydrogenase	H3K4Me ↑	*In vitro* 1 day treatment	Rat hepatocyte	Metabolism	[61]
ALDH1L2	Aldehyde dehydrogenase 1 family, member L2	Hypomethylated	Human alcohol dependent	Peripheral blood	Metabolism	[62]
ALDH3B2	Aldehyde dehydrogenase 3 family, member B2	Hypomethylated	Human alcohol dependent	Lymphocyte	Metabolism	[60]
CYP21A2	Cytochrome P450, family 21, subfamily A, polypeptide 2.	Hypomethylated	Human dose dependent	Lymphoblast	Metabolism	[63]
CYP2A13	Cytochrome P450, family 2, subfamily A, polypeptide13	Hypomethylated	Human alcohol dependent	Lymphocyte	Metabolism	[60]
CYP2C11	Cytochrome P450-2, c11	H3K9Me ↓	*In vitro* 1 day treatment	Rat hepatocyte	Metabolism	[61]
GAD1	Glutamate decarboxylase 1	Hypomethylated	Human alcohol dependent	Peripheral blood	Metabolism	[62]
DBH	Dopamine beta-hydroxylase	Hypomethylated	Human alcohol dependent	Peripheral blood	Neural	[62]

The cell type/tissue of these analyses, as well as the alcohol exposure model and reference to original articles are provided as well. These genes observe a hypomethylation. Though fewer, some genes observe histone modifications in response to alcohol exposure as well.

to DNA and histone methylation, one that also has a large impact on biological processes like development.

Cofactors of Methyl Metabolism and Alcohol

Thus far, we have focused heavily on the two pathways of methyl donation, without expanding on some lesser-known contributors that may ultimately play a role in DNA and histone methylation (Figure 23.1). Vitamin B-12, for example, is a cofactor in the conversion of homocysteine to methionine, while riboflavin influences MTHFR activity. Halsted et al. reviewed the role of B-vitamin cofactors in the pathogenesis of methyl-metabolism related alcoholic liver disease. Namely, patients with alcoholic liver disease have exhibited B6 serum depletion in conjunction with folate deficiency, likely due to the degradation of B6 by the metabolite acetaldehyde.[54] Though B12 serum levels are actually increased in alcoholic liver disease patients, hepatic levels are compromised,[55] likely due to alcohol-induced malabsorption.[56] Chronic ethanol feeding also reduced the presence of riboflavin in the liver of hamsters, and another study found that riboflavin deficiency results in decreased MTHFR activity.[57] Taken together, there is ample evidence that alcohol, to some extent, acts on the cofactors of methionine metabolism, as well as major methyl sources in alcohol-related disease pathogenesis.

Alcohol on Methyltransferase Enzymes and SAM Production

We previously described how alcohol impacts the enzymes involved in methionine metabolism, but alcohol reaches further into the transmethylation pathway via its impact on methionine-transferring enzymes , which can impact not only expression, but regulatory epigenetics in a bidirectional manner (i.e., hyper and hypomethylation). (Figure 23.1 and Table 23.2). After choline and folate sources have culminated in homocysteine–methionine conversion, methionine must then be transformed to its "active" donor state (SAM) by the enzyme methyl adenosyltransferase (MAT). Alcohol–induced oxidative stress has been shown to impair SAM production[58] and dramatically inhibit both the expression and the activity of MAT *in vitro*, via the hypoxic production of NO.[59] More direct studies of alcohol on MAT have described that chronic alcohol exposure initially elevates the expression of MAT genes, though, over time, alcohol appears to decrease MAT activity in the liver. This is concurrent with decreased SAM and DNA methylation (~40%).[25]

Ultimately, the interference of alcohol and alcohol metabolites on methyl metabolism results in the dysregulation of active methyl donor reserves for DNA and histone methylation (as well as other methylation processes outside the scope of this chapter). Meanwhile, ethanol mediated decreases in the ratio of SAM, and the demethylated SAM byproduct SAH, result in the production of NO synthase and NO.[64] This aligns with the finding of a negative correlation between SAM:SAH ratios, and the production of CYP2E1-mediated oxidative stress.[65] The propagation of oxidative stress by alcohol's disruption of methionine metabolism, and the active methyl donor pool, overlap to some extent with the oxidative stress markers released at the metabolism of alcohol (see Section "Alcohol Metabolism and Epigenetic Acetylation"). These interactions add a level of complexity to the investigation of alcohol's role in methyl metabolism, and DNA and histone methylation, but warrant further study nonetheless.

Alcohol and DNA- and Histone-Methylation

DNA and histone methylation are heavily dependent on methyl metabolism. As presented here, there are ample steps on the road to transmethylation whereby alcohol has been shown to act as a deregulator. In effect, alcohol can indirectly influence the DNA methylation profile of cells and tissues, in both adult and fetal alcohol populations. Most notably, hepatocytes and neurons have been extensively profiled, though there is reason to believe the effect of alcohol on aberrant DNA methylation and the histone codes are more far-reaching than we currently understand. Additionally, alcohol metabolism, known to brandish its own array of micromolecular consequences, feeds into methyl metabolism (and thus into DNA and histone methylation), at various places along the pathway. In other words, the effect of alcohol on SAM and epigenetic methylation is an integrative approach, recruiting at least two biochemical pathways –alcohol metabolism and methyl metabolism. By and large, the consensus is that alcohol impairs the production/availability of the active methyl donor required for DNA and histone methylation, though adaptive mechanisms appear to provide at least some level of initial compensation in response to different exposure paradigms.

Though apparently straightforward, the dynamics of alcohol and methyl donor availability are actually more complex. Many high-throughput and smaller-scale analyses have described that the epigenetic effect of alcohol is quite bivalent, either decreasing (Table 23.2) or increasing (Table 23.3) gene methylation, in the presence of a single alcohol exposure model.[66,67] Additionally, alcohol seems to preferentially exert itself on both hypo- and hypermethylated genes[67] (Figure 23.2). This begs the question of whether other factors are at work in the face of environmental pressures seeking to alter epigenetic profiles, and what the exact bearing of

TABLE 23.3 Genes Epigenetically Affected by Alcohol, With Hypermethylation or Inactive Histone Marks, are Listed in Functional Clusters

Epigenetic marks related to gene suppression						
Gene	**Description**	**Epigenetic effect**	**Model**	**Cell type**	**Function**	**References**
GABRP	Gamma-aminobutyric acid (GABA) A receptor, pi	Hypermethylated	Human alcohol dependent	Peripheral blood	Neural	[62]
GABRB3	GABA A receptor B3	Hypermethylated	Human alcohol dependent	Peripheral blood	Neural	[68]
MAOA	Monoamine oxidase A	Hypermethylated	Human dose dependent	Lymphoblast	Neural	[63]
HTR3A	Serotonin receptor 3a	Hypermethylated	Human alcohol dependent	Peripheral blood	Neural	[68]
BLCAP	Bladder cancer associated protein	Hypermethylated	Human dose dependent	Lymphoblast	Cancer	[63]
MBD3	Methyl Cpg binding domain protein 3	Hypermethylated	Human alcohol dependent	Blood	Epigenetic	[68]
DNMT3b	DNA methyltransferase 3b	Hypermethylated	Human chronic alcoholism	Blood	Epigenetic	[27]
SIN3A	Paired amphipathic helix protein 3a	Hypermethylated	Human dose dependent	Lymphoblast	Transcription	[63]
ZNF562	Zinc finger protein 562	Hypermethylated	Human dose dependent	Lymphoblast	Transcription	[63]

(Continued)

TABLE 23.3 Genes Epigenetically Affected by Alcohol, With Hypermethylation or Inactive Histone Marks, are Listed in Functional Clusters *(cont.)*

Epigenetic marks related to gene suppression						
ABR	Active breakpoint cluster region-related	Hypermethylated	Human dose dependent	Lymphoblast	Cell cycle	[63]
ASPM	Asp (abnormal spindle) homolog, microcephaly associated (drosophila)	Hypermethylated	Human alcohol dependent	Lymphocyte	Cell cycle	[60]
NEK6	NIMA-related kinase 6	Hypermethylated	Human alcohol dependent	Lymphocyte	Cell cycle	[60]
CDC23	Cell division cycle 23 homolog	Hypermethylated	Human dose dependent	Lymphoblast	Cell cycle	[63]
LIN37	Lin-37 homolog-cell cycle regulation	Hypermethylated	Human dose dependent	Lymphoblast	Cell cycle	[63]
ERMAP	Erythroid membrane-associated protein	Hypermethylated	Human dose dependent	Lymphoblast	Cell differentiation	[63]
ACTR3C	Actin related protein 3 C	Hypermethylated	Human dose dependent	Lymphoblast	Cell differentiation	[63]

The cell type/tissue of these analyses as well as the alcohol exposure model and reference to original articles, are provided as well. These genes observe a hypermethylation. Though fewer, some genes observe histone modifications in response to alcohol exposure as well.

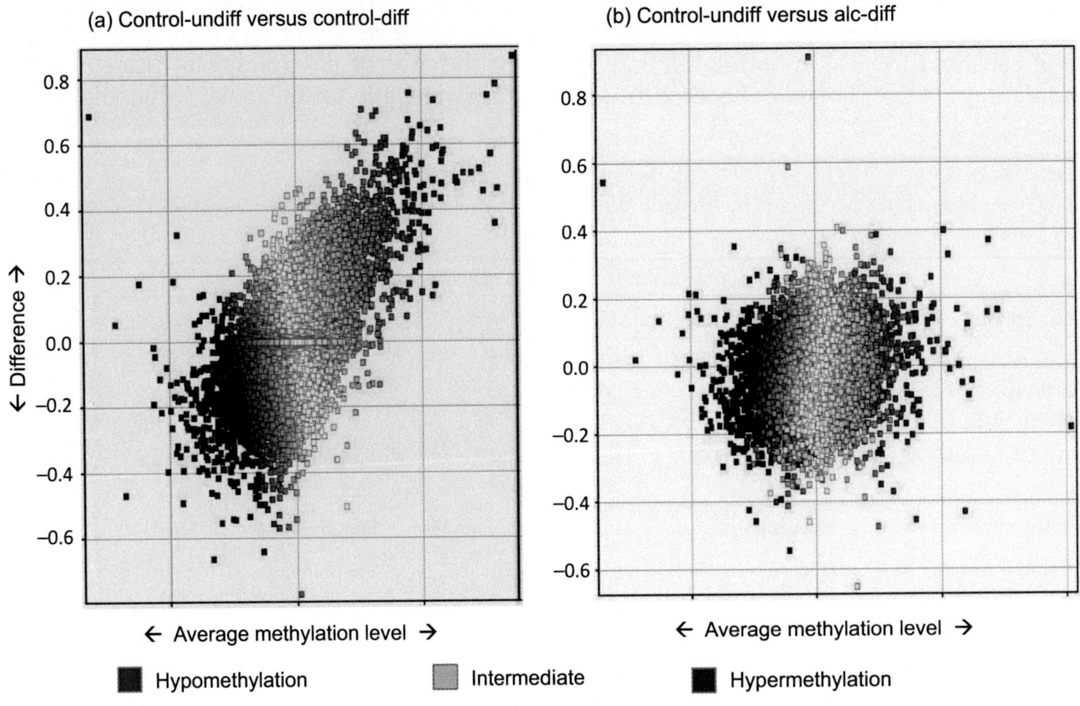

FIGURE 23.2 Genome-wide methylation MvA plot compares DNA methylation during neural stem cell differentiation at absence or presence of alcohol. The difference of promoter (−1300 to +500 bp from Transcription Start Site) methylation between two groups – (a) undifferentiated (undiff) NSCs versus 3-day differentiated (diff) NSCs or (b) undifferentiated NSC versus differentiation of NSC under alcohol treatment (alc-diff) – is plotted on the *Y*-axis, against the average methylation level across epigenome (*X*-axis, also indicated by color, see color legend). In (a), a large number of genes differ in their methylation levels (*Y*-axis) as the differentiation occurred, and prominent differences are seen in genes with hypermethylation (red) and hypomethylation (blue) (*X*-axis). Such differentiation-associated methylation differences were not observed, when treated with alcohol as seen in (b); the number of genes with methylation differences is reduced (*Y*-axis), and the degree of changes on average DNA methylation level are smaller (*X*-axis) between the Undifferentiated and the Alcohol treated group during differentiation. This figure demonstrates that alcohol alters NSC DNA methylation toward an undifferentiated state, preferentially at hypo- and hypermethylated genes. *Adapted from Ref. [67].*

methyl-donor availability is in the grander scheme of epigenetic regulation. At least one item to consider is the role of DNA- and histone-methyltransferases as the catalytic players in the equation of methylation-transfer, and how alcohol may exert influence on this (Section "Alcohol Affects the Epigenetics Converting Enzymes"). Meanwhile, dietary intervention strategies continue to be promisingly investigated as therapies for alleviating alcohol-related disease, whose optimization cannot occur without a more thorough and precise understanding of alcohol pathogenesis of methionine and alcohol metabolism. Next, we profile the effects of alcohol on epigenetic enzymes in order to try and shed greater light on the overarching story of dietary alcohol, and DNA and histone methylation.

ALCOHOL AFFECTS EPIGENETIC CONVERTING ENZYMES

On Histone Code Converting Enzymes

Alcohol can directly affect the efficiency of epigenetic converting enzymes to alter DNA methylation and histone modification (Table 23.3) that, in turn, affect the 3D conformation of DNA and the accessibility of transcriptional machinery. There are several important enzymes that regulate these covalent modifications. Transferring an acetyl group from acetyl-CoA to a lysine residue on histones, histone acetyltransferases (HATs) are associated with transcriptional activation. On the other hand, histone deacetylases (HDACs) remove acetyl groups from lysine residues on histones, and are associated with gene silencing. Alcohol exposure has been shown to increase histone acetylation by several different mechanisms (summarized in Figure 23.3), including modulating the activities of HATs and HDACs. Studies have demonstrated that alcohol exposure can both increase[28] and decrease HDAC activity and expression *in vitro*.[69] This decreased HDAC activity may play a role in the site-specific histone acetylation at lysine 9 on histone 3 (H3K9Ac) that is associated with alcohol consumption.[70] Alcohol treatment has also been directly linked to specific histone acetylation marks, such as increased histone acetylation at H3K9, in primary cortical neurons derived from C57BL/6 mouse fetuses,[71] and increased H4K12 in the mouse nucleus acumbens core.[72] Some of these acetylation marks, including H3K9, have been investigated and found to be preferentially localized on the promoter regions of transcriptionally activating genes.[73] H3K9, along with other histone acetylation marks (i.e., H3K4, H3K27) of this genomic localization, have been implicated in the regulation of developmental genes, such as pluripotency genes.[74,75] Unlike

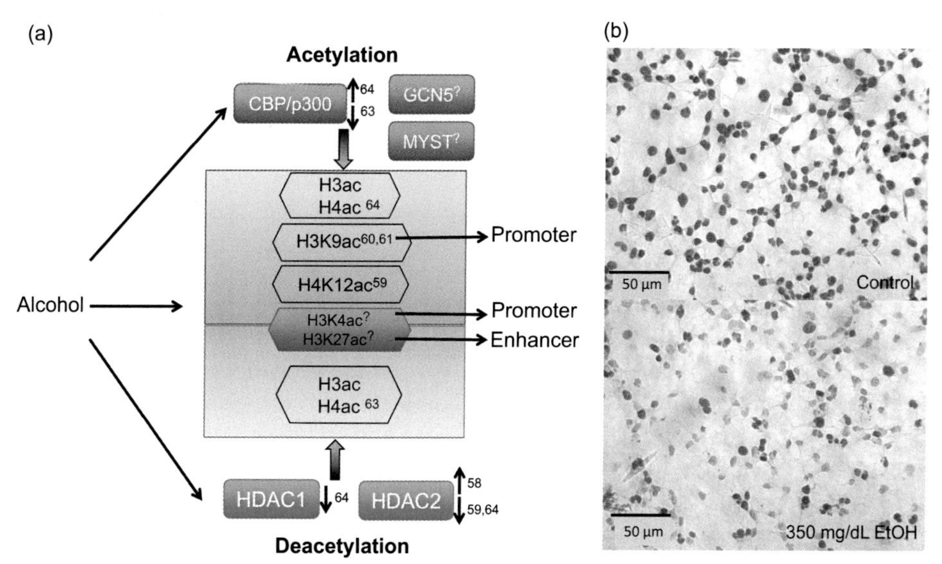

FIGURE 23.3 **Multi-level contribution of alcohol to histone acetylation.** (a) Schematic of alcohol's involvement in histone acetylation via enzymatic or direct contribution. Alcohol alters acetylation and deacetylation process through modification of specific acetylating (CBP/p300) or deacetylating (HDAC1/2) enzymes, though several enzymes remain to be investigated (GCN5, MYST). Interestingly, the alcohol exposure model (i.e., dose, timing) and tissue/cell-type can result in variable histone acetylation, demonstrating the bivalent capacity of alcohol on acetylation. Alcohol has additionally been shown to directly affect histone acetylating marks, independent of acetylation enzymes. For example, H3 and H4 acetylation have both demonstrated sensitivity to alcohol in either the increasing (red) or decreasing (blue) direction. Additionally, some specific lysine residues on the histones 3 and 4 are reportedly affected by alcohol (i.e., H3K9 and H4K12 acetylation). Acetylation marks, such as H3K9, H3K4, and H3K27, are implicated in transcriptional activation, as they tend to be localized at promoter/enhancer sites making the investigation of alcohol's influence of these marks extremely important, as some relationships remain unclear (purple). Question marks indicate areas of unknown alcohol effect requiring deeper investigation. (b) An effect of alcohol on histone acetylase enzyme HDAC1, was shown in 3 day differentiated rat DRG neural stem cells (40×). Top: control, bottom: 2-day treatment of cells with 350 mg/dL EtOH. Alcohol decreases the expression of HDAC1.

H3K9, however, the role of alcohol on these potentially meaningful histone acetylation marks remains to be fully understood.

There are different subfamilies of HATs – among which the GCN5-related *N*-acetyltransferases (GCN5), MOZ, Yeast YBF2, SAS2 and TIP60 (MYST) family HATs, and the E1A binding protein 300/CREB binding protein (p300/CBP) coactivator family have been well studied. While the direct effects of alcohol on GCN5 are yet to be determined, a study showed that GCN5 is involved in alcohol induced histone acetylation, as knock down of *GCN5* reduced ethanol-induced histone acetylation and altered ethanol-induced differential gene expression in human hepatoma cells.[76] Studies on CBP/p300 in an FASD model demonstrated that alcohol exposure decreases CBP expression in the developing rat brain, and that this was associated with decreased H3 and H4 histone acetylation.[30] This finding was different from the effects of acute administration (an anxiolytic dose), a fact that leads to decreased HDAC activity, increased CBP levels, and increased H3/H4 acetylation in the adult rat amygdala.[29] Taken together, these studies suggest that the effects of alcohol on histone acetylation are not unidirectional, but rather dependent on developmental stage, ethanol dosage, and tissue type.

Alcohol also affects histones via histone methylation enzymes. Methylation at histone H3 lysine 9 (H3K9), and lysine 27 (H3K27) is associated with transcriptional silencing, while methylation at H3K4 has been correlated with gene activation. Histone methyltransferases (HMTs) can attach 1–3 methyl groups to an arginine or lysine residue on a histone. Conversely, histone demethylases (HDMs) can remove those methyl groups.[5] Recently, studies showed that alcohol administration at immature (postnatal day 7) time points induced histone methyltransferease G9a activity, leading to increased H3K9 and H3K27 dimethylation and neurodegeneration phenotypes.[31] It has been further reported that alcohol causes a decrease in histone methylation at the H3K9 site, along with decreased expression of at least 9 genes coding for different HMTs.[71] To generalize that histone methylation or demethylation are the lone determinants of alcohol-mediated genomic or phenotypic outcomes would be oversimplistic. Much remains to be uncovered about the effects of alcohol on HDMs that may play just as big a role as HMTs.

On DNA Methylation Enzymes

As mentioned in Section "Alcohol and Methyl Metabolism," alcohol exerts its effects on DNA and histone methylation not only through the deregulation of active methyl donor biogenesis, but also by decreasing the activity of DNA methyltransferase 1 (DNMT1). Similarly, in a human chronic alcoholism study, decreased gene expression of DNMT3a and DNMT3b (the *de novo* methylation enzymes) was found in alcoholic patients, an outcome further associated with global hypermethylation.[27] This result suggests a rearrangement of epigenetic control occurs in alcohol drinkers, though thresholds of consumption for enzymatic alteration, and the longevity and reach of the suppression of DNMTs, is a subject that remains to be fully understood.

A more recent facet of the methyl metabolism pathway that bears much significance on DNA methylation status is 5hmC – the hydroxylated derivative of 5mC. 5hmC occurs when the ten-eleven translocation (TET) family enzymes catalyze the transfer of a hydroxyl group from 2-oxoglutarate (α-ketoglutarate), to the methylated end of the 5-carbon of 5mC. This alteration, static and quite abundant in tissues like the brain, may alter the epigenetic landscape and contribute a complimentary functional role to canonical DNA methylation.

A recent study looked at the interaction between alcohol exposure and aging on 5hmC, and demonstrated that mild chronic alcohol exposure reduced global 5hmC in younger mice, though not altered in older animals.[77] TET1/2/3 expression, levels and the amount of 5mC and unmodified cytosine, were not changed significantly in this study, indicating that the mechanisms underlying the reduction of 5hmC are more complex than the mere regulation of substrate or enzymatic activity. Beside the aforementioned study, the in-depth analysis of alcohol's effect on the TET enzymes is an endeavor yet to be undertaken.

Crosstalk between different epigenetic pathways is a relatively new idea, and not well understood, but it is one that requires more attention. For example, DNA methylation and histone methylation work together during neural stem cell differentiation, where the acquisition of 5hmC triggers the loss of H3K27 trimethylation.[78] Disruption of epigenetic crosstalk by alcohol and many other addictive drugs and environmental factors could ultimately shift cellular patterns from normal to disease states. It is important then to consider several kinds of epigenetic modifiers collectively, in order to better understand the dynamics of epigenetically-mediated disease and to implement effective therapeutic strategies.

EPIGENETIC FOOTPRINT OF ALCOHOL INDUCED DYSFUNCTION AND DISEASE

The influence of alcohol on epigenetic methylation and acetylation, whether consumed in binge or chronic doses, leaves a significant molecular footprint that guides the cellular and physiological onset of alcohol-related diseases. In fact, many altered epigenetic pathways are being increasingly observed and considered as contributors to the mechanisms of disease causality. Profiling disease-state epigenetic signatures may thus prove important in the eventual development and use of epigenetic biomarkers for alcohol-related disease detection. Similarly, the metabolic aspects of the alcohol-epigenetic relationship, as increasingly understood, provide a promising means of alcohol-related disease intervention (see Section "Treatment and Biomarkers of Alcohol Diseases Through Epigenetic Pathway").

Liver Cancer

Among all cancers, liver cancer or hepatocellular carcinoma (HCC) is highly associated with chronic alcohol drinking. Chronic alcoholism is one of the major risk factors for liver-related disease and death,[13] and alcohol induced DNA methylation change has been well studied as a pathogenic mechanism contributing to HCC. Similar to other alcohol related diseases, alcohol related cancer exhibits decreased global DNA methylation, and selected promoter hypermethylation was found in alcohol related-HCC patients. When looking at gene-specific change, Hernandez-Vargas et al. determined that 94 out of 807 cancer-related genes observed altered promoter methylation. Examples of genes that were hypermethylated include *APC, CDKN2A*, and *RASSF1*.[79]

Another gene associated with alcohol induced DNA methylation change in HCC is MAT (recall MAT in Section "Alcohol and Methyl Metabolism"). Alcohol has been shown to reduce MAT activity in hepatic cells, both by decreasing expression of *MAT1A*, and inactivating the protein.[80] Since reduced MAT can lead to a reduction in SAM production, this may at least partially explain the observed global DNA hypomethylation observed in HCC. As told earlier, however, reduced methyl resources cannot account for the bivalent behavior observed in alcohol-induced epigenetic alteration – alcohol related cancer being no exception.

Other than DNA methylation, aberrant histone modifications have also been demonstrated in alcohol induced liver diseases. Ethanol exposure of hepatocytes induced global increases in H3 acetylation, particularly at lysine 9,[70] an effect that is dose and time-responsive. Altered histone H3K4 and H3K9 methylations, and their associated aberrant regulation of genes, have also been observed in ethanol treated hepatocytes. For example, though the expression of CYP2E1 mediates much of the oxidative damage and hepatic toxicity/apoptosis of alcohol[81] and is generally associated with alcohol induced upregulation, increased H3K9 methylation has been interestingly observed at the CYP2E1 gene, and correlated with its downregulation.[61] The likely reality is that alcohol-induced histone modifications are not exclusive, but rather additive, and influenced by multiple facets of alcohol and alcohol metabolism working together to regulate gene expression and disease-states.

Fetal Alcohol Spectrum Disorders

Accumulating evidence supports an epigenetic etiology of fetal alcohol spectrum disorders (FASD), as many FASD dysmorphologies can be attributed to aberrant epigenetic marks affecting gametes during embryonic development, under the insult of alcohol. Alcohol was first shown to reduce fetal DNA methylation in FASD via inhibited DNMT activity.[26] Additionally, our laboratory has shown that alcohol exposure of mouse embryos exhibits DNA methylation changes associated with neural-facial developmental defects and growth retardation analogous to FASD phenotypes in humans.[66]

These alcohol induced changes are not one-directional, rather both hyper- and hypomethylation are observed.[66] Some of the genes associated with altered methylation are imprinted genes and genes involved in cell cycle regulation, apoptosis, cancer, and cell differentiation pathways. In addition, prenatal alcohol exposure has been shown to delay and disrupt the normal progression of DNA methylation (both 5mC and 5hmC) in the developing brain. Both 5mC and 5hmC have been extensively validated as a mechanism guiding temporospatial modulation of genetic expression. Particularly, 5hmC is becoming increasingly accepted not only a byproduct of the demethylation pathway, but also a functional regulator believed to promote hypomethylation and subsequent activation of genes. Recent studies on 5hmC have suggested it is tightly coupled with an important functional role in neural and early developmental systems.[1,82] This adds significance to the finding that alcohol can alter the timely progression of 5mC and 5hmC in neuroepithelial cells, during the differentiation of embryos,[14,67] and in later brain development.[83] Furthermore, from mouse embryonic day 15 (E15) to postnatal day 7 (P7), 5mC, 5hmC, and their respective binding proteins, MBD1 and MBD3, are dynamically distributed in the hippocampus, guiding hippocampal neuronal differentiation and maturation. Unsurprisingly, alcohol disrupts this distribution, altering the chromatic structure correlated with developmental retardation.[83,84]

Further, fetal alcohol exposure studies have revealed a link between epigenetics and the hypothalamic-pituitary-adrenal (HPA) axis, in which proopiomelanocortin (POMC) expression is decreased, due to an alcohol-induced increase of DNA methylation at the promoter of POMC.[85] POMC expressing neurons, located in the hypothalamus, are critical for regulating stress homeostasis, through the HPA axis. Several psychological disorders that are associated with the misregulation of the HPA axis have also been described in FASD models (such as depression and anxiety).[86] The studies in POMC neurons have provided a path for the epigenetic investigation of alcohol's effects on stress regulation associated with FASD.

An interesting and unaccounted mechanism of FASD, with great epigenetic relevance, is that prenatal alcohol exposure is not solely the consequence of maternal intake. Studies have found DNA hypomethylation at the imprinting control region H19 in the sperm of alcoholics,[87] suggesting that alcohol can alter the epigenome of the male germ cells passed to offspring. This paternal effect may be retained in the next generation, as offspring of alcohol exposed sires have demonstrated low birth weights, typically associated with aberrations in chromatin remodeling required for proper development.[88] Epidemiologically, the effects of paternal alcohol drinking on offspring were identified over a decade ago. These corroborate that growth restriction, hyperactivity, mental impairment, and other phenotypes of FASD may originate at least partially from paternal contributions.[89] Germ cells, or more specifically the epigenetic profiles of transmissible germ cells, point to a likely epigenetic mechanism utilized by alcohol to convey molecular and physiological aberrations beyond the realm of the fetal environment. The epigenetic basis of FASD is widely dependent on dose, onset, and length of exposure, as well as tissue/gene of analysis. As in alcohol-related liver disease, the physiological and even transcriptional outcomes can be contributed by variety of epigenetic modifications, arising not just from the maternal environment, but from the epigenetic inheritance that occurs at the earliest formation of the zygote.

Alcoholism/Alcohol Dependence

Given the extensive influence of alcohol on epigenetics, it has been a focus of investigation whether chronic alcohol might alter alcoholic pathology through changing epigenetic paradigms. In other words, can alcohol use epigenetic profiles as a "primer" toward alcohol dependence? Indeed, chronic consumption of alcohol has been shown to alter the methylation of several promoters.[27] The question has driven more and more studies comparing the epigenome of alcoholics and nonalcoholics, mostly through the use of blood samples. Tables 23.2 and 23.3 summarize some of the genes demonstrating epigenetic modification by alcohol exposure. Interestingly, all genes with altered epigenetic modification that code for alcohol metabolism enzymes[60,62] are associated with active epigenetic marks (DNA hypomethylation, increased H3K4 methylation, decreased H3K9 methylation) (Table 23.2). This suggests a consequential activation of ADHs and ALDHs in part can be achieved via epigenetic regulation, in order to metabolize excessive ethanol, accelerate acetaldehyde elimination, and enhance tolerance of alcohol. On the contrary, many hypermethylated genes are related to cell cycle, cell differentiation, transcription, and neural development – suggesting, though not necessitating, that alcohol via epigenetic regulation achieves the repression of critical elements of normal cellular physiology in alcohol-dependent subjects (Table 23.3).

Another group of genes epigenetically altered in alcohol dependent patients are the GABA receptor genes. GABA is the major inhibitory neurotransmitter of the central nervous system and hypermethylation of gamma-aminobutyric acid (GABA) A receptor pi (GABRP) has been identified in alcohol-dependent patients versus their nonalcoholic siblings.[62] A separate study also found that GABA A receptor B3 (GABRB3) was hypermethylated in alcoholic patients.[68] Though further studies are needed to determine whether increased DNA methylation is sufficient to provoke a change in the expression of these GABA genes, alteration of $GABA_A$ receptor expression has been implicated in alcohol tolerance, dependence, and withdrawal.[90] The association of DNA methylation changes in these genes, critical to inhibitory neural circuits, certainly attracts an explanation for the effect of alcohol on addiction.

On the contrary, NMDA receptors initiate excitatory neurotransmission in the brain, and have been implicated in alcohol tolerance and withdrawal symptoms,[91] much like GABA. Studies have shown that the expression of genes encoding NMDA receptors are correlated with changes in DNA methylation. A clinical study, for example, identified altered promoter methylation of the gene encoding NMDA 2B receptor subtype (NR2B) in alcohol dependent patients during withdrawal. Additionally, the level of methylation was inversely correlated with the degree of alcohol addiction.[92] Furthermore, an *in vitro* study using primary neuron culture demonstrated that chronic ethanol exposure can lead to reduced methylation in a regulatory region of the NR2B gene, a phenomenon that was associated with increased gene expression.[93]

Overall, there is abundant evidence that alcohol not only alters the epigenome via modulation of epigenetic enzymes, but also through the modification of the players responsible for regulating the breakdown of the chemical upon ingestion. We recall only some of the vast evidence that alcohol, via DNA methylation, alters the expression of genes affecting alcohol dependence, as well as many associated phenotypes (though a complete resolution of alcohol dependence is by far complete). A positive outcome of the epigenetic study of alcohol dependence is the isolation of epigenetic enzymes as a target of manipulation for therapeutic intervention. In particular, small molecules that inhibit HDAC activity (HDACi), such as trichostatin A (TSA), have been widely applied. For example, Pandey et al. have demonstrated that the anxiety-like behavior observed during alcohol withdrawal (associated with increased HDAC activity, decreased H3K9 and H4K8 acetylation, and decreased expression of neuropeptide Y (NPY) in the amygdala) is reversed with TSA treatment, and that associated epigenetic modifications are normalized.[29]

Transgenerationally Altered DNA Methylation

One of the most attention-grabbing breakthroughs in the field of epigenetics has been the revelation that at least some altered epigenetic marks are rather lasting and germline transmissible.[94] This challenges the widely accepted dogma that epigenetic acquisitions are erased with each generation. This discovery opened up an avenue for the long-term manifestation of environmentally acquired deficit and disease. Additionally, transgenerational epigenetics offers a novel explanation for disease heritability beyond the level of classical genomics (see FASD). Several environmental stressors, such as early life stress and nutritional deficits, may boast a mechanism to bypass germline epigenetic erasure, thereby persisting into future generations. It is worth exploring, then, whether alcohol holds the same capacity as a transducer of abnormal molecular signatures across generations – particularly because of the well-documented heritability of alcohol-related abnormalities.

DNA methylation was found to be similarly patterned in the gametes of male rodents exposed to a fear-odor conditioning model. Gametes of naïve offspring similarly carried the aberrant methylation pattern on the olfactory receptor gene.[95] Chemical pesticides have also been reported to induce the germline transmission of differentially methylated regions to naïve offspring.[96] Similar pesticide studies have extended the transmission of differential methylation through female gamete progenitors, up to the F3 generation.[97]

Whether alcohol holds the ability to pass on epigenetic aberration is a subject of current investigation. Animal models of fetal exposure to alcohol have found that POMC neurons can rewrite promoter methylation and histone codes, and that these can be passed on to future generations through the male germline.[85] Even more interesting was the follow-up study that reported that the alcohol-induced epigenetic modulation could be normalized by choline supplementation.[53] Another study found that parental alcohol-induced hypomethylation of the imprinting gene H19 was passed onto the germline of F1 offspring and the brain of F2 offspring.[88]

It is clear that maternal and paternal germlines can carry epimutations to their offspring – both gonadal and somatic. However, whether subsequent generations can maintain and transmit these to the F2 and F3 generation is a subject of contention.[98] In fact, the very definition of transgenerational epigenetic inheritance is on trial.[99] And though some experts would caution that the incidence of transgenerational inheritance in mammals is actually quite rare,[100] the phenomenon and the potential it holds toward unlocking an untapped mechanism of inherited disease begs for deeper investigation.

TREATMENT AND BIOMARKERS OF ALCOHOL DISEASES THROUGH EPIGENETIC PATHWAY

Hopefully, it has become clear by now that epigenetic mechanisms play an undeniable role in mediating alcohol-related phenotypes. From acute to chronic exposures, during fetal and adult stages, the molecular trails of alcohol are sizable and lasting. Though it will be some time before the precise dynamics are resolved, the manipulation of DNA methylation and histone modification, as a strategy for disease intervention, are moving forward with promising results. Intervention strategies have looked toward the supplementation of folate in alcohol-related folate deficiency. In chronically alcohol-fed rats, for example, folic acid supplementation prevented the oxidative stress of alcohol, likely by reducing the hyperhomocystemia known to induce ER stress responses.[101] Folate supplementation in fetal alcohol models have also been shown to reduce some ethanol-related oxidative stress[102] and alcohol related aberrations.[103] Though the effects of gestational folate supplementation on folate levels themselves appear to be diverse (i.e., they vary depending on how folate levels are analyzed), the epigenetic reach of folate supplementation has been validated via findings such as Wang and coworkers' report of folic acid-induced restoration of miRNA and target gene expression, in a model of prenatal alcohol exposure. In general, methyl supplementation (via folate, choline intervention) has not only demonstrated the ability to normalize serum levels of micronutrients disturbed by ethanol exposure, they have subsequently exhibited the ability to shift global and loci-specific methylation profiles of critical genomic repetitive elements or genes (i.e., LINE-1, IGf2).[104]

As mentioned above, drugs aimed at modifying chromatin structure (i.e., HDAC inhibitors, DNMT inhibitors) have also become increasingly utilized for the treatment of alcohol-dependence and hepatic cancers.[105] Additionally, efforts have begun to better characterize the epigenetic signatures of disease-related genes and regions in an attempt to isolate patient information about alcohol exposure that may not be evident by other means. Cues can perhaps be taken from cancer models, where gene-specific methylation is currently being explored as a diagnostic.

CONCLUSIONS

Alcohol related diseases span a very wide spectrum. Everything from severe liver damage to barely-detectable cognitive abnormalities have been associated with exposure. Alcohol has persisted in human civilization with no signs of erosion. In fact, new forms are being introduced for recreational consumption (i.e., molecular encapsulated (powdered) alcohol). It is important then to fully grasp the molecular consequences of the substance, and to investigate how these beget the phenotypes of alcohol-related disease. Perhaps the key to understanding the molecular effects of alcohol lie in its metabolism and inhibition of other dietary pathways. Substantial evidence has been presented here to precisely implicate alcohol, and alcohol metabolism, in hindering the methyl metabolism pathway guiding DNA and histone modification – potent and lasting agents of transcriptional regulation and, consequently, of cellular and functional change.

Though dynamic and complex, the interactions of alcohol and alcohol metabolites on the biochemical pathways of epigenetic biogenesis have opened a route for the manipulation of dietary factors alcohol appears to dysregulate. In other words, understanding the effects of alcohol on folate, choline, and acetate have allowed the beginning of a thorough investigation of nutritional supplementation as a therapeutic approach, alongside the use of strong and often toxic epigenetic modifying drugs. Theoretically, then, the modification of diet in the presence of alcohol exposure can prevent, alleviate, or even reverse the spectrum of alcohol related phenotypes.

The discussion and opinions herein sought to expand the dimensions of alcohol as a beverage, nutritional substance, and physiological and mental stimulant, to paint the portrait of a commonly accessible molecule reaching into the smallest centers of a cell, where life is directed and epigenetically modified. The commonality of alcohol use in our life and culture, and its intensive capacity to modify DNA methylation and histone-codes regulating gene expression, refresh the contemporary view of alcohol and make apparent the pressing need for a more complete understanding of the dynamic roles it plays in naturally occurring biochemical processes. Alcohol might exert a more profound influence in our lives than we have previously believed. Knowing that its effect on DNA methylation and histone modification may be carried across generations, we should contemplate the possibility that we may, to this day, carry the remnants of the environmental exposures of our ancestors.

Key Facts

- Alcohol metabolites regulate methyl and acetyl donors that are the key fuel for epigenetics.
- Alcohol affects micronutrition (via folate and choline metabolism) that is critical for methyl donor availability in the transmethylation of DNA and histones.
- Alcohol can also affect epigenetic regulatory enzymes directly.
- Epigenetic alteration of alcohol targets is a novel mechanism of alcohol related disease.
- Alcohol-induced epigenetic alterations can be transmitted through the paternal germline, affecting future generations.
- The dietary supplementation of methionine has shown the ability to alleviate alcohol-related outcomes.

Summary Points

- Alcohol and alcohol metabolites modulate acetyl-CoA and methyl donor pools regulating histone modification, and DNA methylation reactions.
- Alcohol can also directly affect the expression and activity of epigenetic modifying enzymes to alter DNA methylation and histone codes, thereby changing the accessibility of chromatin and, ultimately, affecting gene expression.
- The influence of alcohol on epigenetic alteration has been considered a pathogenic mechanism contributing to alcohol-related diseases, including liver cancer, FASD, and alcoholism. These mechanisms may even be germline transmissible.
- The effects of alcohol on epigenetic alteration in disease models are dependent on dose, onset, and length of exposure. Understanding this complexity could provide promising diagnostic and therapeutic strategies.

Acknowledgments

This chapter is supported by the M.W. Keck Foundation and by National Institute of Health AA016698 and P50AA07611 to FCZ. MR is supported by predoctoral training grant 5T32AA007462-29 from the NIAAA. SCLL is supported by P50 and Keck grants listed above. YJC is supported by 103T3040B05 National Taiwan Normal University, Taiwan, during her visiting FCZ's laboratory.

Abbreviations

ADH	Alcohol dehydrogenase
ALDH	Aaldehyde dehydrogenase
BHMT	Betaine-homocysteine methyltransferase
CYP2E1	Cytochrome P450 isoform 2E1
DNMT	DNA Methyltransferease
FASD	Fetal alcohol spectrum disorder
GCN5	GCN5-related-*N*-acetyltransferases
HAT	Histone acetyltransferase
HCC	Hepatocellular carcinoma
HDAC	Histone deacetylase
HDM	Histone demethylase
HMT	Histone methyltransferase
MAT	Methyl adenosyltransferase
MS	Methionine synthase
MTHF	Methyltetrahydrofolate
MYST	MOZ, Yeast YBF2, SAS2 and TIP60 histone acetyltransferases
P300/CBP	E1A binding protein 300/CREB binding protein coactivator
ROS	Reactive oxygen species
SAM	S-adenosylmethionine
TET	Ten-eleven translocation enzyme
5mC	5-Methylcytosine
5hmC	5-Hydroxymethylcytosine

References

1. Lister R, Mukamel EA, Nery JR, et al. Global epigenomic reconfiguration during mammalian brain development. *Science* 2013;**341**(6146):1237905.
2. McGovern PE, Zhang J, Tang J, et al. Fermented beverages of pre- and proto-historic China. *Proc Natl Acad Sci USA* 2004;**101**(51):17593–8.
3. Spooner C. Alcohol: No Ordinary Commodity. Research and Public Policy, Second Edition. *Drug Alcohol Rev* 2011;**30**(1):115.
4. Umulis DM, Gurmen NM, Singh P, Fogler HS. A physiologically based model for ethanol and acetaldehyde metabolism in human beings. *Alcohol* 2005;**35**(1):3–12.
5. Zakhari S. Alcohol metabolism and epigenetics changes. *Alcohol Res* 2013;**35**(1):6–16.
6. Shepard BD, Tuma PL. Alcohol-induced protein hyperacetylation: mechanisms and consequences. *World J Gastroenterol* 2009;**15**(10):1219–30.
7. Ito K, Hanazawa T, Tomita K, Barnes PJ, Adcock IM. Oxidative stress reduces histone deacetylase 2 activity and enhances IL-8 gene expression: role of tyrosine nitration. *Biochem Biophys Res Commun* 2004;**315**(1):240–5.
8. Choudhury M, Park PH, Jackson D, Shukla SD. Evidence for the role of oxidative stress in the acetylation of histone H3 by ethanol in rat hepatocytes. *Alcohol* 2010;**44**(6):531–40.
9. Holownia A, Mroz RM, Wielgat P, et al. Histone acetylation and arachidonic acid cytotoxicity in HepG2 cells overexpressing CYP2E1. *Naunyn Schmiedebergs Arch Pharmacol* 2014;**387**(3):271–80.
10. Zakhari S. Overview: how is alcohol metabolized by the body? *Alcohol Res Health* 2006;**29**(4):245–54.
11. Wang J, Du H, Ma X, et al. Metabolic products of [2-(13) C]ethanol in the rat brain after chronic ethanol exposure. *J Neurochem* 2013;**127**(3):353–64.
12. Siler SQ, Neese RA, Hellerstein MK. *De novo* lipogenesis, lipid kinetics, and whole-body lipid balances in humans after acute alcohol consumption. *Am J Clin Nutr* 1999;**70**(5):928–36.
13. Moghe A, Joshi-Barve S, Ghare S, et al. Histone modifications and alcohol-induced liver disease: are altered nutrients the missing link? *World J Gastroenterol* 2011;**17**(20):2465–72.
14. Resendiz M, Chen Y, Ozturk NC, Zhou FC. Epigenetic medicine and fetal alcohol spectrum disorders. *Epigenomics* 2013;**5**(1):73–86.
15. Soliman ML, Smith MD, Houdek HM, Rosenberger TA. Acetate supplementation modulates brain histone acetylation and decreases interleukin-1beta expression in a rat model of neuroinflammation. *J Neuroinflammation* 2012;**9**:51.
16. Takahashi H, McCaffery JM, Irizarry RA, Boeke JD. Nucleocytosolic acetyl-coenzyme a synthetase is required for histone acetylation and global transcription. *Mol Cell* 2006;**23**(2):207–17.
17. Smith CM, Israel BC, Iannucci J, Marino KA. Possible role of acetyl-CoA in the inhibition of CoA biosynthesis by ethanol in rats. *J Nutr* 1987;**117**(3):452–9.
18. Balusikova K, Kovar J. Alcohol dehydrogenase and cytochrome P450 2E1 can be induced by long-term exposure to ethanol in cultured liver HEP-G2 cells. *In Vitro Cell Dev Biol Anim* 2013;**49**(8): 619–25.
19. Mello T, Ceni E, Surrenti C, Galli A. Alcohol induced hepatic fibrosis: role of acetaldehyde. *Mol Aspects Med* 2008;**29**(1–2):17–21.
20. Brooks PJ, Zakhari S. Acetaldehyde and the genome: beyond nuclear DNA adducts and carcinogenesis. *Environ Mol Mutagen* 2014;**55**(2):77–91.
21. Chrostek L, Tomaszewski W, Szmitkowski M. The effect of green tea on the activity of aldehyde dehydrogenase (ALDH) in the liver of rats during chronic ethanol consumption. *Rocz Akad Med Bialymst* 2005;**50**:220–3.
22. Finkelstein JD. Methionine metabolism in mammals: the biochemical basis for homocystinuria. *Metabolism* 1974;**23**(4):387–98.
23. Barak AJ, Beckenhauer HC, Tuma DJ. Betaine effects on hepatic methionine metabolism elicited by short-term ethanol feeding. *Alcohol* 1996;**13**(5):483–6.
24. Berlin KN, Cameron LM, Gatt M, Miller Jr RR. Reduced *de novo* synthesis of 5-methyltetrahydrofolate and reduced taurine levels in ethanol-treated chick brains. *Comp Biochem Physiol C Toxicol Pharmacol* 2010;**152**(3):353–9.
25. Lu SC, Huang ZZ, Yang H, Mato JM, Avila MA, Tsukamoto H. Changes in methionine adenosyltransferase and S-adenosylmethionine homeostasis in alcoholic rat liver. *Am J Physiol Gastrointest Liver Physiol* 2000;**279**(1):G178–85.
26. Garro AJ, McBeth DL, Lima V, Lieber CS. Ethanol consumption inhibits fetal DNA methylation in mice: implications for the fetal alcohol syndrome. *Alcohol Clin Exp Res* 1991;**15**(3):395–8.
27. Bönsch D, Lenz B, Fiszer R, Frieling H, Kornhuber J, Bleich S. Lowered DNA methyltransferase (DNMT-3b) mRNA expression is associated with genomic DNA hypermethylation in patients with chronic alcoholism. *J Neural Transm* 2006;**113**(9):1299–304.
28. Agudelo M, Gandhi N, Saiyed Z, et al. Effects of alcohol on histone deacetylase 2 (HDAC2) and the neuroprotective role of trichostatin A (TSA). *Alcohol Clin Exp Res* 2011;**35**(8):1550–6.
29. Pandey SC, Ugale R, Zhang H, Tang L, Prakash A. Brain chromatin remodeling: a novel mechanism of alcoholism. *J Neurosci* 2008;**28**(14):3729–37.
30. Guo W, Crossey EL, Zhang L, et al. Alcohol exposure decreases CREB binding protein expression and histone acetylation in the developing cerebellum. *PLoS One* 2011;**6**(5):e19351.
31. Subbanna S, Nagre NN, Shivakumar M, Umapathy NS, Psychoyos D, Basavarajappa BS. Ethanol induced acetylation of histone at G9a exon1 and G9a-mediated histone H3 dimethylation leads to neurodegeneration in neonatal mice. *Neuroscience* 2014;**258**:422–32.
32. Lands WE. A review of alcohol clearance in humans. *Alcohol* 1998;**15**(2):147–60.
33. Zhu B, Buttrick T, Bassil R, et al. IL-4 and retinoic acid synergistically induce regulatory dendritic cells expressing Aldh1a2. *J Immunol* 2013;**191**(6):3139–51.
34. Peng GS, Yin SJ. Effect of the allelic variants of aldehyde dehydrogenase ALDH2*2 and alcohol dehydrogenase ADH1B*2 on blood acetaldehyde concentrations. *Hum Genomics* 2009;**3**(2):121–7.
35. Ehlers CL, Liang T, Gizer IR. ADH and ALDH polymorphisms and alcohol dependence in Mexican and Native Americans. *Am J Drug Alcohol Abuse* 2012;**38**(5):389–94.
36. Liu J, Zhou Z, Hodgkinson CA, et al. Haplotype-based study of the association of alcohol-metabolizing genes with alcohol dependence in four independent populations. *Alcohol Clin Exp Res* 2011;**35**(2):304–16.
37. Edenberg HJ. The genetics of alcohol metabolism: role of alcohol dehydrogenase and aldehyde dehydrogenase variants. *Alcohol Res Health* 2007;**30**(1):5–13.
38. Craciunescu CN, Johnson AR, Zeisel SH. Dietary choline reverses some, but not all, effects of folate deficiency on neurogenesis and apoptosis in fetal mouse brain. *J Nutr* 2010;**140**(6):1162–6.
39. Eichner ER, Hillman RS. Effect of alcohol on serum folate level. *J Clin Invest* 1973;**52**(3):584–91.
40. McGuffin R, Goff P, Hillman RS. The effect of diet and alcohol on the development of folate deficiency in the rat. *Br J Haematol* 1975;**31**(2):185–92.
41. McMartin KE, Collins TD, Bairnsfather L. Cumulative excess urinary excretion of folate in rats after repeated ethanol treatment. *J Nutr* 1986;**116**(7):1316–25.
42. de la Vega MJ, Santolaria F, Gonzalez-Reimers E, et al. High prevalence of hyperhomocysteinemia in chronic alcoholism: the importance of the thermolabile form of the enzyme methylenetetrahydrofolate reductase (MTHFR). *Alcohol* 2001;**25**(2):59–67.
43. Shaw S, Jayatilleke E, Herbert V, Colman N. Cleavage of folates during ethanol metabolism. Role of acetaldehyde/xanthine oxidase-generated superoxide. *Biochem J* 1989;**257**(1):277–80.

44. Barak AJ, Beckenhauer HC, Tuma DJ. Methionine synthase. a possible prime site of the ethanolic lesion in liver. *Alcohol* 2002;**26**(2):65–7.
45. Molloy AM, Kirke PN, Brody LC, Scott JM, Mills JL. Effects of folate and vitamin B12 deficiencies during pregnancy on fetal, infant, and child development. *Food Nutr Bull* 2008;**29**(2 Suppl.):S101–11 discussion S12–S15.
46. Hutson JR, Stade B, Lehotay DC, Collier CP, Kapur BM. Folic acid transport to the human fetus is decreased in pregnancies with chronic alcohol exposure. *PLoS One* 2012;**7**(5):e38057.
47. Lin GW. Maternal-fetal folate transfer: effect of ethanol and dietary folate deficiency. *Alcohol* 1991;**8**(3):169–72.
48. Murillo-Fuentes ML, Murillo ML, Carreras O. Effects of maternal ethanol consumption during pregnancy or lactation on intestinal absorption of folic acid in suckling rats. *Life Sci* 2003;**73**(17):2199–209.
49. Serrano M, Han M, Brinez P, Linask KK. Fetal alcohol syndrome: cardiac birth defects in mice and prevention with folate. *Am J Obstet Gynecol* 2010;**203**(1):e7–e15.
50. Kharbanda KK, Todero SL, King AL, et al. Betaine treatment attenuates chronic ethanol-induced hepatic steatosis and alterations to the mitochondrial respiratory chain proteome. *Int J Hepatol* 2012;**2012**:962183.
51. Wu BT, Dyer RA, King DJ, Richardson KJ, Innis SM. Early second trimester maternal plasma choline and betaine are related to measures of early cognitive development in term infants. *PLoS One* 2012;**7**(8):e43448.
52. Thomas JD, Zhou FC, Kane CJ. Proceedings of the 2008 annual meeting of the Fetal Alcohol Spectrum Disorders Study Group. *Alcohol* 2009;**43**(4):333–9.
53. Bekdash RA, Zhang C, Sarkar DK. Gestational choline supplementation normalized fetal alcohol-induced alterations in histone modifications, DNA methylation, and proopiomelanocortin (POMC) gene expression in beta-endorphin-producing POMC neurons of the hypothalamus. *Alcohol Clin Exp Res* 2013;**37**(7):1133–42.
54. Lumeng L. The role of acetaldehyde in mediating the deleterious effect of ethanol on pyridoxal 5′-phosphate metabolism. *J Clin Invest* 1978;**62**(2):286–93.
55. Kanazawa S, Herbert V. Total corrinoid, cobalamin (vitamin B12), and cobalamin analogue levels may be normal in serum despite cobalamin in liver depletion in patients with alcoholism. *Lab Invest* 1985;**53**(1):108–10.
56. Lindenbaum J, Lieber CS. Alcohol-induced malabsorption of vitamin B12 in man. *Nature* 1969;**224**(5221):806.
57. Bates CJ, Fuller NJ. The effect of riboflavin deficiency on methylenetetrahydrofolate reductase (NADPH) (EC 1.5. 1. 20) and folate metabolism in the rat. *Br J Nutr* 1986;**55**(2):455–64.
58. Chawla RK, Jones DP. Abnormal metabolism of S-adenosyl-L-methionine in hypoxic rat liver. Similarities to its abnormal metabolism in alcoholic cirrhosis. *Biochim Biophys Acta* 1994;**1199**(1):45–51.
59. Avila MA, Corrales FJ, Ruiz F, et al. Specific interaction of methionine adenosyltransferase with free radicals. *Biofactors* 1998;**8**(1–2):27–32.
60. Zhang R, Miao Q, Wang C, et al. Genome-wide DNA methylation analysis in alcohol dependence. *Addict Biol* 2013;**18**(2):392–403.
61. Pal-Bhadra M, Bhadra U, Jackson DE, Mamatha L, Park PH, Shukla SD. Distinct methylation patterns in histone H3 at Lys-4 and Lys-9 correlate with up- & down-regulation of genes by ethanol in hepatocytes. *Life Sci* 2007;**81**(12):979–87.
62. Zhao R, Zhang R, Li W, et al. Genome-wide DNA methylation patterns in discordant sib pairs with alcohol dependence. *Asia Pac Psychiatry* 2013;**5**(1):39–50.
63. Philibert RA, Plume JM, Gibbons FX, Brody GH, Beach SR. The impact of recent alcohol use on genome wide DNA methylation signatures. *Front Genet* 2012;**3**:54.
64. Yu Z, Kone BC. Hypermethylation of the inducible nitric-oxide synthase gene promoter inhibits its transcription. *J Biol Chem* 2004;**279**(45):46954–61.
65. Esfandiari F, Villanueva JA, Wong DH, French SW, Halsted CH. Chronic ethanol feeding and folate deficiency activate hepatic endoplasmic reticulum stress pathway in micropigs. *Am J Physiol Gastrointest Liver Physiol* 2005;**289**(1):G54–63.
66. Liu Y, Balaraman Y, Wang G, Nephew KP, Zhou FC. Alcohol exposure alters DNA methylation profiles in mouse embryos at early neurulation. *Epigenetics* 2009;**4**(7):500–11.
67. Zhou FC, Balaraman Y, Teng M, Liu Y, Singh RP, Nephew KP. Alcohol alters DNA methylation patterns and inhibits neural stem cell differentiation. *Alcohol Clin Exp Res* 2011;**35**(4):735–46.
68. Zhang H, Herman AI, Kranzler HR, et al. Array-based profiling of DNA methylation changes associated with alcohol dependence. *Alcohol Clin Exp Res* 2013;**37**(Suppl. 1):E108–15.
69. Zou JY, Crews FT. Release of neuronal HMGB1 by ethanol through decreased HDAC activity activates brain neuroimmune signaling. *PLoS One* 2014;**9**(2):e87915.
70. Park PH, Miller R, Shukla SD. Acetylation of histone H3 at lysine 9 by ethanol in rat hepatocytes. *Biochem Biophys Res Commun* 2003;**306**(2):501–4.
71. Qiang M, Denny A, Lieu M, Carreon S, Li J. Histone H3K9 modifications are a local chromatin event involved in ethanol-induced neuroadaptation of the NR2B gene. *Epigenetics* 2011;**6**(9):1095–104.
72. Botia B, Legastelois R, Alaux-Cantin S, Naassila M. Expression of ethanol-induced behavioral sensitization is associated with alteration of chromatin remodeling in mice. *PLoS One* 2012;**7**(10):e47527.
73. Millar CB, Grunstein M. Genome-wide patterns of histone modifications in yeast. *Nat Rev Mol Cell Biol* 2006;**7**(9):657–66.
74. Hawkins RD, Hon GC, Yang C, et al. Dynamic chromatin states in human ES cells reveal potential regulatory sequences and genes involved in pluripotency. *Cell Res* 2011;**21**(10):1393–409.
75. Guillemette B, Drogaris P, Lin HH, et al. H3 lysine 4 is acetylated at active gene promoters and is regulated by H3 lysine 4 methylation. *PLoS Genet* 2011;**7**(3):e1001354.
76. Choudhury M, Pandey RS, Clemens DL, Davis JW, Lim RW, Shukla SD. Knock down of GCN5 histone acetyltransferase by siRNA decreases ethanol-induced histone acetylation and affects differential expression of genes in human hepatoma cells. *Alcohol* 2011;**45**(4):311–24.
77. Tammen SA, Dolnikowski GG, Ausman LM, et al. Aging and alcohol interact to alter hepatic DNA hydroxymethylation. *Alcohol Clin Exp Res* 2014;**38**(8):2178–85.
78. Meissner A, Mikkelsen TS, Gu H, et al. Genome-scale DNA methylation maps of pluripotent and differentiated cells. *Nature* 2008;**454**(7205):766–70.
79. Hernandez-Vargas H, Lambert MP, Le Calvez-Kelm F, et al. Hepatocellular carcinoma displays distinct DNA methylation signatures with potential as clinical predictors. *PLoS One* 2010;**5**(3):e9749.
80. Lee SL, Hoog JO, Yin SJ. Functionality of allelic variations in human alcohol dehydrogenase gene family: assessment of a functional window for protection against alcoholism. *Pharmacogenetics* 2004;**14**(11):725–32.
81. Wu D, Cederbaum AI. Ethanol-induced apoptosis to stable HepG2 cell lines expressing human cytochrome P-4502E1. *Alcohol Clin Exp Res* 1999;**23**(1):67–76.
82. Kim M, Park YK, Kang TW, et al. Dynamic changes in DNA methylation and hydroxymethylation when hES cells undergo differentiation toward a neuronal lineage. *Hum Mol Genet* 2014;**23**(3):657–67.
83. Chen Y, Ozturk NC, Zhou FC. DNA methylation program in developing hippocampus and its alteration by alcohol. *PLoS One* 2013;**8**(3):e60503.

84. Chen Y, Damayanti NP, Irudayaraj J, Dunn K, Zhou FC. Diversity of two forms of DNA methylation in the brain. *Front Genet* 2014;**5**:46.
85. Govorko D, Bekdash RA, Zhang C, Sarkar DK. Male germline transmits fetal alcohol adverse effect on hypothalamic proopiomelanocortin gene across generations. *Biol Psychiatry* 2012;**72**(5):378–88.
86. Rachdaoui N, Sarkar DK. Effects of alcohol on the endocrine system. *Endocrinol Metab Clin North Am* 2013;**42**(3):593–615.
87. Ouko LA, Shantikumar K, Knezovich J, Haycock P, Schnugh DJ, Ramsay M. Effect of alcohol consumption on CpG methylation in the differentially methylated regions of H19 and IG-DMR in male gametes: implications for fetal alcohol spectrum disorders. *Alcohol Clin Exp Res* 2009;**33**(9):1615–27.
88. Knezovich JG, Ramsay M. The effect of preconception paternal alcohol exposure on epigenetic remodeling of the h19 and rasgrf1 imprinting control regions in mouse offspring. *Front Genet* 2012;**3**:10.
89. Abel E. Paternal contribution to fetal alcohol syndrome. *Addict Biol* 2004;**9**(2):127–33 discussion 35–6.
90. Enoch MA. The role of GABA(A) receptors in the development of alcoholism. *Pharmacol Biochem Behav* 2008;**90**(1):95–104.
91. Kumari M, Ticku MK. Regulation of NMDA receptors by ethanol. *Prog Drug Res* 2000;**54**:152–89.
92. Biermann T, Reulbach U, Lenz B, et al. N-methyl-D-aspartate 2b receptor subtype (NR2B) promoter methylation in patients during alcohol withdrawal. *J Neural Transm* 2009;**116**(5):615–22.
93. Qiang M, Denny AD, Ticku MK. Chronic intermittent ethanol treatment selectively alters N-methyl-D-aspartate receptor subunit surface expression in cultured cortical neurons. *Mol Pharmacol* 2007;**72**(1):95–102.
94. Skinner MK, Guerrero-Bosagna C, Haque M, Nilsson E, Bhandari R, McCarrey JR. Environmentally induced transgenerational epigenetic reprogramming of primordial germ cells and the subsequent germ line. *PLoS One* 2013;**8**(7):e66318.
95. Dias BG, Ressler KJ. Parental olfactory experience influences behavior and neural structure in subsequent generations. *Nat Neurosci* 2014;**17**(1):89–96.
96. Guerrero-Bosagna C, Settles M, Lucker B, Skinner MK. Epigenetic transgenerational actions of vinclozolin on promoter regions of the sperm epigenome. *PLoS One* 2010;**5**(9):e13100.
97. Nilsson E, Larsen G, Manikkam M, Guerrero-Bosagna C, Savenkova MI, Skinner MK. Environmentally induced epigenetic transgenerational inheritance of ovarian disease. *PLoS One* 2012;**7**(5):e36129.
98. McCarrey JR. Distinctions between transgenerational and non-transgenerational epimutations. *Mol Cell Endocrinol* 2014;**398**(1–2): 13–23.
99. Skinner MK, Anway MD, Savenkova MI, Gore AC, Crews D. Transgenerational epigenetic programming of the brain transcriptome and anxiety behavior. *PLoS One* 2008;**3**(11):e3745.
100. Heard E, Martienssen RA. Transgenerational epigenetic inheritance: myths and mechanisms. *Cell* 2014;**157**(1):95–109.
101. Lee SJ, Kang MH, Min H. Folic acid supplementation reduces oxidative stress and hepatic toxicity in rats treated chronically with ethanol. *Nutr Res Pract* 2011;**5**(6):520–6.
102. Ojeda ML, Nogales F, Jotty K, Barrero MJ, Murillo ML, Carreras O. Dietary selenium plus folic acid as an antioxidant therapy for ethanol-exposed pups. *Birth Defects Res B Dev Reprod Toxicol* 2009;**86**(6):490–5.
103. Xu Y, Li L, Zhang Z, Li Y. Effects of folinic acid and Vitamin B12 on ethanol-induced developmental toxicity in mouse. *Toxicol Lett* 2006;**167**(3):167–72.
104. Downing C, Johnson TE, Larson C, et al. Subtle decreases in DNA methylation and gene expression at the mouse Igf2 locus following prenatal alcohol exposure: effects of a methyl-supplemented diet. *Alcohol* 2011;**45**(1):65–71.
105. Xiao W, Dong W, Zhang C, et al. Effects of the epigenetic drug MS-275 on the release and function of exosome-related immune molecules in hepatocellular carcinoma cells. *Eur J Med Res* 2013;**18**:61.

CHAPTER

24

Molecular Mechanisms of Alcohol-Associated Carcinogenesis

*Helmut K. Seitz, MD, PhD**,**, *Sebastian Mueller, MD, PhD**,**

*Centre of Alcohol Research, University of Heidelberg, Heidelberg, Baden-Württemberg, Germany

**Department of Medicine (Gastroenterology and Hepatology), Salem Medical Centre, Heidelberg, Baden-Württemberg, Germany

INTRODUCTION

According to the International Agency for Research on Cancer (IARC), alcohol is a carcinogen, and its chronic misuse is responsible for the occurrence of various tumors, including those of the upper aerodigestive tract (oral cavity, pharynx, larynx, and esophagus), the liver, colorectum, and female breast.[1] Across the globe, 3.6% of all cancers resulted from chronic alcohol consumption.[2] The fact that alcohol is a risk factor for cancer is not new; more than 100 years ago French pathologists detected that chronic alcohol consumption resulted in an increased risk for esophagus cancer, especially when combined with smoking.[3] It took a long time until epidemiological data were available to give evidence for an association between alcohol and cancer. The National Institute of Health of the United States had three workshops on this issue; one in 1978, a second one in 2004, especially dealing with mechanisms, and the last one in 2010, dealing with general recommendations on the use of alcohol, especially with respect to cancer. Recently a book summarized the knowledge of this workshop.[4]

Various mechanisms may contribute to alcohol-mediated carcinogenesis, including the action of acetaldehyde,[5] the first and most toxic ethanol metabolite, the generation of oxidative stress by various pathways, including the induction of cytochrome P450-2E1 (CYP2E1),[6,7] the effect of alcohol on epigenetics (DNA, histone methylation, and acetylation),[8] the interaction with retinoid metabolism,[4] and special effect of ethanol on intracellular signal transduction pathways.[4] Alcohol also has an immune suppressing effect that may facilitate tumor spread.[4] In addition, tissue specific effects of alcohol are of considerable importance. This is relevant for the liver, where cirrhosis of the liver is a prerequisite for hepatocellular carcinoma (HCC),[9] or the female breast,[10] where estrogens act as a carcinogen and are increased by alcohol consumption,[11] as well as for the esophagus, where gastroesophageal reflux disease, induced by alcohol consumption is an additional risk factor for tumor occurrence. It should also be emphasized that chronic alcohol consumption, even at smaller amounts, may increase the risk for HCC in patients with hepatitis B[12] and C,[13] and patients with hemochromatosis and nonalcoholic steatohepatitis.[14] Alcohol related carcinogenesis may interact with other factors such as smoking,[15] diet, and comorbidities,[12–14] and depends on genetic susceptibility.[16–18]

The administration of ethanol chronically, without any chemical carcinogen, results in an increased incidence of cancer of the upper alimentary tract,[19] liver,[20] breast,[21] and colon.[22]

Various mechanisms of alcohol toxicity have been already discussed in this book series. Some of these mechanisms are also relevant with respect to carcinogenesis. This includes the action of acetaldehyde, as well as its adduct formation, reported by Professors Mei and Muto. In addition, the effect of ethanol on epigenetic modulation, DNA methylation, and one-carbon metabolism has been extensively discussed by Professors D'Addario, Jenssen, and Schernhammer. Therefore, these aspects will only be mentioned briefly. The major emphasis of this review will be on molecular mechanisms involved with oxidative stress due to chronic alcohol consumption. In addition, alcohol and its effects on retinoid metabolism, and estrogen signaling will be discussed.

Molecular Aspects of Alcohol and Nutrition. http://dx.doi.org/10.1016/B978-0-12-800773-0.00024-0

EFFECT OF ETHANOL ON DNA

Chronic ethanol consumption may lead to various DNA adducts: (1) adducts associated with the generation of reactive oxygen species (ROS), whereby some ROS may bind directly to DNA, but others initiate lipidperoxidation and lipidperoxidation products such as 4-hydoxynonenal (4-HNE) or malondialdehyde (MDA) finally bind to DNA forming exocyclic etheno-DNA adducts,[6,23–26] and (2) acetaldehyde DNA adducts.[5,27–29]

The role of ROS, as well as of acetaldehyde in carcinogenesis (genetic, epigenetic, and toxic effects), is illustrated in Figure 24.1.

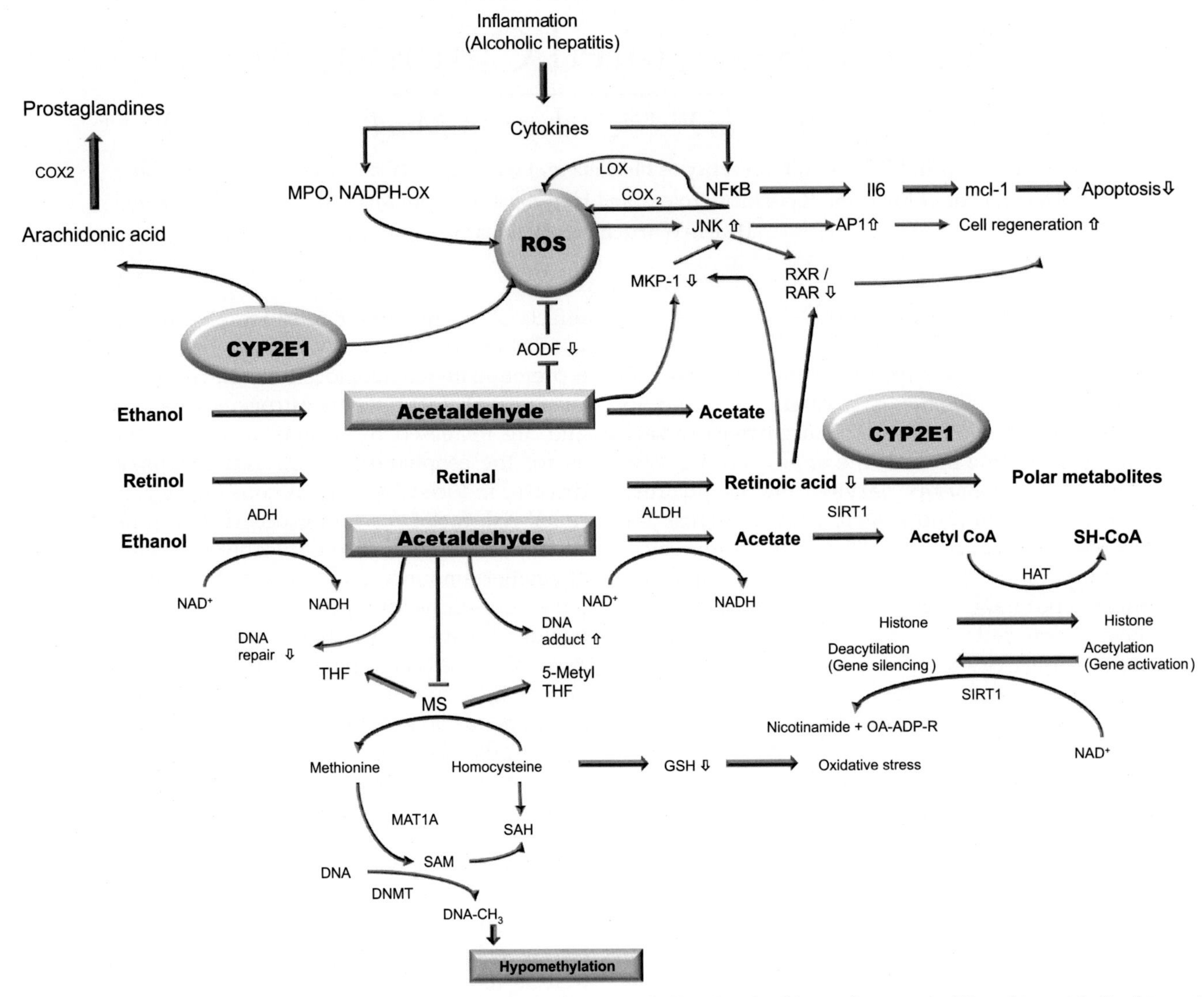

FIGURE 24.1 **Effect of ethanol and acetaldehyde on intermediary metabolism involved in carcinogenesis.** Ethanol is metabolized to acetaldehyde by ADH, cytochrome P450-2E1, and acetaldehyde is further metabolized by ALDH to acetate. The induction of CYP2E1 by ethanol, as well as the liberation of cytokines in inflammatory tissue, leads to the formation of ROS. Acetaldehyde inhibits the antioxidative defense system (AODS) which favors ROS accumulation. ROS leads to lipidperoxidation and finally etheno-DNA adduct formation. CYP2E1 induction decreases retinoic acid (RA) leading together with ROS to hyperregeneration, and decreased apoptosis through various signal pathways. Acetaldehyde leads to DNA adducts, an inhibition of DNA repair, and severe effects on methyl transfer associated with epigenetic changes. Ethanol oxidation also creates NADH, and a change in the cellular redox potential, inhibiting SIRT1 and thus interfering with histone acetylation. For more details see the text. COX-2, cyclooxygenase2; ROS, reactive oxygen species; CYP2E1, cytochrome P450-2E1; Il6, interleukin 6; NFκB, nuclear factor kappa-light-chain-enhanced of activated B cells; NADH, reduced nicotinamide adenine dinucleotide; LOX, lipoxygenase; AP1, activating protein 1; MPO, myeloperoxidase; JNK, c-jun-N-terminal kinase; MKP-1, mitogen-activated protein kinase phosphatase 1; RXR, retinoid X receptor; RAR, retinoid acid receptor; HAT, histone acetylase; SIRT1, silent information regulator1; THF, tetrahydrofolic acid; MS, methionine synthetase; MAT1A, methionine adenosyltransferase 1A; SAH, S-adenosylhomocysteine; SAM, S-adenosylmethionine; ADH, alcoholdehydrogenase; ALDH, acetaldehyde- dehydrogenase; GSH, glutathione; AODF, antioxidative defense system; OA-ADPO-R, O-acetyl-ADP-ribose.

Oxidative Stress and DNA Adduct Formation due to Reactive Oxygen Species

Oxidative stress is an important mechanism in alcohol-associated cancer. ROS following chronic alcohol consumption, may be generated by various mechanisms, including the induction of CYP2E1, the reoxidation of NADH intramitochondrially and the general inflammation, observed in alcoholic hepatitis (AH) due to the increase in inflammatory cytokines, which lead to the generation of ROS via various mechanisms, including an increase in NADPH oxidase, as well as inducible nitric oxide synthase (iNOS), and xanthine oxidase.[30,31] Furthermore, cytokines are released in inflammatory tissue, which not only activate these enzymes, but also activate NFκB, a nuclear transcription factor, which among others stimulates cyclooxygenase 2 (COX-2), lipoxygenase (LOX), and iNOS.[32–34] Upregulation of iNOS COX-2 and LOX also results in the overproduction of ROS.[32] iNOS catalyzes nitric oxide generation, which reacts with oxygen to produce N_2O_3, a strong nitrosating compound that deaminates DNA bases to react with secondary amines, to form *N*-nitrosamines, which are highly carcinogenic.[32]

Another reaction with O_2 leads to peroxynitrite with a formation of 8-nitroguanine. Peroxynitrite also results in a single strand breakage of DNA.[32,35] Such an environmental milieu may be present in AH, with a huge burden of oxidative stress.[36] Some of these ROSs may directly act with DNA, and some others may lead to lipid peroxidation. In addition to AH, other factors due to chronic alcohol consumption may increase ROS, including the induction of CYP2E1,[37,38] iron overload, increase in iNOS with an increased production of nitric oxide, and the generation of a highly reactive peroxynitrite.[30,39] ROS, as well as acetaldehyde, the first metabolite of ethanol activate NFκB,[5,6,32,39] which is an important regulator in carcinogenesis.

Besides inflammation, the induction of CYP2E1 in the liver, and also in other tissues, is of predominant importance in the generation of ROS.[7] We could show that this induction varies in every individual, and occurs at already low alcohol concentrations, such as 40 g/day, and increases significantly with the time of ethanol exposure.[38] Furthermore, CYP2E1 induction by alcohol may also depend on dietary factors, since medium chain triglyceride diminishes CYP2E1 induction, as compared to the application of long-chained triglyceride in animal experiments.[40]

CYP2E1 has a high rate of NADPH oxidase activity, resulting in the generation of large quantities of O_2^-, H_2O_2, and hydroxyethyl radicals.[30,31] All these radicals may lead to lipid peroxidation, with the formation of lipid peroxidation products, such as 4-HNE or MDA. Both chemicals may bind to DNA, forming highly carcinogenic exocyclic etheno-DNA adducts,[7,23–26,41] which have been identified in various tissues including the liver,[42,43] the esophageal,[44] and the colonic mucosa (Seitz, personal communication).

The most important etheno adduct formed with 4-HNE and MDA, and the DNA bases include 1,N^6-etheno-2′-deoxyadenosine (εdA); 3,N^4-etheno-2′-deoxycytidine (εdC); 1,N^2-etheno-2′-deoxyguanosine (1,N^2εdG), and N^2,3-etheno-2′-deoxyguanosine (N^2,3εdG). In addition, also substituted base adducts are formed, such as HNE-dG, carrying a fatty acid chain residue. 2,N^4-etheno-5-methyl-2′-deoxycytidine (ε5mdC), an endogenous, hitherto unknown LPO-derived adduct, was identified in the DNA of human tissue, which could play a role in epigenetic mechanisms of carcinogenesis.[45] Exocyclic etheno-DNA adducts exhibit strong mutagenic properties, producing various types of base pair substitution mutations, and other types of genetic damage in all organisms tested so far.[32,39]

εdA can lead to AT → GC transition, and AT → TA and AT → CG transversions. εdC can cause CG → AT transversions and CG → TA transition, and N^2,3 εdG can lead to GC → AT transition.[45] Incorporation of a single εdA in either DNA strand of HeLa cells showed a similar miscoding frequency island, and was more mutagenic than 8-oxo-dG.[46]

Some etheno adducts are poorly repaired in some tissues and cells supporting their biological relevance.[47] The biological importance of etheno-DNA adducts is further stressed as they are preferentially formed in codon 249 of TP53, which encodes T53, leading to mutation that renders cells more resistant to apoptosis and provides them some growth advantage.[48]

In various animal experiments it has been clearly shown that CYP2E1 has a major input on carcinogenesis and liver disease. Animals overexpressing CYP2E1 under alcohol-containing diet develop severe liver diseases, and an increased load of ROS.[49,50] Knock-out mice for CYP2E1 receiving an ethanol-containing diet have less severe liver lesions, as compared to control animals,[51] and less DNA adducts.[52] Furthermore, the level of CYP2E1 induction correlates significantly with etheno-DNA lesions.[43,44]

We also performed a long-term experiment in rats, with a single dose of the strong carcinogen diethylnitrosamine. The administration of ethanol over 1 month resulted in the formation of preneoplastic lesions[53] and, over 10 months, in the generation of hepatic adenomas, which was not the case when the rats received a control diet.[54] When chlormethiazole, a specific CYP2E1 inhibitor was added to the diet, the formation of liver tumors was completely inhibited, so was the induction of CYP2E1.[54]

We can show in cell culture that HepG2 cells overexpressing CYP2E1 also revealed high levels of etheno-DNA adducts. The formation of these adducts could be completely blocked in the presence of chlormethiazole.[43] In patients with alcoholic liver disease, hepatic CYP12E1 correlated significantly with the levels of hepatic etheno adducts.[43] Finally, a similar observation was made in patients with alcohol initiated esophageal cancer, where CYP2E1 also correlated with the levels of etheno-DNA adducts.[44]

Acetaldehyde Toxicity Associated with DNA Repair and DNA Adduct Formation

Acetaldehyde, the first metabolite of ethanol oxidation, is mutagenic and carcinogenic in animal experiments. IARC has identified acetaldehyde as a carcinogen for human beings, primarily responsible for esophageal cancer.[1]

In addition to acetaldehyde generation from ethanol in various tissues, acetaldehyde can also be formed from ethanol via bacterial oxidation in the mouth, stomach, and colon.[55] Bacterial acetaldehyde production may be of special importance in upper gastrointestinal cancer and in the colon, where acetaldehyde may act as a carcinogen. In addition, the amount of acetaldehyde generated is genetically determined through oxidation by various alcohol dehydrogenases (ADHs) and through its detoxification by acetaldehyde dehydrogenases (ALDHs). Forty percent of Asians have an ALDH gene, which codes for a low activity ALDH enzyme. When these individuals drink alcohol, acetaldehyde accumulates. Such individuals have a striking increase of upper gastrointestinal cancer, demonstrating the carcinogenic effect of acetaldehyde in humans.[16]

A number of *in vitro* and *in vivo* experiments in prokaryotic and eukaryotic cell cultures, as well as in animal models, have identified acetaldehyde as a mutagen and carcinogen.[1] Acetaldehyde causes point mutations in the hypoxanthine–guanosine–phosphoribosyl transferase locus in human lymphocytes, induces sister chromatide exchanges, and gross chromosomal alterations.[56–58] It also induces inflammation and metaplasia of the tracheal epithelium, delays cell cycle progression, and enhances cell injury associated with cellular hyperregeneration.[59] When acetaldehyde is added to the drinking water of rats, the mucosal lesions of the upper alimentary tract observed resembled those following chronic ethanol consumption.[60] When inhaled, acetaldehyde causes nasopharyngeal and laryngeal carcinoma.[61,62] Acetaldehyde interferes with DNA synthesis[5] and repair,[63] injures the cellular antioxidative defense system, especially through binding to glutathione, and results in cellular hyperregeneration of the mucosa of the upper alimentary tract and the colorectum.[5,59,64]

Acetaldehyde also binds to DNA, generating stable adducts with mutagenic properties. The occurrence of stable DNA adducts has been demonstrated in different organs of alcohol fed animals, and in leucocytes of alcoholics.[27,28] The major stable DNA adduct N^2-ethyl-deoxyguanosine (N^2-Et-dG) can be used efficiently by eukaryotic DNA polymerase.[58] In addition to N^2-Et-dG, which primarily serves as a marker for chronic alcohol consumption, another DNA adduct has been identified. The formation of α-methyl-gamma-hydroxy-1,N^2-propano-deoxyguanosine (PdG) occurs in the presence of basic amino acids, histones, and polyamines.[29] PdG has mutagenic properties. It is noteworthy that the generation of this mutagenic DNA adduct is facilitated in hyperregenerative tissue, such as the gastrointestinal mucosa, since polyamines are triggers for cell regeneration.

Acetaldehyde also inhibits DNA repair, by inhibiting O^6-methyl-guanosyltrasferase, an enzyme most relevant for the repair of adducts caused by alkylating agents.[63] Acetaldehyde also binds to various proteins, changing their functional and structural properties, and resulting in the generation of neoantigens with an antibody cascade reaction.[5,30]

EFFECT OF ETHANOL ON EPIGENETICS

Epigenetic changes due to chronic ethanol consumption play a predominant role in the development of cancer. Alcohol can alter the methylation and acetylation pattern of certain DNA regions, as well as of histones.[65] Alcohol also regulates miRNAs, an epigenetic mechanism that controls transcriptional events and the expression of certain genes.[65] In the next paragraph, some of the most important effects of ethanol on epigenetics are briefly described. For more detailed information, it is referred to a recent review article.[8,65,66]

DNA Methylation

ROS produced by CYP2E1 are also involved in hepatic methylation pattern, including DNA methylation.[65] Thus, 8-hydroxy-2-deoxyguanosine decreases DNA methylation during DNA repair.[67] DNA regions rich in the nucleoside cytosine and guanosine (i.e., CpG islands) may incorporate 8-OHdG that inhibits the methylation of adjacent cytosine residues. The reason for this observation is an inhibition of methyltransferase, which results in DNA hypomethylation.[66] 8-OHdG formations can also interfere with a normal function of DNA methyltransferases, and prevent DNA remethylation.[68]

In addition, the availability of the active methyl donor S-adenosine methionine (SAMe) is reduced by alcohol, due to the inhibition of several different methyltransferase reactions.[66] Ethanol inhibits methionine-adenosyltransferase that converts methionine into SAMe, as well as enzymes that help regenerate methionine (betaine-homocysteine methyltransferase and methionine synthase). As a result, SAMe decreases and, therefore, DNA methylation also decreases.[66] It is interesting that SAMe administration to animals inhibits tumor formation in the liver.[69] SAMe content in the liver is decreased in preneoplastic hepatic regions, and SAMe administration blocks the transformation of these lesions into cancer because of its DNA methylation capacity. Subsequently, SAMe administration inhibits the expression of certain cancer inducing genes, such as c-myc, c-Ha-ras, and c-ka-ras.[65]

Histone Modification

Histone modifications include methylation, acetylation, phosphorylation, and ubiquitination. These modifications regulate gene expression, such as, deacetylation, as well as hypermethylation, associated to gene silencing and inactivation of genes that suppress tumor growth (tumor suppressor genes), with consecutive tumor promotion, and stimulation of carcinogenesis.

Histone Methylation

Methylation of histones of one or more lysine amino acids is performed by the action of methyl transferases. One important finding in animals is that the level of methylated histone H3K4 (H3K4me2) and H3K27 (H3K27me3), in the nuclei of liver cells, is found to be increased after alcohol feeding.[70] H3K27 methylation mediates a gene silencing pathway, and is linked to another major silencing pathway (i.e., DNA methylation) via a deacetylase called SIRT1.[71] It is recruited by the PRC2 complex, and contributes also to gene silencing. The SIRT1 levels are increased in alcohol-fed animals.[65]

Histone H3K27me3, together with the histone code writer EZH2, which has intrinsic histone H3K27 methyl transferase activity, regulates stem cell renewal and differentiation progenitor stem cells in the liver. Cells with Mallory–Denk bodies (MDBs), in AH and liver cancer, show a decrease in nuclear H3K27mE3 and an increase pEZH2 (phosphorylated EZH2) in the MDBs.[72] A high expression of EZH is associated with poor survival.[73] MDB-containing cells do not only have progenitor and pluripotent properties, they may also transform to tumor initiating stem like cells. Transformation of these cells to cancer cells involves TLR4 signaling pathway that is upregulated, due to endotoxins and LPS in AH.[65]

Histone Acetylation

In animal experiments, chronic alcohol ingestion increased acetylation of various histones, including H3K18 and H3K9.[74] The nuclear level of β-catenin was increased, indicating the activation of signaling pathway of canonical wnt β-catenin pathway, involved in tumor formation.[65] In addition, it was shown that chronic alcohol fed animals had increased activity of histone acyltransferase (HAT) p300.[74] On the other hand, the activity of deacetylase (i.e., SIRT1) was also found to be increased after alcohol.[65,74] This change was associated by alterations of other molecules, including an increase in RARb and peroxisome proliferator-activated receptor (PPAR) C coactivator 1α (PGC1α) expression, and a decrease in PPAR-γ expression.[65]

HATP300 results in an increase in the signaling molecule p21WAF1/C, p1 (p21),[75] associated with a delay in cycle progression at various stages of the cell cycle, which may prevent the cells from dividing normally. As a result, cell cycle arrest genetic instability and program cell death and oncogenic effects occur.[76,77]

It is important to note that p21 expression is regulated by histone acetylation, and is regulated by a protein complex that is associated with the p21 promoter, including histone deacetylase1, which reduces acetylation.[78] Thus, HDAC1 inhibitors induce p21-expression, and cause cell cycle arrest.[78,79] Such HDAC1 inhibitors are used for cancer treatment. French and coworkers have shown an increase in HDA1C1 levels in alcoholic liver disease, and that HDA1C-inhibitors retard the formation of MDBs in these cells.[80] For a long a time, it was not clear why liver cancer occurs more frequently in the livers of patients who abstained for a longer time from alcohol. However, one reason for that is that the induction of p21 by alcohol with time after alcohol abstinence and that, after that time, cell cycle arrest disappears and hyperproliferation occurs.

ALCOHOL, RETINOIDS, AND CANCER

Retinoic acid (RA) is an important factor regulating cell growth, cell differentiation, and apoptosis. After binding to its receptor, RA regulates the expression of certain genes. Both RA receptors (RARα, RARβ, and RARγ) and retinoid X receptors (RXRα, RXRβ, and RXRγ) function as transcription factors by binding to the RA response element (RARE), and the retinoid X response element (RXRE) located in the 5′ promoter region of susceptible genes. RA binding leads to a conformational change in the retinoid receptors, allowing for the dissociation of corepressors (e.g., NCoR, SMRT, histone deacetylase-containing complexes) and recruitment of coactivators (e.g., CPB/p300, ACTR, DRIP/TRAP), some of which have histone acetyltransferase activity for chromatin decondensation. A decrease of RA results in uncontrolled cell proliferation, inadequate cell differentiation, and altered apoptosis.

Classical studies by Dr Lieber have reported that chronic alcohol consumption results in decreased levels of retinol in the liver, which are claimed to be the cause for night blindness and sexual dysfunction in alcoholics.[81,82] We have clarified the mechanisms for this decrease in hepatic retinol and RA. When hepatic microsomes from ethanol fed rats were incubated with RA, an enhanced degradation of RA was observed, associated with the generation of polar RA metabolites, such as 18-OH-RA and 4-oxo-RA, as compared to microsomes from control rats.[83] Furthermore, RA metabolism could be significantly blocked when chlormethiazole (CMZ), a specific CYP2E1 inhibitor, was added to the *in vitro* system, demonstrating that CYP2E1 is responsible for RA degradation.[84,85]

When rats received ethanol chronically, hepatic RA concentrations decreased, while CYP2E1 was found to be induced. Again, the administration of CMZ prevented the decrease of RA.[86] Thus, CMZ can restore both hepatic retinol and retinyl ester concentrations to normal levels in ethanol fed rats, through blocking enhanced degradation of retinol, and mobilization of vitamin A from the liver into the circulation, indicating that CYP2E1 is the responsible enzyme for ethanol mediated catabolism of retinoids in the liver. It has to be emphasized that not only hepatic concentrations of RA normalized with CMZ, but also cell proliferation, as well as cell cycle behavior.[86]

The impaired retinoid homeostasis results in aberrant retinoid receptor signaling through upregulation of the c-Jun N-terminal kinase (JNK) signaling pathway.[87]

As pointed out CYP2E1 induction by alcohol also results in oxidative stress. Both increased oxidative stress, as well as low RA (via decreased mitogen-activated protein kinase phosphatise-1 (MKP-1)), results in an activation of the JNK-pathway. Cross-talk exists between JNK-pathway and RXR/RAR receptors and, therefore, the activated JNK-pathway activates AP-1 gene, leading to an increase in c-fos and c-jun, which are strikingly elevated in the livers of ethanol fed rats, compared to pair fed controls.[87,88]

Finally, as already pointed out, CYP2E1 mediated catabolism of RA and retinol leads to the generation of various polar metabolites. These polar metabolites revealed high apoptotic properties, since these metabolites resulted in a change of the hepatic mitochondrial membrane potential, a release of cytochrome C, and an activation of the caspase cascade.[83]

Indeed, the intake of vitamin A or β-carotene, a precursor of vitamin A, together with ethanol, results in an increase in hepatic apoptosis and cellular injury.[82] It has also been reported that individuals who smoked and took 30 mg β-carotene for lung cancer prevention, developed more lung cancer than a control population without β-carotene administration. It was finally found that those who smoked and took β-carotene, and developed lung cancer, also had an intake of more than 11 g of alcohol every day,[89] demonstrating that the results elaborated *in vitro* and in animal studies have also implications for the clinical situation.

THE ROLE OF ESTROGENS IN ETHANOL MEDIATED BREAST CANCER

Since chronic ethanol consumption, even at low levels, increase the risk to develop breast cancer,[10,90] and since a strong association exists between circulating estrogens and the risk of breast cancer,[91] the relationship between alcohol and estrogens seems to be of major importance.

Alcohol increases plasma estrogen levels, in controlled feeding studies with human female volunteers.[11,92–94] In postmenopausal women with hormone replacement therapy, 15 or 39 g of ethanol daily, consumed over 8 weeks, resulted in an elevation of serum estrogen sulfate and dehydroepiandrosterone sulfate.[94] In premenopausal women, even small doses of ethanol, equivalent to one drink, resulting in ethanol blood concentrations of not more than 20 mg/100 mL, led to significantly elevated plasma estrogen concentrations.[11]

The mechanisms by which alcohol increases estrogens are not clear but may, however, include an inhibition of their degradation, due to a change in the hepatic redox state after ethanol metabolism, or in an ethanol mediated inhibition of sulfotransferase and 2-hydroxylase, two enzymes involved in estrogen degradation.[10,90] In addition, the ethanol mediated increase in aromatase activity may result in an increased conversion of testosterone to estrogens.[11,95] Finally, ethanol increases the production of luteinizing hormone from the pituitary gland, with a consecutive increase in estrogen release from the ovaries.[11]

Unquestionably, alcohol stimulates estrogen receptor signaling in human breast cancer cells.[96] The activation of this pathway may finally end up in altered cell cycle control, cell proliferation, and apoptosis. Cellular effects of estrogens are mediated via binding to estrogens receptors (ER), either membrane bound or intracellular. Thereafter, estrogens may bind to nuclear ER-α or -β which after dimerization, bind to the estrogen response element of DNA to regulate gene expression. Another pathway involves membrane bound ERs with activation of various protein kinases, such as MAPK and cyclic AMP. There is also evidence that cross-talk exists with pathways downstream of other receptor tyrosine kinases (Figure 24.2).

Ethanol affects estrogens signaling at various levels.

1. Increased circulating estrogens.[11,92–94]
2. Modulation of nuclear receptors.[96,97]
3. Selective stimulation of proliferation of ER+ cultured human breast cancer cells associated with elevated ER-α.[11,97]
4. Stimulation of transcription of estrogen responsive genes in the nucleus, which is ligand dependent and mediated by ER-α, involving the cyclic AMP/protein kinase A (PKA) signaling pathway.[98]

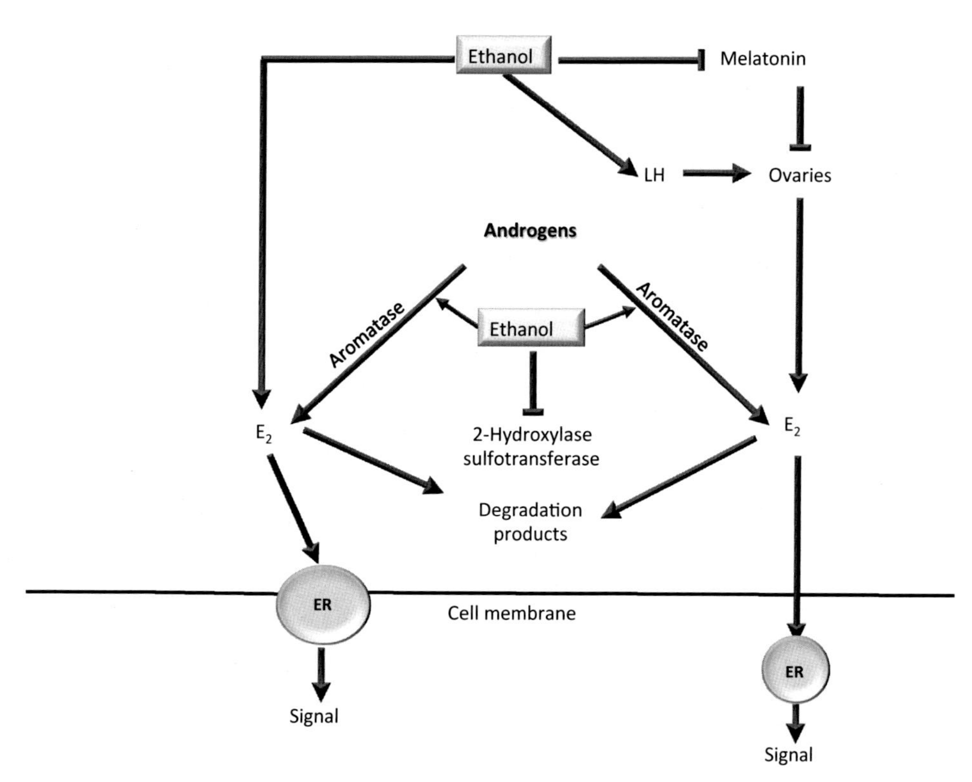

FIGURE 24.2 **Effect of ethanol on estrogens.** Ethanol increases estrogen serum concentrations through various mechanisms, including an increased synthesis from androgens via induced aromatase, and an inhibition of estrogen degradation due to an inhibition of 2-hydroxylase and sulfotransferase. Ethanol also stimulates luteinizing hormone (LH) from the pituitary gland and block melatonin. Both result in an increased secretion of estrogens from the ovaries. Estrogens are bound to their receptors, either at the cell membrane, or intracellularly, with the activation of a signal pathway leading to modulation of cell regeneration and apoptosis. For more details see the text. LH, luteinizing hormone; E, estrogen; ER, estrogen receptor.

SUMMARY AND CONCLUSIONS

According to IARC, alcohol is a carcinogen. Chronic alcohol consumption may exert its carcinogenicity through various mechanisms. During ethanol metabolism, acetaldehyde, as well as ROS, do occur. Since acetaldehyde inhibits the antioxidative defense system, as well as DNA repair, it interacts with methyl transfer, and binds to DNA, ROS leads to lipid peroxidation and the generation of lipid peroxidation products, such as 4-hydroxynonenal, with binding to DNA, and the production of etheno DNA-adducts that are highly carcinogenic.

Besides these effects of ethanol metabolites on DNA, alcohol also has an effect on epigenetics. Alcohol alters methylation, acetylation pattern of certain DNA regions, as well as of histones. Alcohol also regulates miRNAs as epigenetic mechanism that controls transcriptional events in the expression of certain genes.

One of the most important observations is the induction of CYP2E1 through chronic alcohol consumption. CYP2E1 is not only responsible for ROS generation, but also for the activation of various procarcinogens to active carcinogens, and to the enhanced degradation of retinoic acid to polar apoptotic metabolites. A loss in retinoic acid leads to hyperproliferation and dedifferentiation of cells.

Finally, the alcohol-mediated increase in estrogens may have an effect on the alcohol-mediated breast cancer carcinogenesis. Other mechanisms, besides increased estrogens, such as a modulation of nuclear receptors and stimulation of transcription of estrogen responsive genes in the nucleus, may also play a role.

All these mechanisms may play in concert, and one or the other mechanism may prevail, depending on the target organ.

Key Facts

- Alcohol induces cytochrome P450-2E1.
- Cytochrome P450-2E1 is responsible for the increase in ROS, and a decrease in retinoid acid.
- Acetaldehyde, the first metabolite of ethanol, is highly toxic and carcinogenic.
- Epigenetic mechanisms also contribute to the carcinogenicity of alcohol.
- Alcohol increases estrogens that may play a role in breast cancer.

Summary Points

- Chronic alcohol consumption leads to the induction of cytochrome P450-2E1.
- Cytochrome P450-2E1 generates ROS with a consequence of etheno DNA-adducts.
- The first metabolite of ethanol metabolism acetaldehyde damages proteins with structural and functional alterations, and also binds to DNA.
- Epigenetic changes may contribute to alcohol-associated cancer development.
- Chronic alcohol consumption leads to an increased degradation of retinoid acid via enhanced CYP2E1 metabolism, resulting in cellular hyperproliferation and decreased cell differentiation.
- Alcohol results in an increase of estrogens, which may contribute to the increased risk of breast cancer.

Acknowledgment

Original research is supported by a grant of the Dietmar Hopp Foundation.

References

1. Baan R, Straif K, Grosse Y, Secretan B, El Ghissassi F, Bouvard V, et al. Carcinogenicity of alcoholic beverages. *Lancet Oncol* 2007;**8**(4):292–3.
2. Rehm J, Room R, Monteiro R, Gmel G, Graham K, Rehn T. Global and regional burden of disease attributable to selected major risk factors. In: Ezatti M, Murray C, Lopez AD, Rodgers A, Murray C, editors. *Comparative Quantification of Health Risks*. Geneva: World Health Organisation; 2004.
3. Lamu L. Etude de statistique clinique de 131 cas de cancer de l'oesophage et du cardia. *Arch Mal Appar Dig Mal Nutr* 1910;**4**: 451–6.
4. Zakhari S, Vasiliou V, Max Guo Q, editors. *Alcohol and Cancer*. New York, Dordrecht, Heidelberg, London: Springer; 2011.
5. Seitz HK, Stickel F. Acetaldehyde as an underestimated risk factor for cancer development: role of genetics in ethanol metabolism. *Genes Nutr* 2010;**5**(2):121–8.
6. Seitz HK, Stickel F. Risk factors and mechanisms of hepatocarcinogenesis with special emphasis on alcohol and oxidative stress. *Biol Chem* 2006;**387**(4):349–60.
7. Seitz HK, Wang XD. The role of cytochrome P450 2E1 in ethanol-mediated carcinogenesis. *Subcell Biochem* 2013;131–43.
8. Varela-Rey M, Woodhoo A, Martinez-Chantar ML, Mato JM, Lu SC. Alcohol, DNA methylation, and cancer. *Alcohol Res* 2013;**35**(1): 25–35.
9. Seitz HK, Stickel F. Ethanol and hepatocarcinogenesis. In: Watson RR, Zibadi S, Preedy VR, editors. *Alcohol, Nutrition, and Health Consequences*. New York, Heidelberg, Dordrecht, London: Humana Press; 2013 pp. 411–428.
10. Seitz HK, Pelucchi C, Bagnardi V, La Vecchia C. Epidemiology and pathophysiology of alcohol and breast cancer: update 2012. *Alcohol Alcohol* 2012;**47**(3):204–12.
11. Coutelle C, Hohn B, Benesova M, Oneta CM, Quattrochi P, Roth HJ, et al. Risk factors in alcohol associated breast cancer: alcohol dehydrogenase polymorphism and estrogens. *Int J Oncol* 2004;**25**(4):1127–32.
12. Ohnishi K, Iida S, Iwama S, Goto N, Nomura F, Takashi M, et al. The effect of chronic habitual alcohol intake on the development of liver cirrhosis and hepatocellular carcinoma: relation to hepatitis B surface antigen carriage. *Cancer* 1982;**49**(4):672–7.
13. Mueller S, Millonig G, Seitz HK. Alcoholic liver disease and hepatitis C: a frequently underestimated combination. *World J Gastroenterol* 2009;**15**(28):3462–71.
14. Ascha MS, Hanouneh IA, Lopez R, Tamimi TA, Feldstein AF, Zein NN. The incidence and risk factors of hepatocellular carcinoma in patients with nonalcoholic steatohepatitis. *Hepatol* 2010;**51**(6): 1972–8.
15. Tuyns A. Alcohol and cancer. *Alcohol Health Res World* 1978;**2**:20–31.
16. Yokoyama A, Muramatsu T, Ohmori T, Yokoyama T, Okuyama K, Takahashi H, et al. Alcohol-related cancers and aldehyde dehydrogenase-2 in Japanese alcoholics. *Carcinogenesis* 1998;**19**(8):1383–7.
17. Homann N, Konig IR, Marks M, Benesova M, Stickel F, Millonig G, et al. Alcohol and colorectal cancer: the role of alcohol dehydrogenase 1C polymorphism. *Alcohol Clin Exp Res* 2009;**33**(3):551–6.
18. Visapaa JP, Gotte K, Benesova M, Li J, Homann N, Conradt C, et al. Increased cancer risk in heavy drinkers with the alcohol dehydrogenase 1C*1 allele, possibly due to salivary acetaldehyde. *Gut* 2004;**53**(6):871–6.
19. Soffritti M, Belpoggi F, Cevolani D, Guarino M, Padovani M, Maltoni C. Results of long-term experimental studies on the carcinogenicity of methyl alcohol and ethyl alcohol in rats. *Ann NY Acad Sci* 2002;**982**:46–69.
20. Beland FA, Benson RW, Mellick PW, Kovatch RM, Roberts DW, Fang JL, et al. Effect of ethanol on the tumorigenicity of urethane (ethyl carbamate) in B6C3F1 mice. *Food Chem Toxicol* 2005;**43**(1): 1–19.
21. Watabiki T, Okii Y, Tokiyasu T, Yoshimura S, Yoshida M, Akane A, et al. Long-term ethanol consumption in ICR mice causes mammary tumor in females and liver fibrosis in males. *Alcohol Clin Exp Res* 2000;**24**(4 Suppl.):117S–22S.
22. Roy HK, Gulizia JM, Karolski WJ, Ratashak A, Sorrell MF, Tuma D. Ethanol promotes intestinal tumorigenesis in the MIN mouse. Multiple intestinal neoplasia. *Cancer Epidemiol Biomarkers Prev* 2002;**11**(11):1499–502.
23. el Ghissassi F, Barbin A, Nair J, Bartsch H. Formation of 1,N^6-ethenoadenine and 3,N^4-ethenocytosine by lipid peroxidation products and nucleic acid bases. *Chem Res Toxicol* 1995;**8**(2):278–83.
24. Blair IA. DNA adducts with lipid peroxidation products. *J Biol Chem* 2008;**283**(23):15545–9.
25. Chung FL, Chen HJ, Nath RG. Lipid peroxidation as a potential endogenous source for the formation of exocyclic DNA adducts. *Carcinogenesis* 1996;**17**(10):2105–11.
26. Pandya GA, Moriya M. 1,N^6-ethenodeoxyadenosine, a DNA adduct highly mutagenic in mammalian cells. *Biochemistry* 1996;**35**(35):11487–92.
27. Fang JL, Vaca CE. Detection of DNA adducts of acetaldehyde in peripheral white blood cells of alcohol abusers. *Carcinogenesis* 1997;**18**(4):627–32.
28. Wang M, McIntee EJ, Cheng G, Shi Y, Villalta PW, Hecht SS. Identification of DNA adducts of acetaldehyde. *Chem Res Toxicol* 2000;**13**(11):1149–57.
29. Theruvathu JA, Jaruga P, Nath RG, Dizdaroglu M, Brooks PJ. Polyamines stimulate the formation of mutagenic 1,N^2-propanodeoxyguanosine adducts from acetaldehyde. *Nucleic Acids Res* 2005;**33**(11):3513–20.
30. Seitz HK, Stickel F. Molecular mechanisms of alcohol-mediated carcinogenesis. *Nat Rev Cancer* 2007;**7**(8):599–612.
31. Albano E. Alcohol, oxidative stress and free radical damage. *Proc Nutr Soc* 2006;**65**(3):278–90.
32. Bartsch H, Nair J. Chronic inflammation and oxidative stress in the genesis and perpetuation of cancer: role of lipid peroxidation, DNA damage, and repair. *Langenbecks Arch Surg* 2006;**391**(5):499–510.

33. Zha S, Yegnasubramanian V, Nelson WG, Isaacs WB, De Marzo AM. Cyclooxygenases in cancer: progress and perspective. *Cancer Lett* 2004;**215**(1):1–20.
34. Karin M, Greten FR. NF-kappaB: linking inflammation and immunity to cancer development and progression. *Nat Rev Immunol* 2005;**5**(10):749–59.
35. Ohshima H, Bartsch H. Chronic infections and inflammatory processes as cancer risk factors: possible role of nitric oxide in carcinogenesis. *Mutat Res* 1994;**305**(2):253–64.
36. Bautista AP. Neutrophilic infiltration in alcoholic hepatitis. *Alcohol* 2002;**27**(1):17–21.
37. Lieber CS. Microsomal ethanol-oxidizing system (MEOS): the first 30 years (1968–1998) – a review. *Alcohol Clin Exp Res* 1999;**23**(6): 991–1007.
38. Oneta CM, Lieber CS, Li J, Ruttimann S, Schmid B, Lattmann J, et al. Dynamics of cytochrome P4502E1 activity in man: induction by ethanol and disappearance during withdrawal phase. *J Hepatol* 2002;**36**(1):47–52.
39. Bartsch H, Nair J. Lipid peroxidation-derived DNA adducts and the role in inflammation-related carcinogenesis. In: Hiraku Y, Kawanishi S, Ohshima H, editors. *Cancer and Inflammation Mechanisms; Chemical, Biological and Clinical Aspects*. Hoboken, New Jersey: John Wiley and Sons; 2014 p. 61–74.
40. Lieber CS, Cao Q, DeCarli LM, Leo MA, Mak KM, Ponomarenko A, et al. Role of medium-chain triglycerides in the alcohol-mediated cytochrome P450 2E1 induction of mitochondria. *Alcohol Clin Exp Res* 2007;**31**(10):1660–8.
41. Bartsch H, Barbin A, Marion MJ, Nair J, Guichard Y. Formation, detection, and role in carcinogenesis of ethenobases in DNA. *Drug Metab Rev* 1994;**26**(1–2):349–71.
42. Frank A, Seitz HK, Bartsch H, Frank N, Nair J. Immunohistochemical detection of 1,N^6-ethenodeoxyadenosine in nuclei of human liver affected by diseases predisposing to hepato-carcinogenesis. *Carcinogenesis* 2004;**25**(6):1027–31.
43. Wang Y, Millonig G, Nair J, Patsenker E, Stickel F, Mueller S, et al. Ethanol-induced cytochrome P4502E1 causes carcinogenic etheno-DNA lesions in alcoholic liver disease. *Hepatology* 2009;**50**(2): 453–61.
44. Millonig G, Wang Y, Homann N, Bernhardt F, Qin H, Mueller S, et al. Ethanol-mediated carcinogenesis in the human esophagus implicates CYP2E1 induction and the generation of carcinogenic DNA-lesions. *Int J Cancer* 2011;**128**(3):533–40.
45. Linhart KB, Bartsch H, Seitz HK. The role of reactive oxygen species (ROS) and cytochrome P-450 2E1 in the generation of carcinogenic etheno-DNA adducts. *Redox Biol* 2014;**3**:56–62.
46. Levine RL, Yang IY, Hossain M, Pandya GA, Grollman AP, Moriya M. Mutagenesis induced by a single 1,N^6-ethenodeoxyadenosine adduct in human cells. *Cancer Res* 2000;**60**(15):4098–104.
47. Swenberg JA, Fedtke N, Ciroussel F, Barbin A, Bartsch H. Etheno adducts formed in DNA of vinyl chloride-exposed rats are highly persistent in liver. *Carcinogenesis* 1992;**13**(4):727–9.
48. Hu W, Feng Z, Eveleigh J, Iyer G, Pan J, Amin S, et al. The major lipid peroxidation product, trans-4-hydroxy-2-nonenal, preferentially forms DNA adducts at codon 249 of human p53 gene, a unique mutational hotspot in hepatocellular carcinoma. *Carcinogenesis* 2002;**23**(11):1781–9.
49. Butura A, Nilsson K, Morgan K, Morgan TR, French SW, Johansson I, et al. The impact of CYP2E1 on the development of alcoholic liver disease as studied in a transgenic mouse model. *J Hepatol* 2009;**50**(3):572–83.
50. Morgan K, French SW, Morgan TR. Production of a cytochrome P450 2E1 transgenic mouse and initial evaluation of alcoholic liver damage. *Hepatol* 2002;**36**(1):122–34.
51. Lu Y, Zhuge J, Wang X, Bai J, Cederbaum AI. Cytochrome P450 2E1 contributes to ethanol-induced fatty liver in mice. *Hepatol* 2008;**47**(5):1483–94.
52. Bradford BU, Kono H, Isayama F, Kosyk O, Wheeler MD, Akiyama TE, et al. Cytochrome P450 CYP2E1, but not nicotinamide adenine dinucleotide phosphate oxidase, is required for ethanol-induced oxidative DNA damage in rodent liver. *Hepatol* 2005;**41**(2):336–44.
53. Chavez PR, Lian F, Chung J, Liu C, Paiva SA, Seitz HK, et al. Long-term ethanol consumption promotes hepatic tumorigenesis but impairs normal hepatocyte proliferation in rats. *J Nutr* 2011;**141**(6):1049–55.
54. Ye Q, Lian F, Chavez PR, Chung J, Ling W, Qin H, et al. Cytochrome P450 2E1 inhibition prevents hepatic carcinogenesis induced by diethylnitrosamine in alcohol-fed rats. *Hepatobiliary Surg Nutr* 2012;**1**(1):5–18.
55. Salaspuro MP. Acetaldehyde microbes, and cancer of the digestive tract. *Crit Rev Clin Lab Sci* 2003;**40**(2):183–208.
56. Obe G, Jonas R, Schmidt S. Metabolism of ethanol *in vitro* produces a compound which induces sister-chromatid exchanges in human peripheral lymphocytes *in vitro*: acetaldehyde not ethanol is mutagenic. *Mutat Res* 1986;**174**(1):47–51.
57. Helander A, Lindahl-Kiessling K. Increased frequency of acetaldehyde-induced sister-chromatid exchanges in human lymphocytes treated with an aldehyde dehydrogenase inhibitor. *Mutat Res* 1991;**264**(3):103–7.
58. Matsuda T, Terashima I, Matsumoto Y, Yabushita H, Matsui S, Shibutani S. Effective utilization of N^2-ethyl-2′-deoxyguanosine triphosphate during DNA synthesis catalyzed by mammalian replicative DNA polymerases. *Biochemistry* 1999;**38**(3):929–35.
59. Simanowski UA, Suter P, Russell RM, Heller M, Waldherr R, Ward R, et al. Enhancement of ethanol induced rectal mucosal hyper regeneration with age in F344 rats. *Gut* 1994;**35**(8):1102–6.
60. Homann N, Karkkainen P, Koivisto T, Nosova T, Jokelainen K, Salaspuro M. Effects of acetaldehyde on cell regeneration and differentiation of the upper gastrointestinal tract mucosa. *J Natl Cancer Inst* 1997;**89**(22):1692–7.
61. Woutersen RA, Appelman LM, Van Garderen-Hoetmer A, Feron VJ. Inhalation toxicity of acetaldehyde in rats. III. Carcinogenicity study. *Toxicol* 1986;**41**(2):213–31.
62. Feron VJ, Kruysse A, Woutersen RA. Respiratory tract tumours in hamsters exposed to acetaldehyde vapour alone or simultaneously to benzo(*a*)pyrene or diethylnitrosamine. *Eur J Cancer Clin Oncol* 1982;**18**(1):13–31.
63. Espina N, Lima V, Lieber CS, Garro AJ. *In vitro* and *in vivo* inhibitory effect of ethanol and acetaldehyde on O_6-methylguanine transferase. *Carcinogenesis* 1988;**9**(5):761–6.
64. Simanowski UA, Suter P, Stickel F, Maier H, Waldherr R, Smith D, et al. Esophageal epithelial hyperproliferation following long-term alcohol consumption in rats: effects of age and salivary gland function. *J Natl Cancer Inst* 1993;**85**(24):2030–3.
65. French SW. Epigenetic events in liver cancer resulting from alcoholic liver disease. *Alcohol Res* 2013;**35**(1):57–67.
66. Stickel F, Herold C, H.K. S, Schuppan D. Alcohol and methyl transfer: implications for alcohol-related hepatocarcinogenesis. In: Ali S, Friedman SL, Mann DA, editors. *Liver Diseases: Biochemical Mechanisms and New Therapeutic Insights*. New Delhi, Oxford: IBH Publishing Company Pty Ltd; 2005.
67. Weitzman SA, Turk PW, Milkowski DH, Kozlowski K. Free radical adducts induce alterations in DNA cytosine methylation. *Proc Natl Acad Sci USA* 1994;**91**(4):1261–4.
68. Sasaki Y. Does oxidative stress participate in the development of hepatocellular carcinoma? *J Gastroenterol* 2006;**41**(12):1135–48.
69. Hitchler MJ, Domann FE. Metabolic defects provide a spark for the epigenetic switch in cancer. *Free Radic Biol Med* 2009;**47**(2):115–27.
70. Bardag-Gorce F, Oliva J, Dedes J, Li J, French BA, French SW. Chronic ethanol feeding alters hepatocyte memory which is not altered by acute feeding. *Alcohol Clin Exp Res* 2009;**33**(4):684–92.
71. Muntean AG, Hess JL. Epigenetic dysregulation in cancer. *Am J Pathol* 2009;**175**(4):1353–61.

72. French BA, Oliva J, Bardag-Gorce F, Li J, Zhong J, Buslon V, et al. Mallory–Denk bodies form when EZH2/H3K27me3 fails to methylate DNA in the nuclei of human and mice liver cells. *Exp Mol Pathol* 2012;**92**(3):318–26.
73. Gieni RS, Hendzel MJ. Polycomb group protein gene silencing, non-coding RNA, stem cells, and cancer. *Biochem Cell Biol* 2009;**87**(5):711–46.
74. Bardag-Gorce F, French BA, Joyce M, Baires M, Montgomery RO, Li J, et al. Histone acetyltransferase p300 modulates gene expression in an epigenetic manner at high blood alcohol levels. *Exp Mol Pathol* 2007;**82**(2):197–202.
75. Fang Z, Fu Y, Liang Y, Li Z, Zhang W, Jin J, et al. Increased expression of integrin beta1 subunit enhances p21WAF1/Cip1 transcription through the Sp1 sites and p300-mediated histone acetylation in human hepatocellular carcinoma cells. *J Cell Biochem* 2007;**101**(3):654–64.
76. Abbas T, Dutta A. CRL4Cdt2: master coordinator of cell cycle progression and genome stability. *Cell Cycle* 2011;**10**(2):241–9.
77. Serres MP, Zlotek-Zlotkiewicz E, Concha C, Gurian-West M, Daburon V, Roberts JM, et al. Cytoplasmic p27 is oncogenic and cooperates with Ras both *in vivo* and *in vitro*. *Oncogene* 2011;**30**(25): 2458–846.
78. Dokmanovic M, Clarke C, Marks PA. Histone deacetylase inhibitors: overview and perspectives. *Mol Cancer Res* 2007;**5**(10):981–9.
79. Gui CY, Ngo L, Xu WS, Richon VM, Marks PA. Histone deacetylase (HDAC) inhibitor activation of p21WAF1 involves changes in promoter-associated proteins, including HDAC1. *Proc Natl Acad Sci USA* 2004;**101**(5):1241–6.
80. French SW, Bardag-Gorce F, Li J, French BA, Oliva J. Mallory–Denk body pathogenesis revisited. *World J Hepatol* 2010;**2**(8):295–301.
81. Leo MA, Lieber CS. Hepatic vitamin A depletion in alcoholic liver injury. *N Engl J Med* 1982;**307**(10):597–601.
82. Dan Z, Popov Y, Patsenker E, Preimel D, Liu C, Wang XD, et al. Hepatotoxicity of alcohol-induced polar retinol metabolites involves apoptosis via loss of mitochondrial membrane potential. *FASEB J* 2005;**19**(7):845–7.
83. Liu C, Russell RM, Seitz HK, Wang XD. Ethanol enhances retinoic acid metabolism into polar metabolites in rat liver via induction of cytochrome P4502E1. *Gastroenterology* 2001;**120**(1):179–89.
84. Liu C, Chung J, Seitz HK, Russell RM, Wang XD. Chlormethiazole treatment prevents reduced hepatic vitamin A levels in ethanol-fed rats. *Alcohol Clin Exp Res* 2002;**26**(11):1703–9.
85. Chung J, Liu C, Smith DE, Seitz HK, Russell RM, Wang XD. Restoration of retinoic acid concentration suppresses ethanol-enhanced c-Jun expression and hepatocyte proliferation in rat liver. *Carcinogenesis* 2001;**22**(8):1213–9.
86. Wang XD, Seitz HK. Alcohol and retinoid interaction. In: Watson RR, Preedy VR, editors. *Nutrition and Alcohol: Linking Nutrient Interactions and Dietary Intake*. Boca Raton, London, New York, Washington: CRC Press; 2004 pp. 313–321.
87. Wang XD, Liu C, Chung J, Stickel F, Seitz HK, Russell RM. Chronic alcohol intake reduces retinoic acid concentration and enhances AP-1 (c-Jun and c-Fos) expression in rat liver. *Hepatol* 1998;**28**(3): 744–50.
88. Lieber CS. Alcohol and the liver: 1994 update. *Gastroenterol* 1994;**106**(4):1085–105.
89. Albanes D, Heinonen OP, Taylor PR, Virtamo J, Edwards BK, Rautalahti M, et al. Alpha-Tocopherol and beta-carotene supplements and lung cancer incidence in the alpha-tocopherol, beta-carotene cancer prevention study: effects of base-line characteristics and study compliance. *J Natl Cancer Inst* 1996;**88**(21):1560–70.
90. Castro GD, Castro JA. Alcohol drinking and mammary cancer: pathogenesis and potential dietary preventive alternatives. *World J Clin Oncol* 2014;**5**(4):713–29.
91. Yager JD, Davidson NE. Estrogen carcinogenesis in breast cancer. *N Engl J Med* 2006;**354**(3):270–82.
92. Ginsburg ES, Mello NK, Mendelson JH, Barbieri RL, Teoh SK, Rothman M, et al. Effects of alcohol ingestion on estrogens in postmenopausal women. *JAMA* 1996;**276**(21):1747–51.
93. Reichman ME, Judd JT, Longcope C, Schatzkin A, Clevidence BA, Nair PP, et al. Effects of alcohol consumption on plasma and urinary hormone concentrations in premenopausal women. *J Natl Cancer Inst* 1993;**85**(9):722–7.
94. Dorgan JF, Baer DJ, Albert PS, Judd JT, Brown ED, Corle DK, et al. Serum hormones and the alcohol-breast cancer association in postmenopausal women. *J Natl Cancer Inst* 2001;**93**(9):710–5.
95. Etique N, Chardard D, Chesnel A, Merlin JL, Flament S, Grillier-Vuissoz I. Ethanol stimulates proliferation, ERalpha and aromatase expression in MCF-7 human breast cancer cells. *Int J Mol Med* 2004;**13**(1):149–55.
96. Fan S, Meng Q, Gao B, Grossman J, Yadegari M, Goldberg ID, et al. Alcohol stimulates estrogen receptor signaling in human breast cancer cell lines. *Cancer Res* 2000;**60**(20):5635–9.
97. Singletary KW, Frey RS, Yan W. Effect of ethanol on proliferation and estrogen receptor-alpha expression in human breast cancer cells. *Cancer Lett* 2001;**165**(2):131–7.
98. Etique N, Flament S, Lecomte J, Grillier-Vuissoz I. Ethanol-induced ligand-independent activation of ERalpha mediated by cyclic AMP/PKA signaling pathway: an *in vitro* study on MCF-7 breast cancer cells. *Int J Oncol* 2007;**31**(6):1509–18.

CHAPTER

25

Molecular Link Between Alcohol and Breast Cancer: the Role of Salsolinol

*Mariko Murata, MD, PhD**, *Kaoru Midorikawa, PhD**, *Shosuke Kawanishi, PhD**,**

*Department of Environmental and Molecular Medicine, Mie University Graduate School of Medicine, Tsu, Japan
**Laboratory of Public Health, Department of Health Sciences, Faculty of Pharmaceutical Sciences, Suzuka University of Medical Science, Suzuka, Japan

INTRODUCTION

Epidemiological studies show that alcohol intake is an important risk factor for breast cancer.[1–3] Nelson et al. reported recently that alcohol consumption resulted in an estimated 3.2–3.7% of all US cancer deaths, and the majority of alcohol-attributable female cancer deaths were from breast cancer (56–66%).[2] According to data from the WHO Global Burden of Disease project in 2002, 5.2% and 1.7% of all cancers are respectively attributable to alcohol drinking worldwide, in men and women, and, among women, breast cancer comprises 60% of alcohol-attributable cancers.[4] The International Agency for Research on Cancer (IARC)[5] has determined that alcoholic beverages are carcinogenic to humans (Group 1). Understanding the mechanistic basis of this relationship has important implications for breast cancer prevention.

The carcinogenic mechanisms of alcohol-associated breast cancer are not fully understood. A recent review has indicated that the mechanisms of breast cancer roughly comprise a tumor initiation step by cumulative carcinogens, such as acetaldehyde formation and reactive oxygen species (ROS) generation through cytochrome P450 2E1 (CYP2E1) activation.[1] Moreover, as the tumor promoter step, dehydroepiandrosterone sulfate, that is metabolized to estrogen, increases in serum after alcohol drinking in postmenopausal women. The increase of estrogen accelerates the growth rate and induces other characteristics of estrogen receptor-positive cells.[1] In addition to these mechanisms, we suggest here that alcohol-derived salsolinol (1-methyl-6,7-dihydroxy-1,2,3,4-tetrahydroisochinolin) is a novel causative substance of breast cancer, as recently reported.[6]

SALSOLINOL DERIVED FROM ALCOHOL, AS A NOVEL CAUSATIVE SUBSTANCE OF BREAST CANCER

Salsolinol is known as a neurotoxic substance, in relation to chronic alcoholism and Parkinsonism.[7,8] Salsolinol can be endogenously synthesized by a nonenzymatic condensation reaction between acetaldehyde and dopamine in the brain.[9,10] In another pathway, enzymatic enantioselective formation of *R*-salsolinol from acetaldehyde and dopamine was also reported.[9] Salsolinol detected in the brain is likely to be derived from *in situ* synthesis.[11] Several recent studies have demonstrated an increase in salsolinol levels in certain brain areas, after alcohol intake.[12–14]

Salsolinol Detected in Blood and Urine in Relation to Alcohol Drinking

Early studies showed the relationship between alcohol drinking and salsolinol level in plasma and urine. Alcohol is metabolized mainly to acetaldehyde, and acetaldehyde in blood was correlated with urinary salsolinol in healthy men, after ethanol intake.[15] Urinary salsolinol can be highly increased by long-term drinking, even in normal, nonalcoholic subjects.[16] A significant elevation of salsolinol in urine was found after intake of ethanol, and higher urinary salsolinol levels were observed in an aldehyde dehydrogenase (ALDH) deficient group, than the normal ALDH group.[15] Haber et al. showed that the intake of alcohol has an influence of salsolinol levels in urine and plasma, and that different changes in salsolinol output may be affected by a genetic predisposition for alcohol-induced salsolinol formation.[17,18]

Molecular Aspects of Alcohol and Nutrition. http://dx.doi.org/10.1016/B978-0-12-800773-0.00025-2

Salsolinol is contained in alcoholic beverages and a variety of foods, such as cheese, bananas, beef, and milk,[19,20] and Lee et al.[21] suggested that salsolinol from dietary sources such as bananas was the major contributor to plasma salsolinol levels, not from ethanol. Many studies have been published on the possible involvement of salsolinol on underlying ethanol action, but it is still a controversial matter of debate.

Salsolinol Possibly Formed in Mammary Gland

After drinking alcohol, ethanol is metabolized by alcohol dehydrogenase (ADH) to acetaldehyde mainly in the liver. Acetaldehyde is detoxified by ALDH to acetic acid. Alcohol consumers with the ADH polymorphism and/or ALDH deficiency have an increased risk of developing breast cancer.[22–24] The ADH1C allele increased acetaldehyde levels, and women with this genotype were found to be at 1.8-times more at risk of breast cancer than those with other genotypes.[25] In addition to the ADH major pathway, the microsomal ethanol-oxidizing system (CYP2E1) becomes important in the alcohol metabolism with chronic alcohol use.[23] Significant interactions were reported between genetic polymorphisms of CYP2E1 and alcohol intake, for developing breast cancer.[23,26] The CYP2E1 c2 allele containing genotypes carries a 1.9-fold risk for developing breast cancer, compared to nondrinkers.[26] Although multiple mechanisms are involved in alcohol-mediated carcinogenesis, it is well-known that acetaldehyde plays important roles.[27] It is noteworthy that Triano et al. have shown that ADH is highly expressed in the normal mammary epithelium.[28] CYP2E1 is also expressed in normal breast tissues.[29] Castro et al. showed acetaldehyde accumulation in rat mammary tissues after alcohol drinking.[30,31] Genetic factors in low ALDH expression also lead to acetaldehyde accumulation. These reports suggest that generation of acetaldehyde in mammary tissues is one of the participatory factors in breast cancer.

Urinary catecholamine levels in daily life are elevated in women at familial risk of breast cancer.[32] Dopamine is one of catecholamines and is an important neurotransmitter, originating from the central nervous system. Because dopamine does not cross the blood-brain barrier, dopaminergic signaling in the brain is functionally distinct from the peripheral pathways.[33] Collins et al.[34] assumed the reaction of peripheral dopamine with acetaldehyde to be detected as urinary salsolinol, after drinking alcohol. Peripheral dopamine is released from neuronal cells in peripheral tissues, and dopamine released from sympathetic nerves predominantly contributes to plasma dopamine levels.[33] Dopamine is accumulated and compartmentalized by the dopamine transporter and the vesicular monoamine transporter 2 (VMAT2)[35,36] in the brain. Interestingly, Grönberg et al.[37] revealed that specific cells in the epithelium and myoepithelium of mammary glands express neuroendocrine markers, including VMAT2, and suggested that the mammary glands may have neuroendocrine functions, followed by dopamine release. These reports support the idea that dopamine and acetaldehyde may be formed in the mammary glands after drinking and, ultimately, salsolinol might be formed in mammary tissues, as shown in Figure 25.1.

We consider that salsolinol endogenously generated in the mammary glands is a novel causative substance of alcohol-associated breast cancer. In order to examine whether salsolinol can affect tumor initiation and promotion, in relation to mammary carcinogenesis, we investigated the DNA damage and cell proliferation induced by salsolinol.[6]

DNA DAMAGE CAPABILITY OF SALSOLINOL

We have developed an experimental system offering a convenient and useful method to detect the oxidative DNA damage capability of chemicals.[38] Briefly, exon-containing DNA fragments are obtained from the relevant human cancer genes and prepared after subcloning into vectors. The 5′-end-labeled DNA fragments are obtained by dephosphorylation with calf intestine phosphatase and rephosphorylation with [γ-^{32}P]ATP and T_4 polynucleotide kinase, and are further digested with a restriction enzyme to obtain singly labeled fragments. The standard reaction mixture contains salsolinol, ^{32}P-5′-end-labeled DNA fragments, calf thymus DNA (20 μM per base) and a metal ion in 10 mM sodium phosphate buffer (pH 7.8). In certain experiments, endogenous reductant NADH, antioxidant enzyme superoxide dismutase (SOD), or ROS scavenger, are added to the reaction mixture. After incubation at 37°C for 1 h, the DNA fragments are heated at 90°C in 1 M piperidine for 20 min. The recovered and denatured DNA fragments are electrophoresed on an 8% polyacrylamide 8 M urea gel, and autoradiograms are obtained by exposing X-ray film to the gel. Oligonucleotides from ^{32}P-labeled DNA fragments are detected on the autoradiogram as a result of DNA damage. The detailed methods were described previously.[6,38]

Damage to ^{32}P-Labeled DNA Fragments in the Presence of Salsolinol and Metal Ions

^{32}P-labeled DNA fragments and salsolinol were incubated in the presence of various metal ions (Cu(II), Fe(III)EDTA, Fe(III), Fe(III)citrate, Co(II), Ni(II), Mn(II) or Mn(III)). Figure 25.2 shows oxidative DNA damage

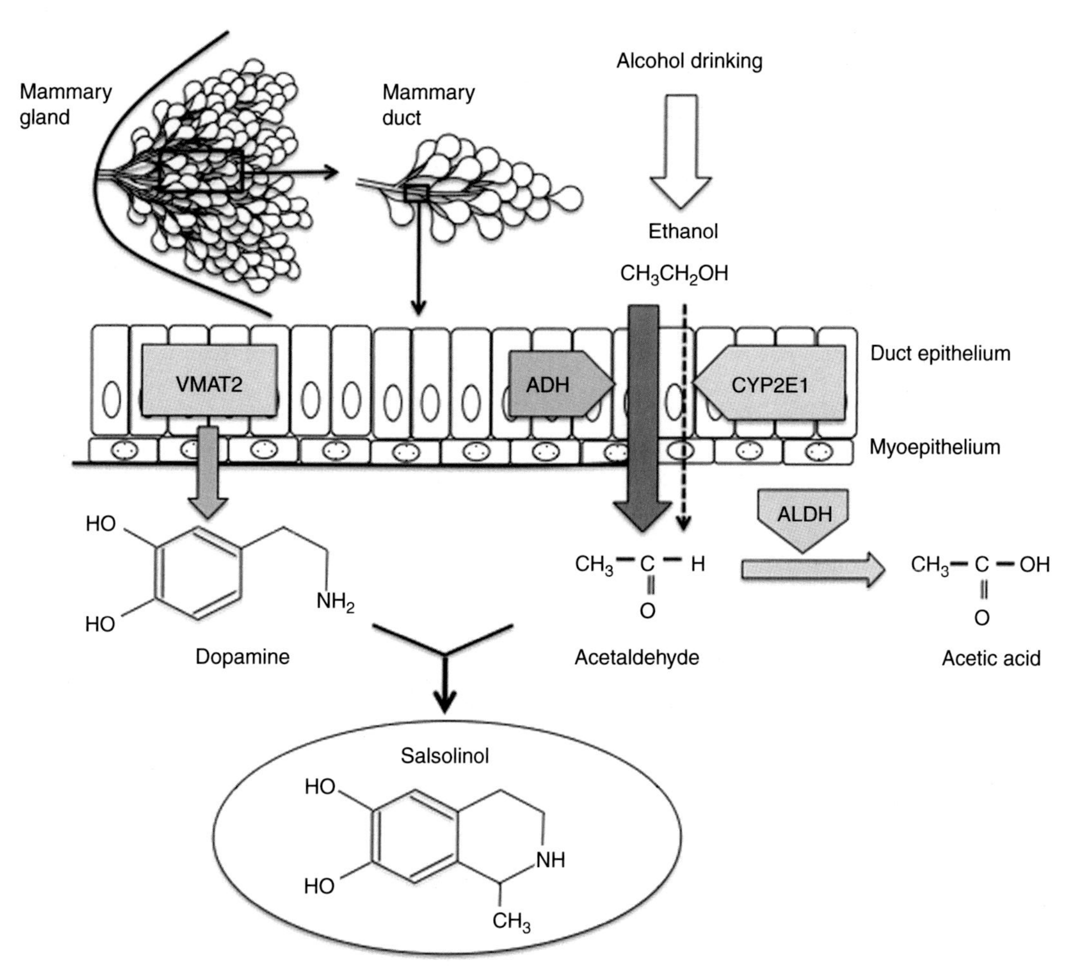

FIGURE 25.1 Formation of salsolinol in mammary tissue. Drinking alcohol may lead to formation of acetaldehyde by alcohol dehydrogenase (ADH) activity and CYP2E1 induction in mammary glands. Acetaldehyde dehydrogenase (ALDH) deficiency results in acetaldehyde accumulation. Vascular monoamine transporter2 (VMAT2)-positive epithelium and myoepithelium of mammary glands may produce dopamine that reacts with acetaldehyde, resulting in the formation of salsolinol.

induced by salsolinol. The autoradiogram (Figure 25.2a) indicated that salsolinol induced DNA damage in the presence of Cu(II) and Fe(III)EDTA in dose-dependent manner. In contrast, salsolinol did not induce DNA damage in the presence of other metal ions (Fe(III), Fe(III)citrate, Co(II), Ni(II), Mn(II), or Mn(III)) under the same conditions (data not shown). Salsolinol induced DNA damage in the presence of Cu(II) much more strongly than Fe(III)EDTA. Even without piperidine treatment, oligonucleotides were formed by salsolinol in the presence of metal ions (data not shown), indicating breakage of the deoxyribose phosphate backbone. The amount of oligonucleotides was increased by piperidine treatment (hot alkaline treatment). Since the altered base is readily removed from its sugar by the piperidine treatment, it is considered that the base modification was induced by salsolinol in the presence of metal ions.

In order to speculate about the reactive species, we performed scavenger experiments on DNA damage by salsolinol (data not shown). In the presence of Cu(II), catalase and bathocuproine, a Cu(I)-specific chelator, inhibited DNA damage but not typical free hydroxyl radical (•OH) scavengers, suggesting the involvement of hydrogen peroxide (H_2O_2) and Cu(I), but not •OH. Therefore, we suppose that the primary reactive species of Cu(II)-mediated DNA damage induced by salsolinol is Cu(I)-hydroperoxo complex [Cu(I)OOH]. In the case of Fe(III)EDTA, typical free •OH scavengers inhibited the DNA damage, indicating the involvement of •OH through the Fenton reaction.

8-OxodG Formation in DNA Treated with Salsolinol

Calf thymus DNA fragments and metal ions were incubated with salsolinol in sodium phosphate buffer at 37°C for 1 h. After ethanol precipitation, DNA fragments were digested to individual nucleosides with nuclease P_1 and calf intestine alkaline phosphatase, and analyzed with an

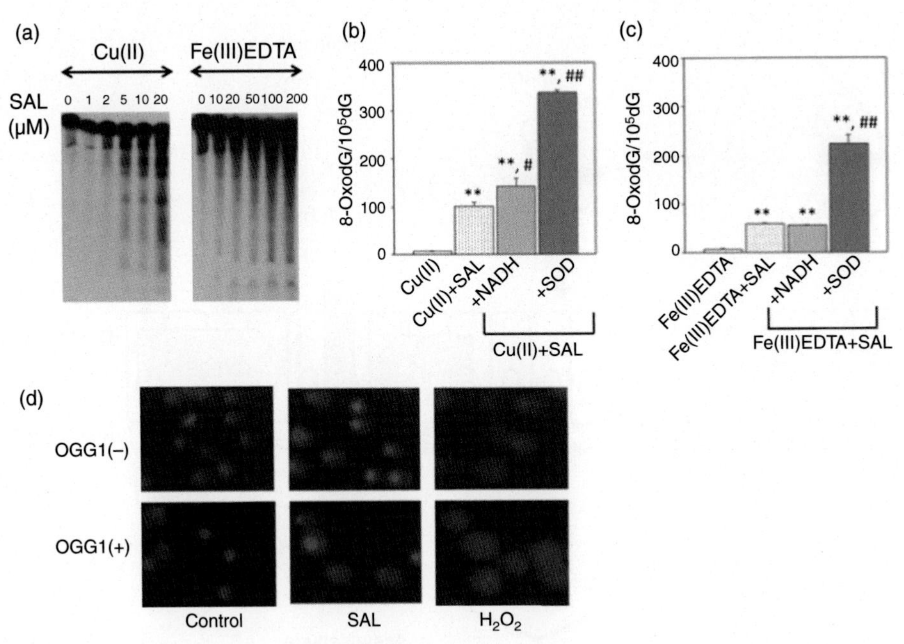

FIGURE 25.2 Oxidative DNA damage induced by salsolinol. (a) The reaction mixture contained the 5′-end ^{32}P-labeled 261-bp fragment from the human c-Ha-*ras*-1 proto-oncogene, 20 μM per base of calf thymus DNA, the indicated concentrations of salsolinol (SAL), and 20 μM $CuCl_2$ in 10 mM sodium phosphate buffer (pH 7.8) containing 5 μM DTPA. The reaction mixture contained 5′-end ^{32}P-labeled 309-bp fragment from the human *p16* tumor suppressor gene, 20 μM per base of calf thymus DNA, the indicated concentrations of salsolinol (SAL), and 20 μM Fe(III)EDTA in 10 mM sodium phosphate buffer (pH 7.8) containing 5 μM DTPA. After incubation at 37°C for 1 h, DNA fragments were treated with 1 M piperidine for 20 min at 90°C, and then electrophoresed on an 8% polyacrylamide/8 M urea gel. The autoradiogram was visualized by exposing an X-ray film to the gel. The detailed methods were described previously.[6,38] (b, c) 8-OxodG formation induced by salsolinol in the presence of metal ions and its enhancement by endogenous compounds. Calf thymus DNA (100 μM per base) was incubated with 100 μM salsolinol (SAL) in the presence of 20 μM $CuCl_2$ (b) or Fe(III)EDTA (c) at 37°C for 1 h. In certain experiments, 100 μM NADH or 30 unit SOD was added to the reaction mixture. After ethanol precipitation, DNA was enzymatically digested to the component nucleosides and analyzed with an HPLC-ECD. The detailed methods were described previously.[6,39] Results are expressed as means and SD of values obtained from three independent experiments. * $P < 0.05$, ** $P < 0.01$: significant difference compared with the metal ion alone by Student's *t*-test. # $P < 0.05$, ## $P < 0.01$: significant difference compared with the condition of salsolinol plus metal by Student's *t*-test. (d) MCF-10A cells were treated with 100 μM salsolinol (SAL) or H_2O_2 at 37°C for 1 h. The alkaline comet assay was used to determine DNA strand breaks and alkaline-labile DNA damage (hOGG1(−)) and 8-oxodG formation (hOGG1(+)).

electrochemical detector coupled to HPLC (HPLC-ECD). The detailed methods were described previously.[6,39]

Salsolinol induced significantly higher 8-oxodG formation in calf thymus DNA, compared with metal ion alone (Cu(II); Figure 25.2b, Fe(III)EDTA; Figure 25.2c). Interestingly, antioxidant enzyme superoxide dismutase (SOD) significantly enhanced salsolinol-induced oxidative DNA damage in the presence of either metal ion. Mn(II) has SOD-mimic activity, and exhibited a similar effect to SOD (data not shown). In contrast, endogenous reductant nicotinamide adenine dinucleotide (NADH) significantly intensified the 8-oxodG formation only in the presence of Cu(II).

Oxidative Damage to DNA of Cultured Human Mammary Epithelial Cells MCF-10A Treated with Salsolinol

Alkaline comet assay without hOGG1 showed that salsolinol slightly increased the number of cells with comet tails, compared to the control, and hOGG1 treatment revealed an increase of damaged cells with 8-oxodG formation (Figure 25.2d). H_2O_2 induced comet tails efficiently, even by the alkaline treatment alone, through the Fenton reaction. In the case of salsolinol, OGG1 exhibited comet tails, suggesting the dominancy of 8-oxodG formation, rather than strand breakage and/or alkaline-labile base modification.

PROPOSED MECHANISMS OF OXIDATIVE DNA DAMAGE INDUCED BY SALSOLINOL

Experiments in our *in vitro* system showed that salsolinol induced oxidative DNA damage in a dose-dependent manner in the presence of Cu(II) or Fe(III)EDTA. Interestingly, an antioxidant enzyme SOD enhanced the DNA damage. An endogenous reductant NADH increased the amount of 8-oxodG formation in the presence of Cu(II),

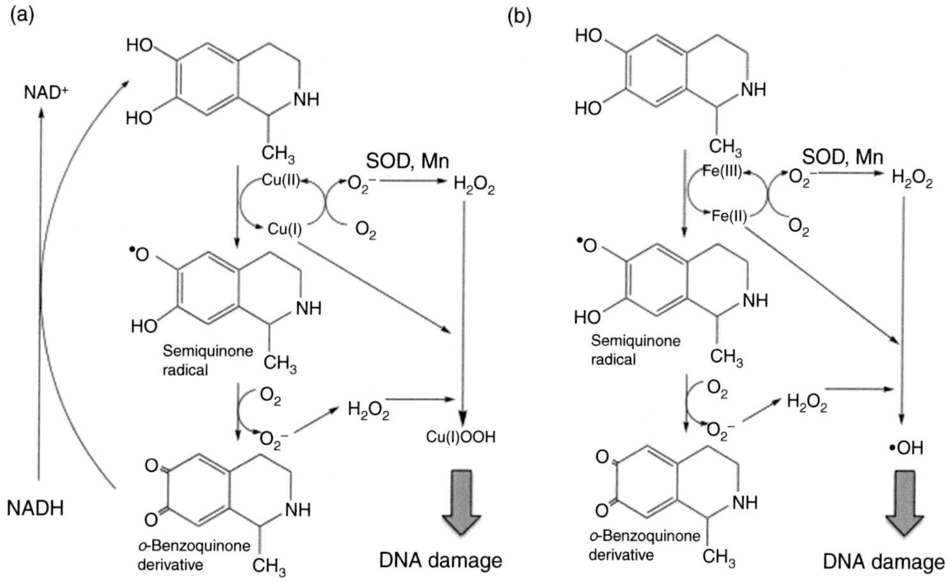

FIGURE 25.3 Proposed mechanisms of oxidative DNA damage induced by salsolinol in the presence of Cu(II) and Fe(III).

but not Fe(III)EDTA. We envision the mechanisms of metal-mediated DNA damage induced by salsolinol to be as shown in Figure 25.3.

Copper-Mediated DNA Damage Induced by Salsolinol (Figure 25.3a)

Salsolinol undergoes Cu(II)-mediated autoxidation to generate Cu(I) and semiquinone radical. Cu(I) reacts with O_2 to generate $O_2^{\bullet-}$ and subsequently H_2O_2. The formed Cu(I) bound to DNA interacts with H_2O_2, resulting in the formation of a Cu(I)-hydroperoxo complex such as DNA-Cu(I)OOH. NADH enhances salsolinol-induced DNA damage in the presence of Cu(II), since the reactive oxygen species may be produced abundantly and nonenzymatically through the reduction of the oxidized form, such as *o*-benzoquinone derivative, by NADH.

Iron-Mediated DNA Damage Induced by Salsolinol (Figure 25.3b)

In the presence of Fe(III)EDTA, salsolinol also undergoes autoxidation to generate Fe(II) and semiquinone radical. Fe(II) reacts with O_2 to generate $O_2^{\bullet-}$ and subsequently H_2O_2. The formation of H_2O_2 by $O_2^{\bullet-}$ dismutation progresses via reduction of Fe(III) to Fe(II). Fe(III)EDTA-mediated DNA damage by salsolinol is caused by •OH generated from the Fenton reaction. In addition, not only Fe(III)EDTA but also Fe(III)ADP, the complex of Fe(III) and adenosine diphosphate (ADP), could mediate DNA damage by salsolinol (data not shown). Pilas et al.[40] demonstrated •OH formation in the presence of a weaker chelator ADP than EDTA. These raise the possibility of a specific metal complex with endogenous compounds that is redox-active transition metal inside the cell.[41] Unlike the case of Cu(II), NADH did not enhance Fe(III)-mediated DNA damage. In the case of Fe(III)-EDTA, enzymes such as NADH-cytochrome reductase may be required to catalyze the formation of reactive oxygen species, as Yang and Cederbaum[42] reported. In an *in vivo* system, NADH may enzymatically enhance Fe(III)-mediated DNA damage induced by salsolinol.

It is believed that most of the metal-related mechanisms of carcinogenesis involve reactive oxygen species.[43,44] Copper ions are an essential component of chromatin, and are known to accumulate preferentially in the heterochromatic regions.[45] •OH is extremely short-lived, and travels a very short distance in water in a cell,[46,47] whereas iron is the most abundant transition metal in chromosome systems.[48] Chronic alcohol intake is associated with increased accumulation of iron.[49] In the presence of metal ions, oxidation of salsolinol to semiquinone radical occurs with the simultaneous reduction of metal ions. As previously reported,[6] we confirmed semiquinone radical formation from salsolinol by the ESR spin-stabilization method using Zn(II).[50] The semiquinone radical is further oxidized to *o*-benzoquinone derivative with the formation of $O_2^{\bullet-}$ that is dismutated to H_2O_2. Reduced metal ions [Cu(I) and Fe(II)] interact with H_2O_2, generated during the redox reaction, to produce Cu(I)OOH and •OH that cause DNA damage, respectively. The antioxidant enzyme SOD enhances the DNA damage, instead of defending against oxidative stress. Our previous papers showed that SOD enhanced oxidative DNA damage, and SOD accelerated the rate of metal-mediated autoxidation of chemicals.[51–53] Mn(II) has the property of catalytically scavenging $O_2^{\bullet-}$[54] similar to SOD. We also observed that Mn(II) enhanced the DNA damage in the same manner as SOD (data not shown). An alkaline comet assay using MCF-10A cells showed that salsolinol induced a slight DNA damage and strand breakage without hOGG1, and a significant increase of cells with the longer comet tail, by hOGG1 treatment. From the results of the alkaline comet assay without hOGG1, H_2O_2 induced DNA strand breakage more efficiently than salsolinol, suggesting that the formation of 8-oxodG may be predominant, rather than strand breakage in the case of salsolinol. 8-oxodG has the ability to pair with adenine, instead of cytosine, during replication, yielding G:C to T:A transversions.[55,56] Salsolinol may play a role in oxidative DNA damage leading to mutation, as an initiation step of carcinogenesis.

CELL PROLIFERATING ABILITY OF SALSOLINOL

We performed a bioassay for measuring cell proliferating activity by the modified method of E-screen assay,[57] using two human cultured cells, normal mammary epithelial MCF-10A cells and estrogen-sensitive breast cancer MCF-7 cells. Briefly, each cell line was trypsinized and plated into 12-well plates at an initial concentration of 3×10^4 cells per well, with growth medium. After the cells were allowed to attach after one day, the growth medium was replaced with phenol red-free experimental medium, in which the sex steroids of the serum were removed by charcoal-dextran treatment.[58] Salsolinol and other compounds were then added at the final solvent concentration in culture medium, so as not to exceed 0.1% DMSO, as this concentration did not affect cell yields. After incubation for 6 days, cells were trypsinized and harvested to count the number. The detailed methods were described previously.[6,58]

Cell Proliferation in Estrogen-Independent MCF-10A Cells Treated with Salsolinol

The human mammary cell line MCF-10A exhibits neither estrogen receptor α nor estrogen receptor β. Salsolinol induced cell proliferation in MCF-10A cells significantly, at the concentrations of 10 μM and 100 μM (Figure 25.4a), compared with the control. 17β-Estradiol did not induce significant cell proliferation (data not shown). An antioxidant, *N*-acetylcysteine (NAC), significantly attenuated cell proliferation (Figure 25.4a), suggesting the involvement of reactive oxygen species in the mechanism of mammary cell proliferation.

Salsolinol induced mammary epithelial cell proliferation in an estrogen receptor-independent manner. Interestingly, NAC inhibited salsolinol-induced cell proliferation in MCF-10A cells. Epidermal growth factor receptor (EGFR) is one of the growth factor receptors most commonly associated with human tumors, including breast cancer, and activated EGFR stimulates various cellular processes, such as cell proliferation.[59] EGFR is known to be affected by reactive oxygen species.[60–62]

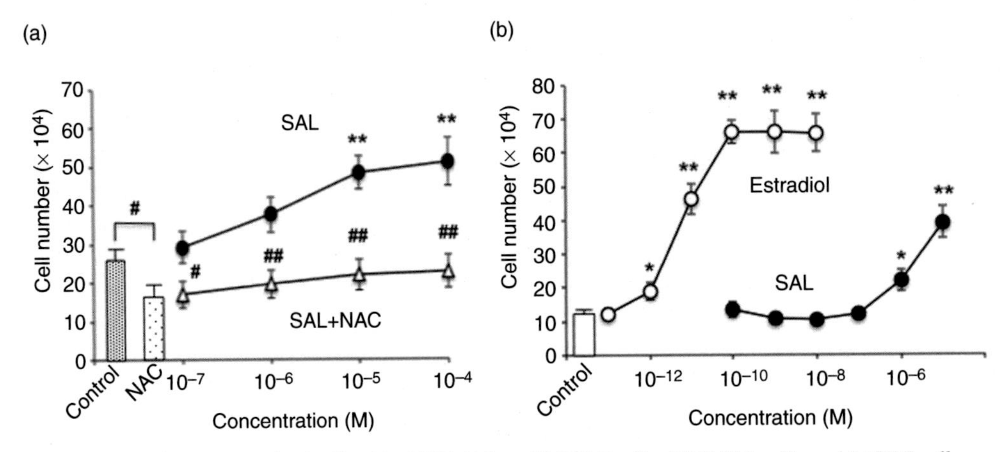

FIGURE 25.4 **Cell proliferating activity of salsolinol in MCF-10A and MCF-7 cells.** MCF-10A cells and MCF-7 cells were trypsinized and plated into 12-well plates at an initial concentration of 3×10^4 cells per well, with seeding medium (MCF-10A cells in DMEM/F12 supplemented with 20 ng/mL EGF, 0.01 mg/mL insulin, 500 ng/mL hydrocortisone, 100 ng/mL kanamycin, 5% horse serum; MCF-7 cells, DMEM supplemented with 100 ng/mL kanamycin and 5% fetal bovine serum). After the cells were allowed to attach for 24 h, the seeding medium was replaced with experimental medium (phenol red-free medium supplemented with 5% charcoal-dextran-serum). (a) MCF-10A cells were incubated with the indicated concentrations of salsolinol (SAL) in the presence and absence of 5 mM *N*-acetylcysteine (NAC). (b) MCF-7 cells were incubated with the indicated concentrations of salsolinol (SAL) or 17β-Estradiol. Control conditions contained 0.1% DMSO. After incubation at 37°C for 6 days, cells were trypsinized, harvested, and then counted. Results are expressed as means and SE of values obtained from 6–8 independent experiments. * $P < 0.05$, ** $P < 0.01$: significant difference compared with the control, and # $P < 0.05$, ## $P < 0.01$: significant difference compared between the conditions with and without NAC by Student's *t*-test. The detailed methods were described previously.[6,58]

We also observed the attenuation of salsolinol-induced proliferation in MCF-10A cells by an EGFR inhibitor, AG1478,[6] suggesting the involvement of EGFR. Therefore, reactive oxygen species derived from salsolinol may participate in cell proliferation via EGFR activation.

Cell Proliferation in Estrogen-Dependent MCF-7 Cells Treated with Salsolinol

To examine the activity of cell proliferation via estrogen receptors, we used the estrogen-dependent human breast cancer cell line MCF-7 that exhibits both estrogen receptors α and β. MCF-7 cells were treated with salsolinol or 17β-Estradiol for 6 days. 17β-Estradiol dramatically induced cell proliferation with maximal proliferating activity at 100 pM, and then plateaued. Salsolinol significantly induced cell proliferation at 1 μM and 10 μM, compared to the control (0.1 % DMSO), and the intensity of maximal cell proliferating ability of salsolinol was about 50%, compared with 100 pM 17β-Estradiol (Figure 25.4b). The estrogen antagonist 4-hydroxytamoxifen significantly inhibited salsolinol-mediated cell proliferation (data not shown), suggesting a cell proliferation mechanism via estrogen receptors.

Salsolinol induced cell proliferation in estrogen-dependent MCF-7 cells, and the treatment with an estrogen receptor antagonist 4-OH-tamoxifen inhibited the cell proliferation, suggesting the involvement of estrogen receptor (ER). A surface plasmon resonance sensor study[6] revealed that the complex of salsolinol and ERα had significantly higher binding ability to estrogen response elements (ERE), than ERβ. Fan et al. reported that alcohol induces a dose-dependent increase in ERα expression levels in MCF-7 cells.[63] Salsolinol may induce mammary cell proliferation through estrogen receptor signaling.

Interestingly, salsolinol can induce mammary cell proliferation via two pathways, estrogen receptor and EGFR. Therefore, salsolinol may contribute to promotion in breast carcinogenesis.

CONCLUSION

We envisage the possible mechanisms of alcohol-associated breast cancer from the viewpoint of salsolinol (Figure 25.5). Alcohol drinking may lead to the formation of salsolinol in mammary glands. Salsolinol has the potential to produce reactive oxygen species (ROS) in the presence of endogenous metals. Salsolinol-derived reactive oxygen species induce both DNA damage, and mammary cell proliferation, via the EGFR pathway. Salsolinol also has estrogenic activity that affects cell proliferation via the ERα pathway. We assume that salsolinol-induced DNA damage and subsequent mutation are the initiation steps, and cell proliferation functions as the promotion step of alcohol-related breast carcinogenesis.

Recently, "acetaldehyde associated with consumption of alcoholic beverages" is classified as Group 1 (carcinogenic to humans),[64] and its carcinogenic mechanism is postulated as DNA adduct formation.[65] In addition, oxidative stress is an essential mechanism of alcohol-associated carcinogenesis.[66,67] Salsolinol may be a key link between acetaldehyde and oxidative DNA damage, and also a link between alcohol and breast cancer.

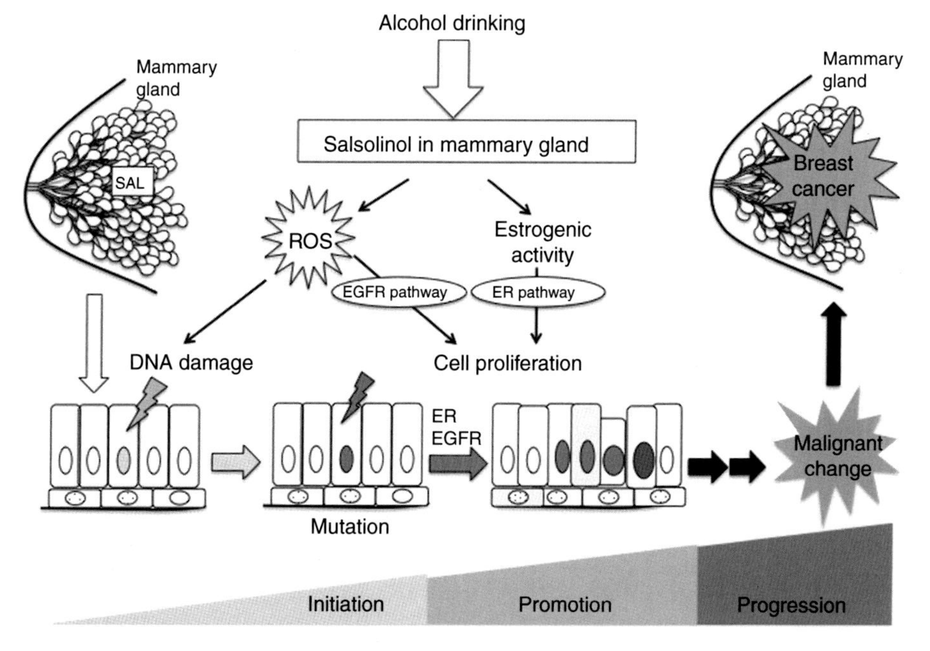

FIGURE 25.5 Possible mechanisms of multistep breast carcinogenesis induced by alcohol-derived salsolinol.

Key Facts

Salsolinol

- Salsolinol is known as a neurotoxic substance, in relation to chronic alcoholism and Parkinsonism.
- Several studies have demonstrated that salsolinol is detected in the brain, blood, and urine, after alcohol intake.
- Salsolinol is endogenously formed by a nonenzymatic condensation reaction between dopamine and acetaldehyde, after drinking alcohol.
- Acetaldehyde is a major metabolite of alcohol by alcohol dehydrogenase (ADH) in the liver.
- Dopamine is accumulated and compartmentalized by the dopamine transporter and the vesicular monoamine transporter 2 (VMAT2).
- Recent studies revealed the presence of both ADH and VMAT2 in mammary glands.
- Salsolinol can be formed in mammary glands.

8-OxodG

- Environmental factors, including infections, chemicals, and some physiological conditions, are involved in the generation of reactive oxygen species (ROS) in relation to carcinogenesis.
- 8-Oxo-7,8-dihydro-2′-deoxyguanosine (8-oxodG) is an oxidized form of guanine residue in DNA, attacked by ROS under oxidative stress condition.
- 8-OxodG is a marker of oxidative DNA damage.
- 8-OxodG is a potentially mutagenic DNA lesion, leading to the transversion of G : C to T : A (G to T transversion).
- Formation of 8-oxodG is a key player of carcinogenesis.

Breast Cancer

- WHO reported that the majority of alcohol-attributable female cancer deaths were from breast cancer.
- Alcoholic beverages are classified as a cause of breast cancer with sufficient evidence.
- Cancer development has multiple steps including (1) initiation: DNA damage followed by mutation; (2) promotion: increase of cell proliferation; and (3) progression: malignant change.
- Estrogen and other estrogenic substances cause proliferation of breast cells that lead to tumor promotion.
- Salsolinol has both abilities of DNA damage and cell proliferation that may contribute to the initiation and promotion steps of alcohol-related breast carcinogenesis.

Summary Points

- This chapter focuses on alcohol-derived salsolinol (1-methyl-6,7-dihydroxy-1,2,3,4-tetrahydroisochinolin) that is a novel causative substance of breast cancer.
- Epidemiological studies show that alcohol intake is an important risk factor for breast cancer.
- Salsolinol is endogenously synthesized by a nonenzymatic condensation reaction between acetaldehyde and dopamine in the brain, and detected in blood and urine after drinking alcohol.
- Recent studies have demonstrated the presence of alcohol dehydrogenase (ADH) and vesicular monoamine transporter 2 (VMAT2) in mammary glands, suggesting the coexistence of acetaldehyde and dopamine, leading to local formation of salsolinol.
- We demonstrated that salsolinol induced metal-mediated oxidative DNA damage and significant mammary cell proliferation that may contribute to multistep breast carcinogenesis.
- Salsolinol could be a molecular link between alcohol and breast cancer.

Acknowledgment

This chapter was partly supported by Grants-in-Aid for Scientific Research from the Ministry of Education, Science, Sports and Culture of Japan.

Abbreviations

hOGG1	Human 8-oxoguanine DNA glycosylase 1
HPLC-ECD	Electrochemical detector coupled to HPLC
IARC	International Agency for Research on Cancer
NAC	*N*-acetylcysteine
8-OxodG	8-Oxo-7,8-dihydro-2′-deoxyguanosine
SOD	Superoxide dismutase

References

1. Brooks PJ, Zakhari S. Moderate alcohol consumption and breast cancer in women: from epidemiology to mechanisms and interventions. *Alcohol Clin Exp Res* 2013;**37**(1):23–30.
2. Nelson DE, Jarman DW, Rehm J, et al. Alcohol-attributable cancer deaths and years of potential life lost in the United States. *Am J Public Health* 2013;**103**(4):641–8.
3. Seitz HK, Pelucchi C, Bagnardi V, La Vecchia C. Epidemiology and pathophysiology of alcohol and breast cancer: Update 2012. *Alcohol Alcohol* 2012;**47**(3):204–12.
4. Boffetta P, Hashibe M, La Vecchia C, Zatonski W, Rehm J. The burden of cancer attributable to alcohol drinking. *Int J Cancer* 2006;**119**(4):884–7.
5. IARC. *Alcohol consumption and ethyl carbamate. IARC Monographs on the evaluation of the carcinogenic risks to humans.* Lyon: IARC Press; 2010 41-1280.
6. Murata M, Midorikawa K, Kawanishi S. Oxidative DNA damage and mammary cell proliferation by alcohol-derived salsolinol. *Chem Res Toxicol* 2013;**26**(10):1455–63.

7. Shukla A, Mohapatra TM, Agrawal AK, Parmar D, Seth K. Salsolinol induced apoptotic changes in neural stem cells: amelioration by neurotrophin support. *Neurotoxicol* 2013;**35**:50–61.
8. Surh YJ, Jung YJ, Jang JH, Lee JS, Yoon HR. Iron enhancement of oxidative DNA damage and neuronal cell death induced by salsolinol. *J Toxicol Environ Health A* 2002;**65**(5–6):473–88.
9. Naoi M, Maruyama W, Dostert P, Kohda K, Kaiya T. A novel enzyme enantio-selectively synthesizes (R)salsolinol, a precursor of a dopaminergic neurotoxin *N*-methyl(R)salsolinol. *Neurosci Lett* 1996;**212**(3):183–6.
10. Naoi M, Maruyama W, Nagy GM. Dopamine-derived salsolinol derivatives as endogenous monoamine oxidase inhibitors: occurrence, metabolism and function in human brains. *Neurotoxicol* 2004; **25**(1–2):193–204.
11. Origitano T, Hannigan J, Collins MA. Rat brain salsolinol and blood-brain barrier. *Brain Res* 1981;**224**(2):446–51.
12. Hipolito L, Sanchez-Catalan MJ, Marti-Prats L, Granero L, Polache A. Revisiting the controversial role of salsolinol in the neurobiological effects of ethanol: old and new vistas. *Neurosci Biobehav Rev* 2012;**36**(1):362–78.
13. Rojkovicova T, Mechref Y, Starkey JA, et al. Quantitative chiral analysis of salsolinol in different brain regions of rats genetically predisposed to alcoholism. *J Chromatogr B Analyt Technol Biomed Life Sci* 2008;**863**(2):206–14.
14. Starkey JA, Mechref Y, Muzikar J, McBride WJ, Novotny MV. Determination of salsolinol and related catecholamines through on-line preconcentration and liquid chromatography/atmospheric pressure photoionization mass spectrometry. *Anal Chem* 2006;**78**(10): 3342–7.
15. Adachi J, Mizoi Y, Fukunaga T, Kogame M, Ninomiya I, Naito T. Effect of acetaldehyde on urinary salsolinol in healthy man after ethanol intake. *Alcohol* 1986;**3**(3):215–20.
16. Matsubara K, Akane A, Maseda C, Takahashi S, Fukui Y. Salsolinol in the urine of nonalcoholic individuals after long-term moderate drinking. *Alcohol Drug Res* 1985;**6**(4):281–8.
17. Haber H, Putscher I, Georgi M, Melzig MF. Influence of ethanol on the salsolinol excretion in healthy subjects. *Alcohol* 1995;**12**(4): 299–303.
18. Haber H, Winkler A, Putscher I, et al. Plasma and urine salsolinol in humans: effect of acute ethanol intake on the enantiomeric composition of salsolinol. *Alcohol Clin Exp Res* 1996;**20**(1): 87–92.
19. Duncan MW, Smythe GA, Nicholson MV, Clezy PS. Comparison of high-performance liquid chromatography with electrochemical detection and gas chromatography-mass fragmentography for the assay of salsolinol, dopamine and dopamine metabolites in food and beverage samples. *J Chromatogr* 1984;**336**(1):199–209.
20. Riggin RM, McCarthy MJ, Kissinger PT. Identification of salsolinol as a major dopamine metabolite in the banana. *J Agric Food Chem* 1976;**24**(1):189–91.
21. Lee J, Ramchandani VA, Hamazaki K, et al. A critical evaluation of influence of ethanol and diet on salsolinol enantiomers in humans and rats. *Alcohol Clin Exp Res* 2010;**34**(2):242–50.
22. Druesne-Pecollo N, Tehard B, Mallet Y, et al. Alcohol and genetic polymorphisms: effect on risk of alcohol-related cancer. *Lancet Oncol* 2009;**10**(2):173–80.
23. McCarty CA, Reding DJ, Commins J, et al. Alcohol, genetics and risk of breast cancer in the Prostate, Lung, Colorectal and Ovarian (PLCO) Cancer Screening Trial. *Breast Cancer Res Treat* 2012;**133**(2): 785–92.
24. Mieog JS, de Kruijf EM, Bastiaannet E, et al. Age determines the prognostic role of the cancer stem cell marker aldehyde dehydrogenase-1 in breast cancer. *BMC Cancer* 2012;**12**:42.
25. Coutelle C, Hohn B, Benesova M, et al. Risk factors in alcohol associated breast cancer: alcohol dehydrogenase polymorphism and estrogens. *Int J Oncol* 2004;**25**(4):1127–32.
26. Choi JY, Abel J, Neuhaus T, et al. Role of alcohol and genetic polymorphisms of CYP2E1 and ALDH2 in breast cancer development. *Pharmacogenetics* 2003;**13**(2):67–72.
27. Seitz HK, Stickel F. Acetaldehyde as an underestimated risk factor for cancer development: role of genetics in ethanol metabolism. *Genes Nutr* 2010;**5**(2):121–8.
28. Triano EA, Slusher LB, Atkins TA, et al. Class I alcohol dehydrogenase is highly expressed in normal human mammary epithelium but not in invasive breast cancer: implications for breast carcinogenesis. *Cancer Res* 2003;**63**(12):3092–100.
29. Iscan M, Klaavuniemi T, Coban T, Kapucuoglu N, Pelkonen O, Raunio H. The expression of cytochrome P450 enzymes in human breast tumours and normal breast tissue. *Breast Cancer Res Treat* 2001;**70**(1):47–54.
30. Castro GD, Delgado de Layno AM, Fanelli SL, Maciel ME, Diaz Gomez MI, Castro JA. Acetaldehyde accumulation in rat mammary tissue after an acute treatment with alcohol. *J Appl Toxicol* 2008;**28**(3):315–21.
31. Fanelli SL, Maciel ME, Diaz Gomez MI, et al. Further studies on the potential contribution of acetaldehyde accumulation and oxidative stress in rat mammary tissue in the alcohol drinking promotion of breast cancer. *J Appl Toxicol* 2011;**31**(1):11–9.
32. James GD, Berge-Landry Hv H, Valdimarsdottir HB, Montgomery GH, Bovbjerg DH. Urinary catecholamine levels in daily life are elevated in women at familial risk of breast cancer. *Psychoneuroendocrinology* 2004;**29**(7):831–8.
33. Rubi B, Maechler P. Minireview: new roles for peripheral dopamine on metabolic control and tumor growth: let's seek the balance. *Endocrinol* 2010;**151**(12):5570–81.
34. Collins MA, Hannigan JJ, Origitano T, Moura D, Osswald W. On the occurrence, assay and metabolism of simple tetrahydroisoquinolines in mammalian tissues. *Prog Clin Biol Res* 1982;**90**: 155–66.
35. Bernstein AI, Stout KA, Miller GW. The vesicular monoamine transporter 2: an underexplored pharmacological target. *Neurochem Int* 2014;**73**:89–97.
36. Hall FS, Itokawa K, Schmitt A, et al. Decreased vesicular monoamine transporter 2 (VMAT2) and dopamine transporter (DAT) function in knockout mice affects aging of dopaminergic systems. *Neuropharmacol* 2014;**76 Pt. A**:146–55.
37. Gronberg M, Amini RM, Stridsberg M, Janson ET, Saras J. Neuroendocrine markers are expressed in human mammary glands. *Regul Pept* 2010;**160**(1–3):68–74.
38. Murata M, Kawanishi S. Mechanisms of oxidative DNA damage induced by carcinogenic arylamines. *Front Biosci (Landmark Ed)* 2011;**16**:1132–43.
39. Murata M, Kawanishi S. Oxidative DNA damage by vitamin A and its derivative via superoxide generation. *J Biol Chem* 2000; **275**(3):2003–8.
40. Pilas B, Sarna T, Kalyanaraman B, Swartz HM. The effect of melanin on iron associated decomposition of hydrogen peroxide. *Free Radic Biol Med* 1988;**4**(5):285–93.
41. Barbouti A, Doulias PT, Zhu BZ, Frei B, Galaris D. Intracellular iron, but not copper, plays a critical role in hydrogen peroxide-induced DNA damage. *Free Radic Biol Med* 2001;**31**(4):490–8.
42. Yang MX, Cederbaum AI. Interaction of ferric complexes with NADH-cytochrome b5 reductase and cytochrome b5: lipid peroxidation, H2O2 generation, and ferric reduction. *Arch Biochem Biophys* 1996;**331**(1):69–78.
43. Desoize B. Metals and metal compounds in carcinogenesis. *In Vivo* 2003;**17**(6):529–39.
44. Kawanishi S, Hiraku Y, Murata M, Oikawa S. The role of metals in site-specific DNA damage with reference to carcinogenesis. *Free Radic Biol Med* 2002;**32**(9):822–32.
45. Agarwal K, Sharma A, Talukder G. Effects of copper on mammalian cell components. *Chem Biol Interact* 1989;**69**(1):1–16.

46. Pryor WA. Oxy-radicals and related species: their formation, lifetimes, and reactions. *Annu Rev Physiol* 1986;**48**:657–67.
47. Tchou J, Grollman AP. Repair of DNA containing the oxidatively-damaged base, 8-oxoguanine. *Mutat Res* 1993;**299**(3–4):277–87.
48. Sissoeff I, Grisvard J, Guille E. Studies on metal ions-DNA interactions: specific behaviour of reiterative DNA sequences. *Prog Biophys Mol Biol* 1976;**31**(2):165–99.
49. Petersen DR. Alcohol, iron-associated oxidative stress, and cancer. *Alcohol* 2005;**35**(3):243–9.
50. Kalyanaraman B, Felix CC, Sealy RC. Semiquinone anion radicals of catechol(amine)s, catechol estrogens, and their metal ion complexes. *Environ Health Perspect* 1985;**64**:185–98.
51. Chen F, Murata M, Hiraku Y, Yamashita N, Oikawa S, Kawanishi S. DNA damage induced by m-phenylenediamine and its derivative in the presence of copper ion. *Free Radic Res* 1998;**29**(3):197–205.
52. Midorikawa K, Kawanishi S. Superoxide dismutases enhance H2O2-induced DNA damage and alter its site specificity. *FEBS Lett* 2001;**495**(3):187–90.
53. Murata M, Nishimura T, Chen F, Kawanishi S, Oxidative DNA. damage induced by hair dye components ortho-phenylenediamines and the enhancement by superoxide dismutase. *Mutat Res* 2006;**607**(2):184–91.
54. Archibald FS, Fridovich I. The scavenging of superoxide radical by manganous complexes: *in vitro*. *Arch Biochem Biophys* 1982;**214**(2):452–63.
55. Shibutani S, Takeshita M, Grollman AP. Insertion of specific bases during DNA synthesis past the oxidation-damaged base 8-oxodG. *Nature* 1991;**349**(6308):431–4.
56. Weiss JM, Goode EL, Ladiges WC, Ulrich CM. Polymorphic variation in hOGG1 and risk of cancer: a review of the functional and epidemiologic literature. *Mol Carcinog* 2005;**42**(3):127–41.
57. Soto AM, Sonnenschein C, Chung KL, Fernandez MF, Olea N, Serrano FO. The E-SCREEN assay as a tool to identify estrogens: an update on estrogenic environmental pollutants. *Environ Health Perspect* 1995;**103**(Suppl. 7):113–22.
58. Murata M, Midorikawa K, Koh M, Umezawa K, Kawanishi S. Genistein and daidzein induce cell proliferation and their metabolites cause oxidative DNA damage in relation to isoflavone-induced cancer of estrogen-sensitive organs. *Biochem* 2004;**43**(9):2569–77.
59. Turtoi A, Blomme A, Bellahcene A, et al. Myoferlin is a key regulator of EGFR activity in breast cancer. *Cancer Res* 2013;**73**(17):5438–48.
60. Burdick AD, Davis II JW, Liu KJ, et al. Benzo(a)pyrene quinones increase cell proliferation, generate reactive oxygen species, and transactivate the epidermal growth factor receptor in breast epithelial cells. *Cancer Res* 2003;**63**(22):7825–33.
61. Drevs J, Medinger M, Schmidt-Gersbach C, Weber R, Unger C. Receptor tyrosine kinases: the main targets for new anticancer therapy. *Curr Drug Targets* 2003;**4**(2):113–21.
62. Valko M, Leibfritz D, Moncol J, Cronin MT, Mazur M, Telser J. Free radicals and antioxidants in normal physiological functions and human disease. *Int J Biochem Cell Biol* 2007;**39**(1):44–84.
63. Fan S, Meng Q, Gao B, et al. Alcohol stimulates estrogen receptor signaling in human breast cancer cell lines. *Cancer Res* 2000;**60**(20):5635–9.
64. IARC. *A review of human carcinogens: personal habits and indoor combustions. IARC monographs on the evaluation of the carcinogenic risks to humans*. Lyon: IARC Press; 2012.
65. Wang M, McIntee EJ, Cheng G, Shi Y, Villalta PW, Hecht SS. Identification of DNA adducts of acetaldehyde. *Chem Res Toxicol* 2000;**13**(11):1149–57.
66. Cahill A, Wang X, Hoek JB. Increased oxidative damage to mitochondrial DNA following chronic ethanol consumption. *Biochem Biophys Res Commun* 1997;**235**(2):286–90.
67. Wright RM, McManaman JL, Repine JE. Alcohol-induced breast cancer: a proposed mechanism. *Free Radic Biol Med* 1999;**26**(3–4):348–54.

CHAPTER

26

Ethanol Impairs Phospholipase D Signaling in Astrocytes

Ute Burkhardt, PhD, Jochen Klein, PhD

Department of Pharmacology, Goethe University Frankfurt, Frankfurt, Hessen, Germany

ASTROCYTES

Glial cells are a heterogeneous group of cells consisting of astrocytes, microglia, and oligodendrocytes. Astrocytes play a crucial role in brain development and in the maintenance of neuronal function. They can have different morphological appearances, but can be identified reliably by staining for the glial fibrillary acid protein (GFAP), an intermediate filament.[2] The branching processes of the astrocytes guide migration and neurite outgrowth of neurons during embryogenesis,[3] and enclose the synapses of neurons in the mature brain. In adult humans, there is a minimum of 100 trillion synapses.[4] Astrocytes also contribute to the blood–brain barrier by surrounding the endothelial cells in blood vessels with the end feet of their processes. Furthermore, they are essential for energy utilization in neurons, for example, by releasing lactate,[5] and they are instrumental for the termination of neuronal signaling by rapid uptake of glutamate and γ-aminobutyric acid (GABA). Another function of astrocytes is the homeostatic maintenance of the extracellular space by importing potassium, and exporting sodium via Na^+/K^+-ATPase. After a brain injury, they build up the glial scar.[6] Astrocytes proliferate throughout all embryonic stages, but the highest proliferation rate is during the brain growth spurt in the third trimester of human pregnancy, or postnatally in rodents,[7] when synaptogenesis occurs.

PHOSPHOLIPASE D

Phospholipase D (PLD) was first identified in plants in 1947, by Hanahan and Chaikoff.[8] Subsequently, many enzymes with PLD activity were described in prokaryotic and eukaryotic organisms, but it took more than 25 years until Saito and Kanfer [9] described PLD in mammalian tissue. Today, we know that PLDs are ubiquitously expressed in all mammalian cells, and they were identified as key players in signal transduction during development and survival of cells. Multiple PLD isoforms have been characterized. Currently, five genes coding for PLD isoforms were described in humans. PLD1 and PLD2 are the most prominent isoforms and have been extensively investigated. Very recently, rare coding variants of PLD3 were studied as risk factors for Alzheimer's disease.[10] PLD4 has no PLD activity, and PLD5 seems to be enzymatically inactive.

Phospholipases belong to the class of hydrolases that catalyze the hydrolysis of ester bonds in lipids. The suffix of the phospholipase characterizes the site of action at the phospholipid molecule (Figure 26.1). While PLAs cleave fatty acid chains at the positions 1 or 2, PLC hydrolyzes phospholipids between the polar head group and the glycerol backbone, producing diacylglycerol (DAG). The most important family of PLCs consists of enzymes that cleave phosphatidylinositol-bisphosphate (PIP_2) producing inositol trisphosphate (IP_3). Under physiological conditions, PLDs catalyze the hydrolysis of PC between the phosphate group and the base (choline), releasing choline, and yielding phosphatidic acid (PA) (Figure 26.2a). The resulting PA is an important lipid signaling molecule, although it makes up only 1–2% of the total lipid amount of the cell,[11] and only 10% of it is used for signaling.

TRANSPHOSPHATIDYLATION

The unique characteristic for PLD is the ability to catalyze a transphosphatidylation reaction in the presence of alcohols. Even at low alcohol concentrations, transphosphatidylation occurs easily because primary alcohols show a higher nucleophilicity than water.[12] During

Molecular Aspects of Alcohol and Nutrition. http://dx.doi.org/10.1016/B978-0-12-800773-0.00026-4

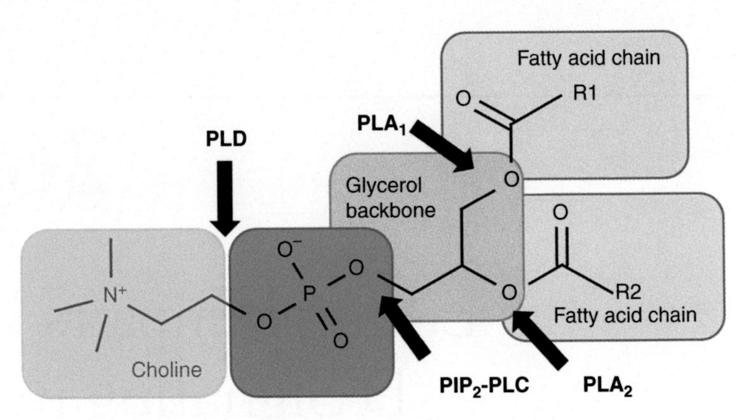

FIGURE 26.1 PC and site of action of phospholipases. The figure shows PC with two fatty acids chains (R1 and R2). Phospholipases A hydrolyze fatty acid chains at C1 (PLA_1) and C2 (PLA_2) of the glycerol backbone. Phospholipase C (PLC) is not known to hydrolyze PC in mammals; the largest family of PLCs preferentially hydrolyzes PIP_2 (phosphatidylinositol-bisphosphate) between glycerol and the polar head group to diacylglycerol (DAG) and inositol trisphosphate (IP_3). PLD cleaves the bond between choline and the phosphate moiety and releases free choline and PA.

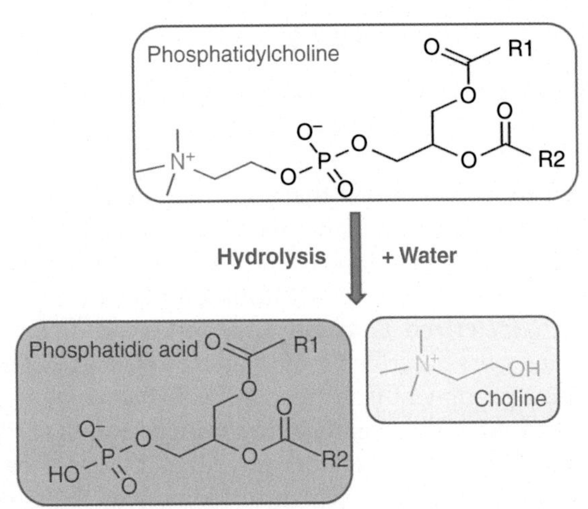

FIGURE 26.2 Hydrolysis of PC by phospholipases. PLDs hydrolyze PC to PA and free choline.

the transphosphatidylation reaction with ethanol, the choline head group is replaced by ethanol, and phosphatidylethanol (PEth) is formed at the expense of PA (Figure 26.2b). PLD is largely selective for primary alcohols; therefore, among the four isoforms of butanol known, only 1-butanol is a good substrate of phospholipase D, whereas t-butanol cannot be transphosphatidylated at all because of steric hindrance.[13] Therefore, different butanols were widely used to study the role of PLD in cell signaling (Figure 26.3).[14,15]

ISOFORMS OF PHOSPHOLIPASE D

In the 1990s, the two most important isoforms of mammalian PLD, PLD1 and PLD2, were cloned. Although both PLDs are ubiquitously expressed, the expression levels of PLD1 and PLD2 differ. For example, both isoforms were found in neurons, but PLD1 is enriched in oligodendrocytes and PLD2 in astrocytes.[16] Although both isoforms share the same substrate specificity, the subcellular localization is different. PLD1 is localized in the perinuclear region including the Golgi apparatus, late endosomes, and the endoplasmic reticulum (ER).[17] In contrast, PLD2 is mainly located at the plasma membrane and colocalizes with β-actin[18] (Figure 26.4). Some publications indicate that certain stimuli trigger translocation of PLD isoforms.[19] Depending on the stimulus, PLD1 translocates to late endosomes and the plasma membrane, upon stimulation with serum (fetal calf serum, FCS), whereas stimulation with the phorbol ester PMA exclusively caused translocation to the plasma membrane.[20] Upon stimulation with receptor agonists and/or desensitization, PLDs translocate to recycling vesicles. Under stimulated conditions, both isoforms can colocalize in the perinuclear region and plasma membrane.[21] PLD isoforms also differ in their activity levels. While PLD1 has a low basal activity, it is known to be activated by numerous factors (receptor agonists, growth factors, etc.). PLD2, in contrast, has a high basal activity and does not usually show a strong activation upon stimulation. In summary, PLD 1 and 2 isoforms differ in their regulation and their role in different signaling pathways.

REGULATION OF PHOSPHOLIPASE D

As mentioned before, PA is a critical lipid-signaling molecule, and small changes in the PA level may have a high impact on cellular processes. As PLD catalyzes the reaction leading to free PA, it is tightly regulated on (1) the transcriptional level, and by binding partners such as (2) small GTPases, (3) protein kinases, and/or (4) structural proteins.

Transcriptional Level

During brain development, PLD expression is ontogenetically regulated. In rats, PLD mRNA was found to increase from embryonal day 19 up to stable expression from postnatal day 19 onward.[22] The rapid increase of PLD activity parallels the rapid increase of glial development and of synaptogenesis. Interestingly, dietary factors can modulate PLD levels. Feeding of dietary oil- or choline-rich diets to the dams increased PLD activity in

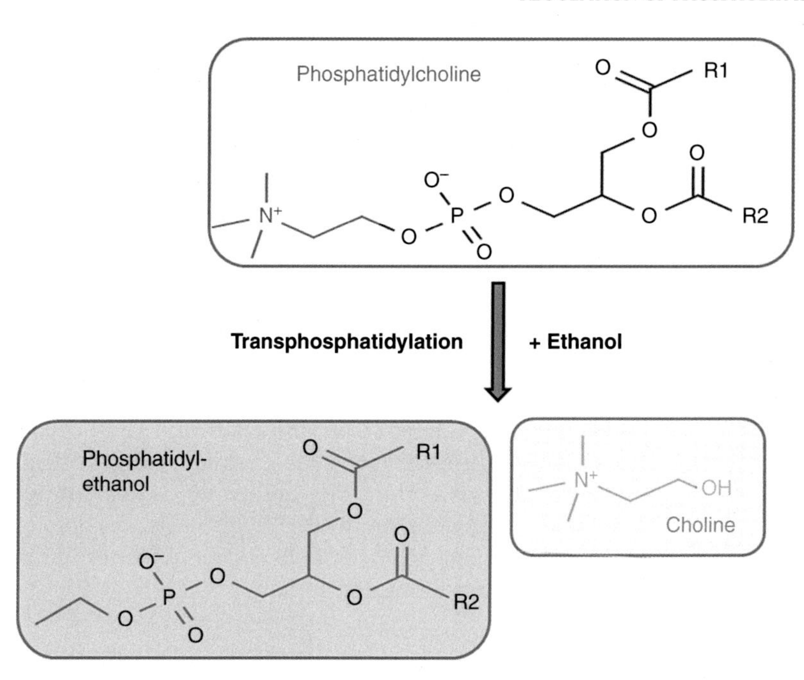

FIGURE 26.3 Transphosphatidylation of PC by phospholipase D. In the presence of ethanol, PLD enzymes catalyze the transphosphatidylation of PC to phosphatidylethanol (PEth) and free choline.

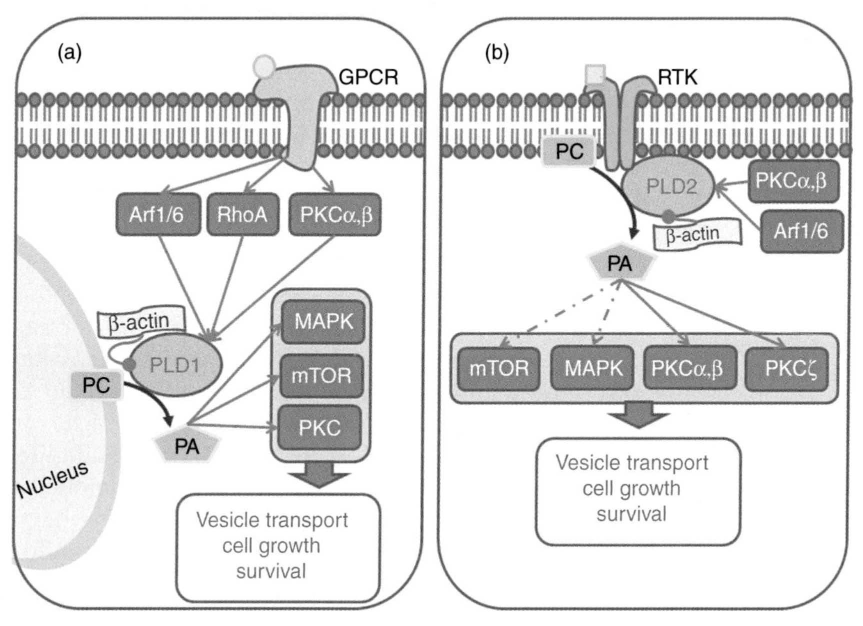

FIGURE 26.4 Regulation of phospholipases D by intracellular pathways. The activities of PLDs are tightly regulated. This figure depicts selected regulators of PLD1 (a) and PLD2 (c). Both isoforms are located at membranes, and are inhibited by β-actin. PLD1 is located in the perinuclear region and can be stimulated by small GTPases such as ADP-ribosylation factors (Arf) and members of the RAS superfamily (Rho, Rac, and Cdc42) that are stimulated by G-protein coupled receptors (GPCR). Members of the protein kinase C (PKC) family interact with both isoforms: PKCα,β and activate both PLDs while PKCζ can be activated by PLD2. PLD2 is located at the cellular membrane and interacts directly with receptor tyrosine kinase (RTK). Both PLDs hydrolyze PC to PA, which mediates mitogenic signals via mitogen-activated protein kinase (MAPK), mammalian target of rapamycin (MTOR), and PKC.

the offspring,[23,24] possibly by modulation of phospholipid synthesis. In adults, the expression of PLD is not prominently modulated. In astrocytes, for instance, phorbol esters cause only a moderate increase in expression.[25]

Small GTPases

Many binding partners of PLD reportedly modulate PLD activity.[19] Small GTPases from the Arf and Rho families have been implicated; Arf1 and Arf6 have been reported to strongly stimulate PLD1 activity and, to a much lower extent, PLD2.[26–28] Arf-stimulated PLD activity is associated with vesicular trafficking, but also with mitogenic effects induced, for example, by PDGF.[29–31] Other PLD-activating members of the Ras superfamily, namely RhoA, Rac1, and Cdc42, are important for organization of the actin cytoskeleton, polarized cell growth, and axonal signaling.[32,33] Members of the Rho family enhanced the catalytic efficiency of PLD1[34] in a process that is important for mitogenic signaling, phagosome formation, and cytoskeletal organization (Figure 26.4a and b). PLD and Rho proteins are also central regulators of neuronal development and neurite outgrowth; for instance, it was observed that neurite outgrowth increases parallel with PLD activation of PDGF-treated hippocampal stem cells. Additionally, inhibition of PLD activity by PLD inactive mutants reduced neurite outgrowth.[35]

Protein Kinases

PLDs are also regulated by protein kinases, for example, protein kinase C (PKC) increases PLD activity after stimulation of G-protein coupled receptors (GPCR).[36] Many protein kinases interact with PLD2, while AMP-activated protein kinase (AMPK) mediates glucose uptake via PLD1 activation.[37] In contrast to PKCα, a major stimulator of PLDs, PKCζ is a downstream target of PLD2[38] (Figure 26.4b). Receptor tyrosine kinases, such as epidermal growth factor receptor (EGFR) phosphorylate PLD2, induce proliferation[35] (Figure 26.4b). Another mitogenic receptor tyrosine kinase (RTK), which activates PLD2 is Janus kinase (JAK3). It plays a critical role in glial cell differentiation,[39] and contributes to the high proliferation rate in breast tumor cells.[40] Nonreceptor tyrosine kinases such as Src also interacts with, and activates, PLD2,[41] for example, PDGF-induced PLD activity could be inhibited by herbimycin A, an Src inhibitor.[42] A third big group of interaction partners are cytoskeletal proteins, such as actins and tubulins. In general, cytoskeletal proteins inhibit PLD activity. Most important is β-actin, which inhibits both PLD1 and PLD2.[43] As mentioned above, PLD2 has a high basal activity, and is negatively regulated by its interaction partners, which includes cytoskeletal proteins, such as β-actin, or neuron-specific proteins, such as synuclein or synaptojanin.[44] Activation of PLD2 activity may be due to dissociation of these inhibitory proteins upon receptor stimulation that was observed in muscarinic receptor-mediated tubulin dissociation.[45] It is also important to mention that PLD activations by neurotransmitters depends on developmental age. For instance, glutamate increased PLD activity rapidly via metabotropic receptors, during the postnatal stage (up to 14 days). In adults, PLD activity was high, and the response to glutamate stimulation was small and delayed.[46] Similarly, muscarinic activation of hippocampal PLD activity could only be demonstrated in immature, not adult tissue.[47] Consequently, neurotransmitter regulation may be restricted to the time of synaptogenesis and glial development.

FUNCTIONS OF PHOSPHOLIPASE D

Analysis of the binding partners of PLD and PA clarified a variety of biological processes in which PLD and PA are involved. PLD participates in vesicle-mediated transport, cytoskeletal organization, cell migration, cell morphogenesis, cell proliferation, and regulation of apoptosis.

Formation of Lipid Messengers

Different downstream signaling pathways are discussed for PA. One possible way is the metabolism of PA to DAG and lyso-phosphatidic acid (LPA). Dephosphorylation by lipid phosphate phosphatase (LPP) converts PA to DAG, which can act as a precursor for the synthesis of phospholipids, such as phosphatidylethanolamine (PEtn) or phosphatidylcholine (PC). The LPPs regulate the level of free lipid phosphates and their dephosphorylated forms, such as DAG, ceramide, and sphingosine. DAG can also activate lipid-dependent kinases such as PKC and, just like PA, is important for cell proliferation and survival. However, it was also suggested that the signaling function of PA and DAG is strongly dependent on the fatty acid composition. According to some data, PA is only active when it is formed from PC, and contains mainly saturated fatty acids; while DAG is only active when it derives from PIP_2 and contains unsaturated fatty acids.[48,49] According to this hypothesis, conversion of PA to DAG, and vice versa, likely serves to terminate the specific, lipid-mediated signals.

Vesicle Formation

An important role of PLD is the regulation of vesicle formation, exo- and endocytosis. The PLD-generated PA is a cone-shaped lipid with negative charge, which causes local distortion of the lipid bilayer. This leads to inward bending and membrane fission, and facilitates the formation of lipid vesicles. Intracellular trafficking to and from the Golgi network is regulated by PLD1 via Arf stimulation.[17,50] Especially in neurons, synaptic plasticity and exo- and endocytosis of neurotransmitter vesicles depend on PLD1- and Arf-regulated membrane fission.[51] PLD can be found in synaptosomes regulating release of neurotransmitters, such as acetylcholine,[51] or gonadotropin-releasing hormone (GnRH), which was inhibited by addition of ethanol.[52] PLD2 is required for receptor internalization, for example, for the angiotensin II receptor.[18] Both PLD isoforms are involved in phagocytosis.[53] Impaired function of both PLD isoforms affect the transport of presenilin-1 and β-amyloid precursor protein (APP), and was suggested to contribute to Alzheimer's disease.[54,55]

Cell Proliferation

Furthermore, PA has been reported to mediate cell proliferation, survival and prevention of apoptosis.[7,48,56–58] Growth factors stimulate PLD activity in astrocytes, and increasing PA levels correlate with induced cell proliferation. A large number of mitogenic signals, such as hormones, growth factors, and lipid ligands can activate PLD via GPCR and RTK. Our own studies demonstrated that serum-, PDGF-, and endothelin-stimulated PLD activation and PA acted as mitogenic second messenger in astrocytes.[7,30] IGF-1 is a signal that promotes growth, development, and differentiation of glial cells during early and late embryonic phase, as well as postnatally.[59] Studies using isoform-specific PLD inhibitors, and RNA interference with isoform-specific siRNA, indicated that PLD (and PLD1, rather than PLD2) contributes to mitogenesis by IGF-1 in astrocytes.[60] In other work, astroglial proliferation was observed with glutamate,[61] muscarinic agonists, such as carbachol,[62] or prolactin.[41] Additionally, exogenous administration of PLD and PA, when introduced into the cell by transient permeabilisation, using SLO-permeabilized astrocytes, resulted in increased cell proliferation.[63]

Besides regulating cell proliferation, PLD prevents apoptosis. In PC12 cells, an anti-apoptotic action of PA was observed by inhibition of proapoptotic sphingomyelinase (SMase).[57] Formation of ceramide (Cer) from sphingomyelin (SM) is regularly observed during apoptosis. Glutamate-induced apoptosis in PC12 cells is accompanied by increased Cer and reduced PA levels. After overexpression of PLD1 or 2, the balance of Cer and PA levels was restored and apoptosis was avoided[64] (Figure 26.8a).

INTERRUPTION OF THE PLD SIGNALING PATHWAY BY ETHANOL

The interference of ethanol with growth factor-related proliferation, differentiation, and migration during brain development has been observed *in vitro* and *in vivo*,[65,66] both in nonneuronal and neuronal tissue. Several molecular mechanisms of ethanol at different stages of development are discussed. Ethanol can (1) interfere on the level of gene expression, (2) form unphysiological phospholids, (3) disturb neuronal development and function, (4) interrupt mitogenic and growth factor-stimulated responses, and (5) induce apoptosis.

Changes of Gene Expression

Altered gene expression by inhibition of retinoic acid (RA) synthesis, homeobox expression, or DNA methylation was suggested as potential mechanism of ethanol toxicity. RA is a signaling molecule during embryonic development, and is synthesized by alcohol dehydrogenase (ADH). In cultured mouse embryos, 100 mM ethanol led to a significant decrease in RA, as ethanol has a high affinity to ADH.[67] Another target on the transcriptional level are the HOX genes, a group of genes that control embryonic development. Ethanol shifted HOX gene expression to apoptosis.[68] However, the relevance of these mechanisms is controversial.

Formation of Unphysiological Phospholipids

Several studies described transphosphatidylation in brain tissue. Phosphatidylethanol (PEth) levels of 12–16 nmol/g brain wet weight were described 2 h after ethanol exposure.[69] Even higher concentrations were found in brain slices (0.4–0.5 μg/g).[70] Due to a slow breakdown (half life 8–10 h), the nonphysiological PEth accumulates during repeated ethanol intake. Secondary effects of PEth accumulation may be changes of physiochemical properties of membranes with characteristic fusion behavior, and transmembrane movement or inhibition of binding of inositol phosphate.[71] However, there is no evidence supporting the role of PEth formation in ethanol toxicity *in vivo*.

Disturbance of Neuronal Function

Besides the antiproliferative effect on astrocytes, ethanol can also disturb neuronal function. As described before, neurite outgrowth depends on Arf-activated PA formation. The neurite outgrowth of cerebellar granule cells was disturbed by addition of ethanol, and overexpression of PLD counteracted the inhibitory effect of ethanol. What is more, transfection with inactive PLD mutants also resulted in reduced neurite outgrowth.[35] However, astrocytes are crucial for the neurite outgrowth during brain development, and the inhibition of neurite outgrowth was also observed in neurons cocultured with ethanol-treated astrocytes.[72]

Other studies demonstrated that ethanol affects exocytosis. In GT1 neurons, ethanol blocked the release of GnRH.[52] In chromaffin cells, ethanol impaired the release of catecholamines upon stimulation with nicotine or calcium,[73] and in ventral striatum dopamine neurotransmission is inhibited by ethanol.[74]

A well-described mechanism for ethanol toxicity is the suppression of neuronal; activity and synaptogenesis by antagonism at NDMA-type glutamate receptors, and agonism at GABAergic receptors.[75] A massive apoptotic neurodegeneration was observed in several brain regions by NMDA antagonists administered to 7-day old rats.[76] The same has been found in rats treated with GABAA-agonists (benzodiazepines and barbiturates).[77] As ethanol has NMDA-antagonistic and GABA-mimetic properties, the apoptotic response in developing rat brain, after ethanol exposure, was even more robust than that observed with NMDA antagonists and GABA mimetics. Furthermore, the ethanol-induced apoptosis was described as Bax-dependent and involved caspase-3 activation.[78]

Interruption of Mitogenic Signaling in Astrocytes

A well-described molecular mechanism is the interruption of mitogenic and growth factor responses. Ethanol is strongly antimitogenic and inhibits neuronal, as well as astroglial, proliferation during brain development.[79]

Acetylcholine is discussed as a mitogen for astrocytes during brain development, and the inhibition of acetylcholine-stimulated astrocyte proliferation by ethanol has been demonstrated. A noncompetitive antagonism of ethanol at the muscarinic receptor M3, as well as an inhibition of the M3-activated intracellular PKC-PLD signaling pathway, were suggested.[80] Ethanol also antagonizes the prolactin-induced JAK/STAT pathway by inhibiting the phosphorylation of JAK2.[81] Additional neurotrophic factors, for example, brain-derived neurotrophic factor (BDNF), are regulators of naturally occurring cell death. Ethanol was found to impair neurotrophic signaling, thereby enhancing cell death by apoptosis.[79] Using comparative studies of ethanol and the isoforms of butanol (1-butanol and t-butanol), it was shown that ethanol inhibits serum-, PDGF- and phorbol ester-induced cell proliferation of rat cortical astrocytes, apparently by disrupting phospholipase D signaling.[14,30] Similar results were obtained in IGF-1-treated cells.[60]

As indicated above, ethanol interacts with PLD in a unique manner. It replaces choline as polar head group and the resulting PEth is formed at the expense of PA in a concentration dependent manner.[30] Already low concentrations of ethanol (0.1/1%) reduced PA formation and astroglial proliferation.[14] The transphosphatidylation reaction is not an inhibition of PLD activity, rather it causes an interruption of the PLD mediated signaling pathway (Figure 26.5a and b). Consequently, downstream targets of PA, such as PKCζ or S6 kinase show reduced activity.[80,82] As described before, exogenous addition of PLD and PA, using transient permeabilization by streptolysin O (SLO), increased cell proliferation of astrocytes. Adding ethanol to PLD-stimulated cells resulted in inhibition of cell growth, while no effect was seen when cells were stimulated by PA. This confirmed that ethanol interrupts the PLD signaling pathway at the level of PLD.[63] This can be explained by formation of PEth at the expense of PA, and the ensuing interruption of PA-induced proliferation. Therefore, proliferation cannot be initiated when the PLD signaling pathway is interrupted, and PA is missing at the right location in the cell. Vesicular trafficking and cytoskeletal organization are also part of cell proliferation processes.

Mimicking the inhibitory effect of ethanol on the PLD signaling pathway, using isoform-specific PLD inhibitors, showed that both PLD inhibitors can interrupt FCS-stimulated cell growth in a dose-dependent manner,[60] and even more effectively than ethanol (Figure 26.6b). Another tool to mimic specific inhibitory effects of ethanol on PLDs is downregulation by transfection with siRNA against PLD. Using PLD isoform-specific siRNA in rat astrocyte cultures, it was shown that PLD1 downregulation results in decreased cell proliferation of astrocytes that were stimulated by FCS and IGF-1[60] (Figure 26.6c; IGF-1 stimulated cells). Recently, PLD1$^{-/-}$ mice, PLD2$^{-/-}$ mice and mice lacking both PLD isoforms were developed.[83,84] Cortical astrocyte cultures were prepared from PLD1/2-deficient mice and PLD activity, and resulting cell proliferation was studied in these cultures.[60] Cell proliferation of these astrocytes was reduced (Figure 26.6d). Comparing the DNA synthesis rate of ethanol-treated wild type cells with the respective PLD-deficient cells, it is evident that ethanol reduces cell proliferation much less effectively in PLD-deficient cells, than in wild type cells (Figure 26.7). We also investigated the effect of ethanol on astrocyte cultures lacking both PLD isoforms, and only a minimal reduction of

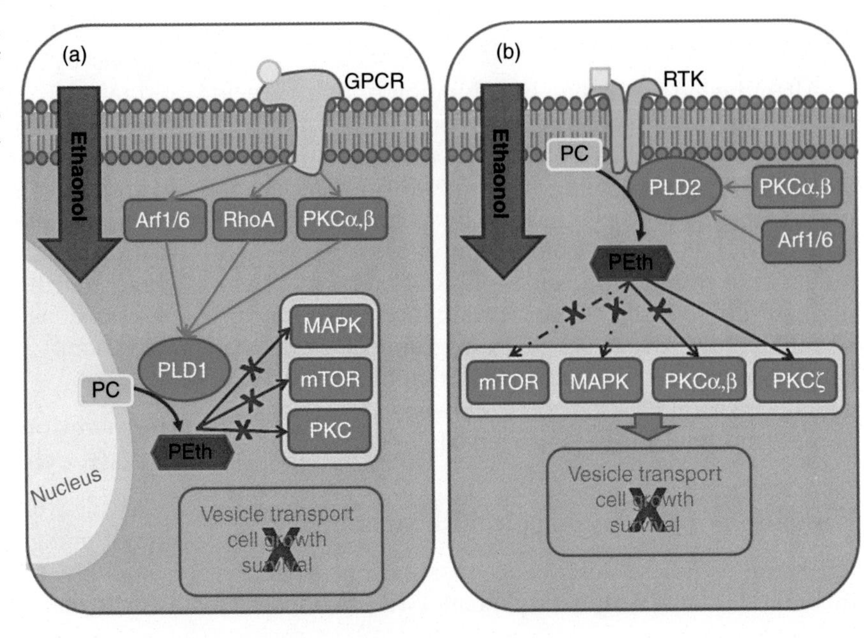

FIGURE 26.5 **Ethanol inhibits the phospholipase D signaling pathway.** In the presence of ethanol, PLDs that are stimulated by growth factors catalyze the transphosphatidylation of PC to phosphatidylethanol (PEth). Thus, further signaling steps through PA formation are interrupted.

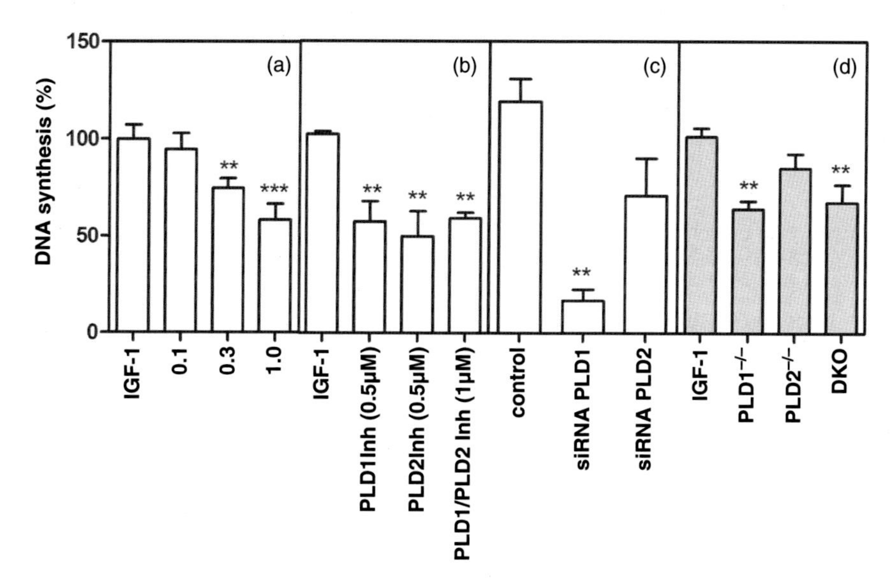

FIGURE 26.6 Cell proliferation in rodent astrocyte cultures stimulated by IGF-1: inhibition by ethanol, PLD inhibitors, siRNA against PLD, and gene deletion. Effects of PLD inhibition in IGF-1-stimulated primary astrocyte cultures. (a) Inhibition of IGF-1-stimulated proliferation by ethanol is dose-dependent. (b) PLD1 and PLD2 inhibitors and the combination of both reduce cell proliferation of rat astrocytes by 50%. (c) Transfection of rat astrocyte cultures with siRNA against PLD1 is more effective than transfection with siRNA against PLD2, when compared to control siRNA. (d) IGF-1-stimulated PLD1$^{-/-}$ mouse astrocyte cultures and astrocytes lacking both PLD isoforms showed reduced cell proliferation, when compared to wild type and PLD2$^{-/-}$ mouse astrocytes. Data are shown as means ± S.D. (n = 5–9). Statistics: $*p < 0.05$, $**p < 0.01$, and $***p < 0.001$ versus untreated and control, respectively (one-way ANOVA + Bonferroni's posttest). *Data taken from Ref. [60].*

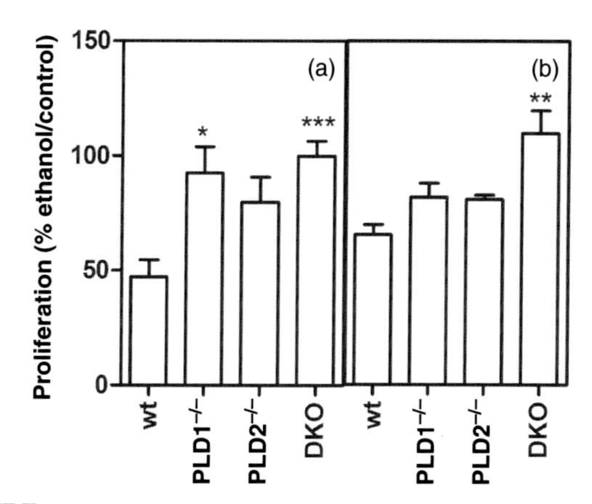

FIGURE 26.7 PLD-deficient mouse astrocyte cultures are resistant to ethanol toxicity. Effects of 0.3% ethanol on (a) IGF-1-stimulated (a) and (b) FCS-stimulated mouse astrocyte cultures. Data are shown as means ± S.D. (n = 5). Statistics: $*p < 0.05$, $**p < 0.01$, and $***p < 0.001$ versus wild type cells (one-way ANOVA + Turkey's Multiple Comparison Test).

DNA synthesis was observed in this experimental setup. Therefore, ethanol evidently does not inhibit cell proliferation when the target PLD is missing from the cells.

Induction of Apoptosis

In addition to its mitogenic effects, PLD is involved in pro- and antiapoptotic processes. Numerous cell culture experiments demonstrated that the balance between PLD and SMase is necessary for the determination of cell fate. SMase hydrolyzes SM, another choline-containing phospholipid of cellular membranes, to Cer. Ceramides induce programmed cell death via caspase activation. In astrocytes, we found that PA controls Cer formation (probably via inhibition of SMase) and therefore inhibits apoptosis. Vice versa, Cer inhibits PLD and keeps apoptotic processes running. Ethanol induced Cer formation and apoptosis in astrocytes.[85] A differential effect of isomeric butanols on Cer formation and programmed cell death was also demonstrated: 1-butanol increased Cer levels and favored apoptosis, while t-butanol had no effect.[15] Additionally, it was found that exogenous administration of PA to SLO-permeabilized astrocytes inhibited ethanol-induced Cer formation (Figure 26.8).

In Vivo Studies

Antimitogenic effects of ethanol, which involve PLD were also demonstrated in transgenic mice. *In vivo* studies with mice lacking PLD1, PLD2, or both PLD isoforms clearly showed that brain development is retarded during the postnatal weeks 1–4 (Figure 26.9);[86] however, PLD-deficient mice catch up as adults. This was also reported for prenatal alcohol exposed (PAE) mice. In a dose regime of 4.0 g ethanol/kg body weight, throughout gestation, PAE mice have a delayed body development, but catch up in adulthood, while a higher dose of ethanol (6.0 g ethanol/kg) resulted in a lifelong smaller body size.[87]

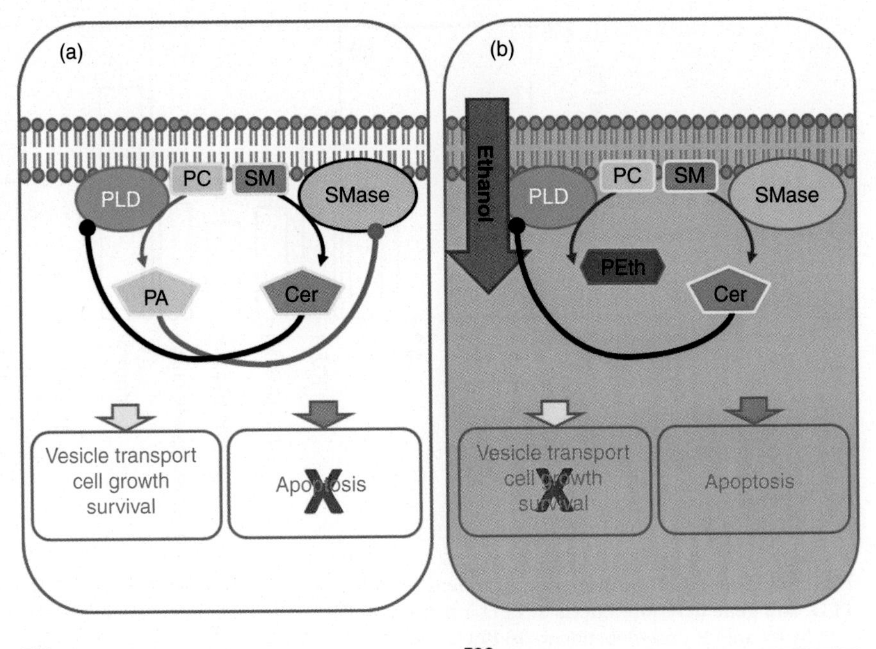

FIGURE 26.8 Regulation of fatty acid second messengers. Fatty acid second messengers are tightly controlled. (a) Activation of PLD induces cell proliferation via PA formation. Additionally, PA controls ceramide (Cer) formation from SM by SMase. Ceramide induces apoptosis and, in a cross-talk, controls PA formation. (b) This regulatory loop is interrupted by ethanol and results in reduced cell proliferation and induced apoptosis.[7,15]

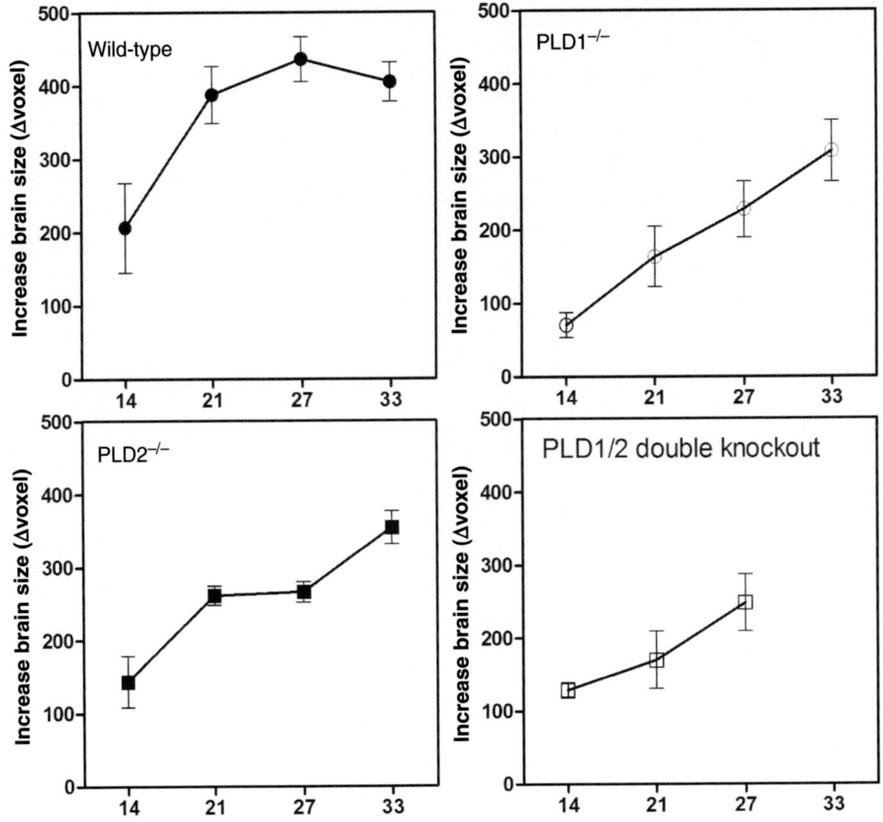

FIGURE 26.9 Postnatal brain developments of mice lacking PLD. Brain of PLD wild type mice grows quickly, while postnatal brain development of mice lacking PLD1, PLD2, or both PLD-isoforms, is delayed. But PLD-deficient mice catch-up to wild type mice as adults. Brain size was measured weekly in MRI scans (voxel) and brain size increase is calculated in Δmean = brain size day x − brain size day 7 ± SD ($n = 4$). *Data published in Ref [86].*

CONCLUSIONS

In summary, the available evidence from cell culture and *in vivo* studies demonstrates that, among other factors, the interruption of growth factor signaling by ethanol, causing reduced PA levels and lipid signaling, is one important part of the pathophysiology of the fetal alcohol spectrum disorder (FASD). This does not exclude the relevance of additional mechanisms, for example, alterations of neurotransmitter release,[82] and suppression of neuronal activity.[75] Nevertheless, there are still open questions about the role of PLD in proliferation, for instance, during the development of cancer. A special interest may also be the role of PLD in neurodegeneration, such as Alzheimer's disease. Since inhibition and downregulation of PLD lead to a strong impairment of brain growth, the development of small molecule stimulators of PLDs may be a future project. To put it all in a nutshell, a better understanding of the function of PLD remains absolutely necessary, because PLDs are so tightly regulated that interventions may have drastic effects.

SUMMARY

This chapter focuses on the role of PLD during development of the brain, and the inhibitory effect of ethanol on PLD signaling pathways. As astrocytes make up for 30% of the volume in the adult mammalian brain cortex,[1] and because astrocytes are more vulnerable to ethanol than neurons, a major focus of this chapter will be on the function of PLD in astrocytes and the effect of ethanol on astrocyte development.

Key Facts

Phospholipase D

The key facts of PLD include tissue distribution, catalytic activity, function, and role of PLD in pathologic mechanisms.

- Species: Bacteria, fungi, plants, and mammals
- Tissue specificity: Ubiquitously expressed in all cells
- Catalytic activity: Hydrolysis/transphosphatidylation
- Substrate: PC
- Products: PA/phosphatidylethanol + choline
- Isoforms: Five in mammals, PLD1, and PLD2 are most important
- Function: Cell proliferation, survival, migration, cytoskeleton reorganization, and endo- and exocytosis
- Relevance: Development, neurodegeneration, cancer, inflammation, and cardiovascular diseases

Summary Points

- This chapter discusses the inhibitory effect of ethanol on the PLD signaling pathway in astrocytes.
- PLDs are ubiquitously expressed enzymes, which hydrolyze PC to PA, a lipid second messenger that leads to cell proliferation, differentiation, and survival.
- In the presence of ethanol, PLDs catalyze the unique transphosphatidylation reaction that interrupts PA signaling, thereby leading to reduced cell proliferation and apoptosis.
- PLD inhibition by PLD inhibitors, downregulation by siRNA, and gene deletion in mice had similar inhibitory effects on cell proliferation; moreover, astrocyte cultures obtained from mice lacking PLD1 and PLD2 were resistant to ethanol toxicity.
- Astrocytes are particularly vulnerable to ethanol toxicity and, as a consequence, brain development is delayed.

Definition of Words

Phospholipids consist of a base (e.g., choline) with a polar phosphate group, which is esterified with a diacylglyceride, and they build up the lipid bilayers of the cellular membranes. Phospholipids are important sources for lipid second messengers, and play a role in the activation of enzymes.

Phospholipase D belongs to the large group of enzymes hydrolyzing phospholipids. PLD is specific for PC. There are five PLD isoforms in humans, but not all are catalytically active. PLDs are involved in proliferation, survival, and migration processes, cytoskeletal reorganization, and endo- and exocytosis.

Transphosphatidylation is the exchange of a primary alcohol (e.g., ethanol) for the polar head group of PC. It is a specific reaction of PLDs in the presence of alcohols. Resulting products are phosphatidylalcohols (e.g., phosphatidylethanol) with no known cellular functions, and choline.

Astrocytes belong to the class of glial cells in the brain, which make up 30% of the brain volume. Mature astrocytes are star shaped and guide neurite outgrowth and synaptogenesis during brain development. Astrocytes tighten the blood–brain barrier and are important for ionic homeostasis, energy utilization of neurons, termination of neurotransmitter signals, and nervous system repair.

Brain growth spurt is the period of synaptogenesis, rapid glial development, and myelination. In humans, the brain growth spurt occurs during the third trimester, while in rodents the highest rate of brain growth is observed after birth, in the neonatal stage, up to postnatal day 14.

Abbreviations

ADH Alcohol dehydrogenase
AMPK AMP-activated protein kinase
APP β-Amyloid precursor protein
BDNF Brain-derived neurotrophic factor
DAG Diacylglycerol
EGFR Epidermal growth factor receptor
ER Endoplasmic reticulum
FASD Fetal alcohol spectrum disorder
FCS Fetal calf serum
GABA γ-aminobutyric acid
GFAP Glial fibrillary acid protein
GnRH Gonadotropin releasing hormone
LPA Lyso-phosphatidic acid
LPP Lipid phosphate phosphatase
IGF-1 Insulin-like growth factor 1
PA Phosphatidic acid
PDGF Platelet derived growth factor
PEth Phosphatidylethanol
PC Phosphatidylcholine
PKC Protein kinase C
PLD Phospholipase D
SLO Streptolysin O
SM Sphingomyelin
SMase Sphingomyelinase

References

1. Bass NH, Hess HH, Pope A, Thalheimer C. Quantitative cytoarchitectonic distribution of neurons, glia, and DNA in rat cerebral cortex. *J Comp Neurol* 1971;**143**(4):481–90.
2. Eng LF, Vanderhaeghen JJ, Bignami A, Gerstl B. An acidic protein isolated from fibrous astrocytes. *Brain Res* 1971;**28**:351–4.
3. Campbell K, Götz M. Radial glia: multi-purpose cells for vertebrate brain development. *Trends Neurosci* 2002;**25**(5):235–8.
4. Matyash V, Kettenmann H. Heterogeneity in astrocyte morphology and physiology. *Brain Res Rev* 2010;**63**:2–10.
5. Magistretti PJ. Neuron-glia metabolic coupling and plasticity. *J Exp Biol* 2006;**209**:2304–11.
6. Michael V Sofroniew, Harry V Vinters. Astrocytes: biology and pathology. *Acta Neuropathol* 2010;**119**:7–35.
7. Klein J. Functions and pathophysiological roles of phospholipase D in the brain. *J Neurochem* 2005;**94**:1473–87.
8. Hanahan DJ, Chaikoff IL. A new phospholipide-splitting enzyme specific for the ester linkage between the nitrogenous base and the phosphoric acid grouping. *J Biol Chem* 1947;**169**(3):699–705.
9. Saito M, Kanfer J. Solubilization and properties of a membrane-bound enzyme from rat brain catalyzing a base-exchange reaction. *Biochem Biophys Res Commun* 1973;**538**(2):391–8.
10. Cruchaga C, Karch CM, Jin SC, Benitez BA, Cai Y, Guerreiro R, et al. Rare coding variants in the phospholipase D3 gene confer risk for Alzheimer's disease. *Nature* 2014;**505**(7484):550–4.
11. Vance JE, Steenbergen R. Metabolism and functions of phosphatidylserine. *Prog Lipid Res* 2005;**44**(4):207–34.
12. Stanacev NZ, Stuhne-Sekalec L. On the mechanism of enzymatic phosphatidylation. biosynthesis of cardiolipin catalyzed by phospholipase D. *Biochim Biophys Acta* 1970;**210**(2):350–2.
13. Ella KM, Meier KE, Kumar A, Zhang Y, Meier GP. Utilization of alcohols by plant and mammalian phospholipase D. *Biochem Mol Biol Educ* 1997;**41**(4):715–24.
14. Kötter K, Jin S, von Eichel-Streiber C, Park JB, Ryu SH, Klein J. Activation of astroglial phospholipase D activity by phorbol ester involves Arf and Rho proteins. *Biochim Biophys Acta* 2000;**1485**:153–62.
15. Schatter B, Jin S, Löffelholz K, Klein J. Cross-talk between phosphatidic acid and ceramide during ethanol-induced apoptosis in astrocytes. *BMC Pharmacol* 2005;**5**:1–11.
16. Saito S, Sakagami H, Kondo H. Localization of mRNAs for phospholipase D type 1 and 2 in the brain of developing and mature rat. *Brain Res Dev Brain Res* 2000;**120**:41–7.
17. Freyberg Z, Sweeney D, Siddhanta A, Bourgoin S, Frohman M, Shields D. Intracellular localization of phospholipase D1 in mammalian cells. *Mol Biol Cell* 2001;**12**(4):945–55.
18. Du G, Huang P, Liang BT, Frohman MA. Phospholipase D2 localizes to the plasma membrane and regulates angiotensin II receptor endocytosis. *Mol Biol Cell* 2004;**15**:1024–30.
19. Jang JH, Lee CS, Hwang D, Ryu SH. Understanding of the roles of phospholipase D and phosphatidic acid through their binding partners. *Cell Signal* 2012;**51**:71–81.
20. Du G, Altshuller YM, Vitale N, Huang P, Chasserot-Golaz S, Morris AJ, Bader MF, Frohman MA. Regulation of phospholipase D1 subcellular cycling through coordination of multiple membrane association motifs. *J Cell Biol* 2003;**162**(2):305–15.
21. Kam Y, Exton JH. Phospholipase D activity is required for actin stress fiber formation in fibroblasts. *Mol Cell Biol* 2001;**21**(12):4055–66.
22. Zhao D, Berse B, Holler T, Cermak JM, Blusztain JK. Developmental changes in phospholipase d activity and mRNA levels in rat brain. *Brain Res Dev Brain Res* 1998;**109**(2):121–7.
23. Holler T, Cermak JM, Blusztain JK. Dietary choline supplementation in pregnant rats increases hippocampal phospholipase D activity of the offspring. *FASEB J* 1996;**10**(14):1653–9.
24. Peng JH, Rhodes PG. Effect of dietary lipids on phospholipase D activity in rat brain. *Neurochel Res* 1999;**24**(8):975–9.
25. Jin S, Schatter B, Weichel O, Walev I, Ryu SH, Klein J. Stability of phospholipase D in primary astrocytes. *Biochem Biophys Res Communic* 2002;**297**:545–51.
26. Oude-Weernink P, Han L, Jakobs KH, Schmidt M. Dynamic phospholipid signaling by G protein-coupled receptors. *Biochim Biophys Acta* 2007;**1764**(4):888–900.
27. Shome K, Nile Y, Romero G. ADP-ribosylation factor proteins mediate agonist-induced activation of phospholipase D. *J Biol Chem* 1998;**273**(46):30836–41.
28. Lopez I, Arnold RS, Lambeth JD. Cloning and initial characterization of a human phospholipase D2 (hPLD2). ADP-ribosylation factor regulates hPLD2. *J Biol Chem* 1998;**273**(21):12846–52.
29. Oude-Weernink PA, López de Jesús M, Schmidt M. Phospholipase D signaling: orchestration by PIP_2 and small GTPases. *Naunyn Schmiedebergs Arch Pharmacol* 2007;**374**:399–411.
30. Kötter K, Klein J. Ethanol inhibits astroglial cell proliferation by disruption of phospholipase D-mediated signaling. *J Neurochem* 1999;**74**:2517–23.
31. Schafer DA, D'Souza-Schorey C, Cooper JA. Actin assembly at membranes controlled by ARF6. *Traffic* 2000;**1**(11):892–903.
32. Perez P, Rincón SA. Rho GTPases: regulation of cell polarity and growth in yeasts. *Biochem J* 2010;**426**(3):243–53.
33. Sit ST, Manser E. Rho GTPases and their role in organizing the actin cytoskeleton. *J Cell Sci* 2011;**125**(5):579–83.
34. Henage LG, Exton JH, Brown HA. Kinetic analysis of a mammalian phospholipase D: allosteric modulation by monomeric GTPases, protein kinase C, and polyphosphoinositides. *J Biol Chem* 2006;**281**(6):3408–17.
35. Sung JY, Lee SY, Min DS, Eom TY, Ahn YS, Choi M, Kwon YK, Chung KC. Differential activation of phospholipases by mitogenic EGF and neurogenic PDGF in immortalized hippocampal stem cell lines. *J Neurochem* 2001;**78**:1044–53.
36. Exton JH. Regulation of phospholipase D. *FEBS Lett* 2002;**531**: 58–61.
37. Kim JH, Park JM, Yea K, Kim HW, Suh PG, Ryu SH. Phospholipase D1 mediates AMP-activated protein kinase signaling for glucose uptake. *PLoS One* 2010;**5**(3):e9600.
38. Kim JH, Kim JH, Ohba M, Suh PG, Ryu SH. Novel functions of the phospholipase D2-Phox homology domain in protein kinase C zeta activation. *Mol Cell Biol* 2005;**25**(8):3194–208.
39. Gomez-Cambronero J. New concepts in phospholipase D signaling in inflammation and cancer. *Sci World J* 2010;**10**:1356–69.
40. Ye Q, Kantonen S, Henkels KM, Gomez-Cambronero J. A new signaling pathway (jak-fes-phospholipase d) that is enhanced in highly proliferative breast cancer cells. *J Biol Chem* 2013;**288**(14):9881–91.
41. Mangoura D, Pelletiere C, Leung S, Sakellaridis N, Wang DX. Prolactin concurrently activates Src-PLD and JAK/Stat signaling pathways to induce proliferation while promoting differentiation in embryonic astrocytes. *Int J Dev Neurosci* 2000;**18**:693–704.
42. Kim BY, Ahn SC, Oh HK, Lee HS, Mheen TI, Rho HM, Ahn JS. Inhibition of PDGF-induced phospholipase D but not phospholipase C activation by herbimycin A. *Biochem Biophys Res Commun* 1995;**212**(3):1061–7.
43. Lee S, Park JB, Kim JH, Kim Y, Kim JH, Shin KJ, et al. Actin directly interacts with phospholipase D, inhibiting its activity. *J Biol Chem* 2001;**276**(30):28252–60.
44. Jenco JM, Rawlingson A, Maniels AJ, Morris AJ. Regulation of phospholipase d2: selective inhibition of mammalian phospholipase D isoenzymes by alpha- and beta-synucleins. *Biochem* 1998;**37**(14):4901–9.
45. Chae YC, Lee S, Lee HY, Heo K, Kim JH, Kim KH, Suh PG, Ryu SH. Inhibition of muscarinic receptor-linked phospholipase D activation by association with tubulin. *J Biol Chem* 2005;**280**(5):3723–30.

46. Klein J, Vakil M, Bergmann F, Holler T, Lovino M, Löffelholz K. Glutamatergic activation of hippocampal phospholipase D: postnatal fading and receptor desensitization. *J Neurochem* 1998;**70**(4):1679–85.
47. Klein J, Lindmar R, Löffelholz K. Muscarinic activation of phosphatidylcholine hydrolysis. *Prog Brain Res* 1996;**109**:201–8.
48. Wakelam MJO, Martin A, Hodgkin MN, Brown F, Pettitt TR, Cross MJ, DeTakats PG, Reinolds JL. Role and regulation of phospholipase D activity in normal and cancer cells. *Adv Enzyme Reg* 1997;**37**:29–34.
49. Hodgkin MN, Pettitt TR, Martin A, Michell RH, Pemberton AJ, Wakelam MJ. Diacylglycerols and phosphatidates: which molecular species are intracellular messengers? *Trends Biochem Sci* 1998;**23**(6):200–4.
50. Sung TC, Roper RL, Zhang Y, Rudge SA, Temel R, Hammond SM, et al. Mutagenesis of phospholipase D defines a superfamily including a trans-golgi viral protein required for poxvirus pathogenicity. *EMBO J* 1997;**16**(15):4519–30.
51. Humeau Y, Vitale N, Chasserot-Golaz S, Dupont JL, Du G, Frohman MA, et al. A role for phospholipase D1 in neurotransmitter release. *Proc Natl Acad Sci USA* 2001;**98**(26):15300–5.
52. Zheng L, Krsmanovic LZ, Vergara LA, Catt KJ, Stojilkovic SS. Dependence of intracellular signaling and neurosecretion on phospholipase d activation in immortalized gonadotropin-releasing hormone neurons. *Proc Natl Acad Sci USA* 1997;**94**(4):1573–8.
53. Corrotte M, Chasserot-Golaz S, Huang P, Du G, Ktistakis NT, Frohman MA, et al. Dynamics and function of phospholipase D and phosphatidic acid during phagocytosis. *Traffic* 2006;**7**:365–77.
54. Liu Y, Zhang YW, Wang X, Zhang H, You X, Liao FF, Xu H. Intracellular trafficking of presenilin 1 is regulated by beta-amyloid precursor protein and phospholipase D1. *J Biol Chem* 2009;**284**:12145–52.
55. Oliveira TG, Chan RB, Huasong T, Laredo M, Shui G, Staniszewski A, et al. Phospholipase D2 ablation ameliorates Alzheimer's disease-linked synaptic dysfunction and cognitive deficits. *J Neurosci* 2010;**30**:16419–28.
56. Foster DA, Xu L. Phospholipase D in cell proliferation and cancer. *Mol Cancer Res* 2003;**1**:789–800.
57. Nozawa Y. Roles of phospholipase D in apoptosis and pro-survival. *Biochim Biophys Acta* 2002;**1585**:77–86.
58. Besson A, Yong VW. Mitogenic signaling and the relationship to cell cycle regulation in astrocytomas. *J Neuro-Oncol* 2001;**51**:245–64.
59. D'Ercole AJ, Ye P. Expanding the mind: Insulin-like growth factor 1 and brain development. *Endrocrinol* 2008;**149**:5958–62.
60. Burkhardt U, Wojcik B, Zimmermann M, Klein J. Phospholipase D is a target for inhibition of astroglial proliferation by ethanol. *Neuropharmacol* 2014;**79**:1–9.
61. Kanumilli S, Roberts PJ. Mechanisms of glutamate receptor induced proliferation of astrocytes. *NeuroReport* 2006;**17**:1877–81.
62. Guizzetti M, Costa P, Peters J, Costa LG. Acetylcholine as a mitogen: muscarinic receptor-mediated proliferation of rat astrocytes and human astrocytoma cells. *Eur J Pharmacol* 1996;**297**:265–73.
63. Schatter B, Walev I, Klein J. Mitogenic effects of phospholipase D and phosphatidic acid in transiently permeabilized astrocytes: effects of ethanol. *J Neurochem* 2003;**87**:95–100.
64. Kim KO, Lee KH, Kim YH, Park SK, Han JS. Anti-apoptotic role of phospholipase D isozymes in the glutamate-induced cell death. *Exp Mol Med* 2003;**28**:38–45.
65. Resnicoff M, Rubini M, Baserga R, Rubin R. Ethanol inhibits insulin-like growth fator-1-mediated signalling and proliferation of C6 and rat glioblastoma cells. *Lab Invest* 1994;**71**:657–62.
66. Guerri C, Renau-Piqueras J. Alcohol, astroglia, and brain development. *Mol Neurobiol* 1997;**15**(1):65–81.
67. Deltour L, Ang HL, Duester G. Ethanol inhibition of retinoic acid synthesis as a potential mechanism for fetal alcohol syndrome. *FASEB J* 1996;**10**(9):1050–7.
68. Wentzel P, Eriksson UJ. Altered gene expression in neural crest cells exposed to ethanol *in vitro*. *Brain Res* 2009;**1305**:S50–60.
69. Lundqvist C, Aradottir S, Alling C, Boyano-Adanez MC, Gustavsson L. Phosphatidylethanol formation and degradation in brains of acutely and repeatedly ethanol-treated rats. *Neurosci Lett* 1994;**179**(1–2):127–31.
70. Henn C, Löffelholz K, Klein J. Stimulatory and inhibitory effects of ethanol on hippocampal acetylcholine release. *Naunyn Schmiedebergs Arch Pharmacol* 1998;**357**(6):640–7.
71. Rodriguez FD, Lundqvist C, Alling C, Gustavsson L. Ethanol and phosphatidylethanol reduce the binding of [3H] inositol-1,4,5-trisphosphate to rat cerebellar membranes. *Alcohol Alcohol* 1996;**31**(5):453–61.
72. Guizzetti M, Moore NH, Giordano G, Costa LG. Modulation of neuritogenesis by astrocyte muscarinic receptors. *J Biol Chem* 2008;**283**:31884–97.
73. Caumont AS, Galas MC, Vitale N, Aunis D, Bader MF. Regulated exocytosis in chromaffin cells. translocation of Arf6 stimulates a plasma membrane-associated phospholipase D. *J Biol* 1998;**273**(3):1373–9.
74. Budygin ES, Mathews TA, Lapa GP, Jones SR. Local effects of acute ethanol on dopamine neurotransmission in the ventral striatum in c57bl/6 mice. *Eur J Pharmacol* 2005;**523**(1–3):40–5.
75. Olney JW. Fetal alcohol syndrome at the cellular level. *Addict Biol* 2004;**9**:137–49.
76. Ikonomidou C, Bosch F, Miksa M, Bittigau P, Vöckler J, Dikranian K, et al. Blockade of NMDA receptors and apoptotic neurodegeneration in the developing brain. *Science* 1999;**283**(5398): 70–4.
77. Olney JW, Wozniak DF, Jevtovic-Todorovic V, Farber NB, Bittigau P, Ikonomidou C. Glutamate and GABA receptor dysfunction in the fetal alcohol syndrome. *Neurotox Res* 2002;**4**(4):315–25.
78. Ikonomidou C, Bittigau P, Ishimaru MJ, Wozniak DF, Koch C, Genz K, et al. Ethanol-induced apoptotic neurodegeneration and fetal alcohol syndrome. *Science* 2000;**287**(5455):1056–60.
79. Guerri C, Pascual M, Renau-Piqueras J. Glia and fetal alcohol syndrome. *Neurotoxicology* 2001;**22**(5):593–9.
80. Costa LG, Vitalone A, Guizzetti M. Signal transduction mechanisms involved in the antiproliferative effects of ethanol in glial cells. *Toxicol Lett* 2004;**149**:67–73.
81. DeVito WJ, Stone S. Ethanol inhibits prolactin-induced activation of the JAK/STAT pathway in cultured astrocytes. *J Cell Biochem* 1999;**74**:278–91.
82. Guizzetti M, Thompson BD, Kim Y, VanDeMark K, Costa LG. Role of phospholipase D signaling in ethanol-induced inhibition of carbachol-stimulated DNA synthesis of 1321N1 astrocytoma cells. *J Neurochem* 2004;**90**:646–53.
83. Elvers M, Stegner D, Hagedorn I, Kleinschnitz C, Braun A, Kuijpers MEJ, et al. Impaired αIIbβ3 integrin activation and shear-dependent thrombus formation in mice lacking phospholipase D1. *Sci Signal* 2010;**3**(103):1–10.
84. Thielmann I, Stegner D, Kraft P, Hagedorn I, Krohne G, Kleinschnitz C, et al. Redundant functions of phospholipases D1 and D2 in platelet α-granule release. *J Thromb Haemost* 2012;**10**(11):2361–72.
85. Pascual M, Valles SL, Renau-Piqueras J, Guerri C. Ceramide pathways modulate ethanol-induced cell death in astrocytes. *J Neurochem* 2003;**87**:1535–45.
86. Burkhardt U, Stegner D, Hattingen E, Beyer S, Nieswandt B, Klein J. Impaired brain development and reduced cognitive function in phospholipase d-deficient mice. *Neurosci Lett* 2014;**572**:48–52.
87. Abel EL, Dintcheff BA. Effects of prenatal alcohol exposure on growth and development in rats. *J Pharmacol Exp Ther* 1978;**207**(3):916–21.

CHAPTER

27

Metabolic Changes in Alcohol Gonadotoxicity

Ganna M. Shayakhmetova, PhD, Larysa B. Bondarenko, PhD, DSc

General Toxicology Department, Institute of Pharmacology and Toxicology of National Academy of Medical Sciences of Ukraine, Kyiv, Ukraine

INTRODUCTION

Alcohol is well documented to affect every organ system in the organism.[1] The researcher's attention to the adverse impact of spirits consumption on male and female fertility is growing. While numerous studies have shown a consistent relationship between heavy drinking and infertility/subfecudity[2,3], additional studies examining moderate consumption are more erratic. Since alcohol is identified as substance, which has a potentially adverse effect on reproductive health in women and men, even a minor effect on fertility are of public and scientific interests.

As it has been reported, ethanol affects all three parts of the hypothalamic-pituitary-gonadal (HPG) axis, a complex system involving feedback from the target organs, the gonads (i.e., the testes and ovaries) to the hypothalamus.[4] A key reproductive hormone, luteinizing hormone releasing hormone (LHRH), is liberated from hypothalamic gonadotropin-releasing hormone (GnRH) neurons into the portal blood system. Upon reaching the pituitary gland, LHRH binds to the specific receptors and activates a complex cascade of biochemical events that results in the synthesis and release of two gonadotropin hormones – luteinizing hormone (LH) and follicle-stimulating hormone (FSH).[4] LH is mainly responsible for gonadal production of androgens (e.g., testosterone). FSH is important for normal development and maturation of sperm in the male and ovarian follicles in the female. The male gonadal hormone testosterone as well as female – estrogens and progesterone, reaches the hypothalamic-pituitary unit to encourage or discourage further release of LHRH, LH, and FSH in a finely tuned system.[4]

The alcohol use in male is known to be associated with low testosterone level and altered levels of other reproductive hormones (e.g., dehydroepiandrosterone, estrogens).[5] Mechanisms of toxic alcohol action on male gonads related with its metabolism, cell damage, and hormonal effects.[6] The alcohol use in female leads to altered levels of estrogens and progesterone, irregularities in the menstrual cycles and ovulation.[7] Together with this it still remains unclear how the volume of consumed alcohol influences on the fertility state.[7]

Ethanol is established to cause disturbances in the main metabolic pathways.[8] Here, we review alcohol-mediated metabolic derangements in male and female gonads that result in fertility disruption.

Ethanol is metabolized mainly by two pathways – oxidative (that take place mostly in the liver) and non-oxidative (that occur mostly in extrahepatic tissues).[9] The major route of ethanol metabolism is the oxidation with cytosolic alcohol dehydrogenase (ADH). The result is a formation of acetaldehyde, a highly reactive and toxic molecule. This reaction is accompanied by the reduction of NAD^+ to NADH. Through such pathway a highly reduced cytosolic environment is generated.[9] However, cytochrome P450 (CYP) system, including CYP2E1, 1A2, and 3A4 isozymes, and catalase also play crucial roles in ethanol oxidation to acetaldehyde.[8,9] Cytochrome P450 isozymes are involved in metabolism of ethanol predominantly after its chronic intake. Alcohol metabolism via CYP2E1 also produces reactive oxygen species (ROS), including hydroxyethyl, superoxide anions, and hydroxyl radicals. They are short lived, unstable and actively reacted with cellular molecules.[9] And at last, another enzyme – catalase (from peroxisomes) also can oxidize ethanol; however, quantitatively this is considered a minor pathway.[9] The acetaldehyde produced due to alcohol oxidation is rapidly metabolized mainly by mitochondrial aldehyde dehydrogenase (ALDH2) to acetate and NADH. Mitochondrial NADH is used by the electron transport chain.[9]

Molecular Aspects of Alcohol and Nutrition. http://dx.doi.org/10.1016/B978-0-12-800773-0.00027-6

The nonoxidative pathway of ethanol metabolism is minor and leads to the formation of fatty acid ethyl ester (FAEE) and phosphatidyl ethanol (PEth). Both PEth and FAEE are known to have intermediate half-life and tendency to accumulate in the liver interfering with cell signaling.[10] A second nonoxidative pathway occurs at high circulating levels of alcohol and involves phospholipase D, which converts phosphatidylcholine to generate phosphatidic acid (PA) and subsequently phosphatidyl ethanol.[8] Phosphatidyl ethanol is poorly metabolized and its effects on the cell are unknown, however it might interfere with the production of PA and disrupt cell signaling.[8]

Considering alcohol biotransformation in the gonads we cannot ignore the existing specificity of metabolic processes characteristic just for these organs and factors influenced on the final outcome (Figure 27.1). First of all, these are: (1) high dependence of all metabolic processes in these organs on hormones production and functioning; (2) age dependence of biotransformation rates; (3) variations in metabolic processes character and rates between sexes.[6,11,12]

It becomes apparent that alcohol adverse effects at each of these levels would cause simultaneous cascade violations in processes of proteins, nucleic acids, lipids and carbohydrates biosynthesis and catabolism.[13,14] In turn specificity of gonads metabolism could modify main streams of ethanol biotransformation in these organs.[15,16]

It should be taken into account that women consume alcohol less frequently and have a lower prevalence of alcohol consumption related problems than men, but they are more sensitive to physiological effects of alcohol than men.[17] The equal doses of alcohol result in a higher blood alcohol level in women than in men and it is thought that this is one reason why women suffer more physical harm from drinking as men.[18]

Because of female reproductive cyclic recurrence the hormonal physiology is more complex than man, and, consequently hormonal feedback on alcohol administration could be more various.[4] Although the differences in the way women and men metabolize alcohol and in the alcohol-related reproductive hormones disturbances have been studied in some depth, research on ethanol-mediated metabolic changes directly in male and females' gonads are limited. Furthermore, in some cases the underlying mechanism of alcohol-induced metabolic injury is clear, but in many others, it remains unknown.

Despite a number of reviews addressed to the effects of ethanol on the balance of reproductive hormones, the changes of proteins, nucleic acids, lipids and carbohydrates metabolism in male and female gonads have not yet been discussed properly. The present review attempts to elucidate alcohol-induced metabolic changes

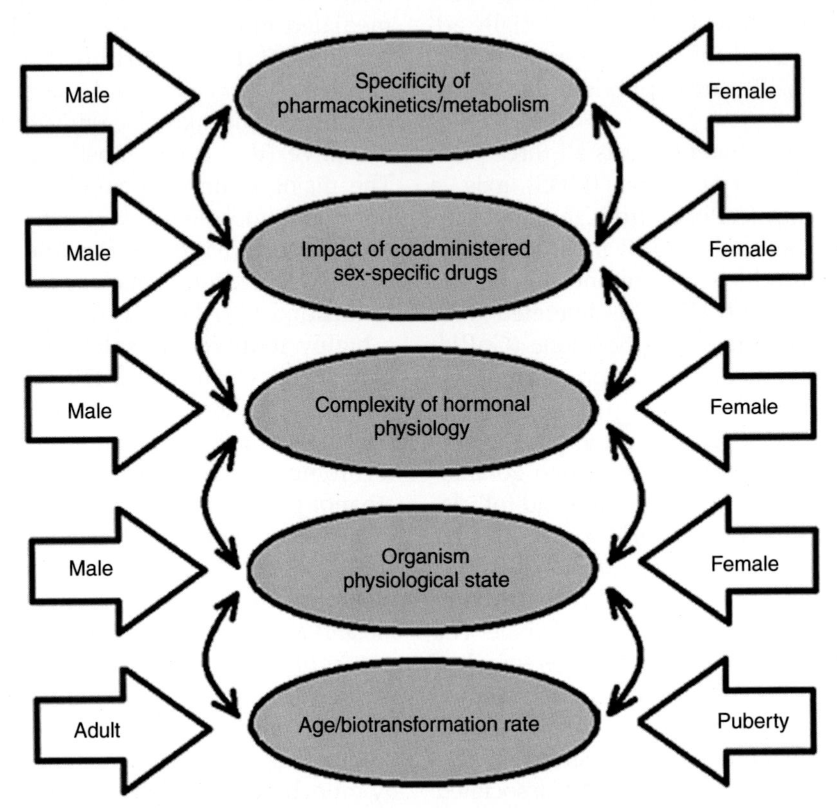

FIGURE 27.1 Factors supposed to influence on ethanol gonadal toxicity. From a mechanistic point of view listed factors can further affect all levels of events leading to the final ethanol-induced male and female gonads injury.

in male and female gonads, which could explain some mechanisms of ovarian and testicular toxicity of ethanol and clear up the prospective direction for the future investigations.

METABOLIC CHANGES IN MALES' ALCOHOL GONADOTOXICITY

Although alcohol is widely used, its impact on the male reproductive function is still controversial. In this chapter we focused our attention on ethanol-induced metabolic changes in male gonads.

As indicated by Wu and Cederbaum,[19] no single process or underlying mechanism can account for all the effects of alcohol on an organism or even on one specific organ. Many mechanisms act in concert, reflecting the spectrum of the organism's response to a myriad of direct and indirect actions of alcohol. One factor that has been suggested as playing a central role in many pathways of alcohol–induced damage is the excessive generation of free radicals, which can result in an oxidative stress. It could be a result of the combined impairment of antioxidant defenses and the production of reactive oxygen species by the mitochondrial electron transport chain, the alcohol-inducible CYP2E1 and activated phagocytes.[19]

The mechanisms by which oxidative stress contributes to alcohol gonadotoxicity are studied extensively from 1980s. It should be noted that in testes reactive oxygen species generation may be beneficial or even indispensable in the complex process of male germ cells' proliferation and maturation, from diploid spermatogonia through meiosis to mature haploid spermatozoa.[20] Conversely high doses, and/or inadequate removal of ROS caused by several mechanisms, that is ethanol metabolism, can be very dangerous, modifying susceptible molecules of testes including DNA, lipids, and proteins.[20]

Many processes and factors could be involved in causing alcohol-induced male gonadal malfunction due to oxidative stress. Taking into account that testicular membranes are rich in unsaturated fatty acids, considering as sensible targets for oxidative damage, it has been supposed that lipid peroxidation is involved into development of gonadal dysfunction with alcohol chronic consumption.[20,21] Indeed, a decreased content of polyenoic acids and a compensatory increase in saturated fatty acids have been shown in rats, which consumed alcohol for 50 days. Such decrease in polyunsaturated fatty acid content was accompanied by an increase in diene conjugates and malondialdehyde (MDA) formation in testicular mitochondria, as well as a decrease in reduced glutathione (GSH) content.[22] Testicular GSH pool was exhausted also in rats chronically treated by alcohol (2.5 g of 50% ethanol/kg body weight (b.wt.) intragastrically during 90 days).[23] In the context of defense against ROS, testicular GSH system plays key functions[24], and its damage following excessive doses of alcohol could be critical for integrity of pro-/antioxidant balance in gonads. In such condition the increase of lipids and proteins oxidation products and decrease of enzymatic antioxidant activity, as well as nonenzymatic antioxidants contents have been reported.[25] A significant depletion in the testicular levels of GSH, protein containing SH groups, tocopherol and ascorbic acid, and an increase in the concentrations of MDA (index of lipid peroxidation) and carbonyl proteins (index of protein oxidation), as well as activity of glutathione peroxidase (GSH-Px) decrease have been shown in male rats consumed alcohol in the mean daily dose 4.05 g/kg (corresponding to the consumption of 41 g of wine (10% alcohol) or 0.71 g of whiskey (40% alcohol) by a man of 70 kg b.wt.).[26] An increase in testicular MDA levels and decrease in the GSH content following long-term ethanol administration in guinea pigs also have been shown.[25] In addition, ethanol administration in male rats led to inhibition of testicular activities of superoxide dismutase (SOD), catalase (CAT), GSH-Px, and xanthine oxidase (XO) with simultaneous increase of MDA and NO levels.[27]

It has been demonstrated that peroxidation injury of testes can be attenuated by dietary vitamin A supplementation.[28] Notably, GSH level were not depleted in the testes of the ethanol-fed rats receiving the vitamin A enriched diet, as well as their testes had a reduced rate of MDA formation. In these animals the signs of testicular atrophy were absent.[28] Such protective effect could be observed due to vitamin A antioxidant action.[28]

In the context of defense against ROS, selenium as the part of GSH system plays key functions.[24] In pups of females exposed to ethanol during gestation and lactation it has been demonstrated decrease in the testicular selenium pool.[29] This element is required for the synthesis of testosterone and the sperm formation and development.[30,31] It is a constituent of selenoproteins GPx1, GPx3, mGPx4, cGPx4, and GPx5 that protect against oxidative damage to spermatozoa throughout the process of sperm maturation and mGPx4 and snGPx4 that serve as structural components of mature spermatozoa.[30] Deficiency of selenium could affect testicular mass with damage to sperm motility, the sperm mid piece, and the shape of the sperm.[32]

Iron plays an important role in various essential cell functions. However, excess iron is toxic and causes lipid peroxidation and tissue damage.[33] Its absorption and transport therefore needs to be tightly regulated.[33] Not only alcohol liver disease frequently exhibit iron overload, but it has been shown that acute ethanol administration (50 mmol/kg) caused iron accumulation in rat testes, which accompanied by lipid peroxidation activation and the content of lipid-soluble antioxidants alpha-tocopherol decrease. These results support the hypothesis concerning iron involvement in paracrine regulation of ethanol-induced testicular disorders development.[34]

Folate and normal activity of one-carbon metabolic pathway enzymes are crucial to nucleotide synthesis, methylation, and maintenance of genomic integrity as well as protection from DNA damage during the course of production of male gamete sperm.[35] Wallock-Montelius et al. reported that ethanol consumption increased folate concentrations in the testes, but not in the epididymis of mini-pigs. On the authors opinion the observed phenomenon can simply reflect a shift in the cell population (i.e., a reduction in germ cells) as testis size becomes smaller. Chronic ethanol consumption at the same time decreased testis methionine synthase activity.[36] These effects may be associated with the reduction in spermatogenesis seen in alcoholics. It must be also taken into account that methionine synthase activity could influence the hormonal regulation of spermatogenesis.[36]

Since oxidative stress can cause inflammation, it is not surprising that alcohol may induce elevated levels of inflammation-promoting cytokines directly in the testes.[37] In ethanol-fed rats increased hypothalamic, pituitary, and testicular tumor necrosis factor α (TNF-α) and interleukin 6 (IL-6) levels have been detected.[38]

Remarkably that acute alcohol administration leads to increase of immunoreactive β-endorphin level in testicular interstitial fluid (TIF) and decrease of TIF testosterone and TIF volume.[39] Sharp decreases of LH and testosterone in serum have been observed in association with acute changes in β-endorphin levels in the pituitary, blood, and testes.[39] On the contrary hypothalamic and testicular immunoreactive β-endorphin levels have been significantly suppressed by chronic alcohol administration.[39]

The capability of ethanol to inhibit testosterone production and cause testicular dystrophy could be associated with metabolic transformation of alcohol into acetaldehyde directly in testes.[40] The subcellular distribution of the enzymes primarily involved in the metabolism of ethanol and acetaldehyde (ADH and aldehyde dehydrogenase (ALDH)) has been analyzed in rats' testes and epididymis. It has been shown that the main localizations of ADH activity were testicular nuclear and cytosolic fractions. Conversely, in the epididymis measurable ADH activity was absent. Testicular ALDH was measurable in all subcellular preparations, that is in the nuclear, mitochondrial, cytosolic, and microsomal fractions.[36] Epididymal ALDH was found evenly distributed between the caput and cauda epididymis.[41] Authors suggest that testicular and epididymal ALDH may play a role in the toxic action of ethanol-derived acetaldehyde on the gonads.[41]

It has been shown that chronic alcohol administration in animals enhanced the activity of testicular NAD^+-dependent ADH.[42,43] By-turn ethanol metabolism to acetaldehyde by ADH activity could reduce, in part, androgen secretion by rat Leydig cells.[44] In these structures increased dihydrotestosterone (DHT) conversion to 3β and 3α-diol was in direct relation to the dose of ethanol and was blocked by 4-methylpyrazole or by saturating NADH concentration, which suggested that this effect was mediated by Leydig cell ADH activity.[45]

Enzymatic NADPH and oxygen-dependent ability of testicular microsomal fraction to transform ethanol to acetaldehyde have been discovered. The authors postulate at least the partial involvement of CYP2E1, P450 reductase, and other enzymes having lipoxygenase-/peroxidase-like behavior into the microsomal transformation of ethanol into acetaldehyde in testes.[40]

Strong evidences concerning localization of microsomal ethanol oxidizing system (MEOS) in Leydig cells were provided.[46] Activity of MEOS in rat Leydig cells has been shown about 47 nmol acetaldehyde per 20 min per mg protein, while activity in crude interstitial cells was about 26 nmol. These results allowed authors to suggest that activity was concentrated in Leydig cells, and it had linear dependence on protein concentration and incubation time. The most effective cofactor was NADPH. Inhibitors of ADH (with 4-methylpirazol) and CAT (with potassium cyanide) activities had no effects on alcohol oxidation by Leydig cells microsomes. MEOS activity here was twice higher than in interstitial cells.[46]

Acetaldehyde can form adducts with reactive residues on proteins or small molecules (e.g., cysteines). These chemical modifications can alter and/or interfere with normal biologic processes, such as signal transduction, and/or be directly toxic to the cell.[8,9] Some investigations have demonstrated that acetaldehyde possessed greater toxicity (then ethanol) as to testosterone production, directly inhibiting protein kinase C – testosterone biosynthesis key enzyme[47], or via causing violations in testis cells pro- and antioxidants balances.[48]

It is important to stress that, in testes CYP2E1 is localized in Leydig cells, where testosterone biosynthesis takes place.[49] In our opinion activation of CYP2E1-dependent metabolizing systems in steroidogenic cells, at least partially, could determine alcohol negative effects in testes. These structures and their microenvironment damages by free radicals, as well as enhanced lipid peroxidation, following of excessive alcohol consumption could be one of the causes of steroidogenesis enzymes inhibition.

Recently, it has been shown, that ROS signaling-mediated c-Jun protein (part of early response transcription factor formation) upregulation suppresses the expression of steroidogenic enzyme genes by inhibiting Nur77 (one of the major transcription factors that regulate the expression of steroidogenic enzyme genes) transactivation, resulting in the reduction of testicular steroidogenesis.[50] It is known that CYP2E1 is an effective generator of hydrogen peroxide.[51] H_2O_2 acts directly on rat Leydig cells to diminish testosterone production by inhibiting cytochrome P450 side chain cleavage enzyme (P450scc) activity and expression of steroidogenic acute regulatory (StAR) protein, which transfers cholesterol to mitochondria in steroid hormone-producing cells.[52,53] Time-depended protein-specific changes in the levels of StAR, the peripheral-type benzodiazepine receptor, and P450scc in Leydig cells of ethanol-exposed rats have been shown.[54] It should be noted that the ability of Leydig cells to release testosterone did not show a simple correlation with changes of above-mentioned proteins.[54]

Changes in the NADH/NAD^+ redox ratio in the testicular cells as a result of alcohol metabolism could play an important role in ethanol-induced disturbances. Series of investigations suggest as mechanism for the inhibition of testicular testosterone synthesis an elevated free NADH/NAD^+ ratio in Leydig cells caused by the metabolism of ethanol. It was demonstrated that the rise in the mitochondrial NADH/NAD^+ ratio rather than in the cytosolic one is connected with the inhibition of testosterone synthesis by ethanol in isolated Leydig cells.[55] The authors suggest that the ethanol-induced high mitochondrial NADH/NAD^+ ratio may deplete mitochondrial oxaloacetate concentrations with following decrease of several transport shuttles activities and interrupting of the flow of mitochondrial citrate into the smooth endoplasmic reticulum. Such events then reflect to decreased rate of steroidogenesis in the presence of ethanol.[55]

Chiao et al. proposed that the major effect of chronic ethanol ingestion upon the enzymes required for steroidogenesis is the reduction of 3β-hydroxysteroid dehydrogenase/isomerase activity, the rate limiting step in sex steroid production from pregnenolone.[56]

There is evidence that ethanol-mediated increase in the NADH/NAD^+ ratio in Leydig cells causes inhibition of reactions catalyzed by 3 beta-hydroxy-5-ene-steroid dehydrogenase/5-ene-4-ene isomerase.[57] This phenomenon can be one of the causes of testosterone synthesis inhibition at the pregnenolone-to-testosterone pathway.[57] On the other hand, the researchers have discovered that ethanol may inhibit testicular steroidogenesis by suppressing at least two steps in the pregnenolone-to-testosterone pathway, the pregnenolone-to-progesterone step catalysed by NAD^+-dependent 5-ene-3β-hydroxysteroid dehydrogenase/isomerase and the 17-hydroxyprogesterone-to-androstenedione step catalysed by the NAD^+-independent C17,20-lyase.[58]

In vitro experiments have demonstrated that 17α-hydroxylase activity of rat testis interstitial cells increased in direct relation to the final concentration of ethanol added, however, 17,20-lyase and 17-ketosteroid reductase (17-KSR) activities were not affected.[59] The authors suggested that ethanol, simultaneously with 17α-hydroxylase activity stimulation, could inhibit the normal coordination of 17,20-lyase activity with the 17-KSR activity.[59] Other authors reported that ethanol-induced inhibition of testosterone biosynthesis was not caused by extratesticular redox increases or by extra- or intratesticular acetaldehyde per se. The inhibition is accompanied by changes in testicular ketone-body metabolism.[60]

Cicero and Bell have discussed major differences between *in vitro* and *in vitro* approaches in understanding of ethanol-induced testicular steroidogenesis reductions.[61] Under *in vitro* conditions ethanol selectively inhibited the conversion of androstenedione to testosterone. This effect was much more general *in vivo*. NAD^+ overcame ethanol's effects on testicular steroidogenesis *in vitro*, but only when labeled or unlabeled pregnenolone was added. In the absence of added pregnenolone, NAD^+ was not effective in preventing ethanol's effects.[61]

As stated below a significant number of ethanol's adverse effect studies in testes have been devoted to its influence on different aspects of energy metabolism, its main ways, sources, messengers, and products.

It is known that the shift in the NADH/NAD^+ ratio increases the rate of fatty acid synthesis and esterification, while simultaneously decreasing mitochondrial β-oxidation of free fatty acids. This change in the redox state can also impair normal carbohydrate metabolism, resulting in multiple effects, including a decreased supply of adenosine 5′-triphosphate (ATP) to the cells.[62] In testicles of rats given drinking water containing 3% ethanol for 8 weeks it has been fixed decrease in the concentration of ATP.[63] A reduction in the testicular phosphodiesters/ATP ratio has been demonstrated in rats after 10 weeks consumption of ethanol-containing liquid diet (36% ethanol of total calories).[64] Likewise decrease of ATP triphosphatase activity, with simultaneous succinate dehydrogenase and sorbitol dehydrogenase activities inhibition, and reduction in fructose content have been detected in testes of long-term alcohol treated guinea pigs.[25] These violations could cause an imbalance in energy production for normal processes of spermatogenesis.

Not numerous clinical investigations on some enzymes activities in male gonads confirm experimental data. It have been demonstrated nearly complete Leydig cells lack of Ca^{2+}-ATPase, decrease of 3β-hydroxy steroid dehydrogenase, and 17β-hydroxy steroid dehydrogenase activity in the testicular biopsies preparations from patients with alcohol induced fertility disturbances.[65]

Novel evidence for a role of lactate as a signaling molecule in the seminiferous tubule was provided. In this regard, lactate considered to be a paracrine factor secreted by Sertoli cells that, in addition to being a source of energy, could regulate germ cell functioning.[66] That's why the ability of ethanol to raise the lactate/pyruvate ratio in the isolated Leydig cell suspension is of special interest.[57]

Effects of ethanol treatment (3.0 g/kg b.wt.) twice daily as a 25% (v/v) on Leydig cell NADPH-generating enzymes have been studied in adult Wistar rats. It has been shown reduction of glucose-6-phosphate dehydrogenase (G-6-PDH), 6-phosphogluconate dehydrogenase (6-PGDH), and NADP-dependent isocitrate dehydrogenase (NADP-ICDH) activities, but malate dehydrogenase (MDH) remained unaltered. Thus, on the authors' opinion, the ethanol treatment impaired Leydig cellular NADPH generation which may be one of the biochemical mechanisms mediating the direct and indirect effects of ethanol resulting in hypoandrogenization.[67] In testes of guinea pigs treated by ethanol (4 g/kg b.wt./day, 90 days) the significant increase of the activity of 3-hydroxy-3-methyl-glutaryl-CoA reductase and decrease in activity of testicular G6PDH and MDH also were fixed. Such changes considered to lead to decreased testosterone level.[42]

There is report on the dose-dependent impairment of Leydig cells glucose oxidative capacity by ethanol. Similarly, ethanol treatment caused significant reduction in LH receptors on the Leydig cell membrane at higher doses (1 and 3 g/kg) whereas no significant change was observed with the lower dose (0.5 g/kg). The authors suggest that the decrease in Leydig cellular LH receptors, glucose oxidation and the activities of 3β-hydroxysteroid dehydrogenase (3β-HSD) and 17-KSR could be possible mechanisms by which ethanol treatment perturbs Leydig cell steroidogenesis.[68]

As far as the NADPH coenzyme is an important factor in regulation of several steps in steroidogenesis, while glucose oxidation acts as a limiting factor on testicular testosterone production[69], abovementioned data demonstrated that the inhibitory effects of alcohol on Leydig cells testosterone production could be mediated through impaired glucose oxidation and defective NADPH generation.

Affecting lipid peroxidation ethanol consumption inevitably causes violations of other lipids metabolism links. It is well known that an increased intake of ethanol leads to hyperlipidemia.[70] In experiments with guinea pigs chronically consumed ethanol significant increase in levels of testicular cholesterol, free fatty acids, phospholipids and triglycerides has been shown.[42] The similar results have been obtained in rats with chronic alcoholism.[23] Another study has demonstrated lipid disturbances in Leydig cells of alcohol-dependent Wistar rats: neutral lipids (esterified cholesterol, triacyl glycerol) were decreased while free cholesterol and diacyl glycerol were increased. At the same time, the reduction in total phospholipids contents was contributed to fractions of phosphatidyl inositol, phosphatidyl serine, phosphatidyl choline and phosphatidyl ethanolamine. Withdrawal of ethanol treatment for 30 days restored these parameters to the normal levels.[67] An increase of the triacyl glycerol concentrations in ethanol-administered groups has been observed.[67]

Ethanol administration can seriously disrupt nucleic acids and proteins metabolism, structure and functions both directly and via its metabolites. ROS produced during ethanol oxidation are the major source of DNA damage, causing strand breaks, removal of nucleotide, and a variety of modifications of the organic bases of the nucleotides.[19] Oxidative stress may play an important role in alcohol-induced DNA disruption in the germ cells.[71]

Our recent investigations have demonstrated that chronic ethanol consumption caused testicular failure along with an over-expression of CYP2E1 mRNA and protein in testes as well as quantitative changes in type I collagen amino acid contents.[72] The profound alcohol-mediated changes in collagen type I amino acid contents may have affected the spermatogenic epithelium state.[72] In testes of alcohol-treated rats the strength of association between the mRNA of CYP2E1 expression levels and numerical indices of spermatogenetic epithelium morphometry was estimated by correlation coefficient. We have found that in alcohol-dependent rats testes mRNA CYP2E1 contents negatively and strongly correlated with spermatogenic index value and positively and strongly correlated with epithelium desquamation occurrence.[73] Correlative links between testicular CYP2E1 mRNA expression levels in alcohol-dependent rodents allow us to suggest the involvement of this isoenzyme in testicular pathology development.

Using Northern blot analysis, Koh et al. elucidated the decrease (46.5%) of pituitary adenylate cyclase activating polypeptide (PACAP) mRNA in rat testes by ethanol administration (3 g/kg i.p., 15% v/v in saline for 10 days). As well as in testicular germ cells, PACAP has been shown to stimulate cAMP production and to contribute to spermatogenesis.[74] On the authors opinion the decrease of testicular PACAP by ethanol administration may contribute to the suppression of male reproductive activity.[74] *In situ* hybridization clearly showed the ethanol-induced decrease in pituitary adenylate cyclase activating polypeptide type I receptor (PAC1) mRNA expression in steroidogenic Leydig cells.[75] This phenomenon accompanied by significant decrease in testosterone serum level.[75] Thus, the decrease in PACAP and PAC1 receptor expression in alcohol-treated rats' testes by ethanol exposure may partly contribute to the male gonadal toxicity of ethanol.

Alcohol-induced violations in testis frequently related with cells severe damage causing testicular germ cell uncontrolled apoptosis [6], which would disintegrate the harmonized microenvironment and the Sertoli/germ cell ratios in the seminiferous epithelium during spermatogenesis.

The importance of caspases in the apoptotic process is well established, expression of Bax (proapoptotic molecule) induces apoptosis in the presence of broad caspase inhibition[76]). *In vitro* experiments with mouse Leydig (TM3) cell line demonstrated ability of ethanol to activate specific intracellular death-related pathways leading to Bax-dependant caspase-3 activation and the induction of apoptosis in Leydig cells.[77] Results of *in vitro* study are in a good accordance with later *in vivo* data, which has confirmed apoptosis induction in the testes of ethanol-treated mice (3 g/kg (15%, v/v), during 14 days, intraperitoneally) through the increased expression of fatty acid syntase (Fas)/ Fas ligand (FasL) and tumor protein p53, upregulation of Bax/Bcl-2 (Bcl-2 is specifically considered as an important antiapoptotic protein) ratio, cytosolic translocation of cytochrome *c* along with caspase-3 activation and GSH depletion.[78] Particularly, Western blot analysis revealed that ethanol repeated administrations decreased the expression of StAR, 3β-hydroxysteroid dehydrogenase (3β-HSD) and 17β-hydroxysteroid dehydrogenase (17β-HSD); increased the expression of active caspase-3, p53, Fas and FasL; and led to upregulation of Bax/Bcl-2 ratio and translocation of cytochrome *c* from mitochondria to cytosol in testes. These events accompanied by upregulation of caspase-3, p53, Fas, Fas-L transcripts; increase in caspase-3 and caspase-8 activities; diminution of 3β-HSD, 17β-HSD, and GSH-Px activities; decrease in the mitochondrial membrane potential along with ROS generation and glutathione pool decrease in the testicular tissue.[78]

The direct evidences that FasL mediates the apoptosis of testicular germ cells induced by acute ethanol administration ((20% v/v), intraperitoneal injection, five times) were provided using FasL transgenic mice.[79] Reverse transcriptase-polymerase chain reaction analysis demonstrated an increase in both Fas and FasL mRNA levels in testes of ethanol-treated rats.[80]

The other findings suggest that ethanol induces apoptotic cell death by suppressing the activation of phosphorylated protein kinase B (pAkt) and phosphorylated extracellular signal-regulated kinase 1/2 (pErk1/2) and the phosphorylation of their downstream targets (the protein Bad, a proapoptotic member of the Bcl-2 family) at two critical sites Ser112 and Ser136 in rat testes.[81]

Thereby, these kinases activities suppression, Fas system upregulation, and elevation of caspases activity in the testes following ethanol administration cause enhanced germ and Leydig cells apoptosis. This process can be involved in pathogenesis of testicular injury resulting in infertility associated with alcohol abuse.

The overall effect of ethanol on the testes affects the metabolism of low-molecular-weight bioregulators. It is well known that alcohol consumption can stimulate NO production and nitric oxide synthases (NOS) expression in different tissues.[82] All three NOS: endothelial NOS (eNOS), inducible NOS (iNOS) and neuronal NOS (nNOS) forms have been identified in testicular tissues.[83] They appear to be involved into the regulation of normal male fertility at multiple levels and as well as pathogenesis of sexual disorders in alcoholics. Particularly in testes NOS has been found in Sertoli cells (eNOS, iNOS), Leydig cells (nNOS, eNOS, iNOS), peritubular cells (iNOS), spermatogenic cells (eNOS, iNOS) and testicular macrophages (iNOS).[83]

The inhibition of the NOS-related NADPH-diaphorase activity in the Leydig cells of chronically alcohol-treated mice has been demonstrated.[84]

The *nNOS* gene produces a testis specific isoform – TnNOS, which has been localized specifically in Leydig cells and is implicated in the control of steroidogenesis.[85] Investigating the alcohol effect, administered intragastrically and delivered via vapors, Herman et al. demonstrated the ability of alcohol to stimulate TnNOS expression, but in these experiments the blockade of NO formation failed to restore a normal testosterone response to human chorionic gonadotropin (hCG).[86]

The involvement of iNOS into ethanol-induced germ cell apoptosis has been demonstrated.[87] Marked increase of iNOS expression has been detected immunohistochemically in Sertoli, germ cells and some interstitial cells in testes of ethanol treated Wistar rats. In addition to that phagocytosed retained elongated spermatids undergoing DNA fragmentation have been revealed.[87] The upregulation of iNOS in the testes may induce apoptosis of germ cells through the generation of excessive NO and androgen suppression.[87] The overproduction of NO due to iNOS activation can play its own role in amplifying testicular injury, but through interaction with superoxide radicals it forms peroxynitrite ($ONOO^-$), another potent oxidizing agent.[88] NO can also react with CO_2 to form nitrogen dioxide ($\cdot NO_2$), a radical of less activity than peroxynitrite but of longer diffusion distance.[88] Peroxynitrite by-turn can modify proteins with thiol groups to generate nitrosothiols disrupting metal-protein interactions and leading to the generation of other metal-derived free radicals.[88]

To avoid imbalance in metabolic processes, cells constantly adjusting their metabolic state based on nutrient availability, using extracellular signaling given by growth factors, hormones, or cytokines. The testicular GnRH and GnRH receptors (GnRH-R) have been found in seminiferous tubules, which were predicted to act as a local regulator of spermatogenesis, but until recently their function was not extensively investigated. Lee et al. demonstrated that GnRH mRNA expression in adult and pubertal rats was dramatically decreased in the testis while no significant change was observed in hypothalamus after both short and long term exposure to ethanol. The pubertal rats showed decrease in testicular GnRH and GnRH-R mRNA expression, whereas GnRH mRNA was increased significantly, while GnRH-R mRNA was further decreased after long term exposure in adults.[89] Authors suggest that the deteriorative effects of ethanol on gonadal activity are more lethal in puberty than adults.[89]

All accumulated data add to the understanding of mechanisms of the testicular metabolic disorders, involved in male alcohol gonadotoxicity. We have summarized them in Table 27.1 and Figure 27.2. Taking into account that majority of experiments have been done with animals or cell cultures, the collection of clinical data is of great importance to reach a general conclusions.

TABLE 27.1 Summary of Alcohol-Mediated Metabolic Changes in Testes

Test system observed	Changes	Possible outcomes
Guinea pigs	↓ GSH ↑ MDA ↓ Ascorbic acid	Pro-/antioxidant balance impairment, oxidative and nitrosative stress leading to disruption of Sertoli–germ cell interactions, disturbances in germ cells differentiation, inhibition of Leydig cells steroidogenic activity, etc.
Rats	↓ GSH ↓ GSH-Px ↓ Protein-containing SH-groups ↓ Tocopherol ↓ Ascorbic acid ↓ SOD ↓ CAT ↓ XO ↓ Polyenoic acids ↑ Saturated fatty acids ↑ MDA ↑ NO	
Rats	↓ Se	Spermatogenesis impairment resulting in abnormal spermatozoa production, which in turn affects semen quality and fertility.
Rats	↑ Fe	Impairment of spermatogenesis paracrine regulation.
Yucatan micropig	↑ Folate ↓ Methionine synthase	Disturbances in nucleotide synthesis, methylation, and maintenance of genomic integrity of germ cells. Abnormalities in the hormonal regulation of spermatogenesis with its following reduction.
Rats	↑ TNF-α ↑ IL-6	Testicular inflammation. Interactions among immune and germ cells resulting in the alteration of spermatogenesis, Abnormalities in the paracrine regulation of spermatogenesis. Apoptosis induction.
Rats	↑ Immunoreactive β-endorphin (acute administration) ↓ Immunoreactive β-endorphin (chronic administration)	Imbalance in regulation of testicular testosterone production. Spermatogenesis impairment.
Guinea pigs Rat Leydig cells	↑ ADH ↓ StAR ↓ PBR ↑ $NADH/NAD^+$	Depletion of mitochondrial oxaloacetate with following decrease of several transport shuttles activities and interrupting of the flow of mitochondrial citrate into the smooth. Impaired carbohydrate metabolism, resulting in multiple effects, including a decreased supply of ATP to the cells. Rise of DHT conversion to 3β and 3α-diol. Inhibition of reactions catalyzed by 3β-hydroxy-5-ene-steroid dehydrogenase/5-ene-4-ene isomerase. Testosterone synthesis inhibition at the pregnenolone-to-testosterone pathway.
Rats	↓ ATP ↓ PDE/ATP ratio ↓ G-6-PDH ↓ 6-PGDH ↓ NADP-ICDH	An imbalance in energy production for normal processes of spermatogenesis. Inhibition of testicular testosterone synthesis. Spermatogenesis impairment.
Guinea pigs	↓ ATP triphosphatase ↓ Succinate dehydrogenase ↓ Sorbitol dehydrogenase ↓ Fructose	
	↓ 3-Hydroxy-3-methyl-glutaryl-CoA reductase ↓ G6PDH ↓ MDH	
Rat Leydig cells	↑ Lactate/pyruvate	

TABLE 27.1 Summary of Alcohol-Mediated Metabolic Changes in Testes *(cont.)*

Test system observed	Changes	Possible outcomes
Human	↓ Ca^{++}-ATPase, ↓ 3β-Hydroxysteroid dehydrogenase ↓ 17β-Hydroxy steroid dehydrogenase	Inhibition of testicular testosterone synthesis. Spermatogenesis impairment.
Rat Leydig cells	↓ 3β -Hydroxysteroid dehydrogenase ↓ 17-Ketosteroid reductase ↓ LH receptors ↓ Glucose oxidation	
Rats	17,20-Lyase, ↑ 17α -Hydroxylase, ↓ 3β -Hydroxysteroid dehydrogenase/isomerase	
Rat testis interstitial cells	↑ 17α-Hydroxylase	
guinea pigs	↑ Cholesterol, ↑ Free fatty acids ↑ Phospholipids ↑ Triglycerides	Changes in structure and function of testicular cells membranes. Disruption of Sertoli–germ cell interactions, disturbances in germ cells differentiation, inhibition of Leydig cells steroidogenic activity, etc.
Rats	↓ Neutral lipids (esterified cholesterol, triacyl glycerol) ↑ Free cholesterol ↑ Diacyl glycerol ↓ Phosphatidyl inositol ↓ Phosphatidyl serine ↓ Phosphatidyl choline ↓ Phosphatidyl ethanolamine	
Rats	↑ CYP2E1 mRNA and protein	Oxidative stress, inhibition of Leydig cells steroidogenic activity
Rats	↓ PACAP mRNA ↓ PAC1 mRNA	cAMP production decrease. Inhibition of testicular testosterone synthesis. Spermatogenesis impairment.
Mouse Leydig (TM3) cell line	↑ Bax-dependant caspase-3	Apoptosis induction. Inhibition of testicular testosterone synthesis. Spermatogenesis impairment.
Rats	↑ Fas mRNA ↑ FasL mRNA ↑ Fas/FasL ↑ Tumor protein p53 ↑ Bax/Bcl-2 ↑ Bax-dependant caspase-3 ↑ caspase-8 ↓ pAkt ↓ pErk1/2	
Mice	↓ NADPH-diaphorase	Imbalance in testicular testosterone synthesis. Spermatogenesis impairment.
Rats	↑ TnNOS (in Leydig cells)	Imbalance in testicular testosterone synthesis. Spermatogenesis impairment.
Rats	↑ iNOS (in Sertoli, germ ↑interstitial cells)	Excessive NO generation, formation of $ONOO^-$ which modifies proteins with thiol groups to generate nitrosothiols disrupting metal-protein interactions and lead to the generation of other metal-derived free radicals, nitrosative stress, apoptosis induction, androgen synthesis suppression. Spermatogenesis impairment.
Puberty rats	↑ GnRH mRNA ↓ GnRH-R mRNA	Spermatogenesis disturbance (more profound in puberty than adults).
Adult rats	↓ GnRH mRNA ↓ GnRH-R mRNA	
Rat Leydig cells	↓ Testosterone synthesis	Spermatogenesis impairment.

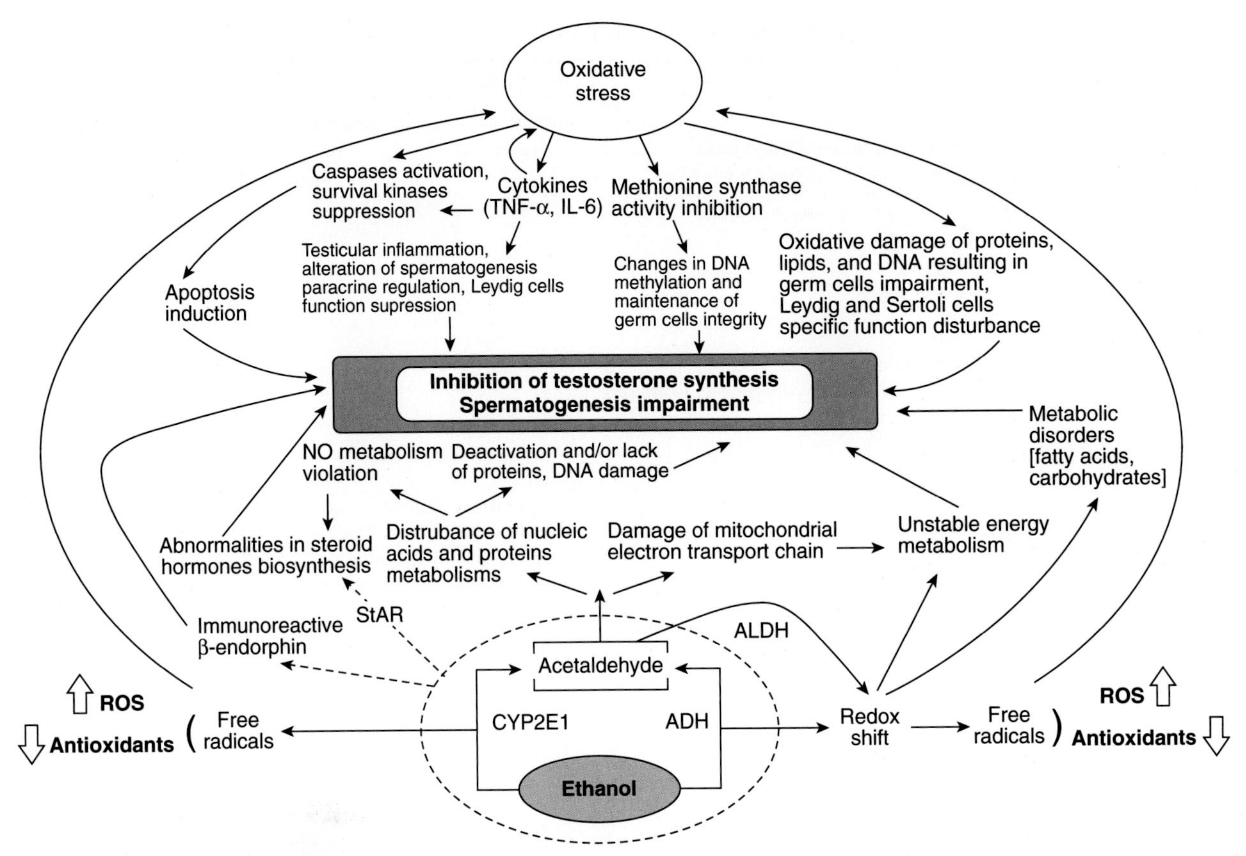

FIGURE 27.2 **Probable testicular metabolic disorders, involved in pathogenesis of ethanol-mediated male gonadal toxicity.** The figure presents a selection of described in presently reported literature ethanol-induced metabolic changes in male gonads (but one must assume that other factors also could be involved, of which many currently remain unknown).

METABOLIC CHANGES IN FEMALES' ALCOHOL GONADOTOXICITY

Female's alcohol gonadotoxicity is of special interest due to its close relationship both with the reproductive organs functioning as themselves, and with ensuing pregnancy or birth of offspring. In female organism ethyl alcohol acts as real gonadal toxin, developing injury of these organs both by direct ethanol effects and by the consequences of ethanol metabolism (e.g., acetaldehyde formation, oxidative stress development, alterations in enzymatic system functioning). This injury occurs at the level of the ovary, the hypothalamus and pituitary. It also can have an impact on general metabolic processes in organism.[4,90]

The general attention of female gonadotoxicity investigation during last decades has been focused on studies of DNA-structure damages and oxidative stress caused by alcohol consumption. For instance, modern methods of investigation (a modification of the standard protocol for comet assay, a combination of alkaline comet version and proteinase-K) have been allowed clarifying exact mechanisms of DNA structure damages by acetaldehyde, induction of DNA-protein crosslinks formation and reduction in DNA migration.[91] But the effects of direct ethanol metabolism to acetaldehyde in ovaries and alcohol-mediated increased susceptibility to oxidative stress in the ovarian tissue did not receive proper attention. In fact it only has been demonstrated, that ethanol direct biotransformation in the rat ovary cells had enzymatic nature, with involvement of XO, NADPH, ADH and ALDH.[92] These processes have been accompanied with cells structure and functioning alterations at the level of the granulose, theca interna and pellucida zones. Investigators observed marked condensation of chromatin attached to the nuclear inner membrane, intense dilatation of the outer perinuclear space, marked dilatation of the rough endoplasmic reticulum, significant detachment of ribosomes from their membranes, swollen mitochondria. Corona radiata cells in the zona pellucida were absent or broken.[92] Thus ethanol metabolite (acetaldehyde) impairs the differentiation of granulosa cells, reduces ovulation and decreases oocyte quality.

The effect of ethanol administration on ovarian function and structure in female rats has been also studied by other authors.[93] It has been showed greatly reduced levels of estradiol and progesterone, which were accompanied (in such conditions) by the reduction of ovarian weight on 60% in comparison with control, disruption of ovarian tissue structure (absence of corpus hemorrhagicum and corpus albicans). Marked estrogen deprivation has been observed as well.[93]

A feedback also exist: progesterone-derived ovarian hormones, via progesterone metabolites effect on γ-aminobutyric acid type A (GABA(A)) receptors, can change effects of ethanol in female rats.[94]

Ethanol negative effects on female reproductive organs could not be limited by alcohol-mediated endocrine disturbances. In the light of Amanvermez et al. findings chronic alcohol administration leading to a significant increase in the levels of protein and lipid oxidation in the ovary [23] possibly has a great impact on the passage of the main metabolic processes in the body as a whole and these organs especially. Particularly oxidative stress accompanied with increased formation of lipid and protein oxidized derivatives can cause many pathophysiologic changes in ovarian functions.[23]

In terms of biochemical transformations successful functioning of the female reproductive organs fully depends on normal rates of nucleic acids, proteins, lipids and hydrocarbons biosynthesis. This in turn demands intensification of starting components and macroergs (such as NADH, NADPH, ATP, guanosine-5′-triphosphate (GTP)) production and replenishment of used energy (in Krebs and β-alanine cycles, glycolysis, β-oxidation of fatty acids, etc.). Besides this, all of the abovementioned processes are subjects to feedback regulation by the low-molecular-weight compounds such as CO_2, NO_2, amines, etc. Character of the metabolic processes changes during gonadotoxic ethanol effects realization depends on many factors. These include doses and terms of ethanol exposure, organism's physiologic state and attendant circumstances (stresses, diseases, environment, and drugs, etc.).

These violations can be very serious. It is proved by the fact, that ethanol may promote carcinogenesis by multiple mechanisms such as production of acetaldehyde (a weak mutagen and carcinogen), induction of CYP system with following oxidative stress development and conversion of procarcinogens to carcinogens. Among this it causes S-adenosylmethionine depletion and induction of global DNA hypomethylation, induction of increased production of inhibitory guanine nucleotide regulatory proteins and components of extracellular signal-regulated kinase-mitogen-activated protein kinase signaling, accumulation of iron and associated oxidative stress, inactivation of the tumor suppressor gene *BRCA1*. All these changes are accompanied with increased estrogen responsiveness (primarily in breast), and impairment of retinoic acid metabolism.[95]

Data showing ethanol inhibitor effect on protein synthesis in female reproductive organs at the transcriptional and translational levels, appeared at 1980s[96,97]: it has been demonstrated that ethanol influenced on the processes of some mRNA transcription and competitively inhibited the interactions of amino acids with amino acyl-tRNA synthetases, whereas acetaldehyde – effected in a non-competitive manner in Chinese hamster ovary cells.

Only state of the art methodology allows obtaining detailed information on ethanol's effects on nucleic acids and proteins metabolisms. In particular, it has been found that ethanol administration during 5 days was capable to cause disturbances in processes of biosynthesis of the prepubertal intraovarian insulin-like growth factor-1 (IGF-1), which was important regulator of ovarian development and functioning.[98] An increase of the expression of mRNAs encoding IGF-1 (both IGF-1a and IGF-1b subtypes) has been demonstrated in the ovaries of ethanol-treated rats by means of an RNAse protection assay. On the other hand, radioimmunoassay data have not shown simultaneous increase of protein levels.[98] On the contrary, ovarian IGF-1 protein contents were decreased in ethanol-treated rats as soon as the expression of Type-1 IGF receptor mRNA and levels of IGF-1 receptor protein, as determined by Western blot analysis. As well as increases in the levels of IGF-binding proteins-3 and -5 have been revealed. Such changes inevitably unbalanced the normal course of ovaries development and functioning and led to their irreversible damage.[98]

Later in experiments on mice the same authors studied ethanol-induced changes of ovarian *IFG-1* gene transcription in dependence on serum growth hormone and its receptor.[99] It has been shown a decrease of *IGF* and insulin-like growth factor-1 receptor (IGF-1R) genes expression.

There is evidence that ethanol in dose 3 g/kg can greatly affect another important regulator of ovary functioning – mitochondrial protein StAR, which plays an essential role in steroid hormones biosynthesis by ovaries via facilitating delivery of cholesterol across the mitochondrial membrane.[100] Northern blot analysis data demonstrated decrease of the levels of two major transcripts of 3.8 and 1.7 kb of ovarian StAR mRNA in ethanol-treated rats, while Western blot analysis showed that ethanol exposure also depressed the basal expression of StAR protein.[100]

Biosynthesis of pregnenolone and estradiol in ovaries were suppressed simultaneously with changes of StAR mRNA content and protein biosynthesis. Thus ethanol-mediated violations in nucleic acids and proteins metabolisms in this study were accompanied with corresponding unbalancing of lipids metabolisms (i.e., steroid hormones biosynthesis). It is quite obvious that such metabolic disorders inevitably affect all cells in female reproductive organs and change their viability and ability to perform specified functions.[100]

NO is intraovarian substance that influences steroidogenesis in opposite direction than StAR protein. Ethanol exposure also greatly changed its metabolism in rhesus monkeys ovaries via regulation of NOS expression.[101] The researchers have been assessed ovarian mRNA encoding three isoforms of NOS by means of real-time polymerase chain reaction. It has been shown, that ethanol exposure caused increase of eNOS and iNOS mRNA expressions levels. Simultaneously increased expressions of eNOS and iNOS proteins have been noted. Such changes, along with inhibition of StAR gene and protein expression, could cause disturbances in biosynthesis of steroid hormones by ovaries.[101]

Thus experiments with females of different mammalian species demonstrate that ethanol simultaneously unbalances gonadal metabolisms of proteins, lipids and biologically active low-molecular-weight compounds such as NO_2, causing multilevel irreversible damage of cells, tissues and organs as a whole.

Acute and chronic ethanol administration possibly causes deleterious effects on HPG axis in female rats by immuno-endocrine interactions with involvement of proinflammatory cytokines: TNF-α and IL-6, which have antireproductive effects.[102] Ethanol exposure stimulated biosynthesis of TNF-α and IL-6 in hypothalamus, pituitaries, and ovaries.[102]

Chronic ethanol consumption greatly changed metabolism of nucleotides and amino acids (starting components of nucleic acids and proteins biosynthesis) in female gonads. Ethanol itself can directly influence on catalytic characteristics of ovarian adenylyl cyclase and luteal adenylyl cyclase.[103,104]

NAD-dependent ADH and ALDH are present both in male and female reproductive tissues.[105] We have nothing to report on $NADH/NAD^+$ ratio in ovaries following ethanol exposure, but on our opinion there is a definite possibility that alcohol can cause violations at the level of NADH and NADPH metabolisms in female gonads, which inevitably leads to negative changes in metabolic transformations energy supply. More studies on this issue are required to understand ethanol female gonadotoxicity mechanisms.

No information is available on ethanol effects on carbohydrates' metabolisms in female gonads too.

In general, despite of evident importance of lipid metabolism and lipid oxidation processes for cell membranes structure and functions, these parameters in female gonads remained insufficiently investigated during ethanol exposure (being in general limited by only steroid hormones metabolism). Few *in vitro* studies allow supposing that ethanol can directly or by its metabolites change membrane ability to fix molecules of hormones and enzymes.[103,104,106] Among this, it has been shown that, administration of 10% ethanol in drinking water during 30 days to immature female mice can alter prostaglandin E production in the oocyte cumulus complexes, accompanied with morphologic abnormalities in the oocytes (high parthenogenetic activated rates).[107] There is also report that ethanol *in vitro* can influence on some processes of phospholipids metabolism.[108]

On the basis of this part of review and the facts summarized in Table 27.2 it seems justified concluding that scientific data on ethanol-mediated metabolic changes in ovaries has been insufficient. However, such investigations are necessary from the point of view of clarifying energy sources for maintaining ovary cells' homeostasis at ethanol exposure. We have colligated all relevant information in the Figure 27.3 and have tried to outline probable

TABLE 27.2 Summary of Alcohol-Mediated Metabolic Changes in Ovaries

Test system observed	Changes	Possible outcomes
Rat ovary	Alterations of normal levels of XO, NADPH, ADH, and ALDH	Cells structure and functioning alterations at the level of the granulosa; theca interna and pellucida zones (condensation of chromatin attached to the nuclear inner membrane, intense dilatation of the outer perinuclear space, marked dilatation of the rough endoplasmic reticulum, significant detachment of ribosomes from their membranes, swollen mitochondria). Impairment of corona radiata cells in the zona pellucida. Impaired differentiation of granulose cells, reduced ovulation, and decreased oocyte quality.
Rats	↑ Oxidative stress ↑ Formation of lipid and protein oxidized derivatives	Alterations of the main metabolic processes in ovaries, decreased oocyte quality.
Chinese hamster ovary cells	↓ The protein synthesis at the transcriptional and translational levels (the processes of some mRNA transcription and the interactions of amino acids with amino acyl-tRNAsynthetases)	
Rats	↑ IGF-1a mRNA ↑ IGF-1b mRNA ↓ IGF-1 protein ↑ IGF-binding proteins-3 and -5	Unbalanced course of ovaries development and functioning, their irreversible damage, decreased oocyte quality.
Mice	↓ IGF mRNA ↓ IGF-1R mRNA	

TABLE 27.2 Summary of Alcohol-Mediated Metabolic Changes in Ovaries *(cont.)*

Test system observed	Changes	Possible outcomes
Rats	↓ 3.8 StAR mRNA ↓ 1.7 kb StAR mRNA ↓ StAR protein	Pregnenolone and estradiol biosynthesis suppression. Metabolic disorders in female reproductive organs, irreversible damage of all cells and change their viability and ability to perform specified functions, decreased oocyte quality.
Rhesus monkeys	↓ StAR mRNA	
Rhesus monkeys	↑ eNOS mRNA and protein ↑ iNOS mRNA and protein	Violations in steroid hormone biosynthesis by ovaries, especially when they are accompanied by inhibition of StAR protein expression processes
Rats	↑ TNF-α ↑ IL-6	Apoptosis induction. Increased antireproductive effects.
Pig ovary	Changes of adenylyl cyclase and luteal adenylyl cyclase catalytic characteristics	Pathophysiologic changes in ovarian cellular functions
Membrane fractions from human corpus luteum		
Mice	Altered prostaglandin E (PGE) production in the oocyte cumulus complexes	Morphologic abnormalities in the oocytes (high parthenogenetic activated rates), decreased oocyte quality
Rats	↓ Estradiol synthesis ↓ Progesterone synthesis	Reduction of ovarian weight, disruption of ovarian tissue structure (absence of corpus hemorrhagicum and corpus albicans). Marked estrogen deprivation.

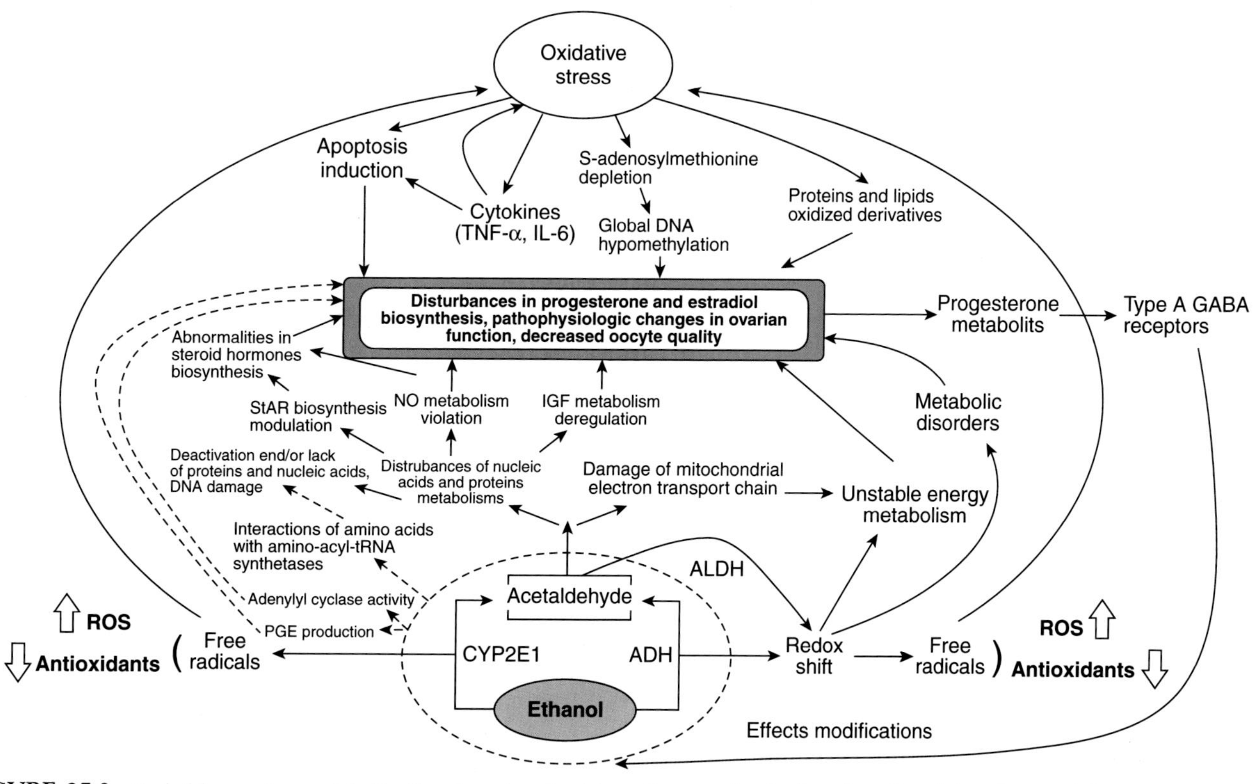

FIGURE 27.3 **Probable ovarian metabolic disorders, involved in pathogenesis of ethanol-mediated female gonadal toxicity.** The figure presents a selection of described in presently reported literature ethanol-induced metabolic changes in female gonads (but one must assume that other factors also could be involved, of which many currently remain unknown).

ovarian metabolic disorders, involved in pathogenesis of ethanol-mediated female gonadal toxicity. Future studies are of particular interest as in this case it's not just about only ovary functions as such but also on the subsequent processes of gestation and offspring birth.

CONCLUSIONS

As it can be clearly seen from the review the effects on gonads of both acute and chronic alcohol administration, as well as their mechanisms, have been studied mainly in male rats. Studies in females are relatively few in number. Ethanol adverse effects on male and female gonads are multiple, serious, and affecting all without exception metabolic processes in these organs. Degree of these effects' relative importance for gonads structure and functions preservation and negative changes reversibility depends on ethanol exposition time, dose, and regime as well as organisms' sex, age, physiological state, and attendant circumstances. At present accumulated numerous data on alcohols' direct and indirect effects on nucleic acids, proteins, lipids, hydrocarbons metabolism in testis and ovary. These violations are accompanied with simultaneous changes in energy metabolism and production of the low-molecular weight-compounds. During gonadotoxic ethanol effects realization arise wide net of systemic transformations exciting the whole complex of metabolic interrelations in testes or ovary (Figure 27.4). A complete picture of violations could be obtained only with state of the art methodological approaches and extensive use of bioinformatics tools for processing huge amounts of collected data, as only such way creates a solid base of mathematical calculations under the justification of real implementation mechanisms of ethanol gonadotoxic effects.

Thus, summarizing the results of studies both of metabolic changes in males' and females' alcohol gonadotoxicity we can state that despite of extensive investigation of ethanol adverse effects these lines of studies require further deepening on the basis of application of new approaches for complex estimation of gonads proteome, lipidome and metabolome changes during ethanol administration.

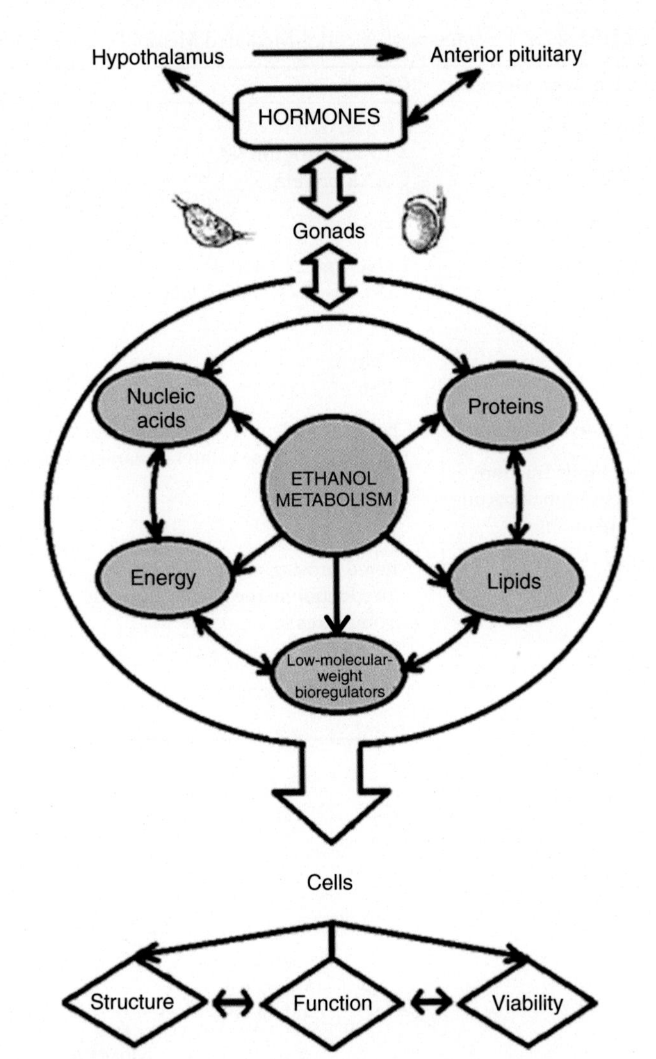

FIGURE 27.4 **Metabolic net potential targets in ethanol gonadal toxicity realization.** Gonadal metabolism of ethanol results in the generation of large quantities of free radicals and other toxic adducts directly in testis and ovary, leading there to disruptions in the normal metabolic processes. Mainly chronic ethanol metabolism in gonads results in impaired nucleic acids, proteins, lipids, hydrocarbons metabolic processes, as well as changes in energy metabolism and production of the low-molecular-weight bioregulators. Such disturbances in turn results in ovary and testis cells changes, leading to the triggering of the gonadal cells apoptosis pathway. The suppression of estrogenic and androgenic functions, disturbance of normal germ cells differentiation, and loss of fertility are the final outcomes.

Key Facts of Metabolic Changes in Alcohol Gonadotoxicity

- Potentially adverse effects of alcohol on reproductive health in women and men are of public and scientific interests.
- Ethanol effects on gonads are multiple, serious, and affecting all without exception metabolic processes in these organs.
- Ethanol-mediated intensification of ROS production can modify structures of DNA, lipids, hydrocarbons and proteins, affect processes of mRNA transcription and interaction of amino acids with amino acyl-tRNA synthetases, cause disturbances in the gonads' metabolism of low-molecular-weight bioregulators.
- Metabolic disturbances in ovary and testis lead to the triggering of the gonadal cells apoptosis pathway.
- The suppression of estrogenic and androgenic functions, disturbance of normal germ cells differentiation, and loss of fertility are the final outcomes of alcohol-mediated metabolic changes in gonads.

Summary Points

- This chapter focuses on alcohol-induced metabolic changes in male and female gonads.
- The effects on gonads of both acute and chronic alcohol administration, as well as their mechanisms, have been studied mainly in male rats.
- Studies in females are relatively few in number.
- Considering alcohol biotransformation in the gonads, the existing specificity of metabolic processes characteristic just for these organs and factors influenced on the final outcome cannot be ignored.
- Analysis of accumulated findings clearly indicates that majority of experiments on alcohol-induced metabolic changes in gonads have been done with animals or cell cultures, the collection of clinical data is of great importance to reach a general conclusions.

Abbreviations

17-KSR	17-Ketosteroid reductase
17β-HSD	17β-Hydroxysteroid dehydrogenase
3β-HSD	3β-Hydroxysteroid dehydrogenase
3β-HSD	3β-Hydroxysteroid dehydrogenase
6-PGDH	6-Phosphogluconate dehydrogenase
ADH	Alcohol dehydrogenase
ALDH	Aldehyde dehydrogenase
ALDH2	Mitochondrial aldehyde dehydrogenase
ATP	Adenosine 5′-triphosphate
CAT	Catalase
DHT	Dihydrotestosterone
eNOS	Endothelial nitric oxide synthase
FAEE	Fatty acid ethyl ester
Fas	Fatty acid syntase
FasL	Fatty acid syntase ligand
FSH	Follicle-stimulating hormone
G-6-PDH	Glucose-6-phosphate dehydrogenase
GnRH	Gonadotropin-releasing hormone
GnRH-R	Gonadotropin-releasing hormone receptor
GSH	Reduced glutathione
GSH-Px	Glutathione peroxidase
hCG	Human chorionic gonadotropin
HPG	Hypothalamic-pituitary-gonadal
IGF-1	Insulin-like growth factor-1
IGF-1R	Insulin-like growth factor-1 receptor
IL-6	Interleukin 6
iNOS	Inducible nitric oxide synthase
LH	Luteinizing hormone
LHRH	luteinizing hormone releasing hormone
MDA	Malondialdehyde
MDH	Malate dehydrogenase
MEOS	Microsomal ethanol oxidizing system
NADP-ICDH	NADP-dependent isocitrate dehydrogenase
nNOS	Neuronal nitric oxide synthase
NOS	Nitric oxide synthase
P450scc	Cytochrome P450 side chain cleavage enzyme
PA	Phosphatidic acid
PACAP	Pituitary adenylate cyclase activating polypeptide
pAkt	Phosphorylated protein kinase B
pErk1/2	Phosphorylated extracellular signal-regulated kinase 1/2
PEth	Phosphatidyl ethanol
ROS	Reactive oxygen species
SOD	Superoxide dismutase
StAR	Steroidogenic acute regulatory protein
TIF	Testicular interstitial fluid
TNF-α	Tumor necrosis factor α
XO	Xanthine oxidase
GABA(A)	γ-aminobutyric acid type A
CYP	Cytochrome P450

References

1. The effects of alcohol on physiological processes and biological development. http://pubs.niaaa.nih.gov/publications/arh283/125-132.htm.
2. La Vignera S, Condorelli RA, Balercia G, Vicari E, Calogero AE. Does alcohol have any effect on male reproductive function? A review of literature. *Asian J Androl* 2013;**15**(2):221–5.
3. Tolstrup JS, Kjaer SK, Holst C, Sharif H, Munk C, Osler M, Schmidt L, Andersen AM, Gronbaek M. Alcohol use as predictor for infertility in a representative population of Danish women. *Acta Obstetricia Et Gynecologica Scandinavica.* 2003;**82**(8):744–9.
4. Emanuele N, Emanuele MA. The endocrine system. Alcohol alters critical hormonal balance. *Alcohol Health Res World* 1997;**21**(1):53–64.
5. Sierksma A, Sarkola T, Eriksson CJ, van der Gaag MS, Grobbee DE, Hendriks HF. Effect of moderate alcohol consumption on plasma dehydroepiandrosterone sulfate, testosterone, and estradiol levels in middle-aged men and postmenopausal women: a diet-controlled intervention study. *Alcohol Clin Exp Res* 2004;**28**(5):780–5.
6. Emanuele MA, Emanuele N. Alcohol and the Male Reproductive System. *Alcohol Res Health* 2001;**25**(4):283–7.
7. Chang G, McNamara TK, Haimovici F, Hornstein MD. Problem Drinking in Women Evaluated for Infertility. *Am J Addict* 2006;**15**(2):174–9.
8. Zakhari S. Overview: how is alcohol metabolized by the body? *Alcohol Res Health* 2006;**29**:245–54.
9. Lieber CS. Metabolism of alcohol. *Clin Liver Dis* 2005;**9**(1):1–35.
10. Maenhout TM, De Buyzere ML, Delanghe JR. Non-oxidative ethanol metabolites as a measure of alcohol intake. *Clin Chim Acta* 2013;**415**:322–9.
11. Seitz HK, Xu Y, Simanowski UA, Osswald B. Effect of age and gender on *in vivo* ethanol elimination, hepatic alcohol dehydrogenase activity, and NAD^+ availability in F344 rats. *Res Exp Med (Berl)* 1992;**192**(1):205–12.
12. Marshall AW, Kingstone D, Boss M, Morgan MY. Ethanol Elimination in Males and Females: Relationship to Menstrual Cycle and Body Composition. *Hepatology* 1983;**3**(5):701–6.
13. Vatsalya V, Issa JE, Hommer DW, Ramchandani VA. Pharmacodynamic effects of intravenous alcohol on hepatic and gonadal hormones: influence of age and sex. *Alcohol Clin Exp Res* 2012;**36**(2):207–13.
14. Frias J, Rodriguez R, Torres JM, Ruiz E, Ortega E. Effects of acute alcohol intoxication on pituitary-gonadal axis hormones, pituitary-adrenal axis hormones, beta-endorphin and prolactin in human adolescents of both sexes. *Life Sci* 2000;**67**(9):1081–6.
15. Ogilvie KM, Rivier C. Gender difference in alcohol-evoked hypothalamic-pituitary-adrenal activity in the rat: ontogeny and role of neonatal steroids. *Alcohol Clin Exp Res* 1996;**20**(2):255–61.
16. Bulfield G, Nahum A. Effect of the mouse mutants testicular feminization and sex reversal on hormone-mediated induction and repression of enzymes. *Biochem Genet* 1978;**16**(7–8):743–50.

17. Mucha L, Stephenson J, Morandi N, Dirani R. Meta-analysis of disease risk associated with smoking, by gender and intensity of smoking. *Gend Med Dec* 2006;**3**(4):279–91.
18. Ely M, Hardy R, Longford NT, Wadsworth MEJ. Gender differences in the relationship between alcohol consumption and drink problems are largely accounted for by body water. *Alcohol Alcoholism* 1999;**34**(6):894–902.
19. Wu D, Cederbaum AI. Alcohol, oxidative stress, and free radical damage. *Alcohol Res Health* 2003;**27**:277–84.
20. Guerriero G, Trocchia S, Abdel-Gawad FK, Ciarcia G. Roles of reactive oxygen species in the spermatogenesis regulation. *Front Endocrinol (Lausanne)* 2014;**5**:56.
21. Coniglio JG, Grogan Jr WM, Rhamy RK. Lipid and Fatty Acid Composition in Human Testes Removed at Autopsy. *Biol Reprod* 1975;**12**:255–9.
22. Rosenblum E, Gavaler JS, Van Thiel DH. Lipid peroxidation: a mechanism for ethanol-associated testicular injury in rats. *Endocrinology* 1985;**116**(1):311–8.
23. Amanvermez R, Demir S, Tunçel OK, Alvur M, Agar E. Alcohol-induced oxidative stress and reduction in oxidation by ascorbate/ L-cys/ L-met in the testis, ovary, kidney, and lung of rat. *Adv Ther* 2005;**22**(6):548–58.
24. Beckett GJ, Arthur JR. Selenium and endocrine systems. *J Endocrinol* 2005;**184**:455–65.
25. Harikrishnan R, Abhilash PA, Syam Das S, Prathibha P, Rejitha S, John F, Kavitha S, Indira M. Protective effect of ascorbic acid against ethanol-induced reproductive toxicity in male guinea pigs. *Br J Nutr* 2013;**110**(4):689–98.
26. Grattagliano I, Vendemiale G, Errico F, Bolognino AE, Lillo F, Salerno MT, Altomare E. Chronic ethanol intake induces oxidative alterations in rat testis. *J Appl Toxicol* 1997;**17**(5):307–11.
27. Uygur R, Yagmurca M, Alkoc OA, Genc A, Songur A, Ucok K, Ozen OA. Effects of quercetin and fish n-3 fatty acids on testicular injury induced by ethanol in rats. *Andrologia* 2014;**46**(4):356–69.
28. Rosenblum ER, Gavaler JS, Van Thiel DH. Vitamin A at pharmacologic doses ameliorates the membrane lipid peroxidation injury and testicular atrophy that occurs with chronic alcohol feeding in rats. *Alcohol Alcohol* 1987;**22**(3):241–9.
29. Ojeda ML, Jotty K, Nogales F, Murillo ML, Carreras O. Selenium or selenium plus folic acid intake improves the detrimental effects of ethanol on pups' selenium balance. *Food Chem Toxicol* 2010;**48**(12):3486–91.
30. Ahsan U, Kamran Z, Raza I, Ahmad S, Babar W, Riaz MH, Iqbal Z. Role of selenium in male reproduction - a review. *Anim Reprod Sci* 2014;**146**(1–2):55–62.
31. Behne D, Hofer T, Von Berwordt-Wallrabe R, Elger W. Selenium in the testis of the rat: studies on its regulation and its importance for the organism. *J Nutr* 1982;**112**(9):1682–7.
32. Garrido N, Meseguer M, Carlos Simon C, Pellicer A, Remohi J. Pro-oxidative and anti-oxidative imbalance in human semen and its relation with male fertility. *Asian J Androl* 2004;**6**:59–65.
33. Swanson CA. Iron intake and regulation: implications for iron deficiency and iron overload. *Alcohol Elsevier Inc* 2003;**30**:99–102.
34. Zhang X, Liu Q, Ha J. Protection of vitamin E against testis lipid peroxidation induced by iron and ethanol. *Wei Sheng Yan Jiu* 1998;**27**(3):184–6.
35. Singh K, Jaiswal D. One-carbon metabolism, spermatogenesis, and male infertility. *Reprod Sci* 2013;**20**(6):622–30.
36. Wallock-Montelius LM, Villanueva JA, Chapin RE, Conley AJ, Nguyen HP, Ames BN, Halsted CH. Chronic Ethanol Perturbs Testicular Folate Metabolism and Dietary Folate Levels in the Yucatan Micropig. *Biol Reprod* 2007;**76**(3):455–65.
37. Emanuele NV, LaPaglia N, Kovacs EJ, Gamelli RL, Emanuele MA. Profound effects of burn and ethanol on proinflammatory cytokines of the reproductive axis in the male mouse. *J Burn Care Res* 2008;**29**(3):531–40.
38. Zhu Q, Emanuele MA, LaPaglia N, Kovacs EJ, Emanuele NV. Vitamin E prevents ethanol-induced inflammatory, hormonal, and cytotoxic changes in reproductive tissues. *Endocrine* 2007;**32**(1):59–68.
39. Adams ML, Cicero TJ. Effects of alcohol on beta-endorphin and reproductive hormones in the male rat. *Alcohol Clin Exp Res* 1991;**15**(4):685–92.
40. Quintans LN, Castro GD, Castro JA. Oxidation of ethanol to acetaldehyde and free radicals by rat testicular microsomes. *Arch Toxicol* 2005;**79**(1):25–30.
41. Messiha FS. Testicular and epididymal aldehyde dehydrogenase in rodents: modulation by ethanol and disulfiram. *Int J Androl* 1980;**3**(4):375–82.
42. Radhakrishnakartha H, Appu AP, Madambath I. Reversal of alcohol induced testicular hyperlipidemia by supplementation of ascorbic acid and its comparison with abstention in male guinea pigs. *J Basic Clin Physiol Pharmacol* 2013;**24**:1–8.
43. Grattagliano I, Vendemiale G, Errico F, Bolognino AE, Lillo F, Salerno MT, Altomare E. Chronic ethanol intake induces oxidative alterations in rat testis. *J Appl Toxicol* 1997;**17**(5):307–11.
44. Murono EP, Fisher-Simpson V. Partial characterization of alcohol dehydrogenase activity in purified rat Leydig cells. *Arch Androl* 1986;**17**(1):39–47.
45. Murono EP, Fisher-Simpson V. Ethanol directly increases dihydrotestosterone conversion to 5 alpha-androstan-3 beta, 17 beta-diol and 5 alpha-androstan-3 alpha, 17 beta-diol in rat Leydig cells. *Biochem Biophys Res Commun* 1984;**121**(2):558–65.
46. Murono EP, Fisher-Simpson V. Microsomal ethanol-oxidizing system in purified rat Leydig cells. *Biochim Biophys Acta* 1987;**918**(2):136–40.
47. Wanderley MI, Udrisar DP. Inhibitory action of *in vitro* ethanol and acetaldehyde exposure on LHRH-and phorbol ester-stimulated testosterone secretion by rat testicular interstitial cells. *Acta Physiol Pharmacol Ther Latinoam* 1994;**44**(4):135–41.
48. Aitken RJ, Roman SD. Antioxidant systems and oxidative stress in the testes. *Oxid Med Cell Longev* 2008;**1**(1):15–24.
49. Jiang Y, Kuo CL, Pernecky SJ, Piper WN. The detection of cytochrome P450 2E1 and its catalytic activity in rat testis. *Biochem Biophys Res Commun* 1998;**246**(3):578–83.
50. Lee SY, Gong EY, Hong CY, Kim KH, Han JS, Ryu JC, Chae HZ, Yun CH, Lee K. ROS inhibit the expression of testicular steroidogenic enzyme genes via the suppression of Nur77 transactivation. *Free Radic Biol Med* 2009;**47**(11):1591–600.
51. Cederbaum AI. CYP2E1-biochemical and toxicological aspects and role in alcohol-induced liver injury. *Mt Sinai J Med* 2006;**73**(4):657–72.
52. Stocco DM. StAR protein and the regulation of steroid hormone biosynthesis. *Annu Rev Physiol* 2001;**63**:193–213.
53. Tsai SC, Lu CC, Lin CS, Wang PS. Antisteroidogenic actions of hydrogen peroxide on rat Leydig cells. *J Cell Biochem* 2003;**90**(6):1276–86.
54. Herman M, Kang SS, Lee S, James P, Rivier C. Systemic administration of alcohol to adult rats inhibits leydig cell activity: Time course of effect and role of nitric oxide. *Alcohol Clin Exp Res* 2006;**30**(9):1479–91.
55. Orpana AK, Orava MM, Vihko RK, Härkönen M, Eriksson CJ. Ethanol-induced inhibition of testosterone biosynthesis in rat Leydig cells: central role of mitochondrial NADH redox state. *J Steroid Biochem* 1990;**36**(6):603–8.
56. Chiao YB, Johnston DE, Gavaler JS, Van Thiel DH. Effect of chronic ethanol feeding on testicular content of enzymes required for testosteronogenesis. *Alcohol Clin Exp Res* 1981;**5**(2):230–6.
57. Widenius TV, Orava MM, Vihko RK, Ylikahri RH, Eriksson CJ. Inhibition of testosterone biosynthesis by ethanol: multiple sites and mechanisms in dispersed Leydig cells. *J Steroid Biochem* 1987;**28**(2):185–8.

58. Akane A, Fukushima S, Shiono H, Fukui Y. Effects of ethanol on testicular steroidogenesis in the rat. *Alcohol Alcohol* 1988;**23**(3):203–9.
59. Murono EP. Direct stimulatory effect of ethanol on 17 alpha-hydroxylase activity of rat testis interstitial cells. *Life Sci* 1984;**34**(9):845–52.
60. Eriksson CJ, Widenius TV, Ylikahri RH, Härkönen M, Leinonen P. Inhibition of testosterone biosynthesis by ethanol. Relation to hepatic and testicular acetaldehyde, ketone bodies and cytosolic redox state in rats. *Biochem J* 1983;**210**(1):29–36.
61. Cicero TJ, Bell RD. Ethanol-induced reductions in testicular steroidogenesis: major differences between *in vitro* and *in vitro* approaches. *Steroids* 1982;**40**(5):561–8.
62. Lieber CS. Alcohol: its metabolism and interaction with nutrients. *Annu Rev Nutr* 2000;**20**:395–430.
63. Grattagliano I, Vendemiale G, Errico F, Bolognino AE, Lillo F, Salerno MT, Altomare E. Chronic ethanol intake induces oxidative alterations in rat testis. *J Appl Toxicol* 1997;**17**(5):307–11.
64. Farghali H, Williams DS, Gavaler J, Van Thiel DH. Effect of short-term ethanol feeding on rat testes as assessed by 31P NMR spectroscopy, 1H NMR imaging, and biochemical methods. *Alcohol Clin Exp Res* 1991;**15**(6):1018–23.
65. Haider SG, Hofmann N, Passia D. Morphological and Enzyme Histochemical Observations on Alcohol induced Disturbances in Testis of two Patients. *Andrologia* 1985;**17**(6):532–40.
66. Galardo MN, Regueira M, Riera MF, Pellizzari EH, Cigorraga SB, Meroni SB. Lactate Regulates Rat Male Germ Cell Function through Reactive Oxygen Species. *PLOS one* 2014;**9**(1):e88024.
67. Srikanth V, Balasubramanian K, Govindarajulu P. Effects of ethanol treatment on Leydig cellular NADPH-generating enzymes and lipid profiles. *Endocr J* 1995;**42**(5):705–12.
68. Rengarajan S, Malini T, Sivakumar R, Govindarajulu P, Balasubramanian K. Effects of ethanol intoxication on LH receptors and glucose oxidation in Leydig cells of adult albino rats. *Reprod Toxicol* 2003;**17**(6):641–8.
69. Kavitha TS, Parthasarathy C, Sivakumar R, Badrinarayanan R, Balasubramanian K. Effects of excess corticosterone on NADPH generating enzymes and glucose oxidation in Leydig cells of adult rats. *Hum Exp Toxicol* 2006;**25**(3):119–25.
70. Baraona E, Lieber CS. Effects of ethanol on lipid metabolism. *J Lipid Res* 1979;**20**:289–315.
71. Emanuele NV, LaPagli N, Steiner J, Colantoni A, Van Thiel DH, Emanuele MA. Peripubertal paternal EtOH exposure. *Endocrine* 2001;**14**:213–9.
72. Shayakhmetova GM, Bondarenko LB, Kovalenko VM, Ruschak VV. CYP2E1 testis expression and alcohol-mediated changes of rat spermatogenesis indices and type I collagen. *Arh Hig Rada Toksikol* 2013;**4**(2):51–60.
73. Shayakhmetova GM, Bondarenko LB, Matvienko AV, Kovalenko VM. Correlation between spermatogenesis disorders and rat testes CYP2E1 mRNA contents under experimental alcoholism or type I diabetes. *Advances in Medical Sciences* 2014;**59**(2):183–9.
74. Koh PO, Won CK, Ho JH. Ethanol decreases the expression of pituitary adenylate cyclase activating polypeptide in rat testes. *J Vet Med Sci* 2006;**68**(6):635–7.
75. Koh PO, Won CK. Decrease of pituitary adenylate cyclase activating polypeptide and its type I receptor mRNAs in rat testes by ethanol exposure. *J Vet Med Sci* 2006;**68**(6):537–41.
76. Xiang J, Chao DT, Korsmeyer SJ. BAX-induced cell death may not require interleukin 1 beta-converting enzyme-like proteases. *Proc Natl Acad Sci* 1996;**93**:14559–63.
77. Jang MH, Shin MC, Shin HS, Kim KH, Park HJ, Kim EH, Kim CJ. Alcohol induces apoptosis in TM3 mouse Leydig cells via bax-dependent caspase-3 activation. *Eur J Pharmacol* 2002;**449**(1–2):39–45.
78. Jana K, Jana N, De DK, Guha SK. Ethanol induces mouse spermatogenic cell apoptosis *in vivo* through over-expression of Fas/Fas-L, p53, and caspase-3 along with cytochrome c translocation and glutathione depletion. *Mol Reprod Dev* 2010;**77**(9):820–33.
79. Hu JH, Jiang J, Ma YH, Yang N, Zhang MH, Wu M, Fei J, Guo LH. Enhancement of germ cell apoptosis induced by ethanol in transgenic mice overexpressing Fas Ligand. *Cell Res* 2003;**13**(5):361–7.
80. Eid NA, Shibata MA, Ito Y, Kusakabe K, Hammad H, Otsuki Y. Involvement of Fas system and active caspases in apoptotic signalling in testicular germ cells of ethanol-treated rats. *Int J Androl* 2002;**25**(3):159–67.
81. Koh PO. Ethanol exposure suppresses survival kinases activation in adult rat testes. *J Vet Med Sci* 2007;**69**(1):21–4.
82. Deng X, Deitrich RA. Ethanol Metabolism and Effects: Nitric Oxide and its Interaction. *Curr Clin Pharmacol* 2007;**2**(2):145–53.
83. Lee NPY, Cheng CY. Nitric Oxide/Nitric Oxide Synthase, Spermatogenesis, and Tight Junction Dynamics. *Biol Reprod* 2004;**70**(2):267–76.
84. Giannessi F, Giambelluca MA, Ruffoli R. The ultrastructural localization of NADPH-diaphorase enzymatic activity in the Leydig cells of mouse. Effects of ethanol. *Ital J Anat Embryol* 1998;**103**:153–65.
85. Wang Y, Newton DC, Miller TL, Teichert AM, Phillips MJ, Davidoff MS, Marsden PA. An alternative promoter of the human neuronal nitric oxide synthase gene is expressed specifically in Leydig cells. *Am J Pathol* 2002;**160**(1):369–80.
86. Herman M, Kang SS, Lee S, James P, Rivier C. Systemic administration of alcohol to adult rats inhibits leydig cell activity: Time course of effect and role of nitric oxide. *Alcohol Clin Exp Res* 2006;**30**(9):1479–91.
87. Nabil E, Yuko I, Yoshinori O. Involvement of inducible nitric oxide synthase in DNA fragmentation in various testicular germ cells of ethanol-treated rats. *J Mens Health* 2011;**8**(S1):S36–S40.
88. Pryor WA, Houk KN, Foote CS, Fukuto JM, Ignarro LJ, Squadrito GL, Davies KJA. Free radical biology and medicine: it's a gas, man! Am. *J Physiol Regul Integr Comp Physiol* 2006;**291**:R491–R511.
89. Lee HY, Naseer MI, Lee SY, Kim MO. Time-dependent effect of ethanol on GnRH and GnRH receptor mRNA expression in hypothalamus and testis of adult and pubertal rats. *Neurosci Lett* 2010;**471**(1):25–9.
90. Gavaler JS, Urso T, Van Thiel DH. Ethanol: its adverse effects upon the hypothalamic-pituitary-gonadal axis. *Subst Alcohol Actions Misuse* 1983;**4**(2–3):97–110.
91. Lorenti Garcia C, Mechilli M, Proietti De Santis L, Schinoppi A, Kobos K, Palitti F. Relationship between DNA lesions, DNA repair and chromosomal damage induced by acetaldehyde. *Mutat Res* 2009;**662**(1–2):3–9.
92. Faut M, Rodríguez de Castro C, Bietto FM, Castro JA, Castro GD. Metabolism of ethanol to acetaldehyde and increased susceptibility to oxidative stress could play a role in the ovarian tissue cell injury promoted by alcohol drinking. *Toxicol Ind Health* 2009;**25**(8):525–38.
93. Van Thiel DH, Gavaler JS, Lester R, Sherins RJ. Alcohol-Induced Ovarian Failure in the Rat. *J Clin Invest* 1978;**61**:624–32.
94. Helms CM, McCracken AD, Heichman SL, Moschak TM. Ovarian hormones and the heterogeneous receptor mechanisms mediating the discriminative stimulus effects of ethanol in female rats. *Behav Pharmacol* 2013;**24**(2):95–104.
95. Purohit V, Khalsa J, Serrano J. Mechanisms of alcohol-associated cancers: introduction and summary of the symposium. *Alcohol* 2005;**35**(3):155–60.
96. David ET, Fisher I, Moldave K. Studies on the effect of ethanol on eukaryotic protein synthesis *in vitro*. *J Biol Chem* 1983;**258**(12):7702–6.
97. Pösö H, Pösö AR. Inhibition of rat ovarian ornithine decarboxylase by ethanol *in vivo* and *in vitro*. *Biochim Biophys Acta* 1981;**658**(2):291–8.

98. Srivastava VK, Hiney JK, Dees WL. Effects of ethanol on the intraovarian insulin-like growth factor-1 system in the prepubertal rat. *Alcohol Clin Exp Res* 1999;**23**(2):293–300.
99. Srivastava VK, Hiney JK, Mattison JA, Bartke A, Dees WL. The alcohol-induced suppression of ovarian insulin-like growth factor-1 gene transcription is independent of growth hormone and its receptor. *Alcohol Clin Exp Res* 2007;**31**(5):880–6.
100. Srivastava VK, Hiney JK, Dearth RR, Dees WL. Acute effects of ethanol on steroidogenic acute regulatory protein (StAR) in the prepubertal rat ovary. *Alcohol Clin Exp Res* 2001;**25**(10):1500–5.
101. Srivastava VK, Dissen GA, Ojeda SR, Hiney JK, Pine MD, Dees WL. Effects of alcohol on intraovarian nitric oxide synthase and steroidogenic acute regulatory protein in the prepubertal female rhesus monkey. *J Stud Alcohol Drugs* 2007;**68**(2):182–91.
102. Emanuele N, LaPaglia N, Kovacs EJ, Emanuele MA. Effects of chronic ethanol (EtOH) administration on pro-inflammatory cytokines of the hypothalamic-pituitary-gonadal (HPG) axis in female rats. *Endocr Res* 2005;**31**(1):9–16.
103. Ekstrom RC, Hunzicker-Dunn M. Opposing effects of ethanol on pig ovarian adenylyl cyclase desensitized by human choriogonadotropin or isoproterenol. *Endocrinology* 1990;**127**(5):2578–86.
104. Rojas FJ, Asch RH. Opposite effects of ethanol on the activation of adenylyl cyclase in human corpus luteum membranes. *Mol Cell Endocrinol* 1985;**40**(2–3):129–36.
105. Mesiha FS. The gender of alcohol and aldehyde dehydrogenases. *Neurobehav Toxicol Teratol* 1983;**5**(2):205–10.
106. Danforth DR, Stouffer RL. 125I-luteinizing hormone (LH) binding to soluble receptors from the primate (Macaca mulatta) corpus luteum: effects of ethanol exposure. *Life Sci* 1988;**43**(7):635–42.
107. Cebral E, Lasserre A, Motta A, de Gimeno MF. Mouse oocyte quality and prostaglandin synthesis by cumulus oocyte complex after moderate chronic ethanol intake. *Prostaglandins Leukot Essent Fatty Acids* 1998;**58**(5):381–7.
108. Yamamoto H, Endo T, Kiya T, Goto T, Sagae S, Ito E, Watanabe H, Kudo R. Activation of phospholipase D by prostaglandin F2 alpha in rat luteal cells and effects of inhibitors of arachidonic acid metabolism. *Prostaglandins* 1995;**50**(4):201–11.

CHAPTER

28

Molecular Effects of Alcohol on Iron Metabolism

*Kosha Mehta, MSc, PhD**, *Sebastien Farnaud, MSc, PhD***, *Vinood B. Patel, PhD**

*Department of Biomedical Sciences, Faculty of Science & Technology, University of Westminster, London, UK

**Department of Life Sciences, University of Bedfordshire, Luton, Bedfordshire, UK

INTRODUCTION

Apart from the hematological disorders, iron deficiency anemia (IDA) and iron overload hereditary hemochromatosis, disturbances in iron metabolism have been implicated in several other diseases such as Alzheimer's, Parkinson's, diabetes, and cancer. Similarly, variations in a number of iron-related proteins have been observed in patients with the classic alcoholic liver disease (ALD). This is anticipated because the pathways of iron and alcohol metabolism share several common features. For example, both iron and alcohol are predominantly absorbed via the duodenum and carried to the liver, the central organ in their metabolism. Here, while the liver secretes the iron–hormone hepcidin to regulate systemic iron levels in the body, it also catabolizes alcohol into carbon dioxide and water, via specific liver enzymes. Moreover, both functional iron overload and alcohol-induced tissue injury, as represented by hereditary hemochromatosis and ALD, respectively, may lead to portal fibrosis followed by cirrhosis, which often terminates in hepatocellular carcinoma. In addition, both alcohol and iron are linked with inflammation and infection.

Owing to these physiological and pathological resemblances, examination of the connection between iron and alcohol metabolism is important, from a clinical perspective. Such an examination can aid in the discovery of novel measures for timely and effective diagnosis and therapeutics for ALD, particularly in a scenario when ALD is one of the major causes of mortality globally. Hence, this chapter highlights the alcohol-induced structural changes in iron-related proteins, and the quantitative alterations in the expression of iron-related genes/proteins; essentially, the molecular effects of alcohol intake on iron metabolism.

OVERVIEW OF IRON METABOLISM

Iron is an essential element required by humans for their survival and development. From 13 mg to 18 mg of heme or nonheme daily dietary iron, only 1–2 mg is absorbed through the duodenum into the circulation and is distributed to different parts of the body, to be used for various metabolic purposes.[1] The three main cell-types involved in iron metabolism are the iron-absorbing duodenal enterocytes, the iron-recycling reticuloendothelial macrophages, and the iron-storing liver hepatocytes. As shown in Figure 28.1, uptake of nonheme iron (mostly ferric iron) begins when a ferric reductase duodenal cytochrome b (Dcytb), located on the apical surface of enterocytes, reduces the poorly bioavailable ferric iron (Fe^{3+}) to ferrous iron (Fe^{2+}). The low pH of proximal duodenum, along with an acidic climate of the brush border membrane, maintains iron in the Fe^{2+} state. The presence of acidic substances in the lumen, such as vitamin C, enhances nonheme iron absorption as it supplies H^+ ions and reduces Fe^{3+} to Fe^{2+}, making it more soluble and thus bioavailable. Conversely, tannins present in tea and fruit juices, phytates present in cereals and polyphenolic compounds found in all plant products, inhibit the absorption of dietary nonheme iron.[2]

The Fe^{2+} iron in the lumen can be transported across the brush border membrane into the enterocyte via the protein divalent metal transporter-1 (DMT1). Heme iron uptake also occurs at the surface of enterocytes via heme carrier protein, and is succeeded by the action of intracellular heme oxygenase that degrades heme to produce the Fe^{2+} iron. Regardless of the type of iron consumed, once inside the enterocytes, iron enters the proposed labile iron pool (LIP) that may act as an iron reservoir for various cellular activities, while excess iron is stored in

Molecular Aspects of Alcohol and Nutrition. http://dx.doi.org/10.1016/B978-0-12-800773-0.00028-8

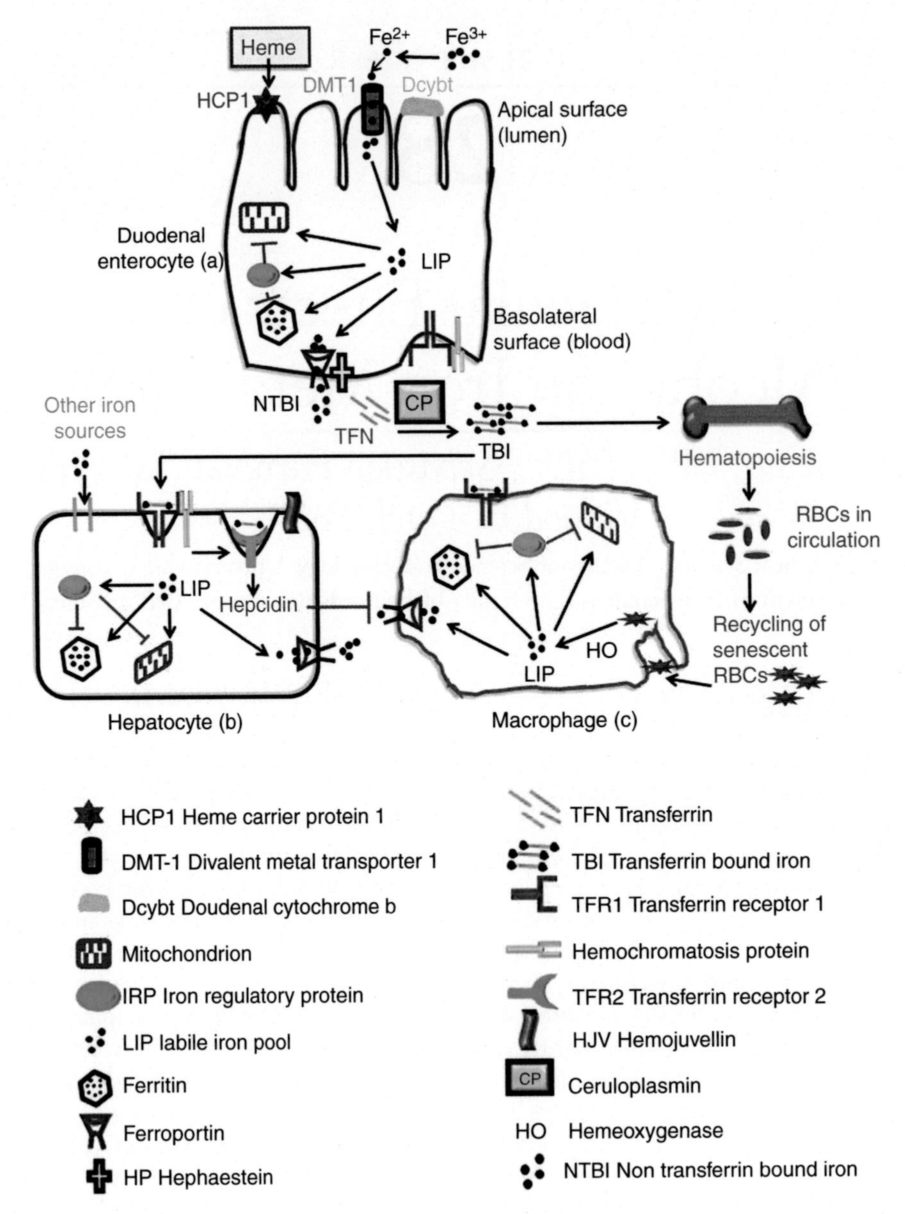

FIGURE 28.1 **Cell types involved in iron absorption and circulation.** The mechanism of iron absorption and circulation, along with the roles of participating proteins, are briefly shown. (a) Mature villus duodenal enterocytes absorb heme and nonheme iron. (b) Hepatocytes store iron as well as produce hepcidin. Here, both TBI and NTBI uptake of iron occurs. (c) Macrophages engulf senescent erythrocytes, particularly in the spleen, and recycle iron back into circulation through ferroportin. Processes in (b) and (c) are common to both heme and nonheme iron transport and utilization. *Figure adapted from Ref. [5].*

the protein ferritin that can hold up to 4500 iron atoms. After 2–3 days, when enterocytes slough from the gut into the lumen, the unabsorbed iron present in these cells is lost via feces.[1,2]

Iron is translocated outside the enterocytes and brought into the circulation through the sole known mammalian transmembrane iron-exporter, ferroportin (encoded by the gene *SLC40-A1*). This is a unidirectional ferrous exporter, and is expressed on all cell types involved in iron transport, including the basolateral surface of the enterocytes, which is in constant access with the circulation.[3] During the exit of iron from the enterocyte, ferroportin is assisted by the ferroxidase hephaestin (encoded by the gene *HEPH*), which converts Fe^{2+} to Fe^{3+}, whereas the ferroxidase ceruloplasmin (encoded by the gene *CP*) assists in the loading of Fe^{3+} onto transferrin. Transferrin, the iron carrier protein in circulation, transports iron throughout the body and binds to the transferrin receptor-1 (TFR1) (encoded by the gene *TFRC*), present on the surface of all cells involved in iron transport. The complex of diferric transferrin-TFR1 is internalized into a vesicle, and taken up within the cell. The low pH of vesicle and intracellular DMT1 on the vesicle surface assist in the release of iron from this complex into the cell cytoplasm.[4] Once the iron is released into the cytoplasm, the complex without iron is then recycled back to the cell surface for further iron uptake. Each TFR1 can undertake approximately 100 such recycling processes in its lifetime, and thereby function to bring iron into the cells. One of the ways, in which the cells maintain intracellular iron levels is by changing the expression of TFR1 on the cell surface, depending on cellular iron-requirement.[4,5] Unlike TFR1, transferrin receptor-2 (TFR2) is expressed mainly on hepatocytes. It has been proposed that TFR2 binds to HFE, the hemochromatosis protein, and may act as a sensor of diferric transferrin in the circulation.[6,7]

Following absorption, the majority of the iron is utilized in the process of erythropoiesis. Approximately 2 million red blood cells (RBCs) are synthesized every second in the bone marrow, which requires 25–30 mg of iron per day for heme synthesis. Where the total amount of iron in the body is 3.0–4.0 g, about 2.5 g is bound to hemoglobin. In addition, iron is utilized by various cell types in several pathways, for example, in the synthesis of myoglobin, an oxygen binding protein found in the muscle tissues of vertebrates, aconitase, which converts citrate to isocitrate in the citric acid cycle, and the membrane bound cytochromes, which conduct redox reactions in electron transport chain in the mitochondria. Furthermore, iron acts as a cofactor for ribonucleotide reductase, which converts ribonucleotides to deoxyribonucleotides, which are eventually used in DNA synthesis.[1,2,5]

Systemic and Cellular Regulation of Iron

Free iron can easily accept and donate electrons, and catalyze the Fenton reaction to generate hydroxyl radicals. These reactive oxygen species (ROS) can severely damage cells and tissues, and both iron and hydrogen peroxide, which participate in the Fenton reaction are capable of oxidizing a wide variety of biological substrates.[8] Under physiological conditions, ROS, such as superoxide anions and hydroxyl radicals, are generated during normal mitochondrial respiration. These contribute to the pathology of cellular damage, and also in the redox signaling to other cell organelles.[9] The resultant oxidative stress is usually combated by cellular antioxidant enzymes like superoxide dismutases and glutathione peroxidases, which catalyze the conversions of superoxide radicals into oxygen and hydrogen peroxide, and the conversion of free hydrogen peroxide to water, respectively.[10] The iron-sequestering proteins such as transferrin, ceruloplasmin and ferritin also provide protection against oxidative stress. These bind to free iron and thereby limit its availability to catalyze the Fenton reaction, and enhance oxidative stress. However, when iron levels exceed beyond the iron-sequestering capacity of the iron-binding proteins, it ultimately renders the inherent antioxidant protection systems insufficient in handling the excess free iron, thereby causing cellular damage that is beyond repair. The damage is particularly accelerated due to the reduction in the levels of cellular antioxidants, such as superoxide dismutase and vitamins A, C, and E under iron overloaded conditions, which normally help to scavenge free radicals.[11,12] Moreover, there is no physiological pathway for the excretion of excess iron, except by blood loss and the natural means of minimal iron loss from the body, occurring as a result of sloughing of mucosal cells and/or menstruation.[1] Thus, although iron is essential for survival, maintaining iron homeostasis is crucial because excess free iron is extremely toxic.

Accordingly, iron is regulated at the systemic as well as cellular level. Systemic iron regulation is mediated by the 25-mer bioactive hepcidin (2.7 kDa), the most predominant isoform of hepcidin that circulates in the human blood,[13] at a concentration of 1.1-55 ng/mL[14] (Figure 28.2). It is produced from an 84-mer preprohormone, the preprohepcidin that is cleaved by signal peptidases to produce the 60-mer prohepcidin, followed by the action of furin-like convertases that cleave prohepcidin to yield the 25-mer bioactive hepcidin.

Typically, hepcidin expression is suppressed by hypoxia and IDA and elevated by lipopolysaccharide (LPS), inflammatory stimuli such as interleukin (IL)1 and IL6, and high iron levels.[15,16] Increased systemic iron levels lead to an increase in hepcidin levels in the

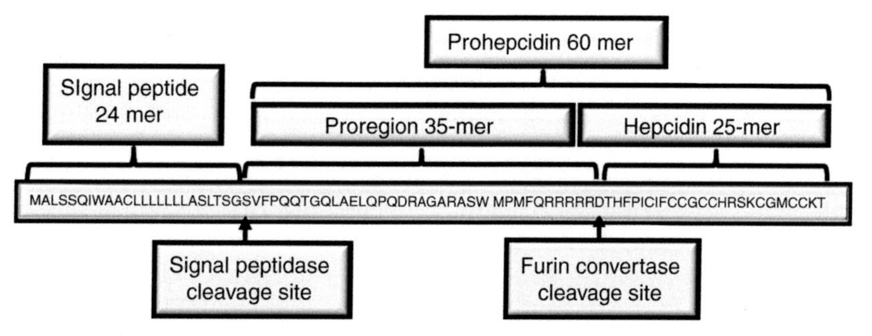

FIGURE 28.2 **Significant sites of cleavage in preprohepcidin (84-mer).** Sites of action of signal peptidase and furin convertase are shown in the figure. Cleavages eventually lead to the bioactive hepcidin-25.

circulation and in the urine, where it offers antibacterial protection.[16,17] Circulatory hepcidin binds to ferroportin, its receptor protein on cell-surfaces, to form a hepcidin–ferroportin complex, which is internalized via endocytosis, followed by proteolytic degradation of both hepcidin and ferroportin in the lysosomes.[18] Since ferroportin functions as an iron exporter, its degradation inhibits iron release from enterocytes, macrophages, and placental cells into the circulation, thus preventing further systemic iron elevation. In contrast, systemic iron deficiency leads to a decrease in hepcidin production,[16] and a resultant increase in iron efflux from cells, which eventually raise systemic iron levels. Thus, hepcidin plays a pivotal role in regulating systemic iron levels in the body by inhibiting duodenal iron absorption, and controlling the release and recycling of iron by macrophages, and iron mobilization by hepatocytes.

Hepcidin (encoded by the gene *HAMP*) transcription, and subsequently systemic iron regulation, is affected by the bone morphogenetic proteins (BMP)-SMAD, Janus kinase/signal transducer and activator of transcription-3 (JAK-STAT) and hypoxia-inducible factor (HIF) pathways, as summarized in Figure 28.3. Briefly, intracellular iron stores influence the binding of BMPs (particularly BMP6) to their receptors on the cell surface, and the membrane-bound hemojuvelin (mHJV) acts as a BMP co-receptor to assist in this process. This leads to phosphorylation and activation of the intracellular SMAD proteins, which transmit a signal to the nucleus to increase *HAMP* transcription in the cells.[19] Also, during inflammation, IL6 induces *HAMP* transcription via the JAK/STAT pathway[19] (Figure 28.3a). When iron stores are low, mHJV is cleaved by furin and/or matripase 2 (a transmembrane protease encoded by the gene *TMPRSS6*, expressed predominantly in liver) to form soluble HJV (sHJV). The sHJV disrupts the BMP-mediated activation and, instead, downregulates *HAMP* transcription[20] (Figure 28.3b). Apart from these pathways, *HAMP* transcription is also regulated by the kinase/mitogen activated protein kinase pathway (ERK/MAPK),[21] the growth differentiation factor (GDF) 15,[22] and twisted grastrulation (TWSG1).[23]

Cellular iron homeostasis is mediated by binding of the two iron regulatory proteins (IRP), IRP1 and IRP2, to iron responsive elements (IREs) on the mRNA of some iron-regulated genes. Binding of IRP1 to IRE in 5′ untranslated regions (UTRs) of transcripts prevents protein synthesis, whereas binding of IRP1 to IREs in the 3′ region provides stability to mRNA, and allows translation into proteins. The binding of the IRPs to the IREs is regulated by the presence of iron (Figure 28.4).

When cellular iron levels are sufficient or in excess, an iron–sulfur cubane (4Fe–4S) is formed, preventing the binding of IRP1 to the IRE in 5′ UTR of ferritin and ferroportin transcripts. This permits translation into proteins that favor more iron storage within cells, and more iron efflux, respectively, and prevent cellular iron overload. In contrast, under cellular iron depletion, IRP1 is iron-free and can therefore bind to the IREs in 3′ UTR on transcripts of genes encoding TFR1 and intestinal DMT1. In doing so, it stabilizes the mRNAs, increases the expression of both proteins, and thereby facilitates increased cellular iron uptake to eliminate cellular iron deprivation. Thus, the IRP–IRE network maintains cellular iron levels, and is not limited exclusively to these exemplified transcripts.[5]

Iron Homeostasis and Disorders

The significance of body iron homeostasis can be realized by reflecting on the diseases and conditions, in which iron homeostasis is disturbed. Iron deficiencies can be broadly classified into absolute iron deficiency, and functional iron deficiency. The former, also referred to as IDA, is characterized by depleted iron stores and low circulating iron levels that limit the availability of iron for RBC synthesis. IDA can occur either due to excessive blood loss, or pregnancy, when there is increased demand for iron, or malnutrition, when there is decreased supply of iron. Approximately one third of the world population is believed to be suffering from IDA, which could manifest in a wide variety of pathophysiological effects, including the developmental retardation in children. Currently, the most commonly prescribed treatment to alleviate IDA is to increase body iron levels by using approximately 200 mg of ferrous salts per day. However, the unabsorbed iron salts cause nausea, abdominal pain, and constipation. To combat these symptoms, low doses of iron salts between 50 mg and 100 mg of iron have been proposed.[24] On the other hand, functional iron deficiency is more complex, as seen during infection/inflammation and cancer, and is referred to as anemia of inflammation, or anemia of chronic disease. This is typically characterized by low serum–iron levels, despite replete iron stores due to iron-restricted erythropoiesis.[5,24]

Iron overload could be acquired or functional. Some examples of acquired iron overload are the iron overdose in children, which is the commonest poisoning in children, and iron overload as a result of repeated blood transfusions.[24] Bantu siderosis is yet another form of acquired iron overload in some people of African descent that occurs due to the consumption of brewed beer made in large iron vessels.[25] Functional iron overload, referred to as hereditary hemochromatosis, is the most frequent genetic disorder in Caucasians, and results from defects in single or multiple genes of iron metabolism, such as *HFE*, *HJV*, *HAMP*, *TFR2*, *SLC40-A1*, *TF*, and *CP*,[26] encoding the proteins HFE, HJV, hepcidin, TFR2, ferroportin, transferrin, and ceruloplasmin, respectively. Under hemochromatosis, intestinal iron absorption can exceed iron loss by approximately 3 mg/day,[26] attaining total

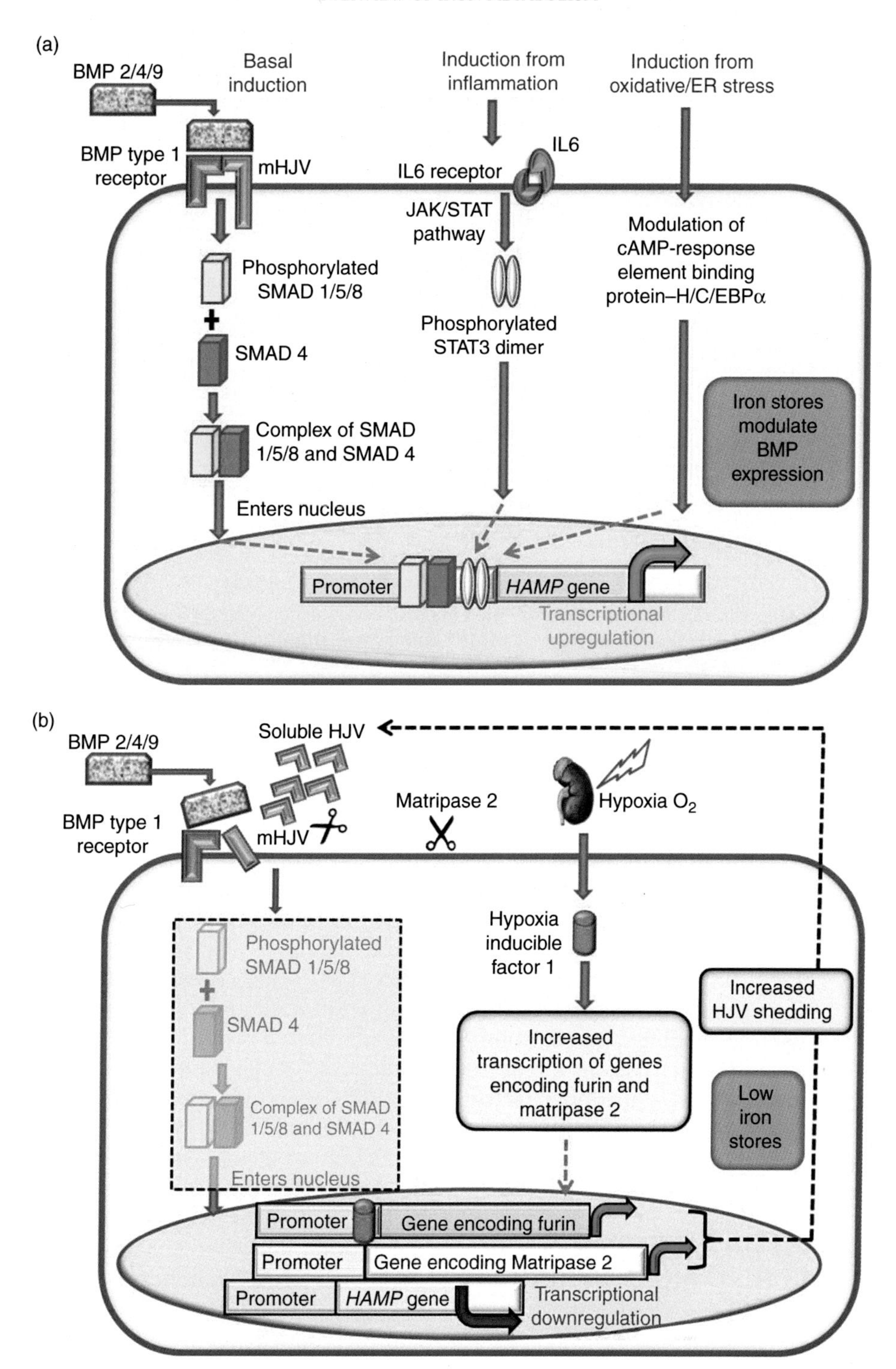

FIGURE 28.3 (a) Pathways leading to transcriptional upregulation of hepcidin. The physiological mechanisms that lead to *HAMP* transcription are shown. The complex of SMAD proteins enter the nucleus, bind to the promoter region of hepcidin gene and stimulate *HAMP* transcription. Also, the STAT3 dimer formed as a result of IL6 binding to its receptor, led to the upregulation of hepcidin. High liver iron stores promote hepcidin synthesis via the BMP pathway. (b) Pathways leading to transcriptional downregulation of hepcidin. The mechanisms that lead to *HAMP* downregulation are shown. Both furin and matripase 2 are able to cleave mHJV to soluble HJV. The soluble HJV does not allow a stable complex to be formed between BMP ligands and receptors, thus inhibiting *HAMP* transcription via the BMP pathway. Hypoxia inducible factor increases the transcription of genes that would cleave the mHJV to soluble HJV. *Adapted from Ref. [27].*

body iron levels of up to 50 g, as compared to the typical 3–4 g. The effect of this can be tissue specific, for example, liver possesses high levels of antioxidants and cytoprotective enzymes, and therefore requires a substantially high level of liver iron overload for observable toxic effects of iron. Alternatively, in the reticuloendothelial cells, for example, macrophages, only about a two- to threefold increase in cellular iron would lead to

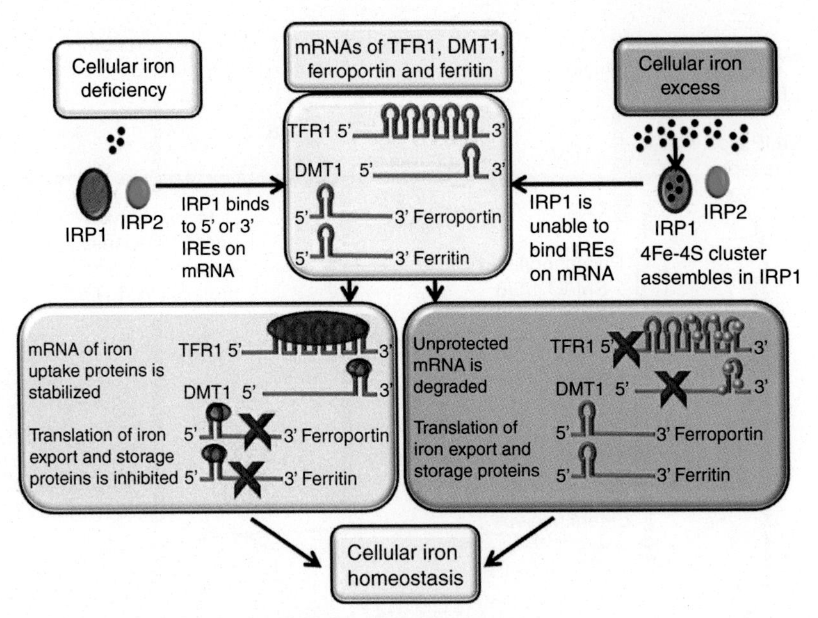

FIGURE 28.4 **Cellular regulation of iron, by iron regulatory proteins (IRPs).** Regulation of intracellular iron levels by IRPs has been shown. Binding of IRPs to IREs in 5′ UTR and 3′ UTR lead to different effects. IRP binding to 5′ IREs result in the inhibition of protein production, whereas binding to 3′ IREs favor protein production.

changes in normal cell functions.[24,28] Accordingly, where hemochromatotic patients show a 10- to 20-fold increase in liver iron stores, before clinical manifestations are observed, patients with Parkinson's disease show only a twofold increase in iron content in substantia *nigra pars compacta* (i.e., a section of the brain),[29] thus demonstrating the tissue specific effect of iron.

In essence, normally, the systemic and cellular iron homeostatic pathways operate in amalgamation to minimize the availability of free iron for the Fenton reaction and, thus, control excess iron-mediated tissue injury. Since alcohol intake is an additional reason, the aforementioned genetic, epigenetic, and nutritional factors that cause digression from the normal iron regulation are not the only reasons for disruption in iron homeostasis.

ALCOHOL AND IRON: CLINICAL OBSERVATIONS AND MOLECULAR EVENTS

Under physiological conditions, alcohol (ethanol being the most commonly consumed alcohol) is eliminated from the body, predominantly by metabolism and minimal (2–8%) by excretion through urine, breath, and sweat. Alcohol uptake occurs in the small intestine by simple diffusion and, via the portal vein, it is carried to the liver where it is catabolized mainly through the oxidative pathway. A series of liver enzymes, namely, alcohol dehydrogenase, cytochrome P450 (CYP2E1), acetaldehyde dehydrogenase, and catalase, collectively degrade alcohol/acetaldehyde into carbon dioxide and water, through the Kreb's cycle.[30] In concomitance with this degradation, several deleterious metabolites, such as acetaldehyde, acetate, and large amounts of ROS and NADH, are generated. Each moiety is capable of deviating the cell from its typical metabolism, directly or indirectly. For example, acetaldehyde is a toxic intermediate, which not only forms protein adducts and disrupts the native functions of several proteins, but also downregulates *HAMP* mRNA expression.[31] Acetate, the oxidative product of acetaldehyde, affects several metabolic pathways,[32] while ROS and NADH alter the cells' redox state and cause cellular damage.

Such molecular and biochemical alternations from the norm, lead to disturbances in iron metabolism and contribute to the alcohol-induced pathophysiology. Particularly in chronic alcohol consumption, evidence indicates that alcohol and iron work together to amplify deleterious effects. Even moderate alcohol consumers show increased hepatic-iron stores, serum transferrin–iron saturation, and ferritin levels, which far exceeds in chronic consumers.[33,34] This high circulating and liver iron content potentiates free radical-mediated toxicity, and exacerbates the pathological progression toward the development of ALD within 10 years, at least in approximately one-third of chronic alcoholics. ALD encompasses of all the hepatic manifestations resulting from alcohol ingestion, beginning from the stage of reversible hepatic steatosis (fatty liver), through hepatitis (liver inflammation), followed by fibrosis, and then cirrhosis, an irreversible stage of severe liver injury.

Although the pH of most alcoholic drinks lies between 2 and 5, and such an acidic pH would favor duodenal iron uptake, the most probable cause of elevated iron levels in alcoholics is the decline in serum hepcidin and prohepcidin levels.[34,35] This was also observed in ethanol-fed mice and rats, and other animal experiments.[36,37] Reduction in hepcidin expression is seen in acute as well as chronic alcohol consumption and could be due to the modulation of one or more of its regulatory pathways, one of which is the previously discussed BMP–SMAD pathway.[19] It was observed that, in alcohol-fed mice, activity of the BMP receptor was inhibited, and the expression of BMP6 and Smad4 was reduced. This was accompanied by attenuation of the binding of Smad4 to the *Hamp* promoter region[38] the process, which normally induces *Hamp* transcription.[19] The consequential drop in *Hamp* transcription, and thereby systemic hepcidin peptide levels, would permit increased duodenal iron absorption and progressively increase systemic iron levels. Not surprisingly, a twofold increase in iron absorption has been observed in alcoholic patients.[39] In addition, it has been observed that hepcidin expression is repressed by hypoxia.[15] Therefore, the alcohol-induced hypoxia, as observed in alcohol-fed *Hfe* knockout mice,[36] may inactivate the BMP/SMAD pathway,[40] and thereby reduce hepcidin expression in alcoholics. However, such mice demonstrate reduced *Hamp* mRNA expression similar to patients with HFE-related and non-HFE-related hereditary hemochromatosis,[41] even in the absence of alcohol feeding. This is the reason why hereditary hemochromatosis, in combination with excessive alcohol consumption, can extensively aggravate liver injury.[42]

The *HAMP* promoter region is strongly activated by the transcription factor CCAAT/enhancer binding protein C/EBPα.[43] Ethanol reduces *HAMP* promoter activity, lowers the levels of C/EBPα protein, and thus decreases the DNA binding action of C/EBPα,[31] thereby interfering with *HAMP* transcription. Since hepcidin expression is also regulated by oxidative stress,[44] the alcohol-mediated oxidative stress may further decrease hepcidin expression.[31] In this perspective, recent findings reveal that although alcohol increases superoxide generation in hepatocytes, mitochondrial superoxide is not responsible for inhibition of *HAMP* transcription.[45] This indirectly implies the involvement of cytosolic or nuclear ROS in regulating *HAMP* transcription.

As both iron and alcohol are associated with infection and inflammation, these mediate some common inflammatory pathways involving *HAMP* transcription. One of these signaling pathways is triggered by the binding of LPS to toll-like receptor (TLR) 4, expressed by various liver cells.[46] Its downstream signaling events involve translocation of the nuclear transcription factor NF-κB into the nucleus, to produce inflammatory cytokines like IL6, which in turn activates another transcription factor, STAT3, that eventually activates *HAMP* transcription.[47] Besides LPS, alcohol also activates this TLR4 signaling pathway.[48] In this perspective, it is surprising to observe that although hepcidin, a type 2 acute phase protein, is normally induced by the inflammatory cytokines IL1 and IL6,[49,50] hepcidin levels in alcoholic patients are not upregulated, despite the presence of inflammation.[51] This was explored in recent investigations that showed the significance of TLR4 in regulating hepcidin expression in the presence of alcohol, via NF-κB.[52] Here, it was observed that *HAMP* mRNA expression was suppressed in alcohol-fed wild type mice as expected, but it wasn't in alcohol-fed TLR4 mutant mice.

Not only hepatocytes, but other cell types of the liver and the duodenum also contribute in alcohol-induced iron-related tissue injury. Hepatic fibrosis precedes cirrhosis, and is a dynamic process that entails inflammatory cytokines produced by Kupffer cells, along with activated collagen-producing stellate cells, and manifests in excessive extracellular matrix.[53] The stellate cells are particularly activated when liver iron exceeds 60 μmol/g and thus play a paramount role in the development of fibrosis.[54] In addition, iron overloaded hepatocytes stimulate procollagen synthesis, which is further enhanced by alcohol, and contributes in the development of hepatic fibrosis and progression toward cirrhosis.[55]

With regards to Kupffer cells, prolonged alcohol consumption increases the permeability of endotoxin through the intestine, leading to its elevation in the plasma,[56] and stimulates the Kupffer cells to release inflammatory cytokines and contribute to the pathogenesis of ALD.[57] However, with respect to hepcidin expression, its downregulation by alcohol was found to be independent of the iron loaded and activated Kupffer cells, indicating that alcohol acts exclusively upon the hepatocytes to reduce hepcidin expression and, thus, affects iron homeostasis in alcoholics.[58]

The overall iron pool is further raised by increased duodenal DMT1 and ferroportin expressions in alcoholics[36,37,59] that enhance iron uptake into the enterocytes, and iron efflux into the circulation, respectively. Moreover, *ex vivo* experiments conducted on HepG2 cells and primary rat hepatocytes showed that the oxidative stress induced by ethanol, increases the activity of IRP proteins.[60] These proteins modulate TFR1 expression on cell surface,[5] as reflected by increased cell-surface TFR1 on hepatocytes in 80% of hepatic tissues of patients with ALD,[61] and thereby promote increased hepatic iron uptake.

It is important to note that both alcohol and iron can independently cause oxidative stress and lipid peroxidation.[62] Regardless of the route of excessive iron accumulation, the resulting unquenched ROS can overwhelm the cellular antioxidant system and increase lipid peroxidation to generate metabolites like malondialdehyde and 4-hydroxynonenal (HNE).[60] Both these moieties have been detected in the liver, and in the circulation of iron overloaded rats.[63] These not only form protein adducts and denature the proteins, but also form DNA adducts leading to mutations and, thus, increased predisposition to liver cancer.[64] Interestingly, adduction of only malondialdehyde or acetaldehyde to a protein leads to the formation of an unstable protein adduct, which has a short half-life. However, the combined binding of malondialdehyde and acetaldehyde in the ratio of 2:1 with the protein, generates a stable hybrid, called the malondialdehyde–acetaldehyde protein adduct that exhibits its deleterious effects in the liver and lungs.[65] Hemoglobin, the predominant iron-harboring protein, is one among the most likely to be modified by adduct formation with aldehydes.[66] While malondialdehyde–hemoglobin adducts in RBCs have been detected in healthy adults in the range of 0.01–10 nmol/g hemoglobin,[67] a specific ELISA result implies threefold higher levels of these adducts in alcoholic patients, than in controls.[68] Moreover, these metabolites stimulate the production of antibodies and generate an immunopathological response, which further contributes to alcoholic liver injury.[69]

Strangely, despite increased iron absorption and high circulating iron levels, one third of chronic alcoholics are diagnosed with anemia.[70] This is predominantly due to the deterioration of the quantity and quality of RBC production. First, heavy alcohol consumption can have a direct toxic effect on the bone marrow and thereby suppresses hematopoiesis, preferentially of the erythroid lineage, resulting in reduced RBC production than normal. Second, the quality of RBCs also gets compromised, as alcohol hinders normal RBC development due to imperfections in mechanisms, such as the improper incorporation of hemoglobin within the RBC precursors. In these cells, iron would mainly be in the stored form of ferritin, and these ferritin-containing cells called ringed sideroblasts cannot mature into functional RBCs, the condition referred to as sideroblastic anemia. Often, alcoholics demonstrate macrocytic anemia, where the RBCs exhibit large vacuoles, and such abnormally structured RBC precursors remain underdeveloped, too. Up to 80% men and 46% women diagnosed with macrocytosis have been found to be alcoholics.[70]

Further complications arise because alcohol consumption reduces folate absorption, which is important for the proper development of RBCs.[71] Reduction in folate absorption has also been observed in combination with vitamin B12 deficiency in alcoholics.[72] Since folate is required for DNA synthesis, as well as for the methylation of DNA, RNA, and proteins,[73] folate deficiency can cause defects in DNA synthesis and prevent nuclear maturation, which leads to the accumulation of enlarged immature RBC precursors, the condition referred to as megaloblastic anemia.[70] Additionally, alcohol inhibits several enzymes of the heme biosynthesis pathway, like coproporphyrinogen oxidase and ferrochelatase,[74] and directly disrupts heme synthesis. Presence of such abnormal RBCs in circulation stimulates hemolysis in the spleen, aiming to remove the dysfunctional RBCs from the circulation. This leads to hemolytic anemia that perpetuates in enlarged spleen (splenomegaly), and overactive spleen (hypersplenism). Although rarely, iron deficiency in alcoholics can occur because of gastrointestinal bleeding, or internal bleeding due to extensive portal fibrosis.[70]

As summarized in Figure 28.5, alcohol-induced reduction in hepcidin expression promotes increased duodenal iron absorption and increased cellular iron efflux into circulation, from iron storing macrophages and hepatocytes. This is accompanied with variations in other iron-related proteins that permit excessive iron accumulation, particularly in the hepatocytes. This excess iron, together with alcohol-induced hypoxia, hepatic lipid peroxidation, and unquenched ROS, exert a synergistic action and lead to tissue injury. Even slightly raised tissue iron levels intensify the toxic effect of alcohol and its metabolites. The availability of free iron may either contribute directly in hepatic injury, or may promote the production of cytokine mediators from Kupffer cells, or may be involved in both ways to accelerate the process of liver degeneration;[75] the latter is more likely.

EXPLORING THE IRON-ALCOHOL LINK FOR DIAGNOSIS AND THERAPEUTICS

In adults, chronic alcohol consumption can result in a myriad of detrimental effects, including adversely affected cardiac output, malfunctioned pancreatic and neuronal functionality, deregulated retinol and homocysteine metabolisms, ketosis, hyperlipidemia, hyperuricemia, lactic acidosis,[30] and increased predispositions to oral[76] and hepatocellular carcinomas.[77] Prenatal exposure to alcohol can deeply affect fetal development and cause atypical mental and physical characteristics, widely referred to as the fetal alcohol syndrome. It can cause oxidative stress and alter somatic growth, lead to facial disfigurations, and inhibit normal neuronal and liver development.[78] Worldwide, 2.5 million people die every year due to alcohol-induced conditions. In the United Kingdom, alcohol is one of the most prevalent causes of death, along with smoking and high blood pressure. "Statistics on Alcohol and illness," published in 2013, indicated that the most common reason for alcohol-related deaths in the United Kingdom was ALD, accounting for 64% of all alcohol-related deaths in 2011, and cost approximately £3.5 billion per year to the National Health Service. In the United States, there are approximately 88,000 alcohol-related deaths every year, as per the latest study published by the "Centers for Disease Control and Prevention" in 2014. Therefore, timely diagnosis of

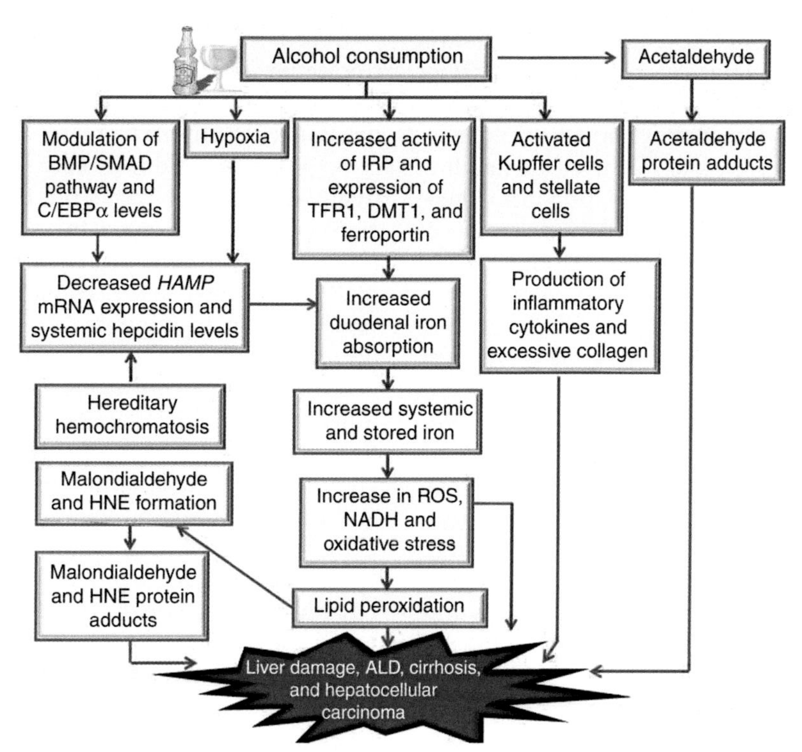

FIGURE 28.5 **Molecular and biochemical alterations following alcohol consumption.** Variations in the expressions of several iron-related genes/proteins following alcohol consumption have been summarized. These alterations mediate biochemical alterations, which cause liver damage, and can lead to cirrhosis as well as hepatocellular carcinoma.

alcohol abuse is important in both clinical and forensic settings.

Presently, several biomarkers of alcoholism and liver injury are used, among which the most common ones are elevated levels of serum gamma glutamyl transferase (GTT), aspartate aminotransferase, alanine aminotransferase (ALT), urinary or hair ethyl glucuronide, and increased mean corpuscular volume (MCV). However, many of these markers are not exclusively alcohol-specific, for example, apart from alcohol consumption, the levels of GTT and ALT increase even in case of increased fat deposition in the liver, due to obesity or diabetes.[79] Assessment of MCV is also not a very specific biomarker of alcohol abuse because enlarged RBCs are detected even in deficiencies of vitamins B_{12} and folate,[80] and in chronic conditions including congenital anemias. Likewise, testing for only ethyl glucuronide in urine samples could be unreliable as glucuronidation can occur in the presence of ethanol, as well as urinary tract infections.[81] Hence, involvement of iron in ALD encourages the exploration of iron-related biomarkers that can be used independently, or in combination with the regular liver function tests for early and more specific detection of alcohol-induced iron overload and liver damage.

Transferrin is one such valuable biomolecule, whose quantitative and structural variations can be indicative of high iron levels, as well alcohol abuse. It is a bilobed glycoprotein, where each lobe can accommodate up to one Fe^{3+} ion and exist in the circulation as monoferric and diferric forms.[82] Therefore, its saturation by iron is often used in the diagnostic laboratory to determine iron status of individuals.[83] Similarly, in order to detect alcohol abuse, a structural property of transferrin has been exploited with respect to its glycosylation status, that is, the presence of sialic acid chains. Transferrin in the human serum exists in several sialylated forms, ranging from monosialo- to octasialo-transferrin, mostly being tetrasialo-transferrin.[84] In earlier studies on chronic alcohol drinkers, serum transferrin with reduced sialic acid content was detected where the proportion of transferrin, with either no sialic acid chains (asialo) or monosialo- and disilo-transferrin, was high.[85] Since then, this carbohydrate-deficient transferrin (CDT) has been used as a biomarker for alcohol abuse, and has now been approved by the United States Food and Drug Administration.[86] Even in the United Kingdom, the Driver and Vehicle Licensing Agency (DVLA) now uses CDT as a unique biomarker to assess alcohol abuse for licensing purposes, and prevent drink–driving, as per the "Secretary of State's Honorary Medical Advisory Panel."

The main advantage of CDT is that, unlike GTT, it reflects alcohol consumption while being independent of alcohol-related or unrelated liver disease.[87] Also, it reports least false positives, compared to other biomarkers, as per the Road Safety Research Report no. 104, (2010). Accordingly, the latest research proposes to test for combination of the biomarkers CDT and hair ethyl glucuronide to obtain optimal information on alcohol usage.[88] In addition to the variation in glycosylation of transferrin, the levels of transferrin and ferritin are elevated in alcoholic fatty liver, and in chronic alcoholics, respectively.[89] This indicates high circulating iron levels, a feature that could be exploited for early detection of the initial stages of ALD. Thus, CDT is a promising and useful "state marker" that reflects a person's alcohol consumption and allows the identification of heavy drinkers, even in the absence of alcohol in the blood. Hence, it can be used to detect relapses in recovering alcoholics, encouraging early intervention and help in alcohol withdrawal.[70]

Both hereditary hemochromatosis and ALD are characterized by high circulating and stored iron levels, accompanied with low hepcidin expression.[90,91] Therefore, the therapies (e.g., phlebotomy) used for hemochromatosis, could also be implemented for ALD. This would lower the body iron content and thereby reduce the risk of free iron mediated tissue injury, or at least decelerate the pathological progression of ALD. Iron-chelating drugs like deferoxamine, deferriprone, and deferasirox are used in patients with iron overload, each presenting their respective advantages and limitations. For example, deferriprone has a better half-life than deferoxamine, it is excreted in urine, and is therefore preferred over deferoxamine. However, caution needs to be executed when using such drugs, due to their inadequate and/or variable chelation, and clinically undetermined lowest safety limits.[24,92] Following further clinical trials, such drugs could be attempted to control the pathological progression in ALD patients, at least in the current scenario when there is no medical treatment available for cirrhosis patients, except for liver transplantation, while they completely abstain from alcohol pre- and posttransplantation.

At the therapeutic forefront, potential routes for evading the pathogenesis of ALD by manipulating iron metabolism in alcoholics have been envisaged. Previously, quercetin, a flavonoid, was shown to efficiently prevent liver injury in mice with iron overload.[93] Also, recently, quercetin has shown promising results in preventing alcohol-induced iron overload in mice by regulating *HAMP* transcription via modulation of the BMP6/SMAD4 pathway.[94] Here, quercetin was shown to alleviate the alcohol-induced suppression of the BMP6/SMAD pathway and consequently promote *HAMP* transcription, thus providing a valuable breakthrough for iron regulation, apart from iron chelation. In an *in vitro* study with the liver-derived HepG2 cells, 4-methylpyrozole, a specific inhibitor of the alcohol-metabolizing enzymes namely alcohol dehydrogenase and cytochrome P450 2E1, was shown to prevent alcohol-induced downregulation of *HAMP* mRNA expression.[31]

In order to regulate iron levels, the effect of direct introduction of hepcidin peptide into the circulation has shown promising results. For example, a strategy has been successfully implemented in rats, where a dosage of ectogenic hepcidin protected rat liver from alcoholic liver damage.[95] In another set of studies, injection of hepcidin peptide prevented the alcohol-induced upregulation of the duodenal proteins DMT-1 and ferroportin in mice,[31] and also decreased ferroportin expression[96] in rats, in order to minimize iron uptake and absorption into the circulation. Equally, a hepcidin-mediated reduction in DMT-1 expression was observed in the human intestinal epithelial cell line, the Caco-2 cells.[97] In a particular case of cirrhosis with alcohol-mediated spur cell anemia, liver transplantation elevated hepcidin levels, and decreased ferritin levels. Such results propose hepcidin replacement as a therapy in these patients.[98]

Vitamins also appear to play a crucial role in regulating ethanol-induced iron overload by modulating iron-related genes/proteins. For example, vitamin C supplementation in alcohol-fed mice improved liver iron content by elevating *HAMP* expression and decreasing *TFRC* expression in the liver, while also reducing ferroportin expression in the intestine.[99] This would simultaneously regulate intestinal iron absorption, iron entry into the circulation, and thereby act as a powerful therapeutic agent to control liver damage. Likewise, mice fed with a vitamin E-rich diet not only prevented the alcohol-induced reduction in *HAMP* expression, but also prevented the upregulation of duodenal DMT-1 protein expression.[31]

Another antioxidant, epigallocatechin gallate (EGCG), found in plenty in green tea, is also a promising natural biomolecule in controlling alcohol-induced iron overload. In mice models of ALD, EGCG caused a drop in serum and hepatic iron content and in hepatic malondialdehyde levels. Moreover, an increase in *Hamp* expression, together with decreased TFR1 protein levels in liver, were observed.[100] This is probably due to the dual functionality of EGCG as an iron-chelator and an antioxidant that would scavenge free radicals and prevent iron-mediated tissue injury.[101] It is ironic that while more than two alcoholic drinks per day increased transferrin–iron saturation and serum ferritin levels, up to two alcoholic drinks per day showed to reduce the risk of IDA without the risk of developing iron overload.[33] However, this does not suggest alcohol consumption as a therapy to alleviate IDA, because alcohol has other deleterious effects, including the different types of noncongenital anemias, as previously discussed. Also, such a therapeutic approach would be controversial and impractical.

Although these diagnostic and therapeutic approaches are promising, many cellular and molecular events remain to be investigated. For instance, although hepcidin expression is induced by inflammation,[15] it is downregulated in alcoholics despite the release of inflammatory cytokines, the reasons for which are yet to be fully understood. Also, it is interesting to note that in hereditary hemochromatosis, insufficient hepcidin expression, despite the iron overload, is due to mutations in one or multiple iron related-genes.[26] However, in ALD, hepcidin expression is decreased, despite the absence of mutations in iron–related genes, and presence of iron overload and inflammation; the factors, which normally induce hepcidin expression. This implies that the regulation of hepcidin in alcoholics is not solely determined by these factors, but it is far more intricate. Additionally, identification and the role of iron-related protein adducts in the development of ALD remains be evaluated.

CONCLUSIONS

Key Facts

Hemochromatosis

- Hemochromatosis is a very common iron overload disease.
- It could be inherited (primary) or acquired (secondary).
- Mutations in the *HFE* gene accounts for 90% of hereditary hemochromatosis.
- The most common mutations in the HFE protein are C282Y, where cytosine is replaced by tyrosine at position 282, and H63D, where histidine is replaced by aspartic acid at position 63.
- Functional HFE protein regulates cellular iron uptake, whereas mutations cause impairment in its regulatory function, and leads to iron overload.

Summary Points

- This chapter focuses on the link between alcohol and iron metabolism.
- Alcohol consumption alters the expression of several iron-related proteins, such as hepcidin, ferroportin, DMT1, TFR1, IRP, transferrin, and ferritin that collectively manifests in high systemic and liver iron levels.
- Contrarily, alcoholics can often be anemic due to the alterations in hematopoiesis that result in reduced and/or abnormal RBCs.
- Thus, alcohol consumption spans both the extremes of iron–related disorders, iron overload, and anemia, and invariably affects the pathways of iron metabolism.
- Since free radical mediated toxicity, due to high iron levels, is common to iron overload, hereditary hemochromatosis, and ALD, the damaging potential of iron is amplified when it acts in conjunction with alcohol.
- The secondary knock-on effects of inflammatory mediators, released by activated Kupffer cells and stellate cells accelerate liver degeneration, and lead to the progression of ALD.
- Understanding the interconnection between alcohol and iron metabolism can aid in the identification of biomarkers for devising better diagnostic and therapeutic strategies to ameliorate liver degeneration caused by alcohol-induced iron overload before it reaches the irreversible stage of liver cirrhosis, if not completely abrogate it.

Definitions of Words and Terms

Cirrhosis An end-stage liver disease characterized by scarring of liver tissue and inflammation that causes loss of the normal functions of the liver.

Erythropoiesis The process of synthesis of red blood cells.

Fibrosis An advanced stage liver disease characterized by excessive collagen production that leads to impairment in normal functions of the liver.

Iron-sequestering proteins Proteins that bind to iron.

Preprohormone A precursor peptide for a hormone that is cleaved by signal peptidases, and other peptidases, to remove a section of amino acids, and yield a shorter biologically active peptide hormone.

Abbreviations

ALD	Alcoholic liver disease
BMP	Bone morphogenetic protein
CDT	Carbohydrate-deficient transferrin
CP	Ceruloplasmin
HFE	High iron
HJV	Hemojuvelin
IDA	Iron deficiency anemia
IRE	Iron responsive element
IRP	Iron regulatory protein
ROS	Reactive oxygen species
TF	Transferrin
TFR	Transferrin receptor

References

1. Miret S, Simpson RJ, McKie AT. Physiology and molecular biology of dietary iron absorption. *Annu Rev Nutr* 2003;**23**:283–301.
2. Sharp PSK. Molecular mechanisms involved in intestinal iron absorption. *World J Gastroenterol* 2007;**13**(35):4716–24.
3. McKie AT, Marciani P, Rolfs A, et al. A novel duodenal iron-regulated transporter, IREG1, implicated in the basolateral transfer of iron to the circulation. *Mol Cell* 2000;**5**(2):299–309.
4. Aisen P. Transferrin receptor 1. *Int J Biochem Cell Biol* 2004;**36**(11):2137–43.
5. Muckenthaler MU, Galy B, Hentze MW. Systemic iron homeostasis and the iron-responsive element/iron-regulatory protein (IRE/IRP) regulatory network. *Annu Rev Nutr* 2008;**28**:197–213.
6. Fleming RE. Iron sensing as a partnership: HFE and transferrin receptor 2. *Cell Metab* 2009;**9**(3):211–2.
7. Rapisarda C, Puppi J, Hughes RD, et al. Transferrin receptor 2 is crucial for iron sensing in human hepatocytes. *Am J Physiol Gastrointest Liver Physiol* 2010;**299**(3):G778–83.
8. McCord JM. Iron, free radicals, and oxidative injury. *J Nutr* 2004;**134**(11):3171S–2S.
9. Venditti P, Di Stefano L, Di Meo S. Mitochondrial metabolism of reactive oxygen species. *Mitochondrion* 2013;**13**(2):71–82.

10. Pigeolet E, Corbisier P, Houbion A, et al. Glutathione peroxidase, superoxide dismutase, and catalase inactivation by peroxides and oxygen derived free radicals. *Mech Ageing Dev* 1990;**51**(3):283–97.
11. Brown KE, Kinter MT, Oberley TD, et al. Enhanced gamma-glutamyl transpeptidase expression and selective loss of CuZn superoxide dismutase in hepatic iron overload. *Free Radic Biol Med* 1998;**24**(4):545–55.
12. Bagchi D, Garg A, Krohn RL, Bagchi M, Tran MX, Stohs SJ. Oxygen free radical scavenging abilities of vitamins C and E, and a grape seed proanthocyanidin extract *in vitro*. *Res Commun Mol Pathol Pharmacol* 1997;**95**(2):179–89.
13. Krause A, Neitz S, Magert HJ, et al. LEAP-1, a novel highly disulfide-bonded human peptide, exhibits antimicrobial activity. *FEBS Lett* 2000;**480**(2–3):147–50.
14. Busbridge M, Griffiths C, Ashby D, et al. Development of a novel immunoassay for the iron regulatory peptide hepcidin. *Br J Biomed Sci* 2009;**66**(3):150–7.
15. Nicolas G, Chauvet C, Viatte L, et al. The gene encoding the iron regulatory peptide hepcidin is regulated by anemia, hypoxia, and inflammation. *J Clin Invest* 2002;**110**(7):1037–44.
16. Pigeon C, Ilyin G, Courselaud B, et al. A new mouse liver-specific gene, encoding a protein homologous to human antimicrobial peptide hepcidin, is overexpressed during iron overload. *J Biol Chem* 2001;**276**(11):7811–9.
17. Park CH, Valore EV, Waring AJ, Ganz T. Hepcidin, a urinary antimicrobial peptide synthesized in the liver. *J Biol Chem* 2001;**276**(11):7806–10.
18. Preza GC, Pinon R, Ganz T, Nemeth E. Cellular catabolism of the iron-regulatory peptide hormone hepcidin. *PLoS One* 2013;**8**(3):e58934.
19. Babitt JL, Huang FW, Wrighting DM, et al. Bone morphogenetic protein signaling by hemojuvelin regulates hepcidin expression. *Nat Genet* 2006;**38**(5):531–9.
20. Finberg KE, Whittlesey RL, Fleming MD, Andrews NC. Down-regulation of Bmp/Smad signaling by Tmprss6 is required for maintenance of systemic iron homeostasis. *Blood* 2010;**115**(18):3817–26.
21. Poli M, Luscieti S, Gandini V, et al. Transferrin receptor 2 and HFE regulate furin expression via mitogen-activated protein kinase/extracellular signal-regulated kinase (MAPK/Erk) signaling implications for transferrin-dependent hepcidin regulation. *Haematologica* 2010;**95**(11):1832–40.
22. Tanno T, Bhanu NV, Oneal PA, et al. High levels of GDF15 in thalassemia suppress expression of the iron regulatory protein hepcidin. *Nat Med* 2007;**13**(9):1096–101.
23. Tanno T, Porayette P, Sripichai O, et al. Identification of TWSG1 as a second novel erythroid regulator of hepcidin expression in murine and human cells. *Blood* 2009;**114**(1):181–6.
24. Munoz M, Garcia-Erce JA, Remacha AF. Disorders of iron metabolism. Part II: iron deficiency and iron overload. *J Clin Pathol* 2011;**64**(4):287–96.
25. Kew MC, Asare GA. Dietary iron overload in the African and hepatocellular carcinoma. *Liver Int* 2007;**27**(6):735–41.
26. Pietrangelo A. Hereditary hemochromatosis. *Biochim Biophys Acta* 2006;**1763**(7):700–10.
27. Andrews NC, Schmidt PJ. Iron Homeostasis. *Ann Rev of Physiol* 2007;**69**:69–85.
28. Crichton RR, Wilmet S, Legssyer R, Ward RJ. Molecular and cellular mechanisms of iron homeostasis and toxicity in mammalian cells. *J Inorg Biochem* 2002;**91**(1):9–18.
29. Crichton RR, Dexter DT, Ward RJ. Brain iron metabolism and its perturbation in neurological diseases. *J Neural Transm* 2011;**118**(3):301–14.
30. Zakhari S. Overview: how is alcohol metabolized by the body? *Alcohol Res Health* 2006;**29**(4):245–54.
31. Harrison-Findik DD, Schafer D, Klein E, et al. Alcohol metabolism-mediated oxidative stress down-regulates hepcidin transcription and leads to increased duodenal iron transporter expression. *J Biol Chem* 2006;**281**(32):22974–82.
32. Israel Y, Orrego H, Carmichael FJ. Acetate-mediated effects of ethanol. *Alcohol Clin Exp Res* 1994;**18**(1):144–1448.
33. Ioannou GN, Dominitz JA, Weiss NS, Heagerty PJ, Kowdley KV. The effect of alcohol consumption on the prevalence of iron overload, iron deficiency, and iron deficiency anemia. *Gastroenterol* 2004;**126**(5):1293–301.
34. Costa-Matos L, Batista P, Monteiro N, et al. Liver hepcidin mRNA expression is inappropriately low in alcoholic patients compared with healthy controls. *Eur J Gastroenterol Hepatol* 2012;**24**(10):1158–65.
35. Jaroszewicz J, Rogalska M, Flisiak R. Serum prohepcidin reflects the degree of liver function impairment in liver cirrhosis. *Biomarkers* 2008;**13**(5):478–85.
36. Heritage ML, Murphy TL, Bridle KR, Anderson GJ, Crawford DH, Fletcher LM. Hepcidin regulation in wild-type and Hfe knockout mice in response to alcohol consumption: evidence for an alcohol-induced hypoxic response. *Alcohol Clin Exp Res* 2009;**33**(8):1391–400.
37. Sozo F, Dick AM, Bensley JG, et al. Alcohol exposure during late ovine gestation alters fetal liver iron homeostasis without apparent dysmorphology. *Am J Physiol Regul Integr Comp Physiol* 2013;**304**(12):R1121–9.
38. Gerjevic LN, Liu N, Lu S, Harrison-Findik DD. Alcohol activates TGF-beta but inhibits BMP receptor-mediated smad signaling and smad4 binding to hepcidin promoter in the liver. *Int J Hepatol* 2012;**2012**:459278.
39. Duane P, Raja KB, Simpson RJ, Peters TJ. Intestinal iron absorption in chronic alcoholics. *Alcohol Alcohol* 1992;**27**(5):539–44.
40. Chaston TB, Matak P, Pourvali K, Srai SK, McKie AT, Sharp PA. Hypoxia inhibits hepcidin expression in HuH7 hepatoma cells via decreased SMAD4 signaling. *Am J Physiol Cell Physiol* 2011;**300**(4):C888–95.
41. Ahmad KA, Ahmann JR, Migas MC, et al. Decreased liver hepcidin expression in the Hfe knockout mouse. *Blood Cells Mol Dis* 2002;**29**(3):361–6.
42. Fletcher LM, Dixon JL, Purdie DM, Powell LW, Crawford DH. Excess alcohol greatly increases the prevalence of cirrhosis in hereditary hemochromatosis. *Gastroenterology* 2002;**122**(2):281–9.
43. Courselaud B, Pigeon C, Inoue Y, et al. C/EBPalpha regulates hepatic transcription of hepcidin, an antimicrobial peptide and regulator of iron metabolism. Cross-talk between C/EBP pathway and iron metabolism. *J Biol Chem* 2002;**277**(43):41163–70.
44. Choi SO, Cho YS, Kim HL, Park JW. ROS mediate the hypoxic repression of the hepcidin gene by inhibiting C/EBPalpha and STAT-3. *Biochem Biophys Res Commun* 2007;**356**(1):312–7.
45. Harrison-Findik DD, Lu S, Zmijewski EM, Jones J, Zimmerman MC. Effect of alcohol exposure on hepatic superoxide generation and hepcidin expression. *World J Biol Chem* 2013;**4**(4):119–30.
46. Kawai T, Akira S. The role of pattern-recognition receptors in innate immunity: update on toll-like receptors. *Nat Immunol* 2010;**11**(5):373–84.
47. Takeda K, Akira S. TLR signaling pathways. *Semin Immunol* 2004;**16**(1):3–9.
48. Hritz I, Mandrekar P, Velayudham A, et al. The critical role of toll-like receptor (TLR) 4 in alcoholic liver disease is independent of the common TLR adapter MyD88. *Hepatology* 2008;**48**(4):1224–31.
49. Nemeth E, Rivera S, Gabayan V, et al. IL-6 mediates hypoferremia of inflammation by inducing the synthesis of the iron regulatory hormone hepcidin. *J Clin Invest* 2004;**113**(9):1271–6.
50. Nemeth E, Valore EV, Territo M, Schiller G, Lichtenstein A, Ganz T. Hepcidin, a putative mediator of anemia of inflammation, is a type II acute-phase protein. *Blood* 2003;**101**(7):2461–3.

51. McClain CJ, Barve S, Deaciuc I, Kugelmas M, Hill D. Cytokines in alcoholic liver disease. *Semin Liver Dis* 1999;**19**(2):205–19.
52. Zmijewski E, Lu S, Harrison-Findik DD. TLR4 signaling and the inhibition of liver hepcidin expression by alcohol. *World J Gastroenterol* 2014;**20**(34):12161–70.
53. Cong M, Iwaisako K, Jiang C, Kisseleva T. Cell signals influencing hepatic fibrosis. *Int J Hepatol* 2012;**2012**:158547.
54. Ramm GA, Crawford DH, Powell LW, Walker NI, Fletcher LM, Halliday JW. Hepatic stellate cell activation in genetic haemochromatosis. Lobular distribution, effect of increasing hepatic iron and response to phlebotomy. *J Hepatol* 1997;**26**(3):584–92.
55. Irving MG, Halliday JW, Powell LW. Association between alcoholism and increased hepatic iron stores. *Alcohol Clin Exp Res* 1988;**12**(1):7–13.
56. Fujimoto M, Uemura M, Nakatani Y, et al. Plasma endotoxin and serum cytokine levels in patients with alcoholic hepatitis: relation to severity of liver disturbance. *Alcohol Clin Exp Res* 2000;**24** (4 Suppl.):48S–54S.
57. Enomoto N, Ikejima K, Yamashina S, et al. Kupffer cell sensitization by alcohol involves increased permeability to gut-derived endotoxin. *Alcohol Clin Exp Res* 2001;**25**(6 Suppl.):51S–4S.
58. Harrison-Findik DD, Klein E, Evans J, Gollan J. Regulation of liver hepcidin expression by alcohol *in vivo* does not involve Kupffer cell activation or TNF-alpha signaling. *Am J Physiol Gastrointest Liver Physiol* 2009;**296**(1):G112–8.
59. Dostalikova-Cimburova M, Balusikova K, Kratka K, et al. Role of duodenal iron transporters and hepcidin in patients with alcoholic liver disease. *J Cell Mol Med* 2014;**18**(9):1840–50.
60. Kohgo Y, Ohtake T, Ikuta K, et al. Iron accumulation in alcoholic liver diseases. *Alcohol Clin Exp Res* 2005;**29**(11 Suppl.):189S–93S.
61. Kohgo Y, Ohtake T, Ikuta K, Suzuki Y, Torimoto Y, Kato J. Dysregulation of systemic iron metabolism in alcoholic liver diseases. *J Gastroenterol Hepatol* 2008;**23**(Suppl. 1):S78–81.
62. Fletcher LM, Bridle KR, Crawford DH. Effect of alcohol on iron storage diseases of the liver. *Best Pract Res Clin Gastroenterol* 2003;**17**(4):663–77.
63. Houglum K, Filip M, Witztum JL, Chojkier M. Malondialdehyde and 4-hydroxynonenal protein adducts in plasma and liver of rats with iron overload. *J Clin Invest* 1990;**86**(6):1991–8.
64. el Ghissassi F, Barbin A, Nair J, Bartsch H. Formation of 1,N_6-ethenoadenine and 3,N_4-ethenocytosine by lipid peroxidation products and nucleic acid bases. *Chem Res Toxicol* 1995;**8**(2):278–83.
65. Wyatt TA, Kharbanda KK, McCaskill ML, et al. Malondialdehyde-acetaldehyde-adducted protein inhalation causes lung injury. *Alcohol* 2011;**46**(1):51–9.
66. Niemela O. Acetaldehyde adducts in circulation. *Novartis Found Symp* 2007;**285**:183–92 discussion 193–197.
67. Tornqvist M, Kautiainen A. Adducted proteins for identification of endogenous electrophiles. *Environ Health Perspect* 1993;**99**:39–44.
68. Lin RC, Shahidi S, Kelly TJ, Lumeng C, Lumeng L. Measurement of hemoglobin-acetaldehyde adduct in alcoholic patients. *Alcohol Clin Exp Res* 1993;**17**(3):669–74.
69. Tuma DJ. Role of malondialdehyde-acetaldehyde adducts in liver injury. *Free Radic Biol Med* 2002;**32**(4):303–8.
70. Ballard HS. The hematological complications of alcoholism. *Alcohol Health Res World* 1997;**21**(1):42–52.
71. Bills ND, Koury MJ, Clifford AJ, Dessypris EN. Ineffective hematopoiesis in folate-deficient mice. *Blood* 1992;**79**(9):2273–80.
72. Maruyama S, Hirayama C, Yamamoto S, et al. Red blood cell status in alcoholic and non-alcoholic liver disease. *J Lab Clin Med* 2001;**138**(5):332–7.
73. Crider KS, Yang TP, Berry RJ, Bailey LB. Folate and DNA methylation: a review of molecular mechanisms and the evidence for folate's role. *Adv Nutr* 2012;**3**(1):21–38.
74. Doss MO, Kuhnel A, Gross U. Alcohol and porphyrin metabolism. *Alcohol Alcohol* 2000;**35**(2):109–25.
75. Pietrangelo A. Iron-induced oxidant stress in alcoholic liver fibrogenesis. *Alcohol* 2003;**30**(2):121–9.
76. Seitz HK, Stickel F, Homann N. Pathogenetic mechanisms of upper aerodigestive tract cancer in alcoholics. *Int J Cancer* 2004;**108**(4):483–7.
77. Stickel F, Schuppan D, Hahn EG, Seitz HK. Cocarcinogenic effects of alcohol in hepatocarcinogenesis. *Gut* 2002;**51**(1):132–9.
78. Jones KL. The effects of alcohol on fetal development. *Birth Defects Res C* 2011;**93**(1):3–11.
79. Lee DH, Jacobs Jr DR, Gross M, et al. Gamma-glutamyltransferase is a predictor of incident diabetes and hypertension: the coronary artery risk development in young adults (CARDIA) study. *Clin Chem* 2003;**49**(8):1358–66.
80. Aslinia F, Mazza JJ, Yale SH. Megaloblastic anemia and other causes of macrocytosis. *Clin Med Res* 2006;**4**(3):236–41.
81. Walsham NE, Sherwood RA. Ethyl glucuronide. *Ann Clin Biochem* 2012;**49**(2):110–7.
82. Makey DG, Seal US. The detection of four molecular forms of human transferrin during the iron binding process. *Biochim Biophys Acta* 1976;**453**(1):250–6.
83. Eknoyan G, Levin N, Nissenson A, Owen Jr W, Levey AS, Bolton K. NKF and RPA collaborating on clinical practice guidelines for chronic kidney disease. *Nephrol News Issues* 2001;**15**(8):13.
84. van Noort WL, de Jong G, van Eijk HG. Purification of isotransferrins by concanavalin A sepharose chromatography and preparative isoelectric focusing. *Eur J Clin Chem Clin Biochem* 1994;**32**(12):885–92.
85. Stibler H, Borg S. Evidence of a reduced sialic acid content in serum transferrin in male alcoholics. *Alcohol Clin Exp Res* 1981;**5**(4):545–9.
86. Fleming MF, Anton RF, Spies CD. A review of genetic, biological, pharmacological, and clinical factors that affect carbohydrate-deficient transferrin levels. *Alcohol Clin Exp Res* 2004;**28**(9):1347–55.
87. Schiff ER, Maddrey WC, Sorrell MF. *Schiff's Diseases of the Liver*. 11th ed. Wiley-Blackwell: Oxford, UK, 2011.
88. Neels H, Yegles M, Dom G, Covaci A, Crunelle CL. Combining serum carbohydrate-deficient transferrin and hair ethyl glucuronide to provide optimal information on alcohol use. *Clin Chem* 2014;**60**(10):1347–8.
89. Moirand R, Lescoat G, Brissot P. Interactions of alcohol and iron proteins. *Ann Gastroenterol Hepatol (Paris)* 1989;**25**(2):51–4.
90. Bridle KR, Frazer DM, Wilkins SJ, et al. Disrupted hepcidin regulation in HFE-associated haemochromatosis and the liver as a regulator of body iron homoeostasis. *Lancet* 2003;**361**(9358):669–73.
91. Bridle K, Cheung TK, Murphy T, et al. Hepcidin is down-regulated in alcoholic liver injury: implications for the pathogenesis of alcoholic liver disease. *Alcohol Clin Exp Res* 2006;**30**(1):106–12.
92. Smith GC, Alpendurada F, Carpenter JP, et al. Effect of deferiprone or deferoxamine on right ventricular function in thalassemia major patients with myocardial iron overload. *J Cardiovasc Magn Reson* 2011;**13**:34.
93. Zhang Y, Li H, Zhao Y, Gao Z. Dietary supplementation of baicalin and quercetin attenuates iron overload induced mouse liver injury. *Eur J Pharmacol* 2006;**535**(1–3):263–9.
94. Tang Y, Li Y, Yu H, et al. Quercetin prevents ethanol-induced iron overload by regulating hepcidin through the BMP6/SMAD4 signaling pathway. *J Nutr Biochem* 2014;**25**(6):675–82.
95. Ji Y, Zhang YN, Kang XX, Xu YQ, Wang C. Protective effect and mechanism of hepcidin in rats with alcoholic liver damage. *Zhonghua Gan Zang Bing Za Zhi* 2011;**19**(4):301–4.
96. Yeh KY, Yeh M, Glass J. Hepcidin regulation of ferroportin 1 expression in the liver and intestine of the rat. *Am J Physiol Gastrointest Liver Physiol* 2004;**286**(3):G385–94.

97. Yamaji S, Sharp P, Ramesh B, Srai SK. Inhibition of iron transport across human intestinal epithelial cells by hepcidin. *Blood* 2004;**104**(7):2178–80.
98. Iqbal T, Diab A, Ward DG, et al. Is iron overload in alcohol-related cirrhosis mediated by hepcidin? *World J Gastroenterol* 2009;**15**(46):5864–6.
99. Guo X, Li W, Xin Q, et al. Vitamin C protective role for alcoholic liver disease in mice through regulating iron metabolism. *Toxicol Ind Health* 2011;**27**(4):341–8.
100. Ren Y, Deng F, Zhu H, Wan W, Ye J, Luo B. Effect of epigallocatechin-3-gallate on iron overload in mice with alcoholic liver disease. *Mol Biol Rep* 2011;**38**(2):879–86.
101. Thephinlap C, Ounjaijean S, Khansuwan U, Fucharoen S, Porter JB, Srichairatanakool S. Epigallocatechin-3-gallate and epicatechin-3-gallate from green tea decrease plasma non-transferrin bound iron and erythrocyte oxidative stress. *Med Chem* 2007;**3**(3):289–96.

Index

A

A1A2 genotypes, 227
AAV8 TUD vector expressing, anti-miR-122, 277
Acetaldehyde dehydrogenases (ALDHs), 308
Acetaldehyde-induced intestinal barrier dysfunction, 177
  direct damage to epithelial cells, 178
  epithelial integrity, 180
  TJs/AJs integrity, 177–178
Acetaldehyde-induced intestinal epithelial barrier function, modulation of, 178–180
Acetaldehyde, on intestinal barrier function, 171, 173, 178, 234, 340
  ADH/ALDH isoenzymes, distribution of, 174
  apical junctional complex, 171
  gastrointestinal tract, generation/metabolism of, 172
    distribution, 172
    elimination, 172
    ethanol absorption, 172
    metabolism, 173, 174
  genetic variations in ethanol, 174
  intestinal epithelial barrier, 171
    integrity, 175–177
  intestinal microbiota, in production/metabolism, 175
  in vitro and *ex vivo* studies, 176
  oxidative ethanol metabolism, in liver cells, 173
  potential mechanisms and pathways, 180
Acetyl-CoA carboxylase (ACC), 9, 101, 190
  in adipocytes, 193
Acetyl-CoA synthesized, 8
Acetyl-coenzyme A (acetyl-CoA), 64
Acetyl-coenzyme A synthetase (AceCS1), 289
Activities of daily living (ADL), 114
Acyl-CoA cholesterol O-acyl transferase (ACAT), 191
Acyl-CoA dehydrogenase (ACAD), 189
Acyl-CoA dehydrogenase (Acadl), 147
Acyl-CoA-diacylglycerol acyltransferase (DGAT), 188
Acyl-CoA synthase (ACS), 7
Adenosine triphosphate (ATP), 86
  adenosine 5′-triphosphate (ATP), 341
*ADH3*1/*1* allele, 234
Adherens junctions (AJs), 171
ADP-ribosylation, 261
Affymetrx, 53
AJs proteins, 178
Alanine aminotransferase (ALT) synthesis, 98, 122, 148, 363
Alcohol abuse, 189, 239
  chronic, 123
Alcohol-associated breast cancer
  carcinogenic mechanisms of, 315
Alcohol-associated carcinogenesis, molecular mechanisms of, 305
  acetaldehyde toxicity with DNA repair, 308
  alcohol toxicity, mechanisms of, 305
  estrogens in ethanol mediated breast cancer, 310–311
  ethanol and acetaldehyde on intermediary metabolism, 306
  ethanol effect on DNA, 306
  ethanol on epigenetics, 308
    DNA methylation, 308
    histone acetylation, 309
    histone methylation, 309
    histone modifications, 309
  ethanol on estrogens, 311
  oxidative stress and DNA adduct formation, 307
  reactive oxygen species, 307
  retinoic acid (RA)/alcohol/cancer, 309–310
Alcohol consumption, 289
  chronic, 225
Alcohol dehydrogenase 4 (ADH4) gene, 248
Alcohol dehydrogenases (ADHs), 17, 143, 234, 288, 308, 316
Alcohol dependence (AD) genes, 227, 247
Alcohol-elevated urinary excretion, 146
Alcohol exposure, 291
Alcohol gonadotoxicity, metabolic changes, 337–339
  ethanol gonadal toxicity
    factors supposed, 338
    realization, net potential targets, 350
  in females, 346–350
  in males, 339–346
  ovarian metabolic disorders, 349
    changes in ovaries, 348
  testicular metabolic disorders, 346
    changes in testes, 344
Alcoholic fatty liver disease (AFLD), 75
Alcoholic hepatitis (AH), 275, 307
Alcoholic liver disease (ALD), 17, 63, 99, 166, 275
  cause of morbidity and mortality, 63
  ethanol-fed micropig model, 101
  fibrosis–pathogenic mechanisms, 64
    adipose tissues/gut, contributions from, 66
    damage associated molecular patterns (DAMPs), 64
    hepatocyte injury-death, 64
    immune cells/chemokines/cytokines, 65
    liver repair, 64
  general mechanisms, 63
    alcoholic liver steatosis, 63
    alcoholic steatohepatitis and fibrosis, 64
    alcohol metabolism/oxidative stress, 64
    gut permeability/lipopolysaccharides (LPS), 64
  hereditary hemochromatosis, 364
  miRNA and extracellular vesicles, 275
  osteopontin (OPN), role, 66–68
    chronic liver diseases, 67
  pathogenic mechanisms, 65
  SAM, effects of, 101–102
  vitamin B regulation. *See* Vitamin B regulation
    methionine metabolism, regulatory effects, 100–101
  zinc deficiency. *See* Zinc deficiency, in ALD
Alcoholic liver injury
  epidemiology and pathogenesis, 99
  hepatic methionine metabolism, 98–99
  methionine metabolism
    to epigenetic regulation of gene expression, 102–103
Alcoholic steatosis, 147
Alcohol induced dysfunction, epigenetic footprint of, 296
  alcoholism/alcohol dependence, 298
  fetal alcohol spectrum disorders (FASD), 297–298
  liver cancer, 297
  transgenerationally altered DNA methylation, 299
  treatment and biomarkers of, 299
Alcohol-induced fatty liver disease, 188
Alcohol-induced hypercholesterolemia
  nutritional therapies, 194–196
  nutritious agents, 195
Alcohol-induced oxidative stress, 133
Alcohol-induced violations, 342
Alcohol intake, 45, 188
Alcohol metabolism, 17–18, 187, 287
  alcohol consumption, 17
  ALD development, 19
  epigenetic changes of, 18–20
  key molecular alterations, 18
  transcriptional genes, 19
Alcohol, overview of, 25
Alcohol use disorder (AUD), 51, 247
Alcohol withdrawal symptom (AWS), 247
  candidate gene studies, 248, 249
    ADH1B/ADH4, 248
    cannabinoid receptor 1 (brain) (CNR1) gene, 248

Alcohol withdrawal symptom (AWS) *(cont.)*
cholecystokinin (CCK) gene, 248
dopamine receptor D2 (DRD2) gene, 249
dopamine transporter (DAT) gene, 249
FK506-binding protein 5 (FKBP5), 250
gamma-aminobutyric acid (GABA) B receptor, 1 (GABABR1) gene, 250
glutamate receptor, ionotropic, *N*-methyl D-aspartate 1 gene (GluN1), 250
5-hydroxytryptamine (serotonin) receptor 1A (5-HT1A) gene, 250
5-hydroxytryptamine (serotonin) receptor 1B (5-HT1B) gene, 250–251
leptin (LEP), 251
methylenetetrahydrofolate reductase (NAD(P)H) (MTHFR) gene, 251
nerve growth factor (NGF) gene, 251
neuropeptide S (NPS) gene, 252
neuropeptide Y (NPY) gene, 251
NR2B receptor, 252
NRH-quinone oxidoreductase 2 (NQO2) gene, 252
opioid receptor, κ1 (OPRK1) gene, 252
solute carrier family 6 (neurotransmitter transporter), member 4 (SLC6A4) gene, 254
type 1 equilibrative nucleoside transporter (ENT1), 250
tyrosine hydroxylase (TH), 253
genome-wide association study, 253–254
limitations, 254
25 SNPs in Genome-Wide Association, 253
ALD. *See* Alcoholic liver disease (ALD)
Aldehyde dehydrogenase (ALDH), 25, 174, 187, 289, 340
ALDH1A3, three transcripts of, 34
ALDH1A2 transcriptional variants, 33
ALDH1L1 protein, 28
ALDH2, mitochondrial activity of, 31
biomedical implications, modifications, 30
acetylation, 31
nitration, 31
phosphorylation, 31
S-nitrosylation, 31
5131-bp ALDH5A1 primary variant, 34
2549-bp of ALDH8A1, 35
3160-bp primary variant ALDH4A1, 34
genomics of superfamily, 31
alternative splicing, 32–35
copy-number variations (CNV), 35–39
human alcohol metabolizing, 25–28
human ALDH1B1, single nucleotide polymorphism (SNP), 39
human transcripts, 32
mitochondrial ALDH2, 40
nonalcohol-metabolizing, 28–30
overview of, 25
Pfam analysis of, 32
polymorphism in alcohol-related diseases, 39–40
protein, alignment of, 29
protein sequences, neighbor-joining dendrogram, 28
proteins, structural/functional features of, 26
sequencing of transcript, 32
superfamily gene, 25
Aldehyde dehydrogenase (ALDH2), 17, 337
ALDH. *See* Aldehyde dehydrogenase (ALDH)
ALDH3A2, 508-amino acid protein, 28
ALDH1A1 protein, 25
ALDH1L1 protein, 28
Aldolase C (ALDOC), 226
Aliphatic methyl esters, 218
Alkaline comet assay, 318
Alkaline phosphatase (ALP), 53
Allium genus, 88
Alzheimers's disease, 113
512-amino acid ALDH1A3, 28
Aminotransferase (AST), 121, 122
AML-12 hepatocytes, 278
AMP-activated protein kinase (AMPK), 4, 9
Anemia, 163
Angiotensin-independent pathways, 113
Ankyrin repeat and kinase domain containing 1 (ANKK1), 227
Antioxidant butylated hydroxytoluene (BHT), 215
Antioxidant treatment, of alcoholism, 121, 126
carotenoids, 122
curcumin (diferuloylmethane), 122
epigallocatechin gallate (EGCG), 122–123
ethanol metabolism/free-radical synthesis, 120
ginsenosides, 123
GSH supplementation, 123–124
importance of, 119–120
metabolism of alcohol, 120
polyenylphosphatidylcholine (PPC), 124
precursors, 123–124
puerarin, 124
quercetin, 124
resveratrol, 124
selenium, 125
silymarin, 125
status in alcoholism, 120–121
sulfur-containing compounds, 125
α-tocopherol, 121–122
vitamin C, 121–122
Antioxidative, action of zinc, 150
Apoptosis, 45–47
cell death through necrosis, 45
cell ethanol exposure, 46
cell stress markers studied in alcohol-exposed neural tissue, 48
ethanol activated apoptosis in CNS, 47
Apoptotic cell death, of hepatocytes, 148
Aryl hydrocarbon receptor (AhR), 75
Aspartate aminotransferase (AST), 53, 98
Astrocytes, phospholipase D signaling, 325
ATP:citrate lyase, 9
Auditory brainstem nucleus, 202
Autologous CD133+ stem cell portal administration, 227
Autologous exosomes, intravenous (IV) administration of, 282
Autosomal recessive cutis laxa type 3A (ARCL3A), 30
AWS. *See* Alcohol withdrawal symptom (AWS)

B

Bax-deficient, 50
Benfotiamine (BF) treatment, male response, 91
Betaine homocysteine methyltransferase (BHMT), 98
Bifidobacterium bifidum, 179, 180
Bile acids (BA), 192
Bioinformatics analysis, 28
Blood alcohol levels (BALs), 49
BNST-projecting neurons, 206
Body mass index (BMI), 66
Bone changes in alcoholics. *See* Vitamin D, bone changes in alcoholics
Bone remodeling, 109
Brain atrophy, 113
Brain-derived neurotrophic factor (BDNF), 329
Brush-border membrane (BBM), 157

C

Caco2-cell monolayers, 149, 175, 177–179, 364
Caenorhabditis elegans, 275
Calcitriol (vitamin D3), 114
Calf thymus DNA fragments, 317
Cannabinoid receptor 1 (*Cnr1*), 269
Capillary electrophoresis (CE), 213, 214
Carbohydrate-deficient transferrin (CDT), 363
Carbohydrate responsive element binding protein (ChREBP), 3, 6, 9
Carnitine palmitoyl transferase 1 (CPT-1), 7, 190
Catalase, 119
Catecholamines, hormones, 5
Cationic liposomes, 281
C57Bl/6J mice, 215
CD36. *See* Cluster of differentiation 36 (CD36)
CD36 protein, 71, 77
expression, 79
inflammatory processes, 79
intracellular trafficking, 73
knockouts, 71
trafficking, 73
CD44 receptor, 67
Cell cycle regulation, 47
Cell signaling pathways, 177
Cellular folate metabolism, 161
Cellular iron homeostasis, 358
Cellular zinc homeostasis, 143
Centrally projecting Edinger–Westphal nucleus (EWcp), 202
Central nervous system (CNS), 45
Ceramide (Cer), 329, 331
Cerebral demyelination syndromes, 86
Chenodeoxycholic acid (CDCA), 192
Child-Pugh score, 151
Chinese hamster ovary cells, 347
Chlormethiazole (CMZ), 309
Chlorzoxazone, *in vivo* studies, 231
Cholesterol ester hydrolase (CEH), 191
Cholesterol esters (CE), 189
Cholesterol ester synthase (CES), 191
Cholesterol regulation
alcohol metabolism, 187
effects on lipid, 187–188
fatty acids, 189
leptin in ALD, 187
triglycerides, 188–189

Cholesteryl ester (CE), 191
Cholic acid (CA), 192
Chronic alcohol intake, 319
Chronic ethanol ingestion, 165
Chronic ethanol vapor, 266
Chronic heavy alcohol consumption, 187
Chronic hepatitis C, 280
Cirrhosis, 99
c-Jun N-terminal protein kinase (JNK)-mediated cell death signaling pathway, 31, 310
Cluster of differentiation 36 (CD36), 71
alcohol-fed WT mice, 75
in brain, 78–79
Ca^{2+} signaling, 76
alcoholic liver disease, 76–77
eicosanoid production/inflammation, 76
FA uptake, 73
alcoholic cardiomyopathy, 74
cardiac CD36 expression, 74
CD36 deficiency protects against alcoholic steatosis, 75–76
hepatic CD36 expression, 74
metabolic consequences, 73
nonalcoholic fatty liver disease (NAFLD), 74–75
trafficking, 73
PAMP-induced inflammation, 78
pattern recognition receptor, 77
alcoholic liver disease, 77–78
TLR4 dimer, 77
TLR2/TLR6 dimer, 77
schematic presentation of, 72
structure-function relationship, 71
functional binding domains, 71–72
posttranslational modification, 73
wild-type (WT) mice, 79
Cobalamin (Cbl), 97
Collaborative Study on the Genetics of Alcoholism (COGA), 247
Collagen-producing myofibroblasts, 64
Colles fracture, 111
Colon folates absorption rate, 157
Corticosterone/cortisol (CORT), 202
Corticotropin releasing factor system
alcohol and brain, 201
CRF system, 201
EtOH drinking, 204–205
EtOH, effects of, 203
genetic knockouts, 202–203
KO effects, 204
nomenclature, 202
stress/EtOH drinking/dependence, 206
HPA-axis, 201
EtOH, effects of, 203
Ucn1/EtOH drinking, 205
urocortins, 201
CpG dinucleotides (CpG sites), 262
^{51}Cr-ethylenediaminetetraacetic acid (^{51}Cr-EDTA), 172
CRF2 agonist Ucn3, 206
CRF-binding protein (CRFBP), 201
CRF-immunoreactivity (IR), 203
CRF receptor antagonist, 203
Crh gene, 205
Crh mRNA, 203
Crohn's disease, 149
Curcumin (diferuloylmethane), 122
CXCR3 chemokines (CXCL9-11), 65
Cyclic-AMP responsive element binding protein (CBP), 265
Cyclopentane perhydrophenanthrene moiety, 107
Cynomologous monkeys, 49
CYP2E1. *See* Cytochrome P450 2E1 (CYP2E1)
*CYP2E1*7C* allele, 233
*CYP2E1*1D* allele, 233
CYP2E1 Dra I polymorphism, 236, 239
CYP2E1 expression, 342
CYP2E1 functional polymorphisms, 239
CYP2E1 gene, 231, 232, 234, 241
CYP2E1 Rsa I/Pst I polymorphism, 232, 233, 239, 241
CYP2E1 Taq I polymorphism, 234
CYP system. *See* Cytochrome P450 (CYP) system
Cystathionine beta synthase (CβS), 99
Cytochrome P450 2E1 (CYP2E1), 149, 173, 187, 231, 275, 305
activation, 315
alcohol dependence/abuse/misuse, terminology of, 233
alcoholic pancreatitis (AP) risks, 239
human epidemiological studies, 239
meta-analysis of, 241
potential links, 239
Rsa I/Pst I polymorphism, 240, 241
CYP450 allele nomenclature, 231
enzyme, catalytic activity of, 232–233
individual susceptibility to alcohol dependence (AD), 233, 234
alcoholic liver disease (ALD), 235–239
human epidemiological studies, 234, 236
meta-analysis of, 234–235, 239
potential links, 236
Rsa I/Pst I polymorphism, 237
isoenzyme, 18
mediated alcohol metabolism, 187
mediated oxidative stress, 293
polymorphisms, 231, 232
population distribution of, 233
Cytochrome P450 (CYP) system, 119, 337, 360
cytochrome P450 side chain cleavage enzyme (P450scc), 340
isoenzymes, 17
isoform 2E1, 288
Cytosin-phosphatidyl-Guanin (CpG), 251
Cytosolic acetyl-CoA, 9

D

DAergic neurons, 205
Death-inducing signaling complex (DISC), 46
Delta-1-pyrroline-5-carboxylate synthetase (P5CS), 30
Dendritic cells (DCs), 277
De novo deoxythymidine monophosphate (dTMP), 160
De novo lipogenesis (DNL), 4, 8
Deoxyuridine monophosphate (dUMP), 159
Diagnostic and Statistical Manual of Mental Disorders (DSM-IV), 233, 247
Dibutyryl cyclic AMP (Bt_2cAMP), 50
Dietary fatty acids, 9
Dietary Se supplementation, 136
Dietary zinc supplementation, 148
Dihydrofolate (DHF), 157
Dimethylglycine (DMG), 98
Direct infusion (DI), 215
DNA damage, 47
DNA fragmentation, 46, 50–52
DNA hypomethylation, 298, 347
DNA methylation, 18, 261, 262, 264, 289, 295–297, 299, 305
DNA methyltransferase 1 (DNMT1), 296
DNA methyltransferase enzymes (DNMT), 289
DNA methyltransferases (DNMTs), 98, 262
DNA polymerase activity, 119
DNA-protein crosslinks formation, 346
Dopamine (DA), 201
Dopamine receptor D2 gene (DRD2), 226
Dorsal root ganglion (DRG) neurons *in vitro*, 124
Driver and Vehicle Licensing Agency (DVLA), 363
Duodenal cytochrome b (Dcytb), 355

E

E-cadherin, 179
Either absorbed, 108
Endoplasmic reticulum (ER), 326
of hepatocytes, 100
stress, 17
Endotoxin, lipopolysaccharide (LPS), 148
Enterobacter cloacae, 175
Enzymatic NADPH, 340
Enzyme betaine-homocysteine methyltransferase (BHMT), 291
Enzyme that oxidizes ethanol, 174
Enzyme phosphoribosylglycinamide formyltransferase (E.C. 2.1.2.2), 160
Epidermal growth factor (EGF), 179
Epidermal growth factor receptor (EGFR), 328
Epigallocatechin gallate (EGCG), 122
Epigenetic acetylation, 287
acetaldehyde, 288–289
acetate, 289
alcohol metabolism, 288
Epigenetic footprint of alcohol induced dysfunction and disease, 287
Epigenetic marks, 261
Epigenetic modulations, 261
acetate produced by ethanol metabolism, 269
alcohol effects, 264
diagram of, 267
on miRNAs pathways, 266
alcohol metabolism influences nutritional status, 268
chromatin, 261
concept of, 262
DNA methylation, 261, 262
epigenetic marks of, 265
future directions, 269
genomic DNA methylation, schematic representation of, 263

Epigenetic modulations *(cont.)*
hepatic NADH/NAD+ ratio, 268
histone modifications, alcohol effects, 262–265
histone acetylation, 265–267
microRNAs (miRNAs), 264
expression, alcohol effects on, 268
pathways, alcohol effects on, 266
nucleosomes, schematic diagram of, 263
open/condensed chromatin, epigenetic marks of, 265
SAM deficiencies, 268
schematics, 264
Epigenetic regulators, 264
Epigenetics converting enzymes, 295
DNA methylation enzymes, 296
histone code converting enzymes, 295–296
Epithelial MCF-10A cells, 320
Epithelial-to-mesenchymal transition (EMT), 177
ERK1/2 signaling pathways, 76
Escherichia coli, 175
ET-fed rats, 53
Ethanol administration, on ovarian function, 346
Ethanol, antimitogenic effects of, 331
Ethanol consumption, 53, 55, 56
affects oxidative balance, 136
apoptosis-related genes, 57
body weight, ethanol consumption, and liver function tests, 54
chronic, 133, 348
effects in adolescent rats
molecular profiling of, 53–58
gene ontologies, 56
transcriptional effects, molecular activity prediction, 58
Ethanol-enhanced lipid oxidation, 51
Ethanol, estrogens signaling, 310
Ethanol exposure, 47
Ethanol feeding, 102
Ethanol-induced CYP2E1 activation, 178
Ethanol-induced health problems, risks of, 231
Ethanol-induced metabolic perturbations, 213
Ethanol-induced steatosis, 101
Ethanol intake biomarkers, 213
ethanol exposure studies
on animal models, 215–219
in man, 219–220
global metabolic profiling, analytical technologies for, 213–214
^{1}H-NMR spectroscopy, 214
HPLC-MS/UPLCMS mouse urine metabolic profiles
3D chromatograms, 214
metabolic profiling, 213
alcohol consumption in light drinkers (LD), 220
principal components analysis plot, 217
relative sensitivity and selectivity, 214
mouse plasma samples, 3D PCA scores plot of, 218
Ethanol-mediated metabolic changes, 348
Ethanol metabolism, in liver, 269
Ethanol-oxidation system, 133
Ethanol-related toxicity, 213
Ethanol supplementation, 192
Ethanol vulnerability, 52
Ethyl glucuronide (EtG), 215
EtOH exposure reduces CRF system, 203
EtOH-induced miR-199 downregulation, 275
EtOH-seeking, stress-induced reinstatement of, 206
EtOH, voluntary consumption, 204
EV-associated miRNAs profile, 280
EWcp-Ucn1 neurons, 202
messenger ribonucleic acid (mRNA), 202
protein expression, 203
Exogenous miRNA-155 mimic, 283
Exosomes, 282
Exposure, to alcohol, 261
Extracellular signal-regulated kinase (ERK), 19
Extracellular vesicles (EVs), 278
Extrahypothalamic CRF System, 205

F

FASD models, 49
Fas system upregulation, 343
Fatty acid binding proteins (FABP), 7
Fatty acid ethyl esters (FAEEs), 338
nonoxidatively, 173
Fatty acids (FAs), 71, 189
uptake, CD36-mediated, 73, 74
Fatty acid synthase (FAS), 101
Fatty acid transport proteins (FATP), 7
Fatty acyl-CoA, 7
into fatty acid ethyl esters (FAEE), 17
Female gonadotoxicity, 346
Fenton reaction, 357
Fetal alcohol exposure, 47
adolescent ethanol models, 51–52
adult models, 52
first trimester models, 50–51
stem cell/slice culture models, 51
third trimester models, 49
Fetal alcohol spectral disorders (FASD), 136, 297, 332
etiology of, 47
Fetal alcohol syndrome (FAS), 226
Fetal calf serum (FCS), 326
Fetal neural tube defects (NTD), 161
Fibroblast growth factor (FGF), 108
Fibrogenic liver repair, 64
Firmicutes, 66
FK506-binding protein 5 (FKBP5), 250
Flavin-adenine dinucleotide (FAD), 7
Fluorescein isothiocyanate-labeled dextran 4 KD (FITC-D4), 175
Fluorescence recovery after photobleaching (FRAP), 172
Folate absorption, 163
Folate binding protein (FBP), 95, 157, 165
Folate cycle independent, 161
Folate metabolism, 159–163
Folate receptors (FR), 157
Folates. *See* Folic acid
Folates absorption, 157–159, 166
ethanol effect, 166
on folate metabolism in the liver, 164–165
on folate transport to colon, pancreas, and fetus, 165–166
on intestinal folate, 163–164
on renal folate reabsorption and excretion, 164
folate-mediated one-carbon metabolism, 160
folic acid, chemical structure of, 158
histidine metabolism, 162
historic background, 163
intracellular folate metabolism pathway, 161
methylation cycle, 162
oxidative stress, 166
Folic acid, 157
supplementation, 194
Folic acid-binding protein (FABP), 163
Follicle stimulating hormone (FSH), 337
Formiminoglutamate (FIGLU), 161
Fourier transform ion cyclotron resonance mass spectrometry (FTICR-MS), 215
Free fatty acids (FFA), 189

G

GABA A receptor B3 (GABRB3), 298
Gamma-aminobutyric acid (GABA), 201, 215
Gamma-glutamyl transferase (GTT), 363
Gas chromatography (GC), 213
Gastrointestinal tract (GIT), 171
GCN5-related *N*-acetyltransferases (GCN5), 296
Gene delivery vehicles (GDVs), 281
Gene expression, in alcoholism, 225
brain serotonin, 225
fetal alcohol syndrome (FAS), 226
genetic risks, 227
microRNAs (miRNA), 227–228
NADH/NAD+ ratio, 264
N-methyl-D-aspartate (NMDA) receptor, 226
stem cell therapy, 227
sweet preference, 226–227
Gene therapy vehicle (GVD), 281
Genome-wide association study (GWAS), 247
Germ cells, 298
γ-Glutamyl hydrolase (GGH), 157, 164
Ginsenoside-free molecules (GFM), 123
Ginsenosides, 123
Glial cells, 325
Glial fibrillary acidic protein (GFAP), 227, 325
Global Assessment of Functioning (GAF) scale, 89
Glucagon, hormones, 5
Glucagon, inactivation of pyruvate kinase, 6
Glucokinase (GK), 3
Glucokinase regulatory protein (GKR), 3
Gluconeogenesis, 3, 6
Glucose, 13
Glucose–alanine cycle, 12
Glucose-6-phosphate, 3
in hepatocytes, 4
synthesis, 4
Glucose-6-phosphate dehydrogenase (G-6-PDH), 342
Glutamate and aminobutyric acid (GABA), 325
Glutamyl carboxypeptidase (GCPII), 95

Glutathione (GSH), 144, 166
 antioxidant, 99
 content, 339
 in numerous organs, 137
Glutathione peroxidase (GSH-Px), 119, 339
 endogenous antioxidant defense enzymes, 133
Glutathione reductase (GR), 133
GLUT 2 glucose transporters, 3
Glyceraldehydes 3-phosphate, 9
Glycine-*N*-methyltransferase (GNMT), 98
Glycogenesis, 3
Glycogenolysis, 3
Glycogen phosphorylase (GP), 5
Glycogen phosphorylase action, 5
Glycogen synthase kinase 3 (GSK3)
 phosphorylate GS, 4
Glycolysis, 4
Glycolytic metabolic pathway, 86
GnRH mRNA expression, 343
Golgi apparatus, 9
Golgi network, 328
Gonadotropin-releasing hormone (GnRH), 328, 337
Green tea, antioxidants, 122
Growth arrest and apoptotic DNA damage inducible gene (GADD153), 102
GSH. *See* Glutathione (GSH)
Guanosine-5′-triphosphate (GTP), 347

H

HAMP expression, 364
 mRNA expression, 361
 promoter region, 361
HBV-induced cirrhosis patients, 219
HDACs. *See* Histone deacetylases (HDACs)
Hedgehog (Hh) ligands, 65
Helicobacter pylori, 175
Hematological disorder, 355
Hepatic amino acid metabolism, 11
 amino acid metabolism, regulation of, 13
 hepatic amino acid catabolism, 11–13
Hepatic carbohydrate metabolism, 3
 blood glucose levels, regulation of, 6
 glycogenesis/glycogenolysis, 3, 6
  absorptive state, 3–5
  postabsorptive state, 5–6
Hepatic CD36 expression, 74
Hepatic CYP7A1 expression, 192
Hepatic fat metabolism, 7
 fatty acid
  oxidation, regulation, 8–9
  regulation of, 9–11
  synthesis, 9
  uptake and oxidation, 7–9
 lipid metabolism, dietary regulation of, 11
Hepatic folate metabolism, 164
Hepatic HMG-CoA reductase activity, 194
Hepatic lipid metabolism, 189
 AMP-activated protein kinase (AMPK), 190
 bile acids, 192
 cholesterol, 191
  effect of ethanol, 191
 PPAR-α, 189–190
 sterol regulatory element-binding proteins (SREBPs), 190
Hepatic pteroylpolyglutamate, 164
Hepatic reduction, 163
Hepatic steatosis, 63
Hepatic stellate cells (HSC), 19, 101, 189
Hepatitis C virus (HCV) replication, 277
Hepatocellular carcinoma (HCC), 98, 101, 218, 297, 305
Hepatocyte nuclear factor 4-α(HNF-4 α), 143
Hepatotoxicity, 120
Hepcidin expression, 357
Hepcidin transcription, 358
Hexose monophosphate shunt (HMS), 86
High-density lipoprotein cholesterol (HDL-C) concentrations, 194
High mobility groupbox 1 (HMGB1), 65
Histone acetylation, 289
Histone acetylation at lysine 9 on histone 3 (H3K9Ac), 295
Histone acetyl transferase (HAT) activity, 289
Histone deacetylases (HDACs), 289, 295
 alcohol-preferring rats, 265
 trichostatin A (TSA), 265
Histone H3 lysine 9 (H3K9), 289, 296
Histone H3 lysine 27 (H3K27), 296
Histone methyltransferases (HMTs), 296
Histone modification, 102
H3K4me3, 267
H3K27me3, 267
H3K27 trimethylation, 296
HMG-CoA reductase, 190, 191
4-HNE. *See* 4-Hydroxynonenal (4-HNE)
Holotranscobalamin (holoTC), 97
HTR3B SNP (rs3782025), 225
5-HT3 receptor (5-HT3R), 225
5-HTT-linked promoter region (5-HTTLPR) polymorphism, 225
Human plasma glucose concentration, 6
4-Hydoxynonenal (4-HNE), 306
Hydrochloric acid, 145
5-Hydroxymethylcytosine (5hmC), 262
4-Hydroxynonenal (4-HNE), 31, 51, 144, 361
 formation of, 51
4-Hydroxyphenylacetate, 219
Hyperglycemia stimulates insulin secretion, 6
Hyperlipidemia, 190
Hypogonadism, 111
Hypothalamic-pituitary-adrenal (HPA) axis, 297
Hypothalamic-pituitary-gonadal (HPG), 337

I

Inducible nitric oxide synthase (iNOS), 100, 307
Inducible transcription factors (ITFs), 204
Inflammatory bowel diseases (IBD), 171
Ingenuity® IPA software, 53
Insulin-like growth factor 1 (IGF-1), 50, 151
Insulin-like growth factor binding protein 3 (IGFBP3), 110
Insulin-like growth factor-1 receptor (IGF-1R) genes expression, 347
International Agency for Research on Cancer (IARC), 305, 315
International Classification of Diseases (ICD-10), 233
Intestinal cell line monolayers, 172
Intestinal epithelial cells (IECs), 171
Intestinal mucosal cells, 12
Intrinsic factor (IF), 97
Iodothyronine deiodinase (DIOs), 133
Iron-chelating drugs, 364
Iron deficiency anemia (IDA), 355
Iron–hormone hepcidin, 355
Iron metabolism, molecular effects of, 355
 absorption and circulation, 356
 alcohol consumption, biochemical alterations, 363
 cellular regulation of, 360
 clinical observations, 360–362
 diagnosis and therapeutics, 362–365
 homeostasis and disorders, 358–360
 iron deficiency anemia (IDA), 355
 overview of, 355–357
 preprohepcidin (84-mer), cleavage, 357
 systemic and cellular regulation, 357–358
 transcriptional upregulation of hepcidin, 359
Iron overload, 358
Iron-related protein, 364
Irritable bowel syndrome (IBS), 171
Isolating EVs, limitation of using differential centrifugation, 280

K

KCs. *See* Kupffer cells (KCs)
Keratin 18 (K18), 64
17-Ketosteroid reductase (17-KSR) activities, 341
Klebsiella pneumoniae, 175
Kreb's cycle, 85
Kupffer cells (KCs), 77, 189, 278, 361
 alcohol-exposed, 78
 express TLR4, 77
 LPS-induced activation, 77

L

Labile iron pool (LIP), 355
Lactobacillus plantarum WCFS1, 179
L-Carnitine, 194
LC-MS metabolic profiling, 214
Leptin, 192–193
 enzymes of lipogenesis, lipolysis, and FA oxidation and cholesterol metabolism, 193
 on ethanol-induced hypercholesterolemia, 194
 ethanol on hepatic gene or enzymes, 194
 in regulating cholesterol synthesis in alcohol-induced fatty liver disease, 195
 regulation of lipid metabolism, 193–194
 schematic representation, 188
Leydig cells, of alcohol-dependent Wistar rats, 342
Lieber-DeCarli diet, 216
 alcohol, 275
 ethanol diet, 67
 liquid diet, 75, 121
Lipid peroxidation ethanol consumption, 342
Lipid phosphate phosphatase (LPP), 328
Lipid rafts (LR), 163
Lipid signaling pathways, endocannabinoid system controls, 269

Lipid-soluble provitamin, 88
Lipopolysaccharide (LPS), 64, 77, 98
Lipoprotein receptor-related proteins 5/6 (LRP5/6), 111
Liquid biopsy, 279
Liquid chromatography (LC), 213
Liquid diet, 215
Liver, 3, 13
Liver cancer, 297
Liver disease progression, 65
Liver fibrosis, 64
Liver function test (LFT) enzyme panel, 53
Liver, GPx imbalance, 138
Liver metabolism
  ChREBP/SREBP-1c, transcription factors, 10
  functions of liver, 4
  hepatic amino acid metabolism, 11
    overview of, 12
  hepatic carbohydrate metabolism, 3
    overview of, 5
  hepatic fat metabolism, 7
  hepatic lipid metabolism
    overview of, 8
Liver pyruvate kinase, 9
Livers, lipid profiling of, 216
Liver-specific DGAT2, overexpression of, 188
Liver-specific miRNA-122, 280
Liver type fatty acid binding protein (L-FABP), 189
Liver uptake, 159
Locus coeruleus (LC), 202
Long-chain fatty acids (LCFAs), 9
Luteinizing hormone (LH), 337
Luteinizing hormone releasing hormone (LHRH), 337
Lyso-phosphatidic acid (LPA), 328
Lysophosphatidylcholines (LPCs), 219

M

Macrophage inflammatory protein (MIP-3), 66
M3-activated intracellular PKC-PLD signaling pathway, 329
Maddrey's prognosis score, 66
Malate dehydrogenase (MDH), 342
Malondialdehyde (MDA) formation, 306, 339
Malonyl-CoA prevents simultaneous oxidation, 8
MAPK pathway, 123
Mass spectrometry (MS), 213
MAT1A deletion, 101
Matricellular protein, 66
Mean corpuscular volume (MCV), 363
Medial prefrontal cortex (mPFC), 268
Membrane-bound hemojuvelin (mHJV), 358
Metabolic profiling, 213
Metallothioneins (MTs), 144
Methionine adenosyl transferase (MAT), 98
Methionine metabolism, with alcoholic liver disease, 101
Methionine synthase (MS), 161, 268
Methyl adenosyltransferase (MAT), 292
Methylated histone K residues, 262
5-Methylcytosine (5mC), 287
N-methyl-D-aspartate (NMDA), 50
1-Methyl-6, 7-dihydroxy-1,2,3, 4-tetrahydroisochinolin, 315
Methylenetetrahydrofolate reductase (MTHFR) gene polymorphism, 161, 166
5,10-Methylene-THF, 160
Methylmalonate semialdehyde dehydrogenase deficiency (MMSDHD), 29
Methyl metabolism, 288, 289
  alcohol and choline/betaine, 291–292
  alcohol and folate, 291
  alcohol, cofactors of, 292
  alcohol/DNA-and histone-methylation, 293–295
  alcohol to histone acetylation
    multi-level contribution of, 295
  DNA methylation, genome-wide methylation *vs.* alcohol plot, 294
  epigenetic marks related to gene activation, 292
  epigenetic marks related to gene suppression, 293
  genetic variants, affecting alcohol metabolism, 290
  metabolic enzymes, alcohol affects, 290
  methyltransferase enzymes and SAM production, 292–293
5-Methyltetrahydrofolate (5-MTHF), 157, 164, 289
Methyltransferase enzymes, 292–293
Microphthalmia isolated 8 (MCOP8), 28
MicroRNAs (miRNA), 227
Microsomal ethanol oxidizing system (MEOS), 17, 231
Microsomal triglyceride transport protein (Mttp), 147
Microvesicles (MVs), 278
miR-122 expression, 277
  hepatic levels of, 277
miRNA-214
  in LX-2 cells, 279
miRNA expression, alcohol induces alterations, 268
miRNA expression profile, in human blood EVs, 280
miRNA expression, SHSY5Y cells, 268
miRNA/extracellular vesicles, 275
  abnormal miRNA expression, mechanism of, 278
  alcohol, cell-specific effects of, 278
  alcoholic liver disease, 275
    biomarkers of, 280
    differential EV, 281
    miRNA-targeted therapies, 281
    role of, 277
    therapeutic delivery, 282
  biogenesis of, 276
  conventional gene delivery vehicles, 281
  extracellular vesicles (EVs), 278–279
    ALD diagnosis, future direction, 281
    as biomarkers in alcoholic hepatitis, 279–280
    isolation/characterization methods to increase diagnostic accuracy, 280–281
    as liquid biopsies, 279
    as natural carriers, 279
  history of, 276
  miRNA-21, 278
  miRNA-122, 277
  miRNA-155, 277
  miRNA-217, 278
  miRNA-34a, 278
  opportunities and suitable potentials of exosomes, 282–284
  target therapies, 284
    using exosomes, 282
  therapeutic delivery, workflow of, 283
Mitochondrial permeability transition opening (MPT), 18
Mitogen-activated protein kinase phosphatise-1 (MKP-1), 310
Molecular activity predictor tool, 53
Monocyte chemoattractant protein-1 (MCP-1), 148, 268
  cytotoxic effects of, 87
mPFC apoptosis, 52
^{99m}Tc-diethylenetriaminepentaacetic acid (^{99m}Tc-DTPA), 172
MTHFR C677T-polymorphism, 251
MT-knockout (MT-KO) mice, 144
MT transgenic (MT-TG), 144
Muscle activity, 111
MyD88-independent pathways, 64
Myeloid differentiation factor 88 (MyD88), 77

N

N-acetylcysteine, 124
N-acetyltaurine (NAT), 218
NAD hydrogenase (NADH), 17
NADP-dependent isocitrate dehydrogenase (NADP-ICDH), 342
National Council on Alcoholism and Drug Dependence (NCADD), 233
National Institutes of Health (NIH), 201
NDMA-type glutamate receptors, 329
1, N^6-etheno-2′-deoxyadenosine (εdA), 307
Neuropeptide Y (NPY), in amygdala, 298
Nicotine adenine dinucleotide phosphate (NADPH), 86
Nitric oxide synthases (NOS) expression, 343
N-methyl-D-aspartate (NMDA), 226
nNOS gene, 343
Nonalcoholic fatty liver disease (NAFLD), 74, 280
Nonalcoholic steatohepatitis (NASH), 66, 191, 279
Nonenzymatic antioxidants, 119
Nonesterified fatty acids (NEFA), 7
Nonoxidative ethanol metabolism, 17
Nontoxic compound, 11
NPY-like immunoreactivity, 251
Nuclear magnetic resonance (NMR) spectroscopy, 213
Nucleus accumbens (NAc), 268
Nutrients, metabolism of, 3

O

Obesity, 192
Omega-3 FAs, 194
OPN. *See* Osteopontin (OPN)
Organic anion transporters (OAT), 159

Osteopontin (OPN), 19, 66
knockout mice, 68
in liver disease, 63
NASH and NASH-fibrosis, 67
neutralization, 67
transgenic mouse, 68
Ovarian metabolic disorders, 348
Oxidative stress, 178
Oxidized low density lipoproteins (ox-LDL), 71
8-oxodG formation, 318

P

Panax notoginseng, 123
Parathyroid hormone (PTH), 110
Paraventricular nucleus of hypothalamus (PVN), 202
Parkinson's disease, 358
Partek Genomics Suite software, 53
Patatin-like phospholipase domain-containing 3 *(PNPLA3)*, 19
Pathogen-associated molecular patterns (PAMPs), 71
PDGF-treated hippocampal stem cells, 327
Pentose-phosphate shunt, 85
Perosisome proliferator receptor-alpha (Ppar-α), 102
Peroxisomal β-oxidation, 7
Peroxisomal proliferator-activated receptors (PPARs), 3, 9, 63, 309
Peroxisome proliferator-activated receptor alpha (PPAR-α), 216
Phosphatidic acid (PA), 338
Phosphatidylcholine (PC) synthesis, 100, 101, 328
Phosphatidyl ethanol (PEth), 213, 325, 329, 338
Phosphatidylethanolamine (PEtn), 328
Phosphatidylethanolamine methyltransferase (PEMT), 98
Phosphatidylinositol-bisphosphate (PIP_2), 325
6-Phosphogluconate dehydrogenase (6-PGDH), 342
Phosphoinositide 3-kinase (PI3K) pathway, 4
Phospholipase D (PLD) signaling, 325
cell proliferation in rodent astrocyte cultures, 331
ethanol inhibits, 330
fatty acid second messengers, regulation of, 332
functions of, 328
cell proliferation, 328–329
lipid messengers, formation of, 328
vesicle formation, 328
interruption by ethanol, 329
apoptosis, induction of, 331
gene expression, changes of, 329
in vivo studies, 331–332
mitogenic signaling in astrocytes, 329–331
neuronal function, disturbance of, 329
unphysiological phospholipids, formation of, 329
intracellular pathways, regulation, 327
isoforms of, 326
mice lacking PLD, postnatal brain developments, 332
PC and site of action, 326
PC, hydrolysis of, 326
PC, transphosphatidylation, 327
PLD-deficient mouse astrocyte cultures, 331
regulation of, 326
protein kinases, 328
small GTPases, 327
transcriptional level, 326
transphosphatidylation, 325
Phospholipases, 325
Phosphorylates glucose, 3
PIDDosome, 47
Plasmodium falciparum, 72
Platelet-derived growth factor (PDGF), 66
PLD signaling. *See* Phospholipase D (PLD) signaling
Polyenylphosphatidylcholine (PPC), 124
Polypeptide type I receptor (PAC1), 342
Polyunsaturated fatty acids (PUFA), 9, 189
POMC expressing neurons, 297
Postnatal exposure, chronic, 51
Posttranslational modifications (PTM), 30
Potassium channel (BK), 227
Ppara-null mice, 216
Pregnenolone, biosynthesis of, 347
Prenatal exposure, to alcohol, 362
Probiotics, 180
Prodynorphin (PDYN) gene, 267
Proenkephalin (PENK), 269
Prostaglandins (PGs)
via cyclooxygenases, 76
Proteinase-K, 346
Protein-coupled folate transporter (PCFT), 157
Protein divalent metal transporter-1 (DMT1), 355
Protein kinase A (PKA), 4
Protein kinase C (PKC), 328
Protein phosphatase 2A (PP2A), 9
Protein tyrosine kinase (PTK), 177
Protein tyrosine phosphatase (PTP), 177
P90RSK, enzymatic protein, 19
P450scc in Leydig cells, 340
$PteGlu_n$ forms, 98
Pteroylmonoglutamate (PteGlu), 95
Puerariae radix, 124
Pueraria lobata, 124
Pyridoxine vitamin B6, 95

R

Radical-scavenging enzymes, 138
RAIDD proteins, 47
Random oligonucleotide primed synthesis (ROPS) assay, 50
RA response element (RARE), 309
Rat hepatocytes, 266
Reactive oxygen species (ROS) formation, 45, 236, 306, 321, 357
Receptor tyrosine kinase (RTK), 328
Recommended dietary allowance (RDA), 143
Reduced folate carrier (RFC), 163
Regulation of Hepatic Fatty Acid Synthesis, 6
Restriction fragment length polymorphism (RFLP) analysis, 231
Retinoic acid (RA) synthesis, 329
Retinoid X receptor (RXR), 189
Retinoid X response element (RXRE), 309
Reverse cholesterol transport (RCT), 192
Reversible heritable epigenetic mechanisms, 262
Rho kinase, 177
Ribosomal S6 kinase family, 19
RP-HPLC-ToF-MS-based metabolic profiling, 219

S

S-adenosine methionine (SAMe), 308
S-adenosylhomocysteine (SAH), 98
S-adenosyl-menthionine (SAM), 18, 98, 262, 289
chronic administration, 262
Salsolinol, breast cancer, 315
alcohol drinking
blood and urine, 315–316
cell proliferating ability, 320
estrogen-dependent MCF-7 cells, 320, 321
estrogen-independent MCF-10A cells, 320–321
DNA damage capability, 316
oxidative damage, 318
8-OxodG formation, 317–318
^{32}P-labeled DNA fragments, 316–317
epidemiological studies, 315
in mammary gland, 316
mammary tissue, formation, 317
in multistep breast carcinogenesis, 321
novel causative substance, 315
oxidative DNA damage, proposed mechanisms of, 318
copper-mediated DNA damage, 319
iron-mediated DNA damage, 319–320
Salsolinol, mammary epithelial cell proliferation, 320
SAM production, 292–293
SAM:SAH methylation ratio, 100
S-analog amino acids, 134
Sarco/endoplasmic reticulum Ca^{2+}-ATPase (SERCA), 76
Se. *See* Selenium (Se)
Secreted phosphoprotein-1 (Spp1), 66
Selenium (Se), 133
deficient offspring, 135
dietary supplementation, 133
elimination routes, 137
supplemented diets, 135, 138. *See also* Selenium supplementation
tissue distribution, 135–137
Selenium supplementation
alcohol and malnutrition, 134–135
alcohol and oxidative balance, 137–139
oxidation/antioxidant balance in kidney, 139
oxidation/antioxidant balance in liver, 138
Selenocysteine (Se-Cys), 133
Selenomethionine (Se-Met), 134
Se-Met absorption, 135
Ser626, 9
Serotonin (5-hydroxytryptamine (5-HT)), 225
Serotonin transporter, 225
Sestatus, 135
Short allele (s-allele), 225

Short chain fatty acids (SCFAs), 179
Silybum marianum, 125
Single nucleotide polymorphisms (SNPs), 247
Sjoegren–Larsson syndrome (SLS), 28
Skin-synthesized, 108
SLC40-A1 gene, 356
SNCA gene, 227
Solute carrier family 29 member 1 (SLC29A1), 250
Sphingomyelin (SM), 329
Sphingosine 1-phosphate (SIP) degradation, 28
Splenomegaly, 362
SREBP-1c gene deletion, 9
SREBP-2 protein expression, 194
Staphylococcus aureus, 72
  CD36-deficient mice, 77
  CD36-mediated internalization, 72
STAT3-SOCS3 pathway, 66
Stearoyl-CoA desaturase-1, 9, 193
Stearoyl-coA desaturase (SCD), 101
Sterol regulatory element-binding protein (SREBP)-1, 100, 190
Sterol regulatory element-binding protein 1c (SREBP-1c), 3, 6
Streptococcus viridans, 175
Sulfur-containing compounds, 125
Superoxide dismutase (SOD), 119, 316, 339
Systemic acetaldehyde levels, 175
Systemic iron regulation, 357

T

Testicular interstitial fluid (TIF), 340
Tetrahydro folate (THF), 157
Thiamine (vitamin B1), 125
Thiamine analogs, 88–89
Thiamine deficiency
  average daily alcohol consumption, 90
  benfotiamine trial, 88–91
  heavy alcohol drinking, 90
  linear mixed model analyses, 90
  logistic mixed model analysis, 90
  mood and behavior, 87
  oxidative stress and inflammation, 87
  secondary effects of, 87
  treatment response in alcohol dependent, 91
Thiamine-dependent enzymes, classification/function of, 86
Thiamine pyrophosphate (TPP^+), 85
  biochemistry of, 85–86
Thiamine tetrahydrofurfuryl disulfide (TTFD), 88
Thiobarbituric acid reactive species (TBARS), 122
Thioredoxin reductase (TrxRs), 133
Thymidine synthetase (TS), 98
Tight junctions (TJs), 171
Tissue inhibitors of metalloproteinase-1 (TIMP-1) levels, 152
TLR4. *See* Toll-like receptor 4 (TLR4)
Toll-like receptor (TLR), 77, 110, 361
Toll-like receptor 4 (TLR4), 64, 100
  signaling, 77
Total radical-trapping antioxidant potential (TRAP), 121
Toxic effects, of iron, 358
Transepithelial electrical resistance (TEER), 172
Transferrin, 363
Transferrin receptor-1 (TFR1), 356
Transferrin receptor-2 (TFR2), 356
Transforming growth factor (TGF), 19
Treatment and biomarkers of alcohol diseases through epegenetic pathway, 291
Triacylglycerol (TAG), 7, 9
Tricarboxylic acid cycle (TCA), 4, 8
Triglycerides (TGs), 73, 187–189
Tumor necrosis factor-α (TNF-α), 120
Tumor suppressor-like properties, 28
TUNEL staining, 52
TwinsUK adult bioresource studies, 219

U

Ucn1 KO mice, 205
Ucns/CRF2 signaling, 202, 206
Ultracentrifugation, 280
Ultraviolet light, 7-dehydrocholesterol for, 107
Uridine diphospate-glucose (UDP-glucose), 4
Urinary catecholamine levels, 316
Urine, UPLC-MS analysis, 216

V

Ventromedial hypothalamus (VMH), 201
Very long chain fatty acids (VLCFAs), 7
Very low density lipoprotein (VLDL), 7, 9, 63, 147, 189
  secretion, 77
Vesicular monoamine transporter 2 (VMAT2), 316
Viral hepatitis, chronic, 67
Vitamin B regulation
  alcoholic liver disease
    SAM, effects of, 101–102
  alcoholic liver injury, epidemiology/pathogenesis, 99–100
  ethanol-fed micropig model, 101
  folate, 95
    enhanced renal excretion, 97
    homeostasis, 96
    intestinal folate malabsorption, 97
    molecule, 96
    reduced hepatic uptake and storage, 97
  methionine metabolic pathways
    chronic alcohol exposure, effects of, 99
    to epigenetic regulation of gene expression, 102–103
    hepatic, 98–99
    regulatory effects, 100–101
  vitamin B6, 98
  vitamin B12, 95, 97
    methionine synthase (MS), 98
1,25 $(OH)_2$ vitamin D, 107, 108
Vitamin D, bone changes in alcoholics, 107–109, 114
  alcoholic individuals with lowest total lean mass values, 112
  bone-resorbing effect of, 110
  deficiency of, 107, 109
    brain alteration, 113
    direct effects, 108, 109
    effects on muscle, 110
    falls and fractures, 111
    pathogenesis of, 109
  effects of therapy
    abstinence, 114
    preclinical studies/clinical trials, 114
  heavy alcoholic men, handgrip strength, 111
  metabolism of, 107–108
  rib fractures, 113
  supplementation in animal models of alcoholism, 114
Vitamin E, 121

W

Wernicke–Korsakoff syndrome, 85–87
Western blot analysis, 347
White adipose tissue (WAT), 149
Wild-type (WT) littermate control, 202
Wistar rats, 218
World Health Organization (WHO), 17

X

Xenograft breast cancer tumors, 282

Y

Yeast YBF2, 296
Yin Chen Hao Tang (YCHT), 215

Z

Zinc deficiency, in ALD
  beneficial effects of, 147
    adipose tissues/adipokines, 149
    alcoholic steatosis, attenuation of, 147
    alcoholic toxicity, extrahepatic actions of, 148
    alcohol-induced hepatic cell death, abrogation of, 148
    alcohol-induced hepatic inflammation, suppression of, 148
    intestinal permeability/endotoxemia, 148
  metabolism and function, 143–144
  occurrence of, 144–145
  zinc dyshomeostasis, mechanisms of, 145
    disturbed zinc metabolism in liver, 146
    inadequate dietary zinc intake, 145
    increased zinc excretion from urine, 146
    reduced zinc absorption from intestine, 145
Zinc deprivation, 148
Zinc dyshomeostasis, in ALD
  mechanisms of, 145
Zinc protection, molecular mechanisms of, 149
  oxidative stress, inhibition of, 149
    antioxidant molecules, upregulation of, 150
    PPAR-α/HNF-4α, 150, 151
    ROS generation suppression, 149
Zinc reabsorption, 146
Zinc supplementation, 147
  in ALD, 151–152
  mouse model, 178
Zinc transporters, 143, 145, 146
ZIP family, 143